A History of
Modern Chemistry

A History of
Modern Chemistry

by

Noboru HIROTA

Kyoto University Press

First published in 2016 jointly by:

Kyoto University Press
69 Yoshida Konoe-cho
Sakyo-ku, Kyoto 606-8315, Japan
Telephone: +81-75-761-6182
Fax: +81-75-761-6190
Email: sales@kyoto-up.or.jp
Web: http://www.kyoto-up.or.jp

Trans Pacific Press
PO Box 164, Balwyn North, Melbourne
Victoria 3104, Australia
Telephone: +61-3-9859-1112
Fax: +61-3-9859-4110
Email: tpp.mail@gmail.com
Web: http://www.transpacificpress.com

Copyright © Kyoto University Press and Trans Pacific Press 2016.

Distributors

Australia and New Zealand
James Bennett Pty Ltd
Locked Bag 537
Frenchs Forest NSW 2086
Australia
Telephone: +61-(0)2-8988-5000
Fax: +61-(0)2-8988-5031
Email: info@bennett.com.au
Web: www.ennett.com.au

USA and Canada
International Specialized Book Services (ISBS)
920 NE 58th Avenue, Suite 300
Portland, Oregon 97213-3786
USA
Telephone: (800) 944-6190
Fax: (503) 280-8832
Email: orders@isbs.com
Web: http://www.isbs.com

Asia and the Pacific (except Japan)
Kinokuniya Company Ltd.

Head office:
38-1 Sakuragaoka 5-chome
Setagaya-ku, Tokyo 156-8691
Japan
Telephone: +81-3-3439-0161
Fax: +81-3-3439-0839
Email: bkimp@kinokuniya.co.jp
Web: www.kinokuniya.co.jp

Asia-Pacific office:
Kinokuniya Book Stores of Singapore Pte., Ltd.
391B Orchard Road #13-06/07/08
Ngee Ann City Tower B
Singapore 238874
Telephone: +65-6276-5558
Fax: +65-6276-5570
Email: SSO@kinokuniya.co.jp

ISBN   978-1-920901-80-6

Contents

Figures

Tables

Acknowledgements

This publication is a translation of the book entitled *Gendai Kagakusi* (*A History of Modern Chemistry*) published in Japanese by Kyoto University Press in 2013. The translation and production of the book was financially supported by a Grant-in Aid from the Japan Society for the Promotion of Sciences (JSPS) (No.266007). We thank the JSPS for its generous support. Translation provided by Forte is also greatly appreciated.

Profile of the Author

Noboru Hirota is emeritus professor of Kyoto University. After graduating from Kyoto University, he did his postgraduate studies at Washington University in St. Louis, USA, where he received his PhD in 1963. From 1965 to 1975 he was a faculty member of the State University of New York at Stony Brook. From 1976 to 2000 he was in the Faculty of Science of Kyoto University. His main research field is physical chemistry, particularly the application of ESR and laser spectroscopy to chemical problems. After his retirement in 2000 he researched the history of chemistry, and published a book in Japanese on the history of modern chemistry in 2013, which formed the basis of this latest publication.

Prologue

What is chemistry?

What exactly *is* chemistry? Answering this question is no easy task. Consulting the Fifth Edition of the *"Iwanami Dictionary of Physics and Chemistry"*, one finds the following entry for "chemistry:"

> The branch of the natural sciences that studies the structure and properties of matter—particularly chemical matter—and the chemical reactions that take place between substances. The field may be subdivided into a number of subfields—such as physical chemistry (also known as theoretical chemistry), inorganic chemistry, organic chemistry, biochemistry, and applied chemistry—based on differences in the methods and objects of study.

After proceeding to give a brief overview of the historical development of chemistry, the entry concludes thusly:

> Several branches of science that lie at the junction between chemistry and other disciplines—including physical chemistry and chemical physics at the interface between chemistry and physics, as well as biochemistry and molecular biology at the interface between chemistry and biology—are currently undergoing rapid development.

This description yields a fully satisfactory common-sense definition: *Chemistry lies between physics and biology—and overlaps with each of these fields—as the discipline concerned with studying the structure and properties of matter.* But this hardly suffices to solve our problem. Scientific disciplines that study the properties of matter include the subdomain of physics known as *solid state physics* and the subdomain of biology known as *molecular biology,* which investigates living organisms and life itself. Needless to say, in nature there are no boundaries; the separation of disciplines is a phenomenon that occurred naturally as the various academic sciences evolved, and thus it is fruitless to attempt a rigorous definition of *any* particular subfield. Even so, the boundaries of the field of chemistry are particularly vague compared to those of physics or biology.

1

Further understanding of the difficulty of answering the question "What is chemistry?" may be gleaned from the following observation. If we look at the Nobel prizes awarded in chemistry since 1960, we see that the prize has frequently been awarded for work that would typically be considered to fall under the domain of molecular biology. It seems the Nobel selection committee casts its net broadly, considering discoveries in molecular biology and biophysics to fall within the realm of chemistry. On the other hand, several recipients of the Nobel Prize in chemistry over the past few decades have commented to the effect that *I am surprised to be receiving the Nobel Prize in chemistry, because I never thought of myself as a chemist!* This suggests that—even among Nobel-caliber scientists—no consensus exists regarding the question of precisely what chemistry *is*. But *why* exactly is it so hard to define our science? To understand this basic difficulty, we must first understand how chemistry developed—and how the conceptual foundations of the subject have evolved over time.

The progress of the natural sciences in the 20th century was nothing short of remarkable. The first half of the century witnessed a revolution in physics—brought about by the development of quantum mechanics and general relativity—while the second half saw major progress in the life sciences, spurred by structural analysis of DNA and the stimulus it gave to molecular biology, amounting to a revolution in the biological sciences. How does the 20th-century development of chemistry compare? The answer is that the development of chemistry experienced *major* strides in the 20th century. By making use of advances in physics, chemists successfully clarified the nature of chemical bonds and came to understand the structure and chemical reactions at the atomic and molecular levels. In addition, 20th-century chemistry stimulated the development of molecular biology by providing a foundation for understanding the phenomena of life processes, making key contributions to the progress of life sciences. Of course, these observations may lead one to question whether chemistry has lost its unique, independent character and become simply another branch of physics or biology. Indeed, a few years ago a well-known scientific journal aired a debate over the proper positioning of chemistry within the natural sciences—suggesting that perhaps chemistry had become merely a support science, existing in the service of other disciplines, with little claim to the status of an essentially independent branch of science.

Chemistry originally arose out of the basic human desire to understand the creation and evolution of matter in the world around us. If we define chemistry as the discipline that studies the structure, properties, and reactions of matter, then it covers a truly unlimited range of territory—from the matter that populates outer space to the substance of life on Earth—and must surely be understood to occupy a central position among the natural sciences. Our understanding of matter experienced dramatic progress in the 20th century, bringing us to the point where now we can observe and control even individual atoms and molecules. Chemistry also boasts the unique feature—not present in other natural sciences—of offering the possibility of creating new materials.

Turning to applied chemistry, the benefits that mankind derives from the fruits of chemistry research are immeasurable and mind-boggling. We live our daily lives surrounded by fibers and plastics based on polymer chemistry and rely on chemical energy sources for over 90% of our energy needs. Medicine and pharmacology have progressed by leaps and bounds thanks to biochemistry and synthetic chemistry. The explosive growth in the world's population during the 20th century was made possible by the use of man-made fertilizers in the agriculture industry. Indeed, chemistry has penetrated every aspect of our lives—from engineering, to medicine, to pharmaceuticals, to agriculture and beyond—and has made possible a materially plentiful human life. Present-day chemistry lies at the heart of the nanoscience that underlies the modern sciences of substances and materials, and its applications continue to have a significant impact on human life. Moreover, at the cutting edges of life-science research, the importance of molecular-level chemistry studies has only grown. Of course, chemistry also has its downsides—such as pollution and environmental destruction—but the domain of modern chemistry research is only continuing to expand.

Why a history of modern chemistry?

Many books discuss the 20th century development of physics and biology, but surprisingly few authors have addressed the contemporaneous progress in basic chemistry. It is certainly true that—if we adopt a traditional notion of what the field of chemistry is—the 20th century did not witness true *revolutions* in basic chemistry comparable to the emergence of quantum mechanics and general relativity in physics or the structural analysis of DNA in biology. However, the historical development of chemistry is replete with intellectual stimulation and abounds with discoveries and inventions that make the pulse race. Moreover, from the perspective of understanding the true nature of matter, the new understanding of material structure and reactions afforded by quantum theory represented a major advance in chemistry, while the elucidation of the structure of DNA was itself a major discovery in chemistry from the perspective of understanding molecular structure. Thus, in studying the historical development of chemistry, it is important not to become ensnared in rigid traditional frameworks separating physics from chemistry and biology, but rather to think more broadly of the science of atoms and molecules.

In today's world, the increasingly fine-grained specialization brought about by the progress of science has made it all too easy for individual researchers to seclude themselves in their own narrow subfields—and all too difficult to obtain an overall perspective on the field as a whole. Yet, a glimpse at the cutting edge of present-day research reveals many interdisciplinary studies that transcend the rigid boundaries between the traditional academic disciplines, and it is precisely from such wellsprings that major new breakthroughs frequently emanate. For this reason, in wide-ranging fields such as chemistry that make frequent contact with other disciplines, it is particularly important to train researchers to adopt a broad-minded perspective; and yet Japan's system of undergraduate and graduate education has a strong tendency to encourage specialization from the earliest stages, resulting all too often in the production of narrow-minded researchers.

One particularly effective means of expanding one's horizons is to look *backward*—to review the history of a branch of science and traverse the pathways along which the subject developed into its present form. Indeed, a look back on the history of a field can help specialized researchers understand how their own research fits into the broader scope of the science as a whole. In recent years, there has been a tendency in Japan to emphasize research with immediate short-term utility; and yet it is by investigating how the most important and influential research was conducted in the past that we learn just what ingredients are crucial for advancing the state of science and technology.

The history of chemistry is discussed in many excellent books. Unfortunately, however, most of these cover only the development of chemistry up to the first half of the 20th century; it is no overstatement to say that there exists *no* written history of chemistry covering the second half of the 20th century. It was precisely in the second half of the 20th century, however, that chemistry developed most rapidly—and the character of the field of chemistry itself underwent its most significant changes. Thus, in thinking about modern chemistry, we need a history that covers developments all the way up to recent progress. Of course, in our current era of scientific specialization, no one individual can possibly master the entirety of any full academic discipline, and thus one might think it a hopeless pursuit for a single author to attempt a comprehensive history of modern chemistry. Indeed, I myself was but a single researcher working in the single subfield of physical chemistry—and even there my research was concentrated in a number of specialized areas: magnetic resonance, spectroscopy, and photochemistry. Thus, until recently, I only knew the faintest outlines of the broader subject of modern chemistry and the history of its development. It has only been since I retired from active research that I have had the time to step back and observe the full breadth of chemistry from a distance, and this opportunity has rejuvenated my interest in understanding the positioning of chemistry within the natural sciences—or, to use perhaps somewhat grandiose language, to think about what chemistry means for mankind. And it finally occurred to me that, as someone who was involved in chemistry research for over 40 years, perhaps my own humble version of the "history of modern chemistry" might be of interest to the world.

Given this background, how should we interpret modern chemistry and write its history? We have sought to write the history of chemistry as simply *the science of atoms and molecules*, rejecting the traditional rigid separations between physics, chemistry, and biology. From this perspective, I have traced the dynamic history of the field—how chemistry evolved in relation to neighboring disciplines such as physics and biology—placing central emphasis on 20th-century developments and leaving ample room to think about the future. Thus, this book is particularly concerned with the development of basic chemistry, including how it has impacted—and been impacted by—society; we survey the history of its evolution while exploring the background factors that explain how the most important discoveries came about.

Because chemistry—like any other domain of culture or art—is ultimately a human endeavor, the story of its progress is inextricably bound up with the stories of the great scientists who developed it. Among these scientists, one finds many examples of fascinating and highly unique individuals; learning their personal stories offers an excellent lesson in the variety of human experience and must surely rank among the most compelling reasons to study the history of chemistry. For this reason, throughout this book we have included brief biographical asides sketching the lives of eminent chemists. We hope these anecdotal tales of the great chemists will help to stoke the reader's enthusiasm for the chemistry they created.

In writing this book, we have referred to a great many sources. From among these, we have singled out those we consider most important for inclusion in the lists of references at the end of each chapter. The main academic journals to which we refer, and their abbreviations, are listed below.

- *Ana. Chem.: Analytical Chemistry*
- *Angew. Chem.: Angewandte Chemie*
- *Ann.: Justus Liebigs Annalen der Chimie*
- *Ann.der Chem.Pharm.: Annalen der Chimie und Pharmacia*
- *Ann. Phys.: Annalen der Physik*
- *Ber.: Chemische Berichte*
- *Ber. Chem. Ges.: Berichte der Deutschen Chemiscen Gesellshaft*
- *Biochem. Biophys. Res. Commun: Biochemical and Biophysical Research Bull.*
- *Bull. Chem. Soc. Jpn.: Bulletin of the Chemical Society of Japan*

- *Bull. Soc.Chem. France: Bulletin de la Societe Chimique de France*
- *Chem. Biochem. Z.: Biochemische Zeitschrift*
- *Chem. Commun.: Chemical communications*
- *Chem. Phys. Letters: Chemical Physics letters*
- *J. Am. Chem. Soc.: Journal of American Chemical Society*
- *J. Biol. Chem. Soc.: Journal of Biological Chemistry*
- *J. Chem. Phys.: Journal of Chemical Physics*
- *J. Chem. Soc.: Journal of the Chemical Society*
- *J. de chim. phys. Journal de Chimie Physique et de Physico-Chimie Biologique*
- *J. Mol. Bio.: Journal of Molecular Biology*
- *J. Org. Chem.: Journal of organic chemistry*
- *J. Phys. Chem.: Journal of Physical Chemistry*
- *Naturwiss.: Naturwissenschaften*
- *Phil. Mag.: Philosophical Magazine*
- *Phys. Rev.: Physical Review*
- *Phys. Rev. Lett. Physical Review Letters*
- *Proc. Chem. Soc. London: Proceedings of the Chemical Society London*
- *Proc. Natl. Acad. Sci.: Proceedings of the National academy of Science*
- *Proc. Roy. Soc.: Proceedings of the Royal society*
- *Trans. Faraday Soc.: Transactions of the faraday Society*
- *Z. Anorg. Chem.: Zeitschrift fur Anorganische und Allgemeine Chemie*
- *Z. Naturforsch : Zeitschrift fur Naturforschung.*
- *Z. Physikal. Chemie: Zeitschrift fur physikalische chemie*
- *Z. Elektrochem.: Zeitschrift fur Elektrochemie*
- *Z. Physik.: Zeitschrift fur Physik*

Part 1

Toward the Formation of Modern Chemistry

Chapter 1: The road to modern chemistry

Chemistry up to the 18th century: The dawn of atomic and molecular science

Lavoisier and his experimental apparatus for the oxidation of mercury

In this book, we trace the historical evolution of chemistry with primary emphasis on the developments in the 20th century. However, because the expansion of human knowledge is a continuous process, in order to understand the full flow of ideas we must first have some knowledge of the situation in earlier centuries. To this end, the first two chapters of this book summarize the history of chemistry up to the 20th century. In this first chapter, we briefly survey progress in chemistry up to the late 18th century, at which point the "chemical revolution" of Lavoisier paved the way for the beginnings of modern chemistry. The history of chemistry throughout this period is treated in detail by many books, to which we refer interested readers for more in-depth treatment.[1–7] The books cited at the end of this chapter on the history of science also include discussions of the history of chemistry.[8–10]

1.1 The Ancient Origins of Chemistry

Humans presumably first became aware of changes in states of matter when they began burning materials to create fire, generating heat and allowing the cooking of food. We can probably date the birth of chemical engineering—technology based on the transformation of matter—to mankind's first use of fire to create earthenware tools or to extract copper and iron from minerals. It was in this way that early humans came to know about matter transformation phenomena, came to put these phenomena to practical use, and to ask the first questions about the formation and properties of matter. Chemistry is the science born out of just

9

this sort of ancient chemical engineering—and the inevitable human curiosity it stoked regarding the nature of matter.

The linguistic origins of the term "chemistry" are traced by some scholars to an ancient Egyptian word meaning "black earth"—an etymological pedigree that traces the beginnings of chemistry to the pottery and metallurgy technologies of ancient Egypt, which later spread to Greece and Rome. Chemical engineering also developed independently—and with distinct, unique perspectives—in ancient China, India, and Arabia. In this chapter, we survey some trains of early thought whose lineage may be traversed continuously from antiquity to the development of modern chemistry, including the views of matter that prevailed in ancient Greece, the science of alchemy—which originated in the Egyptian city of Alexandria, the center of Hellenistic civilization, and is thought to have entered Europe around the 12th century by way of Arabia—and the technologies of iatrochemistry and chemistry that sprouted from it.

Theories of Matter in Ancient Greece

Every ancient civilization had its own understanding of matter, but the views espoused by the philosophers of ancient Greece were to have a particularly pronounced influence on the later development of chemistry. In the sixth century B.C., Thales of Miletus (a Greek city on the Aegean Sea in Asia Minor, part of modern-day Turkey) believed that water constituted the basic component for all matter. Anaximander (610–545 B.C.) believed in the existence of a certain primary substance—present but undetectable in all known substances—that was eternal in duration and constituted the basis of all things. Heraclitus (535–475 B.C.) believed that transformation itself was the one fundamental reality of the universe, and thought of fire, a symbol of transformation, as a primary substance. Separate from these thinkers was a school of thought that believed the world to be composed of atoms moving throughout the vacuum of space. This atomic theory originated with Leucippus and was adopted by Democritus (460–370 B.C.), who believed that atoms were hard and uniform and existed in only a finite number of varieties. Philosophers of the school of Epicurus believed that atoms were the building blocks of the material world, a philosophy clearly articulated by the Roman poet Lucretius (95–55 B.C.) in his poem *"On the Nature of Things."*[11]

But the view of matter that dominated ancient Grecian thought was the four-element theory of Aristotle (384–322 B.C.) (see Figure 1.1). He believed in the existence of shapeless, undefinable primary substances onto which appropriate properties were grafted to constitute elements. All matter on Earth was thought to be composed of the four elements identified by Empedocles—fire, air, water, and earth—while matter in the heavens contained a fifth element known as ether. On earth, the four elements were thought to be combined in various

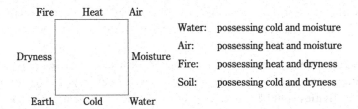

Fire   Heat   Air

Water: possessing cold and moisture

Air: possessing heat and moisture

Dryness   Moisture   Fire: possessing heat and dryness

Soil: possessing cold and dryness

Earth   Cold   Water

Figure 1.1: The four elements of Aristotle and the relationships between them

proportions to comprise the substances around us. The four properties of heat, cold, moisture, and dryness were associated with the four elements, with each element possessing two properties: fire, for example, exhibited heat and dryness, while water exhibited cold and moisture. The Aristotelian elements were not immutable; by discarding one property and incorporating a different property, it was possible to transform one into the other. Although such notions might seem primitive or fantastical to our modern minds, they become somewhat more understandable when we consider the Aristotelian interpretation of the burning of wood in terms of transformations of matter. When a tree branch is burned, the first substance to emerge is smoke, which was deemed to be the element air. As the branch continues to burn, a liquid is produced, which was understood to be the element water. The element fire was thought to dissolve into the atmosphere as a compound of flame and light, ultimately leaving behind only the non-flammable portion of the tree, which was understood to be nothing other than the element earth.[12]

Alchemy

What precisely was alchemy? Colloquially, the term is used to refer to attempts to convert common metals such as lead or tin into precious metals like gold or silver. (The image is captured with particular vividness by the Japanese word for alchemy—renkinjutsu—which quite literally means "the art of refining metals.") However, recent research suggests that the term had a broader significance. Sheppard[1] explains the subject this way:

> Alchemy is a cosmic art by which parts of that cosmos—the mineral and animal parts—can be liberated from their temporal existence and attain states of perfection, gold in the case of minerals, and for humans, longevity, immortality, and finally redemption. Such transformations can be brought about, on the one

1. Quoted from Reference 2, Page 4. This book contains a fairly detailed discussion of alchemy. Sheppard is a British historian.

hand, by the use of a material substance such as "the philosopher's stone" or elixir, or, on the other hand, by revelatory knowledge or psychological enlightenment.

This sort of thinking indicates that the subject of alchemy comprised two types of activities: exoteric, or substantive, pursuits and esoteric, or spiritual, endeavors. These activities could be conducted separately or in tandem. This notion of alchemy encompasses not only the field that originated in Greece and spread to Europe by way of Arabia, but also the fields that burgeoned in India, China, and other civilizations. However, here we will discuss only the former strain of alchemy—the predecessor of the modern chemistry that was later to follow.

This tradition of alchemy originated in the first century B.C. among Greek scholars in Alexandria and quickly spread throughout the entire Mediterranean region. Its birth has been traced to an infusion of Grecian philosophical notions into certain technologies developed by metal-processing tradesmen. The tradesmen, who were attempting to use alloys and plating techniques to make ordinary metals look like precious metals, understood Aristotle's philosophy to justify belief in the possibility of converting ordinary metals to precious metals. These efforts merged with superstitious and religious elements to form the field of alchemy. Many alchemists used alloys of sulfur, mercury, gold, and silver in pursuit of the "philosopher's stone" that would convert dull metals into precious metals, while others attempted to fabricate magical potions known as "elixirs" that would cure all ills and guarantee immortality.

Following the rise of Islam, Greek alchemy spread to parts of the Arabic world under Islamic control, where they blended with elements of Arabic alchemy. Among these was the belief that all metals consisted of sulfur and mercury, and thus that it was possible to turn ordinary metals into precious metals. Later, in the 12th century, alchemy spread from the Arabic world to Northern Europe, where it remained an active focus of research until the 17th century. Even some of the scientists responsible for pioneering the development of modern science—such as Newton and Boyle—devoted considerable time to the study of alchemy.

Needless to say, alchemy proved ultimately fruitless for pursuits such as turning ordinary metals into precious metals and discovering magical elixirs. Nonetheless, the considerable effort devoted to those pursuits brought about progress in methods and equipment for handling chemical substances and made major contributions to the enhancement of mankind's practical knowledge of matter, furnishing a platform atop which modern chemistry was later to develop. Among these, distillation was perhaps the most important technique to emerge from alchemy.

Alchemists used tools—such as experimental dishes, beakers, flasks, mortars, funnels, and crucibles—and a catalog of materials that included, in addition to the

Figure 1.2: A 16th-century distillation apparatus

Cone-shaped metallic distillation chambers were affixed to the upper surface to ensure a wide area of contact with the air, improving the efficiency of the distillation process.

list of seven metals known by the 16th century (iron, copper, gold, silver, mercury, tin, and lead) the new metals zinc, antimony, and arsenic. In addition, alchemists knew of sulfur, sodium carbonate, alum, table salt, and salts including iron sulfate. Observing the progress of alchemy over the centuries, and over the course of its spread from Alexandria to Northern Europe by way of Arabia, one finds major progress in distillation equipment (Figure 1.2), yielding significant advances in techniques for efficiently extracting condensed substances. These advances later enabled the production of acids such as sulfuric acid, nitric acid, and aqua regia.

Iatrochemistry

The practice of medicine systematized by the Greek doctor Galen in the 2nd century A.D. remained dominant across Northern Europe throughout the Middle Ages. However, the situation began to change in the 16th century. The new substances obtained due to progress in distillation methods were seized upon by doctors and changed the face of medicine. Although distillates had been used as medicines already since the 13th century, in the 16th century these were joined by a variety of oil-based and herbal distillation products that were tested for pharmaceutical applications. Their use was systematized by a Swiss doctor generally known as Paracelsus as the field of iatrochemistry. Although Paracelsus spent the majority of his life wandering through various regions of modern-day Germany, he left behind a large number of written contributions to medicine, chemistry, and theology.

Galen (around 129–199)
A doctor in the Roman Empire, Galen rose to fame as the personal physician of Emperor Marcus Aurelius. Using his experience with clinical medicine and autopsy, he systematized Greek medicine based on Hippocrates' theory of *bodily humors.*

Paracelsus (1493–1541)
Paracelsus was a Swiss doctor and alchemist. His full name was Philippus Aureolus Theophrastus Bombastus von Hohenheim, but he was dubbed "Paracelsus" to suggest that his greatness had surpassed even that of the famous Roman doctor Celsus. As a boy, he studied at a mining school in southern Germany, where he was trained as an  analyst in the mineral extraction and processing of various types of metals, and acquired knowledge of metallurgy and chemistry. Starting at the age of 14, he traveled across Europe and spent time at universities in several places, but he ultimately became disillusioned with the academic practice of medicine at universities, believing that traveling and learning various folk remedies constituted a more effective method of medical training. He earned fame as a doctor and in 1527 was appointed as municipal doctor of Basel and a university professor; however, he tended to hold the medical thinking of the time in utter contempt, and his erratic personality and statements did not sit well with medical scholars around him, whereupon he was banished from Basel in less than a year. However, his lectures were popular, his fame spread throughout the world, and he earned many followers. He was the founder of iatrochemistry, which stimulated facets of medicinal chemistry.

Paracelsus believed that alchemy was one of the cornerstones of medicine and that its most important objective—more important than turning lead into gold—should be the creation of medicines. He advocated Aristotle's four-element theory and added the notion of three principles (tria prima): sulfur, mercury, and salt, which was derived from Arabic alchemy. Of course, the three principles of Paracelsus were not the sulfur, mercury, and salt we know today, but rather elements that existed in the core of all matter and which caused the matter to take on various states: Sulfur caused matter to burn, mercury caused matter to liquefy, flow, and evaporate, and salt caused matter to stabilize and solidify. He believed that diseases could be cured by a certain essence known as arcana that could be obtained through methods of alchemy such as

Figure 1.3: Assay balances from the 16th century

distillation and extraction, with particular forms of arcana effective at treating particular diseases; from this hypothesis was born the idea of using inorganic poisons such as sulfur, mercury chloride, and lead acetate as pharmaceuticals. Thus, Paracelsus was the inventor of chemotherapy—the notion that small quantities of poisonous substances could in fact be therapeutic. Although the iatrochemistry of his time never demonstrated adequate results, it promoted cooperation between medicine and chemistry and ultimately stimulated the development of both fields.

The technological legacy of the 16th century

Among the technologies that were to contribute to the development of modern chemistry, refining and metallurgy technologies played a particularly important role. Mining was an important industry in 16th-century Europe, and many books discussing mineral extraction and metallurgy were published; these books tended to be of a practical nature, with little space devoted to theories deemed devoid of practical use. Among the most famous was "*De Re Metallica*", by the German doctor Agricola (1494–1555), who hailed from the region of Saxony. This book includes detailed descriptions of techniques for mineral extraction, refining, metallurgy, and analysis; its copious woodcut illustrations convey an intuitive feel for the technologies of the era. Among the insights we glean from books like this is that significant efforts were devoted to quantitative analysis using balances (Figure 1.3). Of course, the purpose of using balances was a rather mundane one—to determine the quantity of metals that could be extracted from mined rocks—but the pursuit lent the field a quantitative character that had been missing from alchemy. It is believed that the best balances at the time were capable of measurements with a resolution as fine as 0.1 mg. It was also during this era that arsenic, antimony, bismuth, and zinc were recognized as metals.

Technologies developed in the 16th century for handling inorganic substances include methods for fabricating glass, pottery, gunpowder, acid, and salt. For example, the use of cobalt salts to produce blue-tinted glass began in this era—despite the fact that cobalt itself had not yet been discovered. Techniques for producing various kinds of salts—including table salt, soda, alum, and sulfuric acid—were also discussed in the metallurgy books noted above.

Chemistry in the 17th century

From the middle ages through the 16th century, classical notions such as the Aristotelian view of nature and the Ptolemaic picture of the cosmos retained their dominant authority in science; however, a transformation in world views began around the turn of the 17th century. Francis Bacon (1561–1626) distanced himself from the Aristotelian paradigm of syllogism and advocated inductive methods based on facts. Bacon's philosophies served to spur a movement in science away from fruitless methods relying solely on logical thinking. He believed that the role of science was to furnish mankind with new inventions and lives of plenty, a perspective that was adopted by the founders of Britain's Royal Society.[2] On the other hand, René Descartes (1596–1650) insisted, in his "Discourse on Method", that all notions other than obvious truths were to be doubted as a matter of principle. He advocated a deductive method in which complicated problems were to be broken down into the simplest possible constituents; when disparate results were found to follow from identical principles, experiments were required to determine the proper conclusions. Descartes' worldview rejected mysterious or extra-worldly explanations in favor of results based on logical, mechanistic theories; although he espoused a particulate theory of matter, he never arrived at atomic theory due to his denial of the existence of vacuum.

Meanwhile, the atomic theory of the Epicurean school was rejuvenated by the French philosopher Gassendi (1592–1655). In 1649, he proposed that the properties of atoms depended on their shapes, and that in certain cases atoms could bind together to form molecules. He also championed the idea that atoms move through a vacuum devoid of any other matter content. Gassendi's

---

2. The Royal Society was an organization formed, with the permission of King Charles II, to advance the state of science and technology. Although it received the King's blessing, it was a purely citizen-based organization, receiving no financial support from the King or the government.

advocacy of atomic theory was motivated by the successful experiments of Torricelli, who, in 1643, filled a glass tube—with one end closed—with mercury, then inverted this tube in a basin of mercury to create a bubble of vacuum within the tube.

The 17th century witnessed the so-called scientific revolution brought about by the likes of Kepler, Galileo, Newton, and Harvey; while these developments liberated the fields of astronomy, physics, and biology from Aristotelian natural science and Ptolemaic astronomy, as yet there had been no revolution in chemistry. A major reason for this is that, until that point, chemists had worked with such complex mixtures of substances that it was difficult even to identify the component substances they contained. In order for chemistry to make significant progress, it was essential first to develop techniques for isolating pure substances and analyzing them quantitatively, and this process would require a further century's worth of innovation. Nonetheless, mechanistic views of nature, and philosophies that emphasized the importance of experimentation, worked gradually to transform the practice of chemistry.

Although van Helmont (1580–1644) was a student of Paracelsus, he rejected both the four elements of Aristotle and the three principles of Paracelsus, arguing instead that there were two types of matter constituents: (1) water, and (2) a fermenting element that constituted the basis of all activity and organization. The latter gives rise to the various shapes and properties of material bodies. For example, it was believed that the fermenting element acted on water to form soil. van Helmont's major contribution to science was the production, by chemical methods, of gases other than air. He observed that the gas produced by burning coal or alcohol was different from air, and recognized it as identical to the gas produced by fermenting grape juice or pouring vinegar over seashells. He identified the differences between various gases and knew that the same gas could be produced by multiple distinct methods. The science of pneumatic chemistry[3] was also stimulated by von Guericke's invention of the vacuum pump, which heightened interest in gaseous substances. However, the 17th-century scientist who contributed the most to the progress of chemistry was Robert Boyle.

Boyle and the corpuscular philosophy
von Guericke's invention of the air pump in 1654 led to demonstrations of the strength of air pressure and became a topic of widespread interest. Boyle

---

3. Pneumatic chemistry is a term identified to the area of scientific research of the 17th and 18th centuries to understand the physical properties of gases and to study how they relate to chemical reactions.

Robert Boyle (1627–1691)
Boyle was an English natural philosopher, rightly considered a forefather of modern science. Born in Ireland as the 7th son of the Earl of Cork—one of the wealthiest families in England—he was educated at Eton and later studied and acquired experience on the continent of Europe before returning to England in 1644. As a natural philosopher, he followed the pattern of the intellectual elite of the time by being well-versed in many fields of science. From 1655 to 1659, and again from 1664 to 1668, he lived at Oxford, where he mingled with many famous natural philosophers. He was active as a core member of the "Invisible College," the group of scholars that founded the Royal Society.

immediately sought the collaboration of Robert Hooke to build a vacuum pump and conduct research on air pressure and vacuum phenomena. This led to the discovery of Boyle's Law, which held that the pressure of a gas was inversely proportional to its volume. Boyle was a man of extremely devout religious faith; he believed that the purpose of chemistry, in addition to being of service to medicine or industry, was to deepen mankind's understanding of God and the natural world of God's creation. In 1661, he published "*The Sceptical Chymist*", in which he criticized Aristotle's four-element theory and the three principles theory of Paracelsus, and explained his views of elements this way[13]:

> I now mean by elements, [omission], certain primitive and simple, or perfectly unmingled bodies; which not being made of any other bodies, or of one another, are the ingredients of which all those called perfectly mixt bodies are immediately compounded, and into which they are ultimately resolved.

The elements to which Boyle refers here are not the elements of modern chemistry, but rather some sort of primary substance from which he believed all matter to be made. As a believer in corpuscular theory, he attempted to explain the properties of matter in terms of differences in the collective behavior of primary particles. Such a corpuscular theory successfully explained many chemical phenomena and sufficed to refute vague notions such as the four-element theory of the Aristotelian school or the three principles of Paracelsus; however, its ingredients do not correspond directly to our modern chemical elements. It was also during this era that Robert Hooke developed the microscope and used it to demonstrate the existence of a microscopic world invisible to the naked human eye[4]—a backdrop against which it became easier to accept corpuscular theories that assumed the existence of particles too tiny to be seen by eye.

One of Boyle's contributions was to demonstrate the efficacy of chemistry in the service of natural philosophy. Until that point, natural philosophers had frowned upon chemistry as the dubious practice of alchemists; Boyle sought to demonstrate the importance of focusing on chemical phenomena to obtain a proper understanding of mechanistic theories. Although Boyle was a natural philosopher, he was viewed by chemists as a colleague, and natural philosophers came to see him as a chemist worthy of respect.

Another of Boyle's major contributions to chemistry was the research he conducted using vacuum pumps to study the phenomena of burning and breathing. In collaboration with Hooke, Boyle investigated the behavior of combustibles under a variety of conditions and confirmed that burning did not proceed in the absence of air, thus verifying that air was somehow involved in the burning of many substances. He also discovered that heating metals in air caused their mass to increase, which he attributed to absorption of fire particles. To investigate breathing, Boyle conducted experiments involving small birds and mice, from which he concluded that the lungs play the role of extracting impurities from air and expelling them from the body. Boyle believed that air was composed of three types of particles; one of these was the true constituent of air, while the other two were blended with the first in small quantities. Burning and breathing were believed to be caused by these other two particles. Subsequently, Hooke and Mayow conducted independent research on burning and breathing, but their studies were also unable to elucidate the true mechanisms of either phenomenon.

The Phlogiston Theory

While the progress of pneumatic chemistry remained stunted, the phlogiston theory—which developed in the second half of the 17th century in German-speaking lands—offered a powerful framework for explaining the phenomenon of burning. The mining industry was booming in Germanic lands at that time, a backdrop that stimulated interest in the formation of minerals. In 1667, Johann Becher (1635–1682) believed that minerals formed from earth and water. He proposed a theory in which all substances were composed of three types of earth: a liquid earth [responsible for the fluidity, density (or diluteness), and metallic

---

4. Hooke's Micrographia, which revealed to the world the existence of the microscopic world he had gleaned from microscope experiments, was published in 1665. Robert Hooke (1635–1703) is famous for Hooke's law governing the elongation of elastic bodies, as well as for the discovery of plant cells.

nature of materials], an earth with the properties of oil (responsible for oiliness, sulfur-like characteristic, and combustibility), and an earth with the properties of rock (responsible for the ability of materials to mix with each other). This notion later formed the basis of the phlogiston theory proposed in 1718 by Becher's protégé, Georg Stahl (1659–1734). Although Stahl was an atomist, he believed that materials were composed of four types of particles—Becher's three types of earth plus water—and he referred to the oil-like earth as phlogiston. These four types of earth were thought to bind together through affinity for water, or through adhesive forces, to form secondary substances, which corresponded to the elements of modern chemistry. In Stahl's picture, all flammable substances contained phlogiston, which was released to the atmosphere and lost during the burning process. When metals were heated, they would give up their phlogiston, forming metal ash (metallic oxides). Similarly, when metal ash and coal were heated together, the metal ash would transfer its phlogiston to the coal to form metal, while burning sulfur would result in the production of universal acid[5] and phlogiston.

Stahl's phlogiston theory elegantly explained much of what was known at the time about the phenomenon of burning. The assumption that a limited amount of air could only absorb a certain limited amount of phlogiston explained why, in a closed vessel, the burning process would eventually stop: the air would become saturated with phlogiston. Moreover, the assumption that materials only contained a limited amount of phlogiston explained why materials eventually stop burning: their store of phlogiston would be exhausted. Many chemical phenomena were given compelling qualitative explanations by the phlogiston theory. Ignoring the gaseous byproducts generated by the burning of organic matter, it was believed at the time that burning always caused a decrease in mass. Although examples were known in which the burning of a flammable body in air caused its mass to increase, these were not taken particularly seriously. It was not until more quantitative research methods became established in chemistry that the existence of bodies whose masses increase upon burning would come to pose a serious challenge to the phlogiston theory. Thus, for a period of some 60 years—until almost the end of the 18th century—the phlogiston theory remained dominant in chemistry as the most successful available theory of burning.

1.2 The Development of Pneumatic Chemistry

The study of gases made little progress for some 100 years after the second half of the 17th century, largely because methods had not yet been developed for

---

5. Stahl used the term "universal acid" for the acid produced by burning sulfur, which he thought of as a mixture of phlogiston with an acidic primary substance.

preparing or quantitatively analyzing pure samples of various types of gases. Although the dazzling successes of Newtonian mechanics exerted an impact on all branches of science and gave a boost to the mechanistic viewpoint, these developments did not lead immediately to progress in chemistry. Indeed, the phenomena studied by chemists remained so complicated that the pursuit of truth through simple mechanistic theories proved elusive, and efforts to move beyond the phlogiston theory remained unsuccessful until near the turn of the 19th century. Eventually, however, the industrial revolution that began in the 18th century—and particularly the development of steam engines and their applications, which so captivated the attention of scientists and engineers—spurred a phase of active research in pneumatic chemistry.

An experimental technique that contributed significantly to progress in pneumatic chemistry was the pneumatic trough method of capturing gases above liquids, introduced by the English botanist and minister Stephen Hales (1677–1761). Hales, who was educated at Cambridge and was heavily influenced by Newton, believed that the leaves of plants absorb air and retain it in solid form. The solidified air was released by heating, and Hales invented a simple device to capture and measure the volume of this gas. The device consisted of an inverted vessel held in position over a dish of liquid to capture gases above the liquid surface, and using this device, Hales investigated many living organisms in the years around 1730. His main interest was in measuring gas volumes, but his device was to play a major role in advancing research on gases conducted by later chemists; indeed, carbon dioxide, oxygen, and hydrogen were all discovered in short order in the second half of the 18th century.

Black: the discovery of carbon dioxide and quantitative studies

The first major contributor to 18th-century research on pneumatic chemistry was Joseph Black. In the 1750s, Black—who was conducting medical research—became interested in carbonates through studies of the alkaline salts that control stomach acid. Starting with epsom salts ($MgSO_4$) and potassium carbonate (K_2CO_3), he produced magnesia alba ($MgCO_3$) and studied its properties in depth, discovering that the decrease in mass observed upon heating magnesia alba was due to the release of a certain gas. This gas exhibited properties different from those of ordinary air, but was identical to the gas obtained by burning limestone ($CaCO_3$); Black referred to it as "fixed air". The discovery of fixed air was an epochal breakthrough, casting aside the traditional notion that all gases were identical (namely, air). Black also noticed that the lime obtained from burning limestone reacted with water to form lime hydrate [$Ca(OH)_2$], while lime hydrate reacted with potassium carbonate to recover limestone. Moreover, he determined that fixed air was a component of air, and that capturing exhaled

Joseph Black (1728–1799)
Black was an English chemist and doctor who served as Professor of Medicine at Glasgow University. Born in Bordeaux as the son of a Scottish wine merchant, he was educated at the Universities of Glasgow and Edinburgh; it was in Glasgow that he attended lectures given by Cullen and became interested in chemistry. The research he conducted for his thesis at Edinburgh medical school led  him into a subsequent course of research that was eventually to culminate in the discovery of carbon dioxide. His primary contributions to chemistry were the discovery of carbon dioxide and the introduction of quantitative research methods based on mass measurements.

breath in an aqueous solution of lime hydrate caused the liquid to become cloudy. In 1756, Black became a lecturer in both medicine and chemistry at Glasgow, and although his lectures were extremely popular, he made no further outstanding contributions to chemistry. After 1760, he turned to studies of heat and temperature, a shift no doubt motivated by the interests of the times. At that time, steam engines were a focus of intense attention, and many scientists were interested in the properties of heat. Black conducted quantitative studies of dissolution and evaporation phenomena in water; he measured the heats of solution and vaporization, and introduced the notion of specific heat. However, he never presented these results in books or journal articles. Also notable is the fact that the apparatus Black used in his research on vaporization was constructed by none other than James Watt, who was later to earn fame as the inventor of new steam engines. Black and Watt maintained close ties for many years.

Priestley and the discovery of oxygen

The single individual who isolated and studied more new gases than any other scientist in the history of chemistry was an English clergyman and amateur chemist named Joseph Priestley. He conducted systematic studies of nitric oxide, hydrogen chloride, ammonia, sulfur dioxide, silicon tetrafluoride, oxygen, and other gases, investigating properties such as the solubility of these gases in water, their ability to sustain or extinguish flame, their effect on breathing, and their behavior with respect to hydrogen chloride and ammonia. But his most important contribution was the discovery of oxygen. In 1774, he placed red mercury ash (HgO)—obtained by heating mercury in air—in a gas capture vessel filled with mercury, then heated the mixture using a 12-inch lens to focus sunlight. Gas was produced, and the mercury ash reverted to mercury. The gas released did

Joseph Priestley (1733–1804)
The son of a textile craftsman, Priestley lost his mother as a young boy and was raised by his aunt; he was educated at an academy for training Dissenting (that is, not adhering to the Church of England) clergymen. He taught at several schools and even wrote a textbook of English grammar. Later, he became a teacher at Warrington Academy, a school for Dissenters, where he wrote history books. On a trip to London  he met Benjamin Franklin and was inspired to write "*The History and Present State of Electricity*"; he also began to conduct his own chemistry experiments. In 1767, he became Minister of the Unitarian Church in Leeds, and it was here that he began his experiments in pneumatic chemistry (see Column 1).

Carl Scheele (1742–1780)
Scheele was a Swedish chemist. Born the son of a German merchant, he became an apprentice to a pharmacist in Gothenburg, where he became interested in chemistry; later, he continued to conduct chemistry experiments while working as a pharmacist in various regions of Sweden. He  had first-rate analytical techniques, conducting original and ingenious experiments in rather difficult circumstances and leaving behind a wide variety of accomplishments. His most widely known achievements are his discovery of oxygen and chlorine.

not dissolve in water, and candles immersed in it burned more brightly, while the red-hot ash burned brightly as well. To investigate the impact of this gas on breathing, Priestley placed mice in a vessel filled with this gas and found that these field mice lived longer than field mice in a vessel filled with ordinary air. Thus, he had discovered a new gas—oxygen—whose ability to sustain burning and breathing exceeded those of ordinary air. However, Priestley—a firm believer in the phlogiston theory—refused to acknowledge the true nature of this new gas, insisting instead that it was simply "dephlogistonated air," consisting of air from which a large quantity of phlogiston had been removed.

Scheele and the discovery of oxygen

As it happens, there was one other scientist who independently discovered oxygen slightly before Priestley. This was Sweden's Carl Scheele. He believed that air was composed of two types of gas; one of these sustained burning and breathing, while the other did not. By heating metal ash, he obtained a gas that sustained burning (oxygen), which he termed "fire air." Although Scheele wrote

Henry Cavendish (1731–1810)

Although Cavendish was born to a wealthy family, he grew up not knowing his mother, who died when he was just 2 years old—a factor that may have been relevant in shaping his unusual personality. He began research activities under his father, but when his father died, he inherited a massive estate and immersed himself in a life of scientific research. He was extremely reclusive and almost never left the home except to attend scientific meetings; indeed, he almost never spoke even to his servants, preferring instead to conduct all affairs through written notes. His research was quantitative and achieved astonishing degrees of precision compared to the prevailing standards of experimental accuracy in his time.

down this observation in 1773, it was not published in a book until 1777, and thus was not generally known at the time of Priestley's discovery of oxygen. Like Priestley, Scheele was an adherent of the phlogiston theory and lacked a proper understanding of the role of oxygen in burning. Although he obtained chlorine by applying hydrochloric acid to magnesia (magnesium oxide), he thought of this as "dephlogistonated hydrochloric acid." However, using his superior analytical techniques Scheele made major contributions to chemistry, discovering many inorganic substances (including molybdenite and graphite), inorganic acids (including arsenic acid and molybdenum acid), and organic substances (including tartaric acid, oxalic acid, lactic acid, and casein).

Cavendish and the discovery of hydrogen

In 1766, Henry Cavendish discovered hydrogen in the form of a "flammable air." Although flammable gases had been observed before, they were typically mixed with gases such as carbon monoxide or hydrocarbons and thus not clearly identified. Cavendish produced hydrogen by applying dilute sulfuric or hydrochloric acid to zinc and iron, and discovered that its density was far smaller than that of air. Attempting to interpret his experimental results on the basis of the phlogiston theory, he believed the "flammable air" to be phlogiston itself. In 1783, he reported that burning "flammable air" (hydrogen) in Priestley's "dephlogistonated air" (oxygen) resulted in the formation of water. He also pioneered the use of mercury instead of water for the capture of soluble gases.

Although Cavendish achieved many impressive research results spanning a wide range of areas in physics and chemistry, only a few of these were ever made public during his lifetime. Some 70 years after his death, his unpublished

COLUMN 1

Priestley and the quest to fuse science and theology

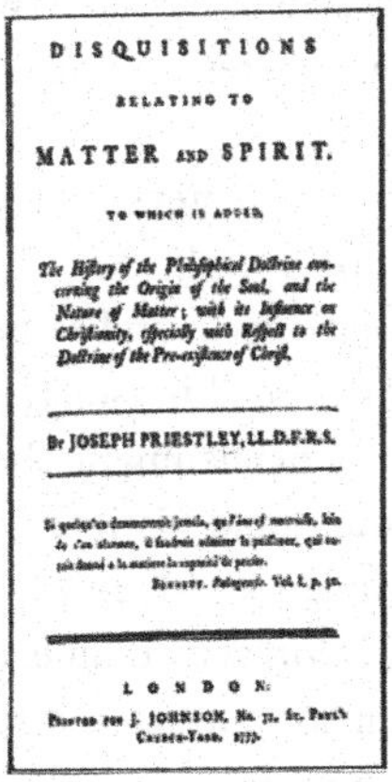
DISQUISITIONS
RELATING TO
MATTER AND SPIRIT.
TO WHICH IS ADDED,
The History of the Philosophical Doctrine concerning the Origin of the Soul, and the Nature of Matter; with its Influence on Christianity, especially with Respect to the Doctrine of the Pre-existence of Christ.

By JOSEPH PRIESTLEY, LL.D.F.R.S.

[illegible]

LONDON:
PRINTED FOR J. JOHNSON, No. 72, St. PAUL'S CHURCH-YARD. 1777.

The cover of Priestley's "Disquisitions Relating to Matter and Spirit."

In the 18th century, there were as yet no professional scientists; instead, all who pursued research in the sciences were amateurs. These amateurs had a variety of motivations for the efforts they devoted to scientific studies. For Joseph Priestley, whose place in history was secured by his discovery of oxygen, science was an essential component of theology, serving as the essential bridge between Christianity and the rationalism of the Age of Enlightenment. Priestley was a fascinating individual who, despite living a tumultuous life in a time of political volatility, managed to remain active in an astonishingly large number of roles: clergyman, theologist, educator, scientist, inventor, natural philosopher, and political theorist.

Priestley, who was fascinated by natural philosophy, first became interested in electricity—a subject in which he conducted experiments and wrote books—before beginning research in pneumatic chemistry around the time he became Minister of the Unitarian Church of Leeds in 1767. Perhaps because he lived next door to a brewery, he began to study the properties of the carbon dioxide gas (CO_2) emitted from the surface of fermenting liquids. The idea of infusing water with carbon dioxide to form soda is also an invention of Priestley's that dates from this time. In 1772, Priestley accepted a position as the personal librarian and family tutor of Lord Shelburne, which offered him more time to pursue research; he carried out a wide range of studies of gases and discovered a number of new gases, of which oxygen was the most important. His activities as a chemist were more productive during these years than at any other time in his life. He also presented a number of important philosophical works during this period.

In 1780, Priestley moved to Birmingham, where he became active in the "Lunar Society," a group of new industrialists and intellectuals. Although he published numerous scientific papers in this intellectually stimulating environment, he was never able to accept Lavoisier's theories and defended the phlogiston theory throughout his life. On the other hand, despite his publication of many works in theology, he was deemed a radical critic of English politics and its discrimination against Dissenters; in 1791, he was the victim of violent attacks sanctioned by Parliament and the King, in which his home and chapel were destroyed and he was forced to flee to London. Later, in 1794, he emigrated to America, where he died in 1804. Although he conducted almost no chemistry research in America, his presence seems to have contributed to stimulating interest in chemistry among Americans. The highest honor awarded by the American Chemical Society—the Priestley medal—commemorates his life's work.

As a chemist, Priestley excelled in the technical inventiveness and creativity of his experiments; in just a short period of time he succeeded in isolating many gases and made major contributions to the advancement of pneumatic chemistry. However, his experiments were of a qualitative nature, and his interpretation of experimental results was somewhat simple-minded and lacking in deep insight. He favored qualitative observations of heat, color, and volume, and was uninterested in quantitative analysis of chemical transformations, a disposition that ensured that he was somewhat unable to keep up with new trends that emerged in chemistry during his lifetime. Although Priestley was radical as a political theorist, he was conservative as a chemist; he remained a staunch believer in the phlogiston theory until the end of his life, and in his final days he became estranged from his chemist friends. But why did Priestley retain such a stubborn insistence on the phlogiston theory? One possible explanation is that, in modern terms, Priestley was more a natural philosopher than a scientist; he understood his mission to be the unification of theology with the dominion of the natural world, and this philosophy prevented him from accepting the new theories of Lavoisier.

Priestley published over 100 works of theology during his lifetime, including "*Disquisitions Relating to Matter and Spirit*", which set out his own personal philosophy of materialism. He believed that everything in the universe was composed of observable matter, but that the human soul was made of a divine substance and thus unobservable by mankind. He also believed that mankind progressed by obtaining a proper understanding of nature, and that these efforts would bring about the Millennial Kingdom of Christ. He rejected the dualism of Descartes—which attempted to separate mind and matter—and attempted to merge theism with materialism and determinism. His ideas on utilitarianism are also thought to have influenced later philosophers such as Bentham, Mill, and Spencer.

papers were organized and published by Maxwell, revealing that he had predated many other scientists in making a number of important discoveries. In 1871, Cambridge University established the "Cavendish Laboratory" in his honor; as of September 2013, this research institution had produced 29 Nobel laureates.

1.3 Lavoisier and the Chemical Revolution

The studies of pneumatic chemistry pursued with such vigor after the mid-18th century had established that air was not elemental, but was rather composed of at least two different types of gases: one of these sustained burning and breathing, while the other did not. It had also become recognized that the gas phase was one of the states in which matter existed, and that there were many types of gases. However, chemists were still attempting to understand chemical phenomena within the framework of the phlogiston theory. Although chemists like Priestley and Scheele were outstanding experimentalists who made many new discoveries and contributed significantly to the development of pneumatic chemistry, they were not particularly gifted in constructing theories or systematizing knowledge. For chemistry to enter the modern era, it was necessary for chemical phenomena to be studied in detail—using the methods of quantitative analysis pioneered by Black—and for the results of these studies to be organized in a systematic fashion. These goals were advanced by Antoine Lavoisier, and the transformation of the chemical sciences he initiated—which came to be known as the "chemical revolution"—marked the birth of modern chemistry.

Lavoisier and new theories of combustion

Lavoisier's interest in the phenomenon of burning appears to have begun sometime around 1770. His first experiments investigated the burning of diamond. When diamond was placed in a closed vessel floating in water and heated by intense light focused by a lens, both the mass of the diamond and the volume of air in the vessel decreased. Next, Lavoisier investigated the burning of phosphorus and discovered that this material, when burned in air, exhibited an increase in mass. It was already known that the calcination[6] of metals caused their mass to increase, and this phenomenon was a major weakness of the phlogiston theory. Lavoisier recognized the seriousness of the theoretical problems posed by combustion and devoted himself wholeheartedly to the pursuit of a resolution. He followed the experiments of Black by reducing red

---

6. Calcination: The process of heating a material in air to remove volatile components, leaving behind an ash-like substance.

Antoine-Laurent de Lavoisier (1743–1794)
Lavoisier was born the son of a wealthy lawyer in Paris. Expecting to enter the legal world, he was educated at the College Mazarin, where he developed an interest in chemistry. He later received a law degree from the University of Paris. While enrolled at the College Mazarin, he learned chemistry from a series of general chemistry lectures delivered by Rouelle in a lecture hall at the Royal Garden. He displayed his formidable abilities during a collaboration in 1766 with the geologist Guettard, while his paper on lighting for large cities exhibited razor-sharp analytical skills and was awarded a prize by the Academy of Sciences. In 1768, he became an assistant at the Academy and was soon recognized as a scientist. He laid the foundations of modern chemistry by organizing the subject on the basis of quantitative experiments. (See Column 2).

lead oxide in coal to obtain "fixed air," recognizing that the burning of tin or lead caused some air to be absorbed. Similarly, when phosphorus was burned in a closed vessel filled with air and floating in water, the volume of the air decreased to 4/5 of its original volume and the mass of the phosphorus increased. On the other hand, when phosphorus was heated in vacuum, it sublimated but exhibited no other phenomena.

In 1774, Lavoisier began experiments on the burning of mercury ash. The gas obtained by heating mercury ash was not "fixed air," but instead exhibited properties similar to that of ordinary air. This gas did not dissolve in water, did not cause birds to suffocate, and did promote burning. At this point, Lavoisier was thinking of this gas as a simpler form of air; however, upon reading Lavoisier's reports, Priestley pointed out that the gas was nothing but Priestley's "dephlogistonated air." Lavoisier then conducted further experiments to produce mercury ash from mercury and air, measuring both the volume of air that was removed and the increase in mass of the mercury. Next, he heated the mercury thus obtained, capturing and investigating the gas produced; he found this gas to exhibit far greater ability than ordinary air to sustain burning and breathing. This gas was clearly identical to the gas termed "dephlogistonated air" by Priestley and "fire air" by Scheele. In regard to this situation, Lavoisier wrote the following[14]:

> The principle which unites with metals during calcination, which increases their
> weight and which is a constituent part of the calx is: nothing else than the healthiest

and purest part of air, which after entering into combination with a metal, [can be] set free again; and emerge in an eminently respirable condition, more suited than atmospheric air to support ignition and combustion.

Burning carbon in this gas produced a weak acid and carbon dioxide; in general, burning non-metals in this gas seemed to produce acidic byproducts. Thus, Lavoisier named this substance "oxygen," meaning a substance that forms acid.

Lavoisier also furthered the studies of burning and breathing, investigating the relationship between the two. After studying breathing by animals, he concluded that inhaled air was either converted to "fixed air" in the lungs or else was absorbed by the lungs and displaced by "fixed air." Later, he proposed a calorimeter that measured the heat produced by chemical transformations; one of his experiments involved placing a guinea pig (a marmot) inside this calorimeter and measuring the quantity of heat produced by allowing the creature to breathe for 10 hours. At the same time, he measured the quantity of carbon dioxide produced by the breathing and compared it to the quantity produced by burning carbon. In this way, Lavoisier demonstrated that breathing was in fact a type of chemical process in which oxygen is burned inside an animal's body to form carbon dioxide. He recognized that air consisted of one component that was active in combustion and breathing and another component that was inactive in these processes; the inactive component, which we know today as nitrogen, he termed azote.

Another of Lavoisier's important contributions was his research into the composition of water. Priestley had known that the combustion of airborne hydrogen via electrical sparks produced mist. Watt and Cavendish also had similarly produced mist from hydrogen and oxygen. Cavendish subsequently demonstrated that this mist exhibited the properties of water. However, as noted above, Cavendish was a firm believer in the phlogiston theory, and he thought of hydrogen as simply water with phlogiston added, while water with phlogiston removed was oxygen. Lavoisier demonstrated that burning hydrogen and oxygen together in a tightly sealed vessel produced water. He also showed that passing water vapor over the surface of red-hot iron caused the water vapor to dissociate, producing hydrogen.

The overturning of the phlogiston theory and the beginnings of new chemical theories

As Lavoisier came to understand the role of oxygen in chemical phenomena such as combustion, metal ash, acids, and the composition of water, he became increasingly suspicious of the phlogiston theory. In a paper sent to the academy in 1783, "Reflections on phlogiston," he advocated for a rejection of the phlogiston picture.

In this paper, Lavoisier analyzes various problems from a logical perspective and demonstrates that, in each case, the underlying phenomena were well explained by oxygen, while explanations based on the phlogiston theory were excessively complicated and even ridiculous. He also criticized the chemists of the time for turning phlogiston into an undefinably vague primary substance:[15]

> Chemists have made phlogiston a vague principle, which is not strictly defined and which consequently fits all the explanations demanded of it. Sometimes it has weight, sometimes it has not; sometimes it is free fire, sometimes it is fire combined with an earth; sometimes it passes through the pores of vessels, sometimes they are impenetrable to it. It explains at once causticity and non-causticity, transparency and opacity, colour and the absence of colours. It is a veritable Proteus that changes its form every instant!

Lavoisier's declaration was immediately accepted by Black, but the famous English chemist Kirwan staunchly defended the phlogiston theory, and Priestley retained his dogged belief in phlogiston until the end of his days. Within France, by around 1787 a number of influential chemists—including Berthollet, Guyton de Morveau, and Fourcroy—had accepted Lavoisier's thesis, and they worked with him to disseminate new theories. One of the supporters of the new theories was Berthollet (1748–1822), a talented chemist with passionate interests in education and industrial applications of chemistry. He invented bleach and contributed to its adoption by the textile industry. He was a colleague of Lavoisier at the Academy. Fourcroy (1755–1809), the youngest of the three, became a professor at the Jardin du Roi (Royal Gardens) in 1784 and became the primary ambassador to the younger generation, responsible for explaining Lavoisier's theories to budding scientists. Later, he wrote the ten-volume *"General System of Chemical Knowledge,"* which contributed significantly to spreading the new theories. The oldest of Lavoisier's allies was Guyton de Morveau (1737-1816), an experimentalist who demonstrated that many metals exhibited increased mass upon calcination in air; he evolved from a devotee of the phlogiston theory to a proponent of Lavoisier's theories. de Morveau was concerned by his observation that, until that time, chemists and pharmacologists had been using different nomenclature; although the terminology of botany and zoology had been revised by Linné, the language used for chemistry was disorganized and chaotic. Thus, in 1782, de Morveau proposed to systematize the jargon of chemistry. In this effort, he was joined by Lavoisier, Berthollet, and Fourcroy, who attempted to codify chemical terminology in the framework of Lavoisier's theories;

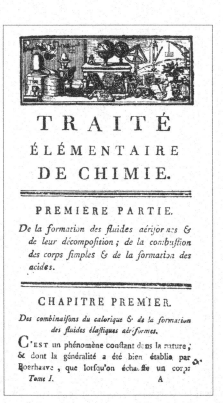

De la formation des fluides aériformes &
de leur décomposition ; de la combustion
des corps simples & de la formation des
acides.

Figure 1.4: The title page of the first edition of Lavoisier's "Elementary Treatise on Chemistry"

the efforts resulted in the publication, in 1787, of the 300-page *"Methode de nomenclature chimique" (Method of Chemical Nomenclature)*. The most important contribution of this dictionary was the set of names it assigned to indivisible substances—elements—a convention that formed the basis of the entire nomenclature system.

The final effort Lavoisier directed toward disseminating the new chemistry was his publication of the textbook *"Traite Elementaire Chimique" (Elementary Treatise on Chemistry)* (Figure 1.4). This textbook, first issued in 1789, remained the standard textbook of chemistry education for the next several decades. It was here that Lavoisier defined chemical elements to be substances that could not be subdivided by chemical techniques. This was a practical and operational definition, which allowed for the possibility that, as analytical techniques progressed, one era's elements would provide divisible—and thus not elements—

Figure 1.5: Lavoisier's table of the elements

In the center column, the elements are classified into four groups, separated by parentheses. The first group consists of elements commonly encountered in the natural world: light, caloric, oxygen, nitrogen, and hydrogen. The next group contains the six non-metals, the third group contains the 17 metals, and the fourth group contains the five types of earth.

at a later time. Lavoisier listed 33 substances as elements (Figure 1.5), although his tally included light and heat (caloric) as well as some substances that were later found not to be elements.

The first part of the "Elementary Treatise on Chemistry" treats heat, gases, and the oxygen theory. The heat and light generated by combustion were interpreted as caloric liberated from oxygen. The book goes on to discuss the formation and properties of elements, including oxides, acids, and salts produced by basic oxides. The principle of the conservation of mass—that the total mass of a system does not change under chemical reactions—is clearly proposed. (Of course, this principle had been implicit in the earlier work of Black and Cavendish, and had been proposed in 1748 by the Russian chemist Mikhail Lomonosov.) The text includes a detailed description of the tools and experimental methods of chemistry. The 33 substances listed as elements include items that Lavoisier suspected to be oxides of metals that were unknown to him—including magnesia, baryta, and alumina—as well as compounds of

COLUMN 2

A Giant of Chemistry—and a Gifted Bureaucrat: Lavoisier and his Remarkable Wife[7]

A portrait of the Lavoisiers

In studying the lives of the great chemists, one finds a surprising number of them to have demonstrated surpassing abilities in other fields entirely unrelated to chemistry, and Lavoisier must without question be counted among their ranks. Although Lavoisier had inherited an enormous estate from his father and was a very wealthy man, in 1768 he attempted to entrench his financial stability further by purchasing a share in the *Ferme générale*, a private-sector institution that collected taxes on goods such as tobacco, salt, and various imported products, and remitted a fixed annual sum to the government. However, this tax-collection scheme was rife with corruption, and its tax collectors were widely despised by the populace; although Lavoisier's motives were purely financial, his decision to participate in the system would eventually prove a matter of fatal significance.

In 1771, the 28-year-old Lavoisier married 13-year-old Marie-Anne Pierrette Paulze, the daughter of a powerful tax collector. The motivation for this 13-year-old girl to marry the much older Lavoisier was to escape a political marriage to an impoverished 50-year old member of the nobility; her father chose Lavoisier—a competent and wealthy subordinate—to be his daughter's groom. Still, the marriage was successful for both parties. Marie-Anne was a highly talented woman; upon coming to appreciate that her husband was a great chemist, she learned chemistry herself and assisted her husband's research as an experimental aide, data recorder, and secretary. She learned English and translated chemistry books in English for the benefit of Lavoisier, who could not read the language; she also studied with painters to learn drawing, and contributed the drawings of experimental apparatus that fill her husband's publications. She upheld masterfully the duties of hostess to the *salons* at which Lavoisier would congregate with other scientist colleagues. She is an

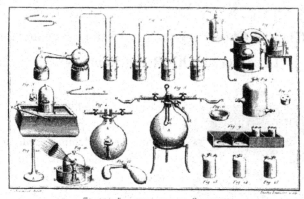

TRAITÉ ÉLÉMENTAIRE DE CHIMIE

A diagram of an experimental apparatus drawn by Marie-Anne Lavoisier and print-ed in the "Elementary Treatise on Chemistry"

outstanding example of a woman who made significant contributions to the advancement of science at a time when women had little opportunity to receive high-level education or perform independent research.

Lavoisier not only left behind a legacy of outstanding achievements in chemistry, but also exhibited superior abilities as a high-level bureaucrat. He was elected to the Academy of Sciences in 1772 and was promoted to full member in 1778. The Academy was a public-sector institution that advised the government in a consulting role; its full members received government stipends and were considered high-level government officials. While a member of the Academy, Lavoisier worked for the state as an administrator of the *Ferme générale* and a supervisor of the Royal Gunpowder and Saltpeter Administration. He would rise each morning at 5 AM, conduct chemistry experiments between 6 and 9 AM, spend the morning in the tax-collection office, spend the afternoon in the Gunpowder Administration, make an appearance at the Academy, then return to his laboratory after dinner to conduct further experiments between 7 and 10 PM. As a member of the Academy, he prepared reports on a variety of topics; among the problems he considered were the Paris water supply, improvements to prisons and hospitals, hypnotism, hydrogen balloons, bleach, ceramics, the manufacture of gunpowder, dyes, and glass production. In the field of agriculture, he operated an experimental farm and pursued methods of increasing crop yields. In his later years, he advocated for adoption of the metric system. As a tax collector, he improved methods of collecting the *gabelle* (a tax on salt) and designed a wall around Paris to prevent unauthorized entry.

Lavoisier also made outstanding contributions as a minister of government finance. When the revolution broke out, the *Ferme générale* was abandoned, and the National Assembly attempted to switch to direct taxes based on land and buildings. However, nobody knew the total amount of government revenue. To address this difficulty, the government asked Lavoisier to design the foundations of a new tax system. He submitted a report—based on population statistics research—titled "The Territorial Wealth of the Kingdom of France," in which he proposed methods of assessing the nation's riches. These proposals involved statistical studies of French agricultural resources and were characterized by a striking degree of originality. Lavoisier applied the same quantitative approach that he used in his scientific studies to solving societal and economic problems.

Nonetheless, despite his major contributions to French society and his peerless achievements in scientific research, Lavoisier was ultimately caught up in the storm of the French Revolution; the Reign of Terror deemed collectors of the *Ferme générale* to be prime targets of popular outrage, with Lavoisier considered no exception. He was falsely accused of mixing tobacco with water and other ingredients to derive illegal interest, thus embezzling funds that should have been remitted to the government, and—together with his tax-collector colleagues—was arrested, charged, convicted, and guillotined in Revolution Square on May 8, 1794. The judge is said to have insisted—in responses to the attempts of Lavoisier's friends to defend him—that "The Republic needs neither scientists nor chemists; the course of justice cannot be delayed"—to which the mathematician Lagrange responded that "It took them only an instant to cut off this head, and one hundred years might not suffice to reproduce its like."

After Lavoisier's execution, Marie too was arrested and held for 65 days; she lost the entirety of her assets, including her land, home, furniture, experimental equipment, farm, and second home. However, after the end of the Terror she launched a successful movement to oust Dupin—a member of the National Convention who had been her husband's accuser—after which she recovered her possessions. Later, working independently, she prepared a second volume of the *Elementary Treatise on Chemistry* and published it in 1803. She continued to hold salons at her home, as she had done previously, and eventually remarried, to Count Rumford, one of the salon participants (see Chapter 2, pages 52, 97). However, the proud Marie insisted on retaining her first husband's name, using the title "Countess Lavoisier de Rumford;" this was a state of affairs the Count found intolerable and which ultimately ensured that the marriage was short-lived.

> **Mikhail Lomonosov (1711–1765)**
> Lomonosov was a gifted Russian scientist and writer who was active in an astonishingly wide range of fields of science and literature. In his scientific work, he proposed the law of conservation of mass, originated the kinetic theory of gas molecules, hypothesized the existence of an atmosphere on Venus, and predicted the existence of a continent at the South Pole. His achievements remained unknown outside Russia for many years after his death.

unknown non-metallic substances, such as radical fluorique[7]. The question of what defined an element remained uncertain in Lavoisier's time, and indeed would continue to puzzle chemists for many years. Nonetheless, the "Elementary Treatise"—in contrast to the chemistry books that preceded it—was written logically and systematically, was easy to understand, and was a major stimulus to the advancement of the new science of chemistry.

1.4 Chemistry and Society in the 18th Century

The beginnings of the chemical industry

The chemical industry began to develop in the mid-18th century, with its first primary objectives being the production of sulfuric acid and alkali. Sulfuric acid was in demand for metal processing, as an ingredient for hydrochloric acid and nitric acid, and as an additive for dye; at the beginning of the 18th century it remained an expensive commodity. In 1736, the English pharmacist Ward burned a mixture of sulfur and saltpeter atop a large glass bottle containing a small amount of water; this produced sulfuric acid at reduced cost. Then, in Birmingham in 1746, Roebuck developed the lead chamber process, in which the glass vessel was replaced by a large container coated with lead; this further advanced the state of sulfuric-acid production technology. The lead chamber process was quickly replicated—and further improved—in France, and by the end of the 18th century sulfuric acid had become inexpensive.

Demand for alkali grew in the second half of the 18th century as an agent for producing glass, soap, paints, and other products, as well as bleach. In particular, France was an active center of glass production and contributed to the increasing demand for lye (potassium carbonate) and soda (sodium carbonate). Increasing

---

7. Radical fluorique: Although the fluoride produced by adding sulfuric acid to fluoric had been known since the 19th century, fluorine itself was not discovered until almost the end of the 19th century.

demand for soap also created the need for inexpensive supplies of the alkali used in its manufacture. In 1775, the French Academy of Sciences offered a prize to anyone who could develop a method for producing high-quality soda from table salt. In response, Leblanc developed a technique in which sodium sulfate—formed from table salt and sulfuric acid—reacts with chalk (calcium carbonate) to form soda. In 1791, Leblanc was granted a patent by King Louis the Sixteenth and built the first soda factory; however, he was soon caught up in the turmoil of the French Revolution, his factory was seized, and his patent became a public asset. Despite this personal tragedy, factories based on Leblanc's method were built in France and England and spurred the development of the soda industry.

The spread of textiles during the industrial revolution created demand for progress in bleaching techniques. The bleaching effects of chlorine were discovered in 1785, but chlorine gas was dangerous and difficult to handle. Tennant, a Scotsman, developed a technique for producing bleaching powder by passing chlorine through lime hydrate. This dramatically reduced the time needed to bleach fabrics and spurred the progress of the textile industry.

The continental blockade imposed by Napoleon after the French Revolution reduced the availability of imported products, stimulating the invention of replacements. In France, the government mobilized elite groups of scientists and engineers to advance the state of science and technology. The École Polytechnique was founded in 1794, with chemistry lectures delivered by Vauquelin, Guyton de Morveau, Berthollet, Chaptal, and others. Although the French chemical industry did experience development, the concentration of the system of technical and advanced education in Paris, and the fact that the system encouraged self-supported and self-motivated workers and was inattentive to costs, eventually caused France to lag behind England.

In England, the education of citizens was handled by private institutions and clubs; typical examples of these include the Literary and Philosophical Society in Manchester—the heart of the industrial revolution—and the Lunar Society in Birmingham. The industrial revolution in England stimulated the development of private-sector corporations and witnessed the emergence of entrepreneurs in the chemical industry. The Leblanc method of soda production was introduced in the 1810s, and the robust growth in the cotton, soap, and glass industries ensured that, in the years after 1830, the Leblanc method formed a cornerstone of the English chemical industry. The hydrochloric acid gas produced as a byproduct of the Leblanc method caused environmental pollution and led to poisonous acid rain, and the processing of industrial waste containing potassium sulfide caused additional problems; nonetheless, England continued to use the Leblanc method as a key pillar on which its chemical industry was built.

References

1. A. J. Ihde, "*The Development of Modern Chemistry*" Dover Publications, Inc., New York, 1984
2. W. H. Brock, "*The Chemical Tree*" Norton, New York, 1993
3. T. H. Levere, "*Transforming Matter: A History of Chemistry from Alchemy to the Buckyball*" Johns Hopkins Univ. Press, 2001
4. B. Bensaude-Vincent, "*A History of Chemistry*" Harvard Univ. Press, 1996
5. F. Aftalion, "*A History of the International Chemical Industry*" Chemical Heritage Foundation, 2005
6. I. Asimov, "*A Short History of Chemistry*" Doubleday and Co., New York, 1965
7. E. Shimao, "*Jinbutsu Kagakushi*" *(Characters in A History of Chemistry)*, Asakura Shoten, 2002
8. J. Gribbin, "*Science: A History*" Penguin Books, 2002
9. J. D. Bernal, "*Science in Histtory*" G.A. Watts & Co. Ltd., London, 1965
10. S. F. Mason, "*A History of the Science*" McMillan,1962
11. T. Lucretius, "*De Rerum Natura*" *(On the Nature of Things)*, A. M. Esolen. trans. John Hopkins Univ. Press, 1995
12. Reference 2, page 56
13. Reference 2, page 68
14. Reference 2, page 106
15. Reference 2, page 112

Chapter 2: The development of modern chemistry

Chemistry in the 19th century: The establishment of the concepts of atoms and molecules and the specialization of the discipline into subfields

A stamp featuring Mendeleev and the first periodic table

In the first half of the 19th century, the modern concept of atoms was proposed and formed the basis for the development of quantitative methods in chemistry, expanding knowledge of chemical substances and chemical phenomena. Atomic weight was introduced to express the relative mass of atoms, and much effort was devoted to their accurate determination. Nevertheless, the distinction between atoms and molecules remained unclear, chemical formulas were frequently inaccurate, and much confusion persisted regarding atomic weights. It was not until around 1860 that the fundamental notions of atoms and molecules were established, whereupon chemistry began to make major progress. In the second half of the 19th century, organic chemistry developed significantly, and advances in physics were incorporated into the new field of physical chemistry. This helped prompt the division of chemistry into many specialized subfields, and the specialization of chemists themselves began in earnest.

Chemistry was a popular science in the 19th century. University-level chemistry education that emphasized the training of researchers began in Germany and lead to the education of many chemists and chemical engineers—a development that provided the backdrop for the development of the chemical industry. In this chapter, we will trace the evolution of these developments in 19th century chemistry. In writing this chapter, we have referred to references[1]–[14] given at the end of the chapter.

Several of the great chemists who were active in the late 19th century went on to win Nobel prizes in the 20th century. To learn about their careers and accomplishments, we have referred to Ref.[9]; we have also used information available on the Nobel Foundation's website, NobelPrize.org.

2.1 Atomic theory and the determination of atomic weights

Lavoisier's Chemical Revolution had given birth to modern chemistry as a quantitative science. However, the definition of "element" in Lavoisier's chemistry was a somewhat operational one that failed to answer the question, posed by humankind since antiquity, of what constituted the true substance of matter. In fact, Lavoisier considered questions such as "how many elements are there?" or "what is the true nature of the elements?" to be fruitless metaphysical pursuits, which he rejected as a waste of time. He explained his ideas as follows[15]:

> If by the term elements we mean to express those simple and indivisible atoms of which matter is composed, it seems extremely probable we know nothing at all about them; however, if instead we apply the term elements or principles of bodies, to express our idea of the last point which analysis is capable of reaching, we must admit as elements, all the substances into which we are capable, by and any means, to reduce bodies during decomposition.

Although chemists—who were influenced by the corpuscular philosophies of the 17th century—believed in the existence of certain particles representing the ultimate constituents of matter, most chemists at the end of the 18th century were satisfied by Lavoisier's definition and pursued practical research in chemistry. However, Lavoisier's chemistry was unable to make quantitative predictions of the products of reactions.

The situation at the turn of the century
Unmistakable progress was made in analytical methods from the end of the 18th century through the beginning of the 19th century. In addition, mineral analysis became an active field due to the growth of the mining industry; consequently, many new metallic elements were discovered, contributing to an expanded knowledge of chemistry. Some chemists who made particularly noteworthy contributions to such discoveries and to advances in analytical techniques were the German Klaproth (1743–1817), the Frenchman Vauquelin (1763–1829), and the Englishman Wollaston (1766–1828). Klaproth, a chemist at the Academy in Berlin, reformed the field of quantitative analysis and made major contributions to the rise of German chemistry. He discovered zirconium, tellurium, titanium, and uranium. Vauquelin, who became a chemist thanks to encouragement from Fourcroy, lectured on analytical methods at a mining school and later became

a professor at the College de France and the Jardin du Roi. Wollaston was a successful English physician who abandoned medicine to immerse himself in chemistry and physics research. He discovered platinum, palladium, and rhodium, while osmium and iridium were discovered by his friend Tennant.

As to the question of what force binds the elements together to form compounds, natural philosophers had been considering this problem, and the notion of "affinity", since ancient times. The Swedish chemist Torbern Bergman attempted to use relative affinities to explain the reaction in which acid and base combine to form salt. French chemist Claude Louis Berthollet retorted that reaction products depend not only on affinities, but also on the quantities of the substances involved in the reaction. Whereas Bergman believed that reactions proceeded in one direction based on affinity alone, Berthollet considered reversible reactions. Berthollet went further by suggesting that all compounds exist in different binding ratios, which put him in conflict with Joseph Proust's "Law of Definite Proportions." This law held that, whenever two or more compounds are formed from a given element, the ratios of the weight of the given element to the weights of the other elements with which it combined would be simple integer fractions. Although this law had been implicitly recognized by many chemists and formed the basis of mass analysis, Proust clarified it through a series of meticulous studies. This set the stage for Dalton's chemical theories of atoms, which provided the crucial link between atoms and Lavoisier's elements and lent solid evidence for the law of definite proportions.

Dalton's atomic theories and the weights of atoms

By the end of the 18th century, some chemists were attempting to use corpuscular theories to explain chemical phenomena. The Irish chemist Higgins envisioned "ultimate particles" and tried to use them to explain chemical phenomena, but he lacked the notion of particles of different weights corresponding to the various individual elements. Lavoisier had understood his indivisible substances (elements) to be composed of atoms with identical properties. The idea that the atoms of different elements had different sizes and properties was first proposed by John Dalton. This offered a clear model for the picture that had been hypothesized by chemists—in which fixed quantities of constituents bound together to form compounds—and opened the door to a quantitative understanding of chemical reactions and chemical binding.

The Lake District of England, where Dalton grew up, experiences wild fluctuations in weather, which spurred Dalton's interest in meteorology and led him to study the amount of water vapor in the atmosphere. He noticed that the amount of water vapor in the air increased with increasing temperature, and that this fact remained true when air was replaced with other gases. In contrast with what was

Torbern Bergman (1735–1784)
Bergman was a Swedish chemist, mineralogist, and taxonomist who became a professor at the University of Uppsala in 1776. He created a table of the chemical affinities of substances and conducted research on metals—particularly nickel—as well as rainbows and the aurora.

Claude Louis Berthollet (1748–1822)
Berthollet was a French chemist, active during the revolutionary and Napoleonic eras, who supported Lavoisier's theory of combustion and helped revise chemistry nomenclature. His accomplishments include the determination of the composition of ammonia and the invention of chloride bleach. He opened up his home laboratory to use by young chemists and supervised their work there.

Joseph-Louis Proust (1754–1826)
Proust was a French chemist who worked as a pharmacist at a Paris saltpeter factory and later directed the Royal Laboratory in Madrid. He conducted quantitative research on the constituent elements of a large number of compounds and contributed to establishing the Law of Definite Proportions.

generally believed at the time, he became convinced that water vapor in air was *not* chemically bound to oxygen or nitrogen. In 1801, he found that adding water vapor to dried air caused the total pressure to increase by precisely the pressure of the water vapor; thus, he had discovered his "law of partial pressures," which states that the total pressure of a gaseous mixture is equal to the sum of the pressures of the individual components. He proposed that particles of the same species repel each other due to repulsive forces acting between them, while particles of different species simply ignore each other; based on this hypothesis, he successfully explained several phenomena such as the partial pressures of gaseous mixtures, diffusion, and uniformity. Although he must surely have believed in the existence of atoms from the beginning, these studies led him to consider a chemical atomic theory, and around 1803 he began developing the concept of atomic weights, publishing "*A New System of Chemical Philosophy*" in 1808.

Dalton's atomic theory was based on the following four assumptions. First, all matter is composed of rigid, indivisible atoms. These atoms are surrounded by a thermal atmosphere, with a varying quantity of heat depending on the state of the collection of atoms. This picture of heat followed Lavoisier's notion of "caloric." Second, atoms could not be destroyed, and in fact remained entirely unchanged by any chemical reaction. Third, there are as many types of atoms as there are elements. Until this point, atomists had believed that there existed just

John Dalton (1766–1844)

Born the son of a poor Quaker textile worker in Eaglesfield, in the central English region of Cambria, Dalton received elementary education at his village school, but later learned mathematics and science on his own. By his teens, he was thoroughly familiar with Newton's "*Principia*". In 1793, he began teaching mathematics and natural philosophy at a dissenting academy in Manchester (the "New College"); after the school moved, he continued his research while earning a living as a home tutor. He was an active member of the Manchester Literary and Philosophical Society, where he presented the results of his research. His greatest accomplishments were his proposal of the chemical theory of atoms, his determination of atomic weights, and his connection of atomic weights to Lavoisier's concept of elements; all of these accomplishments helped lay the foundations for the development of modern chemistry. He is also known for his law of partial pressures and his law of multiple proportions.

a single species of fundamental particle; Dalton's atomic theory differed entirely from this view by linking atoms directly to the very definition of elements.

Fourth, each atom was assigned a relative atomic weight, a quantity that could be determined experimentally. Dalton also made the following assumption regarding the process by which atoms bind to form compound atoms: When just a single type of compound is formed from two elements A and B, he assumed its chemical formula is AB; when two types of compounds are formed, they must be AB_2 and A_2B. Similar rules governed the behavior of three-species and four-species compounds. Thus, Proust's "Law of Definite Proportions" was given a theoretical basis.

With these laws, Dalton was able to use analytical results available at that time to determine the atomic weights. Hydrogen was the lightest of all known substances, so taking its weight to be 1 allowed the relative weights of all other elements to be determined. In Dalton's theories, if the ratios of the numbers of atoms in a compound were incorrect, the atomic weights determined based on those ratios would be in error as well. For example, the only known compound of hydrogen and oxygen was water, but Dalton assumed the chemical formula of water to be HO instead of H_2O. From analytical results on the composition of water, Dalton observed that 12.6 parts of hydrogen reacted with 87.4 parts of oxygen to form water, and thus assigned oxygen an atomic weight of 7. Similarly, he assumed that ammonia (NH_3) consisted of one atom each of nitrogen and hydrogen, and thus concluded that the atomic weight of nitrogen was 6. The

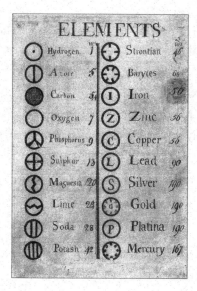

Figure 2.1: Dalton's symbols for the elements and atomic weights.

uncertainty of these atomic weights would cause headaches for chemists for the next several decades.

Atomic symbols

Dalton used symbols to represent atoms (Figure 2.1). For example, hydrogen was ⊙, oxygen was ◯, nitrogen was ⊕, carbon was ●, water was ⊙◯, and ammonia was ⊙⊕. Although Dalton himself had great confidence in these symbols, they led to extra printing costs and were not widely adopted. Nonetheless, the symbols undoubtedly contributed to strengthening belief in the actual existence of chemical atoms and to encouraging chemists to interpret chemical reactions in a visually intuitive way.

The Swedish chemist Jons Jacob Berzelius, who disliked Dalton's symbols, proposed in 1813 a system in which the first one or two letters of the Latin name of an element would be used as the symbol of the element. At first this system represented compounds by joining the symbols of the constituents with a plus sign (such as Cu+O for copper oxide), but later the plus sign was dropped and the constituent symbols simply written aside each other. Numbers of atoms were indicated with superscripts; for example, hyposulfuric acid was S^2O^3. Berzelius also used an abbreviation in which positively charged elements that bind with oxygen were denoted by a dot above the element symbol, but this notation was not generally adopted. The subscript notation used today was introduced in 1834

Jons Jacob Berzelius (1779–1848)
Born the son of a clergyman in Väversunda, Sweden, Berzelius lost his parents at an early age and was raised by relatives. He studied medicine at the University of Uppsala but never became a doctor, instead becoming interested in chemistry. He became a professor of medicine and pharmacology at a college of surgeons in Stockholm, and later served as professor of chemistry for 25 years. He was one of the most influential chemists of the early 19th century. He was a devout believer in atomic theory and analyzed a huge number of inorganic compounds, tirelessly measuring accurate atomic weights. In addition, starting in 1822 and continuing for 27 years, he published an "annual report" in which he introduced European chemistry papers together with commentary.

by Liebig, but Berzelius had laid the foundation for the notation used to represent element symbols, thus allowing chemical theories and analytical results to be presented with clarity and ease.

Gay-Lussac's law of combining volumes and Avogadro's hypothesis

In 1808, Joseph Gay-Lussac reported a study of binding volume ratios in gas-phase reactions. He investigated reactions between hydrogen and oxygen and noted that two parts of hydrogen reacted with one part of oxygen (with parts measured volumetrically). Upon further investigation to determine whether or not other gaseous reactions involved similar integer ratios between the volumes of the reactants, he found that two parts carbon monoxide reacted with one part oxygen to form carbon dioxide, whereas one part ammonia reacted with one part hydrochloric acid to form salt. Based on the results of his research as well as of other researchers, he concluded that the ratios of the volumes of the reactants in gas-phase reactions were always simple integer fractions—the *law of combining volumes.*

The law of Gay-Lussac implied that identical volumes of different gases, at the same temperature and pressure, had identical numbers of reactant molecules, but Gay-Lussac did not emphasize this point. Moreover, although this conclusion supported the law of definite proportions in Dalton's atomic theory, Dalton himself did not believe in Gay-Lussac's law. If—as was believed at the time—individual gas molecules were single atoms, then the results of Gay-Lussac's observations were themselves difficult to understand; indeed, one part CO should react with one part O to form one part CO_2, and the notion that *two* parts CO

Joseph Gay-Lussac (1778–1850)
The son of a judge in Paris, Gay-Lussac graduated from the
Ecole Polytechnique and studied at the École des Ponts et
Chaussées. He was taught chemistry by Berthollet, Guyton
de Morveau, and Fourcroy, and continued his research in
Berthollet's personal laboratory. In 1802, he measured the
thermal expansion coefficients of gases and became famous
for his accurate value of 1.3750; he also worked for many
years as a professor of physics at the Sorbonne. His many accomplishments
include the discovery of the laws of gaseous reactions, the isolation of
potassium, and the discovery of boron and iodine.

Amedeo Avogadro (1776–1856)
The son of well-known lawyer in the city of Turin in the
Kingdom of Sardinia in northern Italy, Avogadro earned a
degree in law but studied mathematics and physics on his
own. In 1806, he became a professor of natural philosophy
at the Royal Institute in Vercelli, and later became a
professor of mathematical physics at the University of Turin.
Avogadro conducted research on electricity, fluid surface
tension, specific heats, and capillary phenomena. In 1811, he presented
Avogadro's Law and proposed that hydrogen, oxygen, and nitrogen were
diatomic molecules; these ideas were not generally accepted until almost
half a century later.

would react with one part O to form two parts CO_2 was hard to believe. Similar
problems afflicted other gas-phase reactions. Another experimental observation
that could not be explained based on this hypothesis was the fact that the density
of carbon monoxide, which was composed of oxygen and carbon atoms, was
lower than the density of oxygen. These problems were ultimately resolved by
the Italian physicist Avogadro.

Avogadro's paper, presented in 1811, resolved the problems mentioned above
based on Gay-Lussac's law and Dalton's atomic theory. He first hypothesized that
identical volumes of gas, at identical temperatures and pressures, contained identical numbers of molecules. He further assumed that the gas molecules involved in
reactions dissociate into half-molecules (atoms)—a statement which signifies that
individual gas molecules were composed of diatomic molecules. These assumptions sufficed to provide elegant explanations of all phenomena. In describing
molecules, Avogadro drew a distinction between two terms: *molecule integrante*
(composite molecule) and *molecule elementaire* (fundamental molecule). Similar

ideas were expounded in 1814 by France's Ampere. However, Avogadro's theory was generally rejected or ignored; in particular, both Dalton and Berzelius found his ideas difficult to accept. In Dalton's atomic theory, repulsive forces—due to heat—existed between atoms, whereupon it was hard to understand how two atoms could bind to form a diatomic molecule. Berzelius believed that electrical forces were the key to chemical binding, and thus that identical atomic species, carrying identical electric charges, should repel each other, again preventing him from accepting the existence of individual diatomic molecules. Thus, Avogadro's hypothesis, notwithstanding its essential correctness, continued to be ignored for almost 50 years. As for the nature of the binding force responsible for the formation of individual diatomic molecules, it would be another 126 years before quantum mechanics was to clarify the character of covalent bonding (see Section 4.3).

The atomic weights of Berzelius

Chemists such as Dalton and Berzelius recognized the importance of the concept of atoms and made efforts to advance it, but they were hampered by major confusion surrounding the determination of atomic weights. Accurate determination of atomic weights required knowledge of the weight ratios of elements bound in compounds and of the proper binding formulas of these compounds. Determination of weight ratios was possible thanks to advances in analytical techniques, but the binding formulas remained problematic. In response to this challenge, Dalton adopted a "Principle of Simplest Combinations," but it gradually became apparent that this principle was not generally valid. Comparison of gas densities was an effective method in a limited number of cases, but not generally applicable as most elements were not gases. To this end, many chemists gave up on determining the atomic weights and contented themselves with measurements of *equivalent weights*: measurements of the quantity of an element that would bind with a fixed quantity of a reference element. It was Berzelius who finally rose to the challenge of atomic weight determination.

Berzelius (see page 45) was an unsurpassed analytical chemist. He prepared high-quality reagents, conducted meticulous experiments using well-conceived methods of weight analysis, and achieved analytical results of high precision for his time. One problem in the determination of atomic weights was the choice of a standard element. Dalton had taken the weight of hydrogen to be 1 and defined the weights of the other elements relative to this choice, but this choice suffers from the drawback that many elements do not form compounds with hydrogen, thus preventing direct determination of binding weight ratios. Moreover, precision analysis of hydrogen compounds was difficult with the experimental methods available at the time. In contrast, oxygen forms stable compounds with many elements and was better suited to accurate analysis; thus, Berzelius defined the atomic weight of oxygen to be 100 and used this reference value to determine many atomic weights.

Berzelius presented the first table of atomic weights in 1814 and a revised version in 1818. He first used Gay-Lussac's law to determine the chemical formulas of water, ammonia, and hydrogen chloride respectively as H_2O, NH_3, and HCl; then, reasoning by analogy, he determined the formula of hydrogen sulfide to be H_2S. Similarly, in view of the close resemblance of selenium (Se) and tellurium (Te) to sulfur, he proposed that the hydrogen compounds formed by those elements were H_2Se and H_2Te. He hypothesized that the chemical formula of a compound formed when one atom of an element A binds with atoms of element B must be either AB, AB_2, AB_3, or AB_4. He noted that oxygen compounds were primarily dioxides such as CO_2 and SO_2, and proposed that this observation would apply to metals as well. Thus, by 1818 he was able to report the atomic weights of 47 elements. In cases where he used the correct chemical formula, many of the atomic weights he determined were accurate even by modern standards, testifying to the extraordinary precision of his analytical techniques. However, because he used incorrect chemical formulas for metal oxides, many of his atomic weights were too large—some 2–4 times the correct values. For example, Berzelius lists the weights of sodium, magnesium, calcium, iron respectively as 92.69, 50.68, 81.93, and 108.55 (in units in which the weight of oxygen is 16), whereas the correct values are 22.99, 24.32, 40.08, and 55.85.

The year 1819 saw two reports that lay the foundation for correcting this error. The French scientists Petit and Dulong discovered that the product of the specific heat and the atomic weight was a constant for many solid elements. Using this law, they pointed out that the values obtained by Berzelius for the atomic weights of lead, copper, tin, zinc, iron, nickel, gold, and many other metals were two times larger than the correct values. In addition, the German chemist Mitscherlich proposed his "law of isomorphism," which stated that materials of similar chemical composition exhibited the same crystal structure. Using this law, it was possible to infer the atomic weights of problematic compounds by comparing them to compounds of known atomic weights. For example, the correct atomic weight of chromium was inferred by comparing chromate to sulfate. In 1826, Berzelius presented a more accurate table of atomic weights that incorporated these revisions. However, this table continued to contain erroneous values for the weights of the alkali metals and other elements.

Prout's hypothesis

Dalton's theory recognized that there were as many atomic species as there were elements; thus, some 50 types of atoms must exist. At the beginning of the 19th

> **William Prout (1785–1850)**
> Proust was an English chemist who studied medicine at the University of
> Edinburgh and conducted research while launching businesses in London.
> In addition to formulating Prout's hypothesis, he boasted many achievements
> in physiology, such as the isolation of urea and the discovery of hydrochloric
> acid in stomach fluids.

century, this remained a difficult proposition for many chemists to accept; it
was hard to believe that God had used more than 50 different types of building
block to construct the world. These chemists continued to wonder if perhaps
Lavoisier's elements were not *true* elements.

In 1815, the Englishman William Prout called attention to the fact that,
if the atomic weight of hydrogen is set to 1, the weights of all other atoms
are nearly integers. Prout wondered if perhaps the "primary element" of the
Aristotelian school was embodied by hydrogen, and proposed that hydrogen
was the essential building block of all matter. For example, if the atomic weight
of chlorine is 36, we imagine that 36 volumes of hydrogen are compressed to
form a single chlorine atom. There were reasonable arguments both for and
against this theory, and much research was conducted to confirm or refute it.
As the determination of atomic weights became more accurate, it became clear
that only some elements had atomic weights which were integer multiples of
the weight of hydrogen, while others did not. The question of *why* this situation
existed would not be resolved until the discovery of isotopes in the 20th century
(see Section 3.3).

Confusion surrounding atomic weights and the concept of equivalent weights

These developments did not put an end to efforts to determine precise values
of the atomic weights. In 1826, the young French chemist Jean-Baptiste-
André Dumas devised an ingenious method for measuring the vapor density
of a substance that is a liquid or solid at room temperature and attempted to
determine atomic weights from vapor densities. Although Dumas succeeded
in accurately measuring the gas densities of many compounds, he was unable
to solve the problem of atomic weights; for example, the value he obtained for
iodine was twice the value determined by chemical analysis, while his weights
for mercury and sulfur were, respectively, one-half and three times the values
obtained by Berzelius. Given the lack of knowledge surrounding the structure of
gas molecules at the time, these results served only to invite further confusion;
many chemists harbored suspicions regarding the atomic theory and attempted

Jean-Baptiste-André Dumas (1800–1884)
Dumas, a French chemist, was born the son of a city administrator in the town of Alès on the lower Rhone. At 15, he
was apprenticed to a pharmacist, but he was not interested
in this work and set off for Geneva to study natural science. Later he went to Paris, where he became a student of
Thenard at the Ecole Polytechnique and then a professor at
the Sorbonne. In addition to the determination of molecular
weights, he made major contributions to analysis and synthesis in organic
chemistry, including quantitative-analytical methods for nitrogen, the isolation of anthracene, and the study of non-cyclic alcohols. Dumas played a key
role in expounding the notion of chemical types and systematizing organic
chemistry.

Leopold Gmelin (1788–1853)
Gmelin was a German chemist who studied at the Universities of Tübingen
and Göttingen and became a professor at Heidelberg. He attempted to
systematize chemical theories using the notion of equivalent weights, and
became the editor of *Handbuch der anorganischen Chemie*.

to use equivalent weights instead of atomic weights.

The notion of equivalent weights had arisen before Dalton's atomic theories
in conjunction with the problem of the neutralization of acids and bases; the
equivalent weight of an element was the weight of the element that formed a
compound with a given quantity of a fixed standard element. Wollaston believed
that equivalent weights were less logical than atomic weights but more useful
in practice. Whereas atomic weights rested on certain hypotheses, equivalent
weights were reliable numbers based on analytical results; in 1814, Wollaston
expanded the concept to include 12 elements and 45 compounds. The German
chemist Leopold Gmelin took the notion further by inventing a system of
chemical formulas based on equivalent weights. In this system, the equivalent
weights of elements were, for example, H=1, O=8, S=16, and C=6, while the
chemical formulas for water and hydrogen sulfide were HO and HS. However,
while equivalent weights were convenient for working with stoichiometry, they
did not help to generalize chemical formulas, and ultimately served only to
deepen the confusion surrounding atomic weights—a confusion that persisted
until the 1850s, when the Italian chemist Cannizzaro clarified the distinction
between atoms and molecules (see Section 2.3).

2.2 The birth of electrochemistry and its impact on chemistry

A major breakthrough in the scientific world came at the beginning of the 19th century with the invention of the battery by the Italian physicist Volta. In the same way that the invention of the air pump spurred the development of pneumatic chemistry, the invention of the battery created the new field of electrochemistry and made a major contribution to the development of chemistry. Since the publication, in the early 17th century, of Gilbert's books on magnetism and electricity,[1] the subject had captivated many scientists, but experiments conducted using static charge buildup in a Leyden jar only lasted for an instant. With the invention of the battery, it became possible for the first time to conduct extended experiments using electric currents. Chemistry was the first branch of science to be majorly impacted by the invention of the battery.

Volta's battery and the birth of electrochemistry

Italian physicist Alessandro Volta was inspired to begin his studies by his friend Galvani's discovery, in 1780, of "animal electricity," in which placing two distinct types of metal in contact with the muscle tissue in a frog's leg caused a flow of electricity. Volta realized that the role of the animal in this experiment was in fact unimportant, and that the flow of electricity actually arose from the contact of the two different metals. In 1800, he reported steady current flow from a galvanic pile consisting of paper or felt dipped in salt water and pressed between two plates made of different metals, such as zinc and silver (Figure 2.2). The invention of the battery had major repercussions; in particular, Berzelius and Davy immediately began using batteries for chemistry research and achieved significant successes.

Berzelius discovered in 1802 that electric current caused salts to dissociate. Working together with his friend and patron Hisinger, he carried out the electrolysis of salt and observed that acids collected near the anode, while bases collected near the cathode. This led him to propose the *electrochemical dualistic theory* (page 53). In 1806, Theodor Grotthuss proposed a theory in which molecules in electrolytes repeatedly alternate between the processes of dissociation and recombination. He believed that, in the electrolysis of water, the negative electrode pulled hydrogen out of water molecules, whereupon the oxygen thus denuded of hydrogen pulled hydrogen out of adjacent water

---

1. In 1600, the English physician and natural philosopher William Gilbert (1544–1603) published an influential book on magnetism, De Magnete, in which he also discussed static electricity.

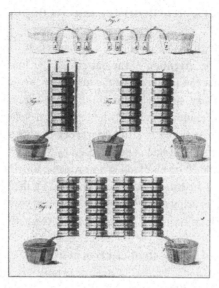

Figure 2.2: Volta's galvanic pile. Felt or paper was dipped in salt water and then pressed between two plates of different metals; this unit was then replicated many times.

molecules, and the process continued indefinitely. This basic picture of proton migration in water—the *Grotthuss mechanism*—remains accepted to this day.

In 1799, under the leadership of such luminaries as the American-born scientist Sir Benjamin Thompson (Count Rumford) and the philanthropic businessman Sir Thomas Bernard, the Royal Institution[2] was founded in London with the objectives of disseminating scientific knowledge and advancing the welfare of the working classes. When Humphry Davy entered the Royal Institution in 1801, he conducted experiments involving the electrolysis of a variety of solutions and discovered that hydrogen, metals, metal oxides, and alkalis tended to collect near the negative electrode, while oxygen and acids collected near the positive electrode. Based on these experiments, Davy surmised that chemical affinity was a type of electrical force, and that if electric currents could be used to overcome chemical affinity, then this would enable the dissociation of compounds that could not be dissociated using standard methods. Continuing to conduct experiments based on this notion, in 1807 he used a strong current to perform the electrolytic decomposition of dissolved caustic potash and caustic

2. Although this institution was established with the blessing of the King, it received no financial assistance from either the King or the national government; it was founded and operated entirely on grants and contributions from wealthy nobility.

Theodor Grotthuss (1785–1822)
Born to a wealthy landowning family from Kurland on the Baltic Sea (present-day Latvia), Grotthuss lived his entire life in the countryside with his mother, except for a 5-year stint in which he studied at universities in Leipzig, Naples, Rome, and Paris. He left behind pioneering accomplishments in electrochemistry and photochemistry.

Humphry Davy (1778–1829)

Born the son of a wood carver in the Cornwall region of England, Davy was apprenticed to a local pharmacist, but became interested in chemistry after reading Lavoisier's *Elementary Treatise on Chemistry*. He became an assistant to the Bristol doctor Beddoes, who was studying the physiological effects of gases, and conducted experiments on a variety of gases he produced. He discovered that dinitrogen monoxide (laughing gas) caused dizziness and had anesthetic effects. In 1801, Davy was hired by the Royal Institution, where he served as a professor from 1802 to 1812. He conducted research on topics such as the isolation of alkali metals and alkali earths by electrolysis and invented safety lights for coal miners. In 1820, he became head of the Royal Institution (Column 3).

soda, and successfully obtained potassium and sodium metals. The following year he succeeded in isolating the alkali earths magnesium, calcium, strontium, and barium.

The electrochemical dualistic theory

Based both on his own electrochemistry experiments and on those of Davy, Berzelius developed an electrochemical dualistic theory to provide a comprehensive explanation of the chemical bond. He proposed that atoms in compounds were bound together by electrical forces and could be dissociated by the application of electric current. All atoms had a *polarity*; some were electrically positive, while others were negative. Oxygen was the most negative of all atoms, while metals were generally positive. Chemical compounds were understood to arise from the mutual attraction between these positive and negative charges, and the resulting compounds had their own polarities. For example, Cu with a positive polarity and O with a negative polarity bound to form CuO, a molecule with a slight positive polarity. Similarly, S with a positive polarity bound with O with a negative polarity to form SO_3, a molecule with a slight negative polarity, and these positive and negative molecules could subsequently interact to form copper sulfate, $CuSO_4$. This dualistic theory explained inorganic compounds quite well and served for some time as the leading theory of the chemical bond. However, as organic chemistry began to develop in the 1830s and thereafter, the dualistic theory was more of an obstruction than an aid to acquiring a proper understanding of chemical bonds in organic compounds. Moreover, in the dualistic theory, it was difficult to understand how a single-component gas could consist of diatomic molecules; this prevented chemists from adopting Avogadro's hypothesis and ensured the persistence of confusion surrounding atoms and molecules.

Faraday and the laws of electrolysis

Michael Faraday[16, 17], who entered the Royal Institution in 1812, produced an impressive range of pioneering accomplishments across an astonishingly wide range of fields in physics and chemistry. Although his research primarily concerned electricity and magnetism, he preferred to think of himself as a natural philosopher, and pursued interesting phenomena throughout the natural sciences, without concern for distinctions between fields such as physics and chemistry. Although his discoveries in the field of electromagnetic induction—which laid the foundations for the development of electromagnetism and electrical engineering throughout the 19th century—are undoubtedly the most important of the many scientific achievements he left to posterity, his accomplishments in chemistry were also important, and we will here discuss his major contributions to that field.

In the first half of the 1820s, Faraday studied carbon compounds and discovered several new substances, including a number of organic compounds such as benzene, isobutene, tetrachloroethylene, hexachloroethane, and naphthalene sulfonate. In 1823, he succeeded in liquefying chlorine through cooling and pressurization, and he later repeated this success with other gases, including CO_2, SO_2, H_2S, and N_2O. He was also the first to note the existence of a critical temperature above which gases could not be liquefied even upon pressurization. In his later years, he conducted pioneering research on gold colloids, a form of matter with direct ties to today's nanoscience. However, his greatest contribution to chemistry was his discovery of the laws of electrolysis.

To make accurate measurements of electrical quantities, he devised a Coulometer that measured the quantity of hydrogen arising from the electrolysis of water, and thus the quantity of electricity that had flowed; he investigated the quantitative relationship between the current that flows in electrolysis and the quantity of substance that is electrolysed. As a result, he discovered two important laws. First, the *chemical effect*—that is, the quantity of substance

Michael Faraday (1791–1867)

Faraday was the son of a blacksmith and received only an elementary education. At the age of 13, he was apprenticed to a bookbinder in London and became knowledgeable on a variety of topics by reading the books he made cover to cover. Among these books were an introduction to chemistry and a portion of an encyclopedia discussing electricity. In 1813, Faraday joined the Royal Institution as an assistant to Davy, where he proceeded to make many important discoveries spanning a wide range of areas in physics and chemistry. In chemistry, he is famous for discovering the laws of electrolysis, the compound benzene, and successfully liquefying chlorine and other gases. In physics, he is famous for discovering electromagnetic induction and for introducing the concept of *lines of force*, which carry the effects of electromagnetic fields through media and vacuum, thus laying the foundations for the future development of electromagnetic theory (Column 3).

that dissociates during electrolysis—is strictly proportional to the quantity of electricity. Second, the quantity of electricity needed to produce 1 g of hydrogen is equal to the quantity of electricity needed to dissociate the chemically equivalent weight of other substances; the electrochemical equivalent weight agreed with the chemical equivalent weight. Thus, he proposed that the electrochemical equivalent weights of hydrogen, oxygen, chlorine, iodine, lead, and tin were respectively 1, 8, 36, 125, and 58. This result was deeply intertwined with the problem of atomic weights, but Faraday did not believe in the existence of atoms and did not attempt to address the determination of atomic weights, which was a major problem at the time. In 1833, in collaboration with Whewell, he proposed many terms that would go on to enjoy widespread adoption, including *electrode, anode, cathode, ion, anion*, and *cation*. However, it is important to note that the meanings he understood for ions, anions, and cations differ from those understood today: in Faraday's understanding, anions and cations were respectively portions of electrolytes discharged at anodes and cathodes. Research on electrochemistry immediately began to yield important practical consequences, and technologies and industries for electroplating began to develop in the late 1830s. However, Faraday's curiosity remained focused on unraveling the mysteries of nature; he was less interested in developing theories based on his experimental results or in pursuing their practical applications, preferring to leave those tasks to others.

COLUMN 3

Davy, Faraday, and the Royal Institution

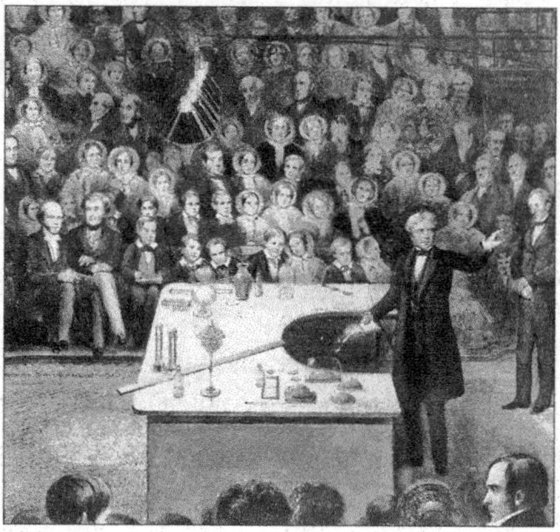

The scene at one of Faraday's Friday Lectures at the Royal Institution. Nearly half of the participants are women.

The Royal Institution in London was originally conceived as an institution to provide technical education to craftsmen and farmers, but it eventually lost sight of this original charitable mission and became a place for its upper-class financial backers to attend lectures and conduct research. Two figures who were particularly active on this stage in the first half of the 19th century were Davy and Faraday. Davy had conducted groundbreaking research on electrochemistry, and as a handsome young man and gifted lecturer, he quickly rose to a position of immense popularity. In an England characterized by a strong tradition of amateurism, he might be called the first professional scientist. Because many of the Royal Institution's sponsors were land-owning nobility, demand for practical research was strong. Davy conducted research related to leather making—an important industry at the time—and devised methods for determining the tannin content of plants, ultimately discovering that catechu from India (a plant in the palm family) had the highest tannin content. He also conducted research on soil analysis and fertilizers, and lectured on agricultural chemistry. Through these efforts, Davy became a star of the Royal Institution and a darling of upper-class society. He was knighted in 1812, married a wealthy widow, became a member of the upper class himself, and enjoyed the life of a gentleman, losing sight of his primary focus on research. It was around this time that Faraday entered the Royal Institution.

Faraday, while a bookbinder's apprentice, attended a series of lectures by Davy with one of his store's customers. Spellbound by what he learned, Faraday compiled the lectures into an illustrated book, which he presented to Davy together with a request to be hired as an assistant. In 1813, this wish was granted. In the fall of that year, the newlywed Davy set off on a continental journey on which he brought Faraday along as an assistant. During this extended sojourn—which lasted a year and half—Faraday not only learned a great deal, but also matured as a scientist through face-to-face meetings with the scientific elite of France and Italy, including Ampere, Chevreul, Gay-Lussac, and Volta. However, Faraday found the attitude he received from Davy's wife—who thought of the young man as a lowly servant—difficult to bear. In 1825, Faraday became laboratory manager and achieved results in a surprisingly wide range of research disciplines. In 1833, he was appointed Fullerian Professor of Chemistry at the Institution. Davy recognized and cultivated Faraday's talent, but the relationship between the two men cooled as Faraday matured into an independent researcher.

A true-to-life depiction of a boisterous scene at the Royal Institution. Davy is holding the bellows, while Count Rumford is standing at the far right.

Through general lectures on science and other events, the Royal Institution made major contributions to English science and culture over a period spanning almost two centuries; these foundations were originally laid by Davy and brought to near-completion by Faraday. The tradition of Friday lectures for the general public was originated in the Davy-Faraday era and lives on to this day in the form of the world-famous "Theatre of Science." Also famous are the Royal Institution's Christmas lectures aimed at children and teenagers, from which the classic "Chemical History of a Candle" was born[18].

2.3 The birth of organic chemistry and the confusion surrounding atoms and molecules

Before the first part of the 19th century, inorganic compounds were the subject of quantitative investigations, but research on *organic* compounds had made little progress. Although a number of organic substances—including alcohols, ether, vinegar, and formic acid—had been known since antiquity, they were treated on a qualitative level. The analysis of organic compounds was vastly more difficult than the analysis of inorganic compounds, and quantitative studies posed a major challenge. Around 1780, Scheele used the fact that the solubility of benzoic acid in water is lower than that of its calcium salt to succeed in producing benzoic acid. In a similar way, he produced other organic acids such as oxalic acid, malic acid, tartaric acid, and lactic acid. Around the same time, the French chemist Rouelle isolated urea from human urine. Lavoisier was also interested in organic compounds and thought of organic chemistry as a branch of chemistry; still, the thinking of the time was dominated by the theory of *vitalism*, which held that organic substances were synthesized by a special "vitality" that existed only in animals and plants.

In 1824, Friedrich Wöhler obtained white crystals from a solution of silver cyanate and ammonium chloride; instead of the ammonium cyanate that had been expected, this substance turned out to be identical to the pure urea $[(NH_2)_2CO]$ obtained from urine. This was an entirely unexpected result, and one that refuted the theory of vitalism then prevalent, which maintained that organic compounds could not be formed from inorganic compounds. Although this result was held in high regard by the chemists of the day, it did not spell the immediate demise of vitalism, which persisted until the late 19th century—although it was successively weakened by the continual progress of organic chemistry.

New organic compounds and the discovery of isomers

As the 19th century progressed, more and more new organic compounds were isolated from plant and animal tissue, paving the way for advances in organic chemistry. Near the beginning of the century, Proust identified glucose, fructose, and sucrose from the sweet juices of plants. He also isolated an amino acid, leucine, from cheese. Crystals of morphine were isolated from poppies by Sertürner in 1805, and by 1835, some 35 different species of alkaloids had been isolated. An important sequence of investigations conducted during this period was the study of fats by Michel-Eugène Chevreul. In 1825, he discovered that fat dissociates into glycerin and fatty acid upon hydrolysis, and he isolated many fatty acids from plants and animals. He also received a patent for a method of fabricating candles using stearic acid; candles made in this way were of higher

Friedrich Wöhler (1800–1882)
Wöhler studied medicine at Marburg and Heidelberg, but became a chemist on the advice of Gmelin. After receiving his medical degree, he studied with Berzelius in Stockholm. After returning to Germany, he taught chemistry at engineering schools in Berlin and Kassel, and became a professor of chemistry at the University of Göttingen in 1836. Although he left behind many achievements in both organic and inorganic chemistry, including the isolation of aluminum, beryllium, and boron, Wöhler's claim to immortality was conclusively sealed by his synthesis of urea from inorganic compounds..

quality than traditional candles made using animal fats. Chevreul also used alcohol and other inactive solvents to isolate substances, and proposed using the melting point as a measure of the purity of a substance.

As the analysis of compounds and the study of their properties progressed, new problems emerged. Until that time, it had been thought that knowledge of the composition of a substance sufficed to identify that substance uniquely. However, in the 1820s, it became clear that there existed substances of identical compositions but distinct physical properties. The composition of the silver salt of cyanic acid used in Wöhler's studies was reported to be 77.23% silver oxide and 22.77% cyanic acid. Meanwhile, Liebig, working in the laboratory of Gay-Lussac, found the composition of silver fulminate to be 77.53% silver oxide and 22.47% cyanic acid. Thus, the analytical results suggested that silver cyanate and silver fulminate, two substances with entirely different properties, nonetheless exhibited identical chemical compositions. Subsequently, Wöhler and Liebig worked together to determine that the compositions of cyanic acid (HOCN) and fulminic acid (HONC) were in fact identical. Various other examples of such *isomers*—substances with identical compositions but distinct properties—were later found, including some inorganic compounds; in 1830, Berzelius proposed the notion of *isomerism* to explain these results.

Liebig and the analysis of organic compounds

At the beginning of the 19th century, the analysis of organic compounds was still in its infancy—but it was soon to make major progress. The first attempts at quantitative analysis of organic compounds were made by Lavoisier. In an effort to determine the volumes of oxygen consumed and carbon dioxide produced by combustion, he placed a known weight of coal in a bell-shaped jar filled with oxygen and burned it above mercury; by absorbing the carbon dioxide released

Michel-Eugène Chevreul (1786–1889)
Chevreul was a French chemist. He was a student of Vauquelin and an assistant at the Gardens in Paris, and he followed his mentor by serving for many years as a professor at the Natural History museum. He is known for his research on fatty acids and on tapestry dyes.

in alkali, he could measure the volume of gas before and after the absorption. He also burned alcohol, fat, and wax in an attempt to determine the volumes of oxygen consumed and carbon dioxide produced, and thus was able to determine the composition of the substances burned. He was aware that these substances consisted of carbon, hydrogen, and oxygen. He also used oxidizing agents, such as mercury oxide, to analyze compounds that were difficult to burn, such as sugar and resin. However, his data regarding constituents such as carbon dioxide and water were inaccurate, and thus the analytical results he obtained were inaccurate as well.

The accuracy with which organic substances can be analyzed depends on the accuracy with which the quantities of carbon dioxide and water produced by the oxidation of hydrogen and carbon in the organic matter can be measured. In 1811, Gay-Lussac and Thenard introduced the use of potassium chlorate as an oxidizing agent—a major innovation in organic analysis—and later replaced the potassium chlorate with copper oxide, which made analyses safer and easier. Later, Berzelius developed an improved apparatus that was simpler and more stable: the water produced by burning was absorbed and weighed by passing the products of the burning process through a tube filled with calcium chloride while the oxygen and carbon gas were captured in a bell-shaped bottle above mercury. Using this apparatus, he was able to determine the composition of simple organic acids. Combustion analysis of carbon and hydrogen was further improved by Justus von Liebig's introduction of methods using the *Kaliapparat* (Figure 2.3); with this device, it was not necessary to capture gases above mercury. The Kaliapparat consisted of a chain of five glass bulbs containing a solution of potassium hydroxide, which absorbed the carbon dioxide produced by burning; the mass of carbon dioxide produced was determined by measuring the weight increase of the bulbs. The quantity of hydrogen in a sample was determined by measuring the increase in weight of a calcium chloride tube placed inside the Kaliapparat; the calcium chloride absorbed the water produced during the reaction. Liebig's improvements made the analysis of organic compounds easier and more reliable.

Classification of organic compounds: The notion of radicals
As more and more organic compounds were analyzed and their composition understood, attention turned to the relationship between composition and

Justus von Liebig (1803–1873)
The son of a pharmacist, Liebig took an early interest in chemistry. Although he studied at the Universities of Bonn and Erlangen, he was dissatisfied with the quality of chemistry education in German universities and dropped out to work with Gay-Lussac. At the age of 23, he became a professor at the University of Giessen, where he reformed the system of chemistry education and launched a program that emphasized experiments. He trained many outstanding organic chemists, and his laboratory developed into a research school. His pupils subsequently moved on to universities in many places, where they made major contributions to the development of organic chemistry and biochemistry. On the research side, Liebig is most famous for improving analytical techniques in organic chemistry. In his later years, he became interested in agricultural chemistry, physiological chemistry, and applied chemistry, and he made significant contributions to all of these fields (Column 6).

Figure 2.3: Liebig's Kaliapparat apparatus for the analysis of organic compounds. The Kaliapparat is at the right, while the calcium chloride tube is at the left.

properties, and efforts began to classify organic compounds based on chemical criteria. The first attempts involved the notion of *radicals*. The word "radical" had been used since the time of Lavoisier to refer to stable components of substances that retained their essential identities even while undergoing a sequence of reactions. In 1815, Gay-Lussac studied hydrocyanic acid and its derivatives and found that the behavior of its radical, CN, was similar to the behavior of chlorine in chlorides or iodine in iodides. Berzelius successfully applied his electrochemical dualisic theory to inorganic compounds, and reasoned that organic compounds were subject to the same laws. Reasoning that organic compounds were analogous to inorganic compounds, he believed that organic acids consisted of the radical of a compound bound with oxygen.

In 1828, Dumas and Boulley studied plant-based organic acid derivatives known as *esters* and were struck by the similarity between esters and ammonium salt, which suggested that esters were compounds of the C_2H_4 radical, termed *etherin*. These two chemists believed alcohol to be $C_2H_4 \cdot H_2O$. Around the same time, Liebig and Wöhler showed that bitter almond oil (benzaldehyde) could transform into a variety of compounds all possessing the same collection of atoms, $C_{14}H_{10}O_2$ (actually it is C_7H_5O), which they termed the benzoyl radical. These types of theories of radicals were favorably embraced by many chemists in the 1830s, who also discussed the equivalents of today's methyl, ethyl, acetyl, and benzoyl groups. However, many problems remained. First, there was still no common standard for the atomic weights of carbon and oxygen. Berzelius used C=12, O=16, but Liebig used C=6, O=8 and Dumas used C=6, O=16, leading to great confusion surrounding chemical formulas. Moreover, Berzelius used silver salts in his analysis of organic acids, but his value for the atomic weight of silver was twice the modern value; consequently, his values for the molecular weights of organic acids were also two times too big. As a result, he thought the formula for acetic acid was $C_4H_6O_3$ (in fact, it is $C_2H_4O_2$)[3]. Second, Berzelius remained a dogged adherent to his electrochemical dualistic theory, but it gradually became clear that it was impossible to apply this theory to the classification of organic compounds.

A development that was to pose major challenges to the electrochemical theory of radicals was the identification, by Auguste Laurent and Dumas, of the problem of *substitution*. Laurent chlorinated many organic compounds and studied their chlorine compounds. He compared the properties of the chlorinated and original compounds and noted that the properties changed little following chlorination, indicating that hydrogen was simply being substituted by chlorine. Dumas was led to study chlorides of fatty acids through his research on the stimulating components of candle smoke, and in the process he investigated the mechanism of alcohol chlorination. He also produced trichloroacetic acid and observed that its properties were similar to those of acetic acid. Thus, he too arrived at the conclusion that hydrogen bound to carbon could be replaced (substituted) by chlorine. However, the substitution of electrically negative chlorine for electrically positive hydrogen could not be explained based on the

---

3. Berzelius took the chemical formula of acetic acid to be $C_4H_6O_3$ and referred to C_4H_3 as an "acetyl group." This is not to be confused with the substance known today as an "acetyl group" (CH_3CO).

Auguste Laurent (1807–1853)
Laurent, a French chemist, was the son of a wine merchant. He was reluctant to pursue his father's line of work; his tutors recommended that he become a scientist, and he studied crystallography and organic chemistry at a mining school. After graduating in 1830, he became a research assistant to Dumas. In 1838, he became a professor of chemistry at the University of Bordeaux, and later conducted research there on coal tar derivatives and other topics. Laurent discovered many organic compounds and attempted to establish a systematic classification scheme for them.

dualistic theory of Berzelius. When substances such as trichloroacetic acid could be formed by substitution, Berzelius attempted to explain them using the theory of "copulae" (conjugators). "Copulae" (conjugators) were electrically neutral portions of molecules that exerted no impact on electrically active radicals; for example, acetic acid $(C_4H_6O_3)$[4] was formed by the binding of anhydrous oxalic acid (C_2O_3) with the methyl copula (C_2H_6), with the substitutions taking place inside the copula.

A notion that arose to replace classification by radicals was the proposal of Dumas to classify organic molecules according to *chemical type*. When multiple organic compounds—such as trichloroacetic acid and acetic acid—contained the same constituents, with the same equivalent weights, bound together in the same ways, and had the same properties and reactivities, they were deemed to belong to the same type. However, precisely what criteria were to be used to define types? As far as Dumas was concerned there *were* no absolute criteria, and ultimately his ideas were to have little impact.

New theories of types

The theory of types of organic compounds was advanced by Laurent and by Charles Frédéric Gerhardt. In 1839, Gerhardt proposed the residue theory, which held that certain inorganic compounds (including water, ammonia, and hydrogen chloride) were extremely stable, allowing them to separate easily from organic compounds; when this happened, the *residue* of the organic compound would bond to form a new organic compound. For example, in the nitration of benzene, hydrogen—a stable substance—is liberated, while the residues, a phenyl group and a nitro group, bond to form nitrobenzene. Gerhardt's residues formed bonds

---

4. Note that, at the time, the formula for acetic acid was not $C_2H_4O_2$ but rather $C_4H_6O_3$.

Charles Gerhardt (1816–1856)
Gerhardt was a French chemist, the son of a producer of lead white in Alsace. He was expected to take over the family business and studied chemistry at the University of Leipzig, but instead of taking over the business went on to study with Liebig and Dumas. After completing his degree in Paris, in 1841 he found work at the University of Montpelier, but left in 1848 to return to Paris. He was later a professor at the University of Strasbourg. Together with Laurent, Gerhardt opposed the dualistic theory of Berzelius and instead proposed the theory of types, serving to classify organic compounds. He was possessed of extraordinary talent, but made many enemies due to his eccentric personality.

with other residues during reactions; they did not simply reside on the original organic compound to give it electrical charge, as in electrochemical dualistic theories. The theory of residues could also be applied to substitution reactions. For example, the substitution reaction of acetic acid by chlorine proceeds when the residues of the reactants—in weights equivalent to the portions removed—react and bind together.

The residue theory also contributed to clarifying the problems of atomic and molecular weights. The chemical formulas used by Berzelius for organic acids were based on silver salts, but because his atomic weight for silver was twice the accepted modern value, the molecular weights of his organic acids was also twice too big. When the residue theory was applied to these organic acids, the molecular weights of the inorganic products liberated was two times larger than the typically used values. Thus, Berzelius proposed that the molecular weights of organic compounds should be halved. This proposal agreed simultaneously with the hypotheses of Avogadro and of Ampere, but it served only to deepen the confusion of the chemists of the time.

In 1846, Laurent adopted Gerhardt's atomic weights and defined the distinction between atomic weights, equivalent weights, and molecular weights as follows. Atomic weights were the smallest weights of the elements present in a compound. Molecular weights were the smallest quantities of substance needed to produce a compound. Equivalent weights varied with the properties of reactions and were either equal to the molecular weights or were integer multiples of them. He also recognized the diatomic nature of molecules such as hydrogen, oxygen, and chlorine. Thus, chemical formulas based on setting the molecular weight of the hydrogen molecule to be $H_2=2$ allowed the creation of a system of chemistry that unified the organic and inorganic branches of the field.

Charles-Adolphe Wurtz (1817–1886)
A French organic chemist, Wurtz studied with Liebig at Giessen and succeeded Dumas as professor at the Ecole de Medecine and later at the Sorbonne. His accomplishments include the discovery of substances such as the alkyl amines, glycol, and aldol, and his establishment of the Wurtz reaction.

Alexander Williamson (1824–1904)
Williamson was an English chemist who studied with Gmelin in Heidelberg and with Liebig in Giessen. From 1855 to 1887, he was a professor at the University of London. He clarified the mechanisms of the production of intermediate products when ether is made from alcohol and sulfuric acid.

Between 1844 and 1845, Gerhardt proposed a systematic classification scheme based on the notion of *homologous series*. This was a further generalization of ideas introduced previously by Schiel and Dumas. In this scheme, compounds belonging to the same "homologous series" had compositions that differed by one CH_2 each, and their melting and boiling points increased in arithmetic progression with each additional CH_2. He also defined *"isologous series"* of compounds, which had similar chemical properties but were not homologous (examples include ethyl alcohol and phenol), and *"heterologous series,"* in which the compounds exhibited different properties, but it was possible to produce one from the other through simple reactions (examples include ethyl alcohol and acetic acid).

This period also witnessed many new discoveries on the experimental front. Hoffman, in Liebig's laboratory, as well as Zinin, at Kazan University in Russia, conducted wide-ranging research on aniline derivatives. Meanwhile, Charles-Adolphe Wurtz, who studied with Liebig and was an assistant to Dumas, embarked on studies of amines. All of these investigations yielded a sequence of new organic compounds related to ammonia, and all these compounds were *basic*, not acidic. Hoffman started from ammonia (1) and synthesized ethylamine (2), diethylamine (3), and triethylamine (4), laying the theoretical foundations for ammonia-type compounds.

$$
\left.\begin{matrix} H \\ H \\ H \end{matrix}\right\} N \qquad \left.\begin{matrix} C_2H_5 \\ H \\ H \end{matrix}\right\} N \qquad \left.\begin{matrix} C_2H_5 \\ C_2H_5 \\ H \end{matrix}\right\} N \qquad \left.\begin{matrix} C_2H_5 \\ C_2H_5 \\ C_2H_5 \end{matrix}\right\} N
$$

$$
\text{(1)} \qquad\qquad \text{(2)} \qquad\qquad \text{(3)} \qquad\qquad \text{(4)}
$$

Laurent proposed that alcohol (2), alcoholate (3), and ether (4) could be described as "analogues" of water (1). However, it was Alexander Williamson who pursued this notion of analogous substances and expanded the idea of classifications based on water type. He attempted to substitute an alkyl group for a single hydrogen atom in ethyl alcohol by reacting potassium ethoxide with ethyl iodide and obtained diethyl ether. Based on this finding, he proposed that both alcohol and ether could be described as the products obtained by replacing one or two hydrogen atoms with ethyl groups.

$$
\left.\begin{array}{l} H \\ H \end{array}\right\}O \qquad \left.\begin{array}{l} C_2H_5 \\ H \end{array}\right\}O \qquad \left.\begin{array}{l} C_2H_5 \\ K \end{array}\right\}O \qquad \left.\begin{array}{l} C_2H_5 \\ C_2H_5 \end{array}\right\}O
$$

$$
\text{(1)} \qquad\qquad \text{(2)} \qquad\qquad \text{(3)} \qquad\qquad \text{(4)}
$$

Williamson described his classification of matter in terms of "water types," and proposed it as a basic model not only for organic substances, but also for inorganic salts and inorganic acids.

Gerhardt further advanced the theory of types by proposing that all organic compounds could be classified in terms of four inorganic types: water, ammonia, hydrogen, and hydrogen chloride[19]. Alcohols, ethers, acids, and esters were of water type; amines, amides, and nitrides were of ammonia type; paraffin, aldehydes, and ketones were of hydrogen type, while halogen compounds were of hydrogen chloride type. However, in Gerhardt's picture, this classification scheme had no *structural* significance; it merely conveyed the information of which substances had the possibility of undergoing which types of reactions.

Cannizzaro's revival of Avogadro's hypothesis

A number of problems remained to be resolved before chemical structure formulas could become reliable. In particular, by the mid-19th century there remained uncertainty surrounding atomic and molecular weights. Indeed, even the atomic weights of fundamental elements such as carbon and oxygen remained murky. Among organic compounds, it was known that there were many isomers with the same empirical formulas, while multimers with the same empirical formulas but different molecular weights also existed. Resolving these conundrums required accurate determinations of molecular weights. A chemist who made significant contributions to addressing these problems was Stanislao Cannizzaro. He realized that, by applying Avogadro's hypothesis, it was possible to solve many of the problems that vexed the chemists of the day, and he presented his ideas in a paper published in an Italian journal in 1858. He attempted to demonstrate the soundness of the assumptions underlying Avogadro's hypothesis..

Cannizzaro first recognized that the atomic weights of elements and molecular weights of compounds could be properly determined by comparing the densities

Stanislao Cannizzaro (1826–1910)
An Italian chemist born in Palermo on the island of Sicily, Cannizzaro studied medicine in Palermo and chemistry at Pisa. He fought in the volunteer army in the 1847 Sicilian revolution; after the battle was lost, he fled to Paris, where he studied with Chevreul. In 1851, he returned to Italy and found a teaching position, and in 1855, he became a professor at Genoa. Five years later, he resigned to rejoin the volunteer corps and agitate for Italian unification. After the Italian political unrest settled, he returned to Palermo. He achieved many successes in organic chemistry, including the discovery of Cannizzaro's reaction in 1858, but he is best known for his proof of the correctness of Avogadro's hypothesis.

of gases and vapor to the density of hydrogen. Because the hydrogen molecule consisted of two atoms, vapor densities determined with reference to hydrogen should be doubled. It was further accepted that—whereas hydrogen, oxygen, nitrogen, and chloride were diatomic molecules—the molecules of sulfur, phosphorous, mercury, and arsenic were *not* diatomic, but rather consisted respectively of 6 atoms, 4 atoms, 1 atom, and 4 atoms. The sulfur molecule consisted of 6 atoms at the temperature of vaporization, but at the higher temperature of 1000°C its molecular weight was 32 times that of hydrogen, from which it followed that the molecule was diatomic at that temperature. One piece of evidence that argued against the use of vapor density to determine molecular weights was that the molecular weights determined by vapor density for substances—such as ammonium chloride and phosphorous pentachloride—were anomalously small. Cannizzaro explained this as resulting from the fact that the vapors of these substances dissociated at high temperatures. In this way, Cannizzaro demonstrated the possibility of using vapor densities and the results of chemical analyses to obtain correct molecular formulas. At first, Cannizzaro's paper went unnoticed; however, this situation was soon to change with the convening of an international conference in the German city of Karlsruhe in 1860.

The objective of this international conference was to share opinions and construct a unified set of criteria to resolve the confusion surrounding atomic weights, equivalent weights, molecular weights, and chemical formulas. The conference was organized by Kekule and Wurtz with help from Weltzien in Karlsruhe, Baeyer in Berlin, and Roscoe in Manchester; it was attended by 140 of the most important chemists in Europe. It was here that Cannizzaro presented his argument that Avogadro's hypothesis should be applied to the determination of atomic and molecular weights; many conference participants were sympathetic to this proposal, but there remained some opposition to the notion of settling scientific questions by vote, and the conference ultimately failed to arrive at a conclusion. However, at the end of the conference Cannizzaro's paper was distributed by Pavesi (from Pavia University), and its impact gradually increased. After reading the paper, Lothar Meyer (see page 82) wrote that "the scales seemed to fall from my eyes. Doubts disappeared and a feeling of quiet certainty took their place."[20]. Meyer himself wrote a book—"*Die Modernen Theorie der Chemie*"—that advanced chemical theories based on Avogadro's hypothesis, and was influential among chemists; he later made important contributions to the establishment of the periodic law. In fact, Mendeleev—who attended the conference—later said that this conference was the first step toward the completion of the periodic law.

Chemists and the reality of atoms and molecules

The revival of Avogadro's hypothesis dispelled the confusion surrounding atomic and molecular weights, but still the chemists of the time did not necessarily believe in the reality of atoms and molecules. Dalton's atoms followed the tradition of Newtonian corpuscular philosophies, holding that atoms had physical reality as the indivisible constituents of matter ("physical atoms"), which bound together to form molecules. However, the subsequent confusion surrounding atomic and molecular weights persuaded some chemists to adopt a view of molecules as the minimal units participating in chemical reactions, leaving aside the question of their physical reality. For these chemists, atoms were "chemical atoms"—the minimal units that were useful for the purposes of explaining chemical phenomena. For many chemists, the question of whether or not atoms were real was an abstract, metaphysical question that could not be decided experimentally and was hence of little interest. In 1867, Kekule defined "chemical atoms" to be "particles of matter that cannot be further subdivided by chemical transformation." Cannizzaro insisted that "physical atoms" and "chemical atoms" were in fact identical, but a significant number of chemists in the late 19th century continued to deny the existence of atoms as real physical entities; indeed, controversy over the reality of atoms survived until the beginning of the 20th century.

2.4 The establishment and development of organic chemistry

As noted in the previous section, by the mid-19th century, the confusion surrounding atoms and molecules had been resolved and knowledge of organic compounds had expanded, setting the stage for major progress in organic chemistry. Until this time, organic chemistry had been treated as just another branch of chemistry, and many of the day's top chemists researched both inorganic

Hermann Kolbe (1818–1884)
A German chemist, Kolbe studied with Wöhler in Göttingen and became a professor at the University of Leipzig in 1865. He is known for the synthesis of acetic acid, for the Kolbe-Schmidt reaction, and for the synthesis of hydrocarbons by electrolysis of carboxylate.

Edward Frankland (1825–1899)
Frankland was an English chemist. After completing an apprenticeship with a pharmacist, he studied with Bunsen at the University of Marburg in Germany; upon returning to England, he became a professor at Owens College in Manchester. He founded the theory of valence based on his studies of organometallic compounds.

and organic compounds. In the second half of the 19th century, however, there arose a new breed of specialists who exclusively studied organic compounds, and organic chemistry began to make major strides toward establishing itself as an independent field of specialty. The first problem that needed to be solved by organic chemistry was the challenge of chemical structure.

Chemical structural formulas and the concept of valence

In order to clarify the notion of chemical structure, it was necessary first to establish the concept of valence. Progress toward this was made in the 1840s and 1850s by Hermann Kolbe, a German, and Edward Frankland. Kolbe, an adherent of Berzelius' theory of radicals, advanced the development of methods to identify the internal structure of the radicals and copulae discussed by Berzelius. From his studies of acetyl (C_4H_3) compounds[4], he was led to believe that these compounds were composed of ethyl conjugated with two equivalent weights of carbon $[(C_2H_3)C_2)]$ and that the carbon acted as the point of action for affinity to the other elements such as oxygen or chlorine. Kolbe and his English friend Frankland attempted to isolate actual alkyl radicals; Kolbe believed he had obtained a methyl group through electrolysis of acetic acid, but in fact, what he had obtained was ethane. Similarly, in 1849, Frankland believed he had obtained an ethyl group from a reaction between zinc and ethyl iodide, but in fact, what he had obtained was butane.

However, as a byproduct of this reaction, Frankland also obtained zinc ethyl $[(C_2H_5)_2Zn]$[21]. This development was important not only because it furnished the first successful example of the synthesis of an organometallic compound, but also because it provided the spur that Frankland needed to progress to the theory of valence. He thought of organometallic compounds as inorganic compounds in which oxygen or other elements were substituted by hydrocarbon groups.

Friedrich August Kekule (1829–1896)
A German organic chemist, Kekule was the son of a public servant in Darmstadt and studied architecture at the University of Giessen, but switched to chemistry on the advice of Liebig. After working in Paris, London, and Heidelberg, he became a professor at the University of Ghent in Belgium in 1858 and at the University of Bonn in 1867. He was a highly imaginative researcher and had a brilliant career as a chemist, determining the structural formula of benzene and laying the foundations of classical organic structure theory.

Archibald Couper (1831–1892)
Born in Glasgow, Couper studied philosophy in Glasgow and Berlin but became interested in chemistry and studied with Wurtz in Paris. Although he started work as a chemist in Edinburgh, his research was cut short by illness.

He also pondered the similarities between organic and inorganic compounds and stated that, in both cases, elements had only a limited number of types of binding force. Although this was not yet a clear notion of valence, it was the first time that atoms in chemical formulas were treated as units with specific individual binding strengths.

In 1858, August Kekule and Archibald Couper independently announced that carbon was tetravalent and could combine with each other to construct carbon chains[22, 23]. In the structural formula proposed by Couper [Figure 2.4(a)], bonds are indicated by solid or dashed lines, a remnant of conventions from the days of type formulas. Kekule introduced the notions of valence and chemical bonds, but did not use lines or symbols to indicate them. He attempted to express chemical formulas in a visually intuitive way using sausage diagrams [Figure 2.4(b)], but this proved inconvenient. Couper's friend Butlerov at Kazan University used the concept of the tetravalence of carbon to develop further the theory of chemical structure. In 1864, Alexander Crum Brown popularized visually intuitive structural formulas in which bonds were denoted by lines and element symbols were circled [Figure 2.4(c)]. After Erlenmeyer dropped the circles surrounding the elements, the use of such diagrams spread rapidly.

The structure of benzene and aromatic compounds
The notion of *unsaturation* had been advanced by Loschmidt, Erlenmeyer, and Meyer, and Crum Brown had used a double bond to explain the bonding of ethylene. However, the structure of benzene and its associated compounds posed

(a)   (b)   (c)

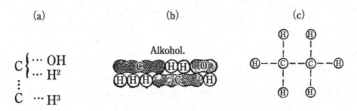

Figure 2.4: The evolution of chemical formulas

(a) Cooper's ethyl alcohol (1858), (b) Kekule's ethyl alcohol (1861), (c) Crum Brown's ethane (1864)

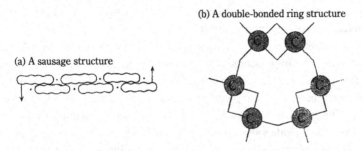

(b) A double-bonded ring structure

(a) A sausage structure

Figure 2.5: Kekule's structural formulas for benzene

In structure (a) eight unsaturated units of affinities are shown by dots and arrows.

a vexing challenge. It was the highly imaginative Kekule who eventually showed the way past this conundrum. From the fact that benzene and its derivatives always contained six or more carbon atoms, he proposed molecules in which the six atoms were bound by single bonds alternating with double bonds. He first conceived of an open-chain structure containing eight unsaturated units of affinity that did not participate in bonding, but he was eventually led to believe that the ends of the chain bound together to form a closed ring structure (C_6A_6, where A represents an unsaturated unit of affinity). Here A would bind to another single atom or multiatomic group. Kekule's first paper employed the sausage structure [Figure 2.5(a)], but by 1865, this had become a hexagonal ring structure [Figure 2.5(b)]. He was aware of the possibility that substitution could lead to a variety of different isomers; for the case of substitutions of halogen atoms, he predicted that (a) substitution of a single atom would lead to one species of isomer, (b) substitution of two, three, or four atoms would lead (in each case) to three species of isomer, and (c) substitution of five atoms would lead to one species of isomer[24]. These predictions were quickly verified by Ladenburg and Korner. Kekule, in his later years, recalled that the idea of the hexagonal

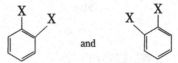

Figure 2.6: Two ortho-substitute structures of benzene that arise if the double bonds are fixed in place.

structure had occurred to him while daydreaming—in front of the fireplace in his home in Ghent, Belgium—of a snake dancing while biting its tail, but the credibility of this anecdote has been a subject of controversy in recent years and is doubted by many science historians[25].

Kekule solved the problem of the tetravalence of the carbon atoms in benzene by proposing that single bonds alternate with double bonds, but many chemists remained unconvinced by this explanation. The reason for their doubt was that the addition reaction seen in typical unsaturated compounds was not observed in benzene. Other structures were proposed to explain this. In addition, Ladenburg pointed out that Kekule's structure should give rise to two distinct types of ortho substitutes, and proposed an alternative structure (Figure 2.6). In response to these objections, Kekule proposed the notion of *resonance structures*, according to which all positions in the benzene ring were equivalent, and there was no reason to expect the existence of two types of ortho substitutes[26]. According to Kekule, atoms in molecules were constantly oscillating around their equilibrium positions and colliding with adjacent atoms, whereupon single bonds and double bonds existed in dynamic equilibrium, and the two benzene ring structures were in fact equivalent. However, the fourth valence of the carbon atoms in benzene remained an unsolved problem and persisted as a source of controversy for decades; its proper resolution had to await the emergence of quantum chemistry (see Section 4.3).

The results of research on aromatic compounds after Kekule's proposal of the structure of benzene did not conflict with Kekule's structure. Although the problem of the fourth valence remained, most chemists took Kekule's structure as a satisfactory working hypothesis and continued their research on this basis. The chemistry of aromatic compounds made rapid progress in the second half of the 19th century. It was quickly proposed that the naphthalene molecule consisted of two condensed benzene rings, and this was confirmed experimentally. Other aromatic compounds, including anthracene and phenanthrene, were also studied. In addition, many hetero-ring compounds such as pyridine and quinoline, in which the ring contains elements other than carbon, were also discovered and studied during this time.

Stereochemistry and the regular tetrahedral theory of carbon

In 1815, the French physicist Biot discovered that many organic liquids have the effect of rotating the polarization planes of light. Because liquids are not crystals in which the direction of the molecules is aligned, this suggested that the ability to change polarization must exist within the structure of the molecules themselves. The solution of this problem by Louis Pasteur in the late 1840s was a major breakthrough in chemistry.

In 1847, working in Balard's laboratory in Paris, Pasteur began research on tartaric acid for his doctoral dissertation. Tartaric acid is a compound obtained from the waste products generated during winemaking, and two types of this substance were known to exist. Solutions of ordinary tartaric acid had the effect of rotating the polarization plane of incident polarized light to the right (it was optically active), but the other type of tartaric acid—known as racemic acid—did not exhibit this effect; it was optically inactive. These two types of tartaric acid were thought to have identical chemical properties and crystal structures, so the origin of the distinct optical behavior was mysterious. Pasteur used a magnifying glass to study the crystal structures of these two types of tartaric acid in detail and showed that, whereas crystals of optically active tartaric acid were asymmetric, racemic acid contained a roughly equal mixture of two types of crystals—which, though asymmetric, were mirror images of each other. Pasteur separated these two types of crystal and showed that a solution of the *first* type rotated polarized light to the *right*, while a solution of the *second* type rotated polarized light to the *left*, and a solution containing an equal mixture of both types was optically inactive (Figure 2.7). Pasteur conducted further research on tartaric acid and developed optical selection techniques to separate racemic mixtures into mirror-image isomers (enantiomers). These included both a method based on differences in the rate of consumption by microorganisms and

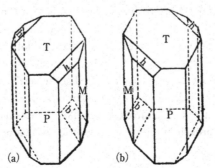

Figure 2.7: Pasteur's sketch of the two optical isomer crystals of tartaric acid

Crystals (a) and (b) are mirror images of each other.

Louis Pasteur (1822–1895)

A French chemist and microbiologist Pasteur was the son of a poor tanner in Dole, in the Jura region of France. He studied with Dumas and Balard at the Ecole Normale Superieure and received a degree in 1857 for research on optical isomers of tartaric acid salts. He served as a professor at the Universities of Strasbourg and Lille and at the Ecole Normale Superieure; in 1867, he became a professor at the Sorbonne, and from 1888 onward directed the Pasteur Institute. After his work on optical isomers, he turned to the study of fermentation; his accomplishments include the discovery of lactic acid bacteria, the development of low-temperature sterilization methods, and his refutation of the natural emergence theory of life. He is known as the father of modern microbiology.

a method in which an optically active base was added to acidic racemic bodies to form diastereomeric salts[5]. Thus, research on optical isomers progressed, but the relationship to the structure of molecules remained unclear. Pasteur later moved away from research on optical activity to study fermentation and the biological decomposition of plant material.

In 1874, Jacobus Henricus van 't Hoff and J. A. Le Bel independently clarified the relationship between optical activity and molecular structure, thus laying the foundations of the field of stereochemistry. At this time, van 't Hoff was just 22 years old and had not received a Ph.D. degree. Le Bel was educated in Paris and served as an assistant to Wurtz. Although lactic acid ($CH_3CH(OH)COOH$) had been identified in many fermentation processes, Berzelius had also found this compound in animal muscle tissue. The lactic acid observed in fermentation processes was optically inactive, but the lactic acid found in muscle tissue was optically active. In 1869, Wislicenus proposed that this distinction arose from differences in the spatial arrangement of atoms. van 't Hoff furthered this notion by treating carbon atoms as regular tetrahedra; in 1874 he showed that, if one assumed that atoms or atomic groups bonded at the vertices of the tetrahedra, then bonds formed between four distinct atoms or atomic groups could exist in two distinct mirror-image structures (Figure 2.8)[27]. That same year, Le Bel took a more abstract approach by proving geometrically that molecules involving

---

5. If two or more asymmetric atoms are present in a molecule, inverting just one of these produces a non-mirror-image isomer. Such isomers are called diastereomers.

Jacobus Henricus van't Hoff (1852–1911)
A Dutch chemist, van 't Hoff was born the son of a physician in Rotterdam, Holland. Against the wishes of his father, he studied chemistry in Delft and Leiden, then studied with Kekule in Bonn and with Wurtz in Paris; he received his degree at Utrecht. In 1876, he joined the faculty of a veterinary school, but after two years became a professor in Amsterdam, where he remained for 18 years before becoming a professor at the University of Berlin. He first studied organic chemistry and laid the foundations of stereochemistry by proposing the regular tetrahedral theory of carbon; later he switched to physical chemistry and achieved successes in research on osmotic pressure and chemical equilibrium. He is famous as a pioneer of physical chemistry. In 1901, he had the honor of receiving the first Nobel Prize in Chemistry.

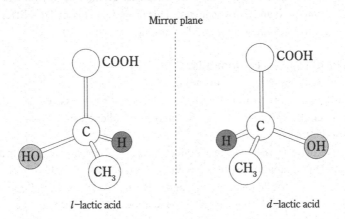

Mirror plane

l−lactic acid   *d*−lactic acid

Figure 2.8: Structures of the d- and l-forms of lactic acid

The two structures are mirror images of each other, and the carbon atoms are asymmetric.

four distinct atoms or atomic groups bound to carbon atoms will exhibit optical activity unless there is internal compensation by a symmetry plane[28].

The isomers of tartaric acid discovered by Pasteur were explained by positing the existence of d-type and l-type structures, each having two asymmetric carbon atoms, and racemic bodies, consisting of a mixture of the two. van't Hoff showed that all known examples of optical activity involved the presence of asymmetric carbon atoms. The notion of asymmetric carbon atoms, put forth by the then-unknown van't Hoff, was not generally accepted at first; the theory initially attracted a fair number of dissenters, but gradually became accepted as

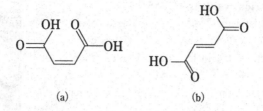

(a)   (b)

Figure 2.9: Structures of maleic acid (a) and fumaric acid (b)

research progressed. Isomers possessing asymmetric carbon atoms were termed *stereoisomers* by Viktor Meyer in 1888.

In an 1874 paper, van 't Hoff proposed the existence of a different type of isomers known as *geometric isomers*. He pointed out that, if a double-bonded carbon molecule contained at least two distinct substituent groups, the double bonds obstruct free rotation within the molecule, giving rise to two species of isomers (termed *cis-* and *trans*-isomers). This proposal explained the differences in the structures of maleic acid and fumaric acid (Figure 2.9).

Analysis of organic compounds

The methods established by Liebig around 1830 for the organic analysis of carbon and hydrogen already exhibited a high degree of completeness, and they did not change much thereafter. However, the analysis of other elements was improved in a variety of ways.

Analysis of nitrogen had to be conducted separately from analysis of carbon or hydrogen. Dumas made two important improvements: (a) he used carbon dioxide instead of air in the combustion tube, and (2) he used a concentrated solution of potassium hydroxide. Using carbon dioxide allowed direct measurement of the volume of nitrogen produced by the sample. Dumas obtained carbon dioxide by heating lead carbonate, but in 1838 Erdman and Marchand invented a carbon dioxide generator that worked by applying acid to carbonate.

In 1841, Varrentrapp and Will introduced a separate method in which the sample was heated together with soda lime; this converted nitrogen into ammonia, which was captured in an acidic solution. In 1883, Kjeldahl improved this technique by using concentrated sulfuric acid, to which permanganate had been added, to process the sample before the soda lime step; this both simplified and improved the accuracy of nitrogen analysis of proteins and other biological substances.

Liebig also developed techniques for analyzing sulfur and halogens in organic compounds. He used nitrate to oxidize organic substances in alkali solutions and analyzed them as precipitates of barium sulfate. The same oxidation method could also be used for organic halides. In this case, the halides produced were analyzed as precipitates of silver salts. More general methods of analysis for sulfur and halogens were proposed by Carius in 1864. In his approach, the sample was oxidized in a closed glass tube using concentrated nitric acid, and the resulting sulfur or halogen hydroxides were analyzed respectively as barium salts or silver salts.

Progress in synthesis techniques

Synthesis of organic compounds was an important and popular field of chemistry in the second half of the 19th century. The reasons for this were many. First, the synthesis of naturally occurring compounds—as well as the synthesis of compounds that do *not* exist in nature—was a challenge for chemists. Moreover, the synthesis of compounds needed to verify the concepts underlying the latest theories and structures in chemistry was a compelling pursuit. In addition, as the chemical industry rose to meet the demand for practical products such as medicines and dyes, the need to synthesize new compounds became more urgent.

A chemist who attained notable success in synthesizing many organic compounds around the mid-19th century was Marcellin Berthelot. He achieved a number of milestones in studies of glycerin and multivalent alcohols. He reacted sulfuric acid with ethylene, and then hydrolyzed the product to synthesize alcohol; he also reversed this process, removing water from alcohol to synthesize olefin. In 1856, he produced methane by high-temperature heating of barium formate, obtaining ethylene, propylene, and acetylene as byproducts. In 1862, he passed hydrogen gas through an electrical arc between carbon electrodes to produce acetylene, while in 1866 he passed acetylene through a red-hot tube to obtain benzene. Thus, he showed that many important organic compounds could be synthesized from inorganic compounds, dealing the final deathblow to the theory of vitalism.

The second half of the 19th century witnessed the discovery of many synthetic reactions that remain in widespread use to this day; many of these are named for their discoverers. A representative sample of such reactions is listed below; the year of the discovery is given in parentheses.

- Wurtz's reaction (1855): $3RX + 3R'X' + 6Na \rightarrow R–R + R'–R' + R–R' + 6NaX$
(Here R and R' denote alkyl groups and X is a halogen)

- Cannizzaro's reaction (1858): $2C_6H_5CHO + KOH \rightarrow C_6H_5CH_2OH + C_6H_5COOK$

- Fittig's reaction (1864): $C_6H_5Br + C_2H_5Br + 2Na \rightarrow C_6H_5C_2H_5 + 2NaBr$

Marcellin Berthelot (1827–1907)

Berthelot was a French chemist and politician. The son of a doctor in Paris, he studied history and philosophy before becoming a chemist. He worked as an assistant to Balard at the College de France and later held joint appointments at the College de France and at a school of pharmacology. Berthelot was a firm believer in the idea that chemical phenomena arose from physical forces, and that  even complex organic compounds could be synthesized without the use of vitalism; he synthesized many compounds—including hydrocarbons, fats, and sugars—and made major contributions to thermochemistry. He was a highly influential chemist in the late 19th century. In his later years, he was active as an educational administrator and politician.

• Friedel-Crafts reaction (1877): $AlCl_3$

$$C_6H_6 + RCl \rightarrow C_6H_5R + HCl$$

(Here R is an alkyl group)

• Perkin's reaction (1868): CH_3COONa

$$C_6H_5CHO + (CH_3CO)_2O \rightarrow C_6H_5CH=CHCOOH + CH_3COOH$$

• Reimer-Tiemann reaction (1876): $C_6H_5OH + CHCl_3 + 3NaOH$

$$\rightarrow HOCC_6H_4OH + 3NaCl + 2H_2O$$

Progress in synthetic methods led to the synthesis of a huge number of compounds, dramatically expanding the catalog of known organic compounds; this created a need to organize what was known and to consolidate the nomenclature of organic compounds. A commission to study nomenclature was formed at an international conference in Paris in 1889, and the recommendations of this commission were accepted in Geneva in 1892, codifying a commonly accepted system of nomenclature for organic compounds. However, some terminology that had been previously used remained in use thereafter due to custom and habit.

Through the first half of the 19th century, Gmelin's "*Handbuch der Chemie (Handbook of Chemistry)*" had played the role of a quick reference manual for chemical information, but this volume proved unable to keep up with the progress of organic chemistry; eventually its section on organic chemistry was discarded and it became a reference on inorganic chemistry alone. To meet the needs of organic chemistry, Beilstein—a Russian-born chemist working in Wöhler's laboratory—compiled a 2,201-page "*Handbuch der Organischen*

Chemie (*Handbook of Organic Chemistry*)" whose first edition was published by a Hamburg publishing house between 1880 and 1882. The second edition, spanning 4,080 pages, was published between 1886 and 1889. In 1866, Beilstein returned to St. Petersburg, where he devoted his full energy to compiling and editing the *Handbuch* from the late 1870s until his death in 1906.

2.5 The periodic law of the elements

The application of Avogadro's Law had eliminated the uncertainty surrounding atomic weights, and the introduction of the notion of valence had solidified the foundations of modern chemistry. The next step was to consolidate the elements into a systematic classification scheme and to understand the relationships among them. Although there had been attempts to classify the elements before 1860, they were doomed to failure by the confusion surrounding atomic weights. Before any meaningful classification of the elements could be constructed, a sufficiently large number of elements had to be known and their atomic weights had to be accurately determined.

In the years between 1790 and 1859, 31 new elements were discovered, bringing the total number of known elements to more than 60; of these, only 5 new elements were discovered between 1830 and 1859. The discovery of new elements was reaching the limits of traditional chemical methods. It was at this point that the new tool of spectroscopy burst onto the stage, allowing the discovery of new elements to proceed apace through the 1860s. This development only furthered awareness of the necessity for a classification system for the elements.

The dawn of spectroscopy and the discovery of new elements

The first scientists to use spectroscopy to discover new elements were Robert Bunsen and Gustav Kirchhoff at the University of Heidelberg in Germany. Bunsen used the color of flames to analyze salt in hard water. When he arrived at Heidelberg, gas-fueled lighting had recently been introduced within the city limits; Bunsen quickly began to use gas in his laboratory and, in an effort to develop a gas burner that produced a colorless flame for use in chemical analysis, invented the Bunsen burner. It was already known that, by using a prism to separate the light arising from the colored flames emitted by salts, one could obtain a unique spectrum characteristic of the elements present. Bunsen and Kirchhoff reasoned that mixtures of salts could be more easily analyzed by using a prism to separate the light emitted by their flames, and thus built a spectrometer (Figure 2.10). They immediately achieved a number of successes with this apparatus, ranging from the analysis of hard water to

Robert Bunsen (1811–1899)

Bunsen, a German chemist, was the son of a philologist in
Göttingen. He graduated from the University of Göttingen,
then became a professor at Kassel in 1836 and at Marburg
in 1839, and in 1852, he succeeded Gmelin as professor
at Heidelberg. He conducted research in many areas
of chemistry and physics. As a chemist, he was highly
inventive; he not only invented the Bunsen burner, but was
also involved in the development of photometers, calorimeters, and other
devices. Bunsen also made major contributions to research on blast furnaces
in England.

Gustav Kirchhoff (1824–1887)

A German physicist, Kirchhoff was the son of a lawyer in
Königsberg. He studied at the University of Königsberg
and served as a professor at the Universities of Heidelberg
and Berlin. In 1849, he discovered Kirchhoff's laws of
electric circuits, and in 1859, he discovered Kirchhoff's
law of radiation. Later, he collaborated with Bunsen in
research on spectral analysis. His studies of spectroscopy
and black-body radiation stimulated later research on atomic structure and
quantum theory.

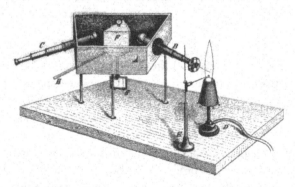

Figure 2.10: The spectrometer of Bunsen and Kirchhoff

the discovery of cesium in 1860 and the discovery of rubidium the following
year. Spectroscopic methods quickly began to demonstrate great promise for
the discovery of other elements. In 1861, thalium was discovered by Crookes
and iridium was discovered by Reich and Richter. Spectroscopic methods also

played a crucial role in the subsequent discovery of gallium, the rare-earth elements, and the noble-gas elements.

First attempts to classify the elements

Attempts to classify the elements continued throughout the first half of the 19th century. In 1817, Döbereiner, a friend of Goethe and a professor at the University of Jena, realized that the alkali earths formed a three-element set (calcium, strontium, and barium). By the end of the 1820s, the concept of three-element groups had been expanded to encompass many more elements, including the alkali metals (lithium, sodium, potassium), the halogens (chlorine, fluorine, iodine), and the sulfur family (sulfur, selenium, tellurium). Gmelin reported many other three-element sets in his *Handbook of Chemistry*.

In 1850, Liebig's friend Pettenkofer, a public health researcher, reported that elements that exhibited similar chemical behavior frequently exhibited equal differences in atomic weight. Dumas realized that, in some elemental series, the pattern of atomic weights followed an arithmetic progression. For example, elements in the series O, S, Se, Te had weights 8, 16 (8+8), 40 (8 + 4×8), and 64 (8 + 7×8). (Dumas used atomic weights in which O=8.) He noted the existence of this sort of numerical sequence for the groups (F, Cl, Br, I), (N, P, As, Sb, Bi), and (Mg, Ca, Sr, Ba, Pb).

In 1862, the French geologist Chancourtois devised a three-dimensional organization scheme for the elements known as the *telluric helix*. In this graph, similar elements were placed in the same vertical row, with points corresponding to the individual elements aligned by 16 gaps. He believed the atomic weights were given by an expression of the form n+16n' where n was 7 or 16 and n' was an integer. However, this report was ignored by chemists and ultimately had little impact.

In 1863, the Englishman Newlands reported that, if one listed the elements in order of atomic weight, then starting from the first element in each group and counting 8 elements down yielded a similar element belonging to the same group. Reasoning by analogy with the notion of an octave in music, he formulated a "law of octaves," in which the elements formed "octaves" in multiples of 8. However, this classification scheme had several apparent problems and was not accepted with much enthusiasm by the chemists of the day. Although he had taken a correct first step by listing the elements in order of atomic weight, his classification scheme yielded too few new insights to be useful. Thus, by 1865, there had been a number of attempts to classify the elements, but none was fully satisfactory.

Dmitri Ivanovich Mendeleev (1834–1907)

Mendeleev was the youngest of 14 children, the son of a teacher in the Siberian city of Tobolsk. After his father lost his eyesight, his mother supported the family by managing a glass factory; the subsequent tragedy of his father's death was followed by further misfortune when the glass factory burned down. Still, his mother, possessed of an indomitable spirit, took Dmitri to Saint Petersburg, where he received a high-school education. After teaching in Odessa, Mendeleev returned to the University of Saint Petersburg, where he researched the physical properties of liquids. He won a traveling fellowship to study abroad in Paris and Heidelberg. In Heidelberg, he studied with Bunsen and Kirchhoff, and before his return to Russia, he attended a conference given in Karlsruhe at which a paper by Cannizzaro made a significant impression on him. Back in Saint Petersburg, he taught at a school of engineering technology and in 1866 became a professor of general chemistry at the University of Saint Petersburg. His immortality in the history of chemistry is secured by his efforts to classify the elements and his publication of a periodic table in 1871.

Lothar Meyer (1830–1895)

Meyer, a German chemist, was the son of a doctor, Meyer was educated to be a doctor himself, but was influenced by Bunsen to become a chemist. He taught medical students at a variety of institutions before becoming a professor of chemistry at the University of Tübingen in 1876. He felt the need to write clear textbooks in a new style, and in 1864 he published *Die modernen Theorien der Chemie* ("Modern Theories of Chemistry"); in the process of revising this book, he proposed a classification of the elements and arrived at a periodic table similar to that of Mendeleev.

The Periodic Laws of Mendeleev and Meyer

The two scientists who succeeded independently in discovering the periodic table were the Russian Dimitri Mendeleev[29] and the German Lothar Meyer. Mendeleev's first paper was published in 1869[30]; although Meyer had independently arrived at similar ideas by that time, his paper was not published until 1870[31]. Mendeleev's periodic table attracted little interest at first, but rose quickly to prominence after the discovery of a new element—gallium—whose existence his paper predicted.

Tabelle II.

Reihen	Gruppe I. — R²O	Gruppe II. — RO	Gruppe III. — R²O³	Gruppe IV. RH⁴ RO²	Gruppe V. RH³ R²O⁵	Gruppe VI. RH² RO³	Gruppe VII. RH R²O⁷	Gruppe VIII. RO⁴
1	H=1							
2	Li=7	Be=9,4	B=11	C=12	N=14	O=16	F=19	
3	Na=23	Mg=24	Al=27,3	Si=28	P=31	S=32	Cl=35,5	
4	K=39	Ca=40	—=44	Ti=48	V=51	Cr=52	Mn=55	Fe=56, Co=59, Ni=59, Cu=63.
5	(Cu=63)	Zn=65	—=68	—=72	As=75	Se=78	Br=80	
6	Rb=85	Sr=87	?Yt=88	Zr=90	Nb=94	Mo=96	—=100	Ru=104, Rh=104, Pd=106, Ag=108.
7	(Ag=108)	Cd=112	In=113	Sn=118	Sb=122	Te=125	J=127	
8	Cs=133	Ba=137	?Di=138	?Ce=140	—	—	—	— — — —
9	(—)							
10	—	—	?Er=178	?La=180	Ta=182	W=184	—	Os=195, Ir=197, Pt=198, Au=199.
11	(Au=199)	Hg=200	Tl=204	Pb=207	Bi=208	—	—	— — — —
12	—	—	—	Th=231	—	U=240	—	

Figure 2.11: Mendeleev's periodic table (1871)

In 1867, Mendeleev began to write a textbook—*Principles of Chemistry*—for use in his lectures. In the process, he attempted to classify elements by valence and atomic weight. Hydrogen, oxygen, nitrogen, and carbon were typical examples of univalent, divalent, trivalent, and tetravalent elements. There were also the groups of alkali metals and halogens. He wrote the names and properties of elements on cards and entertained himself on a long train journey by attempting to classify the cards according to similarities between parameters such as atomic weight, valence, and elemental properties. He realized that, when elements were listed in order of atomic weight, their properties exhibited clear periodicities. He assembled a periodic table and published it in 1869[30]. In this work, Mendeleev introduced a number of bold innovations: He separated hydrogen from the other elements, he left empty spaces for as-yet-undiscovered elements, and he collected elements with similar atomic weights and properties into groups. He also reasoned that, since the atomic weights of the day contained some inaccuracies, it was necessary in some cases to rearrange the order of the atomic weights.

In 1871, Mendeleev published an improved periodic table (Figure 2.11)[32]. In this table, the elements were arranged into 12 horizontal rows and vertical columns labeled I through VIII for groups (families) of elements. Group VIII contained empty entries until one arrived at Fe, Co, and Ni. The treatment of the similarities among transition metals—listed together in Mendeleev's table—and the rare-earth elements, of which only a few had then been discovered, were clearly problematic given the knowledge available at the time. Moreover, the atomic weights contained inaccuracies. For example, the equivalent weight of indium was 38.3, and its atomic weight was thought to be twice this number,

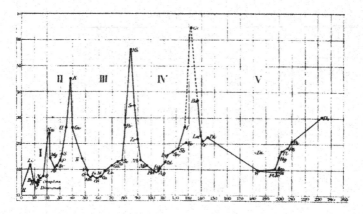

Figure 2.12: Meyer's graph showing periodic law

This is a revised version of the graph he presented in 1872. The horizontal axis indicates the atomic weight, and the vertical axis indicates the atomic volume.

76.6; this would place indium between arsenic and selenium in the periodic table, but there was no open space for it there. Mendeleev instead took the atomic weight to be three times the equivalent weight and inserted indium between cadmium and tin. Similarly, the atomic weight of beryllium was thought at the time to be 14, based on the assumption that its oxide had chemical formula Be_2O_3; when the calculation was redone assuming the chemical formula BeO, one obtained an atomic weight of 9.4, making beryllium the third element. However, the most dramatic consequence of Mendeleev's work was his idea to leave some entries of the table empty and to predict the properties of the elements expected to fill those slots; sure enough, the predicted elements were discovered almost immediately.

Mendeleev predicted the new elements ekaboron (atomic weight 44), ekaaluminium (atomic weight 68), and ekasilicon (atomic weight 72). Here *eka*, the Sanskrit word for "one," was used to mean "in the following slot." The French chemist Boisbaudran, while using spectroscopy to study the rare earths in 1871, identified the spectrum of gallium; he isolated the element and determined its properties. These properties agreed astonishingly well with those predicted by Mendeleev for ekaaluminium, thus confirming the correctness of the periodic table. Ekaboron was discovered in 1879 as scandium, while ekasilicon was discovered in 1871 as germanium; in both cases, the properties of the elements closely matched the predictions of Mendeleev.

Mendeleev's first paper attracted little attention until Meyer presented his paper in 1870. Meyer's *Modern Theories of Chemistry*, published in 1864, was a widely read textbook; in preparing a revised edition, Meyer attempted to classify the elements and ultimately constructed a periodic table similar to that of Mendeleev. However, the publication of this revised edition was unfortunately delayed until 1872, and Meyer presented an outline of his ideas in the form of a paper in 1870[31]. In this paper, Meyer plotted the elements on a graph with atomic weight on the horizontal axis and the atomic volume[6] of the solid elements on the vertical axis (Figure 2.12). Meyer had clearly captured the essence of Mendeleev's period law. Nonetheless, the fair-minded Meyer acknowledged the priority of Mendeleev and confessed that he had personally lacked Mendeleev's courage to predict undiscovered elements.

2.6 Advances in analytical chemistry and inorganic chemistry

Analytical methods were of decisive importance in discovering new elements and correctly determining atomic weights. Chemists such as Klaproth, who discovered several new elements at the beginning of the 19th century, or Berzelius, who accurately determined many atomic weights, were experimentalists with surpassing analytical technique. The introduction of new methodologies, such as spectroscopy, also contributed significantly to the development of chemistry. Thus, analytical chemistry was a crucial field aiding the progress of chemistry as a whole, but in the first half of the 19th century analytical chemistry had not yet emerged as its own specialized subfield. However, the second half of the century witnessed the arrival of many chemists specializing in analytical chemistry, who worked to improve analytical techniques and to develop new methodologies. These developments then contributed to even higher-precision determination of atomic weights.

An important family of specialists in analytical methods was the Fresenius clan. Carl Fresenius (1818–1897) was born in Frankfurt and educated in Bonn. Later he earned a Ph.D. degree working with Liebig in Giessen and became a professor at an agricultural school near Wiesbaden. In 1848, with the help of his father—a wealthy attorney—he established an independent laboratory, where he trained many analytical chemists and conducted a wide-ranging variety of chemical analyses. His sons Heinrich and Theodor, and his grandsons Ludwig and Remigius, were also analytical chemists.

The crowning jewel of 19th century chemistry was organic chemistry, but there were researchers who specialized in the chemistry of elements other than carbon. At the time Mendeleev presented his periodic table, 63 elements were known, but the final 30 years of the 19th century witnessed a continuing succession of new element discoveries—particularly rare earths and noble

---

6. The atomic volume is the volume occupied by 1 mole of individual atoms.

gases—further enriching the world of chemistry and advancing the state of inorganic chemistry. All of this was made possible by advances in analytical methods. As the end of the century approached, coordination chemistry arose as a field fusing the inorganic and organic branches of chemistry; it would subsequently experience rapid progress.

Qualitative analysis, gravimetric analysis, and volumetric analysis

Qualitative analyses to investigate the constituents incorporated in samples had been conducted since antiquity, but these were largely empirical and had never been systematized. The level of systematization remained inadequate even during the time of Fresenius and his grandsons. The use of hydrogen sulfide to separate metal groups was expanded by Carl Fresenius, who demonstrated how to separate a greater number of metals by adjusting the acidity or alkalinity of the solution. The development of a gas-generating device by Kipp in 1862 simplified the use of hydrogen sulfide. However, it would take another 20 years for the systematic foundations of group separation to be solidified and qualitative analysis procedures to be standardized; these developments required an infusion of concepts and knowledge from physical chemistry, including equilibration processes and oxidation and reduction reactions.

The methods of gravimetric analysis that were used so effectively by Bergman, Klaproth, and Berzelius had, by the late 19th century, become popular and reliable analytical techniques. This era did not witness any particular profusion of new ideas or methods, but filter paper, crucibles, and balances were improved and analytical methods attained higher degrees of refinement. Near the mid-19th century, instrument makers began to manufacture and sell new types of balances.

Although methods of volumetric analysis had been used for practical analyses in the past, they were not firmly established as analytical techniques before the 19th century. Their foundations were established by Gay-Lussac. In 1824, he investigated the potency of bleaching powder by observing its color-removing effect on an indigo solution. He then moved on to quantitative titration of acids and bases. However, because the determination of equivalent points was difficult, volumetric analyses were not generally performed. Moreover, proper equipment for volumetric measurements had not yet been developed. It was Mohr who overcame these difficulties to place volumetric analysis on a sound footing.

Karl Friedrich Mohr (1806–1879), who was educated at Bonn, Heidelberg, and Berlin, managed a pharmaceutical business but was interested in chemistry research and conducted studies in analytical chemistry. He was blessed with a particular talent for improving equipment and analytical methods, and he made a number of improvements to pipettes, burettes, flasks, and other tools used for volumetric analysis. Titration methods to quantify chlorides using solutions of silver nitrate were set on firm ground by Mohr's introduction of potassium chromate as an internal indicator. Mohr is also responsible for introducing oxalic acid as a standard reagent for neutralizing titrations and the use of Mohr's salt (ammonium iron sulfate). Later, in 1853, Bunsen introduced iodometry and used it to study reduction-oxidation reactions; this was followed by Schwartz's introduction of the use of sodium thiosulfate in oxidation/reduction titrations.

Instrumental analysis

In the second half of the 19th century, instrumental analysis techniques were first used in analytical chemistry experiments and achieved a number of note-worthy successes. As noted in Section 2.5 above, the spectroscopic methods pioneered by Bunsen and Kirchhoff were to play a decisive role in the discovery of new elements. Spectroscopy was subsequently used by astronomers and enabled the discovery and identification of the elements present in the sun and stars. Indeed, helium was first discovered in the spectrum of sunlight. As spectroscopic methods improved and came to enjoy widespread use, they played important roles in research on rare earths and noble gases in the second half of the 19th century.

Refractometry and polarimetry emerged as spectroscopic methods that exploited the characteristic optical properties of compounds. These techniques became possible due to improvements in the quality of optical lenses that resulted from the demand for better microscopes and telescopes. In 1828, Giovanni Amici developed color-cancelling lenses (achromatizing lens) that could compensate for variations in refractive index due to the wavelength of light; this led to an immediate improvement in the performance of microscopes. Optical tools were further improved by Abbe and Schott in the second half of the 19th century. Abbe, working with a company founded by Carl Zeiss, began to develop and improve optical equipment in 1860, and developed the Abbe refractometer. Because optically active materials rotate plane-polarized light, it was clear that polarimetry could be useful in analysis. Improvements in polarimetry proceeded rapidly, and in the later part of the century, it was widely used for purposes such as the analysis of sugars.

Electrolytic analysis was also important in the second half of the 19th century. In 1864, Oliver Gibbs reported an analysis of copper and nickel using electrolysis. This was followed by many further reports of applications of electrolytic analysis to analytical chemistry.

Determination of atomic weights

Avogadro's hypothesis was accepted after the international conference at Karlsruhe in 1859, and a platform for the accurate determination of atomic weights was established (see page 67). Determining atomic weights accurately was one of the major goals of analytical chemists in the second half of the 19th century. The advent of the periodic table had mandated the reconsideration of several atomic weights, and chemists rushed to measure the atomic weights of the newly discovered elements. In the 1860s, Stas reported accurate atomic weights for 12 elements.

The atomic weight of indium was reconsidered and found to agree with Mendeleev's revision. The atomic weight of platinum was also revised. The atomic weights of the three "eka" elements predicted by Mendeleev were determined by their respective discoverers. The atomic weights of other elements such as vanadium, aluminum, and thorium were also determined during this period.

However, there continued to be no accurate determination of the ratio of the atomic weights of hydrogen and oxygen, despite the obvious importance of this quantity. In 1842, Dumas determined this ratio to be 1:7.98, but suspected that some error remained in this estimate. In the 1880s, this problem was taken up by several chemists, with the ratio of 1:7.9375, reported by Noyes at the U.S. Bureau of Standards, taken to be the most reliable number.

At this time there was still no consensus among chemists as to whether the reference point for atomic weights should be H=1 or O=16; thus, none of the various tables of atomic weights that existed could be considered universal. In 1893, at the request of the American Chemical Society, an official table was constructed by Frank Clarke. The German Chemical Society also released an official table in 1898. In 1903, an international committee regarding atomic weights was formed, with working committees consisting of English, American, French, and German chemists. The internationally recognized table was published from 1903 until the First World War.

In the 1890s, the American chemist Theodore Richards (1868–1928) took up the challenge of accurately determining atomic weights. He had studied in Germany after graduating from Harvard; thereafter he returned to Harvard, where he taught analytical chemistry. Together with many collaborators, he paid careful attention to conducting robust studies, and investigated the atomic weights of a total of 28 elements. For his achievements, he became the first American to be awarded the Nobel Prize in Chemistry, in 1914. Richards noted that metal chlorides or bromides were better suited to the determination of atomic weights than oxides, and used chlorides of barium, strontium, lead, magnesium, nickel, cobalt, iron, calcium, uranium, and cesium to determine atomic weights

Theodore Richards (1868–1928)
Richards was an American chemist who studied at Harvard and received his degree in 1888, then studied abroad in Germany with Viktor Meyer. After returning home, he joined the staff of his alma mater and became a professor in 1901. He made precise measurements of the atomic weights of many elements. In 1913, Richards demonstrated slight differences in the atomic weights of lead taken from various mineral samples, thus confirming the existence of isotopes. For his achievements in the precision measurement of atomic weights, in 1914 he became the first American to win the Nobel Prize in Chemistry.

for these elements. He discovered in 1904 that the value for sodium given by Stas was inaccurate.

Moissan and the discovery of fluorine

By the 1830s, it was known that the three halogens, chlorine, bromine, and iodine, were all elements. At this time, the existence of fluorine was known, but isolating it proved extremely difficult. Fluorite, whose primary ingredient was calcium fluoride, had been used since antiquity as a flux for iron manufacturing and other purposes. Scheele noted that the acid of fluorite could dissolve silicon. Pure hydrofluoric acid was obtained in 1809 by Gay-Lussac and Thenard. Davy tried to obtain fluorine from the electrolysis of fluorides, but he failed in this effort, succeeding only in damaging his health due to the toxicity of the substances he used. In 1854, Fremy at the Ecole Polytechnique noted the release of a gas, believed to be fluorine, from the anode during the electrolysis of anhydrous calcium fluoride, but he was unable to isolate it.

In 1886, Henri Moissan electrolyzed a hydrogen fluoride solution of potassium fluoride at low temperatures and successfully isolated fluorine in a platinum vessel. Moissan was a dominant figure in inorganic chemistry from the end of the 19th century to the beginning of the 20th century. In particular, he advanced the state of research on oxides and fluorides of iron group metals. Using silver fluoride and alkyl iodide, he created a number of alkyl fluorides, and contributed significantly to expanding knowledge of fluorine and fluorine compounds. He developed an electrical furnace in which chemical reactions could be carried out at high temperature and used it in attempts to synthesize artificial diamonds; although he was thought at the time to have succeeded, he is now believed to have failed. Moissan's electrical furnace generated high temperatures by inducing arc discharge between carbon electrodes in a lime block. Using this furnace, he

Ferdinand Henri Moissan (1852–1907)
A French chemist, Moissan graduated from pharmacology
school in Paris, studied chemistry under Fremy at the
natural history museum, then became a professor at a
pharmacology school and at the Sorbonne. He specialized
in inorganic chemistry and succeeded in isolating fluorine.
Moissan laid the foundations of high-temperature chemistry
through achievements such as using an electric furnace to
synthesize many carbides and silicides and successfully dissolving otherwise
indissoluble substances. He is also known for his attempts to synthesize
diamond. He was awarded the Nobel Prize in Chemistry in 1906.

fabricated heat-resistant oxides, metals, carbides, nitrides, silicide, boride, and
conducted wide-ranging studies of their properties. He used carbon to reduce
metal oxides and succeeded in isolating chrome, manganese, molybdenum,
tungsten, vanadium, uranium, zirconium, titanium, and other metals. In so
doing, he laid the foundations for high-temperature chemistry.

Moissan received the 1906 Nobel Prize in Chemistry for his research on
fluorine compounds, chromium compounds, carbides and electrical furnace;
Mendeleev was also a candidate for that year's prize, and in the final vote tally,
the two candidates differed by only a single vote. Both men died in 1907, the
following year. Although Moissan was without question a great chemist, it is a
pity that Mendeleev lost forever the opportunity to be awarded a Nobel Prize.

Isolation of the rare-earth elements
Research on rare-earth elements began at the end of the 18th century, when
Johan Gadolin (1760–1852) discovered that some black minerals found in the
Stockholm suburb of Ytterby contained an unknown type of earth (oxide).
This earth was further investigated by Ekeberg and was named *yttria*. In 1803,
Klaproth, as well as Berzelius and Hisinger, isolated another type of earth,
which Berzelius named *ceria*. These earths were found in the form of oxides;
they were called "earths" for their similarity to oxides of alkali earths, and the
term was qualified to "rare earths" as the substances were thought to exist in
trace quantities in the earth's crust. Research on yttria and ceria was inordinately
challenging, and their discovery marked the beginning of a struggle that was
to challenge analytical chemists for a century. Figure 2.13 summarizes the
historical trajectory of the discoveries of rare-earth elements.

Between 1839 and 1843, the Swedish chemist Mosander discovered that
yttria and ceria were not oxides of a single element, but were rather mixtures.

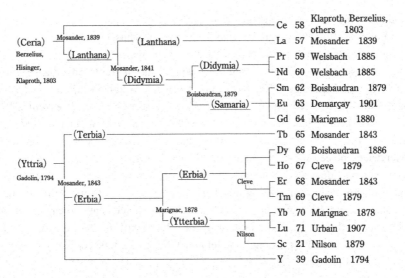

Figure 2.13: History of the isolation of the rare earths elements

Names in parentheses are the names given at the time. Underlined names are mixtures. The column at the right indicates the present-day element symbol and atomic number, the discoverer, and the year of the discovery.

By dissolving these oxides in strong acid and separately crystallizing them as various types of salts, oxides of new elements were isolated from the principal components of yttria and ceria. The elements discovered in ceria were named *lanthana* and *didymia*, while the elements discovered in yttria were named *erbia* and *terbia*. Subsequently it became clear that didymia and erbia were themselves mixtures. These discoveries were puzzling to the chemists of the day. The properties of the oxides of these elements were remarkably similar; they could be distinguished only by atomic weight and by slight differences in solubility. Thereafter there was no major progress in the field until the 1870s.

In 1878, the Swiss chemist Marignac showed that *erbia* could be separated into two components—erbia and *ytterbia*—after which the isolation of the rare earths proceeded rapidly. The following year, Nilson showed that ytterbia could be further separated into *scandia* and ytterbia. On the other hand, erbia could be separated into *three* components [erbia (Er), *holmia*, and *thulia* (Tm)], and holmia was later separated into *dysprosia* (Dy) and holmia (Ho). It was around this time that spectroscopic methods were introduced; by looking at spectra, it was possible to assess the progress of separation of mixtures and to identify new elements. Mosander's didymia was also identified as a mixture; in 1879, it was separated into didymia and *samaria*, and later the elements *praseodymia* (Pr) and *neodymia* (Nd) were isolated from didymia, while samaria (Sm), *europia* (Eu), and *gadolinia* (Gd) were isolated from samaria. Thus, most of the rare-earth elements were known by the end of the 19th century.

The discovery of large numbers of rare-earth elements posed a vexing challenge to Mendeleev. The facts that the atomic weights of these elements were similar to each other, and that their properties were all essentially identical, seemed to destroy the concept of the periodic law. When the periodic law was first formulated in 1869, only 6 rare-earth elements were known, and Mendeleev placed these in Groups III, IV, and V. Meyer placed the rare-earth elements outside the periodic table, and as more and more rare-earth elements were discovered an increasing number of chemists came to view this as the most promising strategy. The problem of the proper placement of the rare-earth elements within the periodic table was not solved until the 20th century with the emergence of Bohr's atomic model and Moseley's studies of X-ray spectra (see Sections 3.2 and 3.6).

The discovery of the noble gases and revisions to the periodic table

After the 1783 experiments of Cavendish, in which he produced nitrous acid (nitrogen oxide)—a nitride—by sending electrical sparks through a mixture of nitrogen (phlogistonated air) and oxygen (dephlogistonated air), Cavendish realized that air contained small quantities of non-reacting gases other than oxygen, nitrogen, carbon dioxide, and water. However, until the 1880s nobody thought that unknown *elements* might be present in air.

Lord Rayleigh (John Strutt), a professor of physics at the Cavendish Laboratory at Cambridge University, was spurred by Prout's hypothesis (see Section 2.1) to believe that the strikingly near-integer values assumed by the atomic weights could not be a coincidence; he embarked on a precise determination of the densities of hydrogen and oxygen. He used a variety of methods to produce oxygen and to measure its density; in 1892, he reported that these values all agreed to within experimental error and yielded a value for the density of oxygen equal to 15.882 times the density of hydrogen. He investigated nitrogen in similar ways and discovered that the nitrogen released from ammonia was lighter than the nitrogen extracted from air by an amount greater than experimental error. He considered various possible explanations for this (such as the existence of N_3), but ultimately could not identify the cause; he reported his results in the journal *Nature* to seek the opinions of others[33].

Around this time, William Ramsay—a professor of chemistry at University College in London—was puzzling over the same conundrum. He discussed the problem with Lord Rayleigh, and—with Rayleigh's encouragement—began a series of experiments to investigate what remained when nitrogen, oxygen,

John Strutt (Lord Rayleigh) (1842–1919)

An English physicist, Strutt was born to a family of nobility in Terling, Essex. He studied at Trinity College in Cambridge and in 1873 became Lord Rayleigh. In 1879, he succeeded Maxwell as the second Cavendish professor, but retired after 5 years to continue research in his private laboratory in Terling. Strutt achieved successes in a wide range of fields of classical physics, including electromagnetic theory, optics, mechanics of continuum solids, and thermal physics. He was awarded the 1904 Nobel Prize in Physics.

William Ramsay (1852–1916)

Ramsay was a British chemist. The son of a civil engineer in Glasgow, Ramsay studied at the University of Glasgow and studied abroad with the organic chemist Fittig at the University of Tübingen. He became a professor at the University of Bristol in 1880 and at University College London in 1887. After discovering several of the noble gases and thus demonstrating the existence of group-0 elements, he turned to research on radioactivity and proposed a theory for the disintegration of radioactive elements. He was awarded the 1904 Nobel Prize in Chemistry.

water vapor, and carbon dioxide were removed from air. By the summer of 1884, he had confirmed that the atmosphere contained unknown, inert gases, and the two scientists proceeded to conduct studies using a division-of-labor scheme: Lord Rayleigh investigated the physical properties of the unknown gases, while Ramsay studied their chemical properties. In August 1894, the two delivered a preliminary report to a meeting of the British Association for the Advancement of Science, where they were exposed to harsh criticism. Nonetheless, they continued to conduct exhaustive experiments and meticulous observations and remained confident in the existence of unknown gases. Ramsay's residual gas exhibited a spectrum not found among any of the existing elements. This gas, which exhibited absolutely no reactivity, was named "argon" after the Greek word for "lazy," and its existence was reported in an 1895 paper[34]. Lord Rayleigh subsequently attempted to exploit differences in diffusion rates to separate nitrogen from argon; he determined the density of argon to be 19.7 (assuming $H_2=1$). The ratio between the specific heat at constant pressure (C_p) and that at constant volume (C_v) indicated that this gas was monatomic. The atomic weight of argon was 40.

The discovery of argon raised the question of where to place this new element in the periodic table. Its atomic weight was larger than that of potassium, but its properties made it impossible to think of it as a successor to potassium. Ramsay demonstrated that argon belonged to a new row of elements positioned between chlorine and potassium, but many chemists refused to believe that argon could be a gas of monatomic molecules. However, the ensuing succession of discoveries of new noble gases left no question as to the need to revise the periodic table.

Ramsay discovered that the inert gas produced by heating the mineral known as cleveite exhibited spectral lines different from those of argon. He realized that this spectrum was identical to that previously observed by Lockyer for the element helium, which was discovered to exist in the sun. Density measurements revealed the atomic weight of this atom to be 4, and it was a monatomic gas like argon[35].

Ramsay also acquired a large quantity of liquid air, which he subjected to fractional distillation, then vaporized the small quantity of liquid remaining at the end of the process and investigated its spectrum. The results indicated new spectral lines, and thus krypton was discovered. By 1898, neon and xenon had also been discovered to be present in air[36]. Thus, 5 inert noble gases had been discovered and the question of where in the periodic table to place atoms of zero valence had been resolved. However, a proper understanding of *why* these elements were inert would have to await the emergence of quantum theory. For his discovery of the noble-gas elements, in 1904 Ramsay became the first Englishman to win the Nobel Prize in Chemistry. The Nobel Prize in Physics that same year was awarded to Lord Rayleigh.

Werner and the birth of coordination chemistry

Chemical structure theory had achieved dazzling successes in the field of organic, but not in inorganic, chemistry. The principle of valence was valid for simple inorganic compounds, but the salts of many inorganic compounds are complicated, and the question of how to represent their structures remained a difficult unsolved problem. The situation was further complicated by the fact that for many metals, the valence is not fixed.

In 1822, Gmelin obtained oxalic acid cobalt salt, $(Co(NH_3)_6)_2(C_2O_4)_3$, from an ammonia solution of cobalt salts. In 1851, Fremy synthesized cobalt salt, $(Co(NH_3)_5Cl)Cl_2$; however, only a portion of the chlorides of this salt precipitated upon the addition of silver nitrate. Thus, interest was spurred in such metal coordination compounds; although coordination compounds of cobalt, chrome, and platinum would continue to be studied for the next 40 years, the problem of their structure remained unresolved. In 1869, Blomstrand proposed that Fremy's cobalt coordination compound was $Co_2Cl_2 \cdot 12NH_3$, and that it formed a

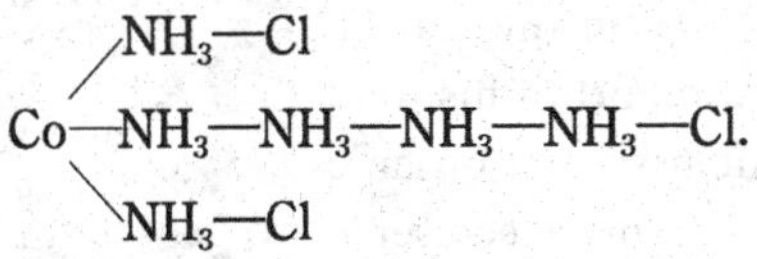

$$NH_3—Cl$$
$$Co—NH_3—NH_3—NH_3—NH_3—Cl.$$
$$NH_3—Cl$$

(a)

1. $[Co(NH_3)_6](NO_2)_3$
2. $[Co(NH_3)_5NO_2](NO_2)_2$
3. $[Co(NH_3)_4(NO_2)_2]NO_2$
4. $[Co(NH_3)_3(NO_2)_3]$
5. $[Co(NH_3)_2(NO_2)_4]K$
6. $[Co(NH_3)(NO_2)_5]K_2$
7. $[Co(NO_2)_6]K_3$

(b)

Figure 2.14: Chemical formulas for cobalt coordination compounds

(a) Jorgensen's chemical formula for $CoCl_3(NH_3)_6$
(b) Werner's chemical formulas for a sequence of coordination compounds

chain structure in which nitrogen was pentavalent. However, when this salt was heated it lost one-sixth of its ammonia, while even with the addition of silver nitrate, one-third of the chloride did not precipitate out of solution. Jorgensen proposed that this coordination compound was $CoCl_3 \cdot 6NH_3$ and that it exhibited the structure shown in Figure 2.14(a).

In response to this problem, young Alfred Werner proposed an entirely original idea[37]. In 1892, he suggested that, in cobalt coordination compounds, ammonia molecules did not form bonds in a chain structure, but rather that cobalt atoms were hexavalent, with ammonia molecules bonded directly to cobalt atoms in the center, with substitutions by nitrous acid or chloride ions. For a sequence of cobalt coordination compounds, he proposed formulas like those shown in Figure 2.14(b).

Werner showed that similar formulas applied to coordination compounds of platinum, chrome, and other metals. He proposed the notions of primary and secondary valence, with the bond between the central metal ion and the ammonia molecules due to secondary valence, which he called the coordination number. To demonstrate the correctness of this theory, Werner applied the theory of electrolytic dissociation, which had only recently been proposed by Arrhenius; from measurements of electrical conductivity, he showed that the negative ions bonded to metal ions were not bound by ionic bonding (Figure 2.15). He thought of chemical effects in terms of an inner region and an outer region,[7]

---

7. In metallic coordination compounds, the region in which ligands are bound directly to the central metal is the inner region, while the region outside of this, in which solvent molecules and counter-ions are weakly bound, is the outer region.

Alfred Werner (1866–1919)
Born in Mulhouse in the Alsace region of France, Werner studied and received his degree from the Zurich Polytechnikum. When he published a report of his joint work with Hantzsch on the stereochemistry of organic nitrogen compounds, he had only just become a lecturer at an engineering school. In 1893, he became an assistant professor at the University of Zurich; he was quickly promoted to professor. He conducted pioneering research on coordination chemistry and proposed new theories of binding in metallic coordination compounds. He was awarded the 1913 Nobel Prize in Chemistry.

(a) *Cobalt Series.*

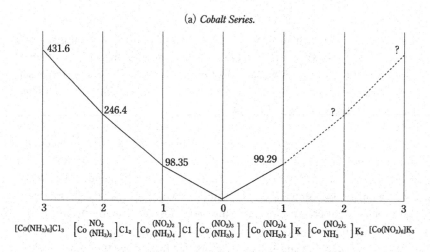

Figure 2.15: Electrical conductivity of cobalt coordination compounds

The vertical axis is the electrical conductivity, while the horizontal axis is the number of ions in the outer region. Taken from Werner's *New Ideas of Complex Cobalt Compounds* (1911).

and he considered the existence of isomers due to the different arrangement of ions in the inner region.

In the case of $Co(NH_3)_2Cl_4$, he envisioned the two isomers shown in Figure 2.16. In fact, the existence of these types of isomers in coordination compounds had been discovered in 1890 by Jorgensen, who found cobalt coordination compounds with two ethylene-diamine substitutions. However, the existence of the two isomers of $Co(NH_3)_2Cl_4$ was not verified until 1907.

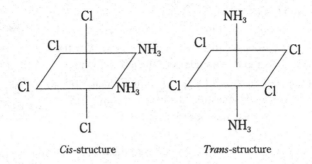

Cis-structure   Trans-structure

Figure 2.16: Stereoisomers of cobalt coordination compounds

Werner's research pioneered many new concepts and created the new field of coordination chemistry[38]. This new field was to play a role in linking organic and inorganic chemistry. Coordination chemistry also played an important role in the field of biochemistry that developed later. His contributions to modern chemistry are difficult to overstate. In 1913, he was awarded the Nobel Prize in Chemistry for his contributions to the understanding of atomic valence in metal coordination compounds.

2.7 Thermodynamics and the kinetic theory of gas molecules[10,39,40]

With the system of atomic weights firmly established and the periodic law of the elements clarified, the groundwork had been laid for major progress in chemistry. Meanwhile, a deeper understanding of heat and energy had been achieved primarily through the work of physicists, and the introduction of these ideas into chemistry contributed significantly to systematizing chemical knowledge.

Since the 17th century, there had existed two competing viewpoints regarding the nature of heat: the *kinetic theory* and the *substance theory*. The kinetic theory held that heat was simply the motion of the atoms or particles that composed material bodies; underlying this theory lay Newton's corpuscular philosophy. The substance theory posited the existence of a massless substance called *caloric* and attempted to explain thermal phenomena in terms of the entry and exit of this substance into matter. This caloric theory exerted a powerful hold on science from the mid-18th century onward. Lavoisier thought of caloric as one of the elements, and many other influential chemists—including Scheele, Dalton, Gay-Lussac, and Avogadro—were believers in the caloric theory. However, in the late 18th century, Count Rumford observed that, in the process of constructing artillery, massive quantities of heat were released by the process of boring holes

Sadi Carnot (1796–1832)

A French physicist and engineer, Carnot was born the son of a famous revolutionary in Paris. He was educated by the finest scientists and mathematicians of the day at the Ecole Polytechnique and became an officer in the French Army. He was interested in the question of how to develop efficient heat engines, and considered the efficiency of steam engines for this purpose. In 1824, he published Reflections on the Motive Power of Fire, in which he proposed the ideas that would eventually be enshrined in the form of the second law of thermodynamics; unfortunately, his ideas were not understood at the time, and he died an early death due to cholera.

Benoit Clapeyron (1799–1864)

Clapeyron was a French physicist and engineer. After studying at the Ecole Polytechnique in Paris, he became involved in the day-to-day business of engineering and taught at the École des Ponts et Chaussées. In 1834, he wrote a report presenting the achievements of Carnot and deriving an equation relating the vapor pressure of a liquid to the temperature

to form cannon barrels; he understood that a close relationship existed between heat and mechanical work, and thus the kinetic theory lived on as well.

Understanding of thermal phenomena was greatly advanced by the industrial revolution of the 18th century, spurred by the steam engine that provided the basic driving force behind the revolution. A major problem facing the engineers of the day was how to build high-efficiency heat engines. In Glasgow, Watt improved Newcomen's steam engine to invent a new type of steam engine that played a major role in the development of the industrial revolution. Heat also became a topic of research interest among scientists. Watt's friend Black, a professor at Glasgow University, introduced the notions of temperature and quantity of heat and laid the foundations of thermal science by establishing methods for measuring them. Studies of heat constituted an active area of research in 19th-century England and France, and it was during this time that the foundations of thermodynamics were established.

Carnot and heat engines

The first major contributions to the process leading to the development of thermodynamics were made by Sadi Carnot. In 1824, he wrote *"Reflections on the Motive Power of Fire"*, which clarified that work is done when heat flows

James Joule (1818–1889)
An amateur English physicist, Joule was born the son of a wealthy brewer in Manchester and received most of his education at home. Dalton, in his later years, tutored him in mathematics. He was uninterested in the brewing business but was passionate about physics research. Throughout his life, he never held a university position in either research or teaching, instead conducting research using homemade apparatus built with his own funds in the brewery. His accomplishments include the discovery of Joule's law relating electric current to heat produced, and measurements of the heat equivalent of work.

from a warm temperature to a low temperature and considered the efficiency with which heat could be converted into work. Carnot's paper was written in the framework of caloric theory, which held that heat was a gas; nonetheless, it correctly derived the conclusions that the efficiency of an idealized heat engine depended only on the temperature of the high-temperature heat source and the low-temperature heat sink, and was independent of the operating substance that drove the engine. However, this paper was so far ahead of its time that it attracted little attention at the time of its publication. Some ten years later, it finally caught the eye of Benoit Clapeyron, who reinforced its content and presented it more clearly with equations; thereafter it would have a major impact on the development of thermodynamics. Unfortunately, in 1832, Carnot contracted cholera and died at the age of 36, and most of the notebooks in which he had recorded his subsequent ideas and developments were lost. From what little remained, it is clear that Carnot had already begun to doubt the caloric theory; he intended to conduct Count Rumford's experiment in water to examine the relationship between the heat evolved and the power used.

The Law of Conservation of Energy and the Mechanical Equivalent of Heat

The pioneering work of Carnot and Clapeyron was not widely known until the 1850s, but there was progress toward an understanding of the nature of heat. In 1840, a German physician named Julius Robert von Mayer (1814–1878), having passed his medical examinations, signed on as a ship's doctor for a Dutch commercial ship bound for Java. He observed that the venous blood of the crew members was significantly redder in warm than in cold regions. He reasoned that, because in warm regions the body needs less oxidation to maintain body temperature, the color of the blood changed less. This observation led Mayer to think deeply about the relationship between human consumption of food and the relationship between heat produced and work done; he became convinced

William Thomson (Lord Kelvin) (1824–1907)

A British physicist, Thomson was born the son of a mathematics teacher in Belfast, Ireland. He was educated as a gifted child and entered the University of Glasgow at the age of 10, and later studied at Cambridge. At the age of 22, he became a professor at Glasgow, where he served as professor of natural philosophy until 1899. He made many important contributions to a wide range of fields of science and technology in the second half of the 19th century. He is known in particular for his contributions to the systematization of thermodynamics, his co-discovery of the Joule-Thomson effect, and his invention of various instruments for measuring electromagnetic phenomena. He is also famous for playing a leadership role in installing the first undersea transatlantic telegraph cable. He is a scientific giant who personifies the scientific and engineering achievements of the British Empire in the Victorian era; he was knighted in 1866.

Hermann von Helmholtz (1821–1894)

A German physiologist and physicist, von Helmholtz learned natural philosophy from his father, studied medicine at the University of Berlin, and became a military doctor while continuing research on physiology and lecturing on physiology at various universities. Later his interest in physics grew, and in 1871, he became a professor of physics at the University of Berlin. He left behind a wide variety of accomplishments spanning a wide range of fields in physics and physiology.

that heat and work must be convertible into each other, and he presented this conclusion in 1842. Although he did not conduct experiments himself, he noted the difference between constant-pressure and constant-volume specific heats of gases and understood that this difference could be ascribed to the work done by the expansion of the gas at constant pressure. The value he obtained using data available at that time was 3.6 J / calorie.

The scientist who made the most important contributions to the law of conservation of energy was James Joule. In 1837, while investigating the effects of heating using an electrical current, he discovered that the quantity of heat arising from the flow of electrical current was proportional to the resistance of the conductor and to the squared magnitude of the current (Joule's law). In 1843, Joule used mechanical work to rotate wheel blades in a water-filled vessel and measured the quantity of heat evolved to determine the mechanical equivalent of heat. He obtained a value of 4.169 J / calorie. Because Joule had not graduated from a university and was not publicly recognized as a scientist, the paper he submitted to the Royal Society was rejected, and his research at first earned a cool reception. However, its importance was recognized by William Thomson at an 1847 conference of the British Association for the Advancement of Science, after which it came gradually to be accepted. Joule and Thomson later became friends and collaborated on studies of the properties of gases; in particular, they discovered the *Joule-Thomson effect*, in which a gas cools due to adiabatic expansion[8].

William Thomson—known as Lord Kelvin—first used the term *thermodynamics* in 1849, and he made major contributions to systematizing this new science. Kelvin realized the essential correctness of Joule's work and was a supporter of his research. Thus, scientists gradually began to accept the notion that heat was not a substance, but was rather a form of motion. When the law of conservation of energy was formalized, it stated that the total change in the energy of a system was equal to the sum of work done on it and the heat supplied to it: the first law of thermodynamics.

A scientist who made major contributions to disseminating the law of conservation of energy was the German physicist and biologist Hermann von Helmholtz. In 1847, he published *"On the Conservation of Force"*, which offered a comprehensive treatment of the problem of energy conservation. He set forth a detailed dynamical theory of heat and showed that energy that appeared to be lost in collisions was in fact converted into heat. The English attorney William Globe (1811–1896) pursued research in physics as a hobby; in 1846, he wrote a book entitled *"On the Mutual Relationship between Physical Forces"*, in which he provided a clear explanation of the conservation of energy.

Entropy and the second law of thermodynamics

Twenty years after the death of Carnot, the importance of his work was realized and further developed by Kelvin. In 1845, Kelvin went to Paris to pursue a research collaboration with Henri Regnault; it was here that he became acquainted with Carnot's work and began to expand on it. As Joule's experiment had demonstrated, work is easily converted into heat; however, clear limitations restrict the conversion of heat into work. When an engine operates, not all the heat it absorbs is converted into work; a portion of the heat is transferred from a high-temperature body to a low-temperature body and a portion of the energy is dissipated. Kelvin pondered this phenomenon

---

8. An adiabatic expansion is one that occurs without transfer of heat between a system and its surroundings.

Henri Regnault (1810–1878)
Regnault was a French chemist who studied with Liebig and achieved his first successes in the field of organic synthesis. Later he turned to the study of the thermal properties of substances; he became a professor of physics at the College de France and contributed to the development of thermdynamics.

in great depth and eventually arrived at what we now know as the second law of thermodynamics: *heat flows irreversibly from warm bodies to cold bodies*.

Based on the Carnot cycle,[9] in 1848 Kelvin proposed the notion of absolute temperature. The definition of absolute temperature was based on the experimental fact that the work done on the outside world by an expanding gas was proportional to its temperature. According to this, the absolute temperature was equal to the temperature in Celsius plus 273.15. Today absolute temperature is denoted by the symbol K in honor of Kelvin. The notion of absolute temperature also demonstrates the existence of a lowest possible temperature: 0 K. In 1851, Kelvin stated in the following way:

> It is impossible (by means of inanimate material agency,) to derive mechanical effect from any portion of matter by cooling it below the temperature of the coldest of the surrounding objects.[10]

This is one way of expressing the second law of thermodynamics.

A contemporary of Kelvin who also developed Carnot's ideas and achieved an even deeper level of understanding was the German physicist Rudolf Clausius. He was a theorist who developed thermodynamics and molecular kinetic theory; he analyzed the Carnot cycle in depth and introduced the concept of entropy. He expressed the second law as follows:

> Heat can never pass from a colder to a warmer body without some other change, connected therewith, occurring at the same time.

The net heat absorbed by a heat engine is the difference between the heat Q_h absorbed from a high-temperature heat bath and the heat Q_l discharged to a low-temperature heat bath, and in an ideal heat engine, this is equal to the amount of work done by the engine. The efficiency η of a heat engine is given by

---

9. The Carnot cycle refers to the ideal heat engine considered by Carnot. In this cycle, an ideal gas used as an operating substance performs work on the outside through a cycle involving isothermal expansion, adiabatic expansion, adiabatic compression, and isothermal compression.

> ### Rudolf Clausius (1822–1888)
> Born in the Prussian city of Koslin (now part of Poland), Clausius studied at the Universities of Berlin and Halle. In 1855, he became a professor of mathematical physics at ETH Zurich, and he later served as professor at Würzburg and Bonn. He is most famous for his continuation of Carnot's research, in which he introduced the concept of entropy and formulated the second law of thermodynamics, but he also left behind outstanding accomplishments in other fields such as molecular kinetic theory and electromagnetism.

$(Q_h - Q_l)/Q_h$. In 1854, Clausius showed that the efficiency η of a Carnot engine operating reversibly between two absolute temperatures T_h and T_l is given by

$$\eta = (Q_h - Q_l)/Q_h = (T_h - T_l)/T_h$$

If the heat engine is non-ideal, its efficiency will be lower than this. From this equation, we obtain the following relationship,

$$Q_h/T_h - Q_l/T_l \geqq 0$$

In this equation the equal sign applies to an ideal engine, and the inequality sign applies to a non-ideal engine. Clausius further advanced the theory by focusing on the quantity Q/T. To represent this quantity, he introduced the notion of entropy (S). When a system underwent an infinitesimal change involving the absorption or emission of heat, the infinitesimal change dS in the entropy of the system was related to the quantity of heat dQ that was absorbed or emitted by the system and to the absolute temperature T according to the relation $dS = dQ/T$ for a reversible reaction. Clausius also considered more general processes involving multiple temperatures and showed that, in an ideal cyclical process consisting entirely of reversible processes, the integral of the quantity $dS=dQ/T$ vanished. For irreversible processes we have $dS > dQ/T$, and in cyclical processes involving irreversible sub-processes the integral of dS is positive. Because spontaneously occurring processes are irreversible, processes that occur naturally will always cause the overall entropy of a system and its surrounding environment to increase; this is the *law of increasing entropy*. Clausius formulated the two laws of thermodynamics as follows:

The energy in the universe is constant. (First law)

The entropy of the universe tends toward a maximum. (Second law)

The concept of entropy was not easy for the scientists of that time to understand. Even Kelvin preferred to think in terms of heat dissipation rather than using entropy. However, the concept of entropy successfully explained many naturally occurring changes and would go on to exert an immeasurably powerful influence over all branches of the natural sciences.

The development of the kinetic theory of gases

Efforts to understand the properties of gases—such as those encapsulated in Boyle's Law or Gay-Lussac's Law—on the basis of molecular theories had been ongoing long before the birth of thermodynamics. In 1738, the Swiss physicist Bernoulli (1700–1782) understood that the pressure of a gas arose from collisions of the gas molecules with the surfaces of the containing vessel, and derived equations corresponding to Boyle's Law. However, this research was so far ahead of its time that it attracted little attention for many years thereafter. In the 19th century, two amateur English scientists—John Herapath (1790–1868) and John Waterston (1811–1883)—presented papers on the kinetic theory of gases to the Royal Society, but both were rejected; it seems the mainstream scientific world of the day was incapable of appreciating new research. Herapath submitted his paper to a different journal, and it was published in 1821, but largely ignored. The case of Waterston is more tragic. The paper he submitted in 1845 was placed in a storage closet at the Royal Society and did not see the light of day until 1892, after his death. By the time his paper was rediscovered, the kinetic theory of gases had already been established by Maxwell.

In 1848, Joule—following the ideas of Bernoulli—calculated the velocity of hydrogen molecules. In 1857 and 1858, Clausius derived the fundamental relationships between the pressure and volume of a gas and its number of molecules, molecular weight, and average velocity; he also derived the *mean free path* representing the average distance traveled by gas molecules between the two successive collisions. This was an effort to resolve doubts surrounding the conclusion that molecules at room temperature travel several hundred meters per second. If molecules indeed move so rapidly, some objected, then why does it take a significant amount of time for an aroma to spread throughout a room? The work of Clausius showed that collisions between molecules ensure that no single molecule can travel long distances in a straight line; instead, the time required for molecules to propagate over a given distance is much longer than would be expected given their high velocities.

Understanding of the kinetics of gas molecules was further advanced by James Clerk Maxwell. In 1865, he resigned his professorship at King's College and retired to the Scottish countryside for a few years, but in 1871, he became

James Clerk Maxwell (1831–1879)
Born to a wealthy family in Edinburgh, from an early age Maxwell was perpetually curious about the world around him. After graduating from the University of Edinburgh, he graduated with honors from Cambridge University in 1854. His first work resulted in pioneering achievements regarding the rings of Saturn, which he showed to consist of many rigid particles executing separate concentric orbits. In 1856, he became a professor at the University of Aberdeen and in 1860 a professor at King's College in London. He is famous for developing the kinetic theory of gases and for deriving Maxwell's equations, which provide a mathematical foundation for electromagnetic phenomena.

the first professor at the Cavendish Laboratory in Cambridge. He made crucial contributions to a wide range of fields of physics. Although his greatest achievement is undoubtedly his formulation of classical electromagnetic theory as summarized in Maxwell's equations, he also made seminal contributions to the kinetic theory of gases.

Maxwell's research on the kinetic theory of gases began in 1859, when he was at the University of Aberdeen, and continued into his time in London. He became interested in the papers of Clausius, and developed them further. He derived the formula for the velocity distribution of gas molecules and presented it in 1860. It was here that he introduced statistical methods to investigate how the velocities of a gargantuan number of particles were distributed. The investigation suggested that the molecular velocities were distributed from the slowest molecules to the most rapid molecules according to the bell-shaped curve known as a Gaussian distribution. He continued his research for several years thereafter and presented a revised version in 1867. This velocity distribution formula became the expression used by Boltzmann to represent the energy distribution of molecules, and today is known as the *Maxwell-Boltzmann distribution* (Figure 2.17). However, Maxwell remained dissatisfied with the understanding of molecular kinetics, because the kinetic theory of gases could not furnish a satisfactory explanation of the ratio γ between the constant-volume and constant-pressure specific heats of gases. Assuming that energy is equally distributed among the translational, rotational, and vibrational degrees of freedom, the value of γ predicted by the kinetic theory of gases for a gas of diatomic molecules was 1.33, but the experimental value was close to 1.4. The resolution of this discrepancy had to await the emergence of quantum theory.

Ludwig Boltzmann (1844–1906)
Boltzmann was born in Vienna and studied at the University
of Vienna; he graduated in 1866 and became a professor
of mathematical physics at the University of Graz in 1869.
In 1873, he became a professor of mathematics at the
University of Vienna, but in 1876 returned to Graz and
became a professor of experimental physics. He later was
a professor of physics at the Universities of Munich and
Vienna. He suffered from manic depression for many years and ultimately
committed suicide in 1906. He founded the science of statistical mechanics,
which unified thermodynamics and the kinetic theory of gases and enabled
an understanding of entropy based on molecular theory.

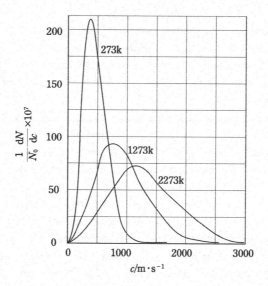

Figure 2.17: Maxwell-Boltzmann distributions for the velocities of molecules

The figure plots the relative probabilities for N_2 molecules at three different temperatures. The horizontal axis is the molecular velocity and the vertical axis is the relative probability.

Boltzmann and entropy

The trail blazed by Maxwell was explored further by the Austrian physicist
Ludwig Boltzmann, who achieved a deeper understanding of the relationship
between probability and the energy distribution and entropy of molecules. In
addition, Boltzmann founded the subject of statistical mechanics, which tied
together the theories of molecular kinetics and thermodynamics. Boltzmann

Ernst Mach (1838–1916)

An Austrian physicist and philosopher, Mach was a professor of the philosophy of science at the University of Vienna. He proposed the notion of the Mach number after studying ultrasonic flows. Mach pursued the true nature of physical insight from the perspective of positivism; he espoused a theory of "economical thinking," in which statements that could not be empirically tested were eschewed as meaningless.

initially attempted to derive the second law of thermodynamics from mechanical principles without using probabilistic interpretations. However, his efforts were unsuccessful; instead, he developed Maxwell's theories and in 1868 provided a more convincing derivation of the velocity distribution formula. He discovered that the probability for a molecule to have an energy E was proportional to the quantity $\exp(-E/k_BT)$, a quantity known today as the "Boltzmann factor." (Here k_B is Boltzmann's constant.[10]) Boltzmann's next efforts were devoted to understanding how systems approach equilibrium. His first objective was to understand how a system approaching equilibrium behaves based on molecular kinetic theory without using probability. However, he found that in fact probabilistic ideas were essential in justifying the approach to equilibrium from molecular theory. Under conditions in which the total energy of a system is constant, the number of possible states that can be assumed by a macroscopic system composed of atoms (or molecules) is colossal; Boltzmann developed methods for counting this number of possible states. He then assumed this number to be proportional to the probability that a state of atoms would be realized, and understood the distribution of the equilibrium state to be the distribution that exhibited the highest probability of realization. In this way, Boltzmann derived the famous relation $S=k_B \ln W$ relating the entropy S to the number of possible states W. This equation not only rendered the concept of entropy more easily understandable, but also allowed the derivation of other thermodynamic properties, opening the door to a theory of statistical mechanics that was capable of calculating thermodynamic quantities from molecular theory. However, Boltzmann's ideas were not generally accepted at first.

The entirety of Boltzmann's research rested on the assumption that atoms were real. Although the majority of chemists in the second half of the 19th century implicitly accepted the reality of atoms, a faction remained that insisted the atom was merely a fictitious notion that conveniently explained natural

10. Boltzmann's constant k_B is the gas constant per molecule, or the gas constant R divided by Avogadro's constant, $k_B = 1.38 \times 10^{-23}$ J/K. For the definition of Avogadro's constant, see p. 176.

phenomena, not an entity that actually existed. In particular, even such scientific luminaries as the physicist and philosopher Ernst Mach and the physical chemist Ostwald argued strongly against atomic theory in the academic societies of the German-speaking world, and Boltzmann was forced to engage in fierce wars of words with their likes. By 1910, the reality of atoms was no longer in doubt, but just before that time Boltzmann tragically died[41]. His gravesite in Vienna's Central Cemetery features not only a bust but also an engraving of his crowning accomplishment, the equation $S=k_{B} \ln W$.

2.8 The birth and growth of physical chemistry

Until the mid-19th century, most scientists whose curiosity tended toward the nature of phenomena involving atoms or molecules were both physicists and chemists simultaneously. Dalton, Faraday, Gay-Lussac, Avogadro, and Bunsen are all characteristic examples. However, physicists and chemists gradually began to follow separate paths. The founders of the fields of thermodynamics and statistical mechanics discussed in the previous section were primarily physicists with little interest in chemistry. On the other hand, in the chemistry of the second half of the 19th century, organic chemistry played a particularly active role, and many organic chemists evinced little interest in theoretical problems with little direct impact on their problems of interest. However, there was a school of chemists who studied physical properties and were interested in the relationship between the chemical composition and the physical properties of a substance. A representative example was Heidelberg University's Hermann Kopp (1817–1892), who measured the specific heats of many compounds in an effort to investigate Neumann's law, which held that the product of the molecular weight and specific heat of a substance was constant. He also researched the melting and boiling points of many substances in an effort to clarify the relationship between molecular structure and boiling and melting. Most of these studies bore little fruit, but did lead to a large accumulation of empirical facts regarding material properties, stimulating efforts to systematize them and understand their theoretical origins. Similarly, in the same way that the regular tetrahedral theory of carbon proposed by van 't Hoff in organic chemistry gave rise to the field of stereochemistry, there emerged researchers who were not satisfied with organic chemistry focusing primarily on synthesis, and sought rather to understand the nature of phenomena by conducting physical chemistry research. Moreover, the fields of thermodynamics and kinetic theory of gases arose in the second half of the 19th century and gradually progressed to a point where they could be used to explain chemical phenomena. It was against this backdrop that physical chemistry became established and developed as a branch of chemistry. In 1887, Ostwald and van 't Hoff founded "*Zeitschrift für physikalische chemie*", beginning the process of securing physical chemistry as a firmly established branch of chemistry.

Properties of gases

In the second half of the 19th century, it became understood that all gases exhibit deviations from ideal-gas behavior at high pressures or low temperatures. In 1873, van der Waals modified the equation of state for a single mole of ideal gas, $PV=RT$ (where R is the gas constant), to read $(P+a/V^2)(V-b) = RT$, a form that agreed well with experiments. This modification takes into account the attractive force between molecules, a/V^2, as well as the actual volume b occupied by the molecules themselves.

The liquefaction of gases was a problem of widespread interest in the second half of the 19th century. The notion of a critical temperature—the temperature above which the gas does not exhibit a phase transition to liquid regardless of how much pressure is applied—was proposed in 1869 by Andrews. This raised the possibility that certain substances previously believed to be eternal gases—such as hydrogen, oxygen, and nitrogen—could be liquefied by cooling below their critical temperature. In 1883, efforts of Dewar in London, and of Wróblewski and Olszewski in Krakow in Poland, made it possible for large quantities of air to be liquefied. Dewar conceived the notion of the Dewar vessel for handling low-temperature liquids. The liquefaction of gases proceeded through the use of the Joule-Thomson effect, by which a gas cools due to adiabatic expansion.

According to the kinetic theory of gases, the ratio γ of the constant-pressure specific heat C_p and the constant-volume specific heat C_v should be 1.67 if one considers only translational motion. Clausius measured this ratio for many gases and found its value to be less than 1.67. He reasoned that the heat absorbed by a gas serves not only to increase the velocity of translational motion, but also to increase rotational and vibrational motion. At the time, a value close to 1.67 had been observed only for mercury vapor; later, after the discovery of the noble gases, their values of γ identified them as monatomic gases.

From thermochemistry to chemical thermodynamics

Since the time of Lavoisier, chemists had been interested in the heat evolved from chemical reactions. In 1840, Henri Hess proposed that the total heat of reaction for a sequence of chemical reactions depended only on the initial and final states and was independent of the intermediate pathways. This was a restricted form of the law of conservation of energy. Precise measurements of the heat evolved

Germain Henri Hess (1802–1850)
A Swiss-born Russian chemist and doctor, Hess was born in Geneva. He emigrated to Russia, and in 1840, he formulated Hess' Law of thermodynamics. He studied with Berzelius and later became a professor at the University of Saint Petersburg.

by chemical reactions became possible from the 1850s onward due to the efforts of Julius Thomsen in Copenhagen and Berthelot in Paris (see page 78). They improved the techniques and equipment for thermochemical measurements and conducted systematic research. The bomb calorimeter developed by Berthelot in 1881 to measure the heat of burning later found widespread use. The initial theoretical backdrop for this type of research was the notion that the heat evolved in a reaction directly reflected the chemical affinities between the reacting molecules, but the incorrectness of this viewpoint gradually became clear.

Applications of thermodynamics to problems in chemistry began in the late 1860s. The first and second laws of thermodynamics showed that reactions were governed by both energy and entropy. For simple phase transitions such as the vaporization of a liquid, the entropy expression derived by Clausius allowed the relation between the vapor pressure p and the temperature T to be expressed by the Clausius-Clapeyron equation, $\mathrm{d}\ln p/\mathrm{d}T = pQ/RT^2$. Here Q is the heat of vaporization. In 1868, the Heidelberg chemist, Horstmann, studied the sublimation of ammonium chloride and discovered that the temperature variations of its vapor pressure could be represented by a similar formula. However, several additional developments were necessary before thermodynamics could be applied effectively to problems in chemistry, such as the direction of reactions.

van 't Hoff, who laid the foundations for stereochemistry with his regular tetrahedral theory of carbon, was inspired by the work of Horstmann to introduce thermodynamics into the study of equilibrium problems[42]. By considering the work done by a system in a reversible process, he related the vapor pressure to the chemical affinity according to $-RT \ln K = A$, where A is the affinity and K is the equilibrium constant; the temperature dependence of the equilibrium constant was given by van 't Hoff's isotherm equation,

$$\mathrm{d}\ln K/\mathrm{d}T = Q/RT^2$$

Here Q is the heat of vaporization and R is the gas constant.

Before thermodynamics could be broadly applied to chemical problems, including reactions, the formal development of thermodynamics needed to progress further. This was achieved between 1874 and 1878 by American

Josiah Willard Gibbs (1839–1903)

Gibbs was born in New Haven, Connecticut, in the U.S., the son of a professor of religious literature at Yale University, and he was educated at Yale himself. In 1863, he received the first doctorate of engineering granted in the U.S., then in 1866 set off for a three-year period of foreign study in Europe, where he was influenced by Kirchhoff and Helmholtz. In 1869, he became a professor of mathematical physics at Yale. He published a paper on equilibrium in multi-component, multi-phase systems in which he introduced the notions of free energy and chemical potential, thus pioneering the application of thermodynamics to chemistry. He also achieved pioneering accomplishments in statistical mechanics. He was the first great American scientist.

physicist Josiah Willard Gibbs at Yale University and in 1882 by Helmholtz[43, 44]. They introduced the new concept of *free energy* given by

Helmholtz free energy: $F = E - TS$

Gibbs free energy: $G = H - TS$

Here E, H, and S respectively denote the energy, enthalpy, and entropy, while T is the absolute temperature. The difference between the free energy of the reactants and the products (ΔF or ΔG) is the driving force behind chemical reactions; reactions proceed in the direction that decreases ΔG at constant pressure or ΔF at constant volume, and the disappearance of these quantities indicates the attainment of the equilibrium state. Because most chemical reactions take place at constant pressure, chemists adopted the Gibbs free energy and began using it to discuss chemical reactions.

Because Gibbs' long paper on thermodynamics was published in the U.S. in the" *Proceedings of the Connecticut Academy of Arts and Sciences*", and because its content was somewhat abstract and difficult to understand, its importance was not appreciated by the European scientific community for some time. In the late 1880s, his papers were finally translated into German by Ostwald and into French by Le Chatelier, and his achievements became widely known. Gibbs went on to achieve fame as a developer of vector analysis and statistical mechanics.

Gibbs' paper included a formula that was to become famous as the *Gibb's phase rule*. This equation related the number of components of a system, C, and the number of phases of the system, P, to the number of degrees of freedom in the equilibrium state, f, according to $f=C-P+2$. This equation is important for understanding equilibrium in non-uniform systems and has useful applications to practical problems such as the composition of alloys; after Gibbs' paper, it became widely used.

One important success of chemical thermodynamics was a formula known as the *Nernst equation* for the electromotive force of a battery. In research conducted between 1882 and 1889, Nernst (see page 231) derived the following equation relating a battery's electromotive force E to its concentration of electrolytes C[45]:

$$E = \text{constant} + (RT/nF)\ln C$$

Here R is the gas constant, F is Faraday's constant, and n is the change in the number of electrical charges. This equation not only yields the electromotive force of a battery, but also forms the starting point for calculations of equilibrium in oxidation-reduction reactions; it remains an extremely important formula to this day.

The origins of chemical kinetics

In 1850, Heidelberg physicist Ludwig Wilhelmy used variations in the degree of optical rotation angles to study the speed of hydrolysis of sucrose in the presence of acid. He found that the rate of the reaction was proportional to the concentration of sucrose, which decreased according to the equation $-dC/dt=kC$. Here C is the concentration, t is the time, and k is the reaction-rate constant. In 1866, Harcourt studied the oxidation of hydrogen iodide by hydrogen peroxide and found that the rate of the reaction varied linearly with the concentrations of both reactants. These were the first steps toward a general treatment of reaction rates.

Norwegian chemists Cato Guldberg (1836–1902) and Peter Waage (1833–1900) conducted detailed studies of equilibrium reactions between 1864 and 1879[46]. They showed that the attainment of equilibrium corresponded to the state in which the reaction ceased, and that equilibrium could be approached either from the side of the products or from the side of the starting materials. They expressed the conditions for equilibrium in terms of concentrations, which they termed the "active masses"; by reasoning that equilibrium corresponded to a balance in the chemical affinities that drive the forward and reverse reactions, they established the concept of the law of mass action. However, it was Berthelot and Pean de Saint-Gilles who first developed clear formulas expressing equilibrium constants; in 1879, they obtained the following equation describing the esterification of ethyl acetate:

Henry Le Chatelier (1850–1936)
A French chemist, Le Chatelier graduated from the Ecole Polytechnique and the Ecole des Mines, then became a professor of mining and later a professor at the Ecole des Mines and the Sorbonne. He is famous for formulating Le Chatelier's principle, but he was also active as an expert on several industries, including cement, glass, and fuel.

$$[H \cdot C_2H_3O_2][C_2H_5OH]/[C_2H_5 \cdot C_2H_3O_2][H_2O]=1/4$$

In 1884, van 't Hoff published a book entitled *"Etudes de Dynamique Chimique"* in which he considered chemical equilibrium from the perspectives of thermodynamics and reversible reactions[42]. To describe dynamic equilibria, he introduced the double-arrow symbol that remains in use to this day; by setting the reaction rates in the forward and reverse directions equal to each other, he derived equations for the equilibrium constants. The experimental examples he had in mind were reactions such as the association / dissociation reaction of nitrogen dioxide, $N_2O_4 \rightleftharpoons 2\ NO_2$, which had been studied by Deville.

Around the same time, Henry Le Chatelier at the Ecole des Mines in Paris proposed the general law now known as *Le Chatelier's principle*: When a change is made to a system in dynamic equilibrium, the system responds in such a way as to minimize that change[47].

In 1889, Arrhenius—while studying the inversion of cane sugar—became interested in the fact that the reaction rate increased extremely rapidly with temperature. He explained this temperature variation in terms of the equilibrium between active molecules and inactive molecules: the more energy needed to activate a molecule, the greater the variation of the reaction rate with temperature. Thus, he arrived at the formula—now known as the *Arrhenius equation*—expressing the temperature dependence of the rate constant; $k=A \exp(-E_A/RT)$[48]. Here E_A is the activation energy.

The concept of a catalyst—a substance whose presence in small quantities could accelerate chemical reactions—was first introduced by Berzelius in 1837, but it was Ostwald who pioneered physical-chemistry research on catalysts[49]. He carried out a wide range of studies emphasizing the notion that catalysts accelerate reactions but do not change their direction. He studied particularly thoroughly the changes in reaction rate due to acidic catalysts. He also used heated iron as a catalyst for the reaction in which ammonia is oxidized to form nitric acid. This turned out to be extremely useful for the industrial production of nitric acid.

Francois-Marie Raoult (1830–1904)
A French chemist, Raoult performed pioneering studies on freezing-point depression and boiling-point elevation, and derived Raoult's Law for the vapor pressure of dilute solutions. He was a professor at the University of Grenoble from 1870 to 1901.

Ernst Beckmann (1853–1923)
Beckmann was a German chemist. After an apprenticeship as a pharmacist, he studied chemistry with Kolbe and was a professor at the Universities of Giessen, Erlangen, and Leipzig. He is known for discovering the Beckmann rearrangement and for devising the Beckmann thermometer, which he used to advance methods of measuring freezing and boiling points.

Hugo de Vries (1848–1935)
A Dutch botanist and geneticist, de Vries was a professor at the University of Amsterdam. He is famous for rediscovering Mendel's Laws and as a chief exponent of the theory of spontaneous mutation.

The idea that light absorption by molecules could induce photochemical reactions was first proposed by Grotthuss (see page 53) in 1819. The reaction of hydrogen and chlorine to produce hydrogen chloride, which had been rediscovered by Draper in 1843[50], was studied in 1860 by Bunsen, and it was observed that the reaction occurred in the presence of light; however, the mechanism of photochemical reactions remained mysterious throughout the 19th century.

Properties of fluids and osmotic pressure

It had been known for some time that the melting points of solutions was lower than those of solvents (freezing-point depression), and that the addition of solute caused the vapor pressure of a solvent to decrease and its boiling point to increase (boiling-point elevation). However, it was not until the second half of the 19th century that quantitative research on freezing-point depression and boiling-point elevation was conducted. In 1882, the French chemist Francois Raoult, at the University of Grenoble, dissolved many organic compounds in solvents and measured the freezing-point depression; he discovered that the product of the molecular weight and the amount of freezing-point depression resulting from the addition of 1 g of solute to 100 g of water was constant. This relationship also held for other solvents. In this way, Raoult showed that freezing-point depression could be used to determine molecular weights[51]. He also conducted detailed studies of boiling-point elevation. Freezing-point depression and boiling-point elevation attracted the attention of chemists as a convenient method for determining molecular weights, and several other laboratories conducted similar experiments. For example, Ernst Beckmann devised a thermometer capable of precision temperature measurement to conduct accurate experiments. However, Raoult's research revealed a curious fact: the extent of freezing-point depression per molecule of inorganic salt was greater than the corresponding quantity for organic compounds. This seemed to suggest that the salts were breaking up into constituent components, which then existed in the solution.

In 1884, van 't Hoff (see page 75) became interested in osmotic pressure through conversations with his botanist colleague, Hugo de Vries. The phenomenon of osmotic pressure had been recognized before, and in the 1870s, the German chemist Wilhelm Pfeffer (1845–1920) had the clever idea of placing a semi-permeable membrane in the walls of a porous fired vessel and used this to conduct a wide range of studies. Upon analyzing the data of Pfeffer and de Vries, van 't Hoff was struck by the similarity in properties between dilute solutions and gases, and he applied the second law of thermodynamics to analyze these properties. The experimental data had already demonstrated that—in analogy to the laws of Boyle and Gay-Lussac—the osmotic pressure was proportional to the density and to the absolute temperature T. van 't Hoff obtained a similar equation for the dependence on the molar concentration C of the solution,

$$\Pi = RTC$$

Here Π is the osmotic pressure and R is the gas constant. Thus, osmotic pressure furnished yet another new method for determining molecular weights. However, although the observed values of the osmotic pressure for many organic compounds agreed with the predictions of this formula, its predictions for inorganic salt solutions were incorrect. To correct for this, an empirical correction parameter i was introduced, and the quantity RTC in the equation above became $iRTC$. The values of i for hydrochloric acid and sodium nitrate were close to 2. These results were announced in 1885, but at that time van 't Hoff did not yet appreciate the significance of the fact that i was close to 2. The situation was clarified by the development of a successful theory of electrolytic dissociation.

Theories of electrolytic dissociation: Arrhenius, van 't Hoff, Ostwald

The term *ion* had been previously introduced by Faraday to refer to substances discharged at electrodes, but the ion envisaged by Arrhenius was an entirely separate entity with an entirely new definition. However, the notion that

Svante Arrhenius (1859–1927)
Born in a suburb of Uppsala in Sweden, Arrhenius studied
chemistry at the University of Uppsala with the rare-earth
researcher Cleve and researched electrochemistry at
the University of Stockholm. After graduating from the
University of Uppsala in 1884 with a thesis on measurements
of electrical conductivity and the theory of electrolytic
dissociation, he collaborated on research with Ostwald,
Kohlrausch, and van 't Hoff; he completed his theory of ion dissociation
and gathered support for it, and is known as one of the founders of physical
chemistry. After receiving the Nobel Prize in Chemistry in 1903, he became
head of the physical chemistry laboratory at the Nobel Institute in Stockholm;
he conducted research in a wide range of fields including astronomy, cosmology, earth science, and physiology (Column 4).

substances dissociate into ions when immersed in solution was not itself a
completely new idea.

The existence of ions was first deduced in the 1850s by Williamson and
Clausius. Further, Hittorf, a German, studied the electrolysis of salt and
hypothesized—and experimentally demonstrated—that the flow of current was
due to the movement of ions, at different speeds, toward the electrodes. In 1874,
Kohlrausch showed that all ions exhibited a characteristic mobility and that the
conductivity of a given salt could be computed from the sum of the mobilities of
its constituent ions. He also discovered that the electrical conductivity increased
as the solution was diluted, approaching an infinite-dilution value. However,
these scientists believed that ions were only produced when current was flowing.

The Swedish chemist Svante Arrhenius chose research on electrochemistry
as the theme of his dissertation, which consisted of two parts: experimental
research and theoretical research. The experimental portion was largely an
extension of the research of Kohlrausch with little new added. In the theoretical
portion, Arrhenius argued that the experimental results of Kohlrausch could be
explained by assuming that salts existed as complicated molecules in solution,
which dissociate as the solution is diluted. However, at this point there was no
clear formulation of the concept of ions. The thesis committee did not grant
a positive assessment to the dissertation Arrhenius submitted in 1883, and he
was unable to find an academic job in Sweden. He sent copies of his thesis to
several prominent European chemists. Willhelm Ostwald received the work
favorably and invited Arrhenius to his laboratory in Riga. Arrhenius conducted
collaborative research with Ostwald in Riga from 1884 to 1886; then, from

Wilhelm Ostwald (1853–1932)
Ostwald was born in Riga, Latvia, to German immigrants; he studied at the University of Tartu. In 1881, he became a professor at the Riga Polytechnicum, and in 1887 a professor at the University of Leipzig. He was active in several fields of research, including chemical equilibrium, reaction rate theory, and catalysts; in addition to achieving major accomplishments in these areas, he also made major  contributions to securing the place of physical chemistry as an established subfield of chemistry. He had close ties to van 't Hoff and Arrhenius, and he contributed to ensuring broad recognition of their work; he discovered and encouraged many promising young chemists. In 1909, he was awarded the Nobel Prize in Chemistry for his work on catalysis, reaction rates and chemical equilibrium. He also worked tirelessly to popularize the science of chemistry; his general textbook "*Grundriss der allgemeinen Chemie*" was read the world over[52].

1886 to 1891, he worked with Kohlrausch in Würzburg and with van 't Hoff in Amsterdam. He completed his ion theory of electrolysis after reading van 't Hoff's paper of 1885.

The theory of electrolytic dissociation developed by Arrhenius elegantly explained both van 't Hoff's results on osmotic pressure and Raoult's results on freezing-point depression. Denoting the number of moles of non-dissociated molecules by m, the number of dissociated molecules by n, and the number of ions arising from the dissociation of a single molecule by k, van 't Hoff's correction factor i was given by

$$i = (m + kn)/(m + n)$$

If we define the degree of dissociation of a molecule, α, according to $\alpha = n/(m+n)$, we find $i = 1 + \alpha(k-1)$. The values of i determined by measurements of electrical conductivity, osmotic pressure, and freezing-point depression all exhibited good agreement. These results were reported in "*Zeitschrift für Physikaslische Chemie*" in 1887[53, 54].

Ostwald used the theory of electrolytic dissociation in his research on acids, and in 1887 derived the famous relation that came to be known as Ostwald's law of dilution[55], which held that the degree of dissociation, α, defined by Arrhenius was given by the following equation.

$$c\alpha^2/(1 - \alpha) = K$$

COLUMN 4

Arrhenius and Global Warming

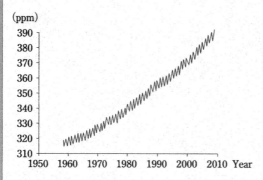

The increase of atmospheric carbon dioxide

Among the great scientists active in the late 19th and early 20th centuries, one finds a number of intellectual giants who studied and were fascinated by a wide range of natural phenomena. Among their ranks, one must without question include Arrhenius. When his research on electrolytic solutions reached a certain state of completion, he turned to research in new fields. He first became interested in problems in physiology, applying methods of physical chemistry to the problem of toxins and antidotes; in 1907, he published a book titled "Immunochemistry" based on lectures he delivered at the University of California. He was also interested in geology, earth science, astronomy, astrophysics, cosmology, and the origins of life, and on these topics, he published both textbooks and surveys for general audiences. His book *"Das werden der welten"*(*Worlds in the Making*) was published in Japanese, in a translation by Torahiko Terada entitled *"Shiteki ni Mitaru Kagakuteki Uchuukan no Hensen"* (*A Historical View of Changing Scientific Pictures of the Universe*)[56]. With regard to the origins of life, he subscribed to the Panspermia theory, which held that life on earth had extraterrestrial origins.

At the end of the 20th century, the problem of global temperature increase due to increased CO_2 attributable to human activity became a cause of serious concern, and it was at this time that research conducted by Arrhenius some 100 years earlier began to attract attention. The so-called greenhouse effect was first recognized by Fourier and Tyndall, but Arrhenius was the first to consider it quantitatively[57]. In 1896, he published an article "On the Influence of Carbonic Acid in the Air on the Temperature of the Ground" in the English journal *Philosophical Magazine*. In this paper, he calculated the extent to which the CO_2 greenhouse effect would cause the Earth's temperature to increase based on information that was known at the time regarding the absorption of infrared radiation by CO_2 and the emission of infrared radiation by material bodies[58]. The results of his calculation predicted that doubling the quantity of CO_2 in the atmosphere would increase the average temperature of the Earth by 4°C—a number close to present-day predictions by the Intergovernmental Panel on Climate Change (IPCC).

But *why* was Arrhenius inspired to calculate the extent of global warming? As a scientist interested in the history of the Earth, his motivation was to understand long-term variations in the planet's temperature, such as ice ages and inter-ice ages. Among the causes of CO_2 increase, natural processes such as the dissolution of limestone were thought to be paramount. Although increased consumption of fossil fuels due to the growth of industry was also considered as a possibility, given the rate of consumption at that time it would have taken 3,000 years for the quantity of atmospheric CO_2 to double. In any case, Arrhenius thought of warming effects due to increased CO_2 as more of a benefit to humankind than a threat to mankind[57]. Indeed, he believed that the warming of the climate might make cold-weather regions into more habitable environments and might further the growth of agriculture by making it easier to supply food to the world's population. In any event, it is astonishing to think that over 100 years ago this giant of chemistry had quantitatively considered the possibility of global warming due to increased CO_2. If recent trends in the growth rate—shown in the figure—continue apace, it will take less than 200 years for the CO_2 concentration to double. What type of world will we then inhabit? The question is worrisome indeed.

Torahiko Terada[1], who visited Arrhenius' physical chemistry laboratory at the Nobel Institute in 1910, described how Arrhenius had installed, on the roof of his home (which doubled as a laboratory), a small telescope, a transit, and other equipment. Arrhenius had explained, with a twinkle in his eye and a slight grin sneaking across his boyish face, that with this he could do "a little astronomy". Terada later wrote that he "had the sense that here, before my very eyes, was the perfect specimen of a person who truly loved learning." On the other hand, Arrhenius had strong influence on the Nobel Prize committee and is said repeatedly to have blocked the prize from being awarded to scientists—such as Nernst or Ehrlich—with whom he had differences of opinion.

A Swedish stamp featuring Arrhenius

---

1. Torahiko Terada (1878–1935): A Japanese physicist and professor at Tokyo University from 1916 to 1935. He achieved fame as an essayist in the school of Natsume Soseki. His essays are still widely read today.

Thomas Graham (1805–1869)
Educated in Scotland, Graham was a professor at London's University College from 1837–1855, then served as Master of the Mint. After studying the diffusion of gases, he switched to research on colloids and is known as the father of modern colloid chemistry.

Here K is a constant and c is the acid concentration. This equation follows from applying the law of mass action to electrolytic equilibrium (ionization equilibrium); for weak acids or weak bases the concentrations of H^+ and OH^- can be immediately determined from this formula, making it extremely useful in practice.

Because Arrhenius' theory of electrolytic dissociation was an entirely new notion to the chemists of the day, it was not generally accepted immediately. It also raised the new and difficult question of *why* salts dissociated in solution. However, it had the support of van 't Hoff and Ostwald, the supreme authorities in physical chemistry at the time, and eventually the electrolytic dissociation theory earned acceptance. The magnitude of its impact on chemistry can hardly be overstated: it systematized and lent solid theoretical evidence to methods of qualitative analysis and volumetric analysis that had previously been based solely on empirically determined facts.

The three great scientists, van 't Hoff, Arrhenius, and Ostwald, all made major contributions to the development of physical chemistry. However, the Leipzig laboratory of Ostwald stands out in particular as something of a global center for physical chemistry research in the late 19th and early 20th centuries, training large numbers of visiting students from several countries, including England, America, and Japan[11]. The journal "*Zeitschrift für Physikalische Chemie*", which Ostwald founded together with van 't Hoff in 1887, also contributed significantly to the establishment of the field. However, Ostwald insisted on explaining all phenomena in terms of energy theories and did not believe in the existence of atoms and molecules until 1909.

The chemistry of colloids and surfaces

Colloidal particles had been known since the time of Davy and Faraday, but it was the English chemist Thomas Graham who began their systematic study. He introduced the word *colloid*, meaning "glue-like," in 1861. He drew a distinction between crystalline substances and colloids: the former, when dissolved, could pass through sheepskin, while the latter could not. He used the term *dialysis*

---

11. Kikunae Ikeda and Yukichi Osaka, two pioneers of physical chemistry in Japan, were foreign-study students in Ostwald's laboratory.

COLUMN 5

Pockels and Lord Rayleigh[59]

Agnes Pockels

In the late 19th century, a time when women were not even afforded the opportunity to go to college, there was one woman who conducted experiments in a corner of her kitchen and achieved groundbreaking advances in the study of surface science, which was reported in the journal "*Nature*". Her efforts to publish her results in this august journal were assisted by none other than Lord Rayleigh, who won the Nobel Prize in Physics for his discovery of argon and produced a dazzling array of achievements spanning a wide range of fields in classical physics.

In 1891, Rayleigh had become interested in physical phenomena at surfaces and was studying monomolecular films. One day he received a letter in German from an unknown correspondent. This turned out to be Agnes Pockels, a woman living in the German city of Brunswick, who for the past 10 years had been conducting experiments in her kitchen to assess the effects on surface tension exerted by oil films on a water surface. Her letter described the results of her research; astonishingly, it seemed she had conducted studies very similar to those Rayleigh was attempting at the time—but had done them some 10 years earlier!

Agnes Pockels (1862–1935) was born in Venice, then part of Austria, to a father serving as an officer in the Austrian army. However, in 1871 her father contracted malaria and was forced to resign his commission, whereupon the family moved to Brunswick. Agnes had a passionate interest in science, but after graduating from the public girls' high school, she was forced to look after her ailing parents. She is said to have become interested, while washing dishes in her kitchen, in the effect of oil and other impurities on the surface tension of water, and began conducting research at around the age of 20. Although she was unable to attend university, her younger brother Friedrich was fortunately studying physics at the University of Göttingen, and through him, she had access to scientific reference material. The homemade apparatus she used to measure surface tension could be said to be a prototype of the device later developed and made famous by Langmuir.

Lord Rayleigh, who was surprised by Pockels' letter, sent an English translation of it to Lockyer, the editor of "*Nature*", together with a letter requesting its publication. Thus it was that the results of her research were published in *Nature* in 1891 and became known to the world as pioneering progress in surface science. Bolstered by Lord Rayleigh's encouragement, Pockels continued submitting papers to "*Nature*" and other journals, but as her parents' illnesses worsened, she was forced to spend increasing amounts of time caring for them, and after the turn of the 20th century, she was largely unable to continue her research. She remained unmarried throughout her life and had no ties to universities or laboratories, but in 1931, she received the Laura Leonard award from the colloid society, and in the same year, she received an honorary doctorate from Braunschweig University of Technology.

Pockels received absolutely no support from the scientists of her native Germany. Rayleigh was a great scientist with commanding authority, but he was also kind-hearted, generous of spirit, and personable, and he never refused assistance to others. He was born the son of rural nobility and was a nobleman himself, becoming the third Lord Rayleigh. Teruhiko Terada wrote that "Rayleigh is a unique product that could arise only from a combination of 'rural nobility' with 'physics,' and that "One gets a sense of warmth, and a lighthearted sense of humor, emanating from the photograph[61]." Incidentally, Pockels' brother Friedrich is also famous for discovering the electro-optical Pockels effect and for inventing the Pockels cell used in laser spectroscopy.

A reproduction of Pockels' experimental apparatus

to refer to his method of separating colloids from crystalline substances using a membrane. He made colloids of a variety of substances—including arsenic trisulfide, silicate, and hydroxides of aluminum and iron; he called solutions of these *sols*, which he distinguished from jelly-like gels. He also discovered that adding a small quantity of salt to a sol caused it to coagulate into hair-like clumps. By the end of the 19th century, the coagulation of colloids by salt had been studied in detail by Hardy and Freundlich, and the effect of the charges of the ions that constituted the salt was appreciated. The fact that colloidal particles carried electrical charge was clear from the observation that colloids placed in an electric field migrate toward the electrodes. The motion of colloids in electric fields was termed *electrophoresis* or *cataphoresis*.

In 1898, Ostwald's assistant Bredig produced a sol of metal colloids by passing an arc current between metal electrodes in water. Using this method, sols of platinum, gold, silver, and other metal colloids were made. Although colloidal particles captured the curiosity of many chemists, it was not until the 20th century that quantitative research on colloids began and came to exert a significant impact on the development of chemistry (see Section 4.4).

Meanwhile, surface phenomena had been attracting the curiosity of scientists since antiquity. In 1757, Benjamin Franklin observed that an oil layer on the ocean surface had the effect of damping waves. He presented his observations in 1774. Theories and measurements of surface tension were subjects of active research by physicists in the second half of the 19th century. In 1891, the young amateur German female scientist Pockels (Column 5) studied oil layers on water surfaces and investigated the relationship between surface tension and the area of the layer; she discovered that the behavior differed significantly above and below a certain critical area, and published her findings in *Nature* thanks to an introduction from Lord Rayleigh[61]. Rayleigh himself was conducting similar research around the same time. In 1899, he showed that indissoluble substances, if given sufficient area, would spread over a liquid surface to form a monomolecular film with the thickness of just a single molecule.

2.9 Organic chemistry of natural products

Organic chemists had long been fascinated by naturally occurring organic compounds. By the second half of the 19th century, chemistry had developed sufficiently to allow study of these complicated organic compounds. In particular, dyes, sugars, proteins, terpenes, and purines were a focus of attention from a practical standpoint. Moreover, many of these substances were of interest for the important role they played in the phenomena of life. The establishment of the organic chemistry of natural products as a major subfield of organic

Emil Fischer (1852–1919)
Fischer, the son of a successful businessman, studied with Kekule at the University of Bonn, then moved to Strasbourg, where he received his degree under Baeyer in 1875 and became Baeyer's assistant. Together with Baeyer, he moved to Munich and became a professor, then went to the University of Erlangen in 1882 and to the University of Würzburg in 1885; in 1892 he succeeded Hofmann as professor at the University of Berlin. Fischer received the second Nobel Prize in Chemistry, in 1902, for his research on sugars and purines. However, despite his towering accomplishments, his later years were not happy ones. He lost two sons in World War I and grew despondent over the human and material loss caused by the war; he contracted cancer, slipped into depression, and committed suicide in 1919.

chemistry not only had important ramifications for the chemical industry, but also laid the groundwork for the explosive growth of biochemistry in the 20th century.

This development was pioneered by the great German organic chemist, Emil Fischer. He placed the chemistry of sugars on solid footing by determining their composition and three-dimensional structure[62]. He also studied enzymes that decompose sugars and demonstrated the relationships among the chemical structures of sugars. He determined the structures of molecules such as uric acid, xanthine, and caffeine, and systematized the chemistry of purine derivatives[62]. Later he turned to the study of proteins, which he hydrolyzed into amino acids, identifying the peptide structure of proteins. In this way, he singlehandedly laid much of the foundations of modern biochemistry. Between the late 19th and early 20th centuries, Fischer's laboratory hosted researchers and trained students from all over the world—including many European nations, America, and Japan—and was a major world center of chemistry research. Of the many students he supervised, 62 became university professors and 6 went on to win Nobel prizes. Among Fischer's protégés, one finds not only Nobel chemistry laureates but also many biochemists who earned prizes in biology and medicine following their exposure to Fischer's teaching (see page 719).

Structure and synthesis of sugars

By 1870, four simple sugars—glucose, fructose, galactose, and sorbose—were known. Hydrolysis of sucrose was known to produce glucose and fructose, while hydrolysis of lactose was known to yield galactose and glucose. The four

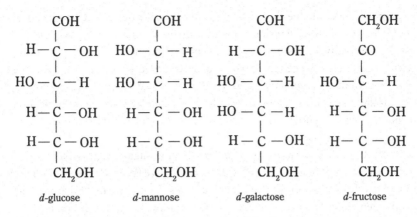

Figure 2.18: The projection formulas introduced by Fischer to represent the structure of sugars

simple sugars were all represented by the empirical formula $C_6H_{12}O_6$ and were known to contain carbonyl groups and to be multivalent alcohols; however, the determination of their structure—including their three-dimensional configuration—proved a thorny challenge. Fischer's success in tackling this problem was unquestionably due to his outstanding abilities and indomitable spirit, but he was also blessed by opportunity and a coincidence of good fortune. The recognition in 1874 of the asymmetry of carbon atoms by van 't Hoff and le Bell (see page 74) proved to be of decisive importance for elucidating the structure of sugars. Moreover, phenylhydrazine, which Fischer discovered in 1875 by studying a sequence of hydrazine derivatives, reacted with sugar to form yellow crystals of osazone, which had a well-defined melting point and was easily separated and refined. These derivatives were to be of crucial importance in studies of sugars.

In previous research, sugars such as glucose that possessed aldehyde groups were believed to have four asymmetric carbon atoms in a structure such as the following:

$$CH_2(OH)-CH(OH)-CH(OH)-CH(OH)-CH(OH)-CHO$$

Similarly, sugars such as fructose that possessed ketone groups were thought to have three asymmetric carbon atoms in a structure such as the following:

$$CH_2(OH)-CH(OH)-CH(OH)-CH(OH)-CO-CH_2(OH)$$

According to the van 't Hoff's law, the aldehyde form should have $2^4=16$ stereoisomers, while the ketone form should have $2^3=8$ stereoisomers. Thus,

Fischer reasoned that the four known isomers of glucose represented just 4 of the 16 possible three-dimensional configurations. In addition to making use of osazone derivatives, he also exploited synthetic methods developed by other German chemists—especially Kiliani—and succeeded not only in clarifying the differences among the sugar isomers known at the time, but also in synthesizing nine of the new isomers he predicted. Fischer pioneered the use of projection formulas like that shown in Figure 2.18 to represent the three-dimensional configuration of sugars.

In 1894, Fischer began to study enzymes using sugars. He discovered that some isomers of glucose were easily fermented by yeast, while others were not fermented despite exhibiting similar structures. He also noticed that maltase hydrolyzed alpha-methyl glucoside but not beta-methyl glucoside. From these observations, he deduced that enzymes were only active when they had a specific configuration that would fit the substrate. In explaining this hypothesis, he used the analogy of a lock and key[63], comparing an enzyme to a key that would fit into only one lock. His research on the two types of methyl-glucoside also demonstrated that these compounds formed ring structures containing oxygen, and that the fifth carbon atom was asymmetric, suggesting the possibility of optical isomers.

Purine and its derivatives

Emil Fischer achieved brilliant success in studies of purine-based compounds. In 1881, he studied caffeine and later investigated related compounds such as uric acid and xanthine. Caffeine had been isolated from coffee and tea in the 1820s. Uric acid was obtained from urine calculus by Scheele in 1776 and was studied

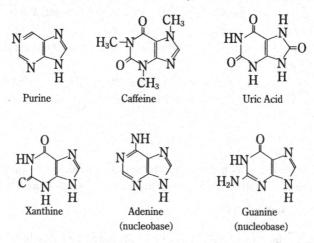

Figure 2.19: Structure formulas for purine compounds

Gerardus Mulder (1802–1880)
Mulder was a Dutch physician and chemist who conducted wide-ranging analytical research on the structural components of living organisms. He is said to have coined the term *protein*, from the Greek word *proteus*, meaning "first", following Berzelius' suggestion. These substances were the primary component of animals and plants nutrients.

by Liebig and Wöhler in the 1830s and by Baeyer and Strecker in the 1860s. Xanthine and the related compounds guanine and adenine had been found in the blood, striated muscle tissue, and urine of animals. Thus, these substances were of great interest, but the relationship among them remained mysterious.

As he had done in his research on sugars, Fischer used synthesis techniques to determine unambiguous structures for these compounds and showed that they all had a purine skeleton (Figure 2.19). By the end of the 19th century, the relationship between purine-based compounds and their dissociation products had been established by Fischer and his collaborators. A small number of animals, including humans, discharged the nitrogen in purine in the form of uric acid, but many animals further decomposed it into allantoin. In birds and reptiles, uric acid is the primary final product of nitrogen metabolism; it is present in large quantities in the guano that accumulates on bird feces.

Proteins and amino acids

Among proteins, one substance that attracted early attention from chemists was albumin. It was well known that substances such as egg whites or blood serum solidified upon heating, and substances exhibiting this sort of condensation were termed albumins. Similarly, the casein obtained by solidifying milk through the addition of acid, and fibrin that segregated from vegetable juices, were known in the first half of the 19th century. It was recognized that all of these substances contained nitrogen, but the challenge of isolating pure substances and identifying their chemical composition proved a daunting one.

The word *protein* was proposed in 1838 by the Dutch chemist Gerardus Mulder[64]. He proposed that egg-white albumin, blood-serum albumin, and fibrin all contained an ingredient with the empirical formula $C_{40}H_{62}N_{10}O_{12}$, and moreover, that in fibrin and egg-white albumin, this unit bound to single atoms of sulfur and phosphorus and could be separated into three common units by alkali processing. Liebig initially supported this hypothesis, but later turned strongly against it. The chemists at that time, who understood the complexity of proteins, strove to identify their composition through analysis; over time a body of knowledge developed regarding the reactions of proteins with heat, acid, and alkali, and meanwhile knowledge of their decomposition products—amino acids—expanded as well.

The first amino acid to be isolated was cystine, discovered in 1810 by Wollaston as one component of urine calculus; however, the fact that this compound originated from proteins was not known until 1899. In 1819, Proust detected leucine as a fermentation product in cheese. Braconnot obtained crystals of sweet-tasting glycine from gelatin, and in the 1850s it was determined that this substance was aminoacetic acid ($NH_2 CH_2 COOH$). However, in general the determination of the structures of amino acids was a difficult problem.

By the 1860s, it was generally accepted that leucine and tyrosine were decomposition products of albumin. In 1866, Ritthausen obtained glutamic acid from the hydrolysis of plant proteins by acid, and two years later, he obtained aspartic acid in the same way and isolated these compounds. By 1890, their structures had been clarified by synthesis. Thus, knowledge of amino acids gradually accumulated, but the giant steps toward clarifying the relationship between proteins and amino acids, and elucidating the structures of proteins, were the results of Fischer's research in the 20th century (see Section 4.9).

The discovery of nucleic acids

In contrast to the other naturally occurring organic substances discussed in this section, nucleic acids were discovered in the context of physiological chemistry. In 1869, Friedrich Miescher, a student of the famous physiological chemist Felix Hoppe-Seyler at the University of Tübingen, began to study white blood cells in an effort to determine the chemical composition of cells. From pus-soaked bandages that he obtained from surgeons at a nearby hospital, he isolated white blood cells and analyzed their chemical properties. He successfully isolated the nucleus from the cellular matter and analyzed its composition; he discovered that the nucleus contained a substance that differed from protein, which he termed *nuclein*[65]. Like protein, this substance contained carbon, hydrogen, oxygen, and nitrogen, but it also contained a significant quantity of phosphorus. He went on to find that this substance was present not only in white blood cells taken from pus but also in a broad family of cells, including yeast, the red blood cells of ducks, and kidney cells. In subsequent research, he observed that salmon sperm cells consisted almost entirely of a nucleus and discovered that nuclein was a huge molecule containing acidic groups; sadly, he contracted tuberculosis and died at the age of 51. In 1889, Altman produced a purified form of nuclein that contained no protein; because this substance had acidic properties, he named it *nucleic acid*.

After Miescher, the next scientist to make major contributions to the chemistry of nucleic acids was Albrecht Kossel. By 1894, he had discovered

Felix Hoppe-Seyler (1825–1895)

A German doctor and physiologist who pioneered the field of physiological chemistry, Hoppe-Seyler became a professor at the University of Tübingen in 1861 and at Strasbourg in 1872. He left behind many accomplishments in the analysis of physiological tissue and in research on metabolism.

Friedrich Miescher (1844–1895)

Miescher was a Swiss biochemist and physiologist. While conducting research under Hoppe-Seyler at the University of Tübingen in 1869, he discovered nucleic acids—which he named nuclein—in white blood cells taken from pus. After discovering nucleic acids, he became a professor at the University of Basel in Switzerland and in 1885 established the first physiological research laboratory in Switzerland. He discovered that breathing is regulated by the concentration of carbon dioxide in the blood.

Albrecht Kossel (1853–1927)

A German biochemist, Kossel studied physiological chemistry with Hoppe-Seyler, worked at the University of Marburg, and became a professor at Heidelberg in 1901. He studied cells, nuclei, and proteins, and discovered adenine and thymine. He was awarded the 1910 Nobel Prize in Physiology or Medicine for research on the chemistry of cells.

the four base constituents of nuclein (adenine, guanine, thymine, and cytosine) and had determined that nucleic acids were composed of these four bases plus sugars and phosphoric acid (Figures 2.19 and 2.20). Uracil was later identified from nucleic acids of yeast. Kossel also conducted excellent studies of proteins—especially histones and other nucleic proteins and protamine—and was awarded the 1910 Nobel Prize in Physiology or Medicine for "establishing the chemistry of cells through research on proteins and nucleic acids." Thus, the importance

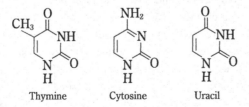

Figure 2.20: Structural formulas for pyrimidine-derivative nucleobases

Otto Wallach (1847–1931)
Wallach was a German organic chemist. He studied with Wöhler and conducted research with Kekule, then became a professor of pharmacology at the University of Bonn and later (in 1889) became head of the chemistry laboratory at the University of Göttingen. Through his research on aromatic oils, he made major contributions to the oil-refining industry and is known for his research on camphor. He received the Nobel Prize in Chemistry in 1910.

of nucleic acids as chemical constituents of the nucleus of cells came to be appreciated; however, none of the chemists of the time, including Miescher and Kossel, could have recognized the genetic significance of nucleic acids. By the end of the 19th century, some cell researchers and geneticists began to develop an inkling that perhaps nuclein had something to do with genes; however, more than half a century elapsed before it became clear that nucleic acids were the carriers of genetic information (Section 6.1).

Terpenes

Aromatic oils obtained by the distillation of plants had been known since the 16th century; they were used for perfumes and pharmaceuticals and attracted the attention of many organic chemists. Among these, the refined oils known as terpenes—and their oxides, camphors—were a subject of active research in the second half of the 19th century. The most famous of the researchers to study terpenes was Otto Wallach, who studied many varieties of terpenes and clarified the relationships between them. Terpenes are organic compounds in which the number of carbon atoms is $5n$ ($n \geq 2$), and in 1887 Wallach proposed that they were composed of isoprene (C_5H_8), a proposal that became known as the isoprene rule. He was awarded the 1910 Nobel Prize in Chemistry for his work on terpenes and camphors.

2.10 The road to the birth of biochemistry

The chemistry of living things and biological phenomena had always been a topic of interest among chemists. Lavoisier conducted pioneering experiments on breathing, and Berzelius and Liebig discussed the causes of fermentation and rotting. Liebig even wrote books on agricultural chemistry and animal chemistry. However, in the second half of the 19th century the term *biochemistry* had yet to emerge; instead, the chemistry of living organisms was largely relegated to the subfield of physiology known as *physiological chemistry*. In 1877, Hoppe-Seyler at the University of Tübingen founded "*Zeitschrift für Physiologische Chemie*", the first specialized journal for the field, and insisted on the independence of the field as an academic discipline; however, it was not until the 20th century that biochemistry became generally recognized as an independent field. Thus, in the second half of the 19th century, biochemical research was conducted by the likes of organic chemists, physiologists, and biologists. As we saw in the previous section, research on the chemistry of naturally occurring organic compounds had gradually made clear the structure of the molecules relevant for living organisms; in the 20th century, this stream merged with the development of physiological chemistry to give rise to the major new academic discipline of biochemistry. In this section, we will trace the 19th century development of the chemistry relevant to living organisms up to the birth of biochemistry.

Agricultural chemistry and plant nutrition

In the world of agriculture, the importance of fecund soil and the nutrients required for plants to grow had been discussed since antiquity, but the discussion was based largely on guesswork. Van Helmont, Bacon, and Boyle believed water to be the most important nutrient for plant growth, while Mayow and Hales emphasized the importance of air. The British physician Woodward believed the soil to be an important nutrient. Although substances such as fertilizer, manure, and lime were known to be useful, there was no firm basis on which to understand why.

Agricultural experimentation began around the end of the 18th century. Lavoisier operated a 300-acre experimental farm with the goal of improving French agriculture, while in Germany, Thaer conducted agricultural experiments on an even more spacious plot of land. Einhof provided analytical data on soil and fertilizer. In 1813, Davy published *"Principles of Agricultural Chemistry"*, which included a discussion of the fecundity of soil. However, the data that could be gathered with the analytical methods of the time had not yet attained a state that could allow these problems to be discussed with any conclusiveness.

The question of which elements were necessary as plant nutrients was an important problem in the first half of the 19th century. In the 1820s, the German botanist C.P. Sprengel (1787–1859) studied the minerals necessary for plant growth; he saw 15 elements (carbon, oxygen, hydrogen, nitrogen, sulfur, phosphorus, chloride, potassium, sodium, calcium, magnesium, aluminum, silicon, iron, and manganese) as essential nutrients for plants, and he noted the possibility that there may be some other elements required only in tiny quantities. He conducted soil analyses and commented on the composition of optimal soil.

In France, Jean-Baptiste Boussingault (1802–1887) conducted experiments in a large botanic garden (farm) and analyses in a laboratory. He analyzed the constituents of plant ash to investigate the utilization of mineral components by plants and the effect of various types of fertilizers. He studied nitrogen sources in the growth of plants and linked the nutritional value of grains to their nitrogen content.

Agricultural chemistry received a major stimulus from the activities of Liebig. In 1840, he published *"Organic Chemistry in its Applications to Agriculture and Physiology"*, a book whose popularity was ensured by his authoritative stature in the field of organic chemistry. Liebig himself had no experience in agricultural chemistry research, and his book was primarily a review based on his opinions of studies conducted up to that time in the field of agricultural chemistry. The first portion of the book treated the nutrition of plants; the ingredients it deemed essential for plant growth were (1) carbon and nitrogen, (2) water and the elements of which it was composed, (3) soil as a source of inorganic elements. Liebig believed that nitrogen could be obtained from the air in the form of ammonia, and thus downplayed the role of soil as a source of nitrogen, a point that became a source of controversy. He recognized the importance of inorganic elements in plant nutrition, and emphasized the need for phosphoric acid, potash, soda, lime, and magnesia, but his incorrect beliefs regarding the equivalence of alkali species became a target of criticism. Liebig was possessed of an incorrigible confidence and tended to directly criticize the theories of others while retaining a dogged insistence on his own ideas. Although his book contained many errors, its emphasis on the importance of inorganic elements spurred the growth of the fertilizer industry. His passion for agricultural chemistry heightened interest in soil chemistry and plant nutrition and clearly played a major role in stimulating research in these fields.

The chemistry of fermentation
The fermentation processes that occur during the production of beer and wine or in the baking of bread from dough are among the biochemical phenomena lying closest to everyday human experience. The processes of fermentation and rotting had always been topics of interest to chemists. Lavoisier and Gay-Lussac attempted to clarify the quantitative relationships involved in the process of sugar fermentation to produce carbon dioxide and alcohol; they showed that a single molecule of sugar dissociates to form two molecules of alcohol plus carbon dioxide. Gay-Lussac believed that oxygen in the air was responsible for inducing fermentation. It was known that yeast was involved in fermentation, but it was thought to be non-biological.

In the 19th century, the factors causing fermentation were a major source of controversy among chemists. Liebig explained that, when sugars come

Willhelm Kühne (1837–1900)
A German physiological chemist, Kühne studied chemistry with Wöhler at Göttingen, then studied physiology at Berlin, Paris, and Vienna. In 1868, he became a professor of physiology at the University of Amsterdam, and later moved to the University of Heidelberg. He studied the chemistry of metabolism and discovered the protein-digesting enzyme, trypsin.

into contact with yeast, the vibrations of the yeast particles destroy the forces that bind molecules together, causing dissociation. Berzelius and Mitscherlich believed that fermentation and rotting proceeded through the effect of catalysts or contact substances. However, as advances in microscopy permitted observations of tiny bodies, it became clear that yeast was a spherical microorganism, and in 1837, the effect of this microorganism on the process of fermentation was independently proposed by three researchers: Cagniard de La Tour, Schwann, and Kützing.

Cagniard de La Tour proposed that yeast multiplied as a spherical microorganism that served to dissociate sugar into alcohol and carbon dioxide. Schwann thought of yeast as a tiny plant with spherical buds and that the fermentation of alcohol was mediated by this tiny plant. Kützing argued similarly that yeast was a microorganism. However, the most influential chemists of the day made little effort to recognize these theories. Liebig insisted stubbornly on his own theory, which held that the microorganisms of yeast were products, not causes, of fermentation. Berzelius also was not favorably disposed toward the microorganism theory. Nonetheless, as research continued from the 1840s through the 1850s, the microorganism theory of yeast gradually gained acceptance.

It was against this backdrop that Pasteur began research on fermentation by yeast. He was inspired to study fermentation by his research on optically active amyl alcohol. He investigated the fermentation of sugar to lactose and discovered that the yeast cells involved in this process were of a variety that differed from the yeast cells involved in the fermentation of sugar to alcohol. He went on to research a variety of fermentation processes, identifying a characteristic microorganism associated with each case. He discovered that, among the microorganisms relevant for fermentation, both aerobic and anaerobic species were present. These studies led him to conclude that *all* yeast cells act as catalysts in the fermentation process: they were "organized ferments."

In the late 18th century, the existence of a number of water-soluble substances—such as diastase, found in malt, which converted starch into sugar, or pepsin, found in stomach acid, which assisted in the digestion of food—gradually became known, and by the end of the 19th century, more than 20 types

Eduard Buchner (1860–1917)
A German biochemist, Buchner studied at the Universities of Munich and Erlangen and assisted Baeyer as an organic chemist; he was also inspired by Nägeli to develop an interest in fermentation phenomena. In 1896, while a professor at the University of Tübingen, he demonstrated that fermentation was a chemical reaction mediated by enzymes, for which work he received the 1907 Nobel Prize in Chemistry. After receiving the Nobel Prize, he became a professor at the University of Breslau and later Würzburg; he volunteered for the German army in World War I, was injured at a battlefield hospital on the Romanian front, and died in 1917 at the age of 57.

of these were recognized. Most of these promoted the hydrolysis of substrates and were considered distinct from ferments, which were thought to involve microorganisms. In 1878, the German physiological chemist Willhelm Kühne proposed that fermentation reactions were carried out not by yeast itself, but by a substance contained in yeast, which he called *enzyme* from the Greek words *en* (in) and *zyme* (yeast). This term later was used to refer to water-soluble reaction-enhancing substances such as diastase and pepsin. Many organic chemists, including researchers such as Traube, wondered if fermentation might not also be caused by water-soluble enzymes similar to these enzymes. However, a long series of attempts to observe such enzymes during the fermentation process yielded nothing but failure. The dominant paradigm among biologists was that of Nägeli, which held that the enzymes involved in fermentation were tied to the cell protoplasm. Thus, the controversy over whether or not fermentation was caused by acellular enzymes continued all the way through the end of the 19th century. It was Buchner who finally put this problem to rest.

In 1897, Eduard Buchner at the University of Tübingen—following a suggestion from his brother Hans, a professor of sanitation at the University of Munich—worked with Hans and with Hans' collaborator Hahn in an attempt to obtain an acellular enzyme solution from yeast. Hans wanted to extract acellular antitoxin from microorganisms and was hoping his younger brother would develop techniques for this purpose. They mixed yeast together with sand and diatomaceous earth in a mortar and ground the mixture into a paste; by compressing this, they obtained an acellular liquid. Upon adding sugar as a preservative to this liquid and letting it sit for a while, they observed that fermentation was occurring. This coincidental discovery was a result of major importance: they had succeeded entirely in severing the link between alcohol fermentation and living cells and put an end to the longstanding controversy surrounding fermentation[66]. Buchner named this enzyme *zymase*. The process of alcohol fermentation was studied in detail in the 20th century, leading not only to the discovery of many intermediate products but also revealing that fermentation was an extraordinarily complex process. Buchner was awarded the 1907 Nobel Prize in Chemistry for his chemical studies of fermentation. Although his success owed much to the assistance of his elder brother, to the techniques developed by Hahn, and to a considerable blessing of good fortune, his discovery was the starting point for modern research on enzymes and was an important step toward the birth of biochemistry.

Breathing and oxidation by living organisms

Lavoisier, in his later years, was fascinated by the relationship between breathing and combustion in animals; in collaboration with Laplace, he developed an ice calorimeter and compared the heat produced by breathing with the heat arising from combustion. They also compared the quantities of heat generated in the production of equal quantities of carbon dioxide via the two processes of burning coal and the breathing of a guinea pig (marmot). In 1789, he and Seguin reached the following conclusion[67]:

> In general, respiration is nothing but a slow combustion of carbon and hydrogen, which is entirely similar to that which occurs in a lighted lamp or candle. Indeed, upon leaving the lung, the air that has been used for respiration no longer contains the same amount of oxygen; it contains not only carbonic acid gas but also much more water that it contained before it had been inspired.

Lavoisier and Seguin concluded that what Lavoisier had termed "animal mechanism" was governed by three processes: breathing, sweating, and digestion. This was the starting point for 19th century animal physiology, which addressed problems such as in which organ oxygen was transformed into carbon dioxide, and in what forms carbon and hydrogen were present. In 1837, Magnus confirmed the presence in blood of dissolved oxygen, carbon dioxide, and nitrogen; he concluded that oxygen was absorbed by the lungs and carried by blood throughout the body, and that oxidation occurred in capillary blood vessels to produce carbon dioxide.

In the mid-19th century, Liebig's ideas on breathing and oxidation in living organisms attracted attention, but this was not grounded on experimental evidence and instead rested primarily on Liebig's authority as a chemist. Liebig's ideas on breathing were little more than modernized versions of the notions of Lavoisier. He believed that oxidation of foods took two different forms—one

George Stokes (1819–1903)
Stokes was an English physicist and mathematician who studied at Cambridge and became a professor there in 1849. He left behind a wide array of accomplishments in areas such as fluid mechanics, vector analysis and spectroscopy, and his name is attached to many laws and phenomena, including Stokes' Law, Stokes' Theorem, and the Stokes shift.

for fats and carbohydrates, which did not contain nitrogen, and another for albumin and casein that did. The former yielded fuel to power animal bodies, while the latter was transformed into blood. Liebig's ideas on the chemistry of living phenomena contained many errors, but efforts to validate his notions did succeed in advancing the state of biochemical research.

Liebig believed that red blood cells contained iron compounds, which were saturated by oxygen and gave up their oxygen in capillaries. By the end of the 1850s, the difference between the color of arterial and venous blood was understood to arise from the oxygenation and deoxygenation of hemoglobin. The spectroscopic research of Hoppe-Seyler and George Stokes in the 1860s further clarified the properties of the iron compounds that bind with oxygen. In 1862, Hoppe-Seyler discovered that the substance known as oxy-hemoglobin had two strong absorption bands in the visible portion of the spectrum; two years later Stokes discovered that the addition of an ammonia solution of iron (II) sulfide caused these two bands to collapse into a single broad absorption band[68]. The research of these two scientists not only laid the foundations of biochemical research on hemoglobin, but was also important for introducing spectroscopic methods into biochemistry. Thanks to research of this type, by 1875 it was understood that the interaction between hemoglobin and oxygen was not an oxidation process but rather a binding of oxygen, and that the role of hemoglobin was to be a carrier of oxygen. In the second half of the 19th century, the idea that oxidation occurs in the cells of animal tissues gradually became accepted; however, a correct understanding of oxidation in living organisms was not achieved until much later.

Digestion and metabolism
Until around the year 1800, the processes of digestion and absorption of nutrients were understood as follows: Food is chewed and dissolved in the stomach and bowels, where it is transformed through the effect of air and water into a liquid that forms the source of blood; the components of blood are then transformed into the components of various tissues. Muscle fiber was believed to arise from solidified fibrin dissolved in blood. The job of chemistry, rooted in the work

Claude Bernard (1813–1878)
A French physiological chemist and professor at the College de France, Bernard achieved many accomplishments in the physiology of metabolism and the nervous system, and is also famous for his contributions to the methodologies used in experimental biology and medicine. He emphasized the importance of hypothesis in experiments.

of Lavoisier and his contemporaries, was to describe these transformation processes in terms of elements found in the bodies of living organisms.

The new physiology began with the task of studying the nutritional roles of the various ingredients of foods. For example, the French physiologist Magendie studied dogs given restricted diets and showed that food containing nitrogen was essential for survival. He also noted that people who ate large quantities of meat, milk products, or other foods with high nitrogen content tended to suffer from ailments such as gout and calculus, and proposed that restricting the intake of such nitrogen-rich foods was an effective remedy. He served as the chairman of a French government committee to investigate the efficacy of gelatin, extracted from bones, as a foodstuff; he reported that gelatin alone did not suffice to sustain life, but that ingredients such as albumin and casein were also unable, on their own, to sustain life. The physiologists of the 19th century believed that digestion and metabolism were the internal workings of the human body to dispose of material that had become unnecessary while replacing it with new material.

In the 1840s, Liebig deduced that urea and uric acid were produced by transformation of the nitrogen content of blood through the effect of water and oxygen, and that these components were identical to the nitrogen content of food. However, it quickly became apparent that digestion and metabolism were vastly more complicated than Liebig had believed. French physiologist Claude Bernard discovered that sugar was present in blood flowing from the livers of animals given food devoid of carbohydrates and concluded that animal liver contained a means of producing sugar[69]. This refuted the notion that the sugars found in the blood and tissue of animals originated from foodstuffs. He also discovered that glucose is always present in blood and is stored as glycogen in the liver. He emphasized the importance of determining the intermediate substances involved in metabolic processes. Bernard was a physiologist who had significant impact in the second half of the 19th century; his *Introduction to the Study of Experimental Medicine*[70] was a classic of the field and was read for many years.

The question of how urea and uric acid were produced in, and expelled from, the bodies of animals was a subject of discussion among many physiologists and chemists in the second half of the 19th century. Dumas and Liebig believed that the degradation of proteins in animal tissue due to oxidation was the source of the urea present in urine, but physiologists countered that only a small portion of the urea in urine originated from proteins in tissues, and the controversy continued. The debate proceeded for some time with various theories proposed on both sides; ultimately, a proper resolution of the problem of the metabolism of proteins would have been impossible to reach with the level of scientific sophistication at the time.

2.11 The education of chemists

The situation at the beginning of the 19th century

Before the 19th century, talented amateurs contributed entirely to advances in chemistry. They were either independently wealthy or had wealthy patrons to support their research. Chemistry had not yet emerged as an independent discipline taught at universities; it was treated as little more than an ancillary branch of medicine. This situation first began to change in post-revolutionary France. In 1794, the Ecole Polytechnique was founded and began to offer excellent scientific education by the finest scholars of the time. Its founding mission was to produce military technologists well versed in matters such as mathematics and mechanics, but chemistry was also taught here; the school's first chemistry professor was Berthollet, and he was succeeded by Guyton de Morveau. There were many outstanding chemists in Paris, and chemistry education also took place at other schools. These schools did not have laboratories, but professors often had private laboratories to which they invited students. Gay-Lussac was one of Berthollet's first students, and his own private students included Dumas and Liebig. In the first quarter of the 19th century, budding scientists interested in chemistry were drawn to Paris, but the chemists of Paris ultimately never succeeded in establishing an influential academic center.

Another center of education at this time was the Stockholm laboratory of Berzelius. However, also in this case, the university did not provide a laboratory, and only students personally selected by Berzelius were accepted. Mitscherlich and Wöhler were educated here.

Germany was not an important center for chemistry research at this time. Liebig and Wöhler, who were later to provide the foundation for German chemistry, were dissatisfied with the state of chemistry in Germany and went abroad to study the cutting edge of the science. Chemistry schools were also absent in England, where individualism was strong; although Thomson, in Scotland, offered systematic chemistry education incorporating experimentation, it had no major impact. These situations were transformed by the reforms in chemistry education brought about by Liebig (see Column 6).

COLUMN 6

Liebig and the remaking of chemistry education

The scene of Liebig's laboratory at the University of Giessen

At the beginning of the 19th century, most universities still did not teach chemistry as a proper academic discipline and did not educate chemists. The individual who transformed this situation and made major contributions to advancing chemistry as a popular discipline was Liebig.

Liebig, who became a professor at the University of Giessen at the young age of 21, invented a completely new style of chemistry education in which many students simultaneously conducted experiments to learn chemistry. At first, he received little support from the university and conducted his efforts as a personal project, but in 1833, his laboratory became an official university facility, and students began flocking not only from across Germany but also from around the world. Liebig's laboratory emphasized analysis, and a systematic approach to chemistry was suited to the research at the time. Above all, Liebig himself was an outstanding analyst; he devised ingenious new experimental equipment to conduct his research, and his passion for science was a major source of inspiration to his students. As his group expanded, Liebig had less time for direct supervision of young students; instead, his senior students assisted in looking after junior students. Senior students were given research projects and delivered daily reports on the progress of their research the previous day; Liebig would discuss these results with the students, ensuring that the students were exposed to a wide variety of problems and had opportunities to teach each other. The University of Giessen became a center of chemistry education and the model for chemistry education in German universities thereafter. The laboratory also developed into a research school, and under Liebig's direction produced large numbers of chemists who would go on to play leading roles in developing organic chemistry and biochemistry in the late 19th and early 20th centuries.

Toward the end of the 1830s, Liebig's primary interest shifted from organic chemistry to agricultural chemistry and physiological chemistry, and he had a major impact on the evolution of these fields. Although he himself did not conduct research in these fields, he wrote books and engaged in controversies (pages 132–133). Liebig may well have played a more significant role than any other chemist in developing and expanding chemistry in the 19th century. But why did he suddenly switch focus from organic chemistry to agricultural chemistry? Germany's severe fall and winter weather ensured that farmlands were barren and that foodstuffs and animal feed were perpetually in short supply. In the mid-19th century, many Germans emigrated to America. It was against this backdrop that Liebig attempted to use his skills as a chemist to develop solutions. Although his book "*Agricultural Chemistry*" contained many errors, it was widely read both in Germany and abroad and was published in multiple editions.

Liebig was impetuous and quick-tempered, and his personality was somewhat dictatorial. In 1832, he became co-editor of the German journal "*Annalen der Pharmacie*", in which he published many papers; in 1837, its name was changed to "*Annalen der Chemie und Pharmacie*" and shortly thereafter to "*Liebigs Annalen*," under which name it continued for 124 years. As an editor, Liebig was fond of battles, and he criticized many of his chemistry contemporaries. A man of extreme self-confidence, he was merciless in his criticism of others, but became enraged if anybody criticized him and tended to clung stubbornly to his own opinions. Because of his difficult personality, he was avoided by most of his fellow chemists, although he remained lifelong friends with the warm-hearted and humble Wöhler.

The final edition of "Liebigs Annalen" (December 1997)

Liebig's educational reforms and their impact

Liebig began a program of chemistry education, centered on experimental instruction, at the University of Giessen. The idea was that students would learn chemistry by performing experiments; the notion has survived to this day and continues to form the template for university chemistry education. Students first learned qualitative and quantitative analysis, then conducted experiments in the synthesis of organic compounds, and finally tackled research problems.

The system of chemistry education established by Liebig was quickly adopted by other German universities. Liebig's friend Wöhler became a professor at the University of Göttingen in 1836, where he introduced educational methods borrowed from Giessen. In contrast to Liebig, Wöhler was an approachable man; he worked with great perseverance to help his students achieve maturity, and many chemists received their training from him. Other centers of research began to develop in Germany. One such center around the middle of the 19th century was Bunsen's laboratory in Heidelberg. Universities in many places began to adopt Liebig's methods; Giessen graduates became professors at universities in many places and introduced the educational methods they learned at Giessen. In some cases, universities offered well-equipped laboratories to attract leading chemists. One example is the University of Berlin, who invited Hofmann after he had just returned from England to the University of Bonn. The University of Berlin went on to become a world center of chemistry research under the leadership of Hofmann and Fischer.[12] The importance of Liebig's educational innovations and his greatness as an educator are vividly illustrated by the large numbers of Nobel laureates in chemistry, and in physiology and medicine, with connections to Liebig (see the charts on page 700, 701)

Thus, chemistry education in German universities became increasingly well established and trained large numbers of chemists; and thus it was that, from the late 19th century through the beginning of the 20th century, Germany laid the groundwork to become the world leader in chemistry and chemical engineering.

The situation in other countries

In France, which had a strong tradition of centralization, university positions were determined by the national government or the Academie de Science. A few prominent chemists such as Dumas and Berthelot exerted strong influence on academic appointments, and Liebig's educational reforms were not adopted. The situation at regional universities outside Paris was dire. As a result, France—

---

12. Nagayoshi Nagai and Yasuhiko Asahina, who made important contributions to organic chemistry and pharmacology in Japan, studied respectively with Hofmann and with Fischer.

August Hofmann (1818–1892)
Born in Giessen, Hofmann first studied law at Göttingen, but later switched to chemistry and studied under Liebig. In 1845, he became the director of London's Royal College of Chemistry, where he taught and conducted research, but in 1864, he returned to Germany and in 1865 became a professor at the University of Berlin. His accomplishments spanned a wide range of areas in organic chemistry. He first researched coal tar, then conducted a long series of studies related to aniline. His name is attached to many laws and reactions, including the Hofmann rearrangement, Hofmann elimination, and Hofmann degradation.

which had flourished as a center of chemistry in the late 18th and early 19th centuries—was unable to preserve its authority.

In England, Liebig was popular, and a movement arose to adopt his reforms. Liebig conducted a speaking tour of England and Scotland in 1837, arguing that chemistry was of use in agriculture and industry. Queen Victoria's husband Prince Albert, a passionate promoter of science education, succeeded in attracting interest from landowners and industrialists in establishing schools of chemistry. In this way, the Royal College of Chemistry was established in 1845 with Liebig's protégé August Hofmann its first director. Hofmann followed the educational methods of Giessen and attracted young scholars interested in chemistry research and education. Hofmann himself was interested in compounds obtained from coal tar and conducted research on them.

Although Hofmann's authority began to attract students, the schools financial backers were dissatisfied with his focus on basic research. English industrialists wanted research that would be of immediate value to industry; they were not interested in supporting research with a long-term focus and were not inclined to start new industries based on science. As time passed, the number of students declined, and the school ultimately merged with the Royal School of Mines; Hofmann resigned in 1864 to return to Bonn as a professor, and one year later became a professor at the University of Berlin, where he continued his research on nitrogen-containing organic compounds.

English universities with long traditions, such as Oxford and Cambridge, were uninterested in chemistry, preferring instead to continue curricula focused on the classics and on mathematics. However, in 1837, Graham took a position at London's University College and began pioneering experiments in colloid chemistry; in 1855, he was joined by Williamson, who studied with both Gmelin and Liebig. In Scotland, the traditions of Cullen and Black lived on in Glasgow and Edinburgh, but the country did not produce any particularly influential organic chemists.

Near the end of the 18th century, leaders in the new world of America, such as Benjamin Franklin and Thomas Jefferson, were fond of science. In 1769, Rush was named the first professor of chemistry at the College of Philadelphia, which had been founded in 1765. From the late 18th century through the early 19th century, chemistry was taught at elite eastern universities as part of natural philosophy or liberal arts; there was no specialized chemistry education. Until the mid-19th century, the primary objective of American universities was training for professions such as legislator or minister, for which science education was considered to be of limited importance. However, in 1846, Silliman proposed the establishment of courses in agricultural chemistry and the physiology of animals and plants at Yale University, and courses in practical chemistry and agricultural chemistry were launched. In 1847, a school of science was founded at Harvard, with Horsford—who had studied with Liebig—serving as headmaster and introducing the educational techniques he had learned at Giessen. Thus, chemistry education was in place at American universities by the mid-19th century, and these schools began training chemists. Meanwhile, as it became increasingly popular to study abroad in Europe after graduation, graduate schools also sprang up. Yale awarded its first science Ph.D. to Gibbs, while 1877 saw the first chemistry degree awarded at Harvard. By the end of the century, graduate schools had also been established at other elite universities such as Johns Hopkins, Pennsylvania, Columbia, Michigan, Chicago, and Wisconsin.

2.12 Chemical industries in the 19th century

Two major events in the 19th-century history of the chemical industry were the development and subsequent abandonment of Leblanc's method for producing soda. The growth of the middle class led to an increase in consumption, increasing demand for the production of chemical-industry products such as soap, glass, and textiles; this created the need for an alkali industry. The spread of the Leblanc method increased demand for sulfuric acid; together with the growth of the fertilizer industry in the second half of the century this created a huge market for sulfuric acid. Moreover, a dye industry based on synthetic organic chemistry emerged, and in Germany, the chemical industry flourished in close connection with university chemistry research. Finally, the end of the 19th century saw the birth of an electrochemistry industry. Thus, the business of chemistry, which had been little more than a small-scale cottage industry at the beginning of the 19th century, had begun by the end of the century to transform into a massive enterprise.

The alkali industry

Although the Leblanc method was developed at the end of the 18th century, it was not until significantly later, well into the 19th century, that a commercial base arose for it to be put into widespread industrial use. In 1823, Muspratt succeeded in the factory production of sodium carbonate using the Leblanc method in Liverpool. This was followed by production using the Leblanc-method at various places across England and France. However, the Leblanc method had a number of problems. The conversion of salt into sodium sulfate produced hydrogen chloride, which was released into the atmosphere and led to rain containing hydrochloric acid; when this rain fell onto neighboring agricultural lands, it threatened the health of local citizens. This resulted in frequent complaints. An additional problem was the danger posed by calcium sulfide, one of the waste products of the process. Moreover, the efficiency with which the process converted raw materials into product was poor. The problem of hydrogen chloride was solved by Gossage's invention of the chloric-acid absorption tower. Technologies were also developed for converting hydrogen chloride into chlorine, which was used in the manufacture of bleach. Finally, calcium sulfide could be oxidized in a combustion furnace and recycled in the form of sulfur. Thus, by 1880 the Leblanc method had been significantly improved. However, the method still involved considerable waste and required unpleasant work, creating a need for a cleaner and more efficient process.

A method for producing soda from ammonium bicarbonate and salt was discovered by Fresnel in France in 1801, but operating this method efficiently proved difficult. Muspratt invested a significant amount of capital in trying, but was unsuccessful. In 1861, Belgian chemist Ernst Solvay (1838–1922) devised a carbonate tower in which water containing ammonia-rich table salt rained down on a rising cloud of carbon dioxide to produce sodium bicarbonate. This method used only readily available ingredients, caused little pollution, and used less fuel than the Leblanc method. The Solvay method gradually replaced the Leblanc method.

Caustic soda had been made by reacting sodium carbonate with calcium hydroxide, but near the end of the 19th century an alternative method was introduced in which the electrolysis of table salt was put to industrial use. A problem with this method was that it required a means of blocking the reaction of sodium and chloride produced between the anode and cathode, but this difficulty was resolved by using mercury electrodes.

The fertilizer industry

Humans had known since antiquity that fertilizer was useful in stimulating the growth of plants. Already in the 15th century, the Incas knew that guano obtained from bird waste was an effective fertilizer. In 1804, Humboldt brought home a sample of guano and determined that it contained nitrogen and phosphorus.

Although it was known that chemical substances could assist plant growth, it was not until the first part of the 19th century that the relationship of elements such as phosphorus, nitrogen, and potassium to plant growth became understood through the research of Boussingault in France, Lawes in England, and Liebig in Germany. Liebig did not understand that soluble fertilizer was necessary for plants to absorb nutrients from soil, but in 1840, he proposed to make fertilizer by treating bone with sulfuric acid. In 1842, Lawes obtained a patent for superphosphate fertilizer made by processing bone powder with sulfuric acid and immediately began factory production. Superphosphate fertilizer was also produced from phosphate minerals. As the fertilizer industry grew, it became the largest consumer of sulfuric acid.

Nitrogen-based fertilizer began to rise to a position of importance around midcentury. Liebig was of the erroneous view that plants capture ammonia from air; he proposed the production of ammonium sulfate. The fact that nitrogen in soil was important for plant growth was shown in 1857 by the American chemist Pugh. In the second half of the 19th century, production of ammonium sulfate became an important industry. In those years, the U.S. and England imported large quantities of guano from Peru and Chile. Chile niter ($NaNO_3$) was also imported and used as fertilizer.

Coal tar compounds and artificial dyes

Hofmann, who had studied with Liebig and earned his academic degree for research on coal tar, became the director of London's Royal College of Chemistry in 1845—thanks to his connections to Liebig—where he continued research on coal tar. Coal tar was a byproduct of the production of coke and coal gas, but it had little use other than as an anticorrosive for wood and was a dangerous waste product. Hofmann and his student Mansfield improved methods of partial distillation to isolate high-quality samples of some 20 substances, including benzene and toluene. This development made it possible to process and study coal tar, opening the door to effective utilization of its components.

In 1856, an 18-year old student, William Perkin, at the Royal College of Chemistry attempted to synthesize quinine ($C_{20}H_{24}N_2O_2$)—an effective malaria

William Perkin (1838–1907)
An English organic chemist, Perkin studied with Hofmann at the Royal College of Chemistry. While attempting to synthesize quinine as an assistant to Hofmann, he stumbled on the coloring agent mauveine, which later became the synthetic dye known as mauve. He received a patent and in 1857 started a synthetic dye factory. He also succeeded in synthesizing tartaric acid, coumalin, alizarin, and other compounds. He is known for Perkin's reaction, in which cinnamic acid is synthesized from aromatic aldehyde.

remedy—from mixtures of allyl chloride (C_3H_5Cl), toluidine (C_7H_9N), and potassium dichromate. However, instead of quinine he obtained brown clumps. Upon carrying out the same reaction using aniline, he instead obtained a black-colored substance, which he discovered could be used to dye silk purple. He had succeeded in stumbling by accident on the first synthetic dye, aniline-dyed mauve. Perkin dropped out of school and started manufacturing mauve with help from his father and brother. As demand for dyes—for room decorations and women's fashions—grew throughout 19th-century Europe, each year brought new marketplace demands for the latest fashionable color varieties, ensuring that the natural dye industry of the time was a large and increasingly profitable business. It was against this backdrop that mauve-dyed silk became all the rage in Paris and quickly caught on in London too. The success of mauve stimulated increasing demand for new synthetic dyes, and new dyes were created from starting materials such as aniline, toluidine, and quinoline. However, the chemical structure of mauve remained unknown, and its synthesis remained an empirical process of trial and error.

In 1858, the triphenyl-methane-based colorant fuchsine was introduced by Freres in France. This was a red dye, which meant it had broader utility than mauve; it was also easy to manufacture. Fuchsine also became the basis for other dyes such as Hofmann violet and Perkin green. The first azo dye—created in 1863—was Manchester brown, consisting of diazonium salt bound to aromatic amine. In the 1880s and thereafter, azo dyes became the most widely produced synthetic dyes. Eventually, natural dyes were entirely eclipsed by artificial dyes and fell largely into disuse.

Synthesis of natural dyes and the synthetic chemical industry
Organic chemistry was a field with strong ties to practical applications from its very beginnings, but progress in synthesis methods during the second half of the 19th century ensured that it made major contributions to the development

Adolf von Baeyer (1835–1917)

Bayer was a German organic chemist. After completing his degree at the University of Berlin, he studied with Bunsen at Heidelberg and later worked with Kekule. In 1875, he became a professor at the University of Munich, where he earned fame for his research on topics such as the synthesis and structural determination of indigo, the discovery of phthalein-based dyes, uric acid derivatives, ring-shaped organic compounds, and the effects of carbon dioxide assimilation. He trained many famous organic chemists, including Fischer and Willstätter. He was awarded the 1905 Nobel Prize in Chemistry.

Figure 2.21: Structural formula of indigo

of industry. The first step in this direction was the development of synthesis techniques for natural dyes, a field with no boundary between basic science and applications.

By the mid-19th century, dyes were being obtained from natural plants and animals. For example, indigo—extracted from the leaves of plants growing on the Coromandel Coast in India—was used to make blue dye, while the madder plant—a plant cultivated in Europe—was used to make red dye. Carmine red was made from the cochineal insect attached to cactus plants in Mexico and Peru. The synthetic dyes pioneered by Perkin were widely used, but the individual who made the greatest contributions to understanding the structure and chemistry of natural dyes and developing chemistry-based technologies for their synthesis was Baeyer.

In 1875, Adolf von Baeyer, who had succeeded Liebig as professor at the University of Munich, succeeded in identifying the structural formula of indigo and developing a method for its complete synthesis (Figure 2.21). Baeyer was also known as an excellent teacher, and his students included three Nobel laureates who laid the foundations of biochemistry—Fischer, Buchner, and Willstätter—as well as many other outstanding organic chemists. Baeyer himself was awarded the 1905 Nobel Prize in Chemistry for his research on organic colorants and hydro-aromatic compounds.

The compound alizarin, known to be a constituent of the natural dye madder, was synthesized from anthracene in the 1860s by Graebe and Liebermann in Baeyer's laboratory. A synthesis method was developed based on the use of fuming sulfuric acid for the oxidation in the first step; this was industrialized by the German chemical company BASF, and synthetic alizarin completely replaced madder. Baeyer began research on indigo in 1870; by 1880, he had succeeded in synthesizing ortho-nitro cinnamic acid as a starting material, and he immediately signed contracts with BASF and Hoechst, but the compound turned out to have no commercial value. Next, in 1882 he synthesized ortho-nitro toluene, and identified its structure in 1883. In 1890, Heumann at BASF developed a manufacturing method based on naphthalene, and the synthesis of indigo was finally commercialized. Hoechst continued to add its own improvements to the process. In this way, universities and the industrial world developed tight collaborative relationships in Germany, and the German chemical industry began to dominate global markets. Many talented chemists had received training in the style of chemical education pioneered by Liebig, and they went on to spearhead research and development efforts in German chemical companies. Under Bismarck, Germany—which had lagged behind other countries during the industrial revolution—advanced ambitious policies to achieve prosperity and military strength, and its industries, with strong ties to chemistry, began to take positions of authority in this process.

Meanwhile, the English dye industry, which had flourished early thanks to the successes of Perkin, proceeded to ignore basic research—notwithstanding Hofmann's exhortations to the contrary—and relied entirely on empirical practice. After Hofmann returned to Germany, many of the German chemists he had trained repatriated as well. Few investors in England were interested in the chemical industry, and the English dye industry gradually faded out of existence.

The pharmaceutical industry
In the first part of the 19th century, the pharmaceutical industry had yet to develop. Pharmacists prescribed medicines derived from plants, mercury derivatives, sulfur, and similar remedies. In 1867, the English surgeon Lister introduced the use of carbolic acid (phenol) as an antiseptic.

In 1853, Kolbe synthesized salicylic acid from phenol and discovered that it was useful for its fever-reducing effects and as a food preservative. Commercial production was initiated 20 years later by his protégé Von Heyden, but it was not widely used due to its side effects on digestion. The Bayer firm sought to develop a fever reducer without side effects; in 1889, it developed phenacetin, and in 1898, it developed acetylsalicylic acid, which was to become world-famous under the

Alfred Nobel (1833–1896)
Nobel was a Swedish chemist, engineer, and businessman.
Born the son of an inventor, he studied chemistry and foreign
languages with a home tutor in St. Petersburg, then joined
his father in the business of manufacturing explosives. In
1866, he invented dynamite, then later invented the first
smoke-free explosives and nitroglycerin-containing ignition
charges, earning a massive fortune. However, he came to
experience remorse over the rampant wartime use of smoke-free explosives
and donated his fortune to the Swedish Royal Academy of Sciences to create
the Nobel Prize Foundation.

name *aspirin*. These products had value as fever-reducers and painkillers and
found widespread use, but were not effective at targeting illnesses themselves.

In the second half of the 19th century, the research of Pasteur and Koch laid
the groundwork for a developing science of bacteriology, and bacteria were
identified as a cause of disease transmission. Through microscope observations
in 1872, Koch discovered that aniline dye selectively colored bacteria. In 1889,
Koch's assistant Ehrlich reasoned that the use of appropriate dyes might be able
to kill germs; in 1891, he discovered that methyl blue was effective against the
insects that carried malaria. This was the birth of chemotherapy and the first of
many subsequent attempts to search systematically for organic compounds that
would selectively kill bacteria without producing side effects. Although it was
not until the 20th century that these efforts would bear fruit—in the form of
pharmaceuticals such as the syphilis drug salvarsan—it was here that the field
of chemotherapy separated from dye chemistry to become an independent field.

The explosives industry

Since the invention of explosives, there had been little change in the ingredients
of black-colored gunpowder: sulfur, charcoal, and saltpeter. After Pelouze, in
France, nitrified cellulose, Schönbein, in Switzerland, discovered that this nitride
was explosive, and in 1846 he produced nitrocellulose from concentrated sulfuric
acid, nitric acid, and cotton. That same year, Sobrero, in Italy, processed the
same acid mixture with glycerin to obtain nitroglycerin. These were the first
synthetic explosives. However, they were all unstable and posed high risk of
explosion; also, nitroglycerin was a liquid, making it unsuitable as an explosive.

In 1866, Alfred Nobel stabilized nitroglycerin by absorbing it into diatomaceous earth. He called the resulting product *dynamite*, obtained patents for it
in several countries, and brought it to market in the form of cylindrical rods.

Dynamite required an igniting agent to explode, making it relatively stable to handle. In 1875, Nobel developed explosive gelatin from a mixture of nitro-glycerin and nitrocellulose. The booming mining and construction industries of the late 19th century created large demand for strong explosives; a military market existed as well. Nobel launched a multinational corporation and sold his product around the world. His business flourished, and by the time of his death, he had amassed an enormous fortune. Upon his death, he donated his assets to establish a foundation of 32 million Kroner to offer Nobel Prizes in five disciplines: physics, chemistry, physiology or medicine, literature, and peace.

Metals and alloys

The progress of the industrial revolution in the late 18th and 19th centuries created rapidly growing demand for iron and steel. The manufacturing methods that had been used until that point were simply unable to keep up with the explosion in demand for steel required to build the steam engines, textile mills, railroads, steel ships, bridges, and other infrastructure underlying the industrial revolution. The development of new manufacturing methods was carried out mainly by engineers.

By the end of the 18th century, it was clear that the difference between pig iron and steel was attributable to their different carbon contents. In 1784, Court developed the "paddle method," in which pig iron was transformed into refined iron and an iron rod was used to stir liquid iron in a reverberatory furnace in which high-temperature gas was reflected from bricks in the roof onto the metal. This method produced high-purity iron because carbon and other impurities burned off in the stirring process; however, it was of limited use for producing steel.

In the 18th century, blast furnaces came to be used to produce pig iron, and in 1828, Neilson improved the blast furnace design by using high-temperature air obtained from waste heat. In 1855, Bessemer conceived of a process in which air was blown into melted pig iron to oxidize impurities and the heat generated by this oxidation process was used to make steel; this process enabled the production of large quantities of steel. However, the Bessemer process had the drawback that it could not be used to convert iron ore containing phosphorus or sulfur. On the other hand, in the 1860s, the Siemens brothers and the father-and-son team of Martin used preheated gas and air to raise reverberatory furnaces to high temperatures and successfully produced steel in an open furnace lined with silicate bricks. This allowed steel to be produced in large quantities even from ore that could not be treated by the Bessemer process. The problem of the Bessemer process—that it was unable to handle ore containing phosphorus—was solved in 1877 by Thomas and Gilchrist, who used bricks composed of basic materials like dolomite to line a converter furnace in which slag would bind with phosphorus. The process of mixing iron with other metals (such as tungsten, chrome, nickel, and manganese) to obtain alloys to lend special properties to steel also started in the 19th century.

One new metal that began to be refined in the 19th century was aluminum. In 1886, Herout in France and Hall in America independently developed methods to produce aluminum through electrolysis. In these processes, oxidized aluminum refined from bauxite was electrolyzed in a solution containing sodium fluoride and cryolite melted in an electrical furnace.

2.13 The introduction of modern chemistry to Japan

Modern chemistry first came to Japan toward the end of the Edo period. In the ensuing 160 years, Japanese chemistry progressed rapidly, catching up with chemistry in the rest of the world and competing today at the cutting edge of global research. A glance at the list of Nobel chemistry laureates through 2014 shows that in the first 50 years of the prize there was not a single Japanese recipient, but in the subsequent 64 years there have been seven. How was such an impressive development accomplished, and how does this story fit into the larger picture of the global evolution of chemistry? In this section, we consider the process by which modern chemistry was introduced into Japan in the 19th century.

The start of modern chemistry education[71, 72]
In the mid-16th century, the nation of Portugal—which was then making inroads in Asia—brought elements of Western culture and equipment to Japan, typified respectively by Christianity and by guns. However, the Tokugawa Bakufu government of Edo-era Japan, fearing the spread of Christianity, imposed a closed-nation policy: Christianity was banned, foreign trade was restricted in 1639 to just the two trading partners of Holland and China, and overseas excursions by Japanese citizens were forbidden. Thereafter, commerce with Dutch merchants at Nagasaki was the sole portal through which Japan was open to the West. When the Tokugawa government negotiated with foreign countries, the task of translating the relevant documents fell for many years to Nagasaki translators.

However, in the second half of the 18th century, ships from Russia, England, and the U.S. began to make frequent appearances in Japanese waters, and it became clear that effective diplomatic negotiations would require a team of translators stationed in Edo (now Tokyo) as well. Thus in 1811, a government office known as the Bansho Wage Goyo (official office for the translation of

Yoan Udagawa (1798–1846)

Udagawa was a Japanese doctor and translator of Dutch texts in the late Edo period. The eldest son of a doctor in the Ogaki *han*, he was adopted by the Dutch-language scholar Genshin Udagawa. In 1816, he became doctor to the Tsuyama *han*, and in 1826 an assistant translator in the *bakufu* government's library of Western books. He was a scholar of Western studies from his early years and introduced, through his translations, many Western books on chemistry, botany, and pharmacology to Japan. He was inspired by a Dutch translation of Lavoisier's *Elementary Treatise on Chemistry* to publish the first systematic Japanese textbook on chemistry, *Seimi Kaiso*.

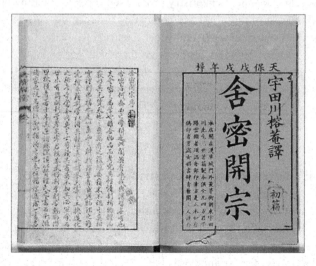

Figure 2.22: Yoan Udagawa's "Seimi Kaiso"

barbarian books) was formed in Tenmon Kata (research office of astronomy and astronomy-based calendar) in the Asakusa district of Edo. However, this office was not always kept fully occupied with diplomatic work, and in their free time the translators undertook projects such as the translation of a Dutch-language encyclopedia (which was itself a Dutch version of a work originally in French) into Japanese.

One of the translators, Yoan Udagawa—who was trained in Western medicine—published between 1837 and 1846 a full 12-volume chemistry textbook titled "*Seimi kaiso*" (Introduction to Chemistry) (Figure 2.22). The Japanese word *seimi* was chosen because it sounded like the Dutch word "chemie." This was the earliest introduction of modern chemistry to Japan. The book was based on a Dutch translation of the textbook "*Elements of Experimental Chemistry*", written by William Henry at the University of Manchester and based on Lavoisier's system of chemistry. Udagawa not only translated this text but also performed experiments. He analyzed water from Atami and other Japanese hot springs, which he found to contain iron, copper, lead, calcium, alum, carbon dioxide, and sulfate ion. In 1856, the *Bansho Wage Goyou* became the *Bansho-shirabe-dokoro*, the School of Western Learning, and in 1863 was renamed the Kaisei-sho (Institute for Translation and Foreign Studies). A chemistry education department known as the *kagaku-sho* (chemistry division) was created in this office. One of the professors was Komin Kawamoto. He translated a Dutch version of the textbook "*Die Schule der Chemie*", by the German chemist Stoeckhardt, and published it as "*Kagaku Shinsho*" ("A New Textbook of Chemistry"). Here Kawamoto replaced Udagawa's term *seimi* with the word *kagaku*, a Japanese pronunciation of characters that had been used to mean "chemistry" in Chinese; this word gradually took root in Japan. Komin Kawamoto is also known for being the first to brew beer in Japan. In this way, modern chemistry was gradually introduced into Japan. Perhaps because the science of chemistry had a certain character that is inextricably linked to practical, real-world pursuits, it seems to have stood out among all traditional Western academic disciplines as a particularly popular subject in Japan.

Modern chemistry education in Japan was initiated by the Dutchman Koenraad Gratama. During the era in which Japan was closed off from the outside world, its only contact with the West was the city of Nagasaki, in which a medical training facility was established in the late 1850s to launch modern medical education in Japan. Chemistry was originally taught here as part of basic education, but in 1865 it was split off from medical education as part of a newly established "Analytical Institute" at which chemistry and physics were taught. Gratama came to Japan in 1866 as a full-time teacher. Gratama was an army doctor who had received degrees in natural science and medicine from the University of Utrecht. The *bakufu* invited him to Japan and planned to establish a new school of science in the *Kaisei-sho*, but the government collapsed before this plan could be enacted. However, the new Meiji government, realizing that science and technology education was essential for prosperity and military strength, moved the science and engineering school in the *Kaisei-sho* to Osaka, where it established an Osaka Bureau of Chemistry (*Osaka seimi kyoku*); it was here that Gratama started teaching in 1869. The Osaka of this time was a vibrant center of Western learning, centered around the *Tekijuku* school founded by Koan Ogata; many students who went on to play leadership roles in the advancement of Japan during the early Meiji era were educated at this school. Gratama's term was complete at the end of 1870, and he returned to Holland the following year. The Osaka Bureau of Chemistry was renamed the Technical School in 1870; thereafter it merged with a school of Western learning to form the Osaka branch of the *Kaisei-sho* Technical School. However, in 1872, the Meiji government instituted a centralizing reform of the education system, under which the Osaka *Kaisei-sho* became the First Middle School of the Fourth University District and lost its role as a school devoted specifically to science and technical education. Nonetheless, this school went on to traverse a meandering historical trajectory that eventually led it to be transformed into Kyoto's Third Higher Middle School, and it was later connected to the founding of Kyoto University.

The new Meiji government transformed the *Kaisei-sho* as the *Kaisei-gakko*. In 1869, *Daigakko* (College) was founded, with *Kaisei gakko* and the medical school respectively renamed *Daigaku nanko* and *Daigaku toko / Daigakko*. In 1871, the Ministry of Education was established, and these universities were renamed simply *Nanko* and *Toko*. Later, in 1875, Nanko was renamed *Tokyo Kaisei Gakko* and Toko was renamed *Tokyo I-Gakko* (Tokyo Medical School); it included the creation of a department of pharmaceutical production. In 1871, the Ministry of Industry established *Kogakuryo* (Institute of Technology) and began preparing to offer engineering education; in 1874, *Kogakkou* (School of Engineering) was formed. An agricultural school was subsequently established at Komaba by the Ministry of the Interior. Thus, by 1875 a system of modern chemistry education had been initiated in five separate disciplines: science, engineering, medicine, pharmacology, and agriculture. These schools invited foreigners to be professors and to develop educational systems; in addition, many students were sent abroad for foreign study, and new academic disciplines were introduced. The first group of foreign-study students included a number of individuals who would go on to play leading roles in the development of Japanese chemistry, including Jyoji Sakurai, Nagayoshi Nagai, and Naokichi Matsui.

The establishment of the university system and the training of chemists
In 1878, Tokyo Kaisei Gakko and Tokyo I-Gakko merged to form Tokyo University. Among its first graduates was Mitsuru Kuhara. Moreover, Japan's Ministry of Industry closed the older *Kogakuryo* (Institute of Technology) and renamed its engineering school *Kobu Dai-Gakko* (College of Engineering). It was in this year that the Chemical Society (renamed the following year to the Tokyo Chemical Society) was formed with Kuhara as its chairman; although this was the birth of the Chemical Society of Japan, at the time its members numbered just 24.

Mitsuru Kuhara (1855–1919)

The son of a doctor in the *han* of *Mimasaka no Kuni Tsuyama* (today the city of Tsuyama in Okayama Prefecture), Kuhara was a member of the first graduating class of the chemistry department of Tokyo University's School of Science in 1877; the following year he became an assistant professor and the first chairman of the Chemical Society of Japan. In 1879, he embarked on a foreign study program at Johns Hopkins University in America, where he obtained a Ph.D. before returning to Japan. In 1886, he became a teacher at the First High School, and later a professor at the School of Science and Engineering at the Kyoto Imperial University, founded in 1900. The primary focus of his research at Kyoto was the Beckmann rearrangement, in which a keto-oxime undergoes an internal molecular rearrangements in the presence of acid; his work was well received overseas.

Jyoji Sakurai (1858–1939)

Sakurai entered *Daigaku Nanko* in 1871 and in 1877 was a member of the second class of Japanese foreign-study students; he traveled to England and studied with Williamson at the University of London. In 1881, he returned to Japan and became the first Japanese professor in the chemistry department of the Tokyo Imperial University School of Science. In 1917, he became the assistant director of RIKEN, and in 1926 became director of the Imperial Academy. He played a central role in the creation of the Japan Society for the Promotion of Science and in academic administration.

At the time of its founding, Tokyo University's Faculty of Science counted among its professors the Englishman Atkinson, who had been present since the days of Tokyo Kaisei Gakko, as well as the American Jewett and others; as Japanese researchers who had been studying abroad returned home, they gradually replaced foreign faculty members. Naokichi Matsui, who had been studying at Columbia University, succeeded Jewett in 1880 as the head of inorganic and analytical chemistry, and in 1881 Jyoji Sakurai, who had been studying with Williamson (see page 65) at London's University College, succeeded Atkinson in charge of organic and physical chemistry.

On the other hand, at the College of Engineering (*Kobu Daigakko*) the Englishman Divers had been a professor since the days of the *Kogakko*; he served until 1899 and trained a large number of chemists and chemical engineers. As an inorganic chemist, he collaborated with many Japanese researchers and published reports in the Journal of the Chemical Society of London; thus, chemistry research of interest to the outside world was already being conducted in Japan in this era.

In 1886, the various academic departments of Tokyo University became the School of Arts and Sciences of the Imperial University system, continuing under names such as the Imperial University Faculty of Science and the Imperial University Faculty of Engineering. Divers moved to the Faculty of Science in 1886, and Matsui moved to the Faculty of Engineering. In 1893, Nagayoshi Nagai returned from his long foreign study in Berlin and became a professor in the Department of Pharmacy in the Faculty of Medicine. In 1890, the agricultural school at Komaba became the Imperial University Faculty of Agriculture; a department of agricultural chemistry was created, and in 1897, Yoshinao Kozai returned from Germany to become a professor there. In 1897, Kyoto Imperial University was established, with its Faculty of Science and Engineering containing departments of pure chemistry and manufacturing chemistry. Courses in medical chemistry were offered by the Faculty of Medicine. Thus, education in modern chemistry began in many schools—including schools of science, engineering, medicine, pharmacology, and agriculture—and the training of Japanese chemists and chemical engineers began in earnest. In the 19th century, Japan busied itself learning the modern chemistry of Europe, and although this period did not boast a great deal of original research destined to live on in the history of world chemistry, a notable exception is Hikorokuro Yoshida's discovery of the oxidizing enzymes of *urushi,* Japanese lacquer (see Column 7).

Modern chemistry education in Japan unquestionably began in a position that lagged behind that of Europe, but this lag was not as severe as one might think. As we saw in Section 2.10, in the 19th century, chemistry was not taught even at the most august English universities such as Oxford or Cambridge. At German universities, the educational reforms of Liebig ensured that chemistry was a popular discipline, but engineering was still not taught at the oldest universities, and instead was taught only at specialized engineering schools (*Technische Hochschule*). America began education in modern chemistry at roughly the same time as Japan. One feature of the Japanese university education system was that chemistry research and education was not confined to science departments, but from the start was incorporated in an applied capacity into many departments, including schools of engineering, medicine, pharmacology, and agriculture. This situation undoubtedly enabled Japanese chemistry to make progress in a wide range of fields. However, in contrast to the West—where science developed amidst longstanding Judeo-Christian cultural traditions—in Japan there was an excessive tendency to emphasize the practical value of science. Chemistry is surely not alone among the sciences in Japan in suffering from a certain weakness of long-term perspective, and in failing to cultivate an atmosphere in

COLUMN 7

Hikorokuro Yoshida and Research on Urushi Japanese Lacquer[73]

Hikorokuro Yoshida

By the 1880s—not long after modern chemistry was first introduced to Japan—Japanese chemist Hikorokuro Yoshida was already performing research destined to earn a permanent place in the international history of chemistry. Yoshida—the fourth son of Toyotoki Yoshida, a top-level samurai advisor to the feudal ruler of the Abe domain in present-day Hiroshima Prefecture—was born in 1859 and enrolled in 1871 at the Daigaku Nanko (also known as Kaisei Gakko), the predecessor of Tokyo University. In 1877, Yoshida began studies at the newly established Chemistry Department at the School of Science in Tokyo University; it was here that he came under the tutelage of a British professor named Atkinson, and conducted research on menthol for his graduation thesis. After graduating in 1880, he took a job in the Analysis Group of the Ministry of Agriculture and Commerce's Imperial Geological Survey, where he began research on the Japanese lacquer known as urushi.

At that time, urushi was already well-known in Western Europe (where it was known simply as "japan"), but its chemistry remained unexplored. The goals of Yoshida's research were ambitious: he sought not only to analyze the chemical composition of urushi, but also to understand the mechanisms responsible for the hardening effect that the substance exhibited—and even to achieve practical improvements in the real-world practice of lacquer craftsmanship. For his research, Yoshida chose urushi trees from the Yoshino region of Japan's Nara Prefecture; he made incisions between the inner and outer layers of tree bark to obtain raw urushi, which he used as the source material for his research, starting with the separation and purification of its components. Yoshida assumed that the primary component obtained from the anhydrous ethanol extraction was an acid, termed this acid *urushiol*, and determined its molecular formula to be $C_{14}H_{18}O_2$. He also discovered the presence of water-soluble components: the protein-like $C_{72}H_{110}N_6O_{24}$ and the acacia-gum-like carbohydrate $C_{12}H_{22}O_{11}$. Next, Yoshida established that the hardening effect of urushi arose from the oxidation of urushiol due to the effects of diastatic substances in humid, oxygen-rich conditions. The "diastatic substances" in question here are in fact enzymes, but the word "enzyme" had only just been coined in 1878 by Kühne, and at the time the only known enzymes were hydrolyzing enzymes. Nobody knew that enzymes could facilitate oxidation reactions, whereupon Yoshida used the word "diastase." Thus, Yoshida may rightfully be considered the discoverer of oxidizing enzymes. Thereafter, the French chemist Bertrand, while researching urushi from Tonkin, assigned the name *laccase* to this oxidizing enzyme, but Yoshida's priority in discovering the enzyme laccase was recognized by Sumner and is noted in Fruton's *"History of Biochemistry"*.[6] Nonetheless, it seems this history is not particularly well known even among Japanese biochemists. Around 1980, when the American biochemist H. S. Mason visited Japan and made a number of inquiries regarding Yoshida's achievements, few Japanese scientists were even familiar with the name.

Structural formula of urushiol

Subsequently, Riko Majima—considered a pioneer of organic chemistry research in Japan—chose structural studies of urushi to be his first research project. Majima showed that urushiol—the acid identified by Yoshida as the primary component of urushi—exhibits the structural formula shown in the figure, with R denoting a blend of several unsaturated alkyl groups. Research on the chemistry of urushi thus amounted to a highly original contribution of Japanese chemistry, conducted at the very dawn of chemical research in Japan and admired the world over.

Thereafter, Yoshida worked as an assistant professor at the School of Science at Tokyo Imperial University and as a professor at Gakushuin University. In

1896, he became a professor at the Third High School, and in 1898 he departed for a two-year foreign-study program in Germany. After returning to Japan, he worked with Mitsuru Kuhara to found the School of Science and Engineering at Kyoto Imperial University, where he taught the third course on chemistry (organic manufacturing chemistry). Yoshida frequently reminded his colleagues that "learning from books alone will make one into an empty skeleton; one has to make contact with the real world!" He stressed the importance of practical education and served as a consultant to a variety of chemical companies. However—in a development perhaps not unrelated to his practical-minded philosophy—he was drawn into the Sawayanagi Incident[1] at Kyoto University and submitted his resignation in 1913 at the age of 54.

The cover of a middle-school textbook authored by Hikorokuro Yoshida

---

1. The Sawayanagi Incident refers to a dispute involving Masataro Sawayanagi, who became President of Kyoto Imperial University in 1913. When Sawayanagi—claiming a need to revitalize the University—attempted summarily to dismiss seven professors, the faculty protested, and Sawayanagi was ultimately forced to resign. The incident played a major role in establishing faculty autonomy at Japanese universities.

which science is seen as a form of culture, whose primary objective is to unlock the fascinating mysteries of nature.[13]

References

1. A. J. Ihde, "*The Development of Modern Chemistry*" Dover Publications, Inc., New York, 1984
2. W. H. Brock, "*The Chemical Tree*" New York, 1993
3. J. Gribbin, "*Science: A History*" Penguin Books, 2002
4. M. J. Nye, "*Before Big Sceience*" Harvard University Press, Cambridge, 1996
5. M. J. Nye, "*From Chemical Philosophy to Theoretical Chemistry*" University of California Press, Berkeley, 1993
6. J. S. Fruton, "*Molecules and Life: Historical Essays on the Interplay of Chemistry And Biology*" John Wiley & Sons, Inc., 1972
7. K. Maruyama "*Seikagaku wo tsukutta Hitobito*" *(Scientists who built Biochemistry)* Shyoukabo, 2001
8. J. S. Fruton "*Protein, Enzymes, Genes*" Yale University Press, 1999
9. L. K. James, ed. "*Nobel Laureate in Chemistry 1901–1992*" Amer. Chem. Soc./ Chemical Heritage Foundation, 1994
10. K. Laidler, "*The World of Physical Chemistry*" Oxford University Press, Oxford, 1993
11. F. Aftalion, "*A History of the International Chemical Industry*" Chemical Heritage Foundation, 1991.
12. R. J. Forbes and J. E. Dijkserhauis, "*A History of Sciences and Technology*" Penguin Books, 1963
13. J. D. Bernal, "*Science in Histtory*" G. A. Watts & Co. Ltd., London, 1965
14. E. Shimao, "*Jinbutsu Kagakushi*" (*Characters in A History of Chemistry*), Asakura Shoten, 2002
15. W. H. Brock, "*The Chemical Tree*" p.129
16. J. M. Thomas "*Michael Faraday and The Royal Institution*" IOP Publishing Ltd., 1991
17. E. Shimao, "*Maikeru Farade-*" (*Michael Faraday*) Iwanami Shoten, 2000
18. M. Faraday, F. J. L. James, D. Phillips, "*The Chemical History of a Candle*" Oxford Univ. Press, 2011
19. M. C. Gerhard, *Ann. Chim. Phys.* **37**, 285 (1852)
20. Ref. 1, Ihde's book, p. 233

---

13. A similar criticism of the Japanese approach to science was voiced pointedly by Erwin Bälz, one of the foreigners who came to Japan early in the Meiji era and who remained in the country for many years[74].

21. E. Frankland, *Phil.Trans.Roy. Soc.*, **142**, 417 (1852)
22. A. Kekule, *Ann. Chem. Pharm.*, **106**, 129 (1858)
23. A. S. Couper, *London, Edingburh and Dublin Phil. Mag. & Science.*, 4th ser. 16, **104** (1858)
24. A. Kekule, *Ann.*, **137**, 158 (1866)
25. J. H. Woitz ed. *"The Kekule Riddle"* Cache River Press, 1993
26. A. Kekule, Ann., **162**, 88 (1872)
27. J. H. van't Hoff, *Archives Neeerlanderes des Sciences exactes et naturarelles*, **9**, 445 (1874)
28. J. A. Le Bel, *Bull. Soc. Chem. France*, **22**, 337 (1874)
29. P. Strathern, *"Mendeleyev's Dream"* Penguin Books, 2001.
30. D. Mendelejef, *Z. Chem.*, **12**, 405 (1869)
31. L. Meyer, *Ann. d. Chem. Phar. u. Suppl.*, VII 354 (1870)
32. D. Mendelejef, *Ann. d. Chem. Phar. u. Suppl.* VIII 123 (1871)
33. Lord Rayleigh, a) *Nature*, 46, 512 (1892), b) *Proc. Roy. Soc.*, **55**, 340 (1894)
34. Lord Rayleigh, W. Ramsay, *Phil. Trans. Roy. Soc.*, **186**, 187 (1895)
35. W. Ramsay, *Proc. Roy. Soc.*, **58**, 81 (1895)
36. W. Ramsay, *Ber.*, **31**, 3111 (1898)
37. A. Werner, *Z. Anorg. Chem.*, **3**, 267 (1893)
38. A. Werner, *"Neue Anschaung der Anorganischen Chemie"* F. Vieweg und Sohn, Braunschweig, 1905
39. S. Tomonaga, *"Buturigaku towa Nandarouka"* (*What is Physics?*) Iwanami shinsho, 1979
40. Y. Yamamoto, *"Netsugakusisou no Shiteki Tenkai"* (*Historical Developments of Ideas in Thermodynamics*) Chikuma Gakugei Bunoko, 2008
41. Though the debate against anti-atomits was not the direct cause of Boltzman's suicide, it has been considered that the fierce debate affected his mental states. David Lindley, *"Boltzmann's Atom"* Free Press, 2001
42. J. H. van't Hoff, *"Etude de Dynamique Chinamique"* Amsterdam, 1884
43. H. Helmholtz, *Sitzber. Kgl. Preuss. Akad. Wiss. Berlin*, 22 (1882)
44. W. Gibbs, *Trans Connecticut. Akad.*, **3**, 108, 343 (1876), *Am. J. Ssci.*, Ser. 3, **16**, 441 (1878)
45. W. Nernst, *Z. Physical Chem.*, **4**, 129 (1889)
46. C. H. Guldberg, W. Waage, *J. prakt. Chem.*, **19**, 69 (1879)
47. H. L. Le Chatelier, *Comptes rendus*, 99, 786 (1884), **106**, 355 (1888)
48. S. Arrhenius, *Z. physikal. Chem.*, **4**, 226 (1889)
49. W. Ostwald, *Z. Elektrochem.*, 7, 995 (1901)
50. J. W. Draper, *Philos. Mag. J. Sci.*, **26**, 465 (1845)
51. F. M. Raoul, *Comptes rendus*, **95**, 1030 (1882)
52. Ostwald also wrote the following introductory text book for novices: W.

Ostwald "*Die Schule der Chemie–erste Einführuing in die Chemie für jederman*" Vieweg, Brunschweig, 1903

53. J. H. van't Hoff, *Z. physical. Chem.*, **1**, 481 (1887)

54. S. Arrhenius, *Z. physikal. Chem.*, **1**, 631 (1887)

55. W. Ostwald, *Z. physical. Chem.*, **2**, 270 (1888)

56. S. Arrhenius, "*Das Werden der Welten*" Leipzig Academische Verlagsgesellsahaft, 1908

57. S. Arrhenius, "*On the Influence of Carbonic Acid in the Air upon the Temperature of the Ground*" Phil. Mag., ser. 5, **41**, 237 (1986)

58. G. Christianson, "*Greenhouse*", Penguin Books, 2000, p 115

59. A. Pockeles, *Nature*, **43**, 437 (1891)

60. Reference 10, p.303–306.

61. Terada Torahiko zenshu Vol.6, "*Reirii Kyou*" (*Lord Rayleigh*) Iwanami shoten, 1997

62. E. Fischer, "Synthesis in the purine and sugar froup" Nobel Lecture, 1902

63. E. Fischer, *Ber. Chem. Ges.*, **27**, 2985 (1894)

64. G. J. Mulder, Ann., **28**, 73 (2838)

65. F. Miescher, *Med. Chem.Unt.* 441and 502 (1871)

66. E. Buchner, *Ber. Chem. Ges.* **30**, 117 (1897)

67. An excerpt from p.233 of reference 6.

68. G. C. Stokes, *Proc. Roy. Soc.*, **13**, 335 (1864)

69. C. Bernard, *Comp. Rend.*, **44**, 578 (1857)

70. C. Bernard "*Introduction a l'etude de la Medecine Experimentale*" Paris, 1865 (English translation: "*An Introduction to the Study of Experimental Medicine*" Dover, 1957)

71. H. Fujita, "*Osaka Seimikyoku no Shiteki Tenkai*" (*Historical Development of Osaka Seimikyoku*) Shibunnkaku Shuppan, 1995

72. M. Imoto, "*Nihon no Kgaku*" *(Chemistry in Japan)* Kagakudoujin, 1978

73. T. Shiba , "*Wakojyunnyaku Tokuhou*", Vol.70, No.3 (2002)

74. T. Berutsu ed. "*Berutsu no Nikki*" *(A diary of Bälz,)*, Translation by S. Suganuma, Iwanami Bunko, 1951

Chronology of modern chemistry history (through the end of the 19th century)

Year	Chemistry	Physics, biology, engineering
1600	Revival of the atomic theory (Gassendi, 1649)	Properties of magnets (Gilbert, 1600) Laws of falling bodies (Galilei, 1604) Invention of the telescope (Lippershey, 1608) Theory of blood circulation (Harvey, 1628) Discovery of vacuum (Torricelli, 1643)
1650	Publication of *The Sceptical Chymist*, definition of elements (Boyle, 1661)	Invention of the vacuum pump (Geuricke, around 1650) Boyle's Law (Boyle, 1662) Observation of cells with microscopes (Hooke, 1665) Dispersion of light (Newton, 1666) Wave theory of light (Huygens, 1678) Laws of motion and universal gravity (Newton, 1687)
1700	Phlogiston theory (Stahl, 1718) Birth of pneumatic chemistry (Hales around 1730) Discovery of zinc (Marggraf, 1746) Discovery of platinum (de Ulloa) 1748	Invention of the steam engine (Newcomen, 1705) Two types of electricity (du Fay, 1733) Watt's steam engine (Watt, 1736)
1750	Discovery of carbon dioxide (Black, 1754) Discovery of hydrogen (Cavendish, 1766) Discovery of nitrogen (Rutherford, 1772) Discovery of oxygen (Scheele, 1772; Priestley, 1774) Discovery of chlorine (Scheele, 1774)	Origins of thunder (Franklin, 1752) Discovery of latent heat and heat capacity (Black, 1761)
1775	New theory of combustion (Lavoisier, 1777)	Animal electricity (Galvani, 1780)

	Chemistry	Physics
	Publication of *Elementary Treatise on Chemistry*; table of the elements; law of conservation of mass (Lavoisier, 1789)	Coulomb's law (Coulomb, 1785–89)
	Discovery of new metallic elements (Klaproth, Vauquelin, Wollaston, et al., 1781–1805)	
	Method of soda production (Leblanc, 1794)	
	Discovery of rare earths (Ekeberg, 1794)	
	Law of definite proportions (Proust, 1799)	
1800	Law of thermal expansion of gases (Gay-Lussac, 1802)	Volta's battery (Volta, 1800)
	Atomic theory; law of multiple proportions (Dalton, 1803)	Interference of light (Young, 1801)
	Henry's law (Henry, 1803)	
	Discovery of alkali metals (Davy, 1807)	
	Laws of gaseous reactions (Gay-Lussac, 1808)	
	Discovery of alkali earths (Davy, 1808)	
	Avogadro's hypothesis (Avogadro, 1811)	
	Prout's hypothesis Prout (Prout, 1815)	Dark lines in the solar spectrum (Fraunhofer, 1815)
		Diffraction and polarization of light (Fresnel, 1816)
	Determination of atomic weights (Berzelius, 1819–1826)	Ampere's law (Ampere, 1820)
	Law of isomorphism (Mitscherlich, 1819)	
	Law of specific heats (Dulong–Petit, 1819)	
	Reform of chemistry education (Liebig, 1824)	Carnot's theorem (Carnot, 1824)
1825	Discovery of benzene (Faraday, 1825)	Ohm's law (Ohm, 1826)
	Synthesis of urea (Wöhler, 1828)	Brownian motion (Brown, 1827)
	Improved methods of combustion analysis (Liebig, 1831)	Electromagnetic induction (Faraday, 1831)
	Laws of gas diffusion (Graham, 1832)	
	Laws of electrolysis (Faraday, 1833)	
	Composition of proteins (Mulder, 1838)	
	Catalytic reactions and fermentation (Berzelius, 1839)	Establishment of the cell theory (Schwann, 1839)
	Hess' law of thermochemistry (Hess, 1840)	Heating effect of electric current (Joule, 1840)
	Applications of chemistry to agriculture (Liebig, 1840)	
	Classification of organic compounds (Gerhardt, 1839–1846; Laurent, 1846)	Mechanical equivalent of heat (Joule, 1843)
	Optical isomers (Pasteur, 1847)	Law of energy conservation (Helmholtz, 1847)

Year	Chemistry	Physics, biology, engineering
1850	Synthetic mauve dye (Perkin, 1856)	Second law of thermodynamics (Clausius, 1850)
	Tetravalent theory of the carbon atom (Kekule, Cooper, 1858)	Second law of thermodynamics (Kelvin, 1851)
	Spectroscopy of atomic spectra (Bunsen, Kirchhoff, 1859)	Joule-Thomson effect (Joule, Thomson, 1854)
	Karlsruhe conference; revival of Avogadro's hypothesis (Cannizzaro, 1860)	Fluorescent effects and bending of cathode rays (Plücker, 1858)
	Concept of colloids (Graham, 1861)	Kinetic theory of gases (Maxwell, 1859)
	Law of mass action (Guldberg, Waage, 1864)	Publication of *On the origin of Species* (Darwin, 1859)
	Structural formula of benzene (Kekule, 1865)	Equations of electromagnetic field (Maxwell, 1861)
	Periodic table of the elements (Mendeleev, 1869)	Principle of increasing entropy (Clausius, 1865)
	Discovery of nucleic acids (Miescher, 1869)	Laws of genetics (Mendel, 1865)
	Theory of asymmetric carbon atoms (van't Hoff, Le Bel, 1874)	Straight-line progress of cathode rays (Hittorf, 1869)
		Equation of state of gases (van der Waals, 1873)
1875	Gibbs free energy, Gibbs equilibrium (Gibbs, 1876)	Statistical interpretation of entropy (Boltzmann, 1877)
	Proposal of "enzymes" (Kühne, 1878)	
	Discovery of ytterbium (Marignac, 1878)	Law of radiation (Stefan, 1879)
	Synthesis of indigo (Baeyer, 1880)	
	Raoult's law (Raoult, 1883)	Concept of free energy (Helmholtz, 1882)
	Synthesis of sugars (E. Fischer, 1884)	
	Temperature variation of equilibrium constants (van't Hoff, 1884)	Spectral series for hydrogen (Balmer, 1885)
	Le Chatelier's principle (Le Chatelier, 1884)	Discovery of anode rays (Goldstein, 1886)
	Discovery of fluorine (Moissan, 1886)	Theory of the microscope (Abbe, 1887)
	Theory of dilute solutions (van't Hoff, 1887)	Photoelectric effect (Hertz, 1887)
	Ionization theory of electrolytic solutions (Arrhenius, 1887)	Experimental demonstration of electromagnetic waves (Hertz, 1888)
	Law of dilution (Ostwald, 1889)	General formula for the spectrum of hydrogen (Rydberg, 1890)
	Expression for the potential of a single electrode (Nernst, 1889)	
	Temperature variation of reaction rates (Arrhenius, 1889)	Wien's law of displacement (Wien, 1893)
	Theory of indicator (Ostwald, 1892)	
	Coordination theory of metal complexes (Werner, 1893)	
	Discovery of noble gases (argon) (Ramsay, Rayleigh, 1894)	
1895	Mechanism of the enzyme action (E. Fischer, 1894)	

Part 2

The Birth and Development of Modern Chemistry

Chapter 3: The revolution in physics from the end of the 19th century to the early 20th century

X-rays, radioactivity, discovery of the electron, and quantum chemistry

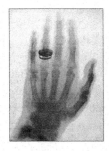

X-ray photo of a hand wearing a ring

Chemistry had made steady advances up to the latter half of the 19th century, but at the end of that century, three startling discoveries were made in succession that fundamentally changed physics and chemistry. These discoveries resulted directly or indirectly from research using vacuum discharge tubes, which in turn depended on technical improvements in vacuum pumps. First, X-rays were discovered by Wilhelm Röntgen in 1895. Next, radioactivity was discovered by Henri Becquerel in 1896, and then the electron was discovered by J. J. Thomson the following year, in 1897. Entering the 20th century, it became clear that an atom consisted of electrons and an atomic nucleus, and that electrons played the main role in most chemical phenomena. The development of chemistry in the 19th century focused on the concept of the atom, whereas chemistry in the 20th century focused on the behavior of electrons. In addition, with the use of X-rays, chemists had a means for precisely understanding the structure of molecules and crystals at the atomic level. Thus, a new chemistry started to develop based on the behavior of electrons and the structure of atoms and molecules. This new chemistry helped promote industrial development by making it possible to synthesize new materials, and opened the way to understanding the complex chemical phenomena related to life (see Section 4.10). In addition, the discovery of radioactivity gave rise to the new fields of nuclear chemistry and radiochemistry. It immediately became clear that classical mechanics and classical electromagnetic theory were inadequate for understanding the behavior of atoms and electrons, and a new branch of physics, namely, quantum theory and quantum mechanics, was born. Quantum mechanics was promptly applied to problems in chemistry, providing insights into fundamental aspects of chemical bonds, and the new field of quantum chemistry was established. Before examining the development of chemistry in the 20th century, we will describe the discovery of the electron, X-rays, and radioactivity and their impact on chemistry, the birth of quantum theory, and quantum mechanics, as well as their first applications to chemistry. In writing this chapter, we often referred to references [1] to [9].

3.1 The Discovery of the Electron

By the end of the 19th century, many chemists had contributed to understanding electrical conduction in solutions, and the theory of electrolytic dissolution was accepted. However, the discovery of the electron resulted from research on electrical conduction in low-pressure gasses: namely, the discharge phenomenon.

Gas Discharge Research

Electrical discharge in a gas is most commonly observed in nature as lightning. Benjamin Franklin (1716–1790) conducted a famous kite experiment in 1752 and demonstrated that a lightning flash is caused by a flow of electricity. Research into electrical discharge in gases was carried out in the 18th century. In 1709, Hauksbee discovered that when the air pressure within a glass vessel was reduced to one sixtieth of atmospheric pressure and the air was then exposed to a source of frictional electricity, light was produced within the vessel. This phenomenon was subsequently investigated by a number of scientists, including Faraday. However, the great advancements in gas discharge research were made in the latter half of the 19th century, after vacuum pumps had improved sufficiently to produce a high vacuum.

In 1855, Johann Geissler (1814–1979) developed a high-performance vacuum pump using a column of mercury as a piston, and was able to reduce the pressure within a glass tube to less than a few thousandths of an atmosphere. Using this pump, Plücker at Bonn University conducted research into the electrical conduction of low-pressure gasses. He observed that when the air in a glass tube was removed, luminescence in most of the tube disappeared, but a green light (a glow) appeared near the cathode. This seemed to indicate that something was emitted from the cathode, flew through the near-vacuum in the tube, and hit the glass tube to emit light. It was also found that the position of this glow could be moved using a magnet, suggesting that the rays that caused the glow carried an electric charge. Hittorf, one of Plücker's students, discovered that when an object was placed near a small cathode, its shadow could be seen on the wall of the tube. Hittorf deduced that the rays emitted from the cathode traveled in a

William Crookes (1832–1919)

Crookes was a British chemist and physicist. He studied chemistry at the Royal College of Chemistry, London, and served as an assistant to Hofmann. He became wealthy due to an inheritance from his father, and constructed an independent laboratory to conduct research. Though he started in organic chemistry, Crookes turned to spectroscopy and discovered the new element thallium from spectrographic analyses. In addition to research related to vacuum discharge, he is known for various scientific activities such as research on rare-earth elements and the discovery of the antiseptic property of phenol.

Heinrich Hertz (1857–1894)
Hertz was a German physicist who was a professor first at the Karlsruhe Institute of Technology, and then at Bonn University. In the laboratory, Hertz created electromagnetic waves and clarified that they undergo reflection and refraction in the same manner as light, and provided the experimental basis for Maxwell's electromagnetic theory of light.

Figure 3.1: A Crookes discharge tube

straight line. These rays were verified by the German physicist Eugen Goldstein and were given the name cathode rays.

The British chemist, physicist, and spiritualist, William Crookes, carried out further detailed research into this phenomenon. He constructed an improved discharge tube called the Crookes tube (Fig. 3.1) to conduct experiments in higher vacuum, and published the results in 1879. When he placed a metallic cross within the discharge tube, he observed that a clear shadow was formed on the wall of the tube when hit with cathode rays, and fluorescence was observed. In addition, he showed that a small windmill could be turned by the impact of cathode rays. This seemed to suggest that cathode rays consist of particles. Furthermore, the demonstration that a thin beam of cathode rays could be bent in a certain direction by a magnet indicated that cathode rays were negatively charged.

However, a different idea was proposed in 1889 by the German experimental physicist, Heinrich Hertz. He found that cathode rays passing between two metal plates, one positively charged and one negatively charged, essentially did not bend, and concluded that cathode rays were not a beam of charged particles. In 1892, he determined that cathode rays easily pass through thin sheets of gold or aluminum. Lenard, one of Hertz's students, found that passing cathode rays through an aluminum-foil window in the discharge tube caused fluorescence in the adjacent air outside the tube. These experimental results led German physicists to conclude that cathode rays were electromagnetic waves and not a charged particle beam. On the other hand, in 1895, the French physicist Jean Perrin (see page 201) showed that cathode rays impart a negative charge to a charge collector placed in a discharge tube, which supported the electrically-charged particle theory. It was in this confusing state of affairs that J. J. Thomson at the Cavendish Laboratory, Cambridge University, started his research to elucidate the true nature of cathode rays.

Thomson's Experiments and the Discovery of the Electron

In 1884, J. J. Thomson was chosen to be the third Cavendish Professor of Physics and the director of the Cavendish Laboratory as the successor to Lord Rayleigh. Thomson was a mathematical physicist, still young, and had no past record as an experimentalist, so his selection surprised British physicists. However, his selection brought about outstanding results. Under his leadership, the Cavendish Laboratory became the world center for experimental physics, and seven of his assistants won Nobel prizes. While Thomson himself was by no means skillful at experimentation, he had an inherent talent for devising experimental methods to solve major problems.

Using magnets to move cathode rays that caused fluorescence, Thomson showed that a negative charge was obtained only when the beam struck the opening of a charge collector, and confirmed Perrin's conclusion. In addition, he conducted experiments that overturned Hertz's conclusion that cathode rays were not bent by an electric field. He thought that the ions produced by the collisions of the cathode rays with gas within the tube were drawn to the electrode to cancel the electrostatic pressure on the electrode. Thompson performed the experiment

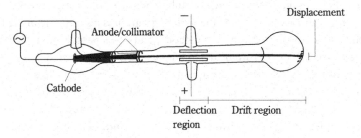

Figure 3.2: Thompson's cathode ray tube

thinking that if the vacuum was sufficiently high, cathode rays should be bent by an electric field, and obtained the expected results. This evidence showed that cathode rays consisted of negatively charged particles, but the next problem was to determine the charge and mass of the particles.

While Thompson could not determine these two values, he could determine the mass/charge ratio by comparing the amount of bending caused by an electric field and by a magnetic field (Fig. 3.2). The average value he obtained was 1.2 × 10^{-11} kg/C (C denotes Coulomb). At the close of the 19th century, the mass/

charge ratio for a hydrogen atom was understood to be 10^{-8} kg/C. Accordingly, the mass/charge ratio for this new particle came to approximately one-thousandth the value for a hydrogen atom. This was a surprising discovery, and implied that this particle had a much smaller mass than an atom, or if not this, then that the particle had an enormous charge. Thomson's value differed from the present day value of 0.56857×10^{-11} C/kg, but the reason for this difference is not known.

The mass/charge ratio did not vary with the gas used in the experiment, nor did it vary with the metal used in the electrode. In addition, Thomson carried out the same experiments using particles obtained by the photoelectric effect and particles obtained from the surface of white-hot metal, but the value did not change. Thus, it became clear that the particles Thomson discovered were universally present. Thomson continued from there to conjecture that this particle was a fundamental component of all matter. He stated it thus:[10]

> Thus on this view we have in the cathode rays matter in a new state, a state in which the subdivision of matter is carried very much further than in the ordinary gaseous state: a state in which all matter—that is, matter derived from different sources such as hydrogen, oxygen, &c.—is of one and the same kind; this matter being the substance from which all the chemical elements are built up.

The top diagram shows the cathode ray tube Thomson used to measure the mass/charge ratio for an electron, and the bottom diagram is a schematic of the tube. Cathode rays are emitted from the cathode, and some pass through the anode and collimator slit and continue down the tube as a narrow beam.

Thomson simply called this fundamental particle a corpuscle and did not use a special name. A number of years previously, the Irish physicist, Stoney, had proposed calling the unit of electricity that is acquired or lost when an atom becomes an ion an "electron." This term became generally accepted, and continues to be used.

Determination of the Charge of an Electron

After the mass/charge ratio for an electron had been determined, the next problem was to determine the mass or charge of an electron. This was first attempted at the Cavendish Laboratory. It was known that a cloud is formed when the temperature of air saturated with water vapor is reduced by expansion. In 1887, C. T. R. Wilson discovered that ions in air produced by X-radiation acted as nuclei for the formation of clouds. Wilson used this phenomenon and built cloud chambers to study elementary particles. Townsend in Thomson's laboratory generated water droplets using ions. The droplets formed condensation nuclei. Townsend measured the velocity at which an individual droplet fell and used

C. T. R. Wilson (1869–1959)
Wilson was a British physicist who studied at Cambridge University and was a professor of natural philosophy there (1925–1934). He discovered that supersaturated water vapor condensed around ions produced by radiation, i.e., the ions acted as nuclei for cloud formation. He also perfected the cloud chamber. Wilson's work contributed greatly to the development of nuclear physics. Wilson was awarded the Nobel Prize for Physics in 1927.

Robert Millikan (1868–1953)
Millikan was an American physicist, born in Illinois. After obtaining a degree at Columbia University and studying in Germany, he became a professor at the University of Chicago in 1910. He determined the charge of an electron by conducting experiments on oil drops. From 1921 to 1945, Millikan was chairman of the California Institute of Technology, and made major contributions that helped this university develop into a world-class science and technology center. He was awarded the 1923 Nobel Prize for Physics.

Stokes' law[1] to determine their radius. The falling water droplets were absorbed by sulfuric acid, and the accumulated charge on the acid was measured. In addition, the amount of water absorbed was measured to determine the charge of a single droplet. The charge obtained was 1.0×10^{-19} C. Wilson improved this method by applying an electric field and measuring the voltage required to stop the fall of water droplets. The electron charge was thus determined, and its mass was shown to be less than one-thousandth of the mass of a hydrogen atom.

Subsequently, Robert Millikan at the University of Chicago carried out similar experiments from 1908 to 1917 using oil drops, and determined a more accurate value for the charge of an electron. Millikan's apparatus allowed observation of the movement of oil drops between two parallel plate electrodes when a voltage was applied. Depending on the voltage applied across the plate electrodes, the charged oil drops rose or fell. Millikan determined the charge on the drops using a method similar to Wilson's. The charge frequently changed, but was always an integer multiple of a small constant value. From this small constant value, the charge of an electron was determined to be 1.592×10^{-19} C. Using this value and the mass/charge ratio of 0.54×10^{-11} C/kg, the mass of an electron was determined to be approximately 9×10^{-31} kg. This implied that the mass of an electron was approximately 1/1850th the mass of a hydrogen atom. Also, since

---

1. Stokes' law: A law indicating that when a sphere of radius a moves at velocity v in a fluid of viscosity coefficient η, and when the sphere is traveling at slow speed, then a resistance proportional to the speed, $6\pi\eta av$, acts on the sphere.

1 F (Faraday constant) was known from electrolysis to be 96,500 C per mole, Avogadro's number $(N_A)^2$ of 6.06×10^{23} was obtained by dividing the Faraday constant by the charge of an electron. After determining the electron charge, Millikan made precise measurements of the photoelectric effect and confirmed Einstein's light quantum hypothesis. In 1923, he was awarded the Nobel Prize for Physics for "research pertaining to the elementary electric charge and the photoelectric effect."

3.2 The Discovery of X-rays and Early Research

The discovery of X-rays was the first of three major discoveries at the end of the 19th century that brought about a revolution in science, and like the discovery of the electron, it arose from experiments using vacuum discharge tubes. While the discovery of the electron was the result of research carried out with the clear objective of elucidating the nature of cathode rays, the discovery of X-rays was an unexpected bounty. The impact on chemistry in the hundred years since X-rays were discovered has been truly enormous. Advances in chemistry described in this book were frequently due to research using X-rays; the circumstances of the discovery of X-rays and early research on X-rays are presented here.

The Discovery of X-rays

Wilhelm Röntgen had been working as a competent physicist at the University of Würzburg without any particularly noteworthy achievements. He had just turned age 50 in 1895, and was conducting research on cathode rays using vacuum discharge tubes. He had been researching fluorescence emerging from vacuum discharge tubes using barium platinocyanide $(BaPt(CN)_4)$ crystals as a detector, which produced fluorescence upon exposure to ultraviolet light. On 8 November 1895, he noticed that the crystals became fluorescent even when the discharge tube was covered with black cardboard. The radiation emitted from the end of the discharge tube struck by cathode rays produced fluorescence on a screen coated with $BaPt(CN)_4$ placed in a location a few feet away. He called this radiation X-rays and investigated the absorption of X-rays using a variety of materials. X-rays passed through paper and wood, but were absorbed by metals, especially lead. A photographic plate wrapped in black paper was easily exposed. Because bones absorbed X-rays well, but X-rays readily passed through soft human tissue,

---

2. Avogadro's number (constant), N_A: The number of elementary particles (atoms, ions, molecules, etc.) in 1 mole of a substance. It is defined as the number of carbon atoms in 12 g of the 12C isotope, atomic mass number 12. N_A=6.022×1023 mol⁻¹

Wilhelm Röntgen (1845–1923)
Röntgen was born in Lennep, Germany, the son of a wealthy textile dealer. He studied mechanical engineering at the Federal Polytechnic Institute in Zurich, and obtained a Ph.D. in physics for his research on gases. When his teacher, Kundt, became a professor at the University of Würzburg, he moved with him to Würzburg, and then became a professor of physics in 1885. After researching electromagnetic phenomena, changes in the physical properties of gasses under pressure, and other topics, Röntgen began to research gas discharges. In 1895, he was catapulted into fame by discovering X-rays. He was awarded the Nobel Prize for Physics in 1901.

it was possible to observe the interior of the human body. Röntgen submitted a paper describing this discovery to the Würzburg Physical-Medical Society, and it was published in January 1896. A preprint of the paper that included an X-ray photograph of his wife's hand, in which the bones were clearly visible, became available 1 January 1896, and news of the discovery appeared in newspapers immediately. An English-language translation of Röntgen's paper appeared in the journal *"Nature"* on 23 January 1896 and in the journal *"Science"* on 10 February 1896.

This discovery created a worldwide sensation in scientific and medical fields. Many scientists set about to do X-ray research, and doctors soon started to use X-rays to diagnosis illnesses. However, in subsequent research it became clear that X-rays also had a harmful effect on the human body. Many researchers were exposed to radiation and suffered from cancer, resulting in over 100 deaths. However, these tragedies led to an awareness of the harm caused by X-rays and knowledge of how to prevent such harm. Röntgen subsequently wrote two more papers, but these were his last contributions to X-ray research. In 1900, he became a professor at the University of Munich. He was awarded the first Nobel Prize for Physics and was given numerous honors. Fame and fortune led Röntgen to largely abandon research, and he died in 1923 at age 78.

The Nature of X-rays and the Structure of Materials

The nature of X-rays was not immediately understood, and for the next ten years or more physicists continued to debate whether X-rays were electromagnetic waves or a beam of particles. Because X-rays were not bent by magnets, they were not believed to be composed of charged particles such as electrons.

Max von Laue (1879–1960)

von Laue was a German theoretical physicist. After studying at various German universities, he became a professor at the University of Zurich in 1912 and a professor at the University of Berlin in 1919. In 1912, he suggested that crystals could be studied using X-ray diffraction, and laid the foundation for crystal physics. He was awarded the 1914 Nobel Prize for Physics.

William Henry Bragg (1862–1942)

After receiving an education at Trinity College, Cambridge, Bragg became a professor at the University of Adelaide, Australia, then in 1909 returned to Britain to become a professor at the University of Leeds. After laying the foundation for X-ray crystallography, he became a professor at University College of London (1915) and director of the Royal Institution in 1923. There he assembled an X-ray research group comprising young, outstanding researchers, promoted research into structure analysis, and mentored talented individuals who became leaders in X-ray research, including Lonsdale, Astbury, and Bernal. He was awarded the 1915 Nobel Prize for Physics.

Lawrence Bragg (1890–1971)

Bragg studied mathematics at Adelaide University, Australia, changed his major subject to physics at Cambridge, and graduated in 1912. In 1912, he derived the Bragg conditions for X-ray diffraction, thus laying the foundation for X-ray crystal structural analysis. With his father Henry, Bragg carried out structural analyses of rock salt, diamond, fluorite, iron pyrite, copper and other substances. Awarded the Nobel Prize for Physics together with his father in 1915, and only 25 years of age at the time, he was the youngest Nobel Prize recipient in science to date. Bragg became a professor at the University of Manchester in 1919, and there promoted research on the X-ray structural analysis of organic compounds and alloys. He became the director of Cavendish Laboratory in 1938.

However, X-rays were not reflected or refracted like ordinary light, and were not diffracted by a diffraction grating comprising closely-spaced lines. The discoverer of X-rays, Röntgen, proposed that X-rays were longitudinal waves of ether, and based his proposal on the elastic body model of ether suggested previously (see page 197). In 1897, Stokes proposed that when a charged cathode

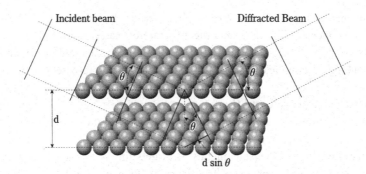

Figure 3.3: Diagram used in the derivation of Bragg's equation for X-ray diffraction

The change in the path length of a beam diffracted from two adjoining planes separated by a distance d is $2d\sin\theta$. The two beams are more intense when this value is an integer multiple of the wavelength.

ray particle collides with metal or glass, ether waves (electromagnetic waves) are generated and emanate spherically from that point.

The penetrating power of X-rays increased with increasing tube voltage, leading Stokes to propose that X-rays were electromagnetic waves with extremely short wavelengths. However, since no diffraction gratings could diffract electromagnetic waves with such short wavelengths, the wavelength of X-rays could not be measured. In 1912, Max von Laue, a physicist at the University of Munich, predicted that the distance between atoms in a solid was suited to X-ray diffraction. His students, Friedrich and Knipping, succeeded in observing an X-ray diffraction pattern using zinc sulfide crystals.[11] Thus, it became clear that X-rays were extremely short wavelength electromagnetic waves.

This finding was soon applied to structural analysis by the Bragg father-son duo in Britain by analyzing the reflections obtained from X-ray diffraction in crystals. In the summer of 1912, the father, Professor Henry Bragg at the University of Leeds, discussed Laue's work with his son Lawrence, who had just become a research student at Cambridge University. Lawrence studied X-ray diffraction in crystals and derived a relational expression (Bragg's equation) between the angle of diffraction (θ), the X-ray wavelength (λ), and the interatomic distance (d):

$n\lambda = 2d\sin\theta$ (n is an integer)

(see Fig. 3.3).[12] They conducted research together during a holiday using a spectrometer designed by Henry at the University of Leeds, and determined the atomic arrangement in several crystals, including diamond.[13, 14] Thus, the path was opened to investigating the structure of materials using X-rays. The two were awarded the Nobel Prize for Physics in 1915, and laid the foundation for making Britain the center for X-ray crystallographic research

Moseley's Research and the Atom

One piece of important early research pertaining to X-rays was carried out by young Henry Moseley in 1913 in Rutherford's laboratory at the University of Manchester. He used various metals as targets for X-rays and measured the wavelengths of the two principal lines generated following impact (characteristic X-rays). The results made it clear that as the atomic number of metals increased, the wavelength of the characteristic X-rays generated decreased. When the square root of the characteristic X-ray frequency was plotted against the atomic number (indicating the order of the element in the periodic table), a linear relationship was obtained, which became known as Moseley's law (Fig. 3.4) [15, 16] This showed unambiguously that the atomic number provided important information about the atom. This research was also effective for confirming the order of the elements in the periodic table. Cobalt had a higher atomic weight than nickel, but the atomic number of cobalt was determined to be smaller than that of nickel. From this linear relationship, Moseley determined the order of

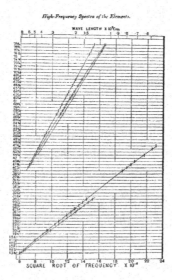

Figure 3.4: Relationship between atomic number and X-ray frequency

The horizontal axis indicates the square root of the frequency, and the vertical axis indicates the atomic number. Reproduced from Moseley's 1913 paper

COLUMN 8

Lawrence Bragg and the Cavendish Laboratory

Cavendish Laboratory

After discovering Bragg's law of X-ray diffraction in 1913 and winning the Nobel Prize at the young age of 25, Lawrence Bragg continued X-ray research at the University of Manchester. Rutherford died suddenly in 1936, and the following year Lawrence Bragg was chosen as his successor to be the director of the Cavendish Laboratory. Because the Cavendish Laboratory had been the world center for nuclear physics during Rutherford's directorship, this decision surprised many in British scientific circles. Disappointment was particularly great among nuclear physicists. Many brilliant young researchers who had conducted research under Rutherford moved elsewhere, and Cavendish lost its leadership position in nuclear physics. Bragg, however, remained calm saying, "We have taught the world very successfully how to do nuclear physics. Now let us teach them how to do something else."

In 1927, a new lecturer position in crystallography was established at the Cavendish, and Bernal's group conducted pioneering research on the X-ray structural analysis of proteins. Bernal left in 1938, but Perutz stayed and continued his research on protein structure. Bragg had an interest in Perutz's research and actively supported it. Shortly afterwards, World War II started, and the reorganization of the laboratory and new developments were postponed until after the war ended.

After the war, Bragg promoted radio astronomy and X-ray crystallography of biological polymers. There was a high level of uncertainty associated with both areas of research, but history proved that this was a brilliant decision. Subsequently, the Cavendish became the birthplace of molecular biology and structural biology. Cavendish also led the world in radio astronomy, and produced numerous Nobel Prize winners.

Bragg achieved great success as the director of Cavendish. What were the reasons why Bragg was so successful? First, he was blessed with a great deal of good luck. Researchers at Cavendish studied radio astronomy, and embarked on the structural analysis of DNA and proteins, at the moment when the technology to pursue these areas became available. It could be said that divine timing provided support. Additionally, Bragg had the ability to make the best use of his good fortune. He also shared a few Golden Rules with new students and researchers. Freeman Dyson, a theoretical physicist who left Cambridge and emigrated to the United States, related the following three rules[17]:

1. Don't try to revive past glories.
2. Don't do things just because they are fashionable.
3. Don't be afraid of the scorn of theoreticians.

Dyson wrote that these three rules formed the backdrop for Bragg's success as director. As director of the Cavendish, instead of trying to regain past glory in nuclear physics, he chose to pioneer new fields. The structural analysis of proteins and radio astronomy were not yet popular fields. At that time, most people considered analysis of the structure of proteins such as hemoglobin to be utterly hopeless research. The world is complicated and full of diversity, outside the scope of many theoreticians. Bragg's Golden Rules were that researchers should be attentive to and get to the core of matters in the study of natural science.

Bragg was an English gentleman who loved sketching, gardening, and bird-watching. One amusing anecdote about him is that after leaving Cavendish, he became director of the Royal Institution in London, and no longer lived in a house with a garden. He therefore became an undercover gardener at a mansion, and once a week enjoyed working in the garden. However, he was eventually found out, and this work came to an end.[18]

"Max Perutz", a Bragg sketch

Henry Moseley (1887–1915)
Moseley was born the son of a professor of botany and anatomy at the University of Oxford; he lost his father when still a child. He received an education at Eton and the University of Oxford, and graduated in 1910. Moseley became a lecturer at the University of Manchester, and conducted research on radioactivity, but turned to research on X-ray spectroscopy and discovered Moseley's law. If he had not died in 1915, it is conjectured that he might have been awarded the Nobel Prize for Physics in 1916.

the rare earth elements, which had been uncertain, and predicted that there were 92 elements spanning hydrogen to uranium. It became clear that the order of the elements in the periodic table should be determined by atomic number and not by atomic weight.

However, if that is the case, just what does atomic number signify? Answering this question was the next big problem. At the time, there was a new way of thinking about the structure of atoms, and Moseley had considered the significance of atomic number in relation to this way of thinking. However, when World War I broke out, Moseley joined the British army engineer corps as a communications officer. In 1915, tragedy struck when Moseley was killed in the landing operations at Gallipoli[3] at the young age of 27. His death was a great loss to physics and chemistry. If he had lived, his achievements unquestionably would have worthy of a Nobel Prize.

3.3 The Discovery of Radioactivity and Isotopes

In February 1896, Henri Becquerel of the Museum of Natural History in Paris became aware of Röntgen's discovery of X-rays. Becquerel had studied materials that fluoresce and phosphoresce, and he wondered if there was a relationship between X-rays and fluorescence and phosphorescence. There were many fluorescent or phosphorescent materials in his laboratory, collected since his grandfather's era. He began experiments with these materials and discovered radioactivity. Becquerel's discovery of radioactivity was also the result of a happy coincidence. Afterwards, enormous advances in radioactivity research

---

3. Gallipoli landing operations: During World War I, these were the landing operations by which the Allied Forces invaded the Gallipoli peninsula, belonging to the German ally, Turkey, with the aim of creating a land link with Russia.

Henri Becquerel (1852–1908)
Becquerel was born into a family of researchers who for three generations studied luminescent materials. His grandfather, Antoine, was an authority on such materials, and a chair in physics was established for him at the Natural History Museum. His father, Alexandre-Edmond, the second generation, was a researcher with an interest in phosphorescent materials. After graduating from École Polytechnique, Henri studied at a civil engineering school and became a civil engineer, but with the death of his father in 1891, he subsequently succeeded him to become professor at the museum. Inspired by the discovery of X-rays by Röntgen, Henri investigated the relationship between fluorescence and X-rays, and discovered radioactivity. He was awarded the 1903 Nobel Prize for Physics.

were made by the Curies and Rutherford, and provided a major driving force that changed 20th century science and technology.

Becquerel's Discovery

In February 1896, Becquerel placed various crystals of fluorescent-phosphorescent materials on top of photographic plates that had been tightly wrapped in thick black paper, exposed these to sunlight for several hours, and then examined whether or not the photographic plates had been exposed. In many cases nothing happened, but in the case of potassium uranyl sulfate, which fluoresces when exposed to light, an image of the crystals appeared on the developed photographic plate. Clearly some kind of radiation had been emitted by the crystals and reached the photographic plate. During several consecutive cloudy days, Becquerel put the photographic plates wrapped in black paper and the uranium salt together in a desk drawer. He assumed that, since the uranium salt was not being exposed to light, the photographic plate should not be exposed. Surprisingly, however, when this photographic plate was developed, he found it had been exposed, clearly demonstrating that the radiation came from the uranium salt. More detailed investigations showed that this radiation was unrelated to phosphorescence or fluorescence, and was peculiar to uranium. It was also discovered that this radiation had penetrating power like X-rays, exposed photographic plates, and ionized air. The most surprising finding was that uranium salt seemed to radiate energy without a source. Becquerel thought that this energy was a form of radiation, but in 1899 it was discovered that these rays were bent by magnets, and thus were

Pierre Curie (1859–1906)
Curie was born in Paris, the second son of a physician. After
receiving an education at home from tutors, his father, and
older brother, he entered the Sorbonne at age 16, graduated
at age 18, and became an assistant for conducting physics
experiments. He discovered the piezoelectric effect in collaboration with his older brother, Jacques. Subsequently,
Curie obtained important results while researching crystals
and magnetism, and built the foundation for modern research on magnetism.
He worked as a low-wage instructor at a private physics and chemistry school
from the age of 24, and was a shy, introverted idealistic dreamer who had no
interest in worldly success. After marriage to Marie, he collaborated with
her to investigate radioactivity, and together they discovered polonium and
radium. He was awarded the 1903 Nobel Prize for Physics (Column 9)

Marie Skłodowska-Curie (1867–1934)
Marie was born in Warsaw, Poland, the youngest of five
children. Her parents were both teachers. She graduated with
excellent grades from gymnasium, but with Poland under
the control of Russia, women had no hope of being students
at a university. Marie diligently studied while working as a
tutor. In order to study at the Sorbonne, at age 24, she turned
to her older sister who was studying medicine and came to
Paris. She worked and studied diligently while enduring great poverty, and
obtained outstanding grades in mathematics and physics. She started research
on radioactivity for her dissertation and discovered polonium and radium.
She was awarded the 1903 Nobel Prize for Physics and the 1911 Nobel Prize
for Chemistry. (Column 9)

clearly not X-rays. However, Becquerel did not pursue research on radiation
much further, leaving subsequent major discoveries to be made by the Curies
and Rutherford.

The Discovery of Radium by the Curies
In September 1897, Marie Curie had just married Pierre Curie in Paris. She
consulted with her husband and began research on the radiation from uranium
salts for her dissertation. Using a crystal electrometer based on the piezoelectric
effect discovered by Pierre's elder brother, Jacques Curie, Marie studied the
ionization of air by radiation. She examined many mineral specimens and
investigated whether minerals other than uranium emitted radiation.

She ascertained that the strength of the radiation emitted from uranium was proportional to the quantity of uranium. It emerged that a pitch blend ore[4] composed of thorium oxide and approximately 80% uranium oxide ionized air much more strongly than uranium salt. In 1898, she used the term 'radioactivity' to express the intensity of this radiation. Marie believed that the strong radiation exhibited by pitch blend was due to the presence of an unknown element in the ore, and together with her husband they started work to isolate that element. They ground pitch blend ore, dissolved it in acid, and performed separations using techniques similar to those used in qualitative analysis. At each step, they verified the presence of radioactive material using an electrometer. After long and tedious effort, they detected the presence of a highly radioactive substance within a sulfur condensate, and in July 1898, published this finding and called the substance polonium, after Marie's native country, Poland.

A different alkaline earth precipitate component also exhibited strong radiation. To separate the element emitting the radiation, it was thought necessary to separate out the barium residue. When the most insoluble portion was isolated from barium chloride, the insoluble material exhibited 60 times the radioactivity of uranium, and a weak, new emission line was observed in spectroscopic analysis. The separation operations were continued until radioactivity 900 times that of uranium was exhibited, and the new emission spectrum was verified. In December 1898, the Curies announced the discovery of a second new element. Because a salt of this new element glowed in the dark, they gave it the name radium, after the Latin word radius (ray). They set about to refine the radium, and by July 1902 had obtained 0.1 g of a chloride of radium from several tons of pitch blend ore. The atomic weight of radium was determined to be 225. Pierre determined that 1 g of radium radiated enough energy in 1 hour to heat 1.33 g of water from 0 to 100°C.

Becquerel, who discovered radioactivity, and the Curies were together awarded the Nobel Prize for Physics in 1903 for "radioactivity research."

Rutherford's Research and the Nature of Radioactivity

As research on radioactivity progressed, it became clear that there was more than one type of radioactivity. The person who made the biggest contribution to understanding the nature of radioactivity was Ernest Rutherford. Rutherford entered the Cavendish Laboratory in 1895, and after investigating the effects of X-rays on the electrical conduction of gases, he actively began research

---

4. Pitch blend: The main uranium ore, the principal component of which is UO_2. It was given this name because it is black, like pitch. Major ore deposits are produced in layers within sedimentary rock.

The Curies[19]

The Curies in the laboratory

Originally from Poland, Marie Curie had rare ability and an indomitable spirit. However, that alone was not sufficient to succeed in the latter half of the 19th century, when female scientists were rare. At the Sorbonne, she met Pierre Curie, who had previously achieved outstanding results as a physicist, and they married in 1895. Marie was fortunate, as Pierre had a gentle and honest personality. Even blessed with talent and the opportunity to study, it was rare for a woman to succeed in a scientific field and to have an understanding husband. There were examples, such as Einstein's first wife (Mileva Maric) and Haber's first wife (Clara Immerwahr) (see page 372), who married great scientists and had unhappy lives.

At the time of meeting Marie in the spring of 1894, Pierre was 35 years old. Talking together about science and society, the two felt that they had many things in common. Pierre was strongly attracted to Marie and passionately proposed marriage. However, Marie had vowed to return to Poland someday and did not respond immediately to the proposal. At long last, Marie accepted Pierre's ardent proposal in July 1895.

Marie began research on radioactivity for her dissertation, and she was the first to discover that ores such as pitch blend have high radioactivity. Thinking that these ores must contain new radioactive elements, Marie set about analyzing pitch blend. Pierre discontinued his research on magnetism and joined Marie in her research. In 1898, they discovered the new radioactive elements polonium and radium, and reported their findings jointly under their names. However, extracting a new element was not enough; it was necessary to determine an atomic weight, and Marie sought to do that. This work required very hard physical labor and persistence. Only 0.1 g of radium salt was purified from several tons of pitch blend.

Marie obtained her doctorate in 1902. The Curies, together with Becquerel, were awarded the Nobel Prize for Physics in 1903 for "radioactivity research," and the Curies immediately became household names. In 1904, Pierre was welcomed into a new professorship established by the Sorbonne, and Marie's laboratory was associated with new position. It appeared as if the long struggles were coming to an end, and that a brilliant future lay ahead. However, a tragedy occurred in the spring of 1906. Pierre slipped and fell when crossing a street, was run over by a horse-drawn cart, and died. After Pierre's death, Marie took over her husband's physics lectures at the Sorbonne, and in 1908, she became the first female professor at the University of Paris.

As a woman and a foreigner, Marie had to struggle against various discriminations and prejudices. In 1911, she was a candidate for an Academy of Science open seat, but because she was a woman and a Pole, and there were members vehemently opposed to the choice of a woman, she was not elected. In November of that year, a newspaper that featured scandal articles published articles on a grand scale about an adulterous affair between Marie and a former student of Pierre, the well-known physicist, Paul Langevin. Marie was tormented daily by attacks from the media. Stones were thrown at her house, and abuse was shouted at her that the foreign woman should go back to her own country. In the middle of this, a report arrived that a second Nobel Prize had been awarded to her: the Nobel Prize for Chemistry. The basis for the award was "the discovery of radium and polonium and investigation of the properties of radium and components thereof." One Nobel Prize committee member went so far as to say that Marie should refuse the prize until it was proven that the scandal was untrue, but Marie replied that the honor was given for her scientific achievements and was not related to personal behavior. She fought poor health to attend the ceremony. However, she suffered for more than a year afterwards from depression and kidney disease thought to be due to radiation damage.

Stamp with design of Madame Curie issued to commemorate the International Year of Chemistry

Ernest Rutherford (1871–1937)
Rutherford was born in Brightwater, New Zealand, the son
of an immigrant. After receiving an education at Nelson
College in New Zealand, he studied physics and mathematics
at Canterbury College in Christchurch, obtained outstanding
grades, and began research on electromagnetism. Rutherford
earned a scholarship and went to the Cavendish Laboratory
in 1895 and began research under J. J. Thomson. After inves-

tigating the effects of X-rays on the electrical conduction of gases, he began
research on radioactivity. He went to McGill University in Canada (1898),
then the University of Manchester (1907), and became director of Cavendish
Laboratory in 1919. He is considered the founder of nuclear physics, and was
awarded the 1908 Nobel Prize for Chemistry.

on radioactivity. In 1898, he found that uranium emitted at least two types of
radiation. One was radiation that was readily absorbed, which he called α (alpha)
rays. The other was radiation that had very strong penetrating power, which
he called β (beta) rays. While α-rays were easily shielded by a thin metallic
sheet, β-rays had a penetrating power 100 times stronger than α-rays. In 1898,
Becquerel and Giesel in Germany bent a component of radiation using magnets,
in a manner similar to how cathode rays could be bent. The mass/charge ratio for
this component was the same as that for an electron. The Curies also investigated
the deflection of radiation by magnets and verified that β-rays were negatively
charged. These observations clearly indicated that a β-ray was an electron ray.

In 1898, Rutherford became a professor at McGill University in Montreal and
actively continued radioactivity research there. Initially, it was believed that
α-rays were not bent by magnets, but in 1903, Rutherford observed that α-rays
were bent by a strong magnetic field and electric field, and also showed that α-
rays were charged particles. From careful measurements, he learned that α-rays
were positively charged and that the mass/charge ratio was approximately twice
that for a hydrogen atom. This showed that α-rays consisted of divalent helium
ions or monovalent molecular hydrogen ions. While helium had been originally
discovered by analysis of the solar spectrum, it had also been discovered by
Ramsay in the uranium ore, cleveite (see page 94). Because helium had been
discovered in association with a radioactive substance, Rutherford thought
that α-rays were probably helium ions. He started work in collaboration with
Frederick Soddy, a young chemist from Oxford at McGill University. Soddy
returned to England after conducting research with Rutherford in Canada in
1900–1902, did research with Ramsay in London in 1903, and found that helium

Frederick Soddy (1877–1956)
After studying at the University of Wales and Oxford University, Soddy went to Canada and started research on radioactivity together with Rutherford at McGill University. He returned to England in 1902 and became a lecturer in physical chemistry at the University of Glasgow in 1904. Soddy continued research on radioactive substances and had many notable achievements. He was a professor at the University of Aberdeen in 1914, and became a professor at the University of Oxford in 1919. He was awarded the Nobel Prize for Chemistry in 1921, but subsequently stopped research on radioactivity. In his later years, he wrote a great deal on topics such as the relationship between science, society and economics.

was produced by radium bromide. In 1907, Rutherford received a professorship at the University of Manchester and returned to England. There, he immediately began research to verify that α-rays were helium ions. He and Royds collected sufficient α-particles radiated from a radium sample and observed that the spectrum of these α-particles was the same as that associated with the helium component in the solar spectrum.

In 1900, Villard in France identified a third type of radiation, one that was not bent by a magnetic field and that had strong penetrating power similar to X-rays. In 1903, Rutherford called this a γ (gamma) ray. The identity of this ray was not immediately clear, but in 1914 Rutherford and Andrade succeeded in observing diffraction by a crystal and confirmed that it was an electromagnetic wave of extremely short wavelength.

Rutherford was awarded the Nobel Prize for Chemistry in 1908. The reason for receiving the award was "research pertaining to the decay of elements and the chemistry of radioactive substances." Although he was an experimental physicist, at this time radioactivity was important to both the fields of physics and chemistry. Since Rutherford's research helped lay the foundations of chemistry, his prize for chemistry was logical.

The Radioactivity of Thorium and the Transmutation of Elements
Rutherford noticed an occasional scatter in the radioactivity from thorium. To investigate this, he blew gas onto the surface of thorium, then collected this gas into a flask to collect the gas produced by this sample. This produced gas exhibited strong radioactivity, and although Rutherford originally thought that this radioactivity came from thorium, it became apparent that

most of the radioactivity was from this produced gas. He called this gas thorium emanation. Rutherford discovered that thorium emanation caused the radioactivity of thorium to decrease, but then to recover over time. This showed that thorium slowly decayed and transformed into a different substance that rapidly decayed. A similar phenomenon had also been observed with radium.

In 1902, Rutherford and Soddy discovered that 54% of thorium radiation was from a radioactive substance produced from thorium, which was called thorium X. Thorium X was separated from thorium salt, and the radioactivity of the residue continued to weaken and did not give rise to thorium emanation. However, when left standing for a few days, the residue regained radioactivity and produced thorium emanation. This phenomenon showed that thorium X was created from thorium, and that thorium emanations came from thorium X.

While the complexity of radioactivity thus gradually became clearer, the most important finding was that a different element was generated by the release of radiation by the original radioactive element. In 1902, Rutherford and Soddy showed that thorium emanation was a new inert gas, and that it belonged to the same family as argon, krypton, and xenon. This new element was initially called "niton", but subsequently came to be called radon. On the other hand, thorium X had different chemical properties than thorium. Thorium X was later understood to be an isotope of radium.

In 1902, in a paper entitled "The Cause and Nature of Radioactivity,"[20] Rutherford and Soddy explained, "the phenomenon of radioactivity is actually the transmutation from one chemical element to another chemical element, and this transmutation is caused by the emission of charge-carrying α-particles or β-particles." This was an example in which the "immutability of elements," which had been believed to be a fundamental principle, did not hold true. The transmutation of elements by radioactivity was also observed in uranium. By 1912, 30 radioactive elements had been reported, and detailed transmutation sequences for uranium, thorium, and actinium had been investigated.

The decrease in radioactivity was shown by Rutherford and Soddy to be exponential

$$I_t = I_0 \exp(-kt)$$

Here, I_0 is the initial radioactivity intensity, I_t is the intensity at time t, k is the decay rate constant, and the half-life $t_{1/2}$, the time for the radioactivity to fall to half its original value, is given by $\ln 2/k$.

In addition, it became clear that the half-lives of radioactive elements ranged from remarkably short to extremely long. The half-life of thorium emanation is 54.5 seconds, and the half-life of radium emanation is 3.82 days. On the other hand, the half-life of radium 226 is 1600 years, and the half-life of uranium 238 is 4.51×10^9 years.

Radioactive Isotopes and Radioactive Transformation Series

In 1911, Soddy pointed out that an element that emits an α-particle transmutates into the element two to the left on the periodic table. At around the same time, Russell discovered that by giving off a β-particle, an element transmutates into the element one to the right on the periodic table. Soddy, Russell, and Fajans generalized this into an overall radioactive transformation process. This meant that atoms of the same element could have different masses. Soddy predicted that while the atomic weight of lead in uranium ore would be 206, the atomic weight of lead obtained from thorium would be 208. This prediction was verified experimentally in 1914 by Richards and Hönigschmidt. A radioactive element occupying the same position in the periodic table but having a different mass was termed an isotope. Soddy limited the use of this term to radioactive elements and did not extend its use to all elements.

Uranium, thorium, and actinium undergo elemental transmutation in a radioactive transformation sequence tied together by α-decay and β-decay. For example, thorium 232 changes to radium 228 by α-decay, with a half-life of 1.41 $\times 10^{10}$ years, and changes to actinium 228 by β-decay with a half-life of 5.77 years. Actinium 228 changes to thorium 228 by β-decay with a half-life of 6.13 years, thorium undergoes α-decay with a half-life of 1.913 years to radium 224 (thorium X), and radium undergoes α-decay to change to radon 220 (thorium emanation). An additional four α-decays and two β-decays provide the stable element lead 208, and the radioactive sequence ends.

Anode Rays and Mass Spectroscopy: Stable Isotopes

In 1886, using a discharge tube with a large diameter cathode pierced with several holes, Goldstein discovered that a particle beam moved from the holes in the opposite direction to the anode. This particle beam produced light, and the color of the light depended on the composition of the low-pressure gas. He called this particle beam a canal ray (anode ray).

In 1897, the German physicist Wien succeeded in bending these canal rays with an electric field and a magnetic field, and from the direction and magnitude of the deflection, he discovered that canal rays were positively-charged particles with a mass/charge ratio several thousand times that of a cathode ray particle. This particle beam was thought to be composed of ionized gas atoms or molecules within the tube, which were generated by the removal of electrons by cathode rays. However, since a portion of the canal rays lost or gained

Francis Aston (1877–1945)
Aston was born in Birmingham, the son of an affluent merchant. He studied organic chemistry under Frankland at the University of Birmingham, but following the discovery of X-rays and radioactivity, he became interested in physics and became J. J. Thomson's assistant at the Cavendish Laboratory. After World War I, Aston developed a mass spectrometer, began research on isotopes, and discovered that they exist for most elements. He was awarded the 1922 Nobel Prize for Chemistry.

charge by collision with gas molecules, it was difficult to investigate canal rays from one type of ion. J. J. Thomson carried out experiments to reduce the probability of collisions of gas molecules with canal ray particles by reducing the gas pressure, and determined the mass/charge ratio for various atoms and molecules with considerable precision. He was thus led to discover isotopes of the same element with different atomic weights.

In 1913, Thomson discovered that the mass/charge ratio for canal rays in neon gas exhibited two different values. One was 20 times larger than that for a hydrogen atom, and the other was 22 times larger. The atomic weight of neon determined by Ramsay was 20.2. Thomson thought that there were two isotopes of neon. Keenly insightful, he focused on the importance of mass spectroscopy to chemistry as early as 1913[21], and pointed out that "this method has astonishingly good sensitivity, extremely small sample quantities are fine, and purity is not particularly required".

However, to realize this expectation, improvements in equipment were needed. Thomson's assistant, Francis Aston, planned to separate gases of atomic weights 20 and 22 by making use of the fact that when a gas is passed through a porous material, the lighter gas diffuses through more quickly, but this research was discontinued with the outbreak of World War I. After the war, Aston developed a mass spectrometer with improved resolution and returned to research. With Thomson's apparatus, ions with the same charge/mass ratio traveled a parabolic trajectory, but Aston made improvements to the equipment so that the ions converged and struck the target as a single point (Fig. 3.5). Thus, Aston could precisely determine the mass of particles within the gas. He showed that 90% of neon had an atomic weight of 20 and that 10% had an atomic weight of 22. The atomic weight of 20.2 determined by Ramsay was explained by the weighted average of the atomic masses of the

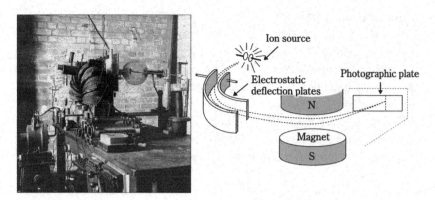

Figure 3.5: Aston's mass spectrometer and a schematic diagram illustrating the principle of mass spectroscopy, in which ions are deflected by an electric field and a magnetic field.

two isotopes. This verified the existence of stable isotopes in stable molecules. Dempster of the University of Chicago built a mass spectrometer at about the same time as Aston and published his findings earlier than Aston, but Aston made many important discoveries pertaining to isotopes and made major contributions to chemistry.

Aston discovered many new isotopes: three isotopes of sulfur (32, 33, and 34), two isotopes of chlorine (35 and 37), and three isotopes of silicon (28, 29, and 30). It became clear that stable, non-radioactive isotopes exist for many elements, and that the presence of isotopes is independent of radioactivity.

The precise values for the atomic weights of isotopes determined by Aston revealed a surprising law between these isotopes and atoms, which he called the "law of whole numbers": "If we take one-sixteenth the atomic weight of oxygen as the basis for atomic weight, the atomic weights of all isotopes are close to a whole number." Actually, this coincided with a hypothesis published by Prout in 1815 that all atoms are constructed from certain elementary particles. Prout thought that this fundamental particle was probably hydrogen, but many elements, such as chlorine, were known that clearly had molecular weights that were not whole numbers of the atomic weight of hydrogen, so this hypothesis had long been disregarded.

However, detailed investigation showed that the precise values for atomic weights deviated slightly from whole numbers. Einstein's theory of special relativity, published in 1905, showed that a (now famous) relational expression holds between mass (m) and energy (E): $E=mc^2$ (c is the speed of light). It was thought that if other atoms consisted of hydrogen atoms, as in Prout's hypothesis, the difference in the actual atomic weight and the value predicted from Prout's hypothesis would indicate the stabilization of the atom. Accordingly, Aston defined a value, called the packing fraction, as the value obtained by subtracting the whole number closest to the atomic weight (mass number) from the atomic weight, then dividing this by the mass number and multiplying by 10,000. When this packing fraction was plotted vs. the mass number, the fraction had positive values for atoms preceding oxygen on the periodic table and negative values for atoms after oxygen; after reaching a minimum value near iron, the packing fraction again increased and became positive. The packing fraction clearly exhibited a correlation with the stability of the atom.

The mass spectrometer that Aston developed did not have sufficient resolution to be immediately useful in other research. However, the resolution of mass spectrometers in the latter half of the 20th century improved dramatically, and mass spectrometers became one of the most important tools in chemistry. Aston started his career as an organic chemist, but following the discovery of X-rays and radioactivity, he began research in physics. He was invited by Thomson to become his assistant at Cavendish and began research on mass spectroscopy. He was awarded the Nobel Prize for Chemistry in 1922. He endeavored subsequently to improve mass spectrometers, and achieved a precision of 1 part per 10,000 in 1927. He also had an interest in the geochemical aspects of isotope separation, and in 1927 created the first table of isotope abundance ratios.

The Discovery of Deuterium

Deuterium (D) is a particularly important stable isotope, and has an atomic weight of 2. In 1931, Harold Urey read a paper in the journal *"Physical Review"* that hinted at deuterium's existence, and planned an experiment to prove it. He thought that hydrogen and deuterium would have different masses, and their atomic spectra would be different. Since the abundance ratio of deuterium in ordinary water is small, the deuterium would have to be concentrated before it could be detected. Theoretical calculations of the vapor pressures of hydrogen and deuterium indicated that deuterium could be concentrated by evaporating liquefied hydrogen. Liquid hydrogen was evaporated at 14 K, and the deuterium was concentrated. Urey identified the presence of deuterium from spectra. His discovery confirmed the existence of deuterium and, importantly, proved that isotopes can be separated by ordinary physical and chemical methods. In 1932, it was shown that deuterium in the form of heavy water (D_2O) could be concentrated by the electrolysis of water. The use of deuterium opened a path to new scientific research. As will be described below, deuterium ions were a useful radiation source in nuclear transmutation, and molecules labeled with

Harold Urey (1893–1981)
Urey was born in Indiana, the son of a clergyman. He studied zoology at the Univrsity of Montana, changed to chemistry, and obtained a Ph.D. under G. N. Lewis at the University of California, Berkeley. Subsequently, Urey studied in Europe with the physicists Bohr and Kramers, and after returning to the US in 1934, he became a chemistry professor at Columbia University. He conducted research in a wide range of fields, including chemical reaction theory, quantum mechanics, and the spectra of molecules, and this background helped him with the discovery of deuterium. He succeeded in separating the isotopes of several elements such as carbon, nitrogen and sulfur, and in his later years had an interest in cosmochemistry and became a pioneer in the field. He was awarded the 1934 Nobel Prize for Chemistry.

deuterium became widely used for investigating the mechanisms of chemical reactions. Urey was awarded the 1934 Nobel Prize for Chemistry for his discovery of deuterium. He served as the first editor of the *"Journal of Chemical Physics,"* and contributed significantly to the development of chemical physics and physical chemistry.

3.4 The Atom as a Reality

Chemists in the 19th century pursued research while presuming the existence of atoms and molecules, and achieved great success. However, Dalton's atom, consisting of an indivisible microparticle, was simplistic and unsatisfactory, and discussions regarding the nature of the atom continued among scientists. In the first half of the 19th century, Prout suggested that atoms are composed of hydrogen as the unit component. Around the middle of the 19th century, the spectra of various elements were obtained, but could not be understood using Dalton's concept of an atom. Physicists developed a different concept of the atom compared to chemists. Ernst Mach of the Viennese school of empiro-criticism was a central philosopher in the latter half of the 19th century who influenced many physicists and physical chemists. Among them there was opposition to theories of the atom that were not based on directly observed phenomena. The end of the 19th century to the beginning of the 20th century saw tense battles of words between atomic theorists and anti-atomic theorists. However, in the first decade of the 20th century, experiment and theory gradually emerged to support the theory of the atom, and by 1910 atomic theory was accepted.

19ᵗʰ Century Physicists and the Atom

Physicists in the first half of the 19ᵗʰ century had an expanded mechanistic view of the world based on Newton's corpuscular philosophy. Classical mechanics was mathematically systemized by Laplace, Lagrange, and others, and was established as the basis for understanding nature in a unified manner. Physicists who received a mathematical education in France, Germany, or Britain began to flourish, and mathematical physics began to develop. These physicists handled particles mathematically using coordinates and mass. Heat, light, and electricity were thought to be fluids comprising massless particles. This fluid, which became increasingly important in the 19ᵗʰ century, was called the "ether."

At the beginning of the 19ᵗʰ century, Fresnel and Young developed a wave theory of light that suggested that light was a transverse wave propagating in the ether. Elucidating the properties of the ether in order to understand matter, heat, light, electricity and magnetism in a unified manner became a major objective of physicists. The ether theory was gradually refined based on the methods of Lagrange and Fourier and became more abstract.

Against this background, William Thomson (later Lord Kelvin) proposed a vortex theory of the atom in 1867. Atoms were thought to be vortices within a perfect fluid such as the ether, allowing the simultaneous existence of discontinuity and continuity. In addition, it was anticipated that the spectra of elements could be explained by oscillations of vortex atoms. Thomson thought that a vortex atom was a completely elastic body, and that a rigorous kinetic theory of gases could be derived to describe such an atom. Based on Thomson's thinking, it should be possible to describe all matter, light, and electricity as movement within the ether.

Physicists focused on the ether largely because of the phenomenon of electromagnetism, discovered by Faraday. Faraday's discovery led to the idea that electromagnetic effects are transmitted along curved lines, and that a charged object induces an electric charge in another object despite the obstacle of a non-conducting shield. Faraday thought that the ether was distributed throughout space and was filled with electric or magnetic lines of force. In fact, using iron powder, magnetic lines of force were shown to be distributed around a magnet, so why should light not be affected by a magnetic force as well? In 1845, Faraday discovered the so-called Faraday effect: when plane polarized light passes through lead borate, the plane of polarization is rotated by a strong magnetic field. Mathematical physicists considered this to be proof that the ether imparts a rotational stress within a magnetic field.

The phenomenon of electromagnetism discovered by Faraday was mathematically systemized by Maxwell. He, too, supported the idea of the ether, and believed the vortex atom to be a possible manifestation of it. Maxwell believed that the vortex atom and the ether were useful hypotheses. However, there was no experimental evidence indicating the existence of the ether. Many physicists did not think it necessary to introduce a discontinuity, such as the existence of an atom, to deal with physical phenomena, and had little interest in a vortex atom. Similarly, chemists of the 19th century explained chemical phenomena based on Dalton's atomic theory, and hence, most chemists had no interest in a vortex atom.

Around the end of the 19th century, Mach strongly attacked the theory of the atom, arguing that the role of a scientist is to observe facts and phenomena, and that theories based on concepts such as the atom, that cannot be observed, are detrimental. Boltzmann, an advocate of the theory of the atom, argued in 1899 that the theory of the atom needed to be developed further, so that it did not contradict the large amount of observational data from physics, chemistry and crystallography. However, others, such as Ostwald, did not agree. To further promote the theory of the atom, experimental evidence was needed that supported the reality of atoms and molecules.

Theory of Brownian Motion and Einstein

In 1827, the English botanist Brown observed the movement of pollen particles under a microscope and saw that the particles exhibited a strong irregular motion. He initially thought that this might be due to the vitality of the pollen, but it became clear that the same phenomenon was observed with any type of microscopic particle. At around the end of the 19th century, the idea arose that particles within a fluid constantly collide with neighboring particles, rebound, and move irregularly. The French physical chemist Jean Perrin thought that this movement indicated the existence of molecules. Papers explaining this phenomenon were submitted independently by Albert Einstein in 1905 and by Marian Smoluchowski the following year.

Einstein derived the following equation regarding particles moving with irregular Brownian motion, focusing on the linear distance linking the starting point and termination point of a particle's path, which is the displacement (x), in a fixed time (t).

$$<x^2> = 2Dt$$

The average of the square of x, $<x^2>$, is proportional to the diffusion coefficient D, and D is related to Avogadro's number N by the Stokes-Einstein equation $D = RT/6N\pi a\eta$. Here, η is the viscosity of the fluid, T is the absolute temperature, and a is the particle radius. The same equation is given for rotational Brownian motion (irregular rotational movement resulting from molecular collisions).

Albert Einstein (1879–1955)
Einstein was born in Ulm, Germany, the son of a Jewish electrical engineer and entrepreneur. He received his primary education in Germany, but spent his boyhood unable to adapt to the strict, militaristic Prussian educational system of the time. Averse to conscription, he dropped out of gymnasium in Germany, but because of excellent marks in physics and mathematics, was permitted to enter the  Swiss Federal Institute of Technology. Here he studied independently and deepened his knowledge of physics and mathematics, but was inattentive to formal classwork and incurred the displeasure of his professors. Because of this, he struggled to obtain a university position, but he was catapulted into fame with the publication of ground-breaking papers in 1905. He successively held professorships at Prague University, The Swiss Federal Institute of Technology, and the University of Berlin. Einstein emigrated to the US in 1933 to escape Nazi Germany, and spent the remainder of his life in academia at the Princeton University Institute for Advanced Study. He was awarded the 1921 Nobel Prize for Physics.

Marian Smoluchowski (1872–1917)
Smoluchowski was an Austrian physicist of Polish descent. He studied at the University of Vienna, and after visiting universities in Paris, Glasgow, and Berlin, became a university professor in 1899 at the University of Lviv and then at the University of Krakow in 1913. He left behind pioneering achievements in statistical physics.

At that time, Einstein had a strong interest in evidence indicating that an atom has a specific size, and conducted research on this topic for his doctoral dissertation. Attempts had been made previously in the latter half of the 19th century to estimate the sizes of atoms and molecules. Based on the kinetic theory of gases, the Austrian physicist, Loschmidt, estimated the sizes of molecules in air and Avogadro's number. Einstein focused on solutions from the perspectives of thermodynamics and statistical mechanics, but his investigations pertaining to the theory of Brownian motion were based on techniques for statistical mechanics, and had a major effect on subsequent physics and chemistry research.

After graduating from university, Einstein obtained a position at the patent office in Bern based on the recommendation of a friend's father, and eked out a living. He had spare time at the patent office and could immerse himself in research. Here he exhibited surprising creative powers and conducted studies that transformed 20th century science. 1905 in particular is called Einstein's Annus Mirabilis (miracle year), in which he published epoch-making papers in three different subjects in the journal *"Annalen der Physik"* that brought about a revolution in science. These papers were on the photoelectric effect,[22] Brownian motion,[23] the special theory of relativity,[24] and the relationship between mass and energy.[25] Any one of these was deserving of a Nobel Prize, but his Nobel Prize for Physics in 1921 was for his achievements related to the photoelectric effect.[5]

Experimental Verification by Perrin[3]

From 1905 to 1912, Jean Perrin experimentally verified the theory of Brownian motion, compared Avogadro's number obtained by various methods, and attempted to prove the atomic theory. He conducted research using mainly emulsions obtained from resins, such as gamboge and frankincense made by drying plant-derived milky liquids. He prepared samples having the same particle size by sorting according to size using centrifugal separation methods, and conducted experiments to determine the density and volume of particles. He observed the motion of these fine particles under a microscope and took numerous photographs to determine the number of particles. Most of his experiments were designed to precisely identify particles that achieved sedimentation equilibrium,[6] and to determine the distribution of the displacement of particles moving with translational and rotational Brownian motion. He conducted detailed experiments for different types and sizes of particles, and varied the temperature and the type and viscosity of the solvent. The experimental results completely agreed with the predictions of Einstein's theory. The value for Avogadro's number (N_A) derived from his results was $6.5–6.9 \times 10^{23}$.

Perrin further compared the values of $N_A/10^{23}$ obtained from analyses of various phenomena in which Avogadro's number is involved. He obtained 13 values for $N_A/10^{23}$ determined from the coefficient of the viscosity of a gas,

---

5. It has been reported that Einstein's Nobel Prize was not awarded for the special theory of relativity because the selection committee could not understand the theory. Despite numerous achievements from 1905 onward, the Nobel Prize was not awarded to Einstein until 1921 because of the war, and perhaps because of the anti-Semitic prejudice of German physicists such as Lenard.

6. Sedimentation equilibrium: During sedimentation, minute particles sink due to the action of gravity. A constant sedimentation rate is achieved when gravity and frictional forces resulting from viscous resistance are balanced. Sedimentation equilibrium refers to the state in which sedimentation and Brownian motion are balanced.

Jean Perrin (1897–1942)
Perrin was a French physical chemist. He studied at the Ecole
Normal Superieure, was a lecturer of physical chemistry at
the Sorbonne in 1897, and was a professor at the Sorbonne
from 1910 onward. After research supporting the idea
that cathode rays were composed of negative particles, he
succeeded in determining Avogadro's number from observations of sedimentation equilibria and Brownian motion, and
proved the existence of molecules. In 1936, he joined the cabinet of the First
Popular Front as Secretary of Research Laboratories. He was awarded the
1926 Nobel Prize for Physics.

Brownian motion, the irregular distribution of molecules (critical opalescence[7]
and the blueness of the sky), black body spectra, the charge on a spherical body,
radioactivity, and other phenomena, and found that all were between 6.0 and
7.5.[3] The fact that a consistent Avogadro's number was obtained from such different phenomena was sufficient to indicate the correctness of the atomic theory
and the actual existence of the atom. By 1909, even the stubborn atomic theory
opponent, Ostwald, came to believe in the existence of the atom. Thus, the atomic
theory was validated. Perrin was awarded the Nobel Prize for Physics in 1926.

3.5 The Advent of the Quantum Theory

Physics in the 19th century was based on the continuity of mass and energy.
After the atomic theory was proven, the continuity of matter was disputed.
Scientists at the end of the 19th century were optimistic that the world could
be fully understood by classical physics based on Newtonian mechanics and
Maxwell's electromagnetism. In 1900, Lord Kelvin stated that two small dark
clouds had appeared on the horizon of classical physics. One of these was that
classical physics could not explain the black body radiation problem. The other
black cloud was that movement within the ether could not be observed.

   At the beginning of the 20th century, doubts were being cast on the continuity
of energy as well. The trigger for this was the problem related to black body
radiation. A body of matter that is heated radiates electromagnetic waves. The
wavelength of the electromagnetic waves depends on the temperature of the

---

7. Critical opalescence: Name given to the phenomenon that when light strikes a liquid near its
   critical point, the liquid appears cloudy due to scattered light.

body, and as temperature increases, electromagnetic waves become shorter. This is demonstrated by the fact that the color of a heated lump of iron changes with increasing temperature, from red to orange and then to white. Quantum theory arose out of attempts to understand this phenomenon.

Planck's Quantum Theory

A black body is an idealized body that absorbs and emits electromagnetic waves of all wavelengths. Black body radiation is approximated by electromagnetic waves emitted from a small hole in a closed container. The spectrum and radiant intensity of this radiation were investigated in detail by German physicists in the latter half of the 19th century.

In 1893, Wien investigated the temperature change of wavelength-dependent radiant energy and discovered a law showing that the product of the wavelength at which the strongest emission occurred and the absolute temperature is constant. Prior to this, in 1879, the Stefan-Boltzmann law had been discovered showing that the total radiated energy is proportional to the absolute temperature to the fourth power. Providing a theoretical explanation for the wavelength dependence of black body radiation was a major topic in physics in the latter half of the 19th century. Lord Rayleigh thought of an electromagnetic field as a collection of oscillators oscillating at different frequencies, and regarded radiation at a frequency v as signifying that an oscillator had been excited and thus oscillated at this frequency. He studied the wavelength dependence of the energy density, which led to a law called the Rayleigh-Jeans law.[8] However, this law only described results obtained at long wavelengths, and completely failed to describe results obtained at short wavelengths. Classical physics clearly could not solve the problem of black body radiation.

The black body radiation problem was solved by the University of Berlin professor, Max Planck. Since 1895, he had been attempting to explain experimental data pertaining to black body radiation from the perspective of the entropy of an electromagnetic oscillator. At a meeting in Berlin on 25 October 1900, Plank proposed an empirical formula for black body radiation. This was a formula devised to fit experimental results, and while no theoretical rationale was provided, it closely matched experimental results. He then concentrated his energy on obtaining theoretical backing for this equation.

Planck focused attention on the fact that solids consist of rows of atoms and that atoms are always oscillating. He hypothesized that the energy of these

---

8. Rayleigh-Jeans law: An equation derived by Rayleigh and Jeans related to thermal radiation. Radiative density $\rho(v, T)$ of frequency v at temperature T is given by $\rho(v, T)=(8\pi v^2/c^3)k_B T$ (where c is the speed of light, and k_B is Boltzmann's constant).

Max Planck (1858–1947)

Plank was born in Kiel, studied at the University of Berlin, and then studied under Helmholtz. He obtained a Ph.D. at the University of Munich, then returned to the University of Berlin in 1889 to become a professor. His initial research was in the field of thermodynamics. He conducted research on topics such as physicochemical applications of thermodynamics and the electrical conductivity of electrolyte solutions, and wrote a textbook that laid the basis for thermodynamics. He was known as a theoretical physicist well versed in physical chemistry. After moving to the University of Berlin, Plank began research on thermal radiation, leading to the proposal of the quantum theory. He was awarded the Nobel Prize for Physics in 1918.

oscillating atoms is not continuous, as thought by classical mechanics, but rather is limited to integer multiples of the energy hv. Here, v is the frequency of the light, and h is a constant that characterizes quantum theory, called Planck's constant.[9] Accordingly, using the same method Boltzmann used to derive an equation for entropy (see page 107), Plank derived the following formula for the frequency dependence of black body radiant energy for an energy density ρ:

$$\rho(v, T) = (8\pi hv^3/c^3)/\{\exp(hv/k_B T)-1\}$$

Here, c is the speed of light and k_B is Boltzmann's constant. In the high-temperature, low-frequency region, $hv/k_B T \ll 1$, and the formula approximates the Rayleigh-Jeans equation. Planck submitted a paper to the German Physical Society on 14 December 1900; this date is recognized as the birthday of quantum theory.[26]

Planck's formula did not attract much attention at the time it was published. At first, it was simply thought to describe black body radiation. Plank himself thought the formula only applied to atoms in solids. However, in a battle of words exchanged in the journal *"Nature"* in 1905, it became obvious that the Rayleigh-Jeans formula was inadequate, and Plank's alternative treatment to the failed classical theory began to attract attention. Einstein's use of quantum theory to describe the photoelectric effect in 1905 completely changed the situation.

---

9. Planck's constant: A constant characterizing the domain of quantum theory, represented as h, where $h = 6.63 \times 10^{-34}$ J·s. If the action of a given mechanical system greatly exceeds h, the system is described by classical mechanics.

Arthur Compton (1892–1962)
Compton was an American physicist who obtained a Ph.D. at Princeton University. He became a professor at the University of Chicago in 1923, and Chancellor of Washington University in St. Louis in 1945. He was awarded the Nobel Prize in 1927 for discovering the Compton effect. He was one of the leaders in the effort to develop nuclear reactors during World War II.

Light quantum hypothesis of Einstein

In the "photoelectric effect", an electron is emitted from the surface of a metal when light is shone on the metal. The effect was discovered by Hertz at the University of Kiel in 1887, and two years later more detailed research was reported by Elster and Geitel. In 1902, Lenard showed that when the light frequency is smaller than a certain fixed threshold v_0, no electrons are emitted regardless of the light intensity, but if the light frequency is larger than v_0, electrons are emitted no matter how weak the light intensity is. The kinetic energy of the emitted electron is proportional to the difference between the frequency of the incident light and v_0. Einstein realized that these observations could not be explained by the wave theory of light and proposed a hypothesis in 1905 to treat light as quantized particles with an energy hv. This theory provided a simple explanation for the photoelectric effect because if hv is greater than a certain value, electrons will be emitted. These particles of light were first referred to as radiant quanta, but the American physical chemist G. N. Lewis proposed the word "photons" in 1926, and this is the term currently used.

The suggestion that light has a particle-like nature was surprising, but Einstein did not deny the wave nature of light. He suggested that light has both a wave nature and particle nature "duality", and in the case of the photoelectric effect or the absorption or emission of light by a molecule, light behaves as a particle. In 1923, the American physicist Arthur Compton discovered that scattered X-rays contain wavelengths longer than those of the incident X-rays (Compton effect). Because X-rays are scattered by electrons in matter, this phenomenon was taken as evidence of light behaving as a particle having a fixed momentum, thus supporting the light quantum hypothesis. Einstein not only dealt with the photoelectric effect of light, but in later years in a different paper also addressed a fundamental problem pertaining to the absorption and emission of light by atoms and molecules. This paper laid the foundation for interpreting spectra, and was fundamental for the development of spectroscopy and lasers.[27]

The applicability of the quantum theory to solving other problems gradually became clear. For example, it was known that the specific heat of a solid decreases

Hantaro Nagaoka (1865–1950)
Nagaoka was born in Nagasaki Prefecture, and graduated from the Department of Physics of Tokyo Imperial University in 1887. He went abroad in 1893 to study with Helmholtz and Boltzmann. He returned to Japan in 1896 and was a professor at Tokyo Imperial University. Nagaoka published a Saturn-type model of the atom in 1903, became a member of the Institute of Physical and Chemical Research in 1917, and engaged in research in the broad fields of geophysics and spectroscopy. He became the president of Osaka Imperial University in 1931 and was awarded the Order of Cultural Merit in 1937.

with decreasing temperature, but this could not be explained by classical theory. In 1907, Einstein showed that it can be explained by quantizing the vibrational energy of atoms in solids. In addition, the specific heat of gases could not be explained by the kinetic theory of gases, and therefore Maxwell doubted the validity of this theory. This problem, too, was immediately solved by the theory of the quantization of energy. In 1911, Einstein provided an explanation for the specific heat of several gases.

In 1913, Einstein and Stern suggested that an oscillating particle would retain residual energy even at absolute zero, and called this energy 'zero-point energy'. Using data for the specific heat of hydrogen gas, they estimated that the zero-point energy of a particle oscillating with a frequency v would be $(1/2)hv$.

3.6 Structure of the Atom and Quantum Theory

At the beginning of the 20th century, with the discovery of the electron, it became clear that Dalton's indivisible atom was incorrect, and alternate models for the structure of the atom began to appear. In 1903, Thomson proposed a model in which "Electrons are embedded in a positively charged continuous structure just like raisins in plum pudding."

At about the same time, Hantaro Nagaoka of Tokyo University proposed a Saturnian model. According to this model, electrons rotated around a central, positively charged body like the rings around Saturn or planets around the sun. Rutherford's group began experiments to investigate the validity of this general model.

Discovery of the Atomic Nucleus by Rutherford

Although Rutherford had conducted outstanding research pertaining to radioactivity in Canada, he returned to England in 1907 and started new

Hans Geiger (1882–1945)
Geiger was a German physicist. After studying at the University of Erlangen and the University of Munich, he worked as Rutherford's assistant. He became a professor at the University of Kiel in 1925, at the University of Tübingen in 1928, and at the Berlin Institute of Technology in 1936. He invented an instrument for measuring ionizing radiation in 1913, known as the improved Geiger-Müller counter.

research at the University of Manchester, exploring the interior of the atom using radioactivity and thus establishing a new field of physics called nuclear physics. He succeeded Thomson in 1919 to become professor of Cavendish Laboratory, and from the 1920's to the 1930's, Cavendish Laboratory became the world center for experimental nuclear physics research. He made innumerable important contributions to physics, and many of his students were awarded the Nobel Prize for Physics. It was natural to presume that he would also win a Nobel Prize in Physics, but for some reason he did not. This is strange, given that Marie Curie was awarded Nobel Prizes in both Physics and Chemistry.

Soon after arriving at the University of Manchester, Rutherford welcomed into the laboratory Hans Geiger, a young researcher from Germany, and Ernest Marsden, originally from New Zealand. They began to study the scattering that occurred when α-particles from a radium source were passed through a slit, collimated into a thin beam, and directed through a thin gold leaf (Fig. 3.6). When α-particles pass close to the atoms that comprise the leaf, their trajectory is bent, and the beam widens. They measured this widening by having the particles impact a sheet of zinc sulfide. Since the zinc sulfide screen gave off a flash of light with each impact of an α-particle, they could count the number of scattered particles. However, counting with the naked eye was difficult work that required patience. Afterwards, Geiger developed a counter tube that came to be known by his name, and this had a major impact on nuclear physics and radiation research.

The number of scattered particles decreased sharply with increasing scattering angle, and when the scattering angle exceeded 2–3°, the number of scattered particles fell to zero. However, one day they made the unanticipated observation that a small number of α-particles were scattered at an angle greater than 90° from the incident direction (backward scattering). To explain this result, Rutherford deduced that the α-particles collided with a small particle within an atom, and that the particle had an enormous electric charge. An analysis of the experimental results using simple calculations indicated that the α-particles were

Charles Barkla (1877–1944)
Barkla conducted research at Cavendish Laboratory, and was a professor at the University of London (1909) and the University of Edinburgh (1913). He estimated the number of electrons in an atom from the scattering of X-rays by gases, and provided clues about the significance of the atomic number. He ascertained that fluorescent X-rays were composed of characteristic K, L, and M series lines. He was awarded the 1917 Nobel Prize for Physics.

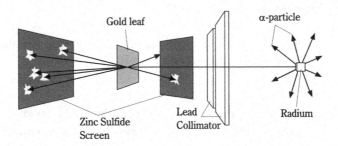

Figure 3.6: Conceptual diagram of the experiment conducted by Geiger and Marsden to demonstrate scattering

scattered backward by impacting with a positively charged core at the center of the atom. Thus, he arrived at an atomic model in which an atom is composed of a small, central nucleus carrying a positive charge, which has almost all the mass of the atom, and is surrounded by electrons.[28] From measurements of the fraction of α-particles scattered at various angles, Geiger and Marsden estimated the value of the positive charge on the nucleus. This value was smaller than the atomic weight but larger than the atomic number, and they found that the charge on the nucleus increased in proportion to the atomic number. They reasoned that the positive charge on the nucleus was half the atomic weight.[29] At about that time, Charles Barkla had come to the conclusion that the number of electrons in an atom was half the atomic weight, based on X-ray scattering experiments. Rutherford immediately recognized that the 1913 X-ray spectra obtained by Moseley correctly represented the number of positive charges in a nucleus. Together, these data revealed the relationship between atomic number, the charge on a nucleus, and the number of electrons around the nucleus.

The Bohr Model of the Atom
Spectroscopic studies of elements began around the middle of the 19th century, but the spectra obtained could not be understood. In the latter half of the 19th

century, a large amount of spectral data had been accumulated, but no progress had been made regarding an explanation of the data. In 1885, Johann Balmer in Switzerland discovered that four lines in the hydrogen spectrum satisfied a certain empirical formula. Rydberg of Lund University, Sweden, introduced the wave number ($1/\lambda$, where λ is wavelength) to represent this with the following formula.

$1/\lambda=R(1/2^2-1/n^2)$

n is a whole number 3, 4, 5, ..., $R=2\pi^2 m_e e^4/h^3$

Here, m_e is the mass of an electron, e is the charge of an electron, and h is Planck's constant. This is called the Balmer series. In a generalized form, this becomes

$1/\lambda=R(1/m^2-1/n^2)$,

and represents the Balmer series when $m=2$. Subsequently, the series for $m=1$ was discovered by Lyman, and the series for $m=3$ was discovered by Paschen, and these series were respectively named after them. However, the origin of these spectra was not at all understood. The person who provided a solution to this problem was the young Danish physicist, Niels Bohr, and major developments further resulted from his insights regarding the structure of atoms.

In 1911, after obtaining his Ph.D. for work on the electron theory of metals, Bohr went to Thomson's laboratory at Cambridge. However, Thomson had no interest in the new physics, so in 1912 Bohr moved to Rutherford's laboratory at Manchester to study the structure of the atom, and examined the theoretical significance of Rutherford's model of the atom. Bohr had recognized during his dissertation research that classical theory was inadequate for explaining the atom, and Rutherford's model of the atom, in which electrons travel around a positively charged nucleus, was unstable when viewed using classical physics. Bohr wondered whether the quantization used by Planck and Einstein might be applicable to addressing the electrons in the atom. He hypothesized that the orbital angular momentum of an electron was quantized, and that an atom existed only in a series of stable steady states of specific energy values. Based on this hypothesis, he constructed a theory for the hydrogen atom that accurately explained the spectrum of the hydrogen atom spectrum.[30]

In Bohr's quantum theory, an emission spectrum was assumed to be the result of electronic transitions from high-energy orbits to low-energy orbits. In this situation, there is a relationship $\Delta E=h\nu$ between the transition energy ΔE and the frequency of the spectrum ν. Thus, he obtained the following formula for the frequency of light ν produced by a transition.

Niels Bohr (1885–1962)
Bohr was born the son of the physiologist, Christian Bohr, of the University of Copenhagen, and was raised in a household with a vibrant academic tradition. He studied at the University of Copenhagen, and after obtaining his Ph.D., studied with Rutherford at Manchester. In 1913, he introduced the quantum hypothesis to Rutherford's atomic model and succeeded in explaining the series of spectral lines for hydrogen. Afterward, Bohr led the efforts to develop old quantum mechanics and later played a leading role in the birth of quantum mechanics, proposed the principle of complementarity, and established the Copenhagen interpretation of quantum mechanics. He was awarded the 1922 Nobel Prize for Physics.

$$v = (2\pi^2 m_e e^4/h^3)(1/m^2 - 1/n^2) \quad n = 3, 4, 5, 6, \ldots.$$

Consequently, if $R = 2\pi^2 m_e e^4/h^3$ and $m=2$, this becomes the Rydberg formula for the Balmer series. This formula reproduced experimentally-obtained formulas extremely well. In addition, by setting $m=1$ or $m=3$, the Lyman series or the Paschen series are described, respectively. This success showed the correctness of the bold hypothesis that atoms in the steady state have discontinuous energy states, and only transitions between states enable the absorption or emission of light. This work was completed in 1913. Bohr further extended the above formula by introducing the atomic number Z, and extended the theory to the single-electron atoms of He^+, Li^{++}, and Be^{+++}. In the case of He^+, the theory solved the mystery of the spectra measured from stars. Subsequently, the spectra of single-electron atoms were obtained by electric discharge for atoms up to oxygen and compared to the predictions of Bohr's theory, and the validity of the theory was established. Bohr's theory was also revolutionary in that it provided solid evidence for the correctness of quantum theory.

Bohr became a professor at the University of Copenhagen in 1916. In 1921, the university established a new theoretical physics research institute for him, and he spent the remainder of his career as its director. Under his excellent leadership, this research institute was an international Mecca for various branches of theoretical physics research. In the period up to the start of World War II, it was a place for research and the interchange of ideas between many young physicists, and Bohr played a leading role in the development of quantum mechanics in the 1920's and 1930's.[31] He was awarded the Nobel Prize for Physics in 1922.

Arnold Sommerfeld (1868–1951)
Sommerfeld was a German theoretical physicist. After
working as an assistant at the University of Göttingen, he
served as a professor at the University of Munich from 1907
to 1940, and played an active role in research in the 1910's
and 1920's. Sommerfeld was known for achievements such
as refining the Bohr model of the atom, introducing the
azimuthal quantum number and magnetic quantum number,
and the electron theory of metals. He was an outstanding teacher during the
developmental stage of quantum theory, and helped train many prominent
physicists, including the Nobel Prize winners Heisenberg, Pauli, Debye,
Bethe, and Pauling.

The Development of Bohr's Theory and the Structure of the Atom

Although Bohr's theory described the structure of single-electron atoms well, it
soon became clear that it was unsatisfactory for handling many-electron atoms,
so new efforts were undertaken to address this shortcoming. In 1915, Arnold
Sommerfeld introduced the concept of elliptical orbits to extend Bohr's theory.
Although research progress was delayed by World War I, further expansion
and correction of the theory were performed by Bohr, Sommerfeld, Wilson,
and others. However, the treatment of many-electron atoms became extremely
complex.

Of the many-electron atoms, the lithium and sodium atoms were comparatively simple and could be handled in a manner similar to single-electron atoms.
This was because the 3rd electron of lithium and the 11th electron of sodium enter
an elliptical orbit far removed from the other electrons. These other electrons
shield the positive charge of the nucleus, allowing the nuclear charge to be
regarding as +1.

Sommerfeld introduced a second quantum number for electrons, called the
azimuthal quantum number, l, in addition to the principal quantum number, n, for
the transverse diameter of the elliptical orbit. This quantum number prescribes
the elliptical characteristics of the orbit as quantized angular momentum. For
an orbit with the same principal quantum number, the larger l is, the higher the
energy. He further introduced a third quantum number, m, to specify the direction
of the plane of the orbit within a magnetic field. This quantum number was related
to the splitting of spectral lines in a magnetic field (the Zeeman effect) and was
called the magnetic quantum number. The Zeeman effect was partially explained
by this, but there were spectral line splittings that could not be explained by the
three quantum numbers, and these were called the anomalous Zeeman effect.

Wolfgang Pauli (1900–1958)
Pauli was a Swiss theoretical physicist. Born in Vienna,
he obtained a Ph.D. studying under Sommerfeld at the
University of Munich. In 1928 he became a professor at the
Swiss Federal Institute of Technology, but emigrated to the
US in 1940, and became a professor at Princeton University.
He made many scientific contributions in addition to the
Pauli exclusion principle, including Pauli paramagnetism
resulting from conduction electrons, and the prediction of the neutrino. A
perfectionist when it came to physics, he was known for his intense criticism
of the work of colleagues and was called the "conscience of physics" by the
physics community. Pauli was awarded the 1945 Nobel Prize for Physics.

Otto Stern (1888–1969)
Stern was a German-American physicist. He studied at the University of
Breslau, was a lecturer at the University of Frankfurt, and became a professor
of physical chemistry at the University of Hamburg in 1923. However, he was
driven by Nazi Germany to emigrate to the US. He helped pioneer research
using atomic and molecular beams. Stern was awarded the 1943 Nobel Prize
for Physics.

Sommerfeld published a book called *"Atombau und Spektrallinien"* (Atomic
Structure and Spectral Lines) in 1919,[32] and this book was like a bible for people
studying the new physics at the time.

In 1922, Bohr further developed his theory of the atom, and suggested a
general model that argued for an electron structure of the atom using a "building
up principle". This principle held that electrons fill an orbital (shell) in order, from
the lowest energy level to the highest energy level.[33] He divided these shells into
the (2), (2, 6), (2, 6, 10), and (2, 6, 10, 14) groups, and thus arranged the elements
of the periodic table up to number 86 in periods of 2, 8, 8, 18, 18, and 32. A correspondence was therefore established between the order of the elements in the
periodic table, that is, the atomic number, and the electronic structure of atoms.
From the second group onward, the shell electrons have different energies and
are further divided into the 2, 6, 10, and 14 subgroups. These were called the s,
p, d and f shells (orbitals) from their connection to spectral lines.[34]

In relation to spectral-line analysis, the "anomalous Zeeman effect", which
had vexed physicists at the time, was solved by the introduction of a new quantum number. In 1924, Wolfgang Pauli added a new quantum number, m_s, to the
electron, and suggested that the state of an electron in an atom is described by

the four quantum numbers n, l, m, and m_s. Pauli introduced the "Pauli exclusion principle" that no two electrons can take the same quantum numbers within an atom.[35] As a result, the number of electrons possible in each orbital was 2 for the s orbital, 6 for the p orbital, 10 for the d orbital, and 14 for the f orbital. With this, quantum theory could explain the order of the elements in the periodic table with no inconsistencies, and provided the foundation for describing the properties of elements based on the electronic structure of an atom. However, Pauli did not provide insights into the physical basis of m_s.

Previously, in 1921, Otto Stern and Walter Gerlach discovered that a heterogeneous magnetic field could split a beam of silver atoms in two. This meant that the silver atom had a magnetic moment. In order to explain this, in 1925, the Dutch physicists Uhlenbeck and Goudsmit proposed the existence of a 4th spin quantum number, m_s, for an electron. They suggested that an electron has an inherent angular momentum called spin, and that the value of this can only be either $\pm 1/2 (h/2\pi)$. Classically, this took the picture of an electron rotating in either a right-hand direction or left-hand direction. Electrons came to be seen as having an inherent magnetic moment associated with the spin angular momentum, and this opened a path to understanding the magnetic properties of matter at the atomic level.

While the Bohr model reasonably described atomic spectra, the periodic table of the elements, and differences in chemical properties, the majority of physics and chemistry problems remained unsolved. Bohr attempted to resolve the situation using a "correspondence principle"[10] to relate classical theory to quantum theory, but this did not yield a fundamental solution. Before long, the new quantum mechanics was introduced by young physicists, and the Bohr model of the atom became outdated (see Section 3.7). However, because his model of the atom was intuitively easy to understand, it continued to be used in chemistry.

The Discovery of the Neutron and the Structure of the Atom

In the 20 years following the discovery of the atomic nucleus by Rutherford, physicists thought that the atomic nucleus was composed of protons and electrons. For example, it was understood that the helium nucleus was formed from 4 protons and 2 electrons and had an overall positive charge of 2. To investigate what an atomic nucleus was actually made of, it was necessary

---

10. Correspondence principle: The principle that physical laws that govern a phenomenon occurring at the atomic level which cannot be explained by classical physics must correspond to classical physics at a certain limit. For example, the energy of light is discontinuous in quantum theory, but in the limit of a large number of photons, the energy must fully reduce to classical physics and be seen as continuous.

James Chadwick (1891–1974)
After studying at Manchester and Cambridge, Chadwick conducted collaborative research with Hans Geiger at the Berlin Institute of Technology. He returned to England in 1919, and joined the Cavendish Laboratory together with Rutherford. In collaboration with Rutherford, Chadwick conducted experiments in which various atoms were smashed with α-particles. In 1932, he showed that the unknown particle emitted during these experiments was a neutron, and was awarded the Nobel Prize in Physics in 1935.

to smash it and examine what was produced. In the period from 1920 to 1932, scientists irradiated various atoms with particle radiation and studied what happened. Rutherford bombarded nitrogen nuclei with α-particles and thought that this produced hydrogen atoms or atoms with a mass of 2. While this meant that nitrogen atoms were destroyed by the impact of α-particles, it did not contradict the thinking at the time that "an atom is composed of protons and electrons."

In 1930, Bothe in Germany discovered that when beryllium was irradiated with α-particles, a new type of radiation was produced. In January 1932, Frederic Joliot and Irene Curie discovered that this particular radiation had no electric charge and could expel protons from paraffin. Bothe and the Joliot-Curies thought that this radiation was a type of γ-ray, but James Chadwick thought that an α-particle had expelled a neutral particle from the beryllium nucleus, and that this particle in turn expelled a proton from the paraffin. Chadwick conducted experiments using boron to verify the presence of a neutral particle, and found that its mass was slightly larger than that of a proton.[36]

Researchers at Cavendish Laboratory in the 1920's had searched on several occasions for a neutral particle in which one proton was tightly bound to one electron, but when Chadwick heard the news of the Joliot-Curies' research, he quickly concluded that the neutron existed. However, Chadwick at first also did not think that the neutron was an elementary particle, and thought that it was a proton-electron complex. However, the spectroscopist, Herzberg, (see page 262) pointed out that the spectrum of a diatomic molecule such as N_2 differs depending on whether the number of elementary particles comprising the atomic nucleus is odd or even, and that the measured spectrum cannot be explained by the notion that a neutron is a proton-electron complex. Experimental results accumulated until about 1934, when it was finally established that the neutron was an elementary particle. Thus, the fact that an atomic nucleus was composed of protons and neutrons was settled, and the number of protons and neutrons came to be determined by the following rules.

Number of Protons = Atomic Number

Number of Neutrons = Atomic Weight – Atomic Number

Chadwick was awarded the Nobel Prize for Physics in 1934.

3.7 The Emergence of Quantum Mechanics and Chemistry

Quantum mechanics emerged in 1925 and could satisfactorily explain the world of atoms and molecules. The emergence of quantum mechanics had an impact on both physics and chemistry. Quantum mechanics allowed many experimental facts concerning chemistry to be reasonably explained for the first time. This section briefly summarizes the development of quantum mechanics, focusing on its connection with chemistry. Although many books have been written concerning the history of the development of quantum mechanics, let me mention here three books.[2, 8, 37]

Heisenberg's Matrix Mechanics[38]
Observational results relating to the atom, such as spectra, could be increasingly explained by Bohr's "correspondence principle," but it was clear that Bohr's theory was incomplete, and that a new theoretical construction was needed. A young German physicist, Werner Heisenberg, proposed the new quantum mechanics in the form of matrix mechanics in 1925. Heisenberg expressed observable physical quantities, such as the frequency of an energy transition, in a matrix form that put indices corresponding to the initial state and the final state of a system into rows and columns. Then, he developed the matrix algebra required to describe the properties of an atom. Heisenberg, Born, and Jordan further developed matrix mechanics. Thus, the characteristics of the atom could be obtained by determining the eigenvalues and eigensolutions for a first-order homogeneous equation. Matrix mechanics was applied to the hydrogen atom problem, and its validity was demonstrated, but complex calculations were required to apply it to most other problems.

De Broglie Waves[39]
There was a line of thinking that was separate from the stream exemplified by Bohr's theory and Heisenberg matrix mechanics. From 1923 to 1924, Louis de Broglie of France proposed a bold idea in opposition to common sense: that particles such as electrons have a wave nature. From analogy with the quantum

Werner Heisenberg (1901–1976)

Heisenberg was born in Würzburg in southern Germany. He studied with Sommerfeld in Munich and Born at Göttingen. He went abroad to study under Bohr, then conducted research under Born, and in 1925 developed matrix mechanics. In 1927 he also discovered the uncertainty principle and made major contributions to the establishment of quantum mechanics. Heisenberg also played a leading role in the development of quantum mechanics, such as predicting the existence of para- and ortho-hydrogen, elucidating the fundamentals of ferromagnetism, and establishing quantum field theory. He held various posts in succession, such as professor at Leipzig University (1927–1941) and director of the Kaiser Wilhelm Institute for Physics (1942–1945). Heisenberg also did a great deal of philosophical writing. He was awarded the 1932 Nobel Prize for Physics.

Louis de Broglie (1892–1987)

de Broglie was born into a French noble family. He studied history and philosophy at the Sorbonne with the aim of becoming a diplomat, but he gradually became interested in physics. Impressed by the minutes of the 1911 Solvay Conference, in which quantum theory was debated, he shifted his studies to physics. Although his research was delayed because of World War I, in 1923 he submitted for his dissertation the concept of a matter wave, in which a wave nature is associated with particles at the atomic level. He was awarded the 1929 Nobel Prize for Physics.

theory of light and the photon hypothesis, he considered the existence of matter waves associated with freely moving particles. Using relativistic arguments, he showed that the wavelength is given by $\lambda = h/mv$. Here, v and m are the particle speed and mass, respectively, and h is Planck's constant. Thus, the particle-wave duality of matter suggested that not only do electromagnetic waves have a particle nature, but particles such as electrons have a wave nature.

De Broglie proposed a theory to explain the interference and diffraction of an electron beam that could be verified experimentally. He attempted to verify the predictions suggested by the theory using data from the diffraction of an electron beam within a crystal, but did not succeed. In 1927, the diffraction of an electron beam was observed by Davisson and Germer at Bell Laboratories in the US and G.P. Thomson (son of J.J. Thomson) in Aberdeen, Scotland,

Erwin Schrödinger (1887–1961)
Schrödinger studied at the University of Vienna, was a
professor at the Universities of Zurich, Berlin, and Graz,
and from 1940 to 1955 was director of the Institute for
Advanced Studies in Dublin. From 1956 onward, he was a
professor at the University of Vienna. In addition to being
one of the founders of quantum mechanics that established
wave mechanics, he also left behind a great deal of work in

a broad range of fields in theoretical physics.[41] In addition, in 1944 he wrote
a book entitled "What is Life?" in which discussed the nature of life from the
viewpoint of a physicist.[42] He was a European man of culture and had a deep
knowledge of both history and philosophy.[43] He was awarded the1933 Nobel
Prize for Physics.

Max Born (1882–1970)
Born was a German theoretical physicist and became a
professor at the University of Göttingen in 1921. He was
initially involved in research on the theory of relativity,
thermodynamics, and solid state properties, but around
1925 made major contributions to establishing the new field
of quantum mechanics together with Heisenberg and other
young researchers. He advanced a statistical interpretation

of Schrödinger's wave function, which became the standard interpretation.
In addition, Born carried out research on several topics, including the Born
approximation for addressing the particle scattering problem. In 1933, he
was driven out of Germany by the Nazis, emigrated to Britain and became a
professor at the University of Edinburgh, but returned to Germany in 1953.
He was awarded the 1954 Nobel Prize for Physics.

and the de Broglie proposition was verified. In the following year, Masashi
Kikuchi at RIKEN in Japan also succeeded to observe the diffraction of an
electron beam.

The Nobel Prize for Physics was awarded to de Broglie in 1929 and to
Davisson and Germer in 1937.

The Schrödinger Wave Equation and the Hydrogen Atom[40]
De Broglie's concept was developed into a wave equation by the Austrian
physicist, Erwin Schrödinger. He applied Hamilton's variational principle to
the de Broglie wave associated with an electron to derive the following wave

Paul Dirac (1902–1984)
Dirac studied electrical engineering at the University of Bristol, but moved to Cambridge University in 1925 and shifted his field of study to theoretical physics. In 1926, he carried out research under Bohr in Copenhagen, published a theory that integrated matrix mechanics and wave mechanics, and created a quantum theory for radiation fields. In 1928, Dirac proposed relativistic quantum mechanics, and suggested the existence of a positive electron. From 1932 to 1969, he was a professor at Cambridge University, and in 1933 was awarded the Nobel Prize for Physics.

equation, familiar to physicists that deal with sound waves and electromagnetic waves.

$$\nabla^2 \psi(x, y, z) + (8\pi^2 m_e/h^2)\{E - V(x, y, z)\}\psi(x, y, z) = 0$$

Here, ∇^2 is the Laplace operator, m_e is the mass of an electron, E is the total energy of an electron, V is the potential energy of an electron, and Ψ is the amplitude of the electron wave. The eigenvalue, E, obtained by solving this equation is the possible energy that an electron may have in the steady state. The time-dependent wave function, $\Psi(x, y, z, t)$, is obtained by multiplying Ψ by $\exp(-2\pi i \nu t)$. By analogy to electromagnetic waves, Ψ^2 is taken to be an index expressing the electron density. However, Max Born argued that Ψ^2 expressed the probability of the existence of an electron in a certain position. Argument among physicists regarding this interpretation of the wave function continued, but the Born interpretation gradually came to be widely accepted.

Matrix mechanics, established by Heisenberg, Born and Jordan, and Schrödinger's wave equation, were developed by completely different methods, but it was soon recognized that both were mathematically the same. In 1927, Paul Dirac published a mathematical form of quantum mechanics that encompassed both of these in an elegant form. The Nobel Prize for Physics was awarded in 1932 to Heisenberg, the founder of quantum mechanics, and to Schrödinger and Dirac in 1933. Max Born, who made major contributions to the development of matrix mechanics, was not awarded a Nobel Prize at this time, but in 1954 was awarded the prize for "Contributions through the statistical interpretation of quantum mechanics." Born was a leading teacher during the developmental period of quantum mechanics, and many leading theoretical physicists studied under him, including six Nobel laureates.

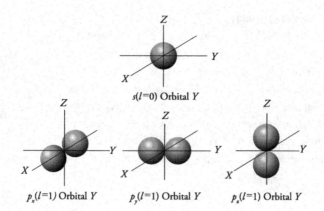

s(l=0) Orbital *Y*

pₓ(l=1) Orbital *Y*   *p_y(l=1)* Orbital *Y*   *p_z(l=1)* Orbital *Y*

Figure 3.7: Diagrams representing the angular dependence of the wave functions for a hydrogen atom

The top illustration shows the s orbital (l=0) and the three bottom illustrations show the p orbitals (l=1)

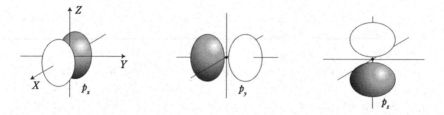

Figure 3.8: Shape of the 2p orbitals

Shape of the 2p orbitals represented by a boundary surface showing the region with a 90% probability of containing an electron. The lightly shaded and darkly shaded areas indicate different signs for the wave function.

Schrödinger's equation is generally used when applying quantum mechanics to chemistry. Schrödinger's equation for hydrogen-like atoms is solved by taking Ze to be the charge on the atom (for hydrogen, Z=1) and r to be the distance of an electron from the nucleus, and setting $V = -Ze^2/r$. If Ψ is expressed as a function in polar coordinates r, θ, and Φ, Ψ is given as the product of the function $R(r)$ of just r and the function $Y(\theta,\varphi)$ of θ and Φ. By using the reduced mass μ,[11] the allowable energy is given by $E_n = -4\pi^2\mu e^2 Z/h^2 n^2$, and is determined by the

---

11. Reduced mass: The reduced mass μ of a system composed of masses m_1 and m_2 is given by $1/\mu = 1/m_1 + 1/m_2$.

principal quantum number n. This energy is the same as the result obtained by the Bohr model. $Y(\theta,\varphi)$ gives the angular dependence of the wave function by a function defined by the two quantum numbers l and m. These wave functions provide the orbital for the hydrogen atom, yielding $Z=1$ (Fig. 3.7). The orbitals for $l = 0, 1, 2, 3$, respectively, are the s, p, d, f orbitals; these orbitals are provided as expressions of the electron probability function using the values of Ψ^2. Thus, the orbitals for the hydrogen atom could be visually represented (Fig. 3.8).

Many-electron Systems and Approximate Solutions

Schrödinger's equation could be solved for the hydrogen atom, but not for many-electron systems in the periodic table (all elements from helium onwards). Attempts to find approximate solutions to Schrödinger's equation for many-electron systems soon began. In 1927, Hartree introduced an approximation to express the wave function, Ψ, of a many-electron system as the product of the wave function $\Psi_i(x_i, y_i, z_i)$ with respect to electron i. Treating Ψ as a trial function, an equation for Ψ_i is introduced using the variational principle, and the field of force with respect to individual electrons is expressed as an average field created by the other electrons. This equation was solved using a self-consistent field.[12] However, this function does not satisfy the Pauli principle that Ψ is anti-symmetric with respect to the exchange of electrons. Consequently, the Hartree-Fock method was devised to address this. With this method, the wave function for all systems, Ψ, is represented approximately by using the determinant $|\Psi_i(x_i, y_i, z_i, \sigma_i)|$ introduced in 1929 by Slater (the Slater determinant), and each Ψ_i is determined by the variational principle. The Slater determinant, represented in the form of a product of Ψ_i, is anti-symmetric with respect to the exchange of electrons. Here, $\Psi_i(x_i, y_i, z_i, \sigma_i)$ is the product of the functions for the spatial coordinates x_i, y_i, z_i of the electron and the spin coordinate functions. The Hartree-Fock method was devised to describe the properties of many-electron atoms, but later it became an important method for handling the electronic state of molecules. (see page 472)

The Heisenberg's Uncertainty Principle[44]

If an electron is represented by a wave equation, what can we know about its precise position? Heisenberg considered the following problem. Extremely short-

---

12. Self-consistent field: The force potential exerted on a particle i by particles other than i is replaced by an average potential V_i created by an electron cloud determined by the wave function of each particle; then, the Schrödinger equation is solved to obtain new wave functions. Using the obtained wave functions, the new potential Vi' is calculated. This operation is repeated until $Vi=Vi'$ is obtained. The Vi thus obtained is called a self-consistent field.

wavelength light (high-energy light) must be used for determining the position of an electron precisely. However, when a high-energy photon collides with an electron, the collision alters the velocity of the electron. If long-wavelength light is used, the collision has little effect on the velocity of the electron, but the position of the electron cannot be precisely determined. Consequently, the uncertainty of the momentum, Δp, is related to the uncertainty in the measurement of the position, Δq. Such considerations led Heisenberg to derive the so-called Heisenberg "uncertainty principle".

$$\Delta p \cdot \Delta q \gtrsim h/2\pi$$

This relational expression for uncertainty is expressed by energy and time. When the intervals for the uncertainty of energy are ΔE and of time are Δt, the relationship $\Delta E \cdot \Delta t \geq h/4\pi$ holds. This shows that to determine the energy with an uncertainty smaller than ΔE, a measurement time longer than $h/4\pi\Delta E$ is needed, and is directly related to the spectral line width observed by spectroscopy.

Around 1927, heated discussions between physicists continued regarding the uncertainty principle, wave nature, particle nature, and other fundamentals of quantum mechanics. Bohr subsequently proposed a viewpoint that came to be called the "Copenhagen interpretation" to describe a phenomenon at the atomic level. Physicists must choose either to take a particle-like viewpoint in discontinuous physics or a wave-like viewpoint in continuous physics. The Heisenberg uncertainty principle implicitly states that the experimentalist's choice inevitably affects the behavior of the system. Bohr's interpretation of quantum mechanics was accepted by most theoretical physicists, but Schrödinger, Einstein and others did not accept it; the controversy continues to the present day. The results of calculations based on quantum mechanics can describe phenomena observed at the atomic level, but doubts remain regarding the basis of quantum mechanics, as indicated by Feynman's statement that "Nobody understands quantum mechanics."[45]

Quantum Mechanics and Chemistry

Quantum mechanics provided chemists with a basic principle for explaining chemical phenomena. In 1929, Paul Dirac, one of the pioneers of quantum mechanics, wrote an introduction to a paper called "Quantum Mechanics of Many-Electron Systems" which include the following:

> The underlying physical laws necessary for the mathematical theory of a large part of physics and the whole of chemistry are thus completely known, and the

difficulty is only that the exact application of these laws leads to equations much too complicated to be soluble.[46]

However, chemistry has not become applied mathematics. Even though classical mechanics was established by Newton and classical electromagnetics by Maxwell, countless physics problems remain to be solved. Although the fundamental laws for understanding atoms and molecules are known, complex chemical phenomena remain poorly understood. If anything, the emergence of quantum mechanics was a new point of departure for chemistry. As will be described in detail in a later chapter, with the emergence of quantum mechanics, the nature of chemical bonds came to be understood for the first time, and the application of quantum mechanics made it possible to understand various chemical phenomena. Quantum chemistry is now a branch of chemistry, and theoretical chemistry, including the application of statistical mechanics, is an important area of chemistry.

Problems in chemistry involve many-electron systems. Their solutions must be approximate because the Schrödinger equation cannot be solved exactly. Consequently, the motion of atomic systems and of electron systems are discussed separately by the Born-Oppenheimer approximation.[47] The mass of an atomic nucleus is large compared to that of an electron, and the movement of a nucleus is extremely slow compared to that of an electron. Therefore, in the approximation, the atomic nucleus is first held fixed, and the energy E of the electron system is determined. E is a function of the position of the atomic nucleus. The point at which E becomes small determines the structure of the molecule, and vibrations in the vicinity of this point are defined as molecular vibrations. The overall rotation of a molecule is handled in accordance with rigid rotation. The movement of electrons, the movement of a molecule, and the rotation of a molecule are determined separately by the Born-Oppenheimer approximation, and the energy of each movement is determined approximately. The foundation is thus established for research into the structure of molecules and the electronic states using spectroscopy (see pages 260–264).

References

1. S. Weinberg, "*The Discoveries of Subatomic Particles* " Cambridge University Press, 2003
2. S. Tomonaga "*Ryousi rikigaku*" (*Quantum Mechanics* (1)) Misuzu Shobo, 1952
3. J, Perrin "*Les Atomes*" Librairie Felix Alcan, Paris, 2nd Ed. 1921

4. A. J. Ihde "*The Development of Modern Chemistry*" Dover Publications, New York, 1984

5. J. Gribbin, "*Science A History*" Penguin Books, 2002

6. M. J. Nye "*Before Big Science*" Harvard Univ. Press, 1996

7. M. J. Nye "*From Chemical Philosophy to Theoretical Chemistry*" Univ. California Press, Berkeley, 1993

8. A. Pais "*Niels Bohr's Time*" Clarendon Press, Oxford, 1991

9. W. J. Moore "*Physical Chemistry*"4th Ed. Prentice-Hall, 1972: many historical episodes are given in this text book.

10. J. J. Thomson, *Phil. Mag.*,**44**, 295 (1897)

11. W. Friedrich, P. Knipping, M. Laue, *Ann. Physik.* **4**, 971 (1912)

12. W. L. Bragg, *Proc. Cambridge Phil. Soc.*, **12**(1), 43 (1912)

13. W. H. Bragg, W. L. Bragg, *Proc. Roy. Soc.* **4**, 88 (1913)

14. W. L. Bragg, *Proc. Roy. Soc., A*, **89**, 248 (1913)

15. H. G. Mosely, *Phil. Mag. Series* **6**, 26, 257 (1913)

16. H. G. Mosely, *Phil. Mag. Series* **6**, 27, 703 (1914)

17. F. Dyson, "*From Eros to Gaia*" p. 1151–154

18. J. M. Thomas, D. Phillips ed. "*The Legacy of Sir Lawrence Bragg*" Science Reviews, p.111.

19. R. Reid, "*Marie Curie*" Saturday Review Press/Dutton, New York, 1974

20. E. Rutherford and F. Soddy, *Phil. Mag.* Series 6, **4**, 561 (1903)

21. J. J. Thomson, *Proc. Royal Soc.*, A **89**, 1 (1913)

22. A. Einstein, *Ann. Phys.*, **17**, 132 (1905)

23. A. Einstein, *Ann. Phys.*, **17**, 549 (1905)

24. A. Einstein, *Ann. Phys.*, **17**, 891 (1905)

25. A. Einstein, *Ann. Phys.*, **18**, 639 (1905)

26. M. Planck, *Ann. Phys.*, **4**, 553 (1901)

27. A. Einstein, *Phys. Z.*, 18, 121 (1917)

28. E. Rutherford, *Phil. Mag. Ser.* 6, **21**, 669 (1911)

29. H. Geiger, E. Marsden, *Phil. Mag. Ser.* 6, 25, 604 (1913)

30. N. Bohr, *Phil. Mag.Ser.* 6, **26**, 1 (1913)

31. Reference 8 gives detailed discussions on Bohr's work and its influences.

32. A. Sommerfeld, "*Atombau und Spektrallinieen*" Friedrich Vieweg und Sohn, Braunschweig, 1919

33. N. Bohr, *Z. Phys.*, *9*, 1(1922)

34. N. Bohr, *Nature*, **112** supplement, 30 (1923)

35. W. Pauli, *Z. Phys.*, **31**, 765 (1925)

36. J. Chadwick, *Proc. Roy. Soc.*,A **136** 692 (1932)

37. T. Takabayashi "*Ryoushiron no Hattenshi*" *(Historical Development of Quantum theory)* Chikuma Gakugei Bunko, 2010

38. W. Heisenberg, *Z. Phys.*, **33**, 879 (1925)
39. L. de Broglie, Thesis (Paris) 1924, *Ann. Phys.*(Paris), **3**, 22 (1925)
40. E. Schrödinger, *Ann. der Phys.*, **79**, 361, 489; **80**, 437; **81**, 109 (1926)
41. W. Moore, "*A life of Erwin Schrödinger*" Cambridge, 1994
42. E. Schrödinger "*What is life*" Cambridge Univ. Press, 1944,
43. E. Schrödinger, "*Nature and the Greeks and Science and Humanism*" Cambridge, 1951
44. W. Heisenberg, Z. Phys., **43**, 172 (1927)
45. R. P. Feynman, "*The Character of Physical Law*" MIT Press, 1967,
46. P. A. M. Dirac, *Proc. Royal. Soc.*, A **123**, 714 (1929)
47. M. Born, R. Oppenheimer, *Ann. Physik.*, **84**, 457 (1927)

Chapter 4: Chemistry in the first half of the 20th century

The maturation and expansion of the science of atoms and molecules

Pauling, with healthy and sickle cell red blood cells in the background, depicted on a stamp

The three major discoveries in physics that occurred at the end of the 19th century—X-rays, radioactivity and the electron—and the subsequent development of quantum theory, greatly transformed 20th century chemistry. Furthermore, the scientific knowledge required to understand and manipulate atoms and molecules expanded enormously and had a huge impact on the way society perceived chemistry. In addition, the methods and approaches to chemistry research greatly changed after World War II. For this reason, this book is divided into two sections, with 1945 as the dividing line. Developments in the first half of the 20th century are summarized in this chapter. The primary documents used in writing this chapter, references [1] to [23], are given at the end of the chapter.

At the start of the 20th century, Germany was the center of chemistry research and the chemical industry. The establishment of chemistry education in universities in the latter half of the 19th century (see Section 2.10) made Germany the leader in all fields of chemistry until the start of World War I. However, excellent research was also carried out in various European countries such as Britain, France, Switzerland, and Sweden, as we saw in Chapter 2. At the start of the 20th century, very high quality research was also emerging from a young country, the United States. Germany lost its overwhelming dominance with its defeat in World War I. Despite this, the Kaiser Wilhelm Institute and similar institutions were established, and German chemistry continued at a high level up to the early1930s. However, as the Nazi regime began to expel Jewish scientists, German chemistry rapidly declined, and in its place, the U.S. gained the upper hand. In Japan, too, which was late to introduce modern chemistry

Henri Poincare (1854–1912)
Henri Poincare was a French mathematician and scientific philosopher. He left behind outstanding work in a broad range of areas in both pure and applied mathematics. Poincare is mainly known as an outstanding scientific thinker and the author of books such as *"Science and Hypothesis," "The Value of Science,"* and *"Science and Method."*

into university curricula, a university system for chemistry education was put in place in the first half of the 20th century, progress was made in the training of chemists and chemical technologists, and the Institute of Physical and Chemical Research (RIKEN) was established; together, these prepared the foundation for world-class research in Japan.

Compared to other areas of the natural sciences, many researchers in chemistry worked in industry. Thus chemists tended to place more stress on the practical value of science compared to researchers in other fields. However, chemists in the academic world were primarily driven by their desire to understand nature; finding practical applications of their discoveries was secondary. At the start of the 20th century, the French mathematician and physicist Henri Poincare placed the highest value on the search for truth, and advocated "science for science".[24] Even in 1949, the biochemist Meyerhof wrote the following.[16]

Biochemistry has an important relationship to the advancement of medical science. Consequently, however, biochemistry must remain a pure science. The impetus to start research in the pure sciences is inspired by a longing for knowledge and is not inspired by anything beyond that.

However, science in the 20th century also provided the foundation for advances in industry, and was strongly connected to national development and warfare. The practical value of science was firmly recognized. Above all, chemistry played a decisive role in World War I.

4.1 Characteristics of Chemistry in the First Half of the 20th Century

Linus Pauling was one of the greatest chemists of the 20th century and was active in wide-ranging fields, from inorganic chemistry to the chemistry of life. He also made important contributions to structural chemistry. In 1951, Pauling summarized the progress of chemistry in the first half of the 20th century as follows.[16]

Linus Pauling (1901–1994)

Linus Pauling was born in a rural area of Oregon and was raised in poverty. He worked his way through school, and after graduating from Oregon Agricultural College, he obtained a Ph.D. from the California Institute of Technology in 1925 for research in structural analysis using X-ray diffraction. Pauling went to Europe in 1926 to study quantum mechanics in Sommerfeld's laboratory. After returning to the US, he became an assistant professor at the California Institute of Technology. Pauling made important discoveries by applying quantum mechanics to problems pertaining to chemical bonds, and he was the world authority on the theory of chemical bonds and structural chemistry. In the mid-1930s, he developed an interest in the application of structural chemistry to biochemistry problems, and conducted pioneering research on topics such as the structure of proteins, the generation of antibodies, antigen-antibody reactions, and molecular genetic diseases. In 1951, Pauling suggested an α-helix structure for proteins, thereby making a major contribution to protein structural chemistry. Pauling was awarded the Nobel Prize for Chemistry in 1954 for "Research pertaining to the nature of chemical bonds and the structure of complex molecules." In his later years, he argued for the benefits of Vitamin C against the common cold and cancer, but this view was not favorably received by many experts. After World War II, he opposed nuclear testing and strove to advance the peace movement, and in 1963, Pauling was awarded the Nobel Peace Prize. In addition to being an outstanding chemist, Pauling was passionate about his causes and boldly pioneered new fields without being afraid of making mistakes.[25]

The half-century we are just completing has seen the evolution of chemistry from a vast but largely formless body of empirical knowledge into a coordinated science. This transformation resulted mainly from the development of atomic physics. After the discovery of the electron and of the atomic nucleus, physicists made rapid progress in obtaining a detailed understanding of the electronic structure of atoms and simple molecules, culminating in the discovery of quantum mechanics. The new ideas about electrons and atomic nuclei were soon introduced into chemistry, leading to the formulation of a powerful structural theory which has welded most of the great mass of chemical facts into a unified system. At the same time great steps forward have been taken through the application of new physical techniques to chemical problems, and also through the continued effective use of the techniques of chemistry itself.

With these words Pauling aptly summarized the progress of chemistry in the first half of the 20th century. Chemistry changed from a science based mainly on empirical knowledge and developed to a logical, exact science.

While the atom played the leading role in 19th century chemistry, the electron was discovered at the end of the century, and it became clear that the electron would play the leading role in many chemical phenomena. The understanding of chemical phenomena based on the behavior of electrons was a major focus of 20th century chemistry. Thus, chemistry and physics were contiguous, and fundamentally they came to be understood based on the same principles. This same trend extended to biology and in the second half of the 20th century much of biology came to be understood at the molecular level. On the one hand, science became specialized and fragmented in the 20th century, while on the other hand, a common basis for understanding matter and life was achieved using the concepts of the electron, atom and molecule. In addition, chemistry began to have a major impact on other sciences, including earth and space science.

Remarkable technological developments were made in chemical industries. Industries that were developed in the 19th century, such as the dye synthesis and alkali industries, laid the foundations for chemical industries, and these industries became much larger in the early 20th century. For example, immediately before World War I, Haber and Bosch developed a method for synthesizing ammonia from atmospheric nitrogen, allowing the production of large quantities of artificial fertilizer. In other ways as well, the applications of chemistry began to have a major impact on society. For example, the polymer chemistry industry was established in the 1920s, and the resulting synthetic fibers and plastics became a part of our lives. Also, chemotherapy made advances in the medical field, and the pharmaceutical industry flourished. However, the poison gases developed by chemists and used in World War I drew attention to the misuse of chemistry.

As seen in Chapter 2, chemistry had become differentiated into specialized fields by the latter half of the 19th century: physical chemistry, inorganic chemistry, analytical chemistry, and organic chemistry, with specialists in each of these fields. Biochemistry was added to this list in the first half of the 20th century, as well as nuclear chemistry and radiochemistry. In this chapter, we will track the progress in each of these fields during the first half of the 20th century. However, as suggested by Pauling's statement given at the beginning of this section, new developments in chemistry became possible by assimilating the results of the physics revolution discussed in the previous chapter, and the development of biochemistry was closely related to progress in biology and medicine. Clearly, collaborations between fields beyond the traditional borders of chemistry were a major driving force behind new developments in chemistry.

4.2 Physical Chemistry (I): Chemical Thermodynamics and Solution Chemistry

The effort to systematically understand chemical phenomena through thermo-dynamics, begun in the 19th century, was largely complete by about 1920 and was formalized in the field of chemical thermodynamics. Tables of thermody-namic functions for chemical substances were created, allowing prediction of whether or not a given chemical reaction was thermodynamically possible. The fundamentals of chemical equilibrium were understood; if the system was at thermodynamic equilibrium, classical physical chemistry could largely describe the macroscopic phenomena of the system. The physical chemistry of solutions, which had been an important field in early physical chemistry research, again became an active field of investigation, and advances were made in understanding electrolyte solutions and acid-base theory. These advances provided a strong foundation for analytical chemistry.

Developments in structural chemistry and theoretical chemistry stand out the most among the advances in physical chemistry in the first half of the 20th century. Understanding chemical bonds was a problem for many years, but following the proposal of the covalent bond concept by Lewis and Langmuir, the nature of chemical bonding was elucidated in parallel with the emergence of quantum mechanics. Quantum mechanics was promptly applied to chemistry and gave rise to the new field of quantum chemistry, and theoretical chemistry, including statistical mechanics, quickly became an important field.

Major advances in structural chemistry were largely the result of dramatic developments in experimental techniques. First, X-ray and electron-beam diffraction provided detailed information on the structure of molecules and crystals of simple organic and inorganic compounds. In addition, spectroscopy advanced and was used to investigate the absorption, emission, and scattering of infrared, visible, and ultraviolet light by molecules, leading to a wealth of information on the structure of molecules and the state of electrons within molecules.

Understanding chemical reaction rates and mechanisms, including details of the course of a reaction at the molecular level and predicting the rate of reaction, remained extremely difficult problems, even though empirical chemical reaction theories and reaction mechanisms for simple gas-phase reactions were being discussed in the 1910s. Transition state theory was proposed in the 1930s, providing a first step toward non-empirical chemical-reaction theory, but an understanding of reaction rates and mechanisms remained a fairly distant goal.

Complex systems of colloids, interfaces, and surfaces became important topics in physical chemistry and quantitative studies advanced. Colloid chemistry also played a major role in the development of biochemistry and polymer chemistry. The emergence of quantum mechanics opened a path to understanding the electrical and magnetic properties of substances at a microscopic level, but chemistry had not yet reached the stage where it was possible to investigate these properties based on quantum mechanics or statistical mechanics. This field developed as a branch of physics known as solid-state physics, and chemists made relatively few contributions.

Germany was the center of physical chemistry in the first twenty years of the 20th century, and Ostwald, Nernst, Haber, and others were active in this field. However, several universities in the US started to graduate excellent researchers, such as G.N. Lewis at the University of California. In addition, Langmuir proved that excellent research could be conducted in corporate laboratories. Quantum mechanics was developed in Germany after World War I, and soon afterwards it was introduced to chemistry by American chemists. Young chemists such as Pauling, Urey, and Mulliken were enterprising, studied the new physics in Europe, and incorporated their findings to develop the new field of physical chemistry. In addition, physicists such as Debye, Heitler, London and Slater successfully addressed problems in chemistry. Consequently, chemical physics arose as a new interdisciplinary field between physics and chemistry. At the end of the 1920s, the center of physical chemistry had already started to move from Germany to the US, and this trend became decisive with the Nazis coming to power and expelling Jewish scholars.

Completion of Chemical Thermodynamics

Various properties of gases, liquids, solutions and solids were studied using the concept of free energy introduced by Helmholtz and Gibbs. The chemical potential $\mu_i = (\partial G/\partial n_i)_{T,p,nj}$ is used to discuss the equilibrium of multicomponent systems such as solutions and alloys. The chemical potential is obtained by partial differentiation of the Gibbs free energy G with respect to the number of moles n_i of component i under constant temperature T, pressure p, and other components. The chemical potential explains colligative properties,[1] phase equilibria, chemical equilibria, and other phenomena such as freezing point depression, boiling point elevation, and osmotic pressure. Additionally, thermodynamic data provide a path to predicting chemical equilibria. In 1901, G.N. Lewis introduced the concept of fugacity[2] in place of pressure to deal with real gases. In 1907,

---

1. Colligative property: Property of dilute solutions determined by the number of solute molecules, not the type of solute molecules. Typical examples of colligative properties are freezing point depression and boiling point elevation.

2. Fugacity: A quantity introduced in place of pressure to thermodynamically treat real gases.

Gilbert Newton (G.N.) Lewis (1875–1946)

G.N. Lewis was born the son of a Boston lawyer, and obtained his Ph.D. under Richards at Harvard. After studying abroad for a year under Ostwald and Nernst, Lewis returned to the US as an instructor at Harvard. Richard's German-like authoritarianism and research policies stressing empiricism caused Lewis to resign from Harvard, and he worked for a year as Superintendent of Weights and Measures for the Bureau of Science in Manila.

Lewis returned to the US, and conducted research on chemical thermodynamics with Noyes' group at the Massachusetts Institute of Technology (MIT), focusing on the systematic study of electrode potentials. In 1912, he was promoted to head of the chemistry department at the University of California, Berkeley. He carried out a large amount of outstanding research until his death in 1946, and helped train many talented scientists. He stressed fundamental principles, and encouraged students to think independently and mature as scientists through open discussions with other students and faculty. His major achievements spanned fields such as the systematization of chemical thermodynamics, chemical bonding theory, acid-base theory, the study of heavy water, isotope research, spectroscopy, and photochemistry. Many leading chemists were trained in his laboratory, including four Nobel Prize winners.

he further introduced the concept of activity[3] for handling real solutions, thus solidifying the foundation of the thermodynamics of non-ideal systems.

The temperature change of the equilibrium constant (K) is given by an expression relating the equilibrium constant K and the heat of reaction according to the van't Hoff equation: $d\ln K/dt = \Delta H/RT^2$. However, in the equation obtained by integration

$$\ln K = -\Delta H/RT + C$$

C is an unknown quantity, and without knowing C, an absolute value for the equilibrium constant cannot be determined.

To solve this problem, Walter Nernst proposed the so-called "Nernst's heat theorem" in 1905.[26] This theory stated that "the change in entropy that

---

3. Activity: A quantity introduced in place of concentration to thermodynamically treat real solutions.

Walther Nernst (1864–1941)

Walther Nernst was born the son of a country judge in Briesen, West Prussia, now a region in Poland. After studying physics and mathematics at the Universities of Zurich, Berlin, Graz, and Würzburg, he began physical chemistry research in Ostwald's laboratory. Here, he focused on electrochemistry and solution chemistry, and discovered a relationship between electromotive force and the change in free energy of a system, leading to the famous Nernst equation. From 1891, he was active as the director of the Institute of Physical Chemistry and Electrochemistry at the University of Göttingen. In 1905, he moved to the University of Berlin, and in 1925 became the director of the Institute of Physical Chemistry. In Berlin, he investigated topics such as the specific heat of solids at low temperature, the density of gases at high temperature, and photo-catalyzed gaseous chain reactions. He also conducted practical research in areas of particular interest to him, such as improving the light bulb. Alongside van't Hoff and Ostwald, he was a leader in physical chemistry in the early 20th century, and was awarded the Nobel Prize for Chemistry in 1920 for "contributions to thermodynamics."

accompanies a physical or chemical change approaches zero as the temperature approaches 0 K." Although experimental evidence was lacking when the proposal was made, Nerst afterwards attempted to obtain experimental data to support this theory, while Einstein showed theoretically that the specific heat of a solid becomes zero at 0 K. Furthermore, Planck expressed in 1910 that "the entropy of a perfect crystal at an absolute temperature of $0°$ is zero," and this generally came to be called the third law of thermodynamics. This is the result predicted from the definition of entropy based on statistical mechanics. According to this, ΔG and ΔH have the same value at 0 K, allowing equilibrium constants to be determined using thermodynamic data.

In the first twenty years of the 20th century, many chemists attempted to collect thermodynamic data. The equilibrium constants obtained from these data were very useful for the chemical industry, and particularly for gaseous reactions, for example, the synthesis of ammonia from nitrogen and hydrogen by the process introduced by Fritz Haber. Haber initially conducted research in electrochemistry, but in 1905 published the book "*Thermodynamik technischer Gasreactionen (The Thermodynamics of Technical Gas Reactions)*", systematizing thermodynamic data so that the yields of industrially important gaseous reactions could be calculated over a wide temperature range.[27] Based on these data, Haber determined a practical method for synthesizing ammonia and, with the support of the BASF Company (Badische Anilin und Soda Fabrik), industrialized the process (ammonia synthesis is described in detail in Section 4.11).

Industrially useful information was also obtained on the stability of hydrocarbons. Calculations based on heat-of-formation data indicated that the passage of hydrogen over carbon heated to 500°C under atmospheric pressure would result in 70% conversion to methane at equilibrium. Large hydrocarbons are thermodynamically unstable, but the rate of decomposition to carbon and methane is slow at ambient temperature. When large unstable hydrocarbons are heated at atmospheric pressure, they decompose to coke and methane by cracking[4], but at high pressure, they are cracked to larger-molecular-weight molecules like those present in gasoline. Thermodynamic data are thus useful for optimizing the experimental conditions for generating the desired hydrocarbons.

Thermodynamic properties were actively investigated by many chemists in the first half of the 20th century, with G.N. Lewis making particularly significant contributions. Lewis and Randall published *"Thermodynamics and the Free Energy of Chemical Substances"*[28] in 1923. This book was read around the world and had a major impact on thermodynamics and its applications to chemistry. The group of Lewis, Gibson, Latimer, and coworkers at the University of California carried out wide-ranging thermodynamics research pertaining to the third law.

Data supporting the third law required that the thermodynamic properties of compounds, especially entropy, be obtained by measurements at very low temperatures. Consequently, technology for achieving very low temperatures was extremely important. At the start of the 20th century, Kamerlingh Onnes of the University of Leiden in Holland achieved a temperature just below 1 K by the evaporation of liquid hydrogen, and succeeded in liquefying helium. In 1924, William Giauque, who studied entropy at very low temperatures at the University of California, proposed an adiabatic demagnetization method to achieve very low temperatures, and together with MacDougall, experimentally demonstrated that temperatures down to 10^{-3} K could be achieved. This method utilizes the relationship between the entropy of a paramagnetic substance and magnetic

---

4. Cracking: Producing low-boiling-point light petroleum by breaking down high-boiling-point heavy petroleum.

Heike Kamerlingh Onnes (1853–1926)

Heike Kamerlingh Onnes was a Dutch physicist. He studied at Groningen and Heidelberg, and in 1882 became a professor at the University of Leiden. He founded a cryogenics laboratory at the University of Leiden, and prepared large quantities of liquid air and liquid hydrogen. In 1908, he succeeded in liquefying helium, and in 1911, discovered superconductivity. In 1913, he was awarded the Nobel Prize for Physics.

William Giauque (1895–1982)

William Giauque was born in Canada, worked for two years after graduating from high school in electric power and chemical companies in Niagara Falls, and entered the University of California intending to become a chemical engineer. Under the influence of G.N. Lewis, he developed an interest in the third law of thermodynamics, and began research on entropy under the guidance of Gibson. His abilities were recognized, and after obtaining a Ph.D., he became a lecturer. He succeeded in developing an adiabatic demagnetization method, and became a professor in 1934. In 1949, he was awarded the Nobel Prize for Chemistry for "contributions to chemical thermodynamics, and particularly, research of various properties of substances at very low temperatures."

field strength;[5] after magnetizing the paramagnetic body under isothermal conditions, the magnetic field is removed under adiabatic conditions. This adiabatic demagnetization method was also an extremely important technique in very-low-temperature physics research.

In the 1930s, statistical mechanics, including quantum statistical mechanics, was actively applied to chemistry problems. Quantum statistical mechanics was used to extract information on thermodynamic functions based on molecular structure data; these structural data were obtained from crystallographic and spectrophotometric research. "*Statistical Thermodynamics*"[29] by Fowler and Guggenheim, published in 1939, was widely read as a classic in the field of statistical thermodynamics.

---

5. Entropy and magnetic field strength: Because the magnetization of ions of paramagnetic substances align in the direction of an external magnetic field, the magnetized state is in a state of lower entropy than the non-magnetized state, and the entropy of the system depends on the strength of the magnetic field.

Chemical thermodynamics progressed to the point where thermodynamic function tables became available for various chemical compounds, and it became possible to predict whether a given chemical reaction would occur under certain conditions or whether it was thermodynamically impossible. Nernst and Giauque were awarded the Nobel Prizes for Chemistry in 1920 and 1949, respectively, for contributions to the development of chemical thermodynamics.

Physical Chemistry of Solutions

Research on solutions developed in connection with research on chemical thermodynamics. The thermodynamic properties, colligative properties, solubility, and phase equilibria of non-electrolyte solutions could be discussed in terms of thermodynamics. A solution obeying Raoult's law was defined as an ideal solution (see Section 2.8), and the divergence from ideal behavior was characterized. Lewis introduced the concept of activity in place of concentration to handle non-ideal solutions. The concept of a "regular solution" was introduced in 1929 by Hildebrand to describe a non-ideal solution in which the entropy of mixing is the same as that of an ideal solution, but the heat of mixing is non-zero. Statistical mechanics was applied to liquids and solutions in the 1930's, but was mainly limited to ideal solutions and regular solutions.

Electrolyte solutions were an important early subject in physical chemistry research. Arrhenius' theory of electrolytic dissociation had the strong support of leaders in physical chemistry, such as van't Hoff and Ostwald, but many chemists remained unconvinced, and some were strongly opposed to it. One outstanding question was: why is an ionic electric charge formed? Ionic bonds were not yet known, and there was no satisfactory answer to this question. Arrhenius' theory of electrolytic dissociation suggested that strong electrolytes should be 100% dissociated, but experiments on electrical conductivity showed that 100% dissociation was only achieved when the solution was extremely dilute; consequently, experimental results differed from predictions, and it was thought that the law of mass action did not hold for strong electrolytes. Arrhenius' research had been limited to aqueous solutions, but research on non-aqueous solutions[6] provided an increasing body of incomprehensible experimental results. There were fierce battles of words between supporters of the electrolytic dissociation theory and those who had doubts about it. New advances in the field arose from this dispute.

Louis Kahlenberg from Wisconsin was one of those at the forefront of the opposition faction.[2] He was a physical chemist who had studied under Ostwald

---

6. Non-aqueous solution: A solution using a solvent other than water.

and initially accepted the theory of electrolytic dissociation. However, he became aware of results concerning non-aqueous solutions[6] which could not be explained adequately by the electrolytic dissociation theory, and changed his mind to become a strong opponent of the theory. He conducted a detailed investigation of the relationship between the dissociation of the solute and the dielectric constant of the solvent. Despite the fact that the dielectric constant of hydrogen cyanide (92) was higher than the dielectric constant of water (80), electrolytes did not dissociate in hydrogen cyanide as much as in water. He conducted detailed research on the electrical conductivity and freezing point depression of non-aqueous solutions, and uncovered a large number of unusual phenomena that could not be explained by the theory of electrolytic dissociation; together, this led him to believe that the theory was unsatisfactory. Armstrong in London also opposed the theory of electrolytic dissociation, and attempted to explain experimental results by the formation of hydrates of the solute and water. However, none of these opponents could offer a theory that satisfactorily explained the observed results.

It was clear that the Arrhenius theory of electrolytic dissociation was applicable only to dilute solutions. In 1909, Bjerrum in Denmark studied the interactions between ions from the concentration dependency of the absorption spectra of chromium salts,[30] but his thinking attracted little attention. Arrhenius thought that strong electrolytes were in a molecular state until they dissolved, but in the 1910s, the ionic character of strong electrolyte crystals became clear from Bragg's X-ray diffraction data, and it became apparent that strong electrolytes were already ions prior to dissolving. In the 1920s, it was discovered that molten salts were good electrical conductors. Attention was focused on the problem of interionic and ion-solvent interactions in solutions.

Peter Debye and Erich Hückel thought that the reason why the electrical conductivity of a strong electrolyte solution was inconsistent with 100% dissociation was due to decreased ion mobility[7] caused by interionic interactions. Ions attract an oppositely charged ionic atmosphere[8] around them, hindering the movement of the ions. In 1923, they introduced a statistical mechanics technique to calculate the strength of the ionic atmosphere, and showed that the reduction in mobility was proportional to the square root of the concentration of the ion, and quantitatively predicted the relationship between conductivity,

---

7. Mobility: In a gas, solution or solid, when charged particles such as ions, electrons, or colloids are subject to a force from an electric field E and move at an average speed v, the coefficient μ defined by $v = \mu E$ is called mobility.

8. Ionic atmosphere: The entirety of ions distributed around a single central ion.

Peter Debye (1884–1966)

Born in Maastricht, the Netherlands, Debye studied at the Aachen University of Technology in Germany. Here, he was noticed by Sommerfeld and became Sommerfeld's assistant in 1904. In 1906, he moved with Sommerfeld to the University of Munich. In 1911, he proposed an equation that improved on Einstein's specific heat equation by taking into consideration the elastic vibration of solids. This advancement made him famous. In 1914, Debye became a professor at the University of Göttingen, and carried out ground-breaking research in areas such as polar molecules, the diffraction of X-rays and electron beams by gases, and powder X-ray diffractometry. He moved to the Swiss Federal Institute of Technology in 1920, and in 1927 he became a professor at the University of Leipzig. From 1934 to 1939 he was the director of the Kaiser Wilhelm Institute in Berlin. In 1923, he proposed the Debye-Hückel theory for electrolyte solutions. In 1940, he left Germany and became a professor in the chemistry department at Cornell University in Ithaca, New York, and continued research in areas such as light scattering by polymer molecules. He was awarded the 1936 Nobel Prize for Chemistry.

Erich Hückel (1896–1980)

Erich Hückel studied physics and mathematics at the University of Göttingen. He was an assistant to Debye in Zurich, and developed a theory for electrolyte solutions jointly with Debye in 1923. Subsequently, he became a teacher at the Technische Hochschule in Stuttgart. In 1931, he led the way for applications of quantum chemistry to organic chemistry by describing the π electron system using molecular orbital theory.

Lars Onsager (1903–1976)

Lars Onsager was an American chemist and physicist born in Norway. He became a professor at Yale University in 1945 and derived the Onsager reciprocal relation from studies of the relationship between irreversible phenomena and fluctuations. He is known for expanding on the Debye-Hückel theory, polar liquid theory, providing an exact solution to the Ising model for ferromagnetic materials, and for his theory describing liquid helium. Onsanger was awarded the 1969 Nobel Prize for Chemistry.

ion concentration, and charge number.[31] The predictions of the Debye-Hückel theory were compared to experimental results and shown to approximately describe the properties of strong electrolyte solutions. In 1926, Lars Onsager

Søren Sørensen (1868–1939)
Søren Sørensen was a Danish chemist. He obtained his Ph.D. at the University of Copenhagen, and was director of the chemistry department of Carlsberg Laboratory, Copenhagen. Research on the ion densities of proteins led Sørensen to conclude that the hydrogen concentration is important, and he proposed the concept of pH.

Nicolaus Brøensted (1879–1947)
Nicolaus Brøensted was a Danish physical chemist. In 1923, he proposed describing acids and bases based on the movement of protons, and he suggested a theory of acid-base catalysts in aqueous solutions. Using the Debye-Hückel theory, in 1924 he related activity coefficients to ion concentrations.

improved the equation by taking Brownian motion into consideration, and proposed a more general equation that applied to both aqueous and non-aqueous solutions.[32] However, the Debye-Hückel-Onsager equation only holds true for dilute solutions, so the problem remained of handling more concentrated solutions and systems experiencing strong ion-ion and ion-solvent interactions. Efforts to develop more general ways of handling electrolyte solutions using statistical mechanics subsequently continued.

Acid-Base Concepts

Acids and bases were important topics related to the theory of electrolyte solutions. The Arrhenius theory of electrolytic dissociation (see page 116) assumed the existence of hydrogen ions (H^+) in acidic solutions and of hydroxide ions (OH^-) in basic solutions. The strength of an acid and base was proportional to the concentration of H^+ and OH^-, respectively. Pure water dissociates slightly to H^+ and OH^-, and since it was known that the ion product of water ($[H^+] [OH^-]$) is 10^{-14} (mol^2/l^2), the hydrogen ion concentration was adopted as an index of the acidity of an acidic or alkaline solution. In 1909, Sørensen introduced the concept of pH ($= -\log[H^+]$) as a convenient index to represent the hydrogen ion concentration.[33] This concept was expanded to express equilibrium constants for acids and bases, and pK_a and pK_b became widely used.

The Arrhenius theory of electrolytic dissociation was useful for handling aqueous solutions, but a more general expansion was desired. In 1923, a definition was proposed by Nicolaus Brøensted and Bjerrum in Denmark, and Lowry in Britain, which focused on the role of the proton in an acid-base system. Brøensted defined an acid as a proton donor and a base as a proton acceptor. Using this concept, new acids and bases could be made via an acid-base reaction. For example, in a mixture of hydrochloric acid and water, $HCl + H_2O = H_3O^+ + Cl^-$, water is a base with respect to hydrochloric acid. The H_3O^+ formed is an acid, and the Cl^- is a base. Thus, acids and bases are relative, and the proton donor is the acid.

In 1923, Lewis further expanded the concept of acids and bases. Lewis defined an acid to be a molecule or ion having an atom that can accept an electron pair from an atom comprising another molecule or atom. A molecule or ion that accepts an electron pair was an acid, and one that donates an electron pair was a base. These were termed a Lewis acid and a Lewis base, respectively. According to this definition, O, HCl, SO_3, BCl_3, and H^+ are Lewis acids, and CN^-, OH^-, tertiary amine, and ether are Lewis bases. He asserted that oxidizing agents should be expanded so as to include molecules that do not contain oxygen, and that acids should not be limited to molecules or ions containing hydrogen. The concept of a Lewis acid and Lewis base came to be used in wide-ranging fields in chemistry.

4.3 Physical Chemistry (II): Chemical Bond Theory and Molecular Structure Theory

G.N. Lewis and the Birth of Chemical Bond Theory

The true nature of chemical bonds was long a mystery for chemists. The electron was discovered at the end of the 19th century, and the structure of atoms began to be revealed by physicists. Consequently, modern chemical bond theory began to focus on electrons. The Arrhenius theory of electrolytic dissociation was a major concern for physical chemists at that time: why did a given substance form ions in solution when another substance did not? What caused the difference in strong electrolytes and weak electrolytes? These problems were closely intertwined with the chemical bond problem. Lewis continued his research in thermodynamics, but also had a strong interest in the chemical bond problem. In 1902, Lewis formulated a hypothesis regarding atomic structure and drew a model of an atom in his notebook in which electrons were arranged at the vertices of a cube, as shown in Figure 4.1, but he did not publish this model. In the interim between this hypothesis and his covalent bond model proposed in 1916, chemists continued trying to establish the structure of atoms and chemical bonds. In 1904, Abegg in Nernst's laboratory noticed that when the individual valences of atoms in a chemical compound were added together, the number came to 8. For example, the valence of Cl in NaCl is 1, but in $HClO_3$ it is 7, and added together these come to 8. Chemists began to realize that 8 electrons had a special significance. In 1908, Ramsay proposed that electrons surround an atom like the skin of an orange, and when a bond is formed, this skin separates to create a layer between the two atoms.

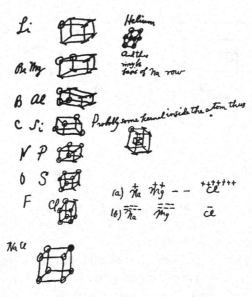

Figure 4.1: Cubic model of an atom in Lewis' notebook (1902)

J.J. Thomson proposed the "plum pudding model" for the atom in 1904, and suggested a theory in which a chemical bond resulted from an electrical force caused by two atoms displacing one or more electrons. This was subsequently accepted by chemists, and some chemists supported the view that bonds had polarity. However, Bray and Branch pointed out that although bonds in many inorganic compounds could be explained by this theory, the bonds in organic compounds could not be explained by polar bonds, and instead required non-polar bonding. Thomson changed his viewpoint following Rutherford's discoveries regarding atomic structure, and in 1914 Thomson proposed that there were two types of chemical bonds: polar and non-polar. This led to Lewis' theory of covalent bonding.

In 1916, Lewis and Kossel in Germany proposed a theory of chemical bonding which focused on an outer shell of 8 electrons (octet theory).[34,35] Using a circular atomic model (Figure 4.2), Kossel described an ionic bond as an electron moving from one atom to another atom and forming a bond through electrostatic interactions. An important point of this theory was that it focused in particular on the electron structure of noble gases in which the outer shell is filled with 8 electrons.

Lewis used a cubic model (Figure 4.2). In atoms on the second row of the periodic table going from Li to F, electrons increasingly occupy the vertices of the cube. The outermost shell of Ne is occupied by 8 electrons, making Ne a

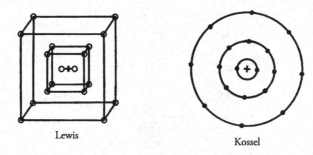

Lewis

Kossel

Figure 4.2: Atomic models of argon from Lewis and Kossel (1916)

stable, inert atom. A chemical bond was thought to arise by two atoms sharing a vertex of the cube that had been occupied by an electron. In this view, in an ionic salt like table salt, a sodium atom gives up 1 electron from its outermost shell to become a positive ion with a structure like that of a noble gas atom, and chlorine receives 1 electron to become a negative ion with a structure like that of a noble gas. This explanation agreed with the results of X-ray analysis. The bond of a diatomic molecule such as F_2 was described as two atoms sharing a cube edge with the two electrons at both ends. The hydrogen atom was described as forming a stable cube and can form a bond with another atom; during bonding, hydrogen shares its one electron with the other atom and the other atom shares an electron with hydrogen (generating an electron pair).

A double-bond in a diatomic molecule like O_2 was described as two cubes sharing a face and sharing four electrons. This model allowed bonds in molecules such as HCl, H_2O, NH_3, CH_4, and CCl_4 to be described. However, the model could not handle a triple bond, and hence the bond in the N_2 molecule could not be described. In addition, while Lewis treated electrons as static, this differed from the mental image physicists had for treating electrons dynamically. He proposed a system using ":" to represent a chemical bond and the element symbol to represent an atom that includes its inner shell electrons. H_2, HCl, H_2O, NH_3, and CH_4 were represented as follows.

$$
\begin{array}{ccccc}
 & & & H & H & H \\
H\!:\!H & H\!:\!\ddot{\underset{..}{Cl}}\!: & :\!\ddot{\underset{..}{F}}\!:\!\ddot{\underset{..}{F}}\!: & H\!:\!\ddot{N}\!: & H\!:\!\ddot{\underset{..}{O}}\!: & H\!:\!\ddot{C}\!:\!H \\
 & & & H & & H
\end{array}
$$

Although the Lewis cubic atom model was short lived, the system for representing bonds was subsequently used widely in chemistry textbooks, and continues to be used. This system made it clear that the essence of a covalent bond is that two atoms share a pair of electrons.

In 1919, Langmuir entered the field of chemical bond theory. He expanded Lewis' theory, and this improved theory found international acceptance.[36] Langmuir introduced the term "covalent bond" and gave lectures across the US and Europe on the theory of chemical bonds. He placed the electrons into shells of 2, 8, 18, and 32 electrons, and attempted to describe chemical properties by assuming stabilization when 8 electrons were in the outermost shell. At the time, Langmuir was already well known because of his pioneering work in surface chemistry (see page 279). Although he respected Lewis' priority in the chemical bond theory, Lewis' theory came to be called the Lewis-Langmuir theory. Lewis did not like this. Relations between the two men deteriorated, and Langmuir withdrew from research on chemical bond theory in 1922. In 1923, Lewis wrote "Valence and the Structure of the Atom,"[37] and this had a major influence, becoming the classic book on chemical bond theory.

At the start of the 20th century, both physicists and chemists conducted research aimed at understanding the structure of the atom, but their mental images of the atom were different, and they were interested in different properties of the atom. As described in the previous chapter, the primary interest of physicists was to explain atomic spectra, and this led to concepts such as the Bohr model of the atom and different atomic orbitals such as s, p, and d. Chemists sought to describe the chemical properties of elements and arrived at the idea of shells containing 8 electrons. In 1921, Bury in Britain proposed that chemical evidence indicated that electrons are arranged in sequence in layers that could hold 2, 8, 18, and 32 electrons, and about the same time Bohr proposed his structure of the atom. When Lewis' book was published in 1923, physicists' and chemists' mental pictures of the structure of the atom were becoming more similar.

An important point of Lewis' theory was that it provided a way for chemists to visually grasp the electronic structures and chemical bonds in atoms and molecules. His concept had a major impact on the subsequent development of electronic theory in organic chemistry (see page 317), and it is widely used even today in introductory books on chemistry. However, his intuitive way of depicting chemical bonding was not highly valued among physicists, who placed more value on rigor.

Lewis' theory was further expanded by Sidgwick at Oxford. Sidgwick showed that in a given compound, the electron pair that participates in the bond is provided by the same atom, and he named this a coordination bond. As will be described later (see page 279), a coordination bond plays a role in understanding the electronic state, particularly in a Werner complex. For example, in a hexammine-cobalt complex, it was proposed that six ammonium molecules each provide a lone pair electron to cobalt, and these 12 electrons in the outer shell form bonds between the cobalt and the six ammonium nitrogen atoms.

COLUMN 10

The Feud between G.N. Lewis and Langmuir[14]

G.N. Lewis conducting an experiment (1944)

A falling out between great scientists often occurs due to competition or differences of opinion. The relationship between the exemplary American physical chemists G.N. Lewis and Langmuir in the first half of the 20th century is an example of such a falling out, and it sparked various conjectures regarding the cause of Lewis' death.

On the afternoon of Saturday, 23 March 1946, the 70-year-old G.N. Lewis was found to have collapsed and died in front of a vacuum line in his laboratory. Hydrogen cyanide gas permeated the area. Officially, it was ruled that he collapsed and died from cardiac arrest during an experiment, and that at that time, a glass vessel that had been in use broke, releasing hydrogen cyanide gas. However, the possibility was raised whether his death was a suicide. On the afternoon of that day, he returned from lunch with a guest in low spirits and went to his laboratory. After a short time, he was found to have collapsed. It was recently discovered that the lunch guest was Langmuir. Langmuir had come to Berkeley that day to receive an honorary doctorate at the University of California, and additionally, had had lunch with a small number of important university officials, including Lewis.

Lewis and Langmuir had been rivals on bad terms since about the end of the 1910s. Lewis was six years senior to Langmuir, and although he had been a pioneer in research on chemical thermodynamics and chemical bond theory and was respected as the doyen of the American chemistry world, he had never won a Nobel Prize. On the other hand, Langmuir had been recognized as a pioneer in surface chemistry at a young age and was awarded the Nobel Prize in 1932. Both Lewis and Langmuir studied in Nernst's laboratory in Göttingen, and their relationship was friendly at first. In 1919, however, Langmuir entered the field of chemical bond theory, and their relationship deteriorated from that time onward. In 1916, Lewis published the paper "*Atoms and Molecules*" in the *Journal of the American Chemical Society*. Afterwards, he temporarily left research due to military service. He had intended to publish his ideas regarding chemical bonds in more detail, but was unable to do so immediately. In the meantime, Langmuir developed and refined Lewis' ideas, disseminating them worldwide. Langmuir had an extroverted, sociable personality, and he was popular as an accomplished lecturer. The theory of covalent bonding came to be called the Lewis-Langmuir theory. However, it was difficult for Lewis to accept Langmuir's name juxtaposed to his on a theory he had first proposed.

Why was a chemist like Lewis, who made many important contributions, not awarded a Nobel Prize? Documents pertaining to the selection process more than 50 years ago have become available and shed light on this question. While nominations for the Nobel Prize are made by scientists from around the world, the decision on who is awarded the prize is made by a selection committee in Sweden. It cannot be denied that the individual subjectivity and preferences of selection committee members have had a major influence on the committee's decisions. In the book "*Cathedrals of Science*," Coffey wrote about why Lewis did not win a Nobel Prize. Lewis was an introvert and did not like to keep company with people outside of his immediate surroundings at Berkeley. He was isolated at Berkeley too, did not have any close acquaintances among European scientists, and had no strong recommenders on the Noble Prize selection committee. He was nominated as a Nobel Prize candidate year after year, but continued to be set aside because of a lack of decisive achievements. Also, he changed his area of research several times, from chemical thermodynamics, to chemical bond theory, to isotopes and then to photochemistry, and this worked against him. When he became a candidate in the 1920s due to his work in chemical thermodynamics, he had already moved away from that field. When he became a candidate due to his contributions on chemical bond theory, quantum mechanical theories had emerged, and the Lewis bond theory was considered out of date. The committee did not value the major impact that Lewis' intuitive electron pair concept had on chemistry, including organic chemistry. A shared prize with Urey for research on isotopes was rumored but did not materialize. His work on photochemistry was set aside as being premature. Therefore, when he saw the young men like Langmuir and Urey win Nobel Prizes, one can easily imagine that Lewis' pride was injured.

In all likelihood, Lewis' death was not a suicide. However, with Langmuir receiving an honorary doctorate on his home grounds of Berkeley, and being forced to sit through a lunch with Langmuir, it would not be surprising if Lewis had been stressed and depressed. It is possible that this stress was sufficient to bring about his cardiac arrest.

Walter Heitler (1904–1981)
Heitler was a German theoretical physicist who studied
at the Universities of Karlsruhe, Berlin, and Munich, and
studied under Sommerfeld. In addition to research on chemical
bonds, Heitler left behind many achievements in fields
such as radiation theory, cosmic rays, and meson theory. He
was employed as a lecturer at the University of Göttingen
but was pursued by the Nazis and moved to Ireland, where
he became a professor at the Dublin Institute for Advanced Studies. After the
war, he became a professor at the University of Zurich.

Fritz London (1900–1954)
London obtained a degree in philosophy from the University
of Munich, but he turned to physics and studied under
Sommerfeld. After working as a lecturer at the University
of Berlin and at a British chemical company, London went
to the US and became a professor at Duke University. In
addition to research on chemical bonding, he is also known
for theories for describing dispersive forces, the phenomenon
of superconductivity, and superfluid helium.

Valence Bond Theory

Understanding the nature of chemical bonds is fundamental to chemistry, but
this was impossible until the advent of quantum mechanics. However, since the
Schrödinger equation has no exact solution even for the hydrogen molecule,
we have to rely on approximation methods. The first quantum mechanical
approximation treatment was a method called the valence bond theory, which
was applied to the hydrogen molecule by the German physicists Walter Heitler
and Fritz London in 1927.[38] They represented the wave function Ψ (see page
217) for the electrons of the hydrogen molecule by the product of the wave
functions Ψ_{1sA} (1) and Ψ_{1sB} (2) for the 1s orbital of the two constituent hydrogen
atoms, A and B. The subscripts 1 and 2 denote the respective electrons of the
atoms. Since the two electrons cannot be distinguished, the electron orbital wave
function Ψ is expressed as

$$\psi = C(\psi_{1sA}(1)\,\psi_{1sB}(2) + \psi_{1sA}(2)\,\psi_{1sB}(1))$$

By taking into account the effect of electron spin, the energy of the hydrogen
molecule is calculated as a function of internuclear distance (see Figure 4.3). Here,
the coefficient C is determined from the fact that the integral over all space covered

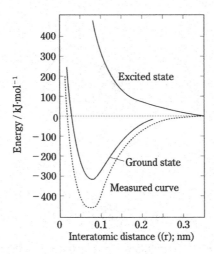

Figure 4.3: Dependence of the energy of the hydrogen molecule on interatomic distance (r)

The energy of a system in which the two atoms are remote from each other is assumed to be zero. The energy of the system is at a minimum at $r = 0.09$ nm. The solid lines show values calculated from the Heitler-London model, and the dotted line shows measured values.

by the probability function for the presence of an electron, Ψ^2, is 1. Using this calculation, Heitler and London obtained a binding energy[9] of approximately 2/3rds the measured value of the ground state in which the electron spins are antiparallel. This was a landmark result that shed light for the first time on the nature of the bond in a diatomic molecule postulated by Avogadro's hypothesis. The binding energy in this calculation consists of two terms. One is called Coulomb energy and is understood by a classical picture of Coulomb interaction between the nucleus and electrons. The other is called exchange energy and cannot be understood using a classical picture. Exchange energy originates from the fact that electrons are indistinguishable and can be exchanged with each other. The magnitude of the exchange energy is determined by the overlap of the wave functions; since the exchange energy is much bigger than the Coulomb energy, it determines the chemical bond strength (Figure 4.4). Accordingly, chemical bonding was shown to be essentially a quantum mechanical phenomenon. The proposal of a quantum mechanical theory for chemical bonding by Heitler and London gave rise to the new field of quantum chemistry. Since then, quantum chemistry has been used to

9. Binding energy: The energy required to dissociate a molecule into individual atoms can be expressed approximately as the sum total of the energy of each bond within the molecule. The energy of each of the bonds is the binding energy.

(a)

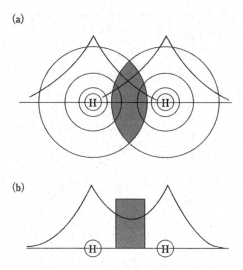

(b)

*Figure 4.4: Conceptual diagram showing the overlap of atomic orbitals in a hydro-
        gen molecule*

(a) Diagram of the overlap of wave functions of two 1s orbitals viewed in a two-dimensional
    plane on the bond axis. There is a large overlap in the space between the atoms (shaded
    area). The concentric circles indicate contour lines for the probability of the presence of an
    electron.
(b) The superposition of two atomic orbitals corresponds to $\Psi = \Psi_{1sA}(1) + \Psi_{1sB}(2)$.

describe many chemistry phenomena, and with the advancement of computers in
the second half of the 20th century, has developed into a major field.

The modern theory of chemical bonding initiated by the Heitler-London
theory was promptly applied to understanding bonding in many compounds.
Linus Pauling made enormous contributions to the development and dissemina-
tion of chemical bond theory. He promoted the investigation of a wide range
of molecular structures using X-ray and electron-beam diffraction, and the
application of quantum mechanics to problems in structural chemistry.

The concept of a partial ionic bond, with properties in between an ionic bond
and a covalent bond, was developed to describe the results of the structural
analysis of many molecules. The ionic bond in metal halides is not 100%
ionic, and the covalent bond in HF or HCl has substantial ionic character. To
understand the properties of various bonds, Pauling introduced the concept of
electronegativity in 1931. Electronegativity reflects the ability of two bonded
atoms to attract an electron. Using Pauling's approach, electronegativity values
for many bonds were estimated from the difference between the bond energies

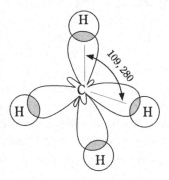

Figure 4.5: Bonding by sp³ hybridized orbitals of the carbon atom in methane

between two atoms, A and B, and the geometric mean of the bond energies of diatomic molecules comprising A and B. In 1934, Mulliken proposed a different approach to electronegativity. Mulliken assumed a measurement scale for electronegativity of ½ the sum of the first ionization energy of an atom[10] and its electron affinity.[11] The values obtained using Pauling's and Mulliken's estimates of electronegativity are generally similar.

In 1931, Pauling and John Slater independently proposed the concept of a hybrid orbital to describe the bonding and regular tetrahedral structure of methane.[39, 40] They proposed that one electron in the carbon 2s orbital jumps to the 2p orbital. This causes one 2s orbital and three 2p orbitals to hybridize, creating four sp³ orbitals. The electron in each sp³ orbital can form a bond by pairing with an electron from hydrogen (Figure 4.5). Linear (sp) and plane-trigonal (sp²) structures were described using the same hybridized orbital

---

10. Ionization energy: Energy required to pull one electron to infinity from an atom or molecule in a vacuum.

11. Electron affinity: Energy released when a neutral atom and an electron bond in a vacuum.

concept. Molecules exhibiting plane-square and octahedral structures were described by hybridized orbitals that additionally included the d orbital. These models agreed well with the results of crystal structure analysis. The theoretical foundation of Pauling's treatment of the concept was further solidified by Slater. In 1933, Pauling and Wheland proposed that aromatic compounds such as benzene "resonate" between different structures. The wave function representing the state of a molecule was expressed as a linear combination of wave functions corresponding to the different resonance structures. This approach had the advantage of appealing to the intuition of chemists and was accepted as the "valence bond method", but proved inadequate for the calculation of excited states and the calculations of complex molecular structures.

In 1935, Pauling, in collaboration with Wilson, published *"Introduction to Quantum Mechanics with Applications to Chemistry,"*[41] and it was widely read as an introductory book on quantum mechanics for chemists. In 1939, Pauling published *"The Nature of the Chemical Bond and the Structure of Molecules and Crystals."*[42] This book has been widely cited as a classic description of chemical bond theory up to the present day. In addition, "General Chemistry"[43] published in 1947 was used worldwide as an introductory chemistry textbook.

Molecular Orbital Method

The molecular orbital (MO) method was developed toward the end of the 1920s as another approximation method for handling the electronic state of molecules. While the electrons that form a bond are represented as pairs in the Heitler-London method, in the molecular orbital method, two electrons fill the molecular orbitals in the molecule in sequence from the lowest-energy orbital, in accordance with the Pauli principle (see page 212). Friedrich Hund and Robert Mulliken independently described the electronic spectra of diatomic molecules by the molecular orbital method in 1927–28.[44, 45] When the internuclear distance of a diatomic molecule becomes small, in the limit, the diatomic molecule becomes a unified atom. The diatomic molecule was described starting with this atom. Hund clarified the differences between σ bonds and π bonds[12] in a diatomic molecule. In 1929, the LCAO (Linear Combination of Atomic Orbitals) method was introduced by Lennard-Jones in Britain to represent molecular orbitals by a linear combination of atomic orbitals (see Section 3.7).[46] In this approach, the wave function Ψ_{AB} corresponding to the molecular orbital of molecule AB

---

12. σ bonds and π bonds: σ bonds are covalent bonds formed when electrons are in an electron orbital (σ orbital) that is symmetric around the bond axis. π bonds are bonds formed by electrons in p orbitals.

Friedrich Hund (1896–1997)
Friedrich Hund was born in Karlsruhe. After majoring in mathematics and physics at the University of Göttingen, Hund became a lecturer at the University of Göttingen in 1925, a professor at the University of Rostock in 1927, and a professor at the University of Leipzig in 1929. In 1927, Hund introduced the concept of molecular orbitals to allow a quantum mechanical interpretation of the spectra of diatomic molecules. He is known for "Hund's rule": if there are several electrons in an atomic orbital, the configuration with maximum spin multiplicity is the most stable.

Robert Mulliken (1896–1986)
After graduating from the Massachusetts Institute of Technology and obtaining a Ph.D. in chemistry from the University of Chicago, Robert Mulliken conducted research on the separation of isotopes and electronic spectra of diatomic molecules. In 1925, he traveled to Europe and studied quantum mechanics, which was being developed at the time. In 1928, he introduced the concept of molecular orbitals to describe the spectra of diatomic molecules. His work on molecular orbital theory contributed to understanding molecular structure and electron configuration. In addition, he proposed the concept of charge transfer interaction between molecules, providing a path to understanding intermolecular interactions and reactions. He was awarded the 1966 Nobel Prize for Chemistry.

is approximated by a linear combination of wave function Ψ_A corresponding to atom A and wave function Ψ_B corresponding to atom B.

$$\psi_{AB} = C_A\psi_A + C_B\psi_B$$

Coefficients C_A and C_B are determined using the variation method.[13]

In the molecular orbital method, the electronic structure of a molecule is obtained by filling electrons in order from the lowest orbital to the higher orbitals. For the hydrogen molecule, linear combinations of the wave functions for the 1s orbitals of the two hydrogen atoms are obtained from $\Psi_g = \Psi_{1sA} + \Psi_{1sB}$ and $\Psi_u = \Psi_{1sA} - \Psi_{1sB}$. Multiplying Ψ_g and Ψ_u by their respective spin function

---

13. Variation method: An important method for obtaining an approximate solution for the Schrödinger equation. The method makes use of the fact that the energy E_1 obtained using an approximate wave function Ψ_1 is greater than or equal to the true energy E_0.

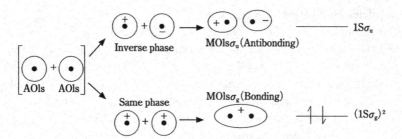

*Figure 4.6: Conceptual diagram describing the creation of a molecular orbital (MO)
from two atomic orbitals (AO)*

The lower orbital is a bonding orbital, with a high electron density between the atoms, and the
stable ground state is the state in which two electrons with antiparallel spins occupy the orbital.
The upper orbital shows an anti-bonding orbital, with a low electron density between the two
atoms.

generates the molecular orbital. The Ψ_g orbital is the stable bonding orbital in
which the electron density overlaps in the region between the two nuclei, and
Ψ_u is the antibonding orbital. These orbitals are symmetric about the molecular
axis. The bonding orbital is called the $1s\sigma_g$ orbital and the antibonding orbital
is called the $1s\sigma_u$ orbital. The ground state electronic configuration of the
hydrogen molecule can be expressed by $(1s\sigma_g)^2$ when the two electrons are in
the $1s\sigma_g$ orbital (Figure 4.6). The simplest treatment of the electronic structure
of a homonuclear diatomic molecule is obtained by extending this approach to
make molecular orbitals by linear combinations of atomic orbitals. With Li_2,
the extension encompasses the molecular orbitals consisting of the 2s atomic
orbitals, and with nitrogen or oxygen molecules, the extension encompasses
the molecular orbitals comprising the 2p orbitals. With heteronuclear diatomic
molecules such as LiH, the molecular orbital is given by an appropriate linear
combination of H and Li atomic orbitals. Molecular orbitals for many-electron
system molecules comprising two or more electrons are affected by electron
interaction effects and are handled in the same manner as many-electron
atoms. An approximate wave function for n electrons is represented by an
n-dimensional Slater determinant (see previous chapter, page 219), and an approximate wave function and approximate energy are obtained by solving the
Slater determinant. Many people conducted research on chemical bonds and
the electron configurations of molecules using the molecular orbital method,
including Hund, Mulliken, Herzberg, Slater, and Coulson.

The molecular orbital method was extended by Hückel in 1931 to approximate
the electronic state of aromatic compounds such as benzene.[47] Hückel
represented the molecular orbital for π electrons by a linear combination of

2p orbitals from the carbons forming the benzene skeleton. This is called the Hückel method, and it allowed application of the molecular orbital method to problems in organic chemistry. Application of the Hückel method shed light on aromaticity and the reactivity of aromatic compounds. In the second half of the 20th century, the molecular orbital method had a large impact on broad-ranging fields in chemistry, but in the first half of the 20th century, prior to the emergence of computers, the ability to handle complex molecules was limited.

Robert Mulliken was awarded the 1966 Nobel Prize for Chemistry for his "fundamental research concerning chemical bonds and the electronic structure of molecules by the molecular orbital method." However, Hund also made major contributions, with pioneering research concerning the electronic structure of molecules. The author feels that the Nobel Prize could have been awarded jointly to Mulliken and Hund.

Hydrogen Bonds and Metallic Bonds

Hydrogen bonds play an important role in chemical phenomena in many areas, such as the association of water and organic acids, and the structure of nucleic acids. Moore and Winmill were likely the first to consider the existence of hydrogen bonds, in 1912. In 1920, Latimer and Rodebush used hydrogen bonding to explain the association of diverse molecules such as water, hydrogen fluoride, ammonium hydroxide, and acetic acid. Pauling explained that hydrogen bonding resulted from an electrical attraction between atoms with high electronegativity, such as fluorine, oxygen, and nitrogen, and hydrogen atoms. Hydrogen bonding explained why small molecules such as water, hydrogen fluoride, and ammonia have high boiling points.

The free electron theory of metals first proposed by Drude in 1900 and extended by Lorentz in 1905 assumed that free electrons moved in the voids of a solid densely packed with atoms. A free electron theory of metals based on quantum mechanics was developed by Sommerfeld in 1927. In typical metals such as sodium and aluminum, the cations occupy densely packed lattice points, and the s and p valence electrons are conduction electrons that move throughout the crystal. In the quantum mechanical treatment of metals, the wave function of a free electron is delocalized over the entire crystal, and the electrostatic interaction between the free electrons and the cations gives rise to the binding force. The contribution of d electrons to bonding was also considered in transition metals.

Polarity of Molecules

When a bond is created between two molecules with different electronegativities, an overall negative charge becomes distributed around the electrically negative atom, and an overall positive charge remains at the electrically positive ion. Consequently, the bond forms an electric dipole moment, μ.[14] The electric dipole moment of a polyatomic molecule can be represented by the vectorial sum of the dipole moments of each bond. In 1914, Peter Debye (see p. 236) derived an expression between the dipole moment and the dielectric constant; based on this expression, he and Lange determined the dipole moment of molecules from measurements of dielectric constants.[48]

Dipole moments provided two pieces of information about the molecular structure: the degree of bond polarity, and the geometric structure of the molecule. The magnitude of the dipole moment is measured in Debye (D) units. The dipole moment when a charge of $\pm e$ is separated by a distance of 0.1 nm is 4.8 D. The interatomic separation distance in HCl is 0.126 nm, and its dipole moment is 1.03D. This indicates that the bond has approximately 1/6th the ionic character of an ionic bond. The dipole moment of CO_2 is zero, which indicates that the molecule is linear. The water molecule has a dipole moment of 1.85D; if it is assumed that the OH bond dipole moment is 1.60D, then the bond angle of the water molecule is close to 105°. Debye wrote a book entitled "*Polar Molecules,*" which is considered a classic masterpiece.[49] Debye trained as a physicist, but he studied important problems in chemistry in the first half of the 20th century and made many contributions to a wide range of areas in physical chemistry and chemical physics. In 1936, Debye was awarded the Nobel Prize for Chemistry for his "investigation of the diffraction of X-rays and electrons in gases."

Intermolecular Forces
Various forces affect atoms and molecules that do not form a chemical bond. These forces are considered in the van der Waals equation of state, and by the 1920s and 1930s were understood in some detail. These forces are essentially due to electrostatic interaction and are described by a gradient of potential energy, U. There are four sources for the potential energy that depend on the distance (r) between interacting atoms or dipoles,

1. The energy resulting from Coulomb interaction between ions carrying a true charge, $U \propto r^{-1}$.
2. The interaction energy between two permanent dipoles, $U \propto r^{-6}$.

---

14. Electric dipole moment: When a pair of positive and negative electric charges, for example, two point charges of $+q$ and $-q$, are separated by a distance l, this charge pair is called an electric dipole, and ql is referred to as the electric dipole moment.

3. The energy between a permanent dipole and a dipole induced by another molecule, $U \propto r^{-6}$.

4. The energy due to forces, called dispersion forces, acting between neutral atoms or molecules, $U \propto r^{-6}$.

An intuitive explanation of dispersion forces is as follows. The nucleus of a neutral atom such as argon is surrounded by a negatively charged cloud. This cloud is spherically symmetric when averaged over time, but it experiences momentary deviations from sphericity, giving rise to non-homogeneous charge distribution, thus causing small, momentary dipoles. These dipoles interact with other dipoles to produce momentary attraction. London developed a quantum mechanical treatment of these dispersive forces in 1930. Forces which act when neutral atoms or molecules condense are called van der Waals forces and are the forces described in 2) through 4) above.

When the distance between nuclei is small, the interaction between the nucleus and electron cloud of one molecule and the nucleus and electron cloud of another molecule becomes repulsive; the magnitude of this repulsive energy acts with a distance dependence of $U \propto r^{-9}$ to r^{-12}. Taking these interactions into account, the Lennard-Jones potential of $U(r) = -Ar^{-6} + Br^{-12}$ is used as an approximation of the intermolecular potential. This sort of weak intermolecular interaction plays a major role in chemical phenomena in wide-ranging fields such as supramolecular chemistry (see page 539) and surface chemistry.

Structural Analysis by Means of X-ray and Electron-beam Diffraction

Information about interatomic distances and bond angles in crystals obtained from X-ray structural analysis provided experimental support for the development of chemical bond theory. In 1912, Laue, in cooperation with Friedrich and Knipping, obtained the X-ray diffraction pattern for zincblende (ZnS) crystals. Laue developed the mathematical techniques necessary to reconstruct the arrangement of atoms within the crystal from the diffraction spots observed on a photographic plate. Henry Bragg, who was aware of Laue's report, promptly devised a spectrometer to measure the strength of an X-ray beam resulting from the amount of ionization caused by X-rays. In 1912, Lawrence Bragg formulated the famous Bragg equation describing diffraction (Chapter 3, page 178).[50] The Bragg father-son duo irradiated crystals with X-rays of fixed wavelength, and measured the intensities of the X-rays scattered by atoms occupying different parallel crystal planes. They determined the distance between layers of atoms in the crystal from the scattering angle from various planes, and then determined the positions of atoms within the crystal.[51] In addition, in 1915 the Braggs introduced a Fourier analysis method to develop a general analytical method.

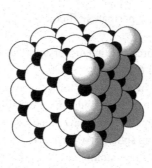

Figure 4.7: Model of the crystal structure of table salt

The white spheres represent Cl⁻ ions and the black spheres represent Na⁺ ions. The radii of the spheres reflect the ionic radius of each ion.

The methods of Laue and Bragg required single crystal samples.[15] In 1916 and 1917, Debye and Scherrer in Göttingen, and Hull at General Electric Company in the US, developed a new method using powdered samples. With this method, film was loaded into a cylindrical camera, and the diffracted X-rays from various planes were observed as concentric circular patterns around the incident X-ray direction.

These X-ray analytical methods provided detailed information regarding crystals and molecules. Table salt was the first structure analyzed using these techniques. In 1914, Lawrence Bragg showed that the Cl⁻ ions and Na⁺ ions in table salt were alternately interlayered to make a face-centered cubic lattice arrangement (Figure 4.7). The distance between the Cl⁻ ion and Na⁺ ion centers was determined to be 0.282 nm. This result showed that NaCl as a crystal exists as ions, providing strong support for the Arrhenius theory of electrolytic dissociation. That same year, diamond crystal was shown to comprise a regular tetrahedral structure of carbon atoms, and the distance between carbon atoms was shown to be 0.154 nm. Subsequently, graphite was revealed to have a sheet-like structure comprising regular hexagonal rings. One carbon atom is bonded to three other carbon atoms with an interatomic distance of 0.142 nm within a sheet, and the sheets lay one over another separated by 0.340 nm. It became clear that graphite and diamond have different properties because they have different crystal structures. The crystal structures of many metals and inorganic compounds were subsequently investigated.

---

15. Single crystal: A crystalline solid in which any section within the crystal has the same orientation with respect to some arbitrary crystal axis.

Shoji Nishikawa (1884–1952)
Shoji Nishikawa graduated from the Department of Physics of Tokyo Imperial University in 1910. He pursued research on the crystal structure of spinel in 1915 under Torahiko Terada and was the first to successfully analyze the structure of crystals with a complex atomic arrangement. In 1922, he became a principal investigator at the Institute of Physical and Chemical Research (RIKEN), and his laboratory achieved internationally acclaimed results in X-ray diffraction and electron-beam diffraction. He was presented the Order of Culture in 1951.

Isamu Nitta (1899–1984)
After graduating from the Department of Chemistry of Tokyo Imperial University in 1915, Isamu Nitta entered Nishikawa's laboratory at RIKEN and started analyzing the crystal structures of organic compounds using X-rays. He continued this pursuit throughout his career. He was a professor at Osaka University from 1933 to 1960 and a professor at Kwansei Gakuin University from 1960 to 1968. In 1970, he determined the structure of the fugu (pufferfish) toxin, tetrodotoxin. He was presented the Order of Culture in 1966.

Compounds such as copper, calcium fluoride (CaF_2), calcite ($CaCO_3$) and iron sulfide (FeS_2) were among the first to be studied using X-ray diffraction. In 1915 in Japan, Shoji Nishikawa at the University of Tokyo analyzed the crystal structure of spinel ($MgAl_2O_4$).[52] Nishikawa later led X-ray diffraction research in Japan at RIKEN, and Isamu Nitta, who studied under Nishikawa, conducted research from the late 1920s through the 1930s to verify the tetrahedral structure of carbon atoms in organic compounds.[53]

The structures of organic compounds such as naphthalene and anthracene, and of Werner complexes such as chloroplatinum complex, were determined at the beginning of the 1920s. In 1928, Lonsdale determined the structure of hexamethylbenzene and found that the C-C bond in the benzene ring was 0.142 nm long, which is in between the length of an ordinary C-C single bond and a C=C double bond.[54] This important result supported the benzene ring resonance theory. The distance between atoms bonded by a covalent bond was a good indicator of the strength and nature of the bond. Consequently, the results of X-ray crystallography provided a detailed understanding of the chemical bonds in inorganic compounds and metal complexes.

Many metals and alloys were crystallographically analyzed in the 1920s. The Debye-Scherrer method was particularly useful for investigating alloys in which metal microcrystals were in a disordered arrangement. Research on silicates in the field of mineralogy began in the 1920s, and the structure of garnet was

John D. Bernal (1901–1971)
A British crystallographer, J.D. Bernal studied at Cambridge University, was a researcher at the Royal Institution, a lecturer at Cambridge University, and a professor at the University of London. He was known for his structural analyses of various compounds by X-ray diffraction, research on the origin of life, and research on the structure of water. Bernal was famous as the author of "The Social Function of Science" and "Science in History." (see Column 11)

William Astbury (1898–1961)
William Astbury studied at Cambridge, and after graduating, pursued research in X-ray crystallography under Henry Bragg at the Royal Institution. In 1928, he moved to the University of Leeds; there, he began X-ray structural analyses of the fibrous proteins keratin and collagen, and became a pioneer in the X-ray structural analysis of biopolymers.

analyzed in 1924. The application of X-ray crystallography in the 1920s and 1930s made major contributions to the development of inorganic chemistry, organic chemistry, metallography, and mineralogy.

X-ray diffraction methods also influenced research on the structure of liquids. In the X-ray analysis of powder crystals by the Debye–Scherrer method, diffraction lines appearing in the photograph broaden as the particle size decreases, and only broad background scattering remains when the diameter of the particle is less than 10 nm. If liquids were completely disordered, no maxima or minima would be expected to appear in X-ray diffraction data, but in fact, maxima are observed. This is because molecules and ions within a liquid are not completely disordered: there is a degree of order between molecules and ions in close proximity. X-ray diffraction thus provides a powerful technique for investigating the structure of liquids. The structure of water was of intense interest, and in 1933, J.D. Bernal (Column 11) and Fowler published a pioneering study on the structure of water based on X-ray diffraction.[55] Subsequently, extensive research was carried out on the structure of liquids using X-ray diffraction.

Structural analyses of complex organic compounds of great biochemical interest also began in the 1930s. Bernal was a pioneer in this area, and in 1937 he analyzed the structure of sterol and pointed out that the structure suggested by organic chemistry techniques was incorrect. In 1945, Hodgkin at Oxford revealed the structure of penicillin. William Astbury of the University of Leeds began analyzing the structure of fibrous biopolymers such as keratin at the beginning of the 1930s, and Michael Polanyi of the Kaiser Wilhelm Institute in

Ernst Ruska (1906–1988)
Ruska studied at the Technical University of Munich and obtained his Ph.D. from the Technical University of Berlin in 1934. He joined the Siemens Company in 1937 to pursue the development of the electron microscope. Ruska served as the director of the Microscopy Institute of the Fritz Haber Institute from 1955 to 1972. He developed the first electron lens in 1931 and developed the first electron microscope in 1933. He was awarded the Nobel Prize for Physics in 1986.

Max Knoll (1897–1969)
Knoll was a German electrical engineer. He studied at the Technical University of Berlin and obtained his Ph.D. at the Institute for High Voltage. In 1927, he became the leader of the electron research group and invented the electron microscope jointly with Ruska in 1931. In 1932, he joined Telefunken and helped develop the television.

Berlin also studied the structure of fibrous proteins by X-ray diffraction from the 1920s through the 1930s. Dorothy Crowfoot (Hodgkin after marriage) (Column 16) and Bernal obtained diffraction images from pepsin crystals at Cambridge in 1934; these were the first diffraction data from a globular protein crystal.[56] In 1937, Perutz at Cambridge began analyzing the structure of hemoglobin. Although it was not possible to conduct X-ray structural analyses of complex biomolecules such as proteins using the X-ray diffraction technology of the time, this bold research effort in Britain led to the birth of molecular biology and structural biology after World War II.

Debye had completed a theoretical treatment of the diffraction of an electron beam by a gas in 1915, and experimental work on this topic was started by the German physicist Wierl in 1930.[57] Maxima and minima were observed in interferograms obtained from samples of disordered molecules, allowing determination of the interatomic distances within molecules. In addition, if the distance between neighboring atoms within a molecule was known, the bond angles could also be calculated. Gas-phase electron beam diffraction was useful for determining the structure of comparatively small molecules.

The negative charge of an electron beam provides some advantages over X-rays for investigating the structure of a substance. Specifically, the appropriate combination of an electric field and magnetic field provides a lens for focusing an electron beam, making possible a microscope that uses an electron beam instead of light. Because the wavelength of an electron beam can be shorter

J.D. Bernal: The Legacy and Complexity of the Sage of Science[58]

Bernal handling an X-ray device

Some scientists show surprising talent and play multifaceted roles. J.D. Bernal (1900–1970) was one such person. His scientific accomplishments are touched on several times in this book. As an X-ray crystallographer, he and Dorothy Crowfoot captured the diffraction pattern of pepsin on photographic film, paving the way for the X-ray crystallographic analysis of globular proteins. In addition, he pointed out that the Windaus structure of cholesterol was incorrect, thereby showing that X-ray analysis was indispensable for analyzing the structures of complex organic compounds. These accomplishments may have been deserving of a Nobel Prize, but many felt that Bernal was interested in too many subjects, and because he could not focus on a single one, his work lacked the depth required for recognition by a Nobel Prize. A voluminous biography of this legendary person was published in 2005[58] and depicts a person with a complex nature, an enormous intellectual capacity, and living a life full of upheaval.

J.D. Bernal was born in Ireland in 1905. His father was a farm manager who converted from Sephardic Judaism to Catholicism, and his mother was American and a graduate from Stanford University. After receiving his primary education in Ireland and completing his secondary education in a British boarding school, he entered Cambridge University where he originally majored in mathematics. However, he changed his major and graduated in natural sciences. His graduate research involved a ground-breaking study of space groups, and his outstanding scientific abilities were recognized. In 1923, he obtained a position at the Royal Institution, headed by Henry Bragg, and worked as a crystallographer. While at Cambridge, he read books in a wide range of fields such as history, art, society, and politics. He absorbed knowledge, exhibited surprising comprehension and an outstanding memory, and enthralled those around him with his extensive knowledge and brilliant intellect. He seemed to know everything about everything, and came to be called by the nickname "Sage." In addition, he went from being a devout Catholic to becoming a Marxist, and he joined activist groups.

Bernal's abilities as a scientist were recognized through his pioneering research in X-ray crystallography from the 1920s to the mid-1930s. In 1938, he went from being a lecturer at Cambridge to becoming a professor at Birkbeck College at the University of London, and at a young age was elected a Fellow of the Royal Society. At the start of World War II, the British government, recognizing his enormous intellectual abilities, appointed Bernal as a member of the Committee for Imperial Defense, whereupon he left science to play a very active role in national defense and operational planning. He participated in the drafting of various operations plans, first as an expert in bombs and air defense, and then from February 1942, as an advisor for Lord Mountbatten, Chief of Combined Operations. Soon, he was taking part in the most important project of all: the Allied Forces' operations to invade the continent. Bernal carefully examined the Allied Forces' planned landing spot and prepared detailed nautical charts to assist landing operations.

Bernal had a wealth of excellent ideas, and he gave these generously to the people around him. His best qualities were that he generated ideas, took the initiative, and motivated and encouraged those around him. The people he influenced—Crowfoot-Hodgkin, whom he charmed with his charisma, Perutz, who began research on hemoglobin under him, Kendrew, who was encouraged by Bernal to become a crystallographer, and others—made major contributions to the birth and development of structural biology and were awarded Nobel Prizes for Chemistry.

However, there are also puzzling facts about Bernal. He became an ardent communist in the 1930s. This in itself was nothing unusual. Disillusioned by World War I and the Great Depression, many intellectuals were attracted to communism. However, as the facts about communism in the Soviet Union became clear, most people became disillusioned and abandoned communism. However, Bernal believed in communism to the end. A dispute arose regarding the genetic inheritance of acquired characteristics in the Soviet Union, and Lysenko, with the support of Stalin, purged the opposing faction. Left-wing scientists in the West opposed the Lysenko hypothesis; in contrast, Bernal suppressed his scientifically critical mind and supported Lysenko. Why would a person with such brilliant intelligence praise the Soviet Union of the inhumane Stalin era and support the Lysenko hypothesis? The author feels that Bernal was a complex and difficult-to-understand human being.

than the wavelength of visible light rays, smaller structures can be observed than is possible with a visible light microscope. The electron microscope was developed by Ernst Ruska and Max Knoll of the Technical University of Berlin in 1931, and the Siemens Company made the first commercial electron microscope. Electron microscopes provided invaluable information in a wide range of scientific fields after World War II. However, Ruska was awarded the Nobel Prize for Physics in 1986, more than half a century after the development of the electron microscope, and after Knoll had died. Given the enormous impact the electron microscope has had on science overall, one has the impression that the Prize was awarded too late.

Molecular Spectroscopy and Structural Chemistry

After Kirchhoff and Bunsen introduced spectroscopy to chemical research (see Section 2.5), large quantities of spectral data were accumulated and used for the identification and analysis of elements. However, prior to the emergence of quantum mechanics, it was impossible to understand the phenomena underlying the spectra. As previously described, efforts to understand atomic spectra were a major factor spurring the development of quantum mechanics. The advent of quantum mechanics provided a means for understanding molecular spectra, and molecular spectroscopy developed largely as a branch of physical chemistry from the 1930s onwards. Measured spectra were analyzed using calculations based on quantum mechanics. In many cases, the spectra could be handled by the Born-Oppenheimer approximation (Chapter 3, page 221) by separating the wave function into two parts: one pertaining to an electron and the other pertaining to the nucleus. Therefore, the wave equation for an electron is calculated using a fixed location for the nucleus, providing information about the electronic configuration. Solving the wave equation for the vibration of the nucleus provides detailed information about the state of molecular vibration and rotation.

The simplest model for molecular vibration is a harmonic oscillator[16] model, which gives equally spaced vibrational energy levels. However, the vibrational energy intervals actually observed become smaller as the energy level increases due to anharmonicity of the vibration. In 1929, Morse introduced a potential function[17] called the Morse function to approximate the anharmonicity. The vibration of polyatomic molecules was interpreted using vibrations called

---

16. Harmonic oscillator: The motion of a particle that is subjected to a force proportional to the distance from a fixed point and that moves linearly is called harmonic motion, and an oscillator that oscillates in such a manner is called a harmonic oscillator.

17. Potential function: Potential refers to potential energy, and a function that indicates this energy is called a potential function.

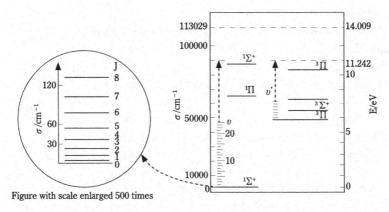

Figure with scale enlarged 500 times

Figure 4.8: The energy levels for carbon monoxide (CO)

The electronic energy levels (right-hand y-axis) and vibrational energy levels (left-hand y-axis) are shown. The circled figure shows the rotational energy levels.

normal vibration.[18] Rotation and vibration were further approximated by separating them. The wave equation for the rotation of a rigid body was solved for symmetric top molecules[19] and asymmetric top non-linear molecules, allowing the rotational energy levels to be obtained. Thus, by the early 1930s, quantized molecular energy levels were understood. For example, the energy levels for CO are shown in Figure 4.8. The energy differences between electronic states, between vibrational states, and between rotational states, correspond respectively to UV-visible light, infrared light, and microwave energy. Electromagnetic absorption or emission occurs when a molecule transitions between quantized energy levels. In 1917, Einstein developed a general theory pertaining to electromagnetic emission and absorption, and derived a function for the probability of absorption, induced emission, and spontaneous emission. [59] With the development of quantum mechanics, this transition probability was approximated by a technique called perturbation theory,[20] and based on this, selection rules were clarified pertaining to transitions between states. Thus, the use of molecular spectroscopy to observe the intensity of electromagnetic wave absorption and emission in relation to electromagnetic wavelength emerged as

---

18. Normal vibration: A pattern of vibration in which all parts of the molecule vibrate simultaneously with the same frequency and phase.

19. Symmetric top molecule: A molecule in which two moments of inertia are the same.

20. Perturbation method: A method for obtaining an approximate wave function when perturbations from electric fields and magnetic fields are small in comparison to the energy of the system.

Gerhard Herzberg (1904–1999)
Gerhard Herzberg was born in Hamburg, Germany. He studied at the Darmstadt University of Technology, and after conducting research under Planck and Born at Göttingen, became a professor at Darmstadt. Hertzberg was pursued by the Nazis, moved to Canada in 1935, and became a professor at the University of Saskatchewan. He was employed as a professor in 1945 for Yerkes Observatory at the University of Chicago, but after three years moved to the National Research Council of Canada in Ottawa. There, he conducted research on the spectra of diatomic molecules and free radicals, and made major contributions by determining the electronic and geometric structures of molecules. In addition, he conducted pioneering spectroscopic studies on interstellar molecules. He was awarded the Nobel Prize for Chemistry in 1971.

a powerful technique to provide detailed information about molecules. These developments in turn led to advances in structural chemistry.

Electronic spectra were interpreted based on the Franck-Condon principle (1926), which assumes that the internuclear distances within a molecule remain constant during electronic state transitions. The electronic spectrum of a gas of diatomic molecules includes detailed information about the electronic state and vibrational energy levels. Analysis of the spectrum allows the interatomic distance and dissociation energy to be obtained. The molecular structures of simple polyatomic molecules comprising several atoms were elucidated from these analyses. Many physicists and physical chemists investigated the electronic spectra of molecules in the 1930s, but the most noteworthy contributions were made by Gerhard Herzberg. In particular, his elucidation of the structures of short-lived free radicals from analysis of their high-resolution spectra had a major impact on chemical reaction theory, photochemistry, and astronomy. Herzberg was awarded the Nobel Prize for Chemistry for this achievement in 1971. The books he authored, "Atomic Spectra and Atomic Structure" (1937) and "Molecular Spectra and Molecular Structure," I (1939), II (1945) and III (1966), are classic books that comprehensively covered the literature of the time in this field.

Infrared spectroscopy and Raman spectroscopy are two types of vibrational spectroscopy useful for observing transitions between vibrational energy levels. Both types of spectroscopy allow detailed analysis of vibrations, providing information on bond distances, bond energies, and intramolecular potentials. In infrared spectroscopy, the sample is illuminated with infrared light and

Chandrasekhara Raman (1888–1970)
Chandrasekhara Raman was born in Tiruchirappalli, India. He received an education in physics at the Madras Presidency College, and obtained a MA degree in 1907. Subsequently, he was employed by the India Ministry of Finance, but continued doing research on the side at the Indian Association for the Cultivation of Science in Calcutta. In 1917, he became a professor of physics at the University of Calcutta, where he continued research on light diffraction and scattering phenomena, leading to the discovery of Raman scattering. He was awarded the Nobel Prize for Physics in 1930.

light absorption by the sample is measured. Transitions between vibrational levels are evident in the near-infrared region, providing a vibration-rotation spectrum, whereas interatomic distances are obtained from analysis of rotational structure. The characteristic vibration frequency of a molecule depends on the structure of the molecule, but because bonds such as OH, NH, CH, C-C, C=C, and C=O within a molecule each show absorption at specific frequencies, infrared spectroscopy could be used to identify organic compounds. In contrast, Raman spectroscopy utilizes light scattered by the sample upon illumination. Most of the scattered light has the same frequency as the incident light, but the frequency of some of the scattered light is shifted (Raman scattering). This phenomenon, known as the Raman effect, was discovered by Chandrasekhara Raman and K.S. Krishnan at the University of Calcutta in India in 1928.[60] Raman and Krishnan created a powerful light source by focusing sunlight with a lens and used this light to illuminate various liquid and gas samples. Using optical filters, they observed scattered light with frequencies different from the incident light. The frequency shift due to Raman scattering differed depending on the atoms or ions in the sample, and information about molecular vibration and rotation could be obtained from differences in the frequencies of the incident light and scattered light. Since the selection rules for infrared spectroscopy and Raman spectroscopy are different, the two techniques are complementary. After the discovery of the Raman effect by Raman and Krishnan, other scientists immediately performed experiments to confirm the effect, and in the 1930's the Raman effect was used to study molecular structures.

Landsberg and Mandelstam in the Soviet Union discovered the same effect using solid samples at about the same time as Raman and Krishnan, but Raman was widely credited for the discovery and was awarded the Nobel Prize for Physics in 1930.

Pure rotational spectra resulting from transitions between rotational energy levels appear in the far infrared region for molecules composed of low-mass elements, and in the microwave region for heavy molecules. Microwave technology advanced due to the development of radar during World War II and was used for scientific research after the war. Microwave spectroscopy was established as a technique at the end of the 1940s, and research on rotational spectra flourished.

The magnetic properties of electrons and nuclei, and magnetic resonance

Electrons have a spin-dependent characteristic angular momentum S, and accordingly, have a magnetic moment of $\mu = -g_e\mu_B S$. Classically, this is imagined as a magnetic momentum produced when an electron rotates on its axis. Here, μ_B is a constant called the Bohr magneton,[21] and g_e is a constant called the g value. Under an external magnetic field, the spin angular momentum is quantized, and the allowable values in the direction of the magnetic field are $\pm(1/2)h/2\pi$. Consequently, the energy of an electron within a magnetic field of strength B is split into two as $\pm(1/2)g_e\mu_B B$ by the interaction with the magnetic field (Figure 4.9). Nuclei with a mass number that is an odd number all have a spin represented by quantum number I, and the value of I is an odd number times 1/2. Accordingly, since nuclei have a nuclear magnetic moment μ_N, the component of μ_N in the direction of the magnetic field within the magnetic field is a value that is quantized in the same manner as for an electron, and the energy is split as 2I+1 with $-g_N\mu_N M_I B (M_I = 1, I-1, ..., -I)$. Magnetic resonance is the observation of transitions due to electromagnetic waves between adjoining levels of spin energy split within a magnetic field. For electrons, this resonance is electron spin resonance (ESR) or electron paramagnetic resonance (EPR), and for nuclei, it is nuclear magnetic resonance (NMR). Microwaves are used to cause an electron spin transition within a magnetic field of a few thousand Gauss, whereas radio waves are used to elicit a nuclear spin transition in a similar magnetic field.

In 1938, Isidor Rabi of Columbia University in New York used the magnetic resonance method with a molecular beam to determine the nuclear magnetic moment of lithium chloride. This was the first successful example of magnetic resonance and marked the beginning of nuclear magnetic resonance spectroscopy[61] Although the magnetic resonance of condensed systems had been predicted in the 1930's by Gorter in the Netherlands, Yevgeny Zavoisky in the Soviet Union was the first to measure the ESR spectra of manganese, chromium and copper salts in 1944.[62]

---

21. Bohr magneton: A unit for quantifying the magnetic moment of an electron. $\mu_B = eh/2m = 9.274 \times 10^{-24}$ J/T

Isidor Rabi (1898–1988)

Rabi was born in Rymanow in an area of present-day Poland to poor Jewish parents who emigrated to New York while he was a child. He received a scholarship and studied chemistry at Cornell University, and obtained a degree in physics at Columbia University in 1927. Subsequently, Rabi studied under Bohr, Heisenberg, Stern and others in Europe at the time quantum mechanics was being developed. After return-
ing to the US, Rabi precisely measured the magnetic moment of atomic nuclei and molecules by combining a magnetic resonance technique with Stern's molecular beam method. He was awarded the Nobel Prize in Physics in 1944.

Yevgeny Zavoisky (1907–1976)

Born in Mogilyov-Podolsk in southern Russia, Yevgeny Zavoisky studied physics at Kazan University. In 1933, he became an associate professor of experimental physics at Kazan University, and investigated the absorption of radio waves by different compounds. In 1944, he measured the electron spin resonance of manganese, chromium, and copper salts.

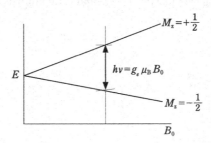

Figure 4.9: Zeeman splitting resulting from the interaction of a magnetic field with electrons of spin quantum number (S) 1/2

The horizontal axis is the magnetic field strength B_0 and the vertical axis is energy E. In the magnetic resonance method, transitions due to absorption of radio waves or microwaves are measured between the two levels split in the magnetic field.

After World War II, a group led by Bleaney at Oxford focused on studying the ESR spectra of paramagnetic salts. The foundation provided by their work allowed ESR to be subsequently widely used in chemistry research. NMR in condensed systems was independently discovered in 1946 by Purcell's group at Harvard University and by Bloch's group at Stanford University. NMR made tremendous advancements in the second half of the 20th century, becoming one of the most important analytical tools in chemistry, as described in detail in the next chapter.

Rene Marcelin (1885–1914)
Rene Marcelin was a gifted French physical chemist who was killed in action during World War I while still a young man. In the Faculty of Science at the University of Paris, Marcelin developed a chemical reaction rate theory based on thermodynamics and statistical mechanics, and in 1914, introduced the concept of a potential surface in the reaction. In addition, he carried out pioneering research in surface chemistry.

4.4 Physical Chemistry (III): Chemical Reaction Theory and the Development of Colloid and Surface Chemistry

Development of Chemical Reaction Theory

There had been little progress in the theory of chemical reactions by the beginning of the 20th century. This changed, first with the emergence of a thermodynamics reaction theory, then with a reaction theory derived from collision theory. This led, in the 1930's, to the emergence of a statistical mechanics reaction theory, called the transition state theory. Together, these theories provided the path to a theoretical understanding of chemical reactions. Pfaundler, van't Hoff, and Arrhenius had suggested in the latter half of the 19th century that reactions required activated molecules.

In 1910 the French physical chemist Rene Marcelin showed that the rate of a reaction could be expressed using affinity (Section 2.8).[63] If Marcelin's affinity is replaced with the Gibbs free energy, the reaction rate v is given by the following equation.

$$v = C[\exp(-\Delta G_1^{o\ddagger}/RT) - \exp(-\Delta G_{-1}^{o\ddagger}/RT)]$$

Here, $\Delta G_1^{o\ddagger}$ and $\Delta G_{-1}^{o\ddagger}$ represent the free energy of activation. The subscript 1 indicates that the reaction proceeds to the activated state, and the subscript −1 indicates the reverse reaction. In the 1920's, the Dutch chemist Scheffer and colleagues expanded on this idea and expressed the reaction rate constant k as

$$k = v\exp(-\Delta S^{o\ddagger}/RT)\exp(-\Delta H^{o\ddagger}/RT)$$

They introduced the concepts of the entropy of activation $\Delta S^{o\ddagger}$ and the enthalpy of activation $\Delta H^{o\ddagger}$. However, they were unable to clearly define the significance of v.

From 1916 to 1918, Trautz in Heidelberg and McCullagh Lewis at the University of Liverpool formulated reaction rates using collision theory by assuming that reactions in the gas phase result from the collision of molecules. [64] Their theory was constructed within the conceptual framework that a unimolecular reaction[22] in the gas phase is initiated by infrared radiation from the walls of the vessel. This idea motivated many experimentalists, but the inadequacy of this theory was gradually recognized and it was discarded around 1920. However, the theory was useful because it predicted that reaction rates were determined by the frequency of molecular collisions. Lewis calculated the number of collisions per unit time and unit volume between two molecules, A and B, which were regarded as rigid spheres. Using Clausius' kinetic theory of molecules, assuming that only collisions with sufficient energy can cause a reaction, and that the fraction of collisions with adequate energy is given by $\exp(-E_A/RT)$, Lewis obtained the reaction rate constant k given by the following equation.

$$k = N_0 d_{AB}^2 \{8\pi\, k_B T(m_A + m_B)/m_A m_B\}^{1/2} \exp(-E_A/RT) = Z_{AB}\exp(-E_A/RT)$$

Here N_0 is the number of molecules per unit volume, d_{AB} is the sum of the radii of A and B, k_B is the Boltzmann constant, and m_A and m_B are the masses of molecules A and B, respectively. Z_{AB} is the number of molecular collisions. Lewis applied this formula to the reaction $2HI \rightarrow H_2 + I_2$ and obtained good agreement with experimental values. Subsequently, however, it was found that the theoretical and experimental values for k for many reactions deviated widely, and it became apparent that reaction rates could not be understood by collision theory. In particular, differences of several orders of magnitude were observed for reactions involving complex molecules. In 1930, Hinshelwood and Moelwyn-Hughes introduced a correction factor p and used pZ_{AB} in place of Z_{AB}. The value of p varies with the individual reaction, and they considered that in order for a reaction to occur, a molecule must collide in a suitable steric configuration. Simple collision theory was clearly insufficient, but in the course of the development of chemical reaction theory, collision theory played an important role, and efforts to explain the differences between the predictions of collision theory and experimental values resulted in more refined theories.

---

22. Unimolecular reaction: A reaction in which only one molecule participates. Typical examples would be reactions in which an excited molecule spontaneously dissociates or isomerizes.

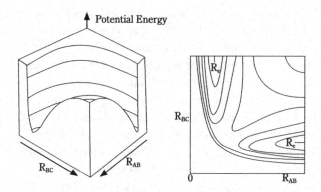

Figure 4.10: Chemical reaction potential surface

Left: The figure shows the energy change of the system for the reaction $H_A + H_BH_C = H_AH_B + H_{C'}$. The energy change for the reaction system assumed to be linear is shown as a function of inter-molecular distances R_{AB} and R_{BC}. The vertical axis is energy.

Right: Figure showing lines of equal energy (equivalent to a contour map) for the potential sur-face shown in the left-hand figure. R_e is the equilibrium internuclear distance.

The most important event in the development of chemical reaction theory in the first half of the 20th century was the emergence of transition state theory. In 1914, Marcelin treated a chemical reaction as the movement of a particle within a phase space and represented the reaction rate using a statistical mechanics technique developed by Gibbs.[65] Sadly, however, Marcelin was killed in action in World War I shortly after proposing this theory. In 1919, Herzfeld applied statistical mechanics to an equilibrium reaction to determine the rate of a dissociation reaction and obtained[66]

$$k = (k_BT/h)[1-\exp(1-(h\nu/k_BT)]_e{}^{-E/RT}$$

Here, ν is the bond vibration frequency, and h is Planck's constant.

In 1931, Henry Eyring and Michael Polanyi at the Kaiser Wilhelm Institute in Berlin calculated a three-dimensional potential energy surface[23] in the course of a reaction for the $H + H_2 \rightarrow H_2 + H$ reaction using a semi-empirical quantum mechanical technique.[67] (Figure 4.10) The following year, Pelzer and

---

23. Potential energy surface: The energy E of the electronic state of a molecule is a function of the internal coordinates indicating the atomic configuration within a molecule. The potential energy surface is a diagram that shows E as a function of internal coordinates. Usually the potential energy surface is drawn with only one or two internal coordinates as variables while the remaining internal coordinates are fixed.

Henry Eyring (1901–1981)

Henry Eyring was born to Mormon parents in Mexico. After studying at the University of Arizona, he obtained his Ph.D. at the University of California, Berkeley in 1927. He became an assistant professor at Princeton in 1931 and the dean of the graduate school of the University of Utah in 1946. Eyring left behind many outstanding contributions in broad areas of theoretical chemistry, but his theoretical research on chemical reactions, and particularly transition state theory, had a major influence on many areas of chemistry. In addition, in 1944 he published *"Quantum Chemistry"*[71] with coauthors Walter and Kimball, and this book long continued to be read as an introductory book on quantum chemistry, along with Pauling's book (see page 248). A second book by Eyring, *"The Theory of Rate Processes"*,[72] published with coauthors Glasstone and Laidler in 1941, was also widely read.

Michael Polanyi (1891–1976)

Michael Polanyi was born in Budapest, Hungary, and studied medicine at the University of Budapest, but he took the opportunity to study chemistry in Haber's laboratory and became a chemist. After World War I, he was head of Haber's section at the Kaiser Wilhelm Institute and was active in a broad range of fields of physical chemistry such as chemical reaction theory, X-ray diffraction, and surface science. In 1933, Polanyi escaped from Nazi Germany, went to England, and continued research at the University of Manchester. After the war, he turned to philosophy and sociology where he advocated the importance of "tacit knowledge" and became a well-known philosopher.

Wigner in Göttingen discussed the reaction system potential surface focusing on the saddle region and obtained a formula describing the reaction rate.[68]

With this background, Eyring at Princeton University and Polanyi and Evans at the University of Manchester in 1935 proposed a transition state theory that could be broadly applied to chemical and physical processes.[69, 70] The essence of this theory was to calculate the concentration of the activated complex by statistical mechanics techniques by assuming that the activated complex shown at a point near the saddle area of the potential surface is in a state of quasi-equilibrium with the reactants. The reaction rate constant k is given by

$$k = (k_B T/h)(q_+/q_A q_B)\exp(-E_0/RT)$$

Here, q_A and q_B are the partition functions[24] for the activated complex q_+ for reactants A and B, and E_0 is the difference in the zero point energies for the activated complex and reactants. In a thermodynamic expression of the transition state theory, k is given by

$$k = (k_B T/h)\exp(-\Delta S^{o\ddagger}/R)\exp(-\Delta H^{o\ddagger}/RT)$$

Here, $\Delta H^{o\ddagger}$ is the enthalpy of activation, and $\Delta S^{o\ddagger}$ is the entropy of activation. It is not possible to precisely calculate the reaction rate constant using transition state theory, but transition state theory provides a framework for understanding how a chemical reaction occurs. For example, the theory provides a qualitative understanding of why reaction rates between polyatomic molecules are much slower than predicted from collision theory. The theory helps to explain experimental results related to the effects of various parameters such as solvents, salt, pressure and isotopes on reaction rates. It was also used widely to understand a variety of organic reactions. Although Eyring and Polanyi did not win a Nobel Prize, it seems to the author that transition state theory was the biggest contribution to chemical reaction rate theory impacting a broad area of chemical research in the first half of the 20th century.

The collision of two molecules in a solution is controlled by molecular diffusion. If two colliding molecules react immediately, the reaction is controlled by diffusion (diffusion-controlled reaction). In the first half of the 20th century, there were no techniques for investigating such fast reactions. Nonetheless, in 1916 Smoluchowski showed during the course of investigating Brownian motion that the rate of a diffusion-controlled reaction, k_d, is given by the following formula[73]

$$k_d = 4\pi(D_A + D_B)$$

Here, D_A and D_B are the diffusion coefficients for molecules A and B.

The Scottish chemist Lapworth suggested in 1908 that activity, and not concentration, should be used to indicate the rate of reactions involving ions, such as an acid-base reaction in a solution. If a reaction proceeds by the scheme A+B → X → reactants and the activity coefficient is represented by γ, then the rate v for a bimolecular reaction can be represented by $v = k[A][B] \, \gamma_A \gamma_B/\gamma_X$.

---

24. Partition function: In a collection of systems in which the probability of energy being E_i is distributed proportional to $\exp(-E_i/kT)$, the partition function is given by $Z = \Sigma_i \exp(-E_i/kT)$.

Following the establishment of the Debye-Hückel theory, Brøensted, Bjerrum and Christiansen used this formula in the 1920s to describe experimental results.

In 1940, the Dutch physicist Kramers set about taking the dynamic influence of solvent on the reaction rate into consideration. Kramers started from the Langevin equation for the Brownian motion of a particle under the influence of a frictional force and a random force. The frictional force is proportional to the velocity in the direction of the reaction coordinate, and the random force results from collisions with solvent molecules. Kramers obtained an expression for the rate constant that a particle goes over a one-dimensional potential barrier.[74] This equation treats Kramers' transmission coefficient κ^{KR} as a factor for multiplying the rate constant from transition state theory k^{TST}. It was shown that when the frictional force increases, the rate constant becomes much smaller than k^{TST}. Kramers' theory attracted the attention of chemists in the latter half of the 20th century.

Understanding Thermal Reactions and Chain Reactions

In 1907, Bodenstein and Lind discovered that the reaction rate v for gas-phase hydrogen (H_2) and bromine (Br_2) is not represented by a simple reaction formula, but rather by the following complex formula.[75]

$$v = k[H_2][Br_2]^{1/2}/(1+[HBr]/m[Br_2])$$

Here, k and m are constants. In 1919, Christiansen and Herzfeld proposed the following reaction mechanism to explain this.

(1) $Br_2 \longrightarrow 2Br$,   (2) $Br+H_2 \longrightarrow HBr + H$,   (3) $H + Br_2 \longrightarrow HBr + Br$
(4) $H + HBr \longrightarrow H_2 + Br$ (5) $2Br \longrightarrow Br_2$

Reaction (4) is the reverse reaction of (2), and HBr produced by reactions (2) and (3) impedes the progress of reactions (2) and (3). Pyrolysis of the Br_2 molecule of (1) proceeds by this mechanism. When the reaction is initiated by photolysis instead of pyrolysis, it was shown that the same reactions occur. A rate equation is obtained for this reaction by using a steady-state assumption in which the concentrations of H atoms and Br atoms participating in the reaction are small,and their time derivatives are given as 0 (d[H]/dt=0 and d[Br]/dt=0). This steady-state assumption was proposed by Chapman and Bodenstein in 1913 and subsequently was used to derive many complex reaction rate equations.

There was much confusion leading up to the elucidation of a mechanism for the decomposition of a single molecule in a gas-phase thermal reaction. In

Cyril Hinshelwood (1897–1967)
Cyril Hinshelwood was born in London and studied at Oxford. He worked in an explosives factory during World War I. In 1921, he became a lecturer at Oxford, and in 1937 became a professor there. In 1922, he published a theory for unimolecular reactions which focused on the role of molecular internal energy. Hinshelwood made many contributions in the 1920s and 1930s towards understanding gas-phase chemical reactions with a focus on chain reactions in the gas phase, such as the reaction to form water from hydrogen and oxygen. Subsequently, he studied liquid phase catalytic reactions and intracellular reactions of bacteria. He was awarded the 1956 Nobel Prize for Chemistry.

1919, Perrin argued that unimolecular reactions occur as a result of absorption of radiation from the walls of the reaction vessel. At that time, radiation theory was accepted by many chemists to describe gas phase reactions in general. According to this idea, reactions that did not depend on pressure were expected to be first order reactions. In 1921, Lindemann in Oxford and Christiansen in Copenhagen proposed that in low pressure regions, unimolecular reactions occur because molecules are activated as a result of molecular collisions. However, in high pressure regions, they proposed that activated A* molecules are deactivated by collision with other A molecules, and at equilibrium, the reaction becomes first-order because the concentration of A* is proportional to A.[76, 77] More specifically, the reaction mechanism can be written as follows.

$$A + A \rightleftharpoons A^* + A, \quad A^* \longrightarrow B + C$$

In low pressure regions, the deactivation of activated molecules does not readily occur, and the reaction becomes a second-order reaction proportional to the rate at which A* molecules are generated. That unimolecular reactions become second-order reactions at low pressure was subsequently confirmed experimentally, and the fundamental correctness of this proposal was acknowledged. However, experimental results and predictions did not agree quantitatively, and various modifications to the theory were proposed.

In 1927, Cyril Hinshelwood pointed out that the energy of an activated molecule is partitioned into molecular vibrational degrees of freedom and corrected the rate of activation.[78] That same year, Rice and Ramsperger at the University of California applied a correction that took into account the fact that the rate at which activated molecules degrade to products depends on the energy.[79] Kassel at the University of Chicago also published the same correction immediately afterwards, and this revised theory became known as the RRK theory of unimolecular reactions.[25][80]

Thermal reactions of organic molecules in the gas phase were also actively investigated in the 1920s and 1930s. In 1925, Taylor suggested the possibility of organic free radicals participating in reactions. He investigated the mercury photosensitized reaction between hydrogen and ethylene and suggested the following mechanism.

$$H + C_2H_4 \longrightarrow C_2H_5, \quad C_2H_5 + H_2 \longrightarrow C_2H_6 + H$$

In this reaction, a hydrogen molecule dissociates upon interaction with a mercury atom that has been excited by ultraviolet light. One of the resulting hydrogen atoms reacts with ethylene to form C_2H_5, which reacts with hydrogen to form H and C_2H_6, and a chain reaction occurs. The presence of a free radical such as C_2H_5 was shown in 1929 by Paneth and Hofeditz at the University of Berlin to trap radicals on a thin film containing vapor-deposited metal such as lead (metal mirror). For example, gaseous tetramethyl lead was identified as a methyl radical that reacted with lead.

In 1934, F.O. Rice and Herzfeld considered the mechanisms by which free radicals participate in the decomposition of many organic molecules and described the reaction rates for these mechanisms. For example, acetyl aldehyde pyrolytically decomposes to methane and carbon monoxide, but the reaction order[26] is 1.5. Rice and Herzfeld described this reaction by a mechanism that included the free radicals CH_3CO and CH_3.

Chain reactions were useful for describing explosions. In 1923, Christiansen and Kramers suggested the possibility of a branched chain reaction. When one atom or radical that is a chain reaction carrier generates two or more carriers, the chain reaction carriers rapidly increase, accelerating the reaction rapidly and giving rise to the possibility of an explosion. A specific experimental example of a branched chain reaction leading to explosion was obtained for the reaction of P_5 and O_2 by Nikolay Semenov of the Soviet Union: he discovered that an explosion occurs above a given pressure limit.[81] In contrast, Hinshelwood investigated the reaction of H_2 and O_2 and discovered the existence of a high pressure limit above which an explosion will not occur.[82] The existence of a high pressure limit was attributed to removal of the chain reaction carriers. Hinshelwood and Semenov

---

25. RRK theory: RRK was derived from the initials for Rice, Ramsperger, and Kassel.

26. Reaction order: The sum of all the exponents of the terms expressing concentrations of molecules or atoms determining the rate of the reaction.

Nikolay Semenov (1896–1986)
Nikolay Semenov was a Soviet chemist. He graduated from Petrograd
University and in 1928 became a professor at the Leningrad Polytechnic
Institute. In 1935, he became the director of the Moscow Chemico-Physical
Institute, and in 1944 concurrently became a professor at the University of
Moscow. Semenov is known for his theory of chain reactions in explosive
reactions and the discovery of the existence of an explosion limit in mixed
gases. He was awarded the 1956 Nobel Prize for Chemistry.

were awarded the Nobel Prize for Chemistry in 1956 "for their research into the
mechanism of chemical reactions, particularly chain reactions."

Photoreactions and excited molecules

The idea that light absorbed by a molecule causes a photochemical reaction
had been proposed in the first half of the 19th century by Grotthuss (1817) and
Draper (1843). Following the light quantum hypothesis by Einstein in 1905 (see
page 204), Stark in 1908 and Einstein in 1912 each introduced the concept of
photons in photochemical reactions.[83, 84] They proposed the principle of photon
activation: in the first stage of a photochemical process, one molecule is activated
by the absorption of one photon and attains an excited state having energy
$E=h\nu$. However, because an activated molecule does not necessarily react, and
because one additional activated molecule produces another activated molecule
or atom to continue the reaction, photoreactions were difficult to understand.
The principle of photon activation led to the concept of the quantum yield of a
photochemical reaction, Φ, that is, the number of reacting molecules consumed
or generated per absorbed photon.

For example, Warburg discovered in 1918 that two molecules of HI decompose
to produce H_2 and I_2 following the absorption of one photon and proposed that
the reaction proceeded as follows.

$$HI + h\nu \longrightarrow H + I, \ H + HI \longrightarrow H_2 + I, \ I + I \longrightarrow I_2$$

In this case, two molecules were produced by the absorption of one photon to
provide a Φ of 2.

In 1913, Bodenstein discovered that in the reaction by which HCl is produced
from H_2 and Cl_2, an extremely large number of molecules react following the
absorption of one photon, and that Φ is between 10^4 to 10^6.[85] In 1918, Nernst
suggested the following chain reaction mechanism to explain this.[86]

$$Cl_2 + h\nu \longrightarrow 2Cl, \; Cl + H_2 \longrightarrow HCl + H, \; H + Cl_2 \longrightarrow HCl + Cl$$

This mechanism showed that the Cl atoms produced reacted successively to cause a chain reaction, and that a large number of HCl molecules were generated by the absorption of one photon.

It had been known for some time that molecules excited by light give off fluorescence or phosphorescence, but the detailed mechanism was not understood. In 1935, Jablonski in Poland proposed a scheme to explain the phosphorescence produced by dye molecules. The scheme involved three levels: a ground state, a fluorescent state, and a metastable state that gives off phosphorescence. Research into the electronic states of molecules using the molecular orbital method was progressing and the excited states of molecules were beginning to be understood. In 1944, G.N. Lewis and M. Kasha suggested that the metastable state of aromatic organic compounds was an excited triplet state. Fluorescence was due to radiative transition between states of the same spin multiplicity (usually between singlet states), and phosphorescence was due to the radiative transition between states of different spin multiplicity (usually from triplet state to singlet state).[87] However, the triplet state theory was not verified until after World War II.

Colloid Chemistry

Colloids and surfaces became subjects for physical chemistry research in the latter half of the 19th century (see Section 2.8), but their nature was not understood. The beginning of the 20th century saw progress in quantitative research and in observational technologies, and the field began to rapidly develop. Concurrently, colloids drew the interest of many researchers. Initially, inorganic colloids such as metal sol and sulfides were the main subjects of interest, but with the development of biochemistry, interest moved to macromolecules such as proteins. At the time it was generally believed that a protein was a colloidal aggregate of small molecules, and this generated interest in colloids. A colloid has a large surface area, and an interest in colloids is inevitably linked to an interest in surfaces and interfaces. Consequently, surface and interface chemistry underwent major development from the beginning of the 20th century.

Understanding colloid particles required determining their size and mass. Colloid particles have been investigated from their light scattering, sedimentation, viscosity, and colligative properties. Tyndall discovered in 1869 that colloid particles scatter light; this phenomenon is known as the "Tyndall effect". Lord Rayleigh conducted detailed research in the 1870s on light scattering by particles much smaller than the wavelength of light, and showed

Richard Zsigmondy (1865–1929)
Richard Zsigmondy was born in Vienna and studied at the universities of Vienna, Munich, and Berlin. After obtaining his Ph.D. at the University of Munich, he spent time at the universities of Berlin and Graz and Schott AG in Jena, and then became a professor at the University of Göttingen. He is known for his accomplishments in areas such as gold colloid and colored glass research, and the elucidation of colloidal states based on light-scattering theory. He and Siedentopf jointly developed the ultramicroscope, allowing Zsigmondy to make major contributions not only to colloid chemistry, but to developments in the fields of biochemistry and medicine. He was awarded the 1925 Nobel Prize for Chemistry.

that the intensity of scattered light was inversely proportional to the wavelength to the 4th power. A theory to describe the case where the size of the scattering particle is large compared to the wavelength was proposed by Mie in 1907. In the 1920s, Staudinger founded the field of macromolecular chemistry and synthetic polymers became a subject of interest in physical chemistry. In 1947, Debye applied Rayleigh's theory to polymer solutions, introducing the use of light scattering for determining the molecular weight of polymers.

The Austrian-born chemist Richard Zsigmondy investigated the use of the Tyndall effect for observing colloid particles under a microscope. He collaborated with the physicist Siedentopf to develop an ultramicroscope in which light is passed through a colloid system and the light scattered at right angles to the incident light rays is observed. Thus, minute particles that cannot be observed directly can be observed as scattered light. The development of the ultramicroscope was reported in 1903 and it was immediately used to observe colloids and confirm the existence of microparticles. Based on this, in 1907 Wolfgang Ostwald defined a colloid as a "state in which a substance is dispersed to a size of about 0.2 to 1 µm". Zsigmondy was awarded the Nobel Prize for Chemistry in 1925 for the development of the ultramicroscope. The ultramicroscope also made major contributions to the fields of biology and bacteriology.

The observations of Brownian motion and sedimentation equilibria by Perrin, and Einstein's and Smoluchowski's theory of Brownian motion at the start of the 20th century, were covered in Chapter 3. This lead to a particularly important relationship from a physical chemistry perspective: the equation relating the diffusion coefficient D to the viscosity of a solution η, and the radius of a particle r pertaining to the diffusion of a particle within a solution.

Theodor Svedberg (1884–1971)
Theodor Svedberg was born in Valbo, Sweden, studied at Uppsala University, and subsequently became a professor at Uppsala University and director of the Physical Chemistry Institute.  He verified the experimental evidence for Brownian motion in regard to colloid particles. He developed ultracentrifugation techniques which he used to understand colloid chemistry and to determine the molecular weights of macromolecules such as proteins. Svedberg made major contributions to the development of biochemistry, biophysics, and macromolecular chemistry. In addition, together with Tiselius, he is known as a pioneer in the development of electrophoresis. He was awarded the 1926 Nobel Prize for Chemistry.

Arne Tiselius (1902–1971)
Arne Tiselius was a Swedish chemist. After graduating from Uppsala University, he continued research as Svedberg's assistant and became a professor at Uppsala University in 1938. He had notable achievements in the electrophoresis of proteins, particularly the development of an electrophoresis apparatus, and adsorption analysis of amino acid and proteolysis solutions. He was awarded the 1948 Nobel Prize for Chemistry.

$$D = k_B T/6\pi r\eta$$

This is called the Stokes-Einstein equation. Here, k_B is Boltzmann constant. Einstein derived the relationship between frictional force f and the diffusion coefficient D for the motion of particles in a solution. Einstein obtained this formula using the frictional forces on a particle of radius r in a solution with viscosity η, ($f = 6\pi r\eta$); this formula had been derived by Stokes in 1851.This equation is still widely used to describe the motion of molecules in a solution.

Perrin's research pertaining to the sedimentation of colloid particles had been limited to the force of gravity and hence was limited to the investigation of large particles and could not be applied to protein molecules. Theodor Svedberg of Uppsala University in Sweden carried out innovative research on the Brownian motion of colloid particles and related phenomena. In 1923, he developed an ultracentrifuge, thus allowing the properties of macromolecules to be investigated using gravitational fields of up to 3×10^5 g. The sedimentation velocity method he developed in 1929 laid the foundation for the determination of the molecular weights of macromolecules, including proteins. Svedberg was awarded the Nobel Prize for Chemistry in 1926 for "colloid research using the ultracentrifuge".

The fact that colloid particles and proteins carry an electric charge was already known at the end of the 19th century. When an electric potential is applied, these particles move toward an electrode in accordance with their electric charge. This is called electrophoresis, and Arne Tiselius in Svedberg's laboratory conducted detailed research on this phenomenon starting around 1930. He developed a U-shaped apparatus to investigate the electrophoresis of macromolecules such as blood proteins. This method subsequently became an extremely important method for the analysis of proteins. Tiselius was awarded the Nobel Prize for Chemistry in 1948 for electrophoresis and adsorption analysis research.

Surface and Interfacial Chemistry

Lord Rayleigh and Pockels (see Section 2.8) conducted research into mono-molecular films from the end of the 19th century to the beginning of the 20th century, and Marcelin continued this research prior to World War I, but it was Irving Langmuir who brought about major developments in this field. In 1917, Langmuir devised a surface pressure gauge to directly measure the surface pressure produced by a film on a liquid to investigate in detail the relationship between surface pressure and surface area. His research revealed the existence of two types of films.[88] In the case of a film of a long-chain fatty acid such as stearic acid on water, there is a critical surface area above which the surface area changes little with respect to increases in pressure. This area corresponds to the area of a monomolecular film created by the fatty acid molecules lining up with their hydrophilic groups in the water and their hydrophobic chain raised into the air. Another type of film was a surface film that behaved like a two-dimensional gas. In 1935, Langmuir and Blodgett developed a technique for making a built-up film of monolayers by transferring a monolayer on the surface of water to a solid substrate such as glass (Langmuir-Blodgett film, LB film). This technique attracted attention in nanotechnology in the latter half of the 20th century.

Langmuir also began innovative research on quantitative investigations related to the adsorption of gases onto solid surfaces.[89] In 1916 he proposed that in addition to physical adsorption in which molecules adsorb onto a surface by van der Waals forces, there is chemical adsorption in which adsorption occurs by the formation of chemical bonds. He introduced an equation called the Langmuir adsorption isotherm equation to show the relationship between the amount of gas adsorbed onto a surface and the gas pressure. This equation was derived based on a model that assumes that there are a fixed number of adsorption sites on the

Irving Langmuir (1881–1957)

Irving Langmuir was born in New York, and after studying metallurgical engineering at Columbia University, he studied under Nernst and obtained his Ph.D. at the University of Göttingen. In 1909, he started working at the research laboratory of General Electric (GE) Co. and continued research there throughout his career. His initial research was on the heat conduction of gases in the presence of a heated metal filament, with the aim of improving the lifetime of light bulbs. He determined that the lifetime of a light bulb depended on the rate of evaporation of the tungsten filament, leading him to recognize the importance of interactions between the filament and gas. He proposed the concept of monolayer adsorption and was a pioneer in solid surface chemical research. In addition, he started research into oil films on the surface of water, and was a trailblazer in monolayer research as well. Together with Blodgett, he pioneered monolayer production methods. He was awarded the 1932 Nobel Prize for Chemistry.

solid surface, that one molecule adsorbs onto one adsorption site, and that there are no interactions between molecules adsorbed onto different sites. According to this model, the relationship between the fraction of the surface occupied by adsorbed molecules θ and gas pressure p is represented by

$$\theta = Kp/(1 + Kp)$$

Here, K is an equilibrium constant for the adsorption process. Accordingly, it was shown that when the pressure is low, $\theta = Kp$, and θ increases in proportion to pressure.

Since the Langmuir isotherm equation is based on a model of uniform adsorption sites, there are many real, non-uniform surfaces which do not conform to the isotherm equation. In these cases, the empirical Freundlich equation, $\theta = Kp^{1/m}$ is often used (m is a number greater than 1).

Langmuir was awarded the Nobel Prize for Chemistry for achievements in surface chemistry in 1932. He joined the research laboratory of General Electric (GE) Co. in 1909. His productive research had the support of the laboratory director, who believed that basic research was important for solving practical problems. Langmuir was the first American industrial researcher to be awarded a Nobel Prize. His career is a representative example of research aimed at solving practical issues that led to important fundamental discoveries.

4.5 The Birth of Nuclear and Radiochemistry

The discovery of radioactivity had an extremely large impact on chemistry during the first half of the 20 century, as seen by the large number of scientists awarded the Nobel Prize in Chemistry up to 1960 who investigated radioactivity, radionuclides, and nuclear reactions.

Rutherford, Marie Curie, Soddy, Frederic Joliot, Irene Joliot, Hevesy, Hahn, and Libby make up a diverse cast of researchers. While the advancement of nuclear research was led by physicists, with Rutherford at Cambridge playing a central role, the study of radioactivity and nuclear reactions developed as an interdisciplinary domain of physics and chemistry. Hahn, who discovered nuclear fission, and Hevesy, a pioneer of radiochemistry, were chemists, but they carried out research in collaboration with physicists, incorporating the findings from new physics and achieved important results. The transmutation of elements had been a major concern of chemists since the days of alchemy, but nuclear chemistry provided new developments that had a major influence on how people viewed the physical world (see Section 3.3).

The discovery of radionuclides proved that many elements transmutate. The next question was "Is it possible to cause transmutation artificially?" Starting with Rutherford, scientists began to investigate whether transmutation of elements could be achieved by bombardment with particle rays. These studies showed that elements can be induced to transmutate and that artificial radionuclides can be made. In addition, it was known that nuclear fission occurred and produced enormous amounts of energy. This subsequently had an extremely large influence on society and civilization.

Thus, "nuclear and radiochemistry" became a leading-edge field of chemistry and an area of active research. In the field of nuclear chemistry, elements heavier than uranium were made by nuclear reactions, and the chemistry of transuranic elements was investigated. The use of radionuclides as tracers became important in analytical chemistry and biochemistry, and made major contributions in a wide range of disciplines such as medicine, geology, astronomy, and archaeology.

Transmutation of Elements

The discovery by Rutherford and others that atoms of one element can spontaneously turn into atoms of a different element destroyed the concept of the immutable element, which had been the foundation of modern chemistry since Dalton. The transmutation of elements, the dream of alchemists, became a reality.

In 1919, Rutherford inserted a radium source into a cylinder and covered one end with metallic foil to shield α-rays. He introduced various gases into the cylinder to investigate the effects of α-ray exposure. When the gas was air, he observed that high-speed hydrogen atoms produced by the collision of α-particles with the gas came out of the cylinder, passing through the metallic foil. This phenomenon did not occur when the gas inside was oxygen or carbon monoxide. Rutherford reported it thusly.[90]

> It is difficult to avoid the conclusion that these long-range atoms arising from the collision of α-particles with nitrogen are not nitrogen atoms, but probably hydrogen atoms. If this is the case, we must conclude that the nitrogen atom is disintegrated by close collision with swift α-particles.

Rutherford and Chadwick discovered that the long-range particle ray comprised positively charged particles that bent in a magnetic field, thus confirming that the particles were protons.

In the 1920s, Rutherford and his research collaborators investigated the effect of α-radiation on a number of elements. Of the elements in the periodic table from boron to potassium, all but carbon, oxygen, and beryllium produced a proton by collision with α-rays. However, to understand how this particle arose, it was necessary to know what happened to the atom following a collision. In 1925, Blackett at Cambridge discovered that out of over 20,000 cloud-chamber photographs, eight α-particle tracks split off into a light particle track and a heavy particle track, and thus obtained evidence that the transmutation of an element occurred by the following nuclear reaction.

$$\,^4_2He + \,^{14}_7N \longrightarrow \,^{17}_8O + \,^1_1H$$

Thus, by 1926 it became clear that a proton was released by the collision of an α-particle with lower atomic number elements, but with heavier elements the α-particles were repelled by the large positive charge of the nucleus and no proton was released.

In 1932, Cockcroft and Walton working in Rutherford's laboratory conducted transmutations by colliding elements with protons, which are lighter than α-particles. The accelerated protons produced by hydrogen discharge were accelerated at high voltage and directed at a lithium oxide target. α-Particles were emitted in the opposite direction from the lithium target. Subsequently, Oliphant and coworkers revealed that ^7_3Li had been transformed to helium.

$$\,^1_1H + \,^7_3Li \longrightarrow \,^4_2He + \,^4_2He$$

These types of nuclear reaction experiments were soon conducted in accelerators.

A voltage of 800,000 electron volts was produced by Cockcroft and Walton's first apparatus, but 1.5 million electron volts was produced by the Van de Graaff accelerator at the Massachusetts Institute of Technology in 1931. In 1929, Wideroe in Germany also devised a linear accelerator. In 1931, Lawrence and Livingston at the University of California designed a cyclotron, which sequentially accelerates positively charged particles by means of a high-frequency electric field within a magnetic field, and achieved 80,000 electron volts. Thus, elements were transmutated using high energy protons, and the properties of the many elemental isotopes obtained were investigated.

In 1932, Urey discovered deuterium (see page 195). Since deuterium has double the mass of hydrogen but the same charge, it became clear that it was more suitable than a proton for many purposes. The neutron was discovered in 1930; since it has no charge, it is not repelled by the nucleus. Neutrons were thus ideal for atomic transmutation experiments and were immediately used in nuclear reaction research. A free neutron is obtained by the collision of a light element and an α-particle. Neutrons were obtained from a mixture of beryllium and α-ray sources such as radium or radon. This is termed an (α,n) reaction, in which a beryllium atom collides which an α-particle and expels a neutron. Neutrons can also be generated using a cyclotron; in this (d,n) reaction, high-energy neutrons are obtained by bombarding beryllium or lithium with deuterium.

The Discovery of Artificial Radioactivity

In 1934, Frederic and Irene Joliot-Curie in Paris investigated the radioactivity produced by irradiating aluminum with α-particles. They found that when they stopped α-ray irradiation, the release of protons and neutrons from aluminum stopped, but the release of positrons continued.[91] This clearly showed that a substance produced by irradiation was radioactive. The same phenomenon was observed with α-particle irradiation of magnesium and boron. The Joliot-Curies thought that the nuclide produced by an (α,n) type nuclear reaction was radioactive and postulated the following scheme.

$$^4_2\text{He} + ^{27}_{13}\text{Al} \longrightarrow ^{30}_{15}\text{P} + ^1_0\text{n}, \quad ^{30}_{15}\text{P} \longrightarrow ^{30}_{14}\text{Si} + ^0_1\text{e}$$

Subsequent experiments confirmed that aluminum was transmuted to radioactive phosphorus. Using a similar approach, two radioactive isotopes with specific half-lives were generated from boron and magnesium by α-ray irradiation and were confirmed to be nitrogen (^{13}N) and silicon (^{27}Si).

Artificial radioactivity immediately attracted major attention both because of scientific interest in the phenomenon itself and because of the possibility of its many applications. Many new radionuclides were created and their proper-

Frederic Joliot-Curie (1900–1958)
Frederic Joliot-Curie studied physics and chemistry at the City of Paris Industrial Physics and Chemistry Higher Educational Institution under Langevin. In 1925, he became an assistant to Marie Curie at the Radium Institute, and in 1926, he married the Curie's eldest daughter, Irene. They collaborated as a couple to conduct experiments on the release of radiation resulting from α-particle collisions and  discovered artificial radioactivity. During World War II, he participated in the resistance movement against Nazi Germany, and after the war, he strove for peaceful uses of atomic energy. He was deeply involved with the peace movement and worked as president of the World Council of Peace. He was awarded the 1935 Nobel Prize in Chemistry.

Irene Joliot-Curie (1897–1956)
Irene Joliot-Curie was the eldest daughter of the Curies. After studying at the University of Paris, she joined the Radium Institute in 1921 to begin research on radioactivity. In 1926, she married her colleague Frederic Joliot, and starting around 1929 they collaborated on research. In 1933, they demonstrated the transmutation of elements using artificial radioactivity. After winning the 1935 Nobel Prize  in Chemistry, she became a professor at the Sorbonne University. She became the director of the Radium Institute in 1947 and worked as a commissioner on France's Atomic Energy Commission from 1946 to 1951.

ties investigated. The discovery of artificial radioactivity coincided with the development of the cyclotron and other accelerators and the improvement of radioactivity detectors such as the Geiger counter and scintillation counter. Research progressed rapidly. Up until about 1930, approximately 40 radionuclides were known, but by around 1960, approximately 900 radionuclides had been generated. The bulk of these were discovered in the 1930s. Frederic Joliot-Curie and Irene Joliot-Curie were awarded the Nobel Prize in Chemistry in 1935 for "artificial radioactivity research."

The Discovery of Nuclear Fission
Many scientists conducted neutron irradiation experiments following the discovery of the neutron. The neutron was considered useful because its lack of charge meant it was not repelled by the nucleus and it could thus penetrate

the interior of an atomic nucleus. Recognizing this, a group led by the physicist Enrico Fermi at the University of Rome bombarded all the elements they could obtain with neutrons and observed whether or not a nuclear reaction occurred. The last element on the periodic table, number 92, uranium (U), produced three types of radioactive compounds, in which β decay occurred with a half-life of 10 seconds, 40 seconds, and 13 minutes. Fermi's group suggested that ^{238}U had captured neutrons to generate ^{239}U by an (n,γ) reaction,[27] and that this then transformed into two new elements, numbers 93 and 94 (transuranic elements). This research created a major sensation, but their findings were not confirmed by other researchers and were considered to be in error.

At the Kaiser Wilhelm Institute for Chemistry in Berlin, the group comprising physicist Lise Meitner and chemists Otto Hahn and Strassmann began collaborative experiments and obtained results different from Fermi's. They thought they had obtained radioactive thorium (^{235}Th) with a half-life of 4 minutes by irradiating uranium with slow neutrons, but this, too, was not confirmed. Irene Joliot-Curie and Paul Savitch in Paris detected radiation with a half-life of 3.5 hours, and discovered that the radioactive compound could be extracted using a carrier consisting of element number 57, lanthanum. They thought they had generated actinium (^{227}Ac). Thus, the results from irradiating uranium with neutrons and the interpretation of the data were confusing, mainly because researchers believed that neutron irradiation of uranium produced an element with an atomic mass close to that of uranium. The concept of nuclear fission ran counter to the common sense of nuclear physicists of that time and was thought to be impossible. In addition, without knowing where the second rare earth elements started from, in which the outer electrons begin to fill the 5f orbital, chemists believed that the product transuranic elements belong to the transition metals. Because of this, attempts were made to extract products as a precipitate using a transition metal salt as a carrier. The inorganic chemist Ida Noddack strongly criticized Fermi's report, and suggested the possibility of nuclear fission, but this was ignored as simple conjecture. Between 1934 and 1938, various experimental data were accumulated, and interpretation of the data puzzled both physicists and chemists.

After the Nazi Germany annexation of Austria in 1938, Meitner feared for her safety and escaped to Sweden. Hahn and Strassmann continued their research in Berlin. They discovered radioactivity in the barium (atomic weight 138) carrier and believed this originated from radium. At the end of 1938, they confirmed that this radioactivity in fact originated from the barium. This result clearly showed that uranium was split by neutron bombardment to produce radioactive barium. In

---

27. (n,γ) reaction: A reaction in which a neutron is captured and an γ-particle is released.

Enrico Fermi (1901–1954)
Enrico Fermi was an American physicist from Italy. In 1927, he published a new statistical method (Fermi statistics) and that same year became a professor at the University of Rome. From 1934 onward he produced many radioactive isotopes by the artificial transformation of elements using neutrons. He moved to the US in 1938 and was engaged in research on nuclear fission chain reactions, and in December 1942 he demonstrated that uranium nuclear fission can be controlled. Fermi was awarded the 1938 Nobel Prize in Physics.

Lise Meitner (1878–1968)
Lise Meitner was born in Vienna and obtained her Ph.D. at the University of Vienna studying under Boltzmann. In 1907, she became a special student of Planck's at the University of Berlin, and there began collaborative research with Hahn. After working as Planck's assistant at the University of Berlin, she moved to the Kaiser Wilhelm Institute for Chemistry, and in 1930 became the head of the physics department. She became involved in nuclear fission research in 1934, and while in exile in Sweden in 1938, proposed a theory showing that nuclear fission of uranium is possible.

Otto Hahn (1879–1968)
Otto Hahn was born in Frankfurt, studied chemistry at the universities of Marburg and Munich, and obtained his Ph.D. in organic chemistry. In 1904, he studied in Ramsay's laboratory in London and discovered radioactive thorium, then continued research on radioactivity under Rutherford at McGill University in Canada. He returned to Germany in 1906, established a radiochemistry laboratory at the University of Berlin, and started collaborative research with Meitner. In 1912, he moved to the Kaiser Wilhelm Institute for Chemistry. In 1934 he participated in experiments, irradiating uranium with neutrons. This led to the discovery of nuclear fission in 1938. After the war, he served for many years as the first president of the Max Planck Society, which was the reorganized Kaiser Wilhelm Society. Hahn was awarded the 1944 Nobel Prize in Chemistry.

January 1939, they reported their results in the journal "Naturwissenschaften".[92] Thinking that physicists would not believe nuclear fission possible, they did not write definitely in the paper that the uranium nucleus had split, but the results clearly indicated nuclear fission had occurred. They believed that elements of around atomic weight 100 were produced by nuclear fission and confirmed the presence of strontium, krypton, and rubidium in the reaction products.

Meitner received word of the Hahn-Strassmann experimental results while in exile in Sweden, and discussed the results with her nephew, the physicist Frisch. Meitner proposed a theory that accepted the concept of nuclear fission. Because of differences in the packing ratio (see page 195) of nuclear fission products such as uranium and barium, Meitner and Frisch believed that a large quantity of energy was released during nuclear fission. Meitner and Frisch wrote a paper for the journal "*Nature*" and suggested that the uranium atom was split by the capture of neutrons, and additionally conjectured that the products of fission would be radioactive.[93]

News of the discovery of nuclear fission was immediately brought to the US by Bohr, and research in nuclear fission promptly started in the US. The existence of products with a large recoil[28] energy was reported by the Joliot-Curies in Paris, Meitner and Frisch in Denmark, and Edwin McMillan in California. The fact that many neutrons were generated during nuclear fission was confirmed immediately. Nuclear fission thus became a central topic of research all over the world. When a review of the field was written in 1940, it already surveyed close to 100 papers.[94] However, with the outbreak of World War II, the possibility of using nuclear fission in weapons was recognized, and research became secret. In 1944, Hahn alone was awarded the Nobel Prize in Chemistry for the discovery of nuclear fission. However, the discovery was clearly the result of joint research with Meitner and Strassmann, and many people thought that Meitner should have been jointly awarded the Nobel Prize in Chemistry.[95]

Transuranic elements[29]

It became clear that Fermi's report of a transuranic element in 1934 was an error, and researchers therefore became cautious in reporting discoveries of new elements. Elements number 93 and 94 were discovered in the 1940's. When Edwin McMillan of the University of California irradiated a thin film of uranium with neutrons, products of nuclear fission were released, but radioactivity with a half-life of 23 minutes and radioactivity with a half-life of 2.3 days remained in the thin film. The material with the 23 minute half-life was uranium-239, and the material with the 2.3 day half-life was a new element with atomic number 93. McMillan and Abelson investigated its

---

28. Recoil: When particle B is emitted from particle A, per the law of conservation of momentum, momentum is imparted to A.

29. Transuranic element: An element having an atomic number greater than uranium.

COLUMN 12

The Contributions of Hahn and Meitner to the Discovery of Nuclear Fission[95]

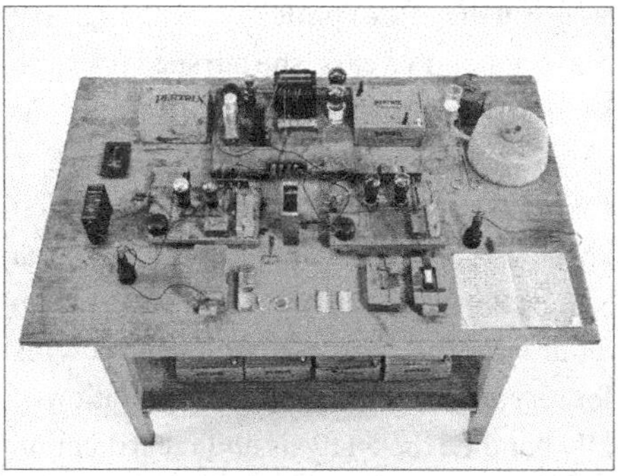

Nuclear fission experimental apparatus of Hahn, Meitner, and Strassmann displayed in the Deutsches Museum

If the discovery of nuclear fission had taken place during a time of peace, one of its discoverers, Lise Meitner, might have become known throughout the world as a great woman of science alongside Madame Curie. Unfortunately, this discovery was made in 1938 when the Nazis had already come to power just before the outbreak of World War II. Meitner was at the mercy of politics, and she was not awarded a Nobel Prize.

Lise Meitner was from Vienna and became a special student of Max Planck's at the University of Berlin. Meitner wanted to conduct experiments. She contacted Rubens, the department head for experimental physics, and was introduced by him to Otto Hahn, who was an assistant in Emil Fischer's chemistry laboratory. Hahn had worked as an organic chemistry assistant at the University of Marburg and studied abroad under Ramsay at the University of London. Hahn entered the field of radiochemistry and extracted radium from barium salt. He produced good results, and after conducting further radiochemistry research under Rutherford at McGill University in Canada, obtained a position in Fischer's laboratory in Berlin.

Hahn and Meitner thus met in 1907 and began a long-term collaboration. At that time, women were prohibited from working in the chemistry laboratory; an exception was made for Meitner, but she was allowed to work only in Hahn's radioactivity test room, which formerly had been a woodworking room. (One year later, women were legally permitted to enter German universities, and Fischer's laboratory was opened to women.) Radioactivity research during this era was a cross-disciplinary domain of physics and chemistry, and Meitner and Hahn were complementary. Meitner excelled at physics and mathematics, and Hahn understood chemistry and excelled at analytical techniques. However, there was no closer relationship between the two of them beyond that of research colleagues. The two achieved results by collaborative research, and Meitner's value as a scientist was established.

In 1912, Hahn became head of radiochemistry at the Kaiser Wilhelm Institute (KWI) for Chemistry, newly established in Berlin. In 1913, Meitner was also given a position at KWI, similar to Hahn's. In 1917, the two of them achieved several important results, such as discovering protactinium. Meitner was recognized for her abilities and accomplishments, and in 1923, she became head of the physics department at KWI and was given the title of professor. Although she had done collaborative research with Hahn for many years, she pursued research and achieved results in nuclear physics independently. Einstein called her "our Marie Curie", and she came to be highly esteemed as a first rate physicist.

In 1934, she learned of Fermi's report describing his neutron irradiation experiments, and she persuaded Hahn to resume their interrupted research collaboration. The young chemist Strassmann joined them, and they began investigating the neutron irradiation of uranium as a three-person team. In this investigation, Meitner took the lead for research planning and interpretation of results, and the two chemists principally handled the analysis and identification of elements produced by nuclear reactions.

Nazi Germany had been expelling scientists of Jewish descent from public office and persecuting them since 1933. One of Meitner's grandfathers was Jewish, but because she was an Austrian citizen, she was not dismissed from the laboratory until 1938. Circumstances changed with Nazi Germany's annexation of Austria in April 1938, and in becoming a German citizen, danger closed in even on her. With the aid of friends, in July she fled to exile in Sweden, while Hahn and Strassmann continued experiments in Berlin. In July, Hahn and Strassmann tried to separate what they thought was radium produced by neutron irradiation of uranium by means of crystallization with a barium bromide carrier, but they repeatedly failed. They finally concluded that the material produced was barium. Hahn informed Meitner of this finding in a letter and asked her for an explanation of the result. Hahn and Strassmann reported only the experimental results in the journal "*Naturwissenshaften*", and this was the first paper reporting the discovery of nuclear fission. Considering the course of research, the paper should have been rightfully under the three joint names of Hahn, Strassmann, and Meitner, but Meitner was considered to be a Jew, and putting her name on the paper was politically not possible.

Meitner heard of the experimental results from Hahn in Berlin immediately before Christmas. She discussed them over the Christmas holiday period with her nephew, the physicist Otto Frisch, and they set out to explain Hahn and Strassmann's experimental results. They showed theoretically that uranium undergoes fission following capture of a neutron and a large quantity of energy is released, and they reported this in the journal "*Nature*". Thus, the nuclear fission of uranium was explained.

Of the four scientists involved in the discovery and explanation of nuclear fission, only Hahn was awarded a Nobel Prize. No one questions that Hahn merited a Nobel Prize, but many doubts have been raised about the fact that only he was awarded a Prize. In 1944, the chemistry division of the Nobel Prize committee decided in secret to give the chemistry prize to Hahn. This was a careless decision made in war time without sufficient investigation and with an added political element. In 1945, there was a move to reconsider this, but in the end, Hahn received the 1944 Nobel Prize in Chemistry as previously decided, and Meitner and Strassmann's contributions were ignored. Meitner and Frisch were candidates for the 1945 Nobel Prize in Physics on Bohr's recommendation, but their nomination was opposed by Siegbahn (1925 Nobel Prize in Physics) who dominated the physics division of the Nobel Prize committee.

Hahn served as president of the Max Planck Society, which was the post-war reorganized Kaiser Wilhelm Society, and actively focused on rebuilding science in Germany. However, he said nothing regarding Meitner's contribution to the discovery of nuclear fission, and kept the honor of the discovery of nuclear fission to himself. When one thinks of the friendship between Hahn and Meitner and their collaborative research over many years, the author cannot help but feel Hahn's dishonesty on this point.

In 1959, the institute for nuclear research in Berlin was named the Hahn-Meitner Institute, and in 1966, the United States Atomic Energy Commission awarded the trio of Hahn, Meitner, and Strassmann the Enrico Fermi Award. The 109th element discovered in 1982 was dedicated to Meitner and given the name meitnerium.

Meitner (left) and Hahn (right) in the laboratory

Edwin McMillan (1907–1991)
Edwin McMillan was an American physicist. After obtaining his Ph.D. at the California Institute of Technology, he went to the University of California where he collaborated with Lawrence and contributed to the development of the cyclotron. Influenced by Pauling, he also had a strong interest in chemistry and carried out research in collaboration with chemists on radionuclides and transuranic elements. McMillan was awarded the 1951 Nobel Prize in Chemistry.

Glenn Seaborg (1912–1999)

Glenn Seaborg was born in Michigan. He obtained his Ph.D. under G.N. Lewis at the University of California, and continued to conduct research on transuranic elements for much of his career at the Berkeley school. In 1940, he discovered plutonium together with McMillan and coworkers. From 1944 to 1957, together with many research collaborators, he synthesized elements number 95 to number 102, and predicted the chemistry of the actinides. Starting in 1950, he was deeply involved with the regulation of nuclear energy, and from 1961 to 1971 he served as the chairman of the Atomic Energy Commission. Element number 106, seaborgium, was named after him. Seaborg was awarded the 1951 Nobel Prize in Chemistry.

chemical properties and ascertained that it was a new element of valence 4 and valence 6, and they called it neptunium (Np).[96] This name signified going beyond uranium and was derived from the planet Neptune, which is outside the orbit of the planet Uranus. Research on neptunium was continued by Glenn Seaborg's group.

Seaborg's group obtained ^{238}Np by irradiating uranium with deuterium, but it became clear that with β decay, this became a new element with atomic number 94. This element had a half-life of 90 years, and its chemical properties were investigated. In addition, based on experiments that used a large quantity of ^{239}Np obtained by neutron irradiation of ^{238}U, it was found that ^{239}Np became element number 94 via a 2.3-day-half-life β decay. The 94th element was named plutonium after the planet Pluto. The fact that plutonium also exists in natural uranium deposits was discovered by Seaborg and colleagues in 1942.

The discovery of element number 93 and 94 raised the expectation that other transuranic elements would be synthesized. By this time element transmutation techniques and isotope identification techniques had been established. Because the outer shell electrons in transuranic elements are in the 5f orbital, it was presumed

Georg de Hevesy (1885–1966)

Georg de Hevesy was born in Budapest, Hungary. After studying at the University of Budapest and the Technical University of Berlin, he obtained his Ph.D. at the University of Freiburg. Subsequently, he conducted research at the universities of Zurich, Manchester, and Copenhagen, and in 1926, he became a professor at the University of Freiburg. He had many outstanding accomplishments in a broad range of fields, including the use of isotopes as tracers, the discovery of hafnium, research on rare earth elements, and applications of X-ray analysis. He had interchange with many physicists such as Rutherford, Bohr, and others, which helped him to conduct outstanding research in broad-ranging fields of chemistry. De Hevesy was awarded the 1943 Nobel Prize in Chemistry.

that the transuranic series would be similar to the rare earth element series. In 1944, Seaborg, James, and Ghiorso bombarded ^{239}Pu with helium ions using a cyclotron and obtained evidence that the 96th element (Curium, Cm) had been produced. They also obtained evidence of the production of element number 95 (americium, Am) by neutron irradiation of ^{239}Pu. Thus, by 1945 elements up number 96 had been obtained.[97] The synthesis of transuranic elements continued after the war, and the final element in this series, number 103, Lawrencium (Lr), was produced in 1961.

The Chemical Uses of Radionuclides and Radioactivity

Georg de Hevesy was a pioneer in the use of radioactive elements as tracers. In 1913, he and Paneth used radium D (^{210}Pb) to determine the solubility of lead salts. In 1921, Hevesy used thorium B (^{121}Pb) to show that the exchange of Pb in $Pb(NO_3)_2$ and $PbCl_2$ in an aqueous solution was rapid, which supported the Arrhenius theory of electrolytic dissociation.[98] In addition, he also conducted pioneering research in the use of tracers in biochemical research. However, at that time there were few radionuclides that could be used, and biochemical applications were limited. With the 1934 discovery of artificial radioactivity by the Joliot-Curies, the number of radionuclides that could be used dramatically increased, and the chemical applications broadened. In 1941, Ruben and Kamen discovered that ^{14}C was produced by an (n,p) reaction of nitrogen.[99] This isotope had a long half-life of 5600 years. ^{14}C was convenient to use as a tracer, and it became clear that it is extremely useful for the investigation of reactions involving carbon compounds. However, the use of ^{14}C in biochemical, medical, and agricultural research did not become widespread until after World War II. Research utilizing radionuclides as tracers developed rapidly after the war.

Attention turned to chemical phenomena caused by high-energy atoms produced by nuclear reactions (hot atoms). In 1934, Szilard and Chalmers discovered that the irradiation of ethyl iodide with neutrons, followed by extraction with water, allowed radioactive iodine ^{128}I to be concentrated and collected. The basis of this method is that a recoiling atom produced during a nuclear reaction separates from the parent molecule (for example, an organic compound) and can be isolated. This is an important method for separating nuclear reaction products (Szilard-Chalmers method). The chemical reactions caused by high-energy atoms and the properties of the excited molecules produced were investigated in detail after the war.

De Hevesy was awarded the 1943 Nobel Prize in Chemistry "for his research on the use of isotopes as tracers in chemical reactions."

4.6 Analytical Chemistry

Methods for separation, detection, analysis and observation are of the utmost importance to the development of chemistry. Analytical chemistry depended mainly on traditional methods at the beginning of the 20th century, but subsequently changed considerably.

Conventional quantitative analysis and volumetric analysis methods incorporated developments from other fields, such as physical chemistry and coordination chemistry, and were further refined to provide new methods of instrumental analysis. There was astonishing progress in instrumental analysis following World War II, supported by advances in electronics and computers. These new techniques revolutionized the way chemical research was done, but the germ of these developments was already apparent in the 1930s. Infrared / visible / ultraviolet spectroscopy, mass spectroscopy, magnetic resonance, electron spectroscopy, chromatography, polarography, microanalysis, and other techniques which figure prominently in modern chemistry research were developed in the first half of the 20th century and made further dramatic advancements after the war. Most of these were brought about by advances in physics, but there were also advances, such as chromatography and microanalysis, that resulted from the ingenious contrivances of chemists themselves. When radioactivity was discovered, attempts were made to make use of radioactivity in analyses, giving rise to radio-activation analysis. Research using radioactivity and radionuclides was classified during World War II, but research during the war made dramatic advances and was the motive force driving post-war chemistry research in wide-ranging areas. Wartime research accelerated the progress of chromatography for isolating fissionable materials.

The demand for mass spectroscopy increased along with progress in isotope chemistry, and the performance of mass spectrometers improved.

Quantitative Analysis

An understanding of equilibrium in solutions resulted from advances in physical chemistry between the end of the 19th century through the beginning of the 20th century. Quantitative analysis helped chemists understand important phenomena and concepts such as the common ion effect, solubility products, the concept of pH, buffer solutions, complex ion formation, and indicator theory. This allowed analytical chemists to conduct more refined analyses.

This progress led to more effective use of indicators. Chemists in the 19th century had used indicators empirically, but in 1891, Ostwald proposed a theory for indicators, and Hantzsch suggested treating indicators as pseudo-acids and pseudo-bases. Many researchers investigated the relationship between hydrogen ion concentration and the change in the color of an indicator. In 1909, Sørensen introduced the concept of pH (see page 237), and in 1911, Tizard studied the sensitivity of indicators. In 1941, Bjerrum published a book on the theory of indicators. New indicators such as sulfonaphthalenes and thymol blue were introduced.

In 1923, Fajans introduced the adsorption indicator fluorescein and derivatives thereof, bringing about advances in precipitation titration.[30] These indicators were useful for clearly indicating the endpoint when titrating a chloride sample with silver ions. Tartrazine and phenosafranine were shown to be effective in acid titrations. Similarly, new indicators were also introduced for redox titration.[30]

Chugaev discovered the reaction between dimethylglyoxime and nickel salt in 1905, and Branch applied this reaction to gravimetric analysis in 1907. Dibasic acids such as oxalic acid and tartaric acid made complexes with various ions, and hence were useful in analyses.

Microanalysis

At the beginning of the 20th century, a sample of several grams was generally required for the elemental analysis of a compound. Friedrich Emich (1860–1940) at the Technical University of Graz in Austria ingeniously used small glassware and capillary tubes to pioneer analyses using just milligrams of sample.

Fritz Pregl worked with Emich to develop the equipment needed to analyze milligrams of organic compounds. He studied medicine at the University of

---

30. Precipitation titration and redox titration: A precipitation titration is a titration that makes use of a precipitate-forming reaction, and a redox titration is a titration that is performed based on an oxidation-reduction reaction.

Fritz Pregl (1869–1930)
Fritz Pregl was born in Laibach, Austria. He studied
medicine at the University of Graz and became a professor
of physiology and histology in 1903, and at the same time acquired a knowledge of chemistry. Subsequently, he studied in
Germany with Ostwald and Fischer. He continued research
into microanalysis methods as a professor at the Institute of
Medical Chemistry at the University of Graz from 1913 until
his death. In 1923, he was awarded a Nobel Prize in Chemistry for "research
of methods of microanalysis of organic substances."

Graz and started physiological studies of bile acid, but only obtained very small
quantities for analysis. He therefore devised an apparatus for conducting analyses
on very small quantities after the example of Emich. He needed a micro-balance
for microanalyses that could precisely weigh samples of a few milligrams. He
obtained a high-performance balance from Kuhlmann in Hamburg that could
measure loads of 20 g with a precision of 0.01 mg. Liebig's carbon and hydrogen
analysis methods (see page 60) were improved and miniaturized. Analytical
methods for elements such as nitrogen, sulfur and the halogens were similarly
miniaturized to be able to handle very small sample quantities. Thus, organic
compound analyses could be conducted on samples of a few milligrams. In 1912,
he published his results up that point, and in 1917, he published a monograph
entitled *"Die Quantitative Organische Mikroanalyse"*.[100] This book was read
widely and researchers from around the world visited Pregl's laboratory at the
University of Graz to learn microanalysis techniques.

The spot test is a qualitative microanalysis method that is simple, has good
sensitivity, and excellent selectivity. In this test, one drop of reagent is dropped
onto a very small quantity of sample in order to analyze the constituents of the
sample. This method was developed and systematized by Feigl. Feigl obtained his
Ph.D. at the University of Vienna in 1920 based on the spot test and subsequently
devoted himself to the development of this technique in Vienna until 1938. After
the annexation of Austria by Nazi Germany, he stayed temporarily in Belgium,
and afterwards took refuge in Brazil.

Instrumental Analysis
In the first half of the 20th century, instrumental analysis was not widespread and
many types of instruments were used for only specialized research. However,
many methods of instrumental analysis developed after World War II were
introduced in this period.

Jaroslav Heyrovsky (1890–1967)

Jaroslav Heyrovsky was born the son of a professor of Roman Law at the University of Prague. After receiving a primary education in Prague, he diligently studied under Ramsay at the University of London, and studied electrochemistry under Donnan. Subsequently, he returned to Prague and began research on the dropping mercury electrode at Charles University in Prague. While there, he discovered the principles of polarography and developed a polarography apparatus together with Shikata. In 1950, he became the director of the Polarographic Institute, and throughout his career he advanced polarography and taught the technique to others. He was awarded the 1959 Nobel Prize in Chemistry.

Masuzo Shikata (1895–1964)

After graduating from Tokyo Imperial University in 1920 and spending time at the Institute of Physical and Chemical Research, Masuzo Shikata became a professor at Kyoto Imperial University. While studying abroad at Charles University in Prague, he developed an automatic recording polarography instrument together with Heyrovsky. He made major contributions to the development of analytical chemistry and electrochemistry. In 1942, he moved to the Mainland Science Institute of China, and in 1953 returned to Japan and became a professor at Nagoya University.

The beginning of the 20th century ushered in great expectations for electrochemical methods, but they often were not all that useful in practice. Potentiometric devices lacked stability, conductivity measurements were tedious, and generally, such methods were not any better than conventional titration methods. Advances in potentiometric methods depended on advances in pH measurement methods. Because of this, there was a need to develop standard electrodes that were stable and easy to use. A glass electrode proposed by Haber and Klemensiewicz in 1909 evolved by the 1930s to withstand actual use. Subsequently, easy-to-use pH meters were developed, and potentiometric titrations became easy and widely used.

In 1922, Jaroslav Heyrovsky of Charles University in Prague published the principles of polarography,[101] and in 1925, he announced the invention of automatic recording polarography jointly with Masuzo Shikata, who was then studying abroad.[102] This was a type of voltammetry that examined the relationship between the potential difference between an indicator electrode and a reference electrode and the flow of electric current between the two electrodes. However, because the relationship between applied voltage and electric current was measured by conducting electrolysis using a dropping mercury electrode

Figure 4.11: Beckman DU spectrophotometer

Marketed in 1941, this spectrometer made spectroscopic analysis with high precision remarkably easy.

for the indicator electrode and a mercury electrode for the reference electrode (anode), the electric current was observed to increase in a step-wise fashion when the solution contacting the indicator electrode contained a reduced ion or base. Polarography was subsequently improved by many researchers and became widely used in qualitative analysis, quantitative analysis, and studies on electrode reactions. Heyrovsky devoted his career to expanding polarography research. He was awarded the 1959 Nobel Prize in Chemistry for the invention of polarographic methods of analysis.

Attempts to use the absorption of visible light for analysis began in the latter half of the 19th century. The relational expression today called the Lambert-Beer law, which describes the change in light intensity due to absorption by a medium, is based on the observations of the 18th century scientist Lambert and the 19th century physicist Beer. This law states that $I=I_o e^{-al}$, where I is the intensity of light passing through a medium with path length l, and a is an absorption coefficient proportional to the concentration of analyte. The Lambert-Beer law is currently used to conduct quantitative analyses from absorbance measurements, but until about 1930, light absorption was used primarily for qualitative analysis. Hellmuth Fischer's use of dithizone is an example of pioneering research in spectroscopic analysis. In 1926, he investigated this compound and showed its practical applications to analysis. He carried out detailed research in the 1930's. This reagent forms colored complexes with many cations, and it was useful for detecting small quantities of ions. Quantitative analysis using the Lambert-Beer law became widespread as spectrophotometers came to market in the early 1940s. Absorption in the ultraviolet region was also used for analysis with the spread of commercial spectrophotometers. (see Fig. 4.11)

Absorption in the infrared region is useful for the identification of organic compounds. Until about 1940, however, infrared spectrographs were research instruments used by physical chemists. Raman scattering was discovered in 1928, and Raman spectroscopy was used for structural chemistry research as a spectroscopic method complementary to infrared spectroscopy (see page 263), but like infrared spectroscopy, this, too, was still limited to physical chemistry research.

Analytical methods that made use of light emission were actively used from early on. The methods had good sensitivity and could detect substances at ultratrace levels. Quantitative analysis by light emission could be achieved by standardizing photographic plates, instruments and procedures. The amount of analyte present was determined from a comparison of the intensity of a specific emission line to that of a standard of known concentration. Emission spectrometry was particularly useful in application areas requiring a large number of analyses.

The mass spectrometer developed by Aston was also still a research instrument in the first half of the 20th century. However, with the availability of stable isotopes as tracers, applications for mass spectrometers expanded and demand increased. Mass spectroscopy became widely used for the analysis of hydrocarbon compounds in the petroleum industry in the 1940's.

Chromatography

Chromatography is an important technique for the separation and purification of compounds, invented by the Russian botanist Mikhail Tsvet.[103] In 1903, Tsvet reported a method for the separation of plant pigments using a column of calcium carbonate to absorb liquid. In 1906, he reported the separation of chlorophyll pigment dissolved in petroleum ether by passing the solution through a column of calcium carbonate. Two pigments, carotene and xanthophyll, appeared as two bands on the column and were separated and extracted with alcohol. He termed this technique 'chromatography'. However, his paper remained largely ignored, in part because Willstätter, an influential figure in chlorophyll research at the time, attempted separations using a different adsorbent. Since he was unsuccessful, he was dismissive of Tsvet's results.

In the 1930s, renowned organic chemists such as Kuhn, Karrer, and Windaus studied biochemically interesting organic compounds such as chlorophyll, carotenoids, vitamins and similar compounds, and the true value of chromatography as a separation and analysis technique was recognized immediately. Interest in chromatography increased, and research aimed at new applications was under-

Mikhail Tsvet (1872–1919)

Mikhail Tsvet was born in Italy to an Italian mother and a Russian father. He was raised in Switzerland and studied physics and mathematics at the University of Geneva, but he turned to botany and received his Ph.D. in this field. In 1896, he moved to Russia, and after conducting research at the Biological Laboratory of the Russian Academy of Sciences in St. Petersburg, he became an assistant at the Institute of  Plant Physiology of Warsaw University in Poland in 1902, and the following year became an assistant professor. In 1906, he invented chromatography. Amid the chaos of the outbreak of the World War I and the outbreak of the Russian Revolution, he died of illness at age 47.

Archer Martin (1910–2002)

The son of a physician, Archer Martin studied at Cambridge University reading biochemistry and obtained his Ph.D. with research on Vitamin E. In 1938, he joined the Wool Industries Research Association in Leeds, and together with his colleague Synge pioneered ground-breaking new techniques such as partition chromatography and paper chromatography in the early 1940s. He conducted research  at the National Institute for Medical Research from 1948 onwards, and developed gas chromatography at the beginning of the 1950s. He was awarded the Nobel Prize in Chemistry together with Synge in 1952.

Richard Synge (1914–1994)

Born in Liverpool, Richard Synge studied at Cambridge University, and obtained his Ph.D. in 1941 with research on the separation of acetylamino acids. Subsequently, he joined the Wool Industries Research Association and developed partition chromatography together with Martin. From 1943 to 1948, he carried out research at the Lister Institute for Preventive Medicine where he developed techniques for separating antibiotic peptides and invented two-dimensional paper chromatography. He was awarded the Nobel Prize in Chemistry together with Martin in 1952.

taken. For example, partition chromatography was introduced by Archer Martin and Richard Synge in Britain. In 1941, they proposed a new method using two immiscible liquids as a stationary phase and a mobile phase.[104] Acetylamino acids were separated using water-saturated silica gel as the stationary phase,

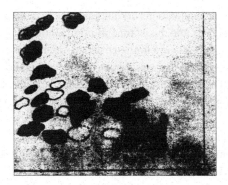

Figure 4.12: An early example of two-dimensional paper chromatography
Separation of potato juice. Stained with ninhydrin. (from Synge's 1952 Nobel Prize lecture)

chloroform as the mobile phase, and methyl orange as an indicator. This method was also used for the separation of organic acids and fatty acids. Consden, Gordon, Martin, and Synge discovered that filter paper saturated with water was a suitable stationary phase for the separation of amino acids, leading to the development of paper chromatography. In 1944, a method was introduced that used two solvents to develop the filter paper in two dimensions (Figure 4.12). This was an inexpensive and efficient method that became an important technique for effectively separating compounds in biochemistry, such as amino acids and antibiotics. Chromatography that made use of ion exchange materials was also developed in the 1940s. By that time many synthetic ion-exchange resins had been developed, and ion-exchange chromatography utilizing these resins was useful for separating the products from nuclear fission reactions during World War II.

Martin and Synge were awarded the 1952 Nobel Prize in Chemistry for devising methods for analyzing amino acids using chromatography. Both were biochemists educated at Cambridge University who conducted research at the Wool Industries Research Association in Leeds. Their results are an important example of research that had a major impact using simple apparatus.

Analysis using Radioactivity
Shortly after the discovery of radioactivity, radionuclides were used in analytical chemistry research. As described previously, Hevesy and Paneth used radium D (^{210}Pb) to determine the solubility of lead salts in 1913. However, the number of radioactive nuclides was limited, as was their application in chemistry. Many radionuclides became available from about 1940, and their application expanded, but their widespread use in analytical chemistry did not occur until after World

War II. Two important applications of radionuclides in analytical chemistry are radio-activation analysis and the isotope dilution method.

Radio-activation analysis is a high-sensitivity analytical method to identify and quantify the radioactive nuclides produced when a sample is bombarded with high-energy radiation or a high-energy particle beam. Radioactive nuclides are identified from the half-life and the energy they emit, and the amount of nuclide required for analysis is determined by the amount of radioactivity emitted. Radio-activation analysis is usually used following the bombardment of a sample with a neutron beam and was developed by Hevesy and Levi in 1936. They discovered that certain rare-earth elements became radioactive when subjected to a neutron beam and used this to identify the element.

The isotope dilution method is a method for determining the quantity of an element X in a sample. In this method, a fixed quantity of a different isotope of that element concerned is added to the sample and the isotope ratio is measured after chemical separation of element X. The analysis result is not affected even if some quantity of the element is lost during subsequent chemical operations, making it a convenient, highly accurate and sensitive method.

Radiometric Dating

Radiometric dating has had a major impact on many fields of science. The dating of minerals using radioactivity was started by the American chemist Boltwood in 1905. He proposed that the age of a mineral could be determined from the half-life of uranium and the amount of helium trapped in the mineral. In 1907, he recognized that the final product of the decay of ^{238}U was ^{206}Pb, and he estimated the age of rocks from the abundance ratio of uranium and lead. His estimated age of the oldest rock was 4.5×10^9 years.

In 1937, Grosse suggested that the impact of cosmic rays produces radio-nuclides. In 1946, Willard Libby of the University of Chicago showed that a certain quantity of radioactive ^{14}C present in living organisms is generated by the collisions of neutrons produced by cosmic rays with atmospheric nitrogen. He used this to establish C-14 dating. ^{14}C has a half-life of 5730 years. ^{14}C becomes incorporated into carbon dioxide ($^{14}CO_2$) and passes through the carbon cycle of organisms through photosynthesis and other biological processes to enter plants and animals. The absorption of ^{14}C stops after the death of the plant or animal, and because ^{14}C decreases with a half-life of 5730 years, the age of the plant or animal can be determined from a measurement of the amount of ^{14}C in the sample. Libby developed a high-precision apparatus to detect minute amounts of ^{14}C. This method became important for dating fossil and archeological samples between 500 and 50,000 years old.[105] Libby was awarded the Nobel Prize in Chemistry in 1960.

Willard Libby (1908–1980)
Willard Libby was born in Colorado, and after obtaining his Ph.D. at the University of California, Berkeley, he remained at Berkeley as an instructor. He participated in the Manhattan Project during World War II, and became a professor at the University of Chicago in 1945. Early in his career he detected and quantified faint radiation in the natural world. In 1946, he developed a method using ^{14}C to determine the age of samples, and made major contributions to the development of a broad range of academic disciplines such as archaeology, anthropology, and geology. In addition, he contributed to biochemical research using isotopes as tracers. He was awarded the Nobel Prize in Chemistry in 1960.

4.7 Inorganic chemistry

Throughout the 19th century, the greatest challenge in inorganic chemistry was discovering new elements and elucidating their properties (see Section 2.5). By the beginning of the 20th century, nearly all of the elements found in nature had been identified, and the periodic table was nearing completion, but there were still incomplete sections. Many of the remaining elements were identified following the development and application of 'new physics'. The chemistry of individual elements was researched in detail and a great deal of empirical knowledge was accumulated. With the periodic table nearing completion and atomic structures clarified, attempts were begun to categorize and understand this vast amount of empirical knowledge.

The Nernst equation shed light on the relationship between redox potential and Gibbs free energy, and the oxidation-reduction potentials of the various oxidation states of each of the elements were measured. The Latimer diagram summarizes these, and makes it possible to predict whether or not a given reaction can take place (Fig. 4.13).

The study of solid inorganic compounds, including metals and alloys, progressed by applying the Gibbs phase rule (see page 111). The area of coordination chemistry, initiated by Werner at the end of the 19th century, was an important field in inorganic chemistry even after the start of the 20th century. Overall, however, the first 30 years of the 20th century were a less active period for inorganic chemistry. Inorganic compounds have complex bonds and structures that are difficult to explain rationally, and there are many irregularities in the behavior of the individual elements, making it impossible to arrive at an adequately unified understanding of the periodic table. Inorganic chemistry

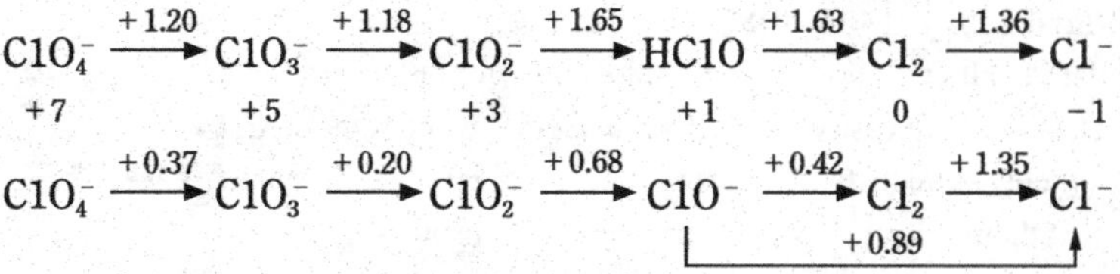

$$\text{ClO}_4^- \xrightarrow{+1.20} \text{ClO}_3^- \xrightarrow{+1.18} \text{ClO}_2^- \xrightarrow{+1.65} \text{HClO} \xrightarrow{+1.63} \text{Cl}_2 \xrightarrow{+1.36} \text{Cl}^-$$

$$\quad +7 \qquad\qquad +5 \qquad\qquad +3 \qquad\qquad +1 \qquad\qquad 0 \qquad\qquad -1$$

$$\text{ClO}_4^- \xrightarrow{+0.37} \text{ClO}_3^- \xrightarrow{+0.20} \text{ClO}_2^- \xrightarrow{+0.68} \text{ClO}^- \xrightarrow{+0.42} \text{Cl}_2 \xrightarrow{+1.35} \text{Cl}^-$$

$$\underset{+0.89}{\rule{8cm}{0.4pt}}$$

Figure 4.13: Latimer diagram

A Latimer diagram shows the redox potentials between oxidation states of an element. The upper series is for chlorine in an acidic solution, and the lower series is for chlorine in a basic solution.

was considered a tedious academic discipline that begins and ends with the compilation of facts.

This situation began to change in the 1930s. The introduction of X-ray crystallography and spectroscopy brought about a new level of accuracy in knowledge about the structures of inorganic compounds and crystals. The development of chemical bond theory by Pauling (see pages 246–248) opened a path to rationally describing the chemical bonds in inorganic compounds. Coordination chemistry (see Section 2.6) began to develop as a field at the intersection between inorganic chemistry, analytical chemistry, organic chemistry, and biochemistry. The discovery of radioactivity (see Section 3.3) also led to the discovery of new elements, and gave birth to the new field of the chemistry of isotopes. The importance of inorganic compounds as catalysts in polymer chemistry was also recognized. Research and development of nuclear weaponry during the Second World War also provided a major stimulus for the study of inorganic chemistry. Inorganic chemistry thus entered a renaissance when World War II came to an end.

A vast amount of research in inorganic chemistry took place during this era, but this chapter gives an overview only of the discovery of new elements, advances in coordination chemistry, advances in understanding the structures of solids and hydrides of boron and silicon, and progress in deducing the chemistry of Earth and the cosmos.

Discovery of new elements and completion of the periodic table

At the end of the 19th century, the periodic table clearly had missing portions. Confusion surrounding the rare earth elements had yet to be resolved, and the question of how many rare earth elements there are had not been settled. Differences between the transition metal elements and other elements also were unclear. With uncertainty regarding how many elements exist, there was lingering doubt about the last remaining portions of the periodic table.

The 20th century brought the discovery of radon (Rn; atomic number 86) in 1900, europium (Eu; atomic number 63) in 1901, and lutetium (Lu; atomic number

71) in 1907. No new elements were discovered for some time thereafter, but in 1917 the teams of Hahn & Meitner and Soddy & Cranston each independently discovered the 91st element, protactinium (Pa; atomic number 91).

Moseley's research on X-ray spectroscopy and Bohr's atomic theory (see Chapter 3) contributed significantly to the completion of the periodic table and the discovery of new elements. Moseley's research determined the atomic numbers of the rare earth elements and brought a resolution to the question of the number of rare earth elements. His research in 1913 on the elements from aluminum (atomic number 13) to gold (atomic number 79) showed that the 61st to 72nd elements were as yet undiscovered. Bohr's periodic table in 1920 had uranium as the 92nd element, and left the 43rd, 61st, 72nd, 75th, 85th, and 87th elements as undiscovered.

Urbain, a rare earth scholar in France who knew of Moseley's research, later collaborated with Moseley and examined his rare earth samples using X-ray spectroscopy, allowing him to confirm the existence of erbium (Er; atomic number 68), thulium (Tm; atomic number 69); ytterbium (Yb; atomic number 70), and lutetium (Lu; atomic number 71). After World War I, Urbain attempted to discover the 72nd element and looked at rare earth samples using X-ray spectroscopy, but his efforts failed. Based on quantum theory, Bohr thought that the 72nd element would be a tetravalent element instead of trivalent, and believed it was not a rare earth. He thought that the rare earths ended with the 71st element, and that the 72nd element was homologous to zirconium. He proposed to his colleague Hevesy that they examine zirconium ore. As Bohr expected, a new element was discovered by Hevesy and Coster using X-ray spectroscopy in 1923, and they named it hafnium (Hf), in honor of "Hafnia", which is the Latin name for Copenhagen, where Bohr had had his research center.[106]

In 1925, Walter Noddack, Ida Tacke (who later married Noddack) and Otto Berg in Berlin reported discovering the elements with atomic numbers 43 and 75. The 75th element was obtained by repeated extraction and concentration from gadolinite, and identified from X-ray spectra. They named it rhenium (Re) in honor of "Rheinland", the part of Germany where Tacke was born. They also reported having discovered the 43rd element, masurium, but later believed they had made a mistake. Later research made it clear that the 43rd element does not have a stable isotope, and does not exist in the natural world in detectable quantities. The discovery of the 43rd element has been attributed to Italy's Segre and Perrier. In 1937, Lawrence, the inventor of the cyclotron, provided Segre and Perrier with samples of molybdenum that had been exposed to deuterium irradiation for several months in a cyclotron; by studying the radioactivity in the sample, they isolated and identified a new radioactive isotope. This element was named technetium (Tc), reflecting the fact that it had been artificially created. In 1908, Masataka Ogawa of Tohoku University reported discovering the 43rd element, which he named nipponium, but

Masataka Ogawa and nipponium[107]

Masataka Ogawa
(1865—1930)

From the end of the 19th century to the beginning of the 20th, the discovery of new elements was one of the major goals being pursued by inorganic chemists. At the time, they still did not know how many elements exist on Earth. However, most of the rare earth elements and noble gas elements had been found by the end of the 19th century, and few stable elements remained to be discovered at the start of the 20th century. This period saw contentious competition to discover new elements. Although he succeeded in isolating new elements, Masataka Ogawa was one chemist who has sadly not been recognized for discovering new elements.

Masataka Ogawa was born in Edo (now Tokyo) on 26 January, 1865, the eldest son of Masahiro Ogawa, a samurai who was a member of the Matsuyama domain and stationed in Edo. The family withdrew to Matsuyama during the Meiji Restoration; with the early loss of her husband, his mother struggled to raise her children. After graduating from Matsuyama Junior High School, he received a scholarship from the former Matsuyama feudal lord and moved to Tokyo, where he completed university preparatory school and attended the Department of Chemistry in the Faculty of Science at the Tokyo Imperial University, graduating in 1889. He advanced to graduate school and studied under Divers, a foreign teacher, but the next year worked at a junior high school in Shizuoka. He was recommended for principal at age 31, but resigned from his position and moved to Tokyo, where he became a research assistant at the Imperial University and resumed studying under Divers. At age 34, he was appointed to be First High School Professor, and traveled to the United Kingdom in 1904 at the age of 39 to study under Ramsay at the University of London. Famous for having discovered several of the noble gases, Ramsay asked Ogawa to discover new elements in the mineral thorianite, which had just been discovered in Ceylon (Sri Lanka).

By repeated dissolution, evaporation, extraction and precipitation—the classical method of separation—Ogawa obtained spectra of what he believed to be a new element present as a trace component. At Ramsay's suggestion, he named it "nipponium (Np)", in honor of Japan's endonym "Nippon". However, during his stay in Britain, he failed to achieve definitive data for the new element, and purchased thorianite to bring back to Japan so that he could continue his studies. His struggle ended when he obtained 0.1 mg of the oxide of a new element from

1 kg of thorianite. Taking its atomic weight as 100, together with the results of his analyses, he determined that this was the 43rd element, which was unknown at the time. His results were reported in the UK chemistry journal "*Chemical News*" as "Preliminary note on a new element". This study, however, was not confirmed by other researchers, and the reliability of the identification was not validated.

Ogawa became a professor of the newly established Tohoku Imperial University in 1911, and in 1918 he was elected president of the university, a position he held until 1928. He diligently continued his study of nipponium, and did not let his new position as president prevent him from continuing his experiments alone in the laboratory provided, next to the president's office. However, his ability to confirm the new element was limited by the conventionally used classical methods available to him. Confirmation required the use of X-ray spectroscopy, which had proven effective in the discovery of hafnium. Unfortunately, no X-ray spectroscopic equipment was available in the 1920s in Japan. In 1930, Tohoku University procured X-ray spectroscopy equipment, making it possible for Japanese scientists to study their samples using this technique. However, Ogawa died that year, and his measurement results were not published. In 1937, Segre discovered the 43rd element as an artificial radioactive element, and decided that this element was not found in nature. This resulted in Ogawa's nipponium being regarded as an "illusory element".

As the 20th century drew to a close, a professor emeritus of Tohoku University, named Kenji Yoshihara, discovered that Ogawa had left behind his X-ray spectral half plates of nipponium and attempted to interpret them. He was consequently able to show that Ogawa's samples of nipponium were of rhenium, the 75th element. Therefore, what Ogawa had thought was the 43rd element was actually the 75th element, which belongs one row down in the periodic table. Rhenium was discovered in 1925 by Noddack and Tacke, but in fact Ogawa had discovered rhenium 17 years earlier. Because Ogawa's analytical competence and perseverance were unrivaled, no one had been able to reproduce his research, but the truth was that he had succeeded in isolating a new element. What regrettably happened was that because he made an error when he postulated the atomic valence, he incorrectly estimated the atomic weight and thought that the 75th element was the 43rd element. Nipponium had not been an illusion, and Ogawa's error was truly a pity.

Morita and his research team at Japan's Institute of Physical and Chemical Research (RIKEN) in 2004 used a linear accelerator to irradiate a ^{209}Bi target with ^{70}Zn ions and succeeded in synthesizing the 113th element. The process of verifying the new element and deciding on its official name has been time-consuming, and no official name has yet been decided for the 113th element. The elemental symbol Np had already been used for the 93rd element, neptunium, so it seems that nipponium will not be listed on the periodic table of the elements.

recent research indicates that was instead the 75th element, rhenium.[107]

The existence of the 61st element was predicted by Brauner in 1902 and was researched in Moseley's laboratory, but it was not actually discovered until 1945. That year, Marinsky, Glendenin, and Coryell, at the Oak Ridge National Laboratory in the United States, obtained this element following the isolation and analysis of fission products in uranium raw fuel for nuclear reactors. This element was named promethium (Pm) in honor of Prometheus, who in Greek mythology stole fire from Mt. Olympus and gave it to mankind.

Following Moseley's research efforts, many attempts were made to discover the 85th and 87th elements. In particular, several groups reported discovering the 87th element, but these were later found to have been errors. In 1939, Perey, at the Curie Institute in Paris, obtained evidence of the 87th element in the decay product of actinium-227. She confirmed that this element had the properties of an alkali metal, as expected, and named it francium (Fr).

Corson, MacKenzie, and Segre at the University of California in 1940 obtained the 85th element by bombarding bismuth with alpha particles in a cyclotron. This element belongs to the halogens, but has metallic properties. It was given the name astatine (At), from the Greek "astatos", meaning "unstable", reflecting the fact that it is a halogen with no stable isotope.

The existence of the transuranic elements, with atomic numbers 93 and higher, became a certainty with the 1940 discovery of neptunium, the 93rd element, by McMillan and Abelson, and the 1941 discovery of plutonium by Seaborg, as described earlier; by 1945, all the elements up to the 96th element had been discovered.

Advances in coordination chemistry

Beginning with his first paper published in 1893, Werner's groundbreaking research on coordination compounds continued past the turn of the century until around 1915 (see page 95). In 1905, he published his "Neuere Anschauungen auf dem Gebiete der Anorganischen Chemie" (New Ideas in Inorganic Chemistry), which provided a detailed and systematic discussion of the field.[108] His research on amine complexes led Werner to introduce the concepts of primary valence and secondary valence for central metal atoms. The most direct evidence for his coordination theory was provided by the existence of the geometric isomers of amine complexes (see page 96). In 1911, Werner also synthesized asymmetric carbon-free enantiomers and succeeded in resolving them optically. His "secondary valence" represented the valence of the bond between a central metal atom and a binding molecule or ion, but the nature of this bond was not clear.

During his initial research on coordination compounds, Werner thought that 4-coordinate complexes of bivalent nickel, platinum, and palladium adopt a planar structure, but this was difficult to prove. In 1922, however, X-ray diffraction data made it clear that in compounds such as K_2PdCl_4 and K_2PtCl_4, the metal atoms are located in the center of the plane of a square formed by four chloride ions, and the existence of cis/trans isomers in these compounds was also confirmed.

An understanding of the nature of coordination bonds would have to await advances in chemical bond theory (see also Section 4.3 in this chapter). In 1920, Kossel proposed an electrostatic model of coordination complexes. The theory stated that the atoms in a complex bind through electrostatic attractions between a charged central ion and a polar molecule, and form a highly stable complex. It was apparent to researchers, however, that this way of thinking is too simple. Fajans focused on the deformation of ion and electron clouds and attempted to explain the colors of complexes.

On the basis of Bohr's atomic theory and the notion of covalent bonds proposed by Lewis and Langmuir, Sidgwick published *"The Electronic Theory of Valency"*[109] in 1927 and tried to explain coordination bonds. He described coordination bonds as one type of covalent bond, where a shared electron pair is donated from one of the atoms. He recognized that with a coordination bond, the atom donating the electrons becomes partially positive, and the atom that accepts the electrons becomes negative, thus creating an electrostatic interaction.

The effort to describe the bonds and physical properties of metal complexes using quantum mechanics began in 1929. The German physicist Hans Bethe submitted a paper on crystal field theory in 1929, and discussed how crystal fields affect the energy of electrons in the d orbitals of metal ions using symmetry.[110] In Bethe's paper, ligands that provide a crystal field are regarded as charged ionic molecules and are handled as a point charge. The paper showed how the energy of the d orbitals is split by the symmetry and intensity of the crystal field.[111] This paper was an important starting point that led to the later development of ligand field theory.

Pauling, in turn, described coordination bonds in 1931 using the idea of hybrid orbitals.[111] For example, in the case of cobalt amine complexes, as illustrated in Fig. 4.14, four unpaired electrons of the trivalent cobalt ion form pairs and occupy the three 3d orbitals along with a lone pair of electrons. The remaining two 3d orbitals, the 4s orbitals, and the 4p orbitals form d^2sp^3 hybrid orbitals, which are thought to be occupied by lone pairs of electrons of, for example, the nitrogen atoms from six ammonia molecules. The dsp^2 hybrid orbitals can be similarly considered as a 4-coordinate complex with a planar structure, such as $Ni(CN)_4$. Pauling discussed this idea in detail in his famous "The Nature of the Chemical Bond". His ideas were intuitive and easy for chemists to accept, and have been used to explain many phenomena in coordination chemistry.

Crystal field theory had successfully been used to explain the magnetic properties of coordination compounds by the American physicist John van Vleck in 1932.

Hans Bethe (1906–2005)
Hans Bethe was an American physicist born in Germany. He studied at the University of Munich, and after graduating went to Cambridge University and the University of Rome to study further and conduct research. Although he took an academic position in Germany, he moved to the United States in 1933 and became a professor at Cornell University. He had major achievements in various field of theoretical physics, and in 1938 he showed that solar energy was created by the nuclear fusion of hydrogen into helium. In 1967, he earned the Nobel Prize in Physics.

John van Vleck (1899–1980)
John van Vleck was an American physicist. After graduating from Harvard in 1922, he became a Harvard University professor in 1934 after working at the University of Minnesota and the University of Wisconsin. He made pioneering achievements in the quantum theory of magnetism and ligand field theory of metal complexes, and was awarded the Nobel Prize in Physics in 1977.

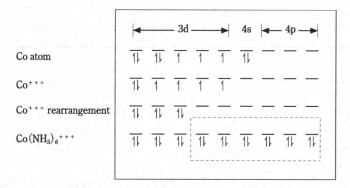

Figure 4.14: Pauling's description of coordinate bonds by hybrid orbitals

NH_3 has coordinate bonds to the six d^2sp^3 hybrid orbitals of Co^{+++}

He further proposed a more satisfying ligand field theory, which integrates crystal field theory and Mulliken's molecular orbital theory.[112] Ligand field theory regards electron delocalization as overlap between metal ion and ligand orbitals. The magnetic properties and absorption spectra of a coordination compound could thus be rationalized on the basis of the energy levels of the orbitals of the d electrons. For example, it was shown that when a crystal field is strong, Hund's rules are broken and the lower levels split by the crystal field are occupied by the electrons in accordance with Pauli's exclusion principle, resulting in a low-spin state. Ligand field theory has since been widely used, even among inorganic chemists.[113]

Chelate compounds, where central metals are connected to organic molecules through two or more bonds, exist as stable molecules. For example, $Ni(NH_2CH_2CH_2NH_2)_3$ is more stable than $Ni(NH_2CH_3)_6$. Chelate compounds have become increasingly important in a broad field of chemistry. Standard reagents for making a chelate used in analytical chemistry include dimethylglyoxime for Ni, 2,2'-dipyridyl for iron, and 8-hydroxyquinoline for aluminum.

Research on chelating molecules important in biological systems has attracted attention in the fields of organic chemistry and biochemistry. In chlorophyll and hemin, magnesium and iron are coordinated in a porphyrin ring, and chelate compounds have been discovered to be involved in many enzymatic reactions. The chemistry of coordination compounds thus began to develop as a domain of interdisciplinary research relevant to nearly all fields of chemistry, and this trend increased throughout the second half of the 20th century.

Inorganic compounds of interest: Boron and silicon hydrides

Many new inorganic compounds were studied in the first half of the 20th century, in particular hydrides of boron and silicon. Hydrides of boron were synthesized from magnesium boride and hydrochloric acid by Jones in 1879. Boron hydrides are molecules that are difficult to handle because they are highly reactive and ignite in air; pure compounds could not be studied until around 1912, when the German chemist Stock developed a technique for synthesizing and purifying them in a vacuum. Stock and his co-workers made a major contribution by using vacuum techniques for conducting research on boron hydrides over an extended period of time.

The simplest boron hydride is diborane (B_2H_6), but determining the structure of this molecule was long a source of frustration for chemists. Stock proposed an ethane-like structure, but the problem was that there were insufficient valence electrons. The possibility of diborane being an ionic structure had also been ruled out. A number of different proposals were made through the 1920s and 1930s, but all were inadequate in one way or another. In 1943, Oxford undergraduate student Longuet-Higgins, working with Bell, proposed a bridge structure for hydrogen. [114] (Fig. 4.15) In 1945, however, Pitzer proposed a model for protonated double bonds, where two protons are present in the electron cloud of the double bond between the borons. Longuet-Higgins explained the hydrogen bridge structure with an electron-deficient "banana orbital", where one electron has entered into the molecular orbitals composed of the 1s orbital of hydrogen and the atomic orbitals of the two borons (Fig. 4.15). Electron diffraction data obtained by Bauer in the

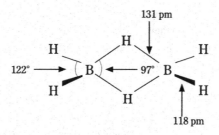

Figure 4.15: Structure and bridging bond of diborane

1930s seemed to support an ethane-like structure, but were inconclusive. Finally, in 1951, electron diffraction studies by Hedberg and Schomaker proved that Longuet-Higgins's structure was correct and put an end to the long debate about the structure of diborane. The bridge structure of diborane had been proposed by Dilthey in 1921. The structures of boranes were extensively researched in the 1950s by Lipscomb and coworkers using X-ray diffraction, and the general properties of the bridge structure were recognized. Similar bridge structures have been found in the structures of many other molecules, such as $Al_2(CH_3)_6$.

In the 1930s, Schlessinger of the University of Chicago synthesized $LiBH_4$, $BeBH_4$, and $AlBH_4$. $LiBH_4$ is stably soluble in water, and the Li^+ and BH_4^- are connected though an ionic bond. $AlBH_4$ is the most easily vaporized of the three compounds. H. C. Brown, a student of Schlessinger, began to apply the hydrides of boron, lithium, and aluminum to organic chemistry in the latter half of the 30s, and discovered that $LiAlH_4$ and $NaBH_4$ are extremely effective at reducing organic compounds. His research on the reduction of organic compounds by borohydride reagents proceeded rapidly thereafter. In 1979, Brown was awarded the Nobel Prize in Chemistry for these important developments in the organic synthesis reactions of boron compounds (see page 522).

Silane (SiH_4) was first synthesized by Wöhler in 1858. He made Mg_2Si from Mg powder and sand (SiO_2), then reacted it with hydrochloric acid to obtain a hydride of silicon in the form of a gas composed mainly of silane. Since carbon and silicon are homologous in the periodic table, hydrides of silicon had also attracted interest. However, silicon hydrides vaporize easily and ignite spontaneously in air, so are difficult to handle; consequently, research on silicon hydrides did not advance. In 1910, Stock used vacuum techniques to study hydrogenated silicons such as SiH_4 and Si_2H_6. These compounds are extremely reactive, and have limited similarity with hydrocarbons. Silane became an important material in the semiconductor industry in the second half of the 20th century, and research on silicon compounds flourished following World War II.

Structures and physical properties of solids

As already mentioned, following the X-ray structural analysis of table salt by Bragg and his son in 1914, the structures of inorganic compounds were extensively studied by X-ray diffraction. Advances in X-ray diffraction techniques brought about a progressively better understanding of complex structures. The bonds in many inorganic compounds are primarily ionic bonds. It thus follows that the size of an ion is an important factor that determines the structure of a crystal. In 1926, Goldschmidt in Norway determined the internuclear distances between ions in various crystals, and published a table of ionic radii.[31] This table of ionic radii was later modified by Pauling, and used in the study of the crystal structures and bonds of inorganic compounds.

The cohesive energy of an ionic crystal is calculated by considering the repulsive forces acting at close range for the sum of electrostatic interactions between ions in the crystal. The sum of electrostatic interactions was calculated by Madelung in 1918 and is given by the following equation.

$$U_M = - MN_A Z^2 e^2 / 4\pi\varepsilon_0 \, r_0$$

Here, M is the Madelung constant[32], N is Avogadro's number, Z is the largest common divisor of the charge of the ions, and r_0 is the distance between ions of opposite charge. The lattice energies of crystals had been determined empirically using a thermochemical cycle, called the Born-Haber cycle; values determined from the Born-Haber cycle were compared with the values obtained from the above equation.

Modern research on the electrical, magnetic, and other physical properties of solids was mainly initiated by physicists. Quantum mechanics was applied to problems in physics and chemistry in the 1930s, but since chemists were primarily interested in molecules, research on solids was carried out by physicists. Here, we touch briefly on the development of research on the physical properties of solids. In 1900, Drude proposed his free electron model, where the electrons in metals are treated as a gas that moves freely inside three-dimensional potential wells. According to this way of thinking, a large heat capacity from the translational energy[33] of the electrons should have been observed. However, the heat capacity

---

31. Ionic radius: Refers to the radius of an ion when the ion is regarded as a sphere.

32. Madelung constant: In this formula, r_0 is the distance between opposing ions and Z is the largest common denominator of the charges of the ions.

33. Translational energy of electrons: If an electron is thought of as a particle, then it has an energy of translation, just like other particle.

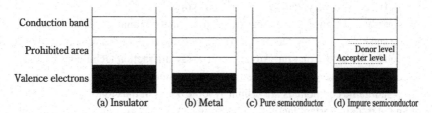

Conduction band

Prohibited area

Valence electrons

(a) Insulator   (b) Metal   (c) Pure semiconductor   (d) Impure semiconductor

Figure 4.16: Band model describing how solids have different electronic properties;
electrons occupy the portion shown in black

of metals could be explained by the vibration of the atoms, and heat capacity
from the electrons was not observed. This discrepancy was resolved in 1928 by
Sommerfeld, who proposed that the electrons are fermions[34] and follow a Fermi-
Dirac distribution rather than a Boltzmann distribution. The electrons in metals
move in a periodic electric field generated by ion core[35], and the electric field is
determined by the crystal structure. In 1928, Bloch (see page 455) introduced the
following Bloch equation to describe the wave function of electrons in such solids,
and calculated the energy of the electrons:

$$\Psi(x) = U_k(x)e^{ikx}$$

where e^{ikx} represents the wave of electrons moving in the x-direction at a wave
number k, and $U_k(x)$ is a function representative of the periodic potential created
by the ion core.

The cohesive energy of a metal is qualitatively understood to be the electrostatic
attraction between the positive ion core of the metal atom and the negative fluid
of highly mobile electrons. The energy levels of the electrons in a solid form
densely-packed bands when atoms assemble. The electrical properties of solids
were qualitatively understood depending on how these energy bands are occupied
by electrons. In the 1930s, band models revealed differences between conductors,
semiconductors, and insulators (Fig. 4.16). With a conductor, such as a metal,
only some parts of the bands of valence electrons are packed with electrons, but
in an insulator, the low energy bands are filled with electrons and there is a large
difference in energy between the packed bands and the empty bands. It was also
understood that a semiconductor is produced when electrons are thermally excited

---

34. Fermions: Particles that have a half-integer spin (1/2, 3/2, ...) and where one state can be occupied
    by only one particle at a time.

35. Ion cores are the cations of metal atoms composed of the nuclei and the electrons excluding
    valence electrons.

from packed valence electron bands to empty conduction bands that differ only slightly in energy level. This model also explains a mechanism by which the addition of impurities generates a semiconductor.

The study of the magnetic properties of substances has a long history, but systematic research began with Pierre Curie's study in 1895. He discovered a law whereby the magnetic susceptibility[36] of a paramagnetic body is inversely proportional to the absolute temperature T. This law was derived theoretically by Langevin in 1905. In 1907, Weiss proposed the Curie-Weiss Law, which states that at temperatures above the transition temperature T_c, the susceptibility of a ferromagnetic body is inversely proportional to T-T_c. Weiss derived this law for a group of magnetic moments that interact with one another. In 1928, Heisenberg submitted an atomic level theory of ferromagnetism in his Heisenberg Model. This model states that in a magnetic body comprising atoms with a magnetic moment derived from electrons, for example, the iron group or rare earths, the positive exchange interactions acting between ions are the cause of ferromagnetism. In 1932, Neel submitted a theory of antiferromagnetic bodies where negative exchange interactions act between adjacent magnetic atoms. These models served as the basis for discussions of the magnetic properties of solid metals and inorganic compounds.

The fact that atoms diffuse easily through a solid had been known since the end of the 19th century, but in the 1920s various models for the defects in solids were proposed, and the process by which this diffusion occurred was revealed. A type of defect where atoms (ions) move to positions outside the lattice points in a crystal was proposed in 1926 by the Russian Frenkel, and defects where there are vacant lattice points were proposed by Schottky in Germany in 1930. Taylor, Orowan, and Polanyi showed in 1934 that there are defects that they called dislocations[37]. The existence of these defects is closely related to the mechanical strengths and reactivities of metals. Elucidating these defects was expected to help solve practical problems in materials sciences.

Chemistry of the Earth and the cosmos[115]

At the beginning of the 19th century, Berzelius and colleagues strove to analyze many different minerals, resulting in the birth of geochemistry. The German chemist Schönbein in 1838 introduced the term "geochemistry" to describe the field exploring the Earth's origins and minerals, but geochemistry did not develop as an established field in the 19th century.

---

36. Magnetic susceptibility: Refers to the χ in $M=\chi H$, the relationship between the magnetization (M) of a substance and the intensity (H) of the magnetic field.

37. Dislocation: A type of lattice defect in a crystal, and a term for the displacement of atoms in a series occurring with linear connections due to deviations in the crystal.

One of the main goals of geochemistry in the late 19th century was to determine the relative quantities of each of the elements that constitute the Earth's crust. F. W. Clark of the United States Geological Survey conducted extensive chemical research on the minerals of North America and published his results as "geochemical data" in 1908.[116] His data showed that the abundance of the elements in the Earth's crust decreases as their atomic weight increases. Chemical analysis of meteorites in the late 19th century revealed that their chemical composition is different from that of terrestrial rocks, so it was believed that the composition of meteorites could provide clues about the chemical composition of the cosmos. This led to efforts to better understand the natural phenomena of both the Earth and the entire cosmos from a chemical point of view. One product of this effort was "Chemical Cosmology" by the German physical chemist, Baur.

Following the establishment of spectroscopy as a method of analysis in the 1860s, it became possible to analyze the elements present in the sun and stars. Scientists such as Crookes and Lockyer grappled with questions such as whether the same elements found on the Earth are present in the stars, and whether the same chemical phenomena take place on Earth and in the stars. Physical chemistry developed significantly between the late 19th century and the early 20th century. Geological/cosmological chemistry was of significant interest to Arrhenius, Nernst, Lewis, and other great chemists responsible for the development of early physical chemistry. Harkins, a physical chemist at the University of Chicago, discovered in 1917 that elements with even-numbered atomic numbers in meteorites were much more abundant than elements with odd-numbered atomic numbers. He discussed this finding in connection with the relative stability of elements. Aston, who developed the mass spectrometer, was also interested in the geochemical significance of isotope separation, and in 1922 was the first to assemble a table indicating the relative abundance ratios of isotopes; he proposed that the abundance ratios could reflect the relative stability of the atomic nuclei during the course of evolution of the elements. It was Victor Goldschmidt in Norway, however, who played the greatest role in the birth of modern geochemistry and the evolution of geochemistry into cosmochemistry. Goldschmidt is regarded as the father of modern geochemistry.

At the beginning of the 20th century, Norway became a center for the study of geochemistry under the leadership of Vogt and Brogger. Swiss-born Goldschmidt applied the Gibbs phase rule based on chemical equilibrium to study the thermal metamorphism of meteorites found near Oslo for his dissertation at the University of Oslo. His research brought new developments to geology. In the 1920s, Goldschmidt and his collaborators used X-ray crystallography to systemically study the relationship between crystal structures and atoms or ions in order to shed light on the general laws relating to the distribution of elements

Victor Goldschmidt (1888–1947)
This Swiss-born mineralogist and crystallographer served as a professor at the University of Oslo (then Kristiania) in Norway starting in 1914, and from 1929 to 1935 as a professor at the University of Göttingen, but he was pursued by the Nazis and returned to Norway. He systematically elucidated the crystal structures of oxides and fluorides, and laid the foundations for inorganic crystal chemistry. His research on the thermal metamorphism of rocks using thermodynamics led to a new field in geology. Goldschmidt studied the distribution of elements in rocks, and is regarded as the father of modern geochemistry. His goal was to place geochemistry on a sound thermodynamic and crystallographic footing, but he ultimately did not achieve this goal.

in important minerals. They also introduced spectroscopic techniques in order to study the elements present in trace amounts, investigated the relative distributions of, for example, rare earth elements, and established that elements with odd-numbered atomic numbers are also less abundant among the rare earth elements. [117] Geochemical research was also very active in the Soviet Union from the late 1910s, with Vernadsky and Fersman playing central roles.

Goldschmidt also included data about the sun, stars, and meteorites in his efforts to determine the relative amounts of elements in the cosmos; in 1938, he published the abundance ratios of the elements as a function of the neutron number of an element. His data later played an important role in the development of the shell model of nuclear structure by Goeppert-Mayer and Jensen, and in the description of the process by which atoms are generated in the cosmos.

Whether or not molecules exist in space had long been a question, but the presence of several diatomic molecules was verified from spectral observations from 1937 to 1941. Those molecules were CH, NH, CN, and CH^+. However, the chemistry of interstellar molecules only made major strides after the birth of radio astronomy following the Second World War (see page 569).

4.8 Organic chemistry (I): The birth of physical organic chemistry and polymer chemistry, and the development of synthetic chemistry

In the late 19th century, the regular tetrahedral theory of the valence of carbon had been established, as had the structure of benzene; together, these provided the basis for developments in classical organic structure theory. New techniques of analysis and synthesis were utilized for synthesizing many natural materials and new materials in the laboratory. This led to the development of organic industrial chemistry, which involves manufacturing useful substances such as dyes and pharmaceuticals (see Section 2.12). Progress in this area saw even greater advances in the early 20th century, leading to the development of physical organic chemistry, natural product organic chemistry, and polymer chemistry.

The development of structural chemistry allowed bond angles and the distances between atoms to be determined, and the various properties of molecules were gradually revealed. Consequently, the structure theory of organic molecules became more accurate and useful. The advent of electronic theory led to the appearance in the 1920s of the electronic theory of organic chemistry, which attempted to rationally explain the mechanisms of organic reactions based on the behavior of electrons. This was the birth of physical organic chemistry.

One of the major advances in organic chemistry in the first half of the 20th century was the development of natural product chemistry: the isolation, analysis, structural characterization and synthesis of the complex molecules present in nature. From the end of the 19th century to the beginning of the 20th century, Emil Fischer laid the foundation for the chemistry of monosaccharides, purine compounds, and proteins. This was followed by further advances in analytical and synthetic techniques, making it possible to take on the challenge of more complex molecules and producing remarkable results. These research efforts resulted in several Nobel Prize awards.

The development of new synthetic methods, which had progressed remarkably during the second half of the 19th century, also made major strides in the first half of the 20th century. In particular, research into the synthesis of useful materials from oil led to the development of powerful synthetic methods that utilize effective catalysts. The development of natural product organic chemistry also provided a major stimulus to organic synthetic chemistry. The greatest achievements of this time, however, were the establishment of polymer chemistry and the development of polymer synthesis. This made it possible to synthesize new fibers, artificial rubbers, and plastics, enabling mankind not only to succeed in creating replacements for natural products, but also to create substances with properties superior to those of natural materials. Polymer chemistry evolved significantly as a field in which basic chemistry and applied chemistry were closely related.

Germany had an overwhelming dominance in organic chemistry in the 19th-century and into the 20th century. After World War I, however, with the establishment of chemical bond theory and structural chemistry, the United Kingdom became the center for physical organic chemistry. In synthetic organic chemistry and natural product organic chemistry, Germany's dominance persisted, but important research results started to appear from outside Germany, such as from the United Kingdom, Switzerland, and the United States. Nobel Prize winners in the field of organic chemistry from 1901 to 1925 included Germany's Fischer, Baeyer, Wallach and Willstätter, and France's Grignard and Sabatier, but from 1926 to 1950 the names of Wieland, Kuhn, Windaus, Butenandt, Diels and Alder from Germany were joined by the names of Haworth and Robinson from the United Kingdom and Switzerland's Karrer and Ruzicka. This text handles the developments in organic chemistry during the first half of the 20th century in two sections. This section addresses the development of physical organic chemistry, polymer chemistry, and synthetic organic chemistry.

The birth and development of physical organic chemistry

Attempts to understand the structures and reaction mechanisms of organic compounds began towards the end of the 19th century and the beginning of the 20th century. For example, Lapworth, a physical chemist from the United Kingdom, in 1903 explained that cyanohydrins are generated from ketones when a cyano group is added to the polarized carbonyl group of a ketone and an acidic proton is withdrawn (Fig. 4.17).

Lapworth studied addition reactions to aldehydes and $\alpha\beta$-unsaturated ketones, as well as benzoin condensation reactions[38]. Based on his results, he suggested that ionic properties play an important role in reaction mechanisms. He developed the concept of alternating polarization in molecules, and based his understanding of reactions on this concept. These ideas later led to the concept of the "inductive effect".

Thiele at the University of Strasburg studied how the position of a double bond changes when a hydrogen or halogen is added to a double bond in an unsaturated compound. In 1899, this led Thiele to propose that the carbon atoms of the double bond have a "residual valence". Another research group focused on partial polarization inside the molecule in an attempt to explain the positions of substitution reactions in benzene derivatives.

Chemical bond theory, based on the concept of electron pairs conceived by Lewis and Langmuir, was introduced at this time in discussions on the structures and reactions of organic compounds. In Lewis's theory of covalent bonds, in the C:C bond in a molecule such as $H_3C:CH_3$, the electrons involved in bonding are shared equally by the two carbons. However, in bonds between different atoms, such as $H_3C:NH_2$, $H_3C:OH$, or $H_3C:Cl$, they are not shared equally by the two atoms, thereby producing a polarization of electrons and a dipole moment. Electronic theory of organic chemistry, which focuses on such polarization of electrons, was

---

38. Benzoin condensation: A condensation reaction between two molecules of aromatic aldehyde that allows the preparation of α-hydroxyketone in the presence of an alkali cyanide.

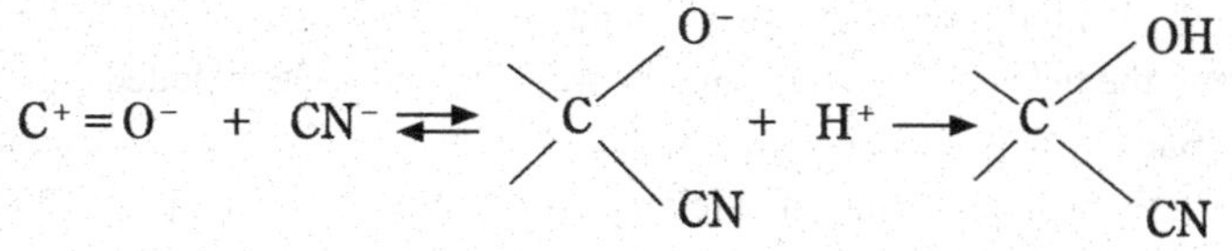

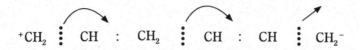

$$C^+ = O^- \; + \; CN^- \; \rightleftharpoons \; \underset{CN}{\overset{O^-}{\diagdown C \diagup}} \; + \; H^+ \longrightarrow \; \underset{CN}{\overset{OH}{\diagdown C \diagup}}$$

Figure 4.17: Mechanism of cyanohydrin synthesis from ketones, proposed by Lapworth

$$^+CH_2 \; \vdots \; CH \; : \; CH_2 \; \vdots \; CH \; : \; CH \; \vdots \; CH_2^-$$

Figure 4.18: Robinson's representation of the mobility of electrons

developed in the United Kingdom by Robinson (see page 328) and Ingold.

Robert Robinson, who was closely associated with Lapworth, in 1922 focused on the mobility of electrons in unsaturated compounds and used a curved arrow to express this mobility (Fig. 4.18). He proposed that with aromatic compounds or conjugated molecules, two electronic effects are important—namely, the transfer of an electron in the electron pairs and the electrostatic inductive effect (I effect)—in an attempt to explain the positions substituted in reactions involving benzene compounds.[118] Christopher Ingold started publishing papers in this field around 1926, and made major contributions to systematizing and developing the electronic theory of organic chemistry. He introduced the concepts of the inductive effect[39] and the mesomeric effect[40], and claimed that these produce changes in the electron density of the molecule and significantly impact the reaction. Ingold examined the effects of adjacent substituents in detail and polarization within the molecule. The formation of bonds in conjugated systems was discussed in terms of whether the reacting atoms were electron-donating or electron-withdrawing. In 1928, he attempted to systematize reaction mechanisms by giving the name "nucleophilic" to reagents that give or share electrons and the name "electrophilic" to those that are the receptors of the electrons.[119] This was an attempt to generally explain the reactive positions in substitution reactions among substituted benzene compounds, which had long been controversial. His electronic theory was sum-

---

39. Inductive effect: An effect where polarization produced by a difference in electronegativity between bonded atoms is transferred to a neighboring bond along the chain of bonds.

40. Mesomeric effect: Refers to a mechanism whereby molecules such as carbonyl compounds are polarized; according to this mechanism, carbons are positively polarized and oxygens are negatively polarized in the carbonyl double-bond.

Christopher Ingold (1893–1970)
A British chemist who studied at Southampton University and the Imperial College of London, and was a professor at the University of Leeds (1924–30) and University College London (1930–61). He developed an electronic theory of organic chemistry, where organic chemical reactions are explained by changes in the charge of substrates and reagents, and contributed greatly to the development of theories in organic chemistry.

marized in the monograph *"Structure and Mechanism in Organic Chemistry"*.[120]

Ingold and Hughes introduced physicochemical research methods, and in particular reaction kinetics-based methods, to the study of reaction mechanisms. From 1933 to 1935, they proposed classifying reaction mechanisms by using the expressions S_N1, S_N2, E1, and E2 (where S refers to a substitution reaction and E refers to an elimination reaction, and where 1 and 2 refer to the number of molecules involved in the step that determines the reaction rate; the subscript N refers to a nucleophilic reaction), thus attempting to systematically understand many different organic reactions. They also advanced research on reaction mechanisms on the basis of the measurement of the dipole moment and dissociation constant, as well as on the basis of isotope effects and thermodynamic quantities. Their description of reaction mechanisms included a detailed discussion involving factors such as solvent polarity, the effects of added salts or catalysts, steric effects, and steric hindrance. Thus, the electronic theory of organic chemistry was developed by an English school of thought, with Ingold at the center.

In 1935, Hammett at Columbia University in New York City published his empirical rule, now called Hammett equation[121], which describes the effect of a substituent on the reaction or equilibrium state of benzene derivatives. According to Hammett, the influence of a substituent group on a given reaction or equilibrium is represented by the formula $\log(k/k_0) = \rho\sigma$. Here, k and k_0 are the reaction rate constant or equilibrium constant when a substituent group is present or absent, respectively, σ is a constant determined by the position and type of substituent, and ρ is a value determined by the conditions of the reaction. In this manner, the electron-withdrawing or -donating effects of a substituent can be discussed quantitatively.

The special properties and stability exhibited by benzene were noted early-on. In 1925, Robinson attributed benzene's unique properties to the presence of six surplus electrons. In 1931, calculations using a molecular orbital approach allowed Hückel to show that six electrons enter the π electron orbitals composed of the p orbitals of the six carbons in the benzene ring to stabilize the molecule. Ingold had by 1922 already proposed that benzene could be expressed as a dynamic equilibrium of two Kekule structures and three Dewar structures; in Pauling's valence bond theory, the stability of benzene was described as a resonance between these structures. Hückel generalized the handling of benzene systems and showed that aromatic properties are obtained when the number of π electrons in a closed ring is represented by 4n+2.

Free radicals

In the second half of the 19th century, it was generally believed that free radicals with unpaired electrons either do not exist, or exist but cannot be isolated. The first stable free radical to be synthesized was triphenylmethyl, by Moses Gomberg of the University of Michigan in 1900. He attempted to obtain hexaphenylethane $(C_6H_5)_3C - C(C_6H_5)_3$ by reacting triphenylmethyl chloride with silver, but the resulting compound did not have the anticipated properties, and gave a highly reactive, yellow solution. His experimental results suggested that it was not $(C_6H_5)_3C - C(C_6H_5)_3$, but rather triphenylmethyl $(C_6H_5)_3C$ radical.[122] Gomberg felt that it would be difficult to arrange six phenyl groups around two carbon atoms, and therefore thought that a free radical had been produced. From a chemical basis, it was speculated that triphenylmethyl would be a stable radical, but obtaining proof for this was difficult in the early 20th century.

Following Gomberg's research, the synthesis of many stable free radicals was attempted. It was discovered that tetraphenyl hydrazine is reversibly decomposed into diphenylamine radical by heating in a toluene solution. 2,2-Diphenyl-1-picrylhydrazyl (DPPH) was confirmed to be a very stable free radical from measurements of its magnetic susceptibility. In the 1930s, many stable free radicals were synthesized and confirmed to be free radicals from measurements of their magnetic susceptibility. For example, Wieland synthesized diphenyl-N-oxide $((C_6H_5)_2NO)$, and Schlenk obtained anion radicals of metal ketyls by reducing aromatic ketones with alkali metals.

Physical chemistry research on reactions in the 1920s suggested the existence of highly reactive free radicals produced by thermal decomposition or photochemical reactions as intermediates, but it was only after World War II that these free radicals became important in the study of the mechanisms of organic chemistry reactions.

Development of stereochemistry

At the start of the 20th century, optical isomers were known to exist not only in organic compounds containing asymmetric carbon atoms, but also to be widely present in metal complexes with two or more chelate rings, as well as other types of organic compounds. van't Hoff had predicted the existence of optical isomers in allenes and cyclic compounds where a C-C bond cannot rotate; optical isomers

Moses Gomberg (1866–1947)
Born in Russia, Gomberg emigrated to the USA at the age of 18 when his family was expelled from Russia. In 1894, he graduated from the University of Michigan and went to Germany to study; after returning home, he became a professor at the University of Michigan. In 1900, he discovered that triphenylmethyl forms a stable free radical.

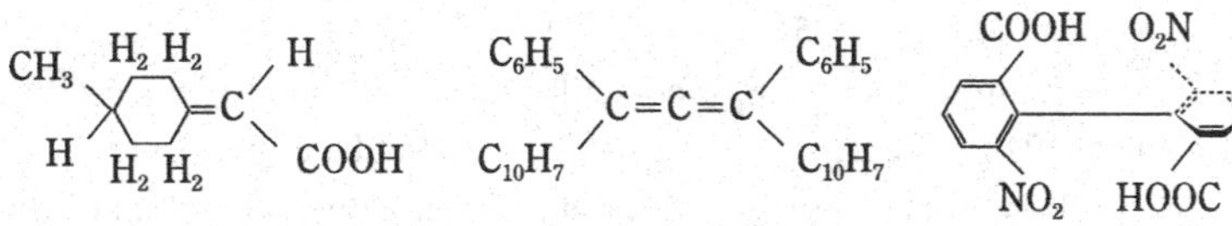

Figure 4.19: Examples of various types of optical isomers

of type (I), shown in Fig. 4.19, were observed in 1909, while optical isomers of type (II) were observed in 1935. It was discovered in the 1920s that optical isomers also are present among compounds such as (III) that have bulky substituents at the ortho position.

Many compounds were also synthesized and studied during research on cis/ trans or syn/anti geometric isomers. These compounds were synthesized by taking advantage of differences in the steric conformation of the two sides of a double-bond or plane of a ring. Cis and trans forms that have C=C double bonds are easily separated if the chemical properties of the two forms differ significantly, as is the case with maleic acid and fumaric acid (see page 76), but there are also many problematic compounds. Generally, it was realized that cis forms more readily dissolve in inert solvents, have lower melting points, and greater heat of combustion, and these differences in physical properties were used for estimating their structures. For simple molecules, measurement of the dipole moment (smaller for trans forms) and the results of X-ray diffraction were also used.

Important developments in stereochemistry in the first half of the 20th century include the analysis of steric conformations. The term "steric conformation" was introduced in 1929 by Haworth as representing a spatial arrangement of atoms within a molecule. When an atomic group connected by a given single bond within a molecule is rotated about this bond, the relative positions of the atoms on both sides of the bonds change, altering the form of the molecule. This spatial arrangement is the steric conformation. For example, in the case of ethane, there are two different arrangements with distinctly different energies: when the anterior C-H bonds are completely overlapped with the posterior C-H bonds (the overlapped structure in Fig. 4.10 (I)), and when the hydrogens on the two carbons are spatially

Sanichiro Mizushima (1899–1983)
Graduating from Tokyo Imperial University in 1923, Mizushima became an assistant professor at the university in 1927 and then a professor in 1938. He advanced the study of molecular structures using radio waves, infrared light, and the Raman effect; he was known in particular for his research on internal rotation in molecules, and discovered gauche rotamers. He was honored with Japan's Order of Culture in 1961.

Yonezo Morino (1908–1995)
In 1931, Morino graduated from the Faculty of Science at the University of Tokyo, and in 1940 he became an assistant professor at the university; between 1943 and 1948, he served as a professor at Nagoya University, and between 1948 and 1969 he was a professor at the University of Tokyo. In conjunction with Sanichiro Mizushima, he discovered the rotamers of ethane derivatives using Raman spectroscopy in 1941; after the war, he advanced the study of molecular structures by electron diffraction and microwave spectroscopy. He was honored with Japan's Order of Culture in 1992.

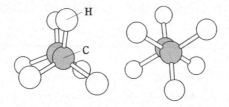

(I) overlapped structure   (II) twisted structure
Figure 4.20: Steric conformation of ethane

staggered (the twisted structure in Fig. 4.20 (II)). In 1937, measurements of entropy enabled Kemp and Pitzer of the University of California to show that the rotation of the methyl groups in ethane is a restricted rotation, not a free one; the energy of the twisted structure is 3 kcal/mole lower than that of the overlapped structure.[123]

The restricted rotation about C-C single bonds was intensively studied beginning in the 1930s, using techniques such as X-ray and electron diffraction, measurement of the dipole moment, and infrared and Raman absorption; detailed research was conducted not only on ethane, but also on compounds such as halogenated ethane. A University of Tokyo group, comprising Sanichiro Mizushima, Yonezo Morino and colleagues, showed that dihalogenated ethane has two twisted rotamer structures: a trans type and a gauche type[41]. The group extensively researched the differences

---

41. Gauche type: When four atoms bond to make A-B-C-D, the term Gauche is used when the dihedral angle between the A-B bond and the C-D bond is 60° in a conformation resulting from rotation

Odd Hassel (1897–1981)

This Norwegian physical chemist studied at the University of Oslo, the University of Munich, and Berlin's Kaiser Wilhelm Society; he served as a professor at the University of Oslo beginning in 1934. Using gas electron diffraction, Hassel showed that cyclohexane adopts a chair-type structure, and also demonstrated the existence of axial and equatorial bonds. He received the 1969 Nobel Prize in Chemistry.

Derek Barton (1918–1998)

After studying at the Imperial College of London and serving as a professor at the University of Glasgow, Barton went on to serve as a professor at Imperial College (starting in 1957) and as director of the Institut de Chimie des Substances Naturelles ("Institute for the Chemistry of Natural Substances") (starting in 1978), among other roles. He laid the foundation for conformational analysis of sesquiterpenes, steroids, and other alicyclic compounds. His research led to the Barton reaction, the total synthesis of aldosterone, and an understanding of the biosynthesis of alkaloids. Barton was a recipient of the 1969 Nobel Prize in Chemistry.

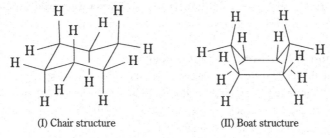

(I) Chair structure   (II) Boat structure

Figure 4.21: Structures of cyclohexane

in energy between the two rotamers.[124]

Research on steric conformations related to saturated cyclic compounds (cycloalkanes) attracted the most attention in organic chemistry. In particular, the structure of cyclohexane was a problem that long attracted the interest of chemists. In the 19th century, cycloalkane was thought to have a planar structure. Baeyer explained that five-membered or six-membered rings exhibit little strain because there is comparatively little deviation in the bond angles from 109.5°, the ideal regular tetrahedral angle for carbon bonds. Furthermore, there are many such molecules, whereas compounds with smaller rings do not exist because of the greater strain. In 1890, the German chemist, Sachse, noted that if the cyclohexane

about the B-C bond.

ring adopts a chair-type or boat-type structure (non-planar structures), then the strain is removed (Fig. 4.21).[125] However, this idea only gained traction once Mohr at Heidelberg University in 1918 noted that cyclohexane may be a mixture of chair-type and boat-type isomers that can readily interconvert.[126] In 1930, Odd Hassel at the University of Oslo in Norway realized from X-ray structural analyses that in crystals, cyclohexane adopts a chair-type structure.

Hassel in 1939 tried to determine the structure of cyclohexane using electron diffraction, and demonstrated that cyclohexane is in a chair-type conformation in the gas phase as well. He also showed that some of the carbon-hydrogen bonds are perpendicular to the molecular plane and some are almost parallel.[127] These were later called the equatorial and axial forms, respectively. Pitzer discussed the stability of the chair structure on the basis of the twisted structure of neighboring carbon atoms.

The results of research on steric conformations had a major impact on the later study of natural products organic chemistry and biochemistry. Steroids, triterpenoids, and other natural compounds include many cyclohexane and cyclopentane rings, and the analysis of steric conformations has been important in elucidating their steric chemistry. Applications of this field were developed in particular by Derek Barton after World War II.[128] Hassel and Barton were awarded the 1969 Nobel Prize in Chemistry for their achievements in studying steric conformations.

Development of synthetic organic chemistry

Having made great strides in the latter half of the 19th century, synthetic organic chemistry continued to develop rapidly well into the 20th century. Increasingly diverse methods of synthesis were developed, and the instruments used also advanced. The first 30 years of research in the 20th century, however, followed the patterns that were established in the latter half of the 19th century. It was only after the 1930s that changes began to appear in synthetic organic chemistry. Several factors can be considered as background for why this change was prompted. First, organic chemistry itself matured, and fields requiring the handling of complex natural organic compounds developed, spurring research in the synthesis of complex molecules of interest. This trend grew as applications of organic chemistry in biochemistry, medicine, pharmacology, and agricultural science increased. Next, the rational understanding of chemical phenomena based on physical chemistry advanced, and served as the foundation for systematic approaches to synthesis. Moreover, there were vigorous attempts to synthesize these complex compounds by incorporating spectroscopic and physical techniques of analysis. However, such changes in synthetic organic chemistry had only begun at the beginning of the 20th century, and it was only by the second half of the 20th century that this effort actually produced fruit. Representative important methods of synthesis

Victor Grignard (1871–1935)

A French organic chemist, Grignard studied at the University of Lyon, and in 1900 created Grignard reagents during his dissertation research. Serving as a professor at the University of Nancy starting in 1909 and as a professor at the University of Lyon from 1919, he had an organized approach to developing applications of the Grignard reaction in organic synthesis. Grignard's method of synthesis opened important new fields in synthetic organic chemistry as a new method of generating C-C bonds. During World War I, he served as a corporal, and was involved in the research and development of poisonous gases. He was a recipient of the 1912 Nobel Prize in Chemistry.

Paul Sabatier (1854–1941)

A graduate from École Polytechnique in Paris, Sabatier studied thermochemistry under Marcellin Berthelot at the Collège de France. He became a professor at the University of Toulouse in 1884, where he first studied compounds such as sulfides, chlorides and chromates. In 1897 he discovered a catalytic reduction method based on adding hydrogen to an organic unsaturated compound using reduced nickel; this made it possible to make fish oil and other oils into solid, hardened fats. Sabatier researched catalytic reduction methods with various metals as catalysts across organic compounds in general, and laid the foundations for the oleochemical industry. He was a recipient of the 1912 Nobel Prize in Chemistry.

developed in the first half of the 20th century are outlined below.

Victor Grignard in 1900 in France discovered that the reagent obtained when isobutyl iodide and magnesium are mixed in anhydrous ether is very stable and converts benzaldehyde to phenyl isobutyl carbinol.[129] Organometallic reagents represented by RMgX (where R is a alkyl group and X is a halogen) obtained from an alkyl halide and magnesium were named after him and are now called Grignard reagents. Grignard reagents are very reactive; for example, C_2H_5MgBr reacts as follows with C=O, –CN, and alkyl halide, respectively.

(1) $C_6H_5COC_6H_5 + C_2H_5MgBr \longrightarrow (C_6H_5)_2C(C_2H_5)OH$

(2) $CH_3CN + C_2H_5MgBr \longrightarrow CH_3COC_2H_5$

(3) $CH_3I + C_2H_5MgBr \longrightarrow CH_3C_2H_5$

Grignard reagents react with many functional groups, so they have been utilized to synthesize diverse compounds by researchers around the world. By 1912, more than 700 papers were published on studies utilizing Grignard reagents, and their use has continued to grow.

Hydrogenation reactions using catalysts were the subject of pioneering research by Paul Sabatier and Senderens in France from the end of the 19th century to the beginning of the 20th century. In 1897 they discovered that reduced nickel can be used to readily convert benzene to cyclohexane, and later Sabatier and his collaborators systematically researched the catalytic reduction of many organic compounds using various reducing metals, openings new fields in organic chemistry. Their research results later became industrially important.

Grignard and Sabatier were jointly awarded the 1912 Nobel Prize in Chemistry. This was only the second time the prize was given to Frenchmen, following Moissan in 1906. Grignard had studied Grignard reagents in Lyon, and served as a professor at the University of Nancy and the University of Lyon. Sabatier continued to conduct research at the University of Toulouse throughout his life. In France, where centralization of power was prominent, it is interesting that both scientists conducted their outstanding research in relatively rural areas. France was not awarded a Nobel Prize in Chemistry in the field of organic chemistry again until Lehn in 1987.

The first quarter of the 20th century saw the development of many organic synthetic reactions, of which many were named after the person who developed them. The following reactions are some typical examples.

- Bucherer reaction (1904): Conversion of naphthol to naphthylamine in the presence of ammonia and sodium sulfite.
- Dakin reaction (1909): Conversion of a phenol aldehyde to a bisphenol and carboxylic acid by hydrogen peroxide.
- Clemmensen reduction (1913): Reduction of a ketone or aldehyde to an alkane using a zinc amalgam and acid.
- Schmidt reaction (1924): Conversion of carboxylic acid to amine by an azide.
- Meerwein-Ponndorf-Verley reduction (1924): Conversion of a ketone to a secondary alcohol by an isopropyl aluminum salt catalyst.
- Diels-Alder reaction (1928): Reaction where a six-membered ring is generated from butadiene and maleic anhydride

The Diels-Alder reaction (Fig. 4.22) was discovered by Otto Diels and Kurt Alder at the University of Kiel.[130] This is a very general reaction, where an addition takes place at the 1,4-position of a conjugated double bond and a double bond is formed at the 2,3-position; a conjugated diene readily reacts with a double or triple bond activated with a carbonyl group or carboxyl group. This reaction has proven very useful for synthesizing cyclic compounds, confirming conjugated dienes, and conducting various polymerization reactions.

Otto Diels (1876–1954)

Diels was born the son of a professor of philology at the University of Berlin, and studied under Emil Fischer at the University of Berlin. He served as a professor at the University of Berlin (1914) and the University of Kiel (1916–1945). His accomplishments include the discovery of carbon suboxide, the development of a method for the dehydrogenation of various hydrocarbons using selenium, and contributions to determining the structure of sterols. The 1928 discovery of the Diels-Alder reaction allowed him to synthesize many organic compounds and elucidate their structures, and he demonstrated the importance of this reaction in the manufacture of synthetic rubbers and plastics. Diels was a recipient of the 1950 Nobel Prize in Chemistry.

Kurt Alder (1902–1958)

Born in the city of Chorzow in southern Poland, then a part of Germany, Alder received his primary education there. He studied at the University of Berlin and the University of Kiel, earning a degree under Diels, and after serving as an assistant (starting in 1926) and lecturer (starting in1936) at the University of Kiel, he moved to I.G. Farben and was involved in research on synthetic rubber. In 1940, he became a professor at the University of Cologne, and continued to study organic chemistry. He was a recipient of the 1950 Nobel Prize in Chemistry.

Figure 4.22: Diels-Alder reaction

Diels and Alder were awarded the Nobel Prize in Chemistry in 1950. This was a full 22 years after the publication of their first paper, showing that the importance of the Diels-Alder reaction grew gradually with time. Diels had been a professor at the University of Kiel since 1916, and by 1928 he was already recognized as an organic chemist with an impressive track record in various

Robert Robinson (1886–1975)
Robinson studied at the University of Manchester and embraced an interest in the organic synthesis of natural products due to Perkin's influence; he also took an interest in the theoretical aspects of organic chemistry, due to Lapworth's influence. After serving as a professor at the University of Sydney, the University of Liverpool, St Andrews University, the University of Manchester and the University of London, he served as a professor at the University of Oxford (1930–1955). He made important contributions to the development of the electronic theory of organic chemistry and made outstanding achievements in research on plant pigments and alkaloids. Robinson was awarded the Nobel Prize in Chemistry in 1947.

Figure 4.23: (I) Tropinone, (II) squalene, (III) riboflavin

fields. Alder earned his degree under Diels in 1926. Unusual in an era where the overwhelming majority of Nobel prizes were in the area of the organic chemistry of natural products, they were the first in the 38 years since Grignard and Sabatier to earn their prize for the development of a method of synthesis.

Natural organic compounds, such as the dye indigo, were already being synthesized in the latter half of the 19th century (see page 147). In 1917, Robert Robinson gained attention for successfully synthesizing tropinone ($C_8H_{13}NO$) (Fig. 4.23 (I)), an alkaloid. This synthesis suggested an approach for synthesizing more complex natural products. The synthesis of more complex organic compounds in natural products began around 1930. Representative examples here include research by Paul Karrer (1889–1971) in Zurich, Switzerland, and the group of Richard Kuhn (1900–1967) and colleagues in Heidelberg, Germany.

In 1931, Karrer synthesized squalene ($C_{30}H_{50}$) (Fig. 4.23 (II)), which can be isolated from shark liver oil. In 1934, he succeeded in the more complex total synthesis of vitamin B_2, also known as riboflavin ($C_{17}H_{20}N_4O_6$) (Fig. 4.23 (III)). His methods of synthesis were also put to use in industrial production. Kuhn, who around the same time had been working on isolating riboflavin and studying its structure, also succeeded in synthesizing riboflavin. Karrer in 1937 started with the synthesis of vitamin E, moved on to tocopherols ($C_{29}H_{50}O_2$), and in 1938 he succeeded in synthesizing a tocopherol. These studies were brilliant achievements in organic synthetic chemistry in the 1930s, but were essentially a prologue to the dramatic development of natural products synthesis after World War II.

The birth and development of polymer chemistry

At the beginning of the 20th century, it was known that substances such as rubbers, cellulose, resins, and proteins had large molecular weights, but it was believed that these were generally colloids comprising aggregates of smaller molecules. Colloid chemistry at this time was an area of significant interest as a new branch of physical chemistry. Leading organic chemists such as Emil Fischer and Wieland believed that no single organic compound had a molecular weight in excess of 5000, and many renowned polymer researchers also thought that polymer compounds were aggregates of smaller cyclic compounds and had colloidal or elastic properties. Hermann Staudinger corrected this common misbelief and established polymer chemistry as a discipline.

Staudinger started out as a typical organic chemist, first making a name for himself with his research on ketenes, but around 1920 he began to concentrate on studying polymer compounds. He measured the molecular weights of rubbers, and in 1917 first published the idea that such high-molecular-weight molecules are macromolecules of long chains connected by covalent bonds; in 1920, he expanded this idea.[131] However, this idea differed from the generally accepted belief that polymer compounds are colloidal, and therefore it was met with fierce opposition. He advanced his research by creating polymers of, for example, polyoxymethylenes (polymers of formaldehyde) and polystyrene with different degrees of polymerization. In 1929, he summarized his research results on poly-oxymethylenes and showed evidence supporting the idea that these polymers are long-chain macromolecules. By researching polystyrenes, he also discovered that polymers created under different conditions can be fractionated into components of different molecular weights. These achievements gradually weakened opposition to his theories.

Staudinger had made full use of the various physical means available at the time to determine the molecular weights of polymers, but in particular he used measurements of the viscosities of solutions to conduct detailed studies. In1932 he proposed the formula $[\eta]=KM$ (where K is a proportionality constant) representing the relationship between $[\eta]$, the intrinsic viscosity of a solution, and M, the molar mass. $[\eta]$ is a value obtained by dividing $((\eta/\eta_0)-1)$ by c, the mass concentration by volume, where η is the viscosity of the solution and η_0 the viscosity of the solvent.

There had been a long-standing controversy between Staudinger and researchers who believed in the colloid theory. However, in the early 1930s his macromolecular theory came to be generally accepted. Unlike other organic molecules, polymers can consist of molecules of various sizes, in which case their molecular weight is the average molecular weight of the constituent molecules. The form of a macromolecule affects the physical and chemical properties of substances made from it. This represents a departure from the concept of typical organic molecules, but Staudinger believed that polymer organic chemistry had great potential. Staudinger was awarded the Nobel Prize in Chemistry in 1953, but he was 72 years old at the time. Many famous organic chemists who were his contemporaries had received a Nobel prize much earlier. Given the importance and societal impact of developments in polymer chemistry, the awarding of this prize seemed surprisingly delayed.

Staudinger developed the study of polymer chemistry from the standpoint of organic chemistry. It was Herman Mark who contributed greatly to the development of polymer chemistry from a physical chemist's standpoint. Mark was involved in X-ray crystallography with Polanyi at the Kaiser Wilhelm Society in Berlin, and conducted structural analyses of cellulose fiber; in 1926, he developed his own doctrine supporting Staudinger's macromolecular theory. Later, Mark worked at the BASF laboratory and conducted further research on the structure of polymeric materials by X-ray structural analysis along with Kurt Meyer, and laid the foundations for the industrial production of polystyrene plastics, polyvinyl chloride, polyacrylics, and synthetic rubbers. Moving to the University of Vienna in 1932, he advanced research on mechanisms of polymerization, rubber elasticity, the viscosity of polymer solutions, and research on a wide range of polymer compounds. Forced to flee because of the annexation of Austria by the Nazis, he worked at a cellulose company in Canada and then served as a professor at the Polytechnic Institute of Brooklyn in New York starting in 1940, which he developed into a center for polymer research and education. He made a major contribution to developing the study of polymeric substances from a field of organic chemistry to the major discipline of "polymer science", which encompassed physical chemistry and physics.[132]

In this manner, the processes controlling the polymerization of macromolecules came to be better understood from the 1920s to the 1930s, as did the

Hermann Staudinger (1881–1965)
Born in Hesse, Germany, he studied at the universities of
Halle, Munich, Darmstadt, and Strasbourg, and served as a
professor first at Karlsruhe and then in 1912 as Willstätter's
successor at the Swiss Federal Institutes of Technology.
Beginning in 1926, Staudinger was a professor at the
University of Strasbourg. Staudinger started out as an organic chemist, with achievements such as the discovery of
ketenes and research on their auto-oxidation mechanisms, but from research
on polymers of isoprene he moved on to investigating the polymerization
of compounds such as styrene and vinyl acetate. He laid the foundations of
polymer chemistry and was recognized by being awarded the 1953 Nobel
Prize in Chemistry.

Herman Mark (1895–1992)
Mark was born in Vienna and studied at the University of Vienna, following
which he conducted pioneering research on the foundations and applications
of polymer chemistry, as well as the application of X-ray structural analysis
to polymer chemistry. He emigrated to America by way of Canada as he fled
the Nazis, and promoted the research and study of polymer science at New
York's Polytechnic Institute of Brooklyn.

structures of polymers, opening the door to major innovations in the manufacture
of plastics, fibers, and synthetic rubbers in the 1930s. Applications arising from
these developments in polymer chemistry soon impacted everyday life, as
discussed in Section 4.11.

4.9 Organic chemistry (II): Organic chemistry of natural products and foundation of biochemistry

Natural products organic chemistry, which was pioneered by Emil Fischer from
the late 19th century to the early 20th century, saw huge advances in the first
half of the 20th century and became the most active area in organic chemistry.
One reason for this is that advances in organic chemistry made it possible to
handle more complex molecules, and it became an appealing field appropriate
for competent, ambitious chemists willing to take on challenges. Another
reason is that this field has close ties to biochemistry, physiology, medicine,
pharmacology, and agricultural science, and these developing fields stimulated
research into natural products organic chemistry and vice versa. In particular,
advances in biochemistry were stimulated by developments in natural products

organic chemistry. The number of chemists who were involved in this field and won the Nobel Prize in Chemistry in the first half of the 20th century gives an understanding of how important this field was. The chain of Nobel recipient chemists comprises Willstätter (1915), Wieland (1927), Windaus (1928), Hans Fischer (1930), Karrer (1937), Haworth (1937), Kuhn (1938), Butenandt (1939), Ruzicka (1939), and Robinson (1947). This section looks at developments in the first half of the 20th century in natural products organic chemistry from the perspective of their achievements.

Carbohydrates

Emil Fischer posited a cyclic structure for methyl glucoside and correctly explained the existence of two kinds of isomers (see page 126), but he did not expand the cyclic structure to glucose itself. Therefore, he was unable to recognize the significance of the fact, discovered by Tanret in 1895, that there are two forms of glucose, with one having an optical rotation of +113° and the other +19°, and that the optical rotation of both of these isomers approaches the equilibrium value of +52.5° when in an aqueous solution. In 1903, Armstrong revealed the relationship between the two kinds of glucose and Fischer's methyl glucoside. In subsequent studies on sugar chemistry that followed, the central challenges were elucidating the cyclic structures of sugars and studying polysaccharides.

In 1925, Walter Haworth of Durham University in the UK (who moved to the University of Birmingham in 1926) and his collaborators proposed a model for the structure of glucose in which a six-membered ring composed of five carbon atoms and one oxygen atom has another carbon atom bonded to it. The Haworth Projection (Fig. 4.24), which was proposed as a method of representing the three-dimensional structure of a carbohydrate, later came to be widely used. The ring structures of important monosaccharides were revealed by around 1930 through the efforts of many researchers, including Haworth and Irvine in the UK and Hudson in the US; the structural elucidation of polysaccharides became the next challenge. Haworth and his collaborators revealed the structures of important disaccharides such as maltose, cellobiose, and lactose, and further

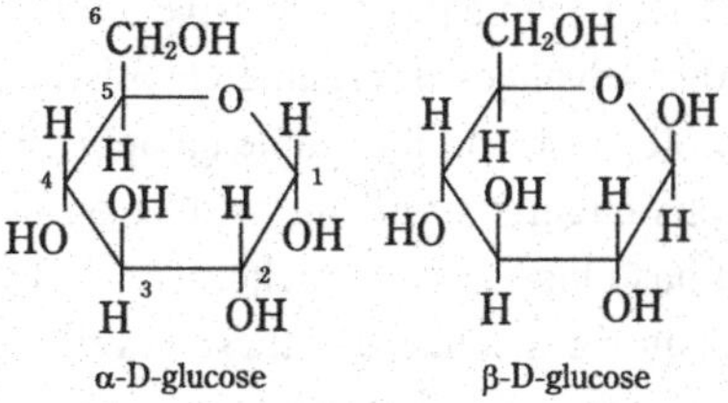

Figure 4.24: Haworth's projection formula

Walter Haworth (1883–1950)
Haworth was a British chemist who studied at the University
of Manchester and the University of Göttingen before
serving as a professor at Durham University starting
in 1920 and at the University of Birmingham starting
in 1925. Haworth confirmed the cyclic structures of
monosaccharides and proposed a formula that came to be
known as the Haworth Projection. In 1932, he determined
the chemical structure of vitamin C, and named it ascorbic acid. He succeeded
in synthesizing ascorbic acid in 1934. He was the recipient of the 1937 Nobel
Prize in Chemistry.

studied the structures of many polysaccharides. Many chemists participated in
research on polysaccharides, including research on substances that are industrially important, such as starches and cellulose.

Haworth in 1937 received the Nobel Prize for his achievements in the study
of carbohydrates and vitamin C. He was the first British Nobel laureate in the
field of organic chemistry.

Proteins and amino acids

By the end of the 19th century, many proteins had been purified and identified,
and a number of amino acids had been discovered. However, elucidating the
structures of the proteins themselves proved an extremely challenging task.
It was Emil Fischer who used techniques of organic synthesis to tackle this
challenge. He thought that a substance like a protein is created when amino
acids are connected by amide bonds to make a polypeptide. Around this time,
Curtius had already removed alcohol from the ester of glycine to produce a dimer.
Fischer and his collaborators put previously developed synthetic techniques to
use to synthesize diverse polypeptides. In 1907, they succeeded in synthesizing
a peptide composed of 18 amino acids.[133] Such peptides provide a color change
in the biuret test[42], as proteins do, and are degraded by pancreatic enzymes.
However, the molecules prepared synthetically were much smaller than naturally
occurring protein molecules.

Molecular weights were determined from measurements of the osmotic
pressure and freezing point depression; the molecular weights of ovalbumin and

---

42. Biuret test: A test where a protein or polypeptide is coordinated with copper ions in an alkaline
    solution, producing a color change from red-purple to blue-violet. It is one of the methods used
    to detect proteins.

hemoglobin were reported to be 14000 and 48000, respectively. Discovering the essence of molecules with such large molecular weights was a major task at the beginning of the 20th century. Around 1916, Fischer published his idea that proteins are composed of 30 to 40 amino acids at most, and that proteins with a molecular weight of 5000 or higher should not be considered to be pure molecules. In 1917, Sørensen reported that ovalbumin has a molecular weight of 34000, and Adair in 1925 reported that the molecular weight of oxyhemoglobin is 66800. In the 1920s, an ultracentrifuge technique developed by Svedberg was used to measure the molecular weights of many purified proteins, and the existence of high-molecular-weight proteins was confirmed. However, the prevailing thought at the time was that these macromolecules were not single molecules, but rather aggregates of much smaller molecules. Towards the end of the 1920s, polymer chemistry became established and the notion that proteins are macromolecules was finally accepted.

In the first half of the 20th century, it was not yet possible to elucidate the structures of proteins. It was known by the end of the 19th century that albumin crystallizes. In the early 1930s, crystals of several proteins were obtained. In 1934, Bernal and Hodgkin (née Crowfoot) at Cambridge obtained the first X-ray diffraction image of a protein crystal, using pepsin, but the technology at the time did not allow determination of the three-dimensional structure of globular protein by X-ray crystal analysis.

Bjerrum made the important finding that amino acids readily form a zwitterion of the form $^+NH_3RCHCOO^-$, but this property of amino acids made their analysis challenging. This was addressed using the method of chromatography (see page 298) developed by Martin and Synge in 1941. Chromatography proved quite useful for analyzing amino acids produced by the hydrolysis of proteins. In particular, the introduction of paper chromatography in 1945 facilitated the separation and identification of individual amino acids.

Nucleic acid

Kossel first demonstrated that nucleic acid is composed of four kinds of bases and phosphate (see page 129), then Phoebus Levene at New York's Rockefeller Institute expanded on this research. Levene studied nucleic acid from 1908 to 1929. First, he confirmed that ribose is the sugar in nucleic acid obtained from yeast. After many years of effort, Levene confirmed that nucleic acid derived from animal thymus contains deoxyribose. It was thus shown that there are two kinds of nucleic acids: ribonucleic acid (RNA) and deoxyribonucleic acid (DNA). Levene thought that the four kinds of bases would be present in the same compositional ratio, and advocated the theory (tetranucleotide theory) that nucleic acid has a structure in which the four kinds of nucleotides (units of

Phoebus Levene (1869–1940)

Levene was born in Lithuania at the time of the Russian Empire, but emigrated to the United States in 1893 and studied at Columbia University. In 1905, he became director of the biochemistry laboratory at the Rockefeller Institute of Medical Research, and researched the structure and functions of nucleic acid. He identified the constituent components of DNA and proposed the tetranucleotide theory of DNA structure.

Alexander Todd (1907–1997)

Todd studied at the University of Glasgow and earned a degree at Oxford University, served as a professor at the University of London and the University of Manchester, and then at Cambridge University from 1944 until 1971. His outstanding achievements included researching the structure and synthesis of nucleotides, the structural determination and synthesis of vitamins, researching alkaloids, and other research on biochemically important organic compounds. He was knighted, and was a recipient of the 1957 Noble Prize in Chemistry.

bonded sugars, phosphates, and bases) are linked in chains in equal numbers. This theory was based on the fact that the compositional ratio of the bases in the yeast-derived nucleic acid with which he had first worked was close to one, but he believed that this ratio was generally true. This theory, however, did not explain how nucleic acid could adopt diverse structures and therefore carry complex information such as genetic information. Levene was an authoritative biochemist at the time, and his theory had considerable influence. In the first half of the 20th century, many biochemists thought that proteins, which are far more complicated in structure than nucleic acid, were the primary bearers of genes, and few chemists paid attention to nucleic acid. At the beginning of the 1940s, the research group headed by the microbiologist Avery, who had researched Diplococcus pneumoniae for many years at the Rockefeller Institute (the same institute as Levene), reported important research results that suggested that DNA was the bearer of genetic information. (See Chapter 6)

It was Alexander Todd of the UK who furthered the study of the chemical structure of nucleic acid from the latter half of the 1930s into the 1940s. He used organic synthesis to determine the structures of nucleotides.[134] In a polynucleotide, the 3'-terminus of a ribose is linked to the 5'-terminus of the next ribose through a phosphate diester bond, and the base is linked to the ribose at the 1'-position (Fig. 4.25). Todd succeeded in synthesizing nucleotide coenzymes such as flavin adenine dinucleotide (FAD) and adenosine triphosphate (ATP), and made major contributions in determining the structures of the vitamins B_1, E, and B_{12}. He was awarded the Nobel Prize in Chemistry in 1957.

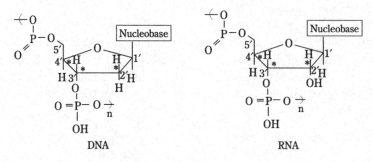

Figure 4.25: Structure of DNA and RNA nucleotide

Chlorophyll and hemin

The pigments in plant leaves, flowers and blood have been of interest to chemists since the 19th century. In 1873, Sorby used a solution to separate and extract two of the pigments for green and yellow from leaves. In the early 20th century, Tsvet used chromatography to perform separations. Richard Willstätter was the leader in this field in the first quarter of the 20th century. He researched pigments from various organisms, as well as photosynthesis and enzymes, at the Swiss Federal Institutes of Technology in Zurich and at Berlin's Kaiser Wilhelm Institute for Chemistry. He and his collaborators discovered in 1906 that chlorophyll, the green pigment in leaves, is composed of two components, a and b, present in a ratio of 3:1. They determined that the molecular formula for chlorophyll-a is $C_{55}H_{72}N_4O_5Mg$, whereas chlorophyll-b contains one more O and two fewer H's. He further researched the structures and properties of the two chlorophylls, and

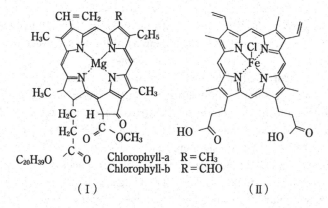

Figure 4.26: Chlorophyll (I) and hemin (II)

Richard Willstätter (1872–1942)

Willstätter was born into a Jewish family in Karlsruhe and studied under Baeyer at the University of Munich. He earned a degree with his studies on the structure of cocaine, and researched the structure of alkaloids as an assistant to Baeyer. In 1905, Willstätter became a professor at the Swiss Federal Institutes of Technology in Zurich, and began to research the structure of chlorophyll. He showed the  similarity in structure between chlorophyll and hemin. In 1912 he became a laboratory director at the University of Berlin and the Kaiser Wilhelm Institute for Chemistry, and in 1916 he became a professor at the University of Munich. In the 1920s, Willstätter researched enzymatic reactions, and claimed that enzymes are not proteins. In 1924, in protest against the rise of anti-Semitism, he resigned from the University of Munich. He was awarded the Nobel Prize in Chemistry in 1915.

Hans Fischer (1881–1945)

Fischer studied chemistry and pharmacology at the University of Lausanne and the University of Marburg, and worked in a hospital in Munich before becoming a researcher at the Berlin Chemical Institute under Emil Fischer. In 1918, he became a professor at the University of Vienna, and in 1921 became a professor at the Technical University of Munich. Fischer made outstanding research contributions on the structures of hemin and chlorophyll, and on the synthesis of porphyrins. His laboratory was destroyed in the war in 1945, and he fell into depression and committed suicide. He received the 1930 Nobel Prize in Chemistry.

found that there is similarity between chlorophyll and blood pigment. Nencki discovered in 1901 that the decomposition of hemoglobin or chlorophyll produces substitution products of pyrrole; Willstätter identified these pyrroles and determined how to synthesize porphyrin from them.[135] Willstätter also researched compounds such as carotenes and the flower pigments called anthocyanins.

Willstätter's research was continued and expanded by Hans Fischer from the 1920s through the 1930s. Fischer began studying the pigment hemin obtained from hemoglobin in 1921 at the Technical University of Munich, and found that removal of iron from hemin produces a porphyrin. He also found that pyrroles are products of hemin decomposition; he then proceeded to study the chemistry of porphyrins and pyrroles.[136]

Fischer confirmed that the skeletal structure of hemin is composed of four pyrrole rings to which methyl, ethyl, vinyl, and propionic acid groups are connected; in 1926, he synthesized porphyrin from pyrrole. In 1929, he went on to synthesize hemin. Later, Fischer studied chlorophyll and its pyrrole decomposition products, determined the structure of chlorophyll, and attempted (unsuccessfully) to synthesize chlorophyll (Fig. 4.26).

Willstätter received a Nobel Prize in Chemistry in 1915 for his research on chlorophyll, and Hans Fischer received the Nobel Prize in Chemistry in 1930 for his research on porphyrin and the synthesis of hemin.

Steroids and hormones

A steroid is a general term for a compound whose structure comprises a steroid skeleton containing three linked cyclohexane rings and one cyclopentane ring; they are found widely in plants and animals as bile acids, cholesterol, and hormones. Steroids have attracted the interest of chemists since the 19th century. In the first half of the 20th century, organic chemistry had advanced to the point of being able to handle such complex compounds, and leading contemporary organic chemists like Adolf Windaus of Göttingen and Heinrich Wieland of Munich sought to elucidate these structures. Windaus studied cholesterol, whereas Wieland studied cholic acid, which is a component of bile acid. In 1919, Windaus converted cholesterol to cholic acid, and showed that the two compounds are structurally closely related. It was recognized that the two have a common scaffold composed of four rings, where cholesterol has a secondary alcohol group and an isooctyl group side chain and cholic acid has a $-CH(CH_3)$ CH_2 CH_2COOH group. However, it was nearly impossible to determine the structures of the scaffolds using chemical methods that rely mainly on oxidative degradation. The cholesterol structure proposed by Windaus is shown in Fig. 4.27(a); however, Bernal conducted X-ray diffraction studies in 1932 that showed that this structure was not correct. Wieland cooperated with Rosenheim and

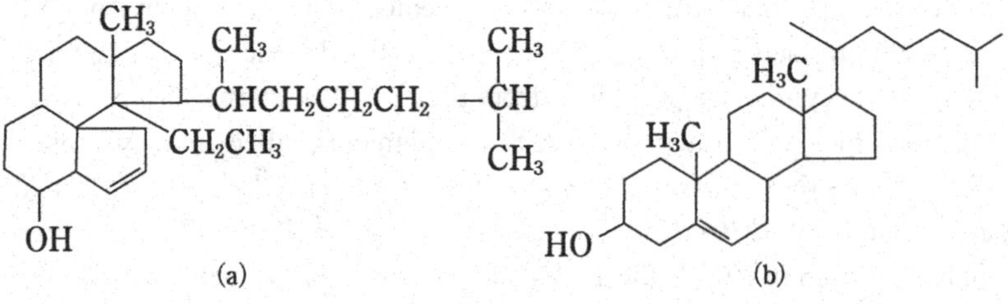

Figure 4.27: Structure of cholesterol

(a) Initial structure given by Windaus, and (b) modified structure given by Wieland.

Adolf Windaus (1876–1959)

Windaus was a German organic chemist who studied at the University of Freiburg and the University of Berlin, and served as a professor at the University of Innsbruck before serving as a professor at the University of Göttingen starting in 1915. Windaus is known, among other things, for his research on the structures of sterols, especially cholesterol, and on the generation of vitamin D by UV irradiation of ergosterol. He was a recipient of the 1928 Nobel Prize in Chemistry.

Heinrich Wieland (1877–1957)

A German organic chemist, Wieland studied at the universities of Berlin, Stuttgart, and Munich; he served as a professor at the Munich Institute of Technology and the University of Freiburg before his long tenure as a professor at the University of Munich (1925–1953). In addition to his groundbreaking research on the structure of bile acid, he is also famous for his research on stable free radicals, and research showing that in vivo oxidation reactions are a dehydrogenation process. His findings contributed greatly to the development of biochemistry and physiology (see page 352). In 1927, Wieland received the Nobel Prize in Chemistry.

King and proposed a modified structure, shown in Fig. 4.27(b). This example shows how physical methods are indispensable for determining the structures of complex organic molecules.

Wieland was awarded the Nobel Prize in Chemistry in 1927 for his research on bile acid, while Windaus was awarded the prize in 1928 for his research on sterols. However, they left behind major achievements beyond their research on steroids. Wieland's legacy includes considerable important research on various alkaloids, in vivo oxidation-reduction reactions, and stable free radicals. Windaus discovered that UV irradiation converts ergosterol to vitamin D.

Sex hormones constitute one class of steroids. Adolf Butenandt of the University of Göttingen, at around the same time as the American, E. A. Doisy, succeeded in isolating the female sex hormone, estrone, as a pure crystal in 1929. In 1931, he crystallized and isolated the male sex hormone, androsterone, and in 1939 obtained testosterone from androsterone. Butenandt extensively studied sex hormones and sterols. Leopold Ruzicka of the Zurich Institutes of Technology independently isolated androsterone. Ruzicka also investigated terpenes, saponins, and sterols. Butenandt and Ruzicka were jointly awarded

Adolf Butenandt (1903–1995)

Butenandt was a German organic chemist and biochemist who studied under Windaus at the University of Göttingen; in addition to professorships at the Technical University of Danzig (1933–36), the University of Tübingen (1944–56), he was also a director of the Kaiser Wilhelm Institute of Biochemistry (1936–1945). In 1956 he became a professor at the University of Munich, and in 1960 president of the Max Planck Gesellschaft. Butenandt was successful in isolating and crystallizing sex hormones and determining their chemical structures. For this achievement, he was chosen for the 1939 Nobel Prize in Chemistry, but he declined because the Nazis forbade him from receiving the award; Butenandt received the award after World War II, in 1949. His later successes included isolating ecdysone, a molting and metamorphosis hormone in insects, and the pheromone, bombykol.

Leopold Ruzicka (1887–1976)

A Swiss organic chemist born in Croatia, Ruzicka studied at the universities of Karlsruhe and Zurich, and became a professor at the Swiss Federal Institutes of Technology in 1929. He is known for his research achievements on polymethylene, terpenes, saponins, and sterols, and the structural determination of the sex hormone, testosterone. He was a recipient of the 1939 Nobel Prize in Chemistry.

the 1939 Nobel Prize in Chemistry for their achievements in the study of sex hormones.

Vitamins and carotenes

By the 20th century, there was gradual recognition that many types of compounds are essential to maintaining life and growth, including proteins, fats, carbohydrates, and inorganic salts, as well as organic compounds present in trace amounts; these trace compounds came to be called vitamins. In the first half of the 20th century, research on vitamins had attracted the attention of many nutritionists, biochemists, and organic chemists. Organic chemists with major achievements in vitamin research include Paul Karrer of the University of Zurich and Richard Kuhn of the Kaiser Wilhelm Society in Heidelberg, Germany.

In 1926, Karrer began studying plant pigments, and researched anthocyanins, flavins, and carotenoids. He revealed that carotenoids form a conjugated double bond system where eight isoprene units are bonded, with both ends having the

Paul Karrer (1889–1971)

Karrer was born in Moscow to Swiss parents, but he moved to Switzerland as a young child and studied chemistry under Werner at the University of Zurich, where he earned his degree. In 1917, he became a professor at the University of Zurich; his achievements include the structural determination of plant pigments, in particular carotenoids, and he discovered that carotenes are precursors of vitamin A.

Karrer also made important contributions in elucidating the structures of vitamin C, vitamin B_2, vitamin E, and other vitamins. He was awarded the Nobel Prize in Chemistry in 1937.

Richard Kuhn (1900–1967)

Kuhn was born in the suburbs of Vienna and received his secondary education in Vienna; after studying medical chemistry at the University of Vienna, he studied under Willstätter and earned a degree at the University of Munich. After the University of Munich and the Swiss Federal Institutes of Technology, Kuhn became a professor at the University of Heidelberg in addition to his duties

as Principal (1929) and later Director of the Institute for Chemistry at the Kaiser Wilhelm Institute for Medical Research. His achievements include researching carotenoids and enzymes, isolating and synthesizing vitamin B2 crystals, and synthesizing vitamin A. Due to Nazi interference, he declined the Nobel Prize in Chemistry in 1938, but received a medal and certificate of merit after the war.

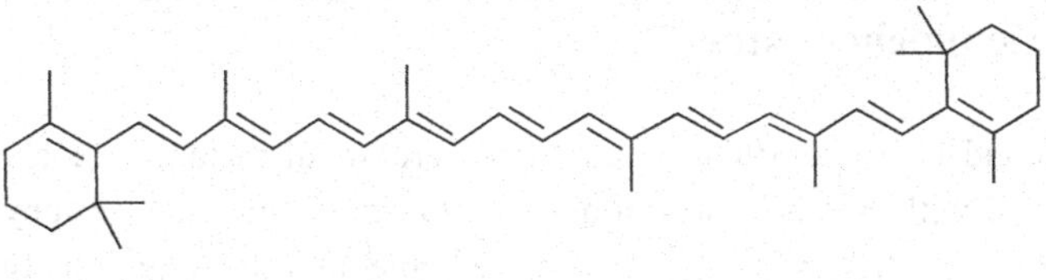

β-carotene

Figure 4.28: Structure of β-carotene

same structure. By 1930, Karrer had determined the structures of carotenes and lycopene (Fig. 4.28).

Karrer went on to discover that the addition of water converts carotene into two molecules of vitamin A. This was the first example of the determination of the molecular structure of a vitamin and represented a major breakthrough because many vitamin researchers at the time did not believe that vitamins were specific molecules.

Karrer later continued to study vitamins, and confirmed Szent-Györgyi's proposal for the structure of vitamin C; in 1934, at around the same time as Kuhn, he succeeded in the total synthesis of vitamin B_2, riboflavin. In 1937, he began to study vitamin E, tocopherols, and in 1938 synthesized the first tocopherol. Karrer also made major contributions to the structural determination and synthesis of tocopherols. In 1930 he published his "*Lehrbuch der Organischen Chemie*", a textbook on organic chemistry, which was translated into many languages in 13 different editions and was widely used as a textbook all over the world between the 1930s and the 1950s.[137] Karrer was awarded the Nobel Prize in Chemistry in 1937 for his research on carotenoids, flavins, and vitamins A and B.[138]

Between 1926 and 1929, Kuhn worked at the Swiss Federal Institutes of Technology in Zurich and studied the structure and synthesis of polyenes and their light absorption. He later continued his studies of carotenoids in Heidelberg and discovered a new carotene isomer; he isolated carotenoids from many natural products and determined their structures. From 1933 to 1934, he went on to isolate, crystallize, and synthesize vitamin B_2, and in 1937 succeeded also in synthesizing vitamin A. He was active in introducing new methods of analysis and physical techniques, and was an organic chemist with wide interests in theoretical issues such as stereochemistry and optical properties, as well as in biochemistry. Kuhn was awarded the Nobel Prize in Chemistry for the synthesis of vitamin B_2 in 1938. Doisy and Dam were awarded the Nobel Prize in Physiology or Medicine in 1943 for their research on vitamin K.

4.10 Establishment and development of biochemistry: Dynamic biochemistry

In the second half of the 19th century, there were two directions in biochemistry, which deals with the chemistry of biological phenomena. One was physiological chemistry, a field of physiology in medicine, and the other was the study of the molecules that make up organisms, a field of organic chemistry. These two directions coalesced in the 20th century, culminating in the establishment of one large, independent discipline of study. Research on the molecules that make up organisms, led by Emil Fischer, continued to grow in the 20th century and revealed the structures of sugars, amino acids, peptides, and purines; later, as was described in the previous section, the structures of diverse organic molecules involved in life phenomena, such as vitamins and hormones, were also elucidated. These developments laid the groundwork for biochemistry to shift gradually from the study of the structure of biologically relevant molecules to a dynamic biochemistry that aims to elucidate the reactions in the body.

Reactions in vivo are controlled by enzymes, so the chemistry of enzymes became a central issue in biochemistry. At the end of the 19th century, Buchner discovered that the enzyme chymase, isolated from cells, could ferment sugar into alcohol, thus setting the foundation for modern biochemistry. However, the process of alcohol fermentation entails many intermediate products, and it became increasingly apparent that the process of fermentation is extremely complex. Biochemistry in the first half of the 20th century revealed the details of processes such as respiration, digestion, and metabolism; revealing the enzymes that mediate these processes was central to the major advances in fields spanning chemistry, biology, medicine, pharmacology, and agriculture. In 1926, the enzyme urease was crystallized, and was shown to be a protein, but shedding light on the complex metabolic processes in which enzymes are involved proved to be a very difficult task. In the second half of the 1930s, radionuclide tracers were introduced to metabolic studies, thus beginning major developments in biochemical research, but it was only after World War II that this approach produced significant results.

Biochemists who were active at this time included some who started as organic chemists, and some who became researchers in biochemistry after being biologists or medical scientists. By 1950, Nobel Prizes in Chemistry had been awarded to Harden and Euler-Chelpin, who contributed to elucidating the existence of coenzymes and their fermentation processes, to Sumner and Northrop, who crystallized enzymes and showed that they are proteins, and to Stanley, who crystallized the tobacco mosaic virus. There were also many achievements in the field of biochemistry, resulting in Nobel Prize awards in Physiology or Medicine. Several biochemists awarded the Prize in Physiology or Medicine in the first half of the 20th century were Kossel for nucleic acid and protein research, Hill and Meyerhof for showing that the contraction of muscles comes from the energy released by glycolysis, Warburg for research on respiration and in vivo oxidation/ reduction, Szent-Györgyi for discovering vitamin C and studying muscle contraction, and Hopkins, Doisy, and Dam for their research on vitamins. The close relationship between the biochemical study of metabolism and medicine seemed to destine biochemical advances to be rewarded with the Prize in Physiology or Medicine, but the distinction between the Prize in Chemistry and the Prize in Physiology or Medicine is in fact not quite so clear.

Development of enzyme research

The study of enzymes proved to be key for biochemical research in the first half of the 20th century. Modern enzyme chemistry has its origins in Buchner's

Arthur Harden (1865–1940)
Harden was a British chemist who was born in Manchester
and studied at the University of Manchester before traveling
to study and earn a degree in organic chemistry at the
University of Erlangen in Germany. He served as a lecturer
at the University of Manchester, and later served as a director of the biochemistry department in the Jenner Institute
and then as a professor at the University of London (1912).

He studied sugar fermentation for over 20 years, beginning in 1897, and
clarified the role of phosphoric acid in fermentation and showed the existence
of coenzymes. Harden also made pioneering contributions to the study of
bacterial enzymes. He was a recipient of the 1929 Nobel Prize in Chemistry.

Hans von Euler-Chelpin (1873–1964)
Born in Augsburg, Germany, Euler-Chelpin first studied art
but turned to science and studied at the University of Berlin.
After two years of study under Nernst in Göttingen, he
became an assistant to Arrhenius in Stockholm and in 1906
became a professor at Stockholm University. His interest
shifted from physical chemistry to biochemistry, and he
researched sugar fermentation and enzymes. He confirmed

the existence of coenzymes and clarified their properties. Euler-Chelpin also
showed how sugars and phosphoric acid behave in the body. He was awarded
the 1929 Nobel Prize in Chemistry.

discovery of cell-free alcohol fermentation (see Section 2.10). Arthur Harden and
Young of the University of London in 1904 dialyzed yeast juice to divide it into
a protein and non-protein fraction in an attempt to ferment sugar; they reported
that neither component resulted in fermentation independently, but when the two
were mixed, fermentation did occur. This led to the discovery that the dialyzed
fraction included a component that complemented the enzyme, a "coenzyme"
or "prosthetic group". This component was stable to heat. Harden and Young
named this substance cozymase.[139] They went on to discover that when potassium phosphate is added to yeast juice, the generation of CO_2 by fermentation
increases markedly, and the phosphoric acid binds with and is fixed to the sugar.
However, it became apparent that the process of fermentation is extremely
complex, and it would take many years until it became fully understood.

One person who advanced Harden and Young's research was Hans von Euler-
Chelpin of Stockholm University. He and his collaborators obtained samples of
concentrated cozymase and studied its chemical properties; they revealed that

James Sumner (1887–1955)

Sumner was born near Boston, graduated from Harvard University, and studied biochemistry at Harvard Medical School. After earning his degree, he researched the purification and separation of the enzyme urease at Cornell University. He succeeded in crystallizing urease in 1926, and claimed that enzymes are proteins, but this view was not accepted immediately. By the 1930s, his claims were accepted, supported by research from Northrop's group. He was awarded the 1946 Nobel Prize in Chemistry. (See column 14)

John Northrop (1891–1987)

Born in New York City, Northrop researched enzymatic chemistry and earned a degree at Columbia University. During World War I, he investigated the process of fermentation for the industrial production of acetone and ethyl alcohol, and later at the Rockefeller Institute was engaged in researching fermentation. He succeeded in crystallizing many pure enzymes, such as pepsin and trypsin, and showed that enzymes are proteins. In 1938, he also successfully isolated bacteriophages for the first time. He was a recipient of the 1946 Nobel Prize in Chemistry.

cozymase has a molecular weight of 490 and had properties similar to nucleotides (compounds where a sugar, a base, and phosphoric acid are bound together).[140] Harden and Euler-Chelpin were jointly awarded the Nobel Prize in Chemistry for their research on alcohol fermentation in 1929.

Since Harden and Young's discoveries, coenzymes and prosthetic groups have also been discovered and identified during research on many enzymatic reactions. A coenzyme is the non-protein moiety released when an enzyme comprises a protein and a prosthetic group that reversibly dissociates from the protein. This prosthetic group is a small-molecular compound and is easily dialyzed. Compounds identified as coenzymes include pyrimidines and purines, pentasaccharides such as ribose and deoxyribose, phosphoric acid compounds, and vitamin B complexes.

At the start of the 20th century, many chemists, including Fischer and Buchner, began to hold the belief that enzymes are proteins, but it would take time until this could be established.

Willstätter, an organic chemist who was renowned in the 1920s, claimed that since enzymes do not undergo color reactions specific to proteins, proteins

Wendell Stanley (1904–1971)

Stanley was born in the state of Indiana in the U.S., and studied locally at Earlham College where he was a well-known football player. He studied organic chemistry under Roger Adams, a renowned organic chemist, in graduate school at the University of Illinois; after traveling to study at the University of Munich, in 1931 he joined the Rockefeller Institute for Medical Research. Beginning in 1932, Stanley studied the tobacco mosaic virus. He successfully crystallized it in 1935, and proved that the virus has properties intermediate between those of a pure substance (in that it can be crystallized) and the properties of a living organism (in that it can proliferate). He was awarded the Nobel Prize in Chemistry in 1946.

were merely enzyme vehicles, and that enzymes are unknown small molecules that are adsorbed onto colloidal proteins. At the time, many scholars were like-minded with the famous Willstätter and doubted that enzymes were proteins.[141] In 1926, at the end of nine years of patient research, James Sumner (column 13) of Cornell University succeeded in obtaining the enzyme urease, which breaks urea down into CO_2 and ammonia, by extracting it from bean flour and then crystallized it.[142] This crystal showed high enzyme activity, and exhibited the properties of a protein. However, Sumner's samples included impurities, and the notion that enzymes are proteins was not immediately accepted because it went against the dominant way of thinking at the time. Later, the group led by John Northrop of the Rockefeller Institute worked on crystallizing various enzymes. Between 1930 and 1935, he himself succeeded in crystallizing pepsin, and his collaborators crystallized one enzyme one after another, including trypsin, chymotrypsin, and ribonuclease; all of these enzymes were shown to be proteins.[143] It was thus established that Willstätter's way of thinking was wrong and that enzymes are proteins. In 1936, Wendell Stanley of the Rockefeller Institute attracted attention by crystallizing plant viruses, such as tobacco mosaic virus, in the same way as enzymes.[144] It was surprising that a virus, which behaves similar to an organism, had been crystallized. The 1946 Nobel Prize in Chemistry was shared by Sumner, for his discovery that enzymes could be crystallized, and Northrop and Stanley for their purification and preparation of enzyme and virus protein.

With the progress made in the study of physicochemical reaction kinetics at the start of the 20th century came the start of attempts to understand the factors that determine the rate of enzyme reactions. Enzyme reaction kinetics was developed on the basis of Fischer's thinking of the enzyme/substrate complex as an intermediate. In 1913, Leonor Michaelis and M. Menten showed that the initial rate of a reaction (v_0) can be given by the following equation when the reaction involves an enzyme (E) and a substrate (S) that form an enzyme/substrate complex, ES, and the reaction follows E+S $\rightleftharpoons$ ES$\rightarrow$E+P (product).

$$v_0 = v_{max}(S_0)/[K_m + (S_0)]$$

Here, (S_0) is the initial concentration of the substrate, K_m is an empirical parameter called the Michaelis constant, and v_{max} is the reaction rate when the enzyme is saturated with the substrate. Many researchers would later revise and expand this equation, but its importance has not diminished, even today, as a fundamental equation for enzyme kinetics.

Physicochemical techniques also proved useful when introduced into the experimental methods of enzyme research. Barcroft of the U.K. was the first to use pressure measurements in enzyme research (1902), but Germany's Warburg made major contributions in biochemistry by using improved pressure gauges to measure the amount of gas absorbed and generated during microbial respiration and fermentation, the respiration of animal and plant tissue sections, and by enzymatic reactions occurring in solution. His pressure gauge is called the Warburg manometer, and was used widely throughout the world from the 1920s to the 1940s. Warburg's studies incorporated spectroscopy into enzyme research. He showed that enzymes that contain iron play very important roles in in vivo oxidation reactions; to do so, he skillfully utilized the absorption spectra and action spectra of enzymes that have specific absorption bands. These advances in research methods brought major developments in the study of the chemical properties of enzymes.

In 1923, Hevesy and his collaborators introduced the use of radioactive isotopes as tracers for biochemical research by using radium-D (^{210}Pb) and thorium-B (^{203}Pb). Later, the use of isotopes as tracers made it possible to reveal

COLUMN 14

Sumner's indomitable fighting spirit and controversy over the nature of enzymes[23]

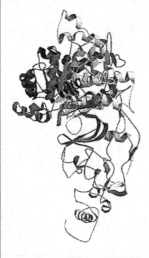

Urease from H. pylori

At the end of the 19th century, it was generally believed that enzymes are proteins. However, in the early 20th century, colloid chemistry was in the spotlight as a new field of physical chemistry, and this gave force to a dual carrier theory, which posited that enzymes are low-molecular weight catalytic materials that have attached to a protein colloid. Strong arguments for this theory came from the prominent organic chemist, Willstätter, and his followers. They attempted to obtain pure enzymes and worked on purifying invertase and other enzymes, but failed to crystallize them. They thus believed that enzymes could not be pure proteins.

It was Sumner, an unknown biochemist from Cornell University, who first succeeded in crystallizing an enzyme. While out hunting with a friend at the age of 17, he suffered a freak accident where his friend's gun misfired and a bullet entered Sumner's left hand, which was his dominant hand. This resulted in amputation of his left arm above the elbow. This would seem to be a handicap that would sink any hopes of becoming an experimental scientist, but he overcome his difficulties with an indomitable fighting spirit and became a biochemist. Furthermore, Sumner was an avid athlete, and excelled at many sports such as skiing, tennis, swimming, and mountain climbing. With this indomitable fighting spirit, he also fought against the mainstream of academia, with long years of challenging research on enzymes. He was the first to crystallize an enzyme, earning him the Nobel Prize in Chemistry in 1946.

Sumner was born into a family that ran a farm and also was engaged in textile fiber spinning on the side, in the small town of Canton, near Boston. He studied chemistry at Harvard University and graduated in 1910, then began to work at his uncle's textile fiber spinning company. Several months later he was asked to substitute for one semester at Mount Allison University in Canada; lecturing there sparked an interest in education. Thinking he would focus on studying biochemistry in graduate school at Harvard, he called on Otto Folin, a professor of biochemistry at Harvard Medical School, and asked to study under him. Folin advised him to study law instead of biochemistry, thinking that a one-armed man had no hope of succeeding in biochemistry, but Sumner insisted that even without one of his arms he could still easily conduct experiments, and begged for his permission to enroll. Eventually, Folin accepted him, unable to deny Sumner's strong will. Folin gave Sumner the topic of showing where urea is synthesized in the body. To quantify urea, a method of quantification was used where urea was converted to ammonia by the action of the enzyme urease. Thus, Sumner was fortunate that he was introduced to urease.

Leaving graduate school, Sumner found work as an assistant professor in a laboratory at Cornell University's Department of Physiology and Biochemistry. However, he was charged with giving lectures and training students, and found that he had limited time to devote to his own research. Then, reading a report that extract of jack beans had high urease activity, he had the idea of purifying urease from jack beans. At that time, the prevailing belief was that enzymes are not proteins, so this did not strike many people as a plan with much hope of success.

Sumner finely ground jack beans with a coffee mill and attempted to extract urease from the powder using various solvents. He made repeated attempts to precipitate urease by adding neutral salts or organic solvents, or by lowering the temperature, but could not concentrate urease activity. After beginning his research in 1917, he spent six years working on purifying urease but had no publishable results. At long last, in 1923 he discovered that most urease activity was precipitated when he cooled a 30% ethanol extract to $-5°C$. Crystallization remained impossible, however. In April 1926, he switched from ethanol solution to acetone solution in his attempts at crystallization. Finally, he succeeded in obtaining crystals with high urease activity. He excitedly reported his results to academia, but there was almost no response. Willstätter's followers did not believe his results, and controversy between them and Sumer continued for several years. In the early 1930s, Northrop succeeded in crystallizing pepsin, and from 1933 to 1934 Northrop and Kunitz crystallized trypsin, chymotrypsin, and other enzymes as the enzyme protein theory gained growing support.

Willstätter and his colleagues, however, refused to back away from their claims, and the controversy lasted throughout the 1930s. They claimed that Sumner and Northrop's crystals still contained impurities and included unknown active substances. There was some truth to this claim, in that many enzymes only function in the presence of a small-molecule coenzyme. One factor that prolonged this debate was that Sumner had chosen urease, which is easy to crystallize, while Willstätter's choice, invertase, is difficult to crystallize.

the pathways by which more complex compounds are synthesized in the body from simpler compounds, as well as the mechanism by which intermediates are formed during metabolic processes. The most important radionuclides in biochemical research are ^{14}C and ^{32}P, but it was only after World War II that these were used widely in metabolic research and produced important results.

Respiration and in vivo oxidation/reduction

Understanding respiration, and oxidation and reduction in the body, were central challenges in biochemistry at the beginning of the 20th century. In the second half of the 19th century, it became apparent that hemoglobin, which contains iron, in the blood was involved in respiration, and it was a generally accepted notion that oxygen bound to hemoglobin moves in the blood to the different tissues, where a chemical reaction takes place inside the cells.

The Danish physiologist Christian Bohr (father of Niels Bohr) in 1904 studied the levels of oxyhemoglobin (HbO_2) in the blood. HbO_2 is the product formed when oxygen binds to hemoglobin (HB). The relationship between the ratio of HbO_2 in the blood is given by $Y=(HbO_2)/\{(Hb)+(HbO_2)\}$ and the partial pressure of oxygen when at equilibrium was studied. Bohr discovered that Y provides an S-shaped curve with respect to the partial pressure of oxygen (Fig. 4.29). Various theories were proposed to explain this result, but in 1911 Hill suggested that hemoglobin units assemble and interact cooperatively, where the first HbO_2 molecule makes it easier for the next Hb and oxygen to bind together. In the 1920s, it became clear that hemoglobin is composed of four subunits.

Chemical thermodynamics developed from the latter half of the 19th century to the beginning of the 20th century, and made it possible to understand equilibrium and oxidation-reduction reactions on the basis of thermodynamics. Despite active discussion in the second half of the 19th century regarding how oxygen is involved in in vivo reactions, all theories lacked confirmation. However, the involvement of enzymes in in vivo oxidation gradually came to be recognized, and by around 1910 the existence of various oxidases was recognized. Under these circumstances, it was Otto Warburg of the Kaiser Wilhelm Society in Berlin who significantly contributed to understanding intracellular oxidation.

It was not until 1910 that the involvement of iron in intracellular oxidation became known. In 1908 Warburg began his research, and except for interruptions during the war, he spent the next 20 years working on intracellular oxidation. He used a manometer to quantitatively study oxygen uptake by red blood cells and sea urchin eggs. Between 1921 to 1925, he studied oxygen uptake and how it is hindered by carbon monoxide and other compounds using model experiments. His experiments used charcoal as a catalyst, made by burning blood, hemin, and other iron-containing components. He concluded that "... molecular

Otto Warburg (1883–1970)
After studying chemistry and earning his degree under
Fischer, Warburg studied medicine in Heidelberg. He
became a professor at the Kaiser Wilhelm Institute for
Biology in 1918, and in 1931 was named director of the
Kaiser Wilhelm Institute for Cell Physiology, making major
contributions in the study of respiration and oxidation, photosynthesis, sugar metabolism, cancer, and other fields. He
contributed significantly by introducing physicochemical techniques to the
study of biochemistry. Warburg was awarded the Nobel Prize in Physiology or
Medicine in 1931 for his achievements in the study of intracellular oxidation.

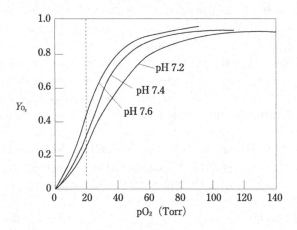

Figure 4.29: Saturation curve of oxyhemoglobin relative to the partial pressure of
oxygen

The horizontal axis is the partial pressure of oxygen, and the vertical axis is the ratio of oxyhemoglobin.

oxygen reacts with divalent iron, whereby there results a higher oxidation state
of iron. The higher oxidation state reacts with the organic substance with the
regeneration of divalent iron …. Molecular oxygen never reacts directly with the
organic substance".[145] Warburg believed that this conclusion applied to in vivo
oxidation as well, but at the time a catalyst that combined with iron in the body
had not yet been described. In carefully designed experiments, Warburg showed
that an iron-containing enzyme comprised an iron-porphyrin compound. He and
Neglein used light irradiation to reactivate respiration in yeast in which the iron-
porphyrin was inhibited by binding with CO. Reactivation of respiration was
plotted, allowing its efficiency to be determined as a function of the excitation

David Keilin (1887–1963)
A British biologist, Keilin was born in Moscow. He studied medicine in
Belgium and researched the life cycle of parasites in Paris before moving to
the University of Cambridge. There, he served as an assistant lecturer, and
starting in 1931, as a professor. In 1924, while studying the life cycle of the fly,
Keilin discovered cytochromes and revealed their role in cellular respiration.

wavelength. This provided a spectrum for the enzyme involved in respiration.
The resulting spectrum showed strong absorption specific to porphyrin at 420 nm
(a Soret band) and weak absorption by the amino acids in proteins, demonstrating
that this enzyme was a heme-containing compound.[146]

In the mid-1880s, MacMunn in the UK discovered that a respiratory pigment
in the muscle tissue of animals gave four absorption bands when in a reduced
state, and that these absorption bands disappear when the tissue is in an oxidized
state. David Keilin, a biologist in Cambridge, in 1925 discovered that the spectra
obtained from insect muscle tissue and yeast suspensions have four absorption
bands, and gave the name "cytochrome" to the pigments producing these bands.[147]
These absorption bands disappear when the suspension is shaken with air, and they
reappear when the suspension is allowed to stand. He showed that these pigments
act as intermediates in oxygen respiration in organisms. Keilin understood that
three of these four absorption bands came from different pigments, so he named
them cytochrome a, b, and c. He assumed the existence of an enzyme, oxidase,
which catalyzes oxidation of the reduced cytochromes.

An organic chemist, Wieland (see page 339) in 1912 proposed a theory for the
oxidation of an aldehyde to an acid, where first water is added to the aldehyde
and then dehydrogenation proceeds catalytically. His focus was on the fact that
when in the presence of a platinum or palladium catalyst, organic compounds
are oxidized even in the absence of oxygen. He attempted to apply this to in
vivo oxidation reactions as well. He reported that bacteria anaerobically oxidize
ethanol or acetaldehyde to acetic acid, provided that there is a hydrogen acceptor present, such as quinone or methylene blue. Wieland's research suggested
that there are reduction enzymes (reductases) that are different from oxidation
enzymes (oxidases).

At the end of the 1910's, Thunberg put finely chopped muscle tissue into a
solution of methylene blue, added deaerated organic acid, and measured the
time required for the methylene blue color to fade. This led to the discovery
that organic acids such as lactic acid and succinic acid promote the reduction of
methylene blue; he concluded that these compounds are dehydrogenated by the
enzyme dehydrogenase, similar to Wieland's view.[148] The Wieland-Thunberg

theory was accepted by many researchers in the 1920s, and in vivo oxidation came to be thought of as involving both oxygen and hydrogen.

In the 1920s, oxidation-reduction electronic theory advanced, and led to a unified manner of understanding that oxidation is the process of losing an electron and reduction is the process of obtaining an electron. The redox potentials of many reactions were measured, and attempts were made to understand in vivo oxidation as a flow of electrons based on the redox potential of the compounds. Redox potentials were also measured for in vivo redox systems. There was a search for compounds other than heme[43] that are involved in electron transfer in in vivo oxidation. Compounds initially considered included glutathione, which has an SH group, and ascorbic acid (vitamin C), which Szent-Györgyi isolated from the adrenal cortex and from oranges. In the 1920s, vitamins were a popular topic of research, and organic chemists rapidly revealed their chemical structures. This was also a time when in vivo oxidation reactions catalyzed by enzymes were beginning to be understood as electron transfer reactions, and it came to be recognized that many vitamins are constituents of the electron transfer system. Research by Warburg in the 1930s played a decisive role in providing a unified approach to grasping in vivo oxidation as a flow of electron transfers, and included Wieland's dehydrogenase and Keilin's cytochromes.

In 1928, the Americans Baron and Harrop discovered that oxygen uptake by red blood cells in the presence of glucose increases markedly with the addition of a small amount of dye such as methylene blue, and the glucose is oxidized. Warburg studied the mechanism that produces this effect in greater detail, and reached the conclusion that provided there is an enzyme that is not stable to heat and is not dialyzed, and a coenzyme that is stable to heat and is dialyzed, then the presence of a heme compound is not essential for the oxidation of glucose phosphate by methylene blue. In 1932, Warbug and Christian isolated a coenzyme and a protein from yeast that together comprised a yellow enzyme. The yellow prosthetic group is a type of flavone, a derivative of vitamin B_2, and was named flavin mononucleotide (FMN) (Fig. 4.30).[149]

Warburg and Christian also conducted research on enzymes and coenzymes that are involved in the oxidation of glucose phosphate. This research was related to the research by Euler-Chelpin on cozymase in alcohol fermentation. They discovered an adenylate system coenzyme, diphosphopyridine nucleotide (DPN) (Fig. 4.31).[150] This coenzyme is formed when adenine, nicotinic amide, pentose, and phosphoric acid are linked in the ratio 1:1:2:2. This molecule is now called nicotinamide adenine dinucleotide (NAD).

---

43. Heme: A complex compound where divalent iron ions are coordinated to porphyrin.

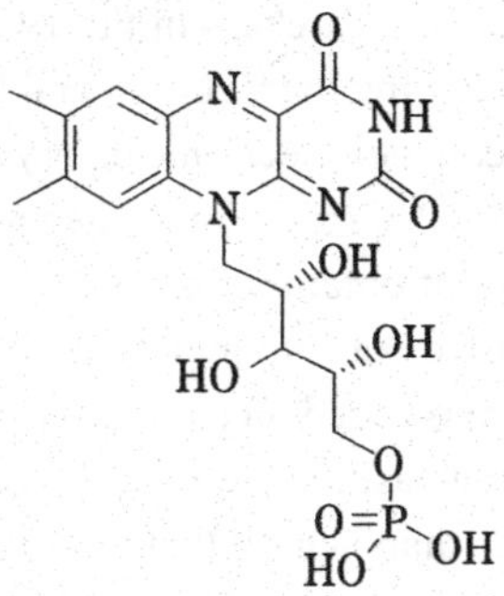

Figure 4.30: Flavin mononucleotide (FMN)

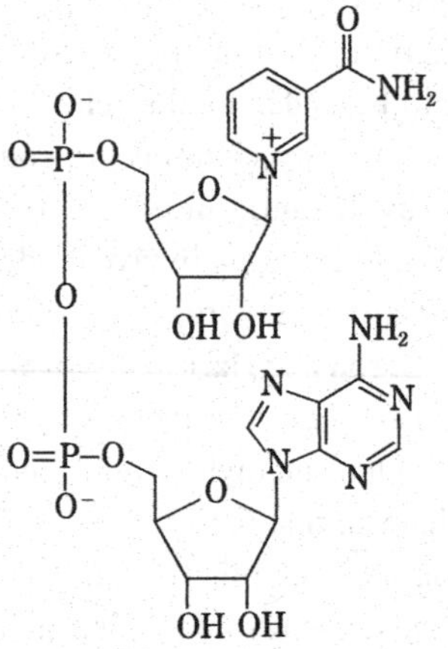

Figure 4.31: Diphosphopyridine nucleotide (DPN)

Thus, by 1945, in vivo oxidation was recognized as a process that involves heme compounds, FMN, and NAD, but it was not until after World War II that the details regarding in vivo redox processes began to be understood.

Elucidation of the glycolytic mechanism and the citric acid cycle

Elucidating the glycolytic process, where glucose is fermented by yeast and converted into ethanol and CO_2, was one of the major topics in biochemical research in the first half of the 20th century. This was an enormous task, requiring the isolation and identification of many intermediates and enzymes, and many researchers contributed to this research. The launching point was when, in the early 20th century, Harden and Young discovered that 1) an inorganic

Otto Meyerhof (1884–1951)
After earning a degree in medicine at Heidelberg University, Meyerhof became a physiologist studying under Warburg. After serving as a professor at the University of Kiel, he became a director of physiology at Berlin's Kaiser Wilhelm Institute for Medical Research in 1924. As a director of medical research at Heidelberg University beginning in 1929, he played a leading role in the major challenges in biochemistry at the time, such as the mechanism of action in glycolysis and the role of ATP. In 1938, Meyerhof fled Nazi Germany to France, and later traveled to the US, where he became a professor at the University of Pennsylvania. He was awarded the Nobel Prize in Physiology or Medicine in 1922 for discovering the correlation between lactic acid generation and oxygen consumption in muscle.

phosphoric acid is involved in alcohol fermentation, and 2) fermentation requires two components, zymase and cozymase, of the cell-free extract of yeast.

It became apparent at the end of the 19th century that glucose breaks down to produce a molecule comprising three carbons, and that this molecule converts into CO_2 and ethanol, but it was not until 1913 that Neuberg proposed his theory that first two methylglyoxals are produced, then one is reduced to glycerol, and the other is oxidized to pyruvic acid, then acetaldehyde, and finally to ethanol. In 1918, Otto Meyerhof discovered that the coenzyme discovered by Harden and Young was also found in muscle tissue, and later discovered that it is needed for glycolysis to occur in muscle.[151] This revealed the similarities between alcohol fermentation and the glycolytic process in muscle. By 1925, pyruvic acid was recognized as an important intermediate in both processes.

In 1906, Harden and Young proposed that glucose can form an ester bond with phosphoric acid, and separated the phophorylated sugar, but it was shown in 1928 that this ester is fructose-1,6-diphosphoric acid. Later, Ohle put forth the idea that glucose first is converted into glucose-6-phosphoric acid, then to fructose-1,6-diphosphoric acid, and finally breaks down to produce a phosphate ester of a C_3 glyceraldehyde and a phosphate ester of dihydroxyacetone. In 1933, Embden proposed the notion that these products are converted to glycerol 3-phosphate and 3-phosphoglycerate, with the latter being converted to pyruvic acid. Meyerhof and Lohmann in 1934 showed that in muscle extract, fructose-1,6-diphosphate breaks down into a triose phosphate. However, no one knew how the esterification of phosphoric acid proceeds in glycolysis, or what its biochemical significance was.

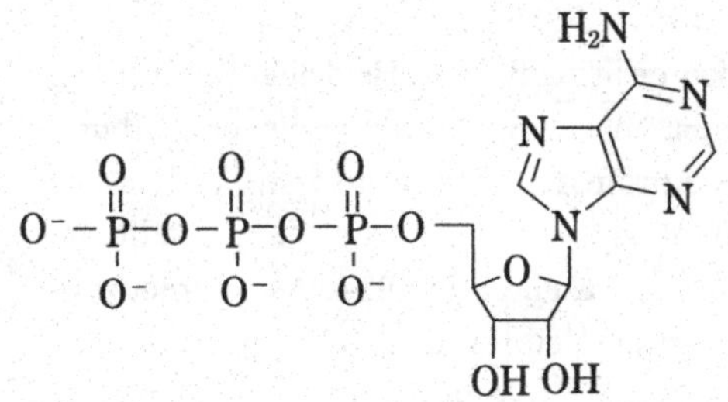

Figure 4.32: Adenosine triphosphate (ATP)

In 1929, Fiske, Subbarow, and Lohmann each independently discovered adenosine triphosphate (ATP) (Fig. 4.32) in muscle or yeast.[152,153] In 1931, Meyerhof and Lohmann showed that ATP is involved in the process of glycolysis. Parnas and colleagues thought that ATP is not involved in the entire process of glycolysis, but rather is involved in some specific step, and engages in the transfer of phosphate groups. Meyerhof's group conducted most of the detailed studies on these processes and identified the structures of the intermediates and enzymes involved in the reaction. It became clear that glucose is broken down by the action of ATP into a phosphate ester of a C_3 compound, which then breaks down into pyruvic acid, acetaldehyde, and ethanol (Fig. 4.33). This process is known as the Embden-Meyerhof pathway.

It was shown that about 30 kJ/mol is released when the bonds between the phosphoric acids in ATP are broken to generate adenosine diphosphate (ADP) and is stored when ADP is converted to ATP. The degradation of glucose is shown by:

$$\text{glucose} + 2\text{ADP} + 2 \text{ phosphoric acids} \rightarrow 2 \text{ ethanol} + 2CO_2 + 2\text{ATP} + H_2O$$

ATP acts like a high-energy currency in all living organisms, and it plays a critical role in energy metabolism, a fact that was pointed out by Fritz Lipmann and Kalckar.

By around 1935, the mechanism of glycolysis was well understood, but the relationship between the glucose oxidation mechanism and cellular respiration remained unknown. In 1935, the Hungarian-born biochemist Albert Szent-Györgyi discovered that cellular respiration increases dramatically when a small amount of succinic acid, fumaric acid, malic acid, or oxaloacetic acid is added to minced pigeon muscle. He showed that these acids convert in the order succinic acid → fumaric acid → malic acid → oxaloacetic acid.[154] Martius and Knoop

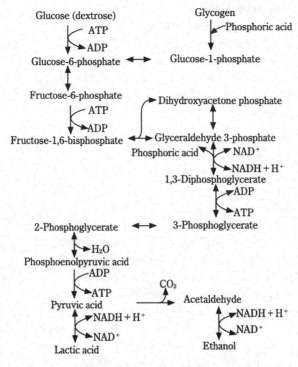

Figure 4.33: Embden-Meyerhof pathway[8]

showed that citric acid changes to cis-aconitic acid and then to isocitric acid, which is then dehydrogenated to produce α-ketoglutaric acid.

α-Ketoglutaric acid breaks down into CO_2 and succinic acid in oxidative decarboxylation, providing a reaction pathway from citric acid to oxaloacetic acid. When citric acid is produced from oxaloacetic acid, the catalytic citric acid cycle is completed. In 1936, Hans Krebs of the University of Sheffield in the U.K. discovered that citric acid is produced enzymatically from pyruvic acid and oxaloacetic acid, and thought this was the completion of the citric acid cycle. The enzymes involved in the cycle explained the speed of breathing that is actually observed. This established the citric acid cycle, where pyruvic acid is oxidized to produce CO_2 (Fig. 4.34)[155] The details of the process by which citric acid is produced from pyruvic acid remained unknown, but in 1945 Kaplan and Lipmann discovered coenzyme A (CoA), and in 1951 Severo Ochoa showed that pyruvic acid reacts with acetyl CoA and then oxaloacetic acid to produce citric acid. This revealed the entire cycle, now called the citric acid cycle or the TCA cycle.

Fritz Lipmann (1899–1986)
After studying medicine at universities in Königsberg, Berlin, and Munich, Lipmann studied under Meyerhof at the Kaiser Wilhelm Society, and in 1939 emigrated to the U.S. to become a professor at Harvard University from 1949 to 1957. He was awarded the 1954 Nobel Prize in Physiology or Medicine for his studies on the roles of coenzyme A and the high-energy phosphate bonds in metabolism.

Albert Szent-Györgyi (1893–1986)
A Hungarian-born biochemist, Szent-Györgyi first studied medicine in Budapest before traveling from one university to another in Europe as he continued his research; in Hopkins's laboratory at Cambridge, he isolated a reducing compound from the adrenal gland in 1927. He accepted the position of professor of medical science at the University of Szeged, where he showed that the isolated reducing  compound is ascorbic acid; he named this antiscorbutic factor vitamin C. Szent-Györgyi also made notable achievements in studying cellular respiration, and was awarded the Nobel Prize in Physiology or Medicine in 1937. Beginning in 1938, he started biochemical studies on muscle and showed that muscle contracts in the presence of ATP as an energy source when two proteins, actin and myosin, assemble together. After the Second World War, he continued studying muscle contraction in the U.S.

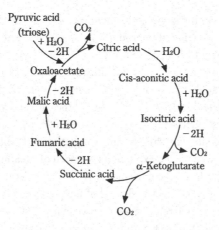

Figure 4.34: Krebs citric acid cycle (1937)

Hans Krebs (1900–1981)

Born in Germany, Krebs studied medicine at universities in Germany, and between 1926 and 1930 he served as an assistant in Warburg's laboratory and conducted biochemical research. Later, he entered the medical profession, but had to flee Nazi Germany to the U.K., where he continued his biochemical studies at Cambridge and the University of Sheffield. In 1945, Krebs became a professor at the University of Sheffield. In 1953, he was awarded the Nobel Prize in Physiology or Medicine for the discovery of the tricarboxylic acid (TCA) cycle, but he had many other achievements, including discovering the urea formation cycle (ornithine cycle) in the liver.

Severo Ochoa (1905–1993)

Ochoa studied medicine at the University of Madrid, and studied under Meyerhof in Heidelberg. In 1940 he emigrated to the U.S., and served as a professor at New York University School of Medicine starting in 1954. He succeeded in synthesizing RNA in vitro in 1959, and for this achievement he was awarded the Nobel Prize in Physiology or Medicine in 1959.

Left: Gerty Cori (1896–1957); Right: Carl Cori (1896–1984)

Classmates at the University of Prague, Gerty and Carl Cori married in 1920 and emigrated to the U.S. in 1922. In 1936, they discovered "Cori ester" at St. Louis's Washington University; in 1942, they discovered an enzyme that catalyzes the reaction for synthesizing this ester, and in 1943 they succeeded in synthesizing glycogen. This established the "Cori cycle", which is a reaction whereby glycogen in the liver is converted to glucose and then to lactic acid, to serve as an energy source for muscle movement. They were awarded the Nobel Prize in Physiology or Medicine in 1947.

In animals, glucose is stored in the liver and muscle as glycogen, which is broken down to glucose in the blood. The husband and wife team of Carl and Gerty Cori of Washington University in St. Louis in the 1930s elucidated each of the pathway steps in the synthesis and degradation of glycogen by hindering the activity of the enzyme required in the subsequent step.[156] They succeeded in synthesizing glycogen by way of glucose-6-phosphoric acid and glucose-1-phosphoric acid from glucose in test tubes in the presence of ATP and enzymes.

ATP was a donor of phosphoric acid in the first step, and was generated from ADP in the last step. They were awarded the 1947 Nobel Prize in Physiology or Medicine for "the discovery of glycogen's catalytic conversion process". Carl and Gerty Cori were educated in Prague, but they emigrated to the United States in the 1920s. They conducted many pioneering studies and educated outstanding researchers at Washington University beginning in 1931 (See Page 639, Chapter 6).

Photosynthesis

Photosynthesis is the reverse of the oxidation of sugar, because it is a process where CO_2 is reduced to sugar by light and H_2O is oxidized to O_2. By the early 20th century, it was thought that light absorbed by photosynthetic pigments directly reduces CO_2, and the reduced CO_2 and water react to form sugar. In this way of thinking, the O_2 produced in photosynthesis derives from water. Dutch-born American microbiologist van Niel, however, showed that anaerobic green sulfur bacteria use H_2S to perform photosynthesis and generate sulfur. In 1931, he used the hydrogen donor H_2A to generalize photosynthesis with the following formula.[157]

$$6CO_2 + 12H_2A \longrightarrow (CH_2O)_6 + 12A + 6H_2O$$

With plants, H_2A is water, whereas it is H_2S with sulfur bacteria. This formula shows that photosynthesis proceeds in a two-step reaction, comprising a light reaction where H_2A is oxidized by light energy and a dark reaction where CO_2 is reduced with the H thus produced.

Two different studies showed that this theory was correct. In 1937, R. Hill isolated chloroplasts and irradiated them with light in the absence of CO_2 but in the presence of an electron acceptor such as quinone or $Fe(CN)_6^{3-}$. Hill observed that O_2 was generated and the electron acceptor was reduced. This is called the Hill reaction, and it showed that CO_2 is not directly involved in the generation of O_2. In the 1940s, various radioactive isotopes started to be used in chemical research as tracers. Ruben and Kamen in 1941 used H_2O and CO_2 labeled with ^{18}O and proved that the O_2 generated in photosynthesis is derived from H_2O.

Little progress had been made in understanding how CO_2 is converted to a sugar since Baeyer's theory at the close of the 19th century, which posited that CO_2 is first converted to formaldehyde, which polymerizes to form hexose. Ruben, Kamen, and Hassid used $^{11}CO_2$ and chlorella to confirm that ^{11}C is incorporated into an organic compound in the dark reaction, but the short half-life of this isotope (22 minutes) made it unsuitable for research on photosynthesis.

In 1940, they conducted a nitrogen (n, p) reaction and discovered the ^{14}C isotope, with a half-life of 5600 years (see page 291), but the Second World War stopped them from using this isotope to study photosynthesis. The end of the war was followed by major advances in metabolism studies using ^{14}C as a tracer, and Calvin's group at the University of California used $^{14}CO_2$ to elucidate the pathway by which plants convert CO_2 to sugar (see Section 6.3.5).

Lipid metabolism

Most lipids are triacylglycerols composed of a fatty acid and glycerol, which are then digested to fatty acids. Fatty acids are converted to CO_2 and H_2O in oxidation metabolism, in the same way as sugars. In-depth study of the degradation of fatty acids began with Knoop at the University of Strasbourg in 1904. He synthesized various fatty acids with a phenyl group attached to a terminus and gave the compounds to dogs. Metabolites with attached phenyl groups were isolated from their urine and studied. He discovered that when a fatty acid with an odd number of carbons was used, hippuric acid (an amide of benzoic acid and glycine) was excreted, whereas phenylaceturic acid (an amide of phenylacetic acid and glycine) was excreted when a fatty acid with an even number of carbons was fed to the dogs. He accordingly inferred that in the degradation of fatty acids, the carbon at the β-position (second) from the carboxyl group is oxidized. Specifically, the degradation of fatty acids takes place according to the following pathway.

$$CH_3(CH_2)_n \, CH_2 \, CH_2COOH \longrightarrow CH_3(CH_2)_n \, COCH_2COOH$$
$$\longrightarrow CH_3(CH_2)_{n-2} \, COOH$$

When the number of carbons is decreased in increments of two, phenyl-substituted fatty acids with an odd number of carbons are ultimately oxidized to benzoic acid. There were pros and cons to Knoop's hypothesis, but there was no alternative explanation of the mechanism of oxidation in the first half of the 20th century. His hypothesis was shown to be correct and the overall mechanism of oxidation was revealed in the 1950s, when CoA and the various enzymes involved in fatty acid oxidation were discovered. It should be noted that in these experiments, Knoop first used labeling techniques to study metabolism. His method of labeling with a phenyl group was not widely applicable, but by the 1930s, when isotopes became available, methods of labeling with isotopes started to provide major advances in the study of metabolism.

Pioneering research on fatty acids using isotope labels was conducted by Rudolph Schoenheimer in 1935.[158] He used fats substituted with deuterium

Rudolph Schoenheimer (1898–1941)
Schoenheimer was born in Berlin, where he studied medicine and earned a degree from the University of Berlin. After studying chemistry at the University of Leipzig, he studied cholesterol and atherosclerosis at the University of Freiburg; it was there that he learned from Hevesy how to use isotope labeling techniques to study reactions. In 1933, Schoenheimer emigrated to the U.S. to flee the Nazis; at Columbia University, he met Urey and established a technique for using deuterium to label biomolecules and track metabolism, thus bringing about a revolution in the study of metabolism. His research clearly showed that the molecules that constitute living organisms are constantly being interchanged and exist in a dynamic equilibrium. However, he committed suicide in 1941, at the pinnacle of his research.

to conduct his research. At the time, it was thought that animals store fat in readiness for energy metabolism, and consumed it as needed. He synthesized deuterated stearic acid and gave it to mice for a defined length of time, then killed the mice and studied the deuterium content of their fatty acids. His research showed that fatty acids are stored and consumed continually. Feeding deuterated stearic acid to mice led to the discovery of deuterated palmitic acid and oleic acid, and feeding deuterated oleic acid led to the discovery of deuterated stearic acid. These studies showed that stearic acid, oleic acid, and palmitic acid are present in a dynamic state in which they interconvert with each other.

The synthesis of fatty acids in the body was later revealed to occur by condensation of C_2 units, contrary to oxidation. In 1945, Rittenberg and Bloch used isotope labels to show that the condensation units derive from acetic acid.

Metabolism of proteins and amino acids
The metabolism of nitrogen-containing compounds was a problem that had attracted interest since the 19th century, but no advances were made until the 20th century. Urea had been isolated from urine by Rouelle in 1773, but its origins remained unknown. In 1904, Kossel and Dakin discovered the enzyme arginase in animal muscle, and observed that arginase hydrolyzes arginine into urea and ornithine. By around 1930, it was believed that arginase was involved in the synthesis of urea in the body, but it was difficult to describe the generation of urea solely by the breakdown of arginine obtained from protein.

In 1932, Krebs and Henseleit showed that the first reaction in the synthesis of urea in the liver is the addition of one molecule of ammonia and one

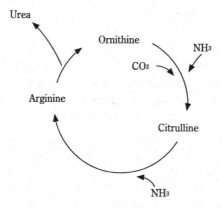

Figure 4.35: Ornithine cycle

molecule of carbonic acid to the δ-amino group of ornithine, resulting in the removal of one molecule of water and the production of citrulline. Krebs and Henseleit thought that citrulline then reacts with another ammonia molecule to produce arginine, and that this is hydrolyzed to produce urea and ornithine. With this reaction pathway, Krebs proposed a "cycle" (the "ornithine cycle") whereby ornithine reacts with ammonia and carbonic acid to synthesize citrulline (Fig. 4.35). This "cycle" not only put forward a new way of thinking about the synthesis of urea, but also was a pioneering force in the systematization of metabolic pathways such as the citric acid cycle. However, it was only in the 1950s that the more detailed steps involved in the ornithine cycle came to be understood.

An understanding of the details of amino acid metabolism would also have to wait until after World War II, but by around 1940 the transfer reactions of amino groups important in amino acid metabolism were known. In 1937, Braunstein and Kritzmann showed that minced pigeon or mouse muscle produces a reaction where the amino group of glutamic acid is transferred to pyruvic acid or oxaloacetic acid to produce alanine or aspartic acid. This was intended to prove that a transamination reaction is catalyzed by enzymes.

In 1937, ^{15}N started being used in the study of metabolism. From 1939 to 1941, Schoenheimer and his collaborators made use of ^{15}N to conduct pioneering research on protein metabolism.[159] They gave amino acids labeled with ^{15}N to animals, and analyzed the amino acids obtained by the hydrolysis of proteins in their tissues and examined the isotope ratios. This research showed that the amino acids in proteins in tissue constantly undergo chemical changes and exist in a dynamic state, mixing with amino acids ingested in the diet or with amino acids from other proteins in the tissue.

Vitamins and hormones

Up to the beginning of the 20th century, it was thought that food is necessary for supplying the materials required for generating energy and constructing the body, and thus the intake of proteins, fats, and carbohydrates is essential. It had not yet been recognized that trace amounts of chemical compounds are required to maintain life. However, it was known empirically that many diseases are due to dietary deficiencies. In 1747, Lind, a surgeon in the Royal Navy, treated sailors suffering from scurvy by giving them oranges or lemons. Kanehiro Takagi, a medical officer in the Japanese Navy, discovered that soldiers do not contract beriberi if they are given more diverse foods in place of white rice. In 1886, Eijkman discovered that rice bran (unpolished rice) has many components that prevent and cure beriberi.

Gowland Hopkins, a biochemist at Cambridge, in the early 20th century raised mice on various diets and studied their lifespans. He concluded that natural foods contain innumerable substances that are necessary to maintain health. He noted that rickets and scurvy are prevented by a proper diet. In 1911, Funk extracted a water-soluble compound from yeast that acts against beriberi and named it a vitamin. In 1910, prior to Funk's publication, Umetaro Suzuki had extracted this component from rice bran and called it "oryzanin"; however, Suzuki's publication was not immediately known to the world because it was written in Japanese. For the discovery of vitamins, Eijkman and Hopkins were awarded the Nobel Prize in Physiology or Medicine in 1929, but it was Suzuki who first extracted vitamin as a substance, and it is regrettable that his achievement went unappreciated. Hopkins made many contributions as a biochemist, and was an eminent scholar who educated numerous Nobel laureates, but his research on vitamins was not verified; consequently, it is questionable whether he should have been awarded the Nobel Prize as the discoverer of vitamins.

In 1913, McCollum in the U.S. discovered that there is a component in butter and egg yolk fat that is essential to the growth of mice, and extracted this component. The component extracted by McCollum had properties unlike those of the compound extracted by Funk. McCollum called this compound "oil-soluble A", while Funk call the compound he extracted "water-soluble B". In 1920, Drummond renamed these two components vitamin A and vitamin B. He also extracted a compound that has an effect against scurvy from citrus fruit, and named it vitamin C. The name vitamin D was given to an anti-rickets compound. Trace components needed for the maintenance of life thus were found one by one and named in alphabetical order. When it was realized that vitamin B encompasses a group of compounds that have similar properties, these became the vitamin B group and were named B_1, B_2, B_3, etc.

Gowland Hopkins (1861–1947)
After studying chemistry at the Royal School of Mines in London, Hopkins became an assistant at a London hospital and studied medicine before graduating from the University of London. He accepted a position as a lecturer at the University of Cambridge and then in 1914 as a professor. Despite his many achievements in the field of biochemistry, his greatest achievement was the leading role he played in educating many young researchers. Hopkins was a recipient of the 1929 Nobel Prize in Physiology or Medicine.

Umetaro Suzuki (1874–1943)
Born the second son of a farmer in the village of Horino-shinden, Haibara-gun (now Makinohara City) in Shizuoka Prefecture, Suzuki graduated from Tokyo Imperial University's Faculty of Agriculture, and specialized in plant physiology in graduate school. In 1900 he became an assistant professor at the same university before traveling to Switzerland and Germany the next year; in Berlin, he studied proteins under Emil Fischer. Returning to Japan in 1906, he became a professor at the Morioka College of Agriculture and Forestry, and from 1907 until 1934 he served as a professor at his alma mater. In 1910 Suzuki succeeded in extracting an anti-berberi compound (vitamin B1) from rice bran, and named it oryzanin. Starting in 1917, Suzuki held a concurrent position at RIKEN.

As stated in the previous section, determining the chemical structures of vitamins was a major topic in organic chemistry from the latter half of the 1920s to the 1930s. When the structures of vitamins were revealed due to the efforts of many organic chemists and biochemists, it became apparent that many vitamins function as coenzymes and are required by enzymes for activity inside the body. It follows that a deficiency in a vitamin would cause dysfunction of the metabolic systems involving enzymes requiring vitamins as coenzymes.

In 1902, the Englishmen Baylis and Starling discovered a secreted substance that promotes the secretion of pancreatic juice from the duodenal membrane, and named it a hormone. Hormones are secreted by glands in different places in the body, and are transported by blood to various places and promote physiological effects. The simplest of these is adrenaline, which Jokichi Takamine and his assistant Keizo Uenaka succeeded in isolating from animal adrenal glands and crystallizing in 1901.

Insulin is an anti-diabetic hormone and is secreted from the islets of Langerhans in the pancreas. Insulin was separated by Canada's Banting and

Jokichi Takamine (1854–1922)
After graduating from the Imperial College of Engineering's
Department of Applied Chemistry, Takamine moved to the
U.K. to study; after returning to Japan, he served as a deputy
director at the patent bureau. He studied superphosphate
fertilizers and opened the door to the manufacture of
artificial fertilizer in Japan. Takamine emigrated to the U.S.
in 1890 and created Taka-diastase; in 1901, he successfully
extracted adrenaline. In 1902 he opened the Takamine Laboratory in New
York and was active in invention, research and development. In 1913 he returned temporarily to Japan, where he participated in the founding of RIKEN.

Edward Doisy (1893–1986)
An American biochemist, Doisy served as a professor at St. Louis University
starting in 1923. In 1929, he succeeded in extracting the sex hormone estrone.
In 1939, he separated two forms of vitamin K (K1 and K2), determined their
structures, and succeeded in synthesizing K1. Doisy was a recipient of the
1943 Nobel Prize in Physiology or Medicine.

Best in 1921. In 1927, the German biochemists Zondek and Aschheim discovered
that male mice become sexually aroused when injected with an extract from the
urine of pregnant women. Two years later, Butenandt (see page 339) and Edward
Doisy succeeded in separating this sex hormone. By the mid 1930's, various
hormones had been discovered and attempts were being made to elucidate their
structures and to synthesize them.

4.11 Development of applied chemistry

Industries based on chemistry, such as the synthesis of dyes and soda, had seen
developments since the 19th century and were economically important. In the
20th century chemical industries developed further, increased in scale and became more widespread. A representative example of this was during World War
I, when Haber and Bosch developed a method for synthesizing ammonia using
atmospheric nitrogen. The development of polymer chemistry in the 1920s gave
rise to superior artificial fibers like nylon, and plastics. Medicine also saw paths
open up for eradicating infectious diseases with advances in chemotherapy, and
the pharmaceutical industry evolved in parallel. Chemical achievements were
also put to use in agriculture, including the generation of fertilizer and pesticides,
and the production of food increased to support the ever-growing population.
Applied chemistry thus began to majorly contribute to making people's lives

richer and more convenient. This section describes the outstanding achievements of applied chemistry, especially in the first half of the 20th century.

Atmospheric nitrogen fixation and high-pressure chemistry

In the second half of the 19th century, Europe after the Industrial Revolution saw a rapid rise in the demand for fertilizer accompanying the increasing amount of land being farmed in order to feed the growing population. The source of nitrogen fertilizer at the time was mainly Chile saltpeter and the ammonia byproduct from the manufacture of gas, but there was concern that this source could not meet the rapidly rising demand, so finding new ammonia sources became a pressing issue. The English physicist and chemist Crookes, worrying about the depletion of natural resources in Chile, in 1898 suggested fixing the nitrogen that accounts for 80% of the atmosphere. In 1903, Norway's Birkeland and Eyde developed a method of making NO by reacting nitrogen and oxygen at the high temperatures obtained in an electric arc, then oxidizing NO into NO_2 and reacting it with limestone as nitric acid to produce calcium nitrate. However, this method was very inefficient and consumed a great deal of electricity, and was not successfully industrialized outside Norway. Frank and Caro were the first to produce ammonia industrially, by hydrolyzing calcium cyanamide in heated water vapor. Cyanamide is made by reacting nitrogen with calcium carbide at high temperatures, and was also used directly as a fertilizer.

At the start of the 20th century, many chemists were intrigued by the possibility of directly reacting nitrogen and hydrogen together to obtain ammonia. In 1900, Le Chatelier established the temperature and pressure conditions at which hydrogen and nitrogen become ammonia in the presence of a catalyst, but discontinued his research because of an accidental explosion. Nernst studied the thermodynamics of this reaction, but it was Fritz Haber who succeeded in synthesizing ammonia. Haber had earned a degree in organic chemistry, but after obtaining work at the Technical University of Karlsruhe, he became interested in physical chemistry and conducted research on electrochemistry and the thermodynamics of gases. He was interested in solving practical problems using physical chemistry, and published his *"Grundriss der technischen Elektrochemie auf theoretischer Grundlage (Outline of technical electrochemistry based on theoretical foundations)"* in 1898 and *"Thermodynamik technischen Gasreaktionen (Thermodynamics of technical gas-reactions)"* in 1905. In 1905, Haber measured the equilibrium constant for the reaction N_2 (gas) $+ 3H_2$ (gas) $\rightleftharpoons 2NH_3$ (gas). However, his measurement results were different from the calculations based on the heat theorem by Nernst, leading to a fierce debate between the two. From the viewpoint of chemical equilibrium, NH_3 is efficiently generated when the pressure is raised and the temperature is lowered, but the

Fritz Haber (1868–1934)

Haber was born the son of a wealthy merchant in Breslau in German Silesia (now Wrocław, Poland). After graduating from the gymnasium, he started working at the family business, but quit and went to study organic chemistry at the University of Berlin and Heidelberg University. After two years of military service, he earned a degree in organic chemistry in 1891 at the Technical University of Berlin. He served as an assistant at the University of Karlsruhe starting in 1894 and began researching topics in physical chemistry. Haber wrote a textbook on electrochemistry and gaseous reactions, earning him a favorable reputation and a promotion. In 1901 he started to work on developing a method based on equilibrium theory for synthesizing ammonia from gaseous nitrogen. Haber was promoted to professor in 1906, and in 1912 he was appointed director of the newly-founded Kaiser Wilhelm Institute for Physical Chemistry and Electrochemistry. In 1918, he was awarded the Nobel Prize in Chemistry (column 15).

Carl Bosch (1874–1940)

Bosch studied organic chemistry at the University of Leipzig, and after earning his degree in 1898, he studied industrial technology and joined BSAF, becoming president of the company in 1919. In 1913, he succeeded in industrializing Haber's method of synthesizing ammonia, and developed the Bosch process of producing hydrogen from water gas. He was a recipient of the 1931 Nobel Prize in Chemistry.

reaction is slower at lower temperatures. For this reason, the catalyst is important for increasing the reaction rate. In 1909, Haber and his assistant Le Rossignol succeeded in synthesizing ammonia at 175 atm and 550°C using osmium as a catalyst. BASF, which had been involved in the synthesis of ammonia, was in a collaborative relationship with Haber, and was emboldened by these results and dispatched the metallurgist Carl Bosch and the catalyst expert Mittasch to Karlsruhe, where they set out to industrialize Haber's method.

For industrialization, it is essential to develop reactors that can withstand high temperature and pressure and catalysts that are inexpensive and efficient. BASF used a special steel that has a low carbon content for its reactors, and developed efficient iron catalysts to overcome the many technical difficulties. The method, called the Haber-Bosch process, saw the industrial production of 8,700 tons of ammonia in 1913. The hydrogen needed for synthesis was obtained

Friedrich Bergius (1884–1949)
Bergius studied at universities in Breslau, Leipzig, and Berlin, and in 1909 became a lecturer at the Technical University in Hannover. In 1914, while at Goldschmidt AG in Essen, he succeeded in liquefying coal by hydrogenation at high pressure, and also showed the possibility of obtaining sugar from wood. He was awarded the Nobel Prize in Chemistry in 1931.

from the water gas ($CO+H_2$) obtained when water vapor is reacted with coke, and by oxidizing carbon monoxide to carbon dioxide using a metal catalyst. Nitrogen was fractionated from liquid air. BASF expanded its production during World War I, and in 1918 Germany was producing 200,000 tons of synthetic ammonia per year. Ammonia was being used not only for fertilizers, but also to produce nitric acid. Nitric acid is obtained by oxidizing ammonia using a high-temperature platinum catalyst (the Ostwald process). Nitric acid is essential for the production of explosives, and at the outbreak of World War I in 1914, demand for nitric acid increased rapidly. In the war, supplies of Chile saltpeter become unreliable, and each country began synthesizing its own ammonia.

The success of the Haber process is one example of basic chemistry's outstanding contributions to industry, and is the product of cooperation among outstanding scientists and engineers. Haber was appointed director of Berlin's Kaiser Wilhelm Society in 1912, and by 1933 had played a leading role in the development of physical chemistry in Germany while in this position. He was awarded the Nobel Prize in Chemistry in 1918 for his achievements in synthesizing ammonia. Haber was a scientist of rare genius, but in spite of his fame, his life was full of tragedy and misfortune (column 15). He was a patriot and made major contributions to Germany, but by virtue of being Jewish, he was forced to exile himself from Nazi-ruled Germany, and passed away in exile in Switzerland in 1934.

The high-pressure chemistry techniques developed for the Haber-Bosch process found other applications as well. In 1921, France's Patard developed a method for synthesizing methanol by adding hydrogen under high pressure to carbon monoxide in the presence of a catalyst. In 1911, Ipatieff, a Russian, began his pioneering research on adding hydrogen to fats and oils under high pressure. As a country with no oil resources, Germany had a vested interest in making synthetic gasoline by reacting abundantly-available coal with hydrogen at high temperature and pressure. In 1913, the German chemical engineer Friedrich Bergius acquired a patent for making gasoline by hydrogenating pulverized coal at high pressure, and two years later he manufactured experimental equipment

COLUMN 15

Haber's glories and tragedies[20, 21]

Haber's Institute for Physical Chemistry and Electrochemistry at the time of its founding

Fritz Haber was an eminent chemist with a rare capacity for genius who developed an efficient process for synthesizing ammonia from atmospheric nitrogen. However, his personal life was as full of tragedy as it was filled with glory. He lived at a time that posed challenges for scientists in terms of ethics and their contributions to their country.

The Kaiser Wilhelm Society, composed of several different laboratories, was founded in 1912 in order to maintain Germany's dominance in science and technology. It was proposed that the Institute for Physical Chemistry and Electrochemistry, with Haber as its director, be set up in Berlin as one of these laboratories. The Institute was opened in 1912, and Haber left Karlsruhe to become its director. Shortly thereafter World War I began, and the war between Germany and France was a long-lasting conflict that could not be resolved across trench lines. Surrounded on all sides by the enemy, Germany was no longer able to obtain the outside resources necessary for war. Haber was a patriot who blindly cooperated with the German state at this time of national crisis, and he devoted himself to assuring a German victory.

One of the greatest problems of the time was the manufacture of nitric acid, which is needed to produce explosives. Haber strove to increase the production of ammonia by synthesis, and succeeded in increasing nitric acid production by means of the catalytic oxidation of ammonia, thus making it possible for Germany to continue the war. Another area of wartime research where Haber was involved was the development of poison gas. To break the stalemate of trench warfare at the front lines, the military considered spraying poison gas at the battlefront. Haber proposed using chlorine gas; on 22 April 1915, chlorine gas was used at the battlefront in Ypres. His institute was in full cooperation with the development of the new poison gas.

The German economy collapsed as a result of the defeat and paying the harsh reparations imposed by the Treaty of Versailles. In an effort to prevent a crisis, Haber spent 8 years attempting to extract gold from seawater. However, the analytical results that he relied upon were wrong; the amount of gold he was able to obtain from seawater was far smaller than he expected, and his plans ended in unmitigated failure.

Ever the patriot, Haber had voluntarily cooperated with the state during the war, and even after the war he tried to avert an economic crisis in Germany. However, the results ended up damaging Haber's reputation. After the war, he was regarded as a war criminal by the Allies. In 1920, he was awarded the Nobel Prize in Chemistry for his synthesis of ammonia, but in the face of considerable protest and opposition. Not only was he complicit in the development of poison gas, but the synthesis of ammonia enabled Germany to prolong the war. Many scientists on the Allied side, including Rutherford, did not and would not forgive him for this.

Haber concentrated on his work, and ignored his family. His first wife, Clara Immerwahr, had specialized in chemistry and was the first woman to earn a doctoral degree at the University of Breslau; she was an intellectual and had a delicate character. She married Haber with the hope of continuing her research, but this hope was shattered immediately. Immersed in his research, Haber became extremely irritable when at home, and overwhelmed by her overbearing and egocentric husband, Clara gradually fell into despair. She was opposed to his research on poisonous gas, and committed suicide with Haber's pistol on 1 May, 1915, immediately after poisonous gas was used at Ypres. Some were of the opinion that her suicide was in protest of Haber's poisonous gas research, but the truth is uncertain.

Following World War I, Haber focused on reconstructing the Institute, which had fallen on difficult times. Haber strove to forge an interdisciplinary institute that covered broad fields of physics, chemistry, and physiology. In the second half of the 1920s, with Polanyi (chemical kinetics), Freundlich (colloid chemistry), Ladenburg (atomic physics) and others as group leaders, in addition to many other outstanding researchers, the Institute entered its prime and contributed many superb advancements to science. In a colloquium held every other week at Haber's Institute (the Haber Colloquium), the latest research results were introduced in a wide range of fields said to range "from the helium atom to the flea", and leading scholars in chemistry, physics, and physiology at the time would participate in lively discussions in a free and interdisciplinary atmosphere that became world-famous. Haber demonstrated outstanding leadership as its director. However, this era of glory did not last long. When the Nazis came to power in 1933, laws that excluded Jewish people from civil service came into effect, and many of his institute's scholars were dismissed. Haber was exempted because of his achievements during World War I, but he resigned in protest. In the summer of that year, he went on a journey of exile and hid temporarily in Cambridge; in 1934, he died in exile in Switzerland while on his way to Palestine. Haber had devotedly served the German nation, but ultimately he fell victim to the Nazis.

for producing 30 tons of gasoline per day. Around 1932, spearheaded by Bosch, IG Farben began to manufacture 300,000 tons of synthetic gasoline annually using the Bergius process.

Germany's Franz Fischer and Tropsch in 1923 discovered that when water gas is passed over a catalyst of iron oxide at medium pressure and 200°C, an aliphatic hydrocarbon that can be used as a fuel for engines is obtained. From 1935 onward, Germany used this method to produce hydrocarbons on a commercial basis.

Bosch and Bergius in 1931 were awarded the Nobel Prize in Chemistry in 1931 "in recognition of their contributions to the invention and development of chemical high pressure methods". This is the first example of the Nobel Prize in Chemistry being given for industrial achievements.

New metals and alloys

The most important metal at the beginning of the 20th century was carbon steel, but demand for other metals was also increasing. Demand for pure copper had increased markedly with the development of the electrical industry; with the popularity of canned food, tin was in demand for tin-plating cans. Demand for zinc also increased for zinc-plated steel sheets.

Changes were underway in the products and methods of iron production. Wrought iron had mostly given way to mild steel, and open-hearth furnaces came to be widely used in the production of steel. Quality also improved as understanding of the steps for manufacturing steel advanced. Steel is essentially an alloy, so control of the alloy components and proper heat and mechanical treatments allowed the production of steel suitable for special applications.

In the second half of the 19th century, metallurgy had progressed and alloys had attracted interest. With the emergence of metallographs and the application of Gibbs' phase rule (Chapter 2, page 111), the study of metals and alloys became a discipline with a scientific basis. In 1863, H. C. Sorby in Britain developed a technology for polishing and etching metal surfaces, and observed metal surfaces with a metallograph. As a result, crystal structures came to be associated with various properties such as the ductility and extensibility of metals. Roozeboom in the Netherlands in 1887 began to apply the phase rule to study alloys, and by the 20th century phase diagrams[44] were being used extensively to investigate alloys. X-ray structural analysis was also introduced to the study of

---

44. Phase diagram: In a system of a designated composition, a phase diagram is a diagram where temperature, pressure, composition, and other factors are coordinates for representing the number of phases that exist in equilibrium with the surroundings, as well as their compositions and respective relative quantities.

alloys. Research using these scientific methods markedly increased knowledge of alloys, and helped with the development of useful alloys.

Improvement of steel first began with the introduction of manganese by Hadfield in 1882. He discovered that manganese improves steel's resistance to shock and abrasion. Manganese was also known to work as a deoxidizer when present in small amounts. Pig iron is iron that contains 20% manganese and 5% carbon, and was widely used as an additive for making steel with the desired manganese content, but gradually steel with higher manganese content came to be used. Manganese steel was also used for ships because of its hardness. Steel containing about 40% nickel has a small coefficient of thermal expansion, and therefore was used to make items such as measuring tapes, and pendulums for timepieces. Stainless steel containing nickel and chromium was introduced by Brearley in 1912, and became important in the food, pharmaceutical, and chemical industries. Chrome steel has both high hardness and strength, and so it has been used in armor, such as for tanks and warships. Iron alloys were studied extensively, and alloys for special applications were developed and made in quick succession.

Aluminum began to be produced at the end of the 19th century, but at first was used only for cookware. After World War I, however, aluminum alloys containing copper, manganese, magnesium, and other elements were made and their applications increased. Aluminum alloys are light and strong, so they became important for the manufacture of railway vehicles, aircraft, and automotive engines.

Plastics

Animal horns had long been used to make combs and jewelry. The rubber-like resin, gutta-percha, from the Malay Peninsula was also used for electrical insulators. Plastics resulted from the need to make such products with synthetic materials. First, semi-synthetic plastic was made using nitrocellulose. In 1862, the Englishman Parkes made many different products by mixing nitrocellulose, alcohol, camphor, and vegetable oil; he sold these products under the trade name Xylonite. In the U.S., Hyatt, who had been looking to replace ivory for billiard balls, developed a material containing only nitrocellulose and camphor, which he commercialized as celluloid in 1872. Celluloid was used for clothing collars, knife handles, photographic film, and billiard balls.

Baekeland, a Belgian, first earned a degree in chemistry at Ghent University before emigrating to the U.S., where he extensively studied synthetic resins obtained by the condensation-polymerization of phenols with formaldehyde in his laboratory in Yonkers, New York. Baekeland showed the conditions under which a liquid resin can become a thermosetting molding powder, and developed Bakelite. First marketed in the U.S. in 1909, Bakelite was a dark, opaque plastic, but it could be easily molded and also was highly insulating, so it found a market immediately in the telephone, automobile, and radio industries. In response to such trends, the plastics industry began to grow more rapidly in the first half of the 20th century.

In 1918, the Czechoslovakian-born chemist John substituted phenol with urea and succeeded in obtaining a transparent amino resin by polycondensation with formaldehyde. The Austrian Pollack made improvements, and amino resin came to be widely used, from bottles to decorating supplies.

In 1901, Otto Röhm had written a doctoral thesis relating to the polymerization of acrylic acid. In the 1930s, resins were being developed by polymerizing acrylic acid and related compounds. In particular, a transparent thermoplastic was developed in 1931 by methyl methacrylate polymerization that melts at 110°C and can be easily molded. This organic glass was lighter than inorganic glass and was produced in Germany, the U.K., and the U.S., becoming a major industrial product. Plastics made by polymerizing styrene, vinyl chloride, ethylene, and other compounds also came to market in succession in the 1930s.

Artificial fibers and nylon

The first artificial fiber was artificial silk made from nitrocellulose. In 1883, the British inventor Swan developed a manufacturing process for making threads of nitrocellulose. This could be carbonized and used as filaments in light bulbs. The Frenchman Count Chardonnet developed a method for reducing flammability by partially hydrolyzing nitrate, and in 1891 the production of artificial silk began. The German Pauly in 1897 developed a method for making yarn by dissolving cellulose in an ammonia solution of copper hydroxide and precipitating it with sulfuric acid; this method was applied to fabrics. The Englishmen Cross and Bevan in 1892 made cellulose fibers from a syrup (viscose) obtained by dissolving cellulose in carbon disulfide and caustic soda. The cellulose fibers made by these methods were marketed first as "artificial silk", but gradually the term "rayon" came to be used for fibers obtained from chemically treated cellulose. Because it can be made inexpensively, viscose became a mainstay of artificial fiber production in the 1920s. The quality of rayon improved and it came to be widely accepted by consumers.

In 1927, the American company DuPont made the decision to establish a laboratory for conducting basic research, and the general manager of the chemical department, Stine, managed to lure Wallace Carothers away from Harvard University to lead their polymer research efforts. At the time, DuPont felt its corporate laboratories should be conducting basic research, even if that research might not have direct industrial applications.

Wallace Carothers (1896–1937)
Born in Iowa, Carothers was educated at Tarkio College in Missouri and went on to earn a master's degree at the University of Illinois in 1921. After serving as a lecturer in chemistry at the University of South Dakota, he returned to the University of Illinois and earned a degree in 1924 studying under Adams. He spent two years lecturing in organic chemistry at Illinois before being noted for his creative approach and becoming a lecturer at Harvard University. In 1928, Carothers was appointed a director in the organic chemistry laboratory at DuPont, where he studied high-molecular-weight polymers and invented products such as synthetic rubber and nylon. However, he suffered from depression and committed suicide in 1937.

In 1928, Carothers began research at DuPont, initially on acetylene polymer. An assistant, Collins, discovered that a rubber-like solid is produced when chloroprene is separated and polymerized. This was the first synthetic rubber, "neoprene". Carothers attempted to make a polymer with a molecular weight of 4,000 or more. Another assistant, Hill, in 1930 synthesized a polyester with a molecular weight of 12,000 and showed that it has superior strength and elasticity. Aliphatic polyester has too low of a melting point and dissolves easily in water, so it is not suitable for fibers; therefore, Carothers began studying polyamide instead of polyester. After a series of tests, he chose adipic acid and hexamethylene diamine and attempted polycondensation. The result, Polyamide 6-6, has resilience, is resistant to water, and its fibers have a high melting point. Thus, nylon was born in February 1935. In 1936 nylon threads were used to make prototype stockings, and in 1939 production of nylon stockings began on an industrial scale. Nylon was introduced to the market alongside an extensive advertising campaign for fibers made from coal, air, and water. I.G. Farben also developed amide fibers made from caprolactam, called Perlon.

Carothers' research at DuPont met with brilliant success, but he personally had long been suffering from depression. Around the time nylon was made, his condition worsened significantly. On 29 April, 1937, without ever seeing the fruits of his own research, he took his own life as a result of his depression.

Synthetic rubber
Rubber did not attract much attention from chemists until the 20th century. The rubber industry had little relationship with chemistry. At the end of the 19th

century, with the spread of bicycles, pneumatic tires started being used and demand for rubber increased. At the turn of the 20th century, with the growth of the automobile tire industry, the U.S. saw a rapid rise in demand and the rubber industry became a major industrial sector. One of the problems faced by the rubber industry was the unreliability of the raw material supply, and efforts were begun to develop synthetic rubbers.

The relationship between the rubber and chemical industries began in 1906 when the American Oenslager discovered that aniline promotes the vulcanization[45] of rubber. Thiocarbanilide, which is easier to handle, later came to be used as a vulcanization accelerant. To prevent oxidation from deteriorating rubber, it was also discovered that aromatic amines and phenols are effect as antioxidants, as are quinones.

Research on the composition of rubber began in 1860 when C. G. Williams identified isoprene as a main degradation product in rubber. The French chemist Bouchardat in 1879 thought that the molecule that forms isoprene binds to form a chain-like molecule by ethylene double bonds. The chemistry of macromolecules, however, advanced little until Staudinger established the concept of macromolecules in the 1920s.

Germany during World War I suffered from a shortage of natural rubber, and was determined to develop synthetic rubber. A chemist at Bayer, a chemical company, attempted to synthesize rubber by polymerizing dimethyl butadiene or methyl isoprene, which were easily obtained from acetone, and produced "methyl rubber". This rubber lost its competitive edge against natural rubber following the war, but the research and development of synthetic rubber continued. In 1930, I.G. Farben developed a copolymer of butadiene with styrene (Buna S) that had excellent abrasion resistance, and a copolymer with acrylonitrile (Buna N) that had excellent corrosion resistance against oil and solvents.

In the U.S., DuPont brought polychloroprene to market under the name "neoprene" in 1933. Neoprene was developed when Carothers' group expanded on research by J.A. Newland at the University of Notre Dame in the 1920s; they added monovinyl acetylene to hydrochloric acid to synthesize 2-chlorobutadiene, which they then polymerized to polychloroprene. Neoprene is superior to natural rubber in its oxidation resistance and corrosion resistance.

Chemotherapy and pharmaceuticals

Paul Ehrlich made major contributions to the development of chemotherapy.

---

45. Vulcanization: An operation to improve rubber's elasticity, strength, and other properties by adding sulfur or other accelerators to raw rubber and heading to strongly bind the rubber molecules together.

Ehrlich, who studied medicine in Breslau, Strasbourg, Freiburg, and Leipzig, had a keen interest in how dyes affect living tissue. In the second half of the 19th century, Germany's dye industry was growing, and many aniline dyes were on the market. People began using these dyes to dye tissue for the purpose of microscopic observation. Ehrlich studied techniques for dyeing bacteria from his nephew Weigert, a histologist. He focused on how specific dyes can selectively dye specific bacteria and tissues, and thought that using the proper dye might make it possible to kill bacteria. In 1891, he found that methylene blue acts against the malaria parasite. This was the birth of chemotherapy, and the development of drugs that selectively act against a pathogen without adverse side effects on the body was important. Ehrlich developed a "side-chain theory" which claimed that side chains of molecules attach to pathogenic bacteria and inhibit their activity.

In the early 1900s, Ehrlich and Kiyoshi Shiga discovered that the azo dye, trypan red, is effective against trypanosomiasis, such as sleeping sickness. They confirmed that atoxyl, an organic derivative of arsenic synthesized by Bechamp in 1863, is effective against trypanosomiasis. Around this time, the spirochete pathogen that causes syphilis was also discovered, and was believed to be closer to trypanosomes than bacteria. Ehrlich systematically experimented with organic arsenic compounds and together with his assistant, Sahachiro Hata, discovered that arsphenamine, compound no. 606 (later sold by Hoechst under the name Salvarsan) is effective in treating syphilis. The success of salvarsan was a major stimulus for the chemical industry, and the synthesis of new molecules was promoted. However, little was known about how to minimize side effects, so there were few advances in chemotherapy from 1910 until 1930. Ehrlich was awarded the Nobel Prize in Physiology or Medicine in 1908.

In 1932, Bayer acquired the patent for the sulfa drug Prontosil. Experiments on mice by Gerhard Domagk showed Prontosil to be effective in controlling streptococcal infection. Later clinical trials confirmed its efficacy and it became a pharmaceutical drug. A group led by Trefouel at the Pasteur Institute revealed that the efficacy of Prontosil does not reside in the azo group-containing dye, but rather in the sulfanilamide produced when the dye is broken down in the body. Sulfanilamide has similar efficacy to Prontosil, discovered in 1908. The patent on Prontosil had expired, allowing the manufacture and use of Prontosil to spread rapidly. Domagk was awarded (but declined) the Nobel Prize in Physiology or Medicine in 1939. The success of sulfanilamide stimulated investigations by pharmaceutical companies into numerous other compounds having a sulfamide group. The British company May & Baker manufactured sulfapyridine and sulfathiazole, which are effective against pneumonia. Sulfamide drugs have been effective against many pathogens, but not effective against those causing diseases such as tuberculosis and leprosy.

Paul Ehrlich (1854–1915)
Born in Silesia, Ehrlich studied medicine at the University
of Leipzig before becoming an assistant at the University
of Berlin, following which he entered Koch's Institute
of Infectious Diseases. There, he developed a method of
determining the antibody titer of diphtheria serum, among
other accomplishments. Becoming head of the immunology
laboratory in 1896, Ehrlich proposed a side-chain theory
for describing the mechanism of biological immunity. With the establishment of the Institute of Experimental Therapy in 1899, he began studying
chemotherapy based on side-chain theory and the affinity of dyes to tissues.
He was a recipient of the 1908 Nobel Prize in Physiology or Medicine.

Gerhard Domagk (1895–1964)
After teaching at the University of Greifswald and the University of Münster,
Domagk became head of the Institute of Pathology and Bacteriology at Bayer.
In 1932, he succeeded in developing Prontosil, which is a sulfa drug. He was
awarded the 1939 Nobel Prize in Physiology of Medicine, but declined it at
the instruction of the Nazis.

A true revolution in chemotherapy was brought about by the discovery of
antibiotics, but this is regarded as a typical example of great serendipity. In
1928, antibiotics were discovered entirely by chance by Alexander Fleming, a
Scottish microbiologist working at St. Mary Hospital in London. At the time, he
had been researching the influenza virus, but he went on vacation and left many
culture dishes unattended. Back from vacation, he noticed that Staphylococcus
culture dishes had been contaminated with mold. He then discovered that the
mold impeded growth of bacteria inoculated into the medium, and believed that
the mold secreted a compound with an antibacterial action (Fig. 4.36). Fleming
extracted the active antibacterial compound from Penicillium notatum, a blue
mold, and named it penicillin. He focused on how this compound was harmless
to animals and white blood cells, studied the bacteria killed by penicillin, and
discovered that it is effective against many bacteria. However, as he was not a
chemist, he was neither able to stabilize nor to determine the structure of the
compound. Fleming announced his discovery in 1929, but it attracted very little
attention, and his research did not advance. Ten years later, Walter Florey and
Ernst Chain at Oxford University started studying penicillin. They confirmed
Fleming's results, and extracted an impure form of a compound that exhibited
strong bactericidal activity against certain bacteria. In 1942, Chain produced

Alexander Fleming (1881–1955)

A British microbiologist, Fleming graduated from the University of London and worked for the bacteriology department in St. Mary's Hospital Medical School, where he discovered lysozymes. In 1928, while researching staphylococcus, he noticed by chance that Penicillium inhibited the growth of bacteria; he extracted the active ingredient and named it penicillin in his 1929 report. Fleming was a professor of bacteriology at the University of London from 1928 to 1948. He was a recipient of the 1945 Nobel Prize in Physiology or Medicine.

Walter Florey (1898–1968)

A British pathologist born in Australia, Florey was a professor of pathology at Oxford from 1935 to 1962. After the successful isolation of lysozymes, he and Chain succeeded in isolating and purifying penicillin in 1939. He was a recipient of the 1945 Nobel Prize in Physiology or Medicine.

Ernst Chain (1906–1979)

Chain was a German-born British biochemist. After graduating from university, he was admitted to the Charité–Universitätsmedizin Berlin, but he defected to the U.K. in 1933 and conducted biochemical research under Hopkins at Cambridge. Chain became a lecturer in pathology at Oxford University in 1935, and he succeeded in separating and purifying penicillin, alongside Florey. He was a recipient of the 1945 Nobel Prize in Physiology or Medicine.

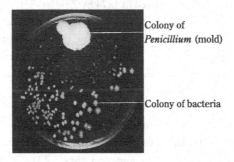

Colony of
Penicillium (mold)

Colony of bacteria

Figure 4.36: Photograph by Fleming showing how bacteria do not grow near the mold

a yellow powder of penicillin, and confirmed its effects in clinical trials, and research on penicillin advanced rapidly. Around this time, with strong support from the federal government, the U.S. began mass production of penicillin; with demand persisting in wartime, penicillin became a well-established therapeutic drug against infectious diseases. Later, many researchers raced to participate in the study of antibiotics, and new antibiotic compounds were discovered in quick succession. The structure of penicillin was determined using X-ray structural analysis by Dorothy Hodgkin in 1945. (See 5.2.1)

The trio of Fleming, Florey, and Chain was awarded the Nobel Prize in Physiology or Medicine for their respective penicillin-related achievements in 1945.

Emergence of pesticides

In the 20th century, agricultural production in developed countries increased dramatically, largely due to synthetic fertilizers, pesticides, and herbicides. Fertilizers were produced by the fixation of atmospheric nitrogen and the discovery and development of new synthetic routes to phosphoric acid and potassium salt. Furthermore, it was confirmed that plants require trace elements such as boron, cobalt, manganese, copper, and molybdenum.

Inorganic products were used as insecticides beginning around 1870. The realization that sulfur is effective against mold led to the widespread use of sulfur and copper sulfate. Most insecticides used before World War II were inorganic products; few organic products were used, and those that were included natural products such as nicotine extracted from tobacco and pyrethrins extracted from Asteraceae plants, but they were more expensive than inorganic products. The development of synthetic organic insecticides brought about a breakthrough in agriculture, a forerunner of which was dichlorodiphenyl trichloroethane (DDT).

DDT had already been discovered in 1874 by Zeidler, but in 1938 Paul Müller at the Swiss company J. R. Geigy discovered its strong and lasting insecticidal effects, and Geigy started selling DDT. In 1943, the U.S. military successfully used this insecticide to ward off malaria, which had become widespread in Casablanca, as well as typhus, which had become widespread in Naples, thus giving DDT worldwide recognition as an insecticide. DDT is effective in combating insects of all kinds, and was thought to be harmless to higher animals. Easy and inexpensive to manufacture, DDT became a popular pesticide and was widely used. It was only after World War II that the problems associated with the widespread use of DDT became apparent. (See 5.7.1)

The success of DDT prompted a search for organochlorine compound-based insecticides. The British company ICI started selling benzene hexachloride (BHC). This had been synthesized by Faraday in 1825; one of its nine isomers is odorless and insecticidal. Organophosphate compounds were also studied, and after World War II insecticides such as parathion and malathion were developed.

> **Paul Müller (1899–1965)**
> Müller was a Swiss chemist who studied at the University of Basel and conducted research into dyes and skin tanning agents at the Geigy laboratories. In 1939 he succeeded in developing DDT, which has strong insecticidal effects against flies, mosquitoes, lice, and other insects. His achievements earned him the 1948 Nobel Prize in Physiology or Medicine.

The discovery of organochlorine-based insecticides earned Müller the Nobel Prize in Physiology or Medicine in 1948.

4.12 Chemistry in Japan

Establishing an environment of chemical education and research in Japan

Japan at the start of the 20th century had just started teaching modern chemistry at two of its universities, and the research environment was very harsh. However, Japanese scientists trained abroad started performing state-of-the-art research. Jokichi Takamine, who had studied under Divers at the Imperial College of Engineering, entered the Ministry of Agriculture and Commerce after studying at the University of Glasgow, but in 1890 he emigrated to the U.S. and in 1901 crystallized adrenaline at a laboratory in New Jersey. Hantaro Nagaoka studied under Boltzmann while in Europe after graduating from Tokyo Imperial University; after returning to Japan, he proposed his Saturnian model of the atom in 1903. Umetaro Suzuki graduated from the Agricultural University in 1896, conducted research under Emil Fischer in Berlin, then returned to Japan, where he was the first to discover vitamins, in 1910.

After the Russo-Japanese War and as part of the modernization of Japan, new universities and higher specialized technical schools were founded, and training of chemists and chemical engineers progressed. 1907 saw the decision to establish Tohoku Imperial University and Kyushu Imperial University in Sendai and Fukuoka, respectively. Tohoku Imperial University formed the College of Science and set up a department of basic chemistry, while Kyushu Imperial University set up a department of applied chemistry in its College of Engineering. The newly-founded universities and technical colleges were the birthplace of new research and education, and the training of chemists and engineers advanced along with research developments. New national universities would follow, including Hokkaido (1918), Taipei (1928), Tokyo Institute of Technology (1929), the University of Tokyo of Literature and Science

(1929), Hiroshima University of Literature and Science (1929), Osaka (1931), and Nagoya (1939), providing the newly educated chemists with a forum for research and education. 1917 saw the founding of RIKEN, largely due to Jokichi Takamine's insistence on promoting basic chemistry. Research at RIKEN led to major contributions to the development of basic research in physics and chemistry in pre-World War II Japan.

Leading chemists of the early 20th century

The early 20th century saw leading chemists traveling to Europe and the U.S. to study the most recent developments in chemistry, then returning to Japan where they established new laboratories and conducted outstanding research while educating many new researchers. The efforts of these leading chemists lead to Japanese chemistry being respected worldwide. It is thanks to these people that Japanese chemistry approached the level in the rest of the world. Here, we discuss some of these chemists.

Kikunae Ikeda worked in the field of physical chemistry as an assistant to Joji Sakurai (see page 155) at Tokyo Imperial University. He went to Ostwald's laboratory to study, and then returned to Japan and became a professor in 1901; in 1907, he discovered that the umami component in kelp is sodium glutamate. Full-scale introduction of physical chemistry in Japan, however, was achieved by the efforts of two students of Sakurai, Yukichi Osaka (1867–1950) and Masao Katayama (1877–1950). Osaka traveled to Germany and studied under Ostwald and Nernst before becoming a professor at Kyoto Imperial University in 1903; conducting his research on chemical equilibria and reaction rates in solutions, he wrote his "Physical Chemistry", the first textbook on physical chemistry in Japan. Katayama studied in Europe from 1905 until 1909; he researched the electromotive force of batteries in Zurich and dissociation equilibria under Nernst in Berlin, before becoming a professor at Tohoku Imperial University in 1911. He published his "Kagaku honron (Chemistry main principle)" in 1915, which represented the first textbook on physical chemistry in Japan based on molecular theory. In 1919, he became a professor at Tokyo Imperial University as a successor to Sakurai, and studied topics such as surface tension while also training many excellent physical chemists, thus laying the foundations of structural chemistry in Japan. These foundations led to the cutting-edge research of San'ichiro Mizushima and Yonezo Morino (see page 322) in infrared and Raman spectroscopy.

Riko Majima made major contributions in organic chemistry.[160] After graduating from Tokyo Imperial University in 1899, Majima began study-ing the components of lacquer around 1903. In 1907, he began studying

Kikunae Ikeda (1864–1936)

Ikeda graduated in 1889 from Tokyo Imperial University's Department of Chemistry in the Faculty of Science, then served as an assistant professor in 1896; he spent 1899 to 1901 in Germany, where he studied catalytic reactions under Ostwald. After returning to Japan, Ikeda became a professor in physical chemistry at his alma mater. In 1908 he invented "Ajinomoto", which has glutamate as its main ingredient. He participated in the 1917 founding of RIKEN, where he continued his research.

Riko Majima (1874–1962)

An 1899 graduate of Tokyo Imperial University's Department of Chemistry in the Faculty of Science, Majima left Japan to study at the Swiss Federal Institutes of Technology in Zurich and the Davy-Faraday Research Laboratory in London. He became a professor at Tohoku University in 1911, and researched the structures and synthesis of natural organic compounds such as the main components of lacquer, alkaloids, indole derivatives, and the main components of *Lithospermi radix*. Majima concurrently served as chief scientist at RIKEN, and contributed to the establishment and operation of Hokkaido University, Osaka University, and the Tokyo Institute of Technology. He served as president of Osaka University.

in Germany, Switzerland, and the U.K. After returning to Japan, he was appointed a professor at the newly-founded Tohoku Imperial University in 1911, and in 1917 he succeeded in the structural determination and synthesis of urushiol, the main component of lacquer. Concurrently, he studied indoles, Aconitum alkaloids, and other compounds, and trained many protégés. In 1932, he was involved in the founding of Osaka Imperial University, and the following year he became a full-time professor there. The next generation of leaders in organic chemistry research in Japan included Tetsuo Nozoe and Shiro Akahori, as well as many former students of Majima. Majima had made a major contribution to laying he foundation for Japanese organic chemistry and to raise its prominence world-wide. In the second half of the 20th century, Japanese organic chemistry was at the cutting edge, largely thanks to the traditions set down by Majima.

Yasuhiko Asahina is one of the pioneers of organic chemistry research in pharmaceuticals. After graduating from Tokyo Imperial University in 1905, he

Tetsuo Nozoe (1902–1996)
Born in Sendai, and a 1926 graduate from Tohoku Imperial University, Nozoe worked at the newly-established Taipei Imperial University and in 1937 became a professor at that university, then in 1948 he became a professor at Tohoku University. While in Taipei, Nozoe discovered hinokitiol, a seven-membered ring carbon compound obtained from the Taiwan cypress. Nozoe was a pioneer of research on non-benzene-based aromatic compounds and was honored with Japan's Order of Culture in 1958.

Shiro Akahori (1900–1992)
Akahori was born in Shizuoka and graduated in 1925 from Tohoku Imperial University before traveling to Germany and the U.S. to study; in 1939, he became a professor at Osaka University, and later became its president. He extensively studied molecules such as amino acids, proteins, and enzymes, and established a leading protein research laboratory at Osaka University. He was honored with Japan's Order of Culture in 1965.

Yasuhiko Asahina (1881–1975)
After graduating from Tokyo Imperial University's Department of Pharmaceutical Sciences in the Faculty of Medicine in 1905, Asahina traveled to Switzerland and Germany to study under Willstätter and Emil Fischer; after returning to Japan, he became a professor at the University of Tokyo in 1918. He was known for his study of the components of East Asian traditional medicines and his study of lichen components. A pioneer of natural products organic chemistry in Japan, Asahina was honored with Japan's Order of Culture in 1943.

went to Switzerland and Germany and studied under Willstätter and Fischer. He then returned to Japan and later served as a professor at Tokyo Imperial University and as a director of the Research Institute for Natural Resources. Asahina advanced the separation and structural determination of plant components, and was particularly known for his research on lichen components and for laying the foundation for natural products organic chemistry as a discipline in pharmaceutical sciences.

In inorganic chemistry, Yuji Shibata helped introduce coordination chemistry to Japan. After graduating from Tokyo Imperial University in 1907, Shibata traveled to Europe and studied under Werner at the University of Zurich and under Urbain at the Sorbonne before returning to Japan and

Gen'itsu Kita (1883–1952)
Kita was born in Nara prefecture and graduated from Third High School before graduating from Tokyo Imperial University's Department of Applied Chemistry in 1906. After serving as an assistant professor there, he became an assistant professor at Kyoto Imperial University's Faculty of Engineering in 1916 and then was a professor there between 1923–1943. He conducted applied research, with a focus on the basics, in a broad range of fields including fermentation, fats and oils, fibers, fuels, and rubbers, and trained many talented chemists. After the war, he served as the first president of Osaka Prefecture University. (See column 16)

Ichiro Sakurada (1904–1986)
Born in Kyoto and a 1926 graduate of Kyoto Imperial University, Sakurada immediately entered RIKEN and started studying cellulose. He studied in Germany between 1928 and 1931, and in 1936 began serving as a professor at Kyoto University. In 1939, he developed polyvinyl alcohol fibers, "Synthetic No. 1", and after World War II he marketed these fibers under the trade name "vinylon". He also made major contributions in fundamental research on the structures and physical properties of polymers. He was honored with Japan's Order of Culture in 1977.

becoming a professor at Tokyo Imperial University in 1919. Shibata conducted research on the colors of complexes, and was a pioneer in coordinate chemistry in Japan. His student, Ryutaro Tsuchida, submitted his spectrochemical series of metal complexes at Osaka Imperial University.

Gen'itsu Kita at Kyoto Imperial University made major contributions to the development of both basic and industrial chemistry in Japan.[161] After graduating from Tokyo Imperial University in 1906, Kita moved to Kyoto Imperial University in 1916 and studied in the U.S. and France before becoming a professor at Kyoto Imperial University in 1921. There, he conducted diverse research on subjects such as fats and oils, rayon, synthetic rubber, synthetic fibers, and synthetic oils, and trained a great deal of talent. One of his students, Ichiro Sakurada, made major contributions to the development of polymer science in Japan. Kita's leadership, which emphasized basic research, also produced leading chemists in the field of basic chemistry. The tradition he established continued to cultivate talent, including the Nobel laureates Ken'ichi Fukui and Ryoji Noyori in later years (see column 16).

COLUMN 16

Gen'itsu Kita and the formation of the Kyoto School

Instructors in the Kita Group in 1937. Oda, Kita, and Sakurada, from right to left.

The level of chemical research in Japan became world-class after World War II, and by 2012 seven scientists had become Nobel laureates. This achievement was largely due to the pioneers that laid the foundations for chemical research before the war. One of these pioneers was Gen'itsu Kita, who contributed significantly to developing applied chemistry founded on basic chemistry at the Kyoto Imperial University, Faculty of Engineering. This section introduces Kita and his influence, with reference to Yasu Furakawa's paper "Gen'itsu Kita and the Formation of the Kyoto School" published in "*Kagakushi kenkyu* (*Research on the history of chemistry*)" in 2010.[161]

Gen'itsu Kita was born in Hirabata Village (now Yamatokoriyama City), Ikoma District, Nara Prefecture in 1883; after studying at Third High School, he graduated in 1906 from the Department of Applied Chemistry in Tokyo Imperial University's Faculty of Engineering. Kita was a graduate student of Michitada Kawakita, then became a lecturer the next year and an assistant professor in 1908. However, Kita and Kawakita differed in personality and their approach to research and education, and Kita was thus unhappy at Tokyo Imperial University. Kita felt that a solid foundation in basic chemistry was essential for researchers in applied chemistry. However, the focus at Tokyo Imperial University was on obtaining a practical industrial education. Tokyo Imperial University was the center of engineering education at a time when Japan was intent on introducing advanced technology from the U.S. and Europe. At Kyoto Imperial University, two professors in the Department of Manufacturing Chemistry were forced to retire because of the "Sawayanagi incident" in 1914, making the recruitment of teachers an urgent task. Yoshio Nakazawa, a classmate of Kita's from Third High School and the Tokyo Imperial University, was hurriedly invited from Kyushu Imperial University as a professor. Nakazawa asked Kita, who had been languishing in Tokyo, if he would come to Kyoto. Kita accepted and moved to Kyoto in 1916, despite the fact that he would not be promoted to professor and would remain an assistant professor.

Two years after moving to Kyoto, Kita was ordered to spend two years conducting industrial chemical research in the West. This was the practice at the time when an assistant professor was in line for promotion to professor. Germany was defeated in 1918 in World War I, so he went to the U.S. and France. In the U.S., Kita studied with the physical chemist Arthur Noyes at the Massachusetts Institute of Technology (MIT). Noyes had introduced the emerging field of physical chemistry to the U.S., and he helped G. N. Lewis at MIT and Pauling at the California Institute of Technology to develop. In France, Kita studied enzymes under Bertrand at the Pasteur Institute. Bertrand was a scholar who had studied the chemistry of lacquer (see column 7). The fact that Kita studied under Noyes during this stay is important; basic courses in mathematics and physics were central to the teaching of chemistry at MIT, and it gave Kita confidence in his educational philosophy that basic knowledge of the sciences is important in engineering. In 1921, immediately after returning to Japan, Kita was promoted to professor. This new standing allowed him to greatly expand on his own approaches to teaching and research based on his philosophy. He also had a connection with RIKEN that had provided him a better environment for research.

The Institute of Physical and Chemical Research (RIKEN) was founded in 1917. Kita was first employed as a "junior-researcher", and then as a senior researcher in 1922 when a senior researcher system was inaugurated. Researchers at this level had discretionary power over human resources, payroll, and research expenses. Laboratories could be established both at the main site in Tokyo and at each of the imperial universities, so the RIKEN Kita Laboratory was established at Kyoto Imperial University, and Kita's group was employed there as research students, assistants, and temporary employees. This enabled Kita to greatly expand research activities in the Kyoto area. Ichiro Sakurada, Shinjiro Kodama, Ryohei Oda, Keiichi Sishido, Haruo Shingu and other later professors at Kyoto University were RIKEN research students or temporary employees in their younger days.

RIKEN at this time was a paradise for researchers, providing a free atmosphere removed from academic cliques and authoritativeness, as captured in a memoir[162] by Sin'ichiro Tomonaga. The "scientism industry" advocated by Masatoshi Okochi, a member of the third generation of RIKEN directors, was said to "promote basic scientific research that is based on researchers' freedom of thought and their results led to various industrial technologies, which at the same time produces a large number of talented researchers and increases academic achievements", and was consistent with Kita's own aims.

In 1918, because of Kita, a fifth chair (laboratory) was founded at the Kyoto Imperial University, focused on the study of fats and oils, petroleum, and fibers. Kita's students made achievements not only in applied chemistry, but also in a broad range of fields that included basic chemistry. His research group developed

RIKEN in the Taisho era (around 1923)

into a large school for chemistry in Japan. Kita welcomed the German organic chemist Lauer as a full-time lecturer in order to strengthen the foundations of synthetic chemistry. Ryohei Oda, an assistant to Kita, later became professor and laid the traditions of organic synthetic chemistry. This tradition led to the training of many excellent organic synthetic chemists, and later to Ryoji Noyori earning a Nobel Prize. Graduating from the fifth laboratory in 1926, Ichiro Sakurada studied abroad for three years at the University of Leipzig and the Kaiser Wilhelm Institute. He studied polymer chemistry under Kurt Hess and laid the foundations for polymer chemistry in Japan. He developed the polyvinyl alcohol-based synthetic fiber vinylon, which was Japan's first synthetic fiber, in 1939. Shinjiro Kodama spent two years, beginning in 1930, studying under Michael Polanyi (see page 269) at the Kaiser Wilhelm Institute, and he introduced the new field of quantum mechanics upon his return to Japan. Ken'ichi Fukui, whom he mentored, was Japan's first Nobel laureate in Chemistry, and was awarded the prize for his theoretical studies of chemical reactions. Fukui educated many excellent quantum chemists. Kodama himself was a leader in the development of artificial petroleum, and conducted both theoretical and experimental research with a focus on basic research. He transferred to Sumitomo Chemical from a professorship at Kyoto University, and served as research director and vice president.

Kita was a scholar who always sensed the latest research trends and predicted where learning was headed. As such, Kita was at his best when he was an "organizer of research" rather than a "researcher". The fields of research promoted by Kita involved the production of materials important to national defense before and during the war, such as fibers, artificial petroleum, synthetic rubber and industrial chemicals. Kita's research supported national policy, he acquired significant research funding, and was able to greatly advance his research. By developing an academic culture that emphasized the basics, he trained many scientists who contributed greatly to developing the related fields of polymer chemistry, quantum chemistry, catalyst chemistry, and synthetic organic chemistry.

4.13 Chemistry and society

Among the various fields of the natural sciences, chemistry has traditionally been the field with the strongest ties with society. Before the 20th century, researchers in other disciplines of the natural sciences were mostly researchers at a university. However, the close ties between basic and applied chemical research meant that advances in chemistry were linked to advances in industry. Therefore, many excellent researchers were working in corporate laboratories by the start of the 20th century, as demonstrated by the chemical industry in Germany. There was also close cooperation between universities and the industrial world. This trend spread to other developed nations after World War I, especially the U.S., and the chemical industry advanced significantly.

The first half of the 20th century was a time when the world suffered two world wars and an economic crisis between the wars. The chemical industry played a critical role in the execution of war; war both generated new demand and stimulated technological development. This chapter will look at changes in the chemical industry in the first half of the 20th century, the relationship between chemistry and society, and war and its impact on chemistry.

Changes in the chemical industry in the first half of the 20th century
The 20th century was a time when cutting-edge technological innovation significantly advanced industry. Innovation was dependent on close ties between science and technology, but the chemical industry had an especially deep relationship with basic chemistry and applied chemistry. At the beginning of the 20th century, chemical industry was rapidly developing, and was prominent in Western Europe, but its level of development varied greatly from country to country.

The German chemical industry towered above that of all other countries at this time. Since Liebig, German universities had enriched their teaching of chemistry with an emphasis on organic chemistry; technical institutes were founded across the country, and teaching emphasized applied chemistry, producing trained and highly competent chemical engineers. The German system did not just focus on organic chemistry. Physical chemists such as Ostwald, Haber, and Nernst were keenly interested in applied chemistry and were actively involved in industry. The employment of competent chemists and engineers in industry was a distinctive feature of chemistry in Germany, as was the close collaboration between universities and industry. In 1911, the Kaiser Wilhelm Society was formed in Berlin, and this society established seven institutes whose directors included Haber and Willstätter. The German government strongly supported chemistry as the basis for the development of the industry.

The chemical industry in Germany benefited from geography. Germany had coal from Ruhr and Silesia, potash ore from Prussia and Alsace, salt from the Neckar valley, and other natural resources. The chemical industry was dominant in the production of sulfuric acid, industrial gas, dyes, and potash, and commanded an overwhelming share of the global dye market. BASF, Bayer, Hoechst and other large chemical companies were the core of a chemical industry in which many companies thrived.

Pre-World War I Britain's chemical education was more academic in its approach and was not able to meet the industrial demands of the times. There were also few ties between universities and industry. The chemical industry in the U.K. at the turn of the 20th century lagged behind much of the rest of the world. The dye industry, which had begun with Perkin's mauve, found itself in decline, and the soda industry and fertilizer industry also relied on outdated techniques and were in decline. Around the beginning of World War I, the U.K.'s chemical industry was far from noteworthy.

In the time since Lavoisier, chemistry had been the science that was the pride of France, but its days of glory had already ended in the 19th century. The educational system was centralized and unable to fully satisfy the needs of the times. In higher education at its universities and Grandes Ecoles, the humanities and mathematics were emphasized more than experimental sciences. Chemical engineering schools, which were supported by local business leaders and the relevant authorities, did produce excellent engineers, but there was no competition with Germany. The chemical industry was in the process of developing, but overall it did not compare to the level of industry in Germany.

In other European countries, distinctive contributions were made in the limited field of chemical industry. The Belgium-based company Société Solvay developed as a multi-national company and produced most of the world's sodium carbonate. In the field of fine chemicals, Swiss companies maintained a certain share in competition with Germany. Ciba and Hoffmann-La Roche were successful with pharmaceutical products, as was Geigy with dyes. The Zurich school of the Swiss Federal Institutes of Technology (Eidgenössische Technische Hochschule, ETH) was training superb chemists and engineers. In Sweden, the Nitro Nobel Company, founded by Nobel, was making profits in the manufacture of dynamite, and developed into a multi-national company.

All aspects of the chemical industry were developing in the United States at the turn of the 20th century. Chemical education in universities had been reformed by young chemists who had studied in Germany at the end of the 19th century, and several universities were graduating eminent chemists and engineers. There were no barriers between universities and the world of industry, and scientists were actively addressing the issues faced by industry. Prior to

World War I, there was no chemical industry with a tradition to be proud of, but for the most part, America was self-sufficient regarding basic chemical products.

World War I, chemistry, and the chemical industry

World War I was the first conflict where science played a major role. The war was an all-out-war between the nations involved, and lasted longer than expected. Consequently, strength in the chemical industry had a major impact on the ability to wage war, while conversely, the war also majorly impacted the chemical industry. The first problem at hand was ensuring nitrogen resources as feedstock for fertilizer and nitric acid, used in the manufacture of explosives. Before the war, most of the world's consumption of nitrogen was for fertilizer, and two thirds of it came from sodium nitrate mined in Chile. Nitric acid, required for the manufacture of explosives, also was almost always made from Chile niter. Germany had difficulty with imports from Chile due to a U.K. naval blockade. Just before the Great War, BASF developed a method for synthesizing ammonia, so the German government energetically assisted with its production. In 1916, synthetic ammonia represented nearly half of the nitrogen produced in Germany. The fact that Germany could continue the war until November 1918 was clearly in part due to the success of the Haber-Bosch process.

The Allied countries had their own nitrogen sources, so securing a source of nitrogen was not a grave problem. Also, they were able to procure explosives from the U.S. immediately after the outbreak of war. DuPont was an American manufacturer capable of producing enormous amounts of smokeless gunpowder, and benefited significantly from the war.

At first, however, the chemical industry in the Allied countries was inadequate for the war effort. The fact that they were unable to procure German chemical goods such as medicines, dyes, glass products and potash products during the war was initially a major blow to the Allies, but the demand from the war stimulated and greatly helped develop the British and American chemical industries. The huge demand for many products promoted the development of new production technologies, and production volume increased substantially.

The use of poison gas as a weapon first appeared during World War I, and the Great War was said to be a "war of chemists". In Germany, lead by Haber, the Kaiser Wilhelm Institute was developing poison gas. Poison gas was first used in the Battle of Ypres on 22 April 1915, where chlorine gas was used. Phosgene was later used by the French army in the Battle of Verdun in February 1916. Afterwards, the German army used mustard gas. An ardently patriotic man, Haber justified the use of poison gas with his belief that war called for any means possible to win. Nearly all German chemists supported this position. For his involvement in the development of poison gas, Allied scientists were strongly opposed to Haber being awarded a Nobel Prize after the war. However, the Allied side likewise had many chemists who believed that all means possible must be used in war. In France, Grignard was involved in the development of phosgene. Poison gas was a chemical weapon that drew a great deal of attention, but it did not play a major role in deciding the outcome of the war.

The chemical industry in World War I

Lasting from 1914 to 1918, the war had a major impact on the economies of the combatant countries in Europe, and the same was also true of the chemical industry. Having lost territory, Germany lost its lead position in synthetic dyes, which had been close to a monopoly before the war. Opportunities for overseas market expansion in dyes and other chemical products opened up for British and American manufacturers. Companies involved with the military were especially prosperous in the U.S. However, when the war ended, the industrialized nations had an over-abundance of chemical products. The situation became even more difficult as economies entered the postwar economic depression. To counter this, industry strove to maintain market share by forming cartels and developing new products and applications.

Huge cartels between manufacturers in Europe were formed in the 1920s and 1930s. This made it possible for the giant chemical corporations to maintain their dominance in the market of key products. In the U.S., antitrust law prevented the formation of cartels. In both Europe and the U.S., however, enterprises formed alliances and centralized both during and after the war. In Germany, under the leadership of Bosch, eight affiliated companies were merged in 1926 to form I.G. Farben. This new company was a giant, accounting for one third of sales in the chemical industry in Germany. I.G. Farben also expanded into new areas such as explosives, rayon, and artificial petroleum.

The birth of I.G. Farben accelerated mergers in the chemical industry in the U.K., and in 1926 the United Alkali Company, British Dyestuffs Corporation, Nobel Enterprises, and Brunner Mond joined to form Imperial Chemical Industries (ICI).

In the U.S., antitrust law made it difficult for one company to occupy a dominant position by merger, but the U.S.'s domestic market was large enough for several leading companies to coexist in parallel. Nonetheless, expansion and integration led to the appearance of huge chemical companies in the U.S., such as DuPont, Union Carbide, and Monsanto. These companies focused on research and development and strove to forge new markets.

Examples included the discovery of new applications for chemical products, finding ways to add value to undeveloped raw materials, the development of quick-drying lacquer, and the use of acetylene as a raw material.

Nitrocellulose-based lacquer had been introduced before the war, and low-viscosity nitrocellulose-based lacquer was developed by DuPont and Hercules. This greatly reduced the drying time for automotive coatings. In the U.S., the spread of automobiles was accompanied by a dramatic increase in the demand for quick-drying lacquer, and research on other types of quick-drying lacquer also thrived; in the second half of the 1920s, alkyd resin lacquer appeared.

Acetylene attracted attention because it is highly reactive and can be easily generated from coal by using inexpensive calcium carbide. Because acetaldehyde is manufactured from acetylene and can be oxidized to produce acetic acid, it was useful as a raw material for the aliphatic chemical industry. Tetrachloroethane and dichloroethane are also produced by the chlorination of acetylene, and appeared on the market as non-flammable solvents. Trichloroethylene was in demand for dry cleaning and metal degreasing. In 1925, Newland discovered divinyl acetylene, which later became a raw material for neoprene synthetic rubber.

In the U.S., the automotive industry flourished in the 1930s, and there was a rapid increase in demand for high-octane gasoline. Most petroleum companies used a thermal decomposition method to purify crude oil, but there was a need to increase the distillate and improve quality. Houdry developed catalytic cracking using a silica/alumina acid-based catalyst in 1931, greatly improving production volume and raising the octane number of gasoline. Thermal decomposition and catalytic cracking techniques were utilized not only to manufacture gasoline, but also to produce gases such as ethylene, propylene, and butylene.

The Great Depression of 1929 to 1931 was a major setback for the chemical industry, but it recovered faster than other industries. Efforts in research and development were fruitful and markets expanded as new products were introduced, such as new fibers, plastics, alloys, and pharmaceuticals. At the time, coal tar was gradually losing its dominant position as a raw material for synthetic compounds and was replaced by aliphatic hydrocarbons obtained from oil and natural gas.

World War II and Chemistry

Increased production of chemistry-related munitions was essential for waging war, and World War I had stimulated the demand for chemical products, providing a major stimulus for industrializing the discoveries of chemists in the 1920s and the 1930s. Germany revoked the Treaty of Versailles in 1935 and immediately began preparations for war. The Nazis forced German industry to work closely with the state. The cooperation of the chemical industry, which manufactured most of the munitions, was essential to Germany's war efforts. At the end of the 1930s, preparations for war advanced and there was a sharp increase in the production volume of synthetic gasoline, synthetic rubber, and rayon. Most of the budget for the chemical industry was allotted to I.G. Farben, which had already grown into a giant, centralized corporation, and I.G. Farben began to fully support the upcoming war effort.

The other confrontational countries in Europe, France and the U.K., were not adequately prepared for war. France had not developed a synthetic gasoline or synthetic rubber industry. The U.K. was manufacturing synthetic gasoline, but not synthetic rubber. The task of providing the Allies with chemistry-related munitions fell on the huge chemical industry in the U.S. American power was dominant in every aspect, but especially in the development of the atomic bomb, synthetic rubber, penicillin, and aviation fuel.

The most significant event during World War II was arguably the development of the atomic bomb. As is well known, the idea of obtaining enormous energy by harnessing nuclear energy came from advances in physics in the first half of the 20th century. The development of the atomic bomb for the Manhattan Project in the U.S. during World War II was the fruit of organized research and development by a group of scientists and engineers. However, competent physicists, led by Robert Oppenheimer, a theoretical physicist from the University of California, were at the core of the project. The fact that the atom possesses tremendous energy was shown by the famous equation $E=mc^2$, derived from Einstein's theory of special relativity, but the possibility that considerable energy could actually be obtained by nuclear fission was shown experimentally by the nuclear chemists Hahn and Strassmann using uranium in 1938. Chemistry and chemists also played a major role in the development of the atomic bomb.

Szilard, Wigner, and Teller were exiled physicists from Hungary who feared that Nazi Germany would develop an atomic bomb. The three persuaded Einstein to write a letter to President Roosevelt telling him of the risk that the Nazis would develop the atomic bomb first. The Manhattan Project began in December 1941, and scientists and engineers were mobilized en masse to develop the atomic bomb. The first atomic bomb used uranium and plutonium. Its development was based on four successes: 1) controlling the chain reaction in the fission of uranium, by the group led by Fermi at the University of Chicago; 2) separating and concentrating uranium-235 by gas diffusion of uranium hexafluoride; 3) producing plutonium-239; and 4) developing and manufacturing the atomic bomb at Los Alamos. Of these, 2) and 3) were made possible by the contributions of chemists and the chemical industry. The concentration of uranium-235 by gas diffusion was developed by Abelson and Urey, and Union Carbide was given the task of separating uranium-235 from large amounts of natural uranium. DuPont was responsible for producing plutonium-239. In the Manhattan Project, leading physicists who had fled Europe made major contributions, but it was the force of the American chemical industry that actually made it possible to produce plutonium-239 and uranium-235 in the quantities necessary to manufacture the bomb. Nazi Germany also had plans to manufacture the atomic bomb, but implementation was difficult and ultimately it did not happen.

The U.S. embarked on plans to increase the production of synthetic rubber immediately after the start of the war against Japan. An agreement to select butadiene styrene elastomer and develop the required technical knowledge was forged by the government, Standard Oil, Dow, and the four major tire manufacturers. Enormous amounts of money and a huge workforce were dedicated to developing a system for producing synthetic rubber. The manufacturing processes improved, and at the conclusion of the war, the production volume of synthetic rubber had grown 100-fold.

As for penicillin, the Oxford group in 1939 succeeded in separating and purifying Penicillium notatum, which had been identified by Fleming. Wartime U.K. was not capable of mass production, however. In the U.S., the group at the Northern Regional Research Lab in Illinois developed a technique for culturing the mold in a mixture of corn syrup and lactose, and mass production became possible. Several companies, including Merck, started production, and the supply of penicillin for war efforts was ensured; after the war, production could meet the general public's demand.

High-octane gasoline was needed for aviation fuel, and was made possible by the catalytic cracking process invented by Eugene Houdry. Houdry's "fixed-bed" catalyst, however, degraded over time due to carbon deposits. The "fluidized bed" catalyst proposed by Odell was studied by a team at the Massachusetts Institute of Technology (MIT), and in 1942 was industrialized by Standard Oil of New Jersey. Thus, the high-octane gasoline required by the United States Air Force could be obtained in high yield, and the problem of supplying gasoline for aviation was solved. This was a decisive factor in the war.

References

1. A. J. Ihde, "*The Development of Modern Chemistry*" Dover Publications, Inc. New York, 1984
2. W. H. Brock, "*The Chemical Tree*" New York, 1993
3. J. Gribbin, "*Science A History*" Penguin Books, 2002
4. M. J. Nye, "*Before Big Sceience*" Harvard University Press, Cambridge, 1996
5. M. J. Nye, "*From Chemical Philosophy to Theoretical Chemistry*" University of California Press, Berkeley1993
6. K. Laidler, "*The World of Physical Chemistry*" Oxford University Press, Oxford, 1993

7. G. Friedlander and J. Kennedy, *"Nuclear and Radiochemistry"* John Wiley & Sons, 1955

8. J. S. Fruton, *"Molecules and Life"* John Wiley and Sons, 1972

9. J. S. Fruton, *"Protein, Enzymes, Genes"* Yale University Press, New Heaven and London, 1999

10. L. K. James, ed. *"Nobel Laureate in Chemistry 1901–1992"* Amer. Chem. Soc./ Chem. Her. Fond., 1994

11. *"Iwanami Rikagaku Jiten"* (Iwanami Dictionary of Physics and Chemistry) 5th ed., Iwanami shoten, 1998

12. R. J. Forbes and J. E. Dijkserhaus, *"A History of Science and Technology"* Penguin Books, 1963

13. F. Aftalion, *"A History of the International Chemical Industry"* Chemical Heritage Foundation, 1991

14. P. Coffey, *"Cathedrals of Science: Personalities and Rivalries that made Modern Chemistry"* Oxford Univ. Press, 2008

15. C. Reinhardt (Ed.) *"Chemical Sciences in the 20th century"* Wiley-VCH, 2001

16. J. R. Oppenheimer Ed. *"The age of Science, 1900–1950"* Scientific American, September 1950

17. W. J. Moor, *"Physical Chemistry"* 4th edition, Prentice-Hall, New Jersey, 1972

18. J. Heine, *"Physical Organic Chemistry"* McGraw-Hill, New York, 1956

19. D. Voet, J. Voet, *"Biochemistry"* 2nd edition, John Wiley and Sons, 1995, 3rd edition, 2004

20. D. Stoltzenberg *"Fritz Harber"* Chemical Heritage Press, Philadelphia, 2004

21. E. Shimao, *"Jinbutsu Kagakushi"* (Characters in the history of chemistry) Asakura Shoten 2002

22. K. Maruyama, *"Seikagaku wo Tsukutta Hitobito"* (Scientists who made biochemistry) Shokabo, 2001

23. H. Shinohara, *"Seimeikagaku no Sennkusya"* (Pioneers in Life Science) Koudansha Gakugeibunnko, 1983

24. H. Poincare, *"La Valueur de la science"* Flammarion, 1905

25. a) A. Serafini, *"Linus Pauling"* Simon & Schuster, London, 1989
    b) T. Hager, *"Force of Nature: The Life of Linus Pauling"* Simon & Scuster, 1995

26. W. Nernst, *Kgl. Ges. Wiss. Nachrichten,Math-Phys. Klasse 1,* 1 (1906)

27. F. Harber, *"Thermodynamik technischer Gasreactionen:Sieben Vortrage"* R. Oldenbourg, Munich, 1905

28. G. N. Lewis and M. Randall, *"Thermodynamics and the Free Energy of Chemical Substances"* McGraw-Hill, New York, 1923
29. R. H. Fowler and E. Guggenheim, *"Statistical Thermodynamics"* Cambridge Univ. Press. 1939
30. N. Bjerrum, *Zeitschr. Anorg. Chem.*, **63**, 140 (1909)
31. P. Debye and E. Hückel, *Physikal Z.*, **24**, 185 (1923)
32. L. Onsager, *J. Chem. Phys.*, **2**, 599 (1934)
33. S. P. Sørensen, *Comt Rend. des travaux du Laboratoire de Carsberg* Vlll 1(1909)
34. G. N. Lewis, *J. Am. Chem. Soc.*, **38**, 762 (1916)
35. W. Kossel, *Ann. Physik*, **49**, 229 (1916)
36. I. Langmuir, *J. Am. Chem. Soc.*, **41**, 868 (1919), 42, 274 (1920)
37. G. N. Lewis, *"Valence and the Structure of the Atom and Molecules"* Chemical Catalogue Co. New York, 1923
38. W. Heitler and F. London, *Z. Physik*, **44**, 455 (1927)
39. L. Pauling, *J. Am. Chem. Soc.*, **53**, 1367 (1931)
40. J. C. Slater, *Phys. Rev.*, **38**, 1109 (1931)
41. L. Pauling and E. B. Wilson, *"Introduction to Quantum Mechanics with Application to Chemistry"* Mcgraw-Hill, 1935
42. L. Pauling, *"The Nature of the Chemical Bond, and the Structures of Molecules and Crystals"* Cornell Univ. Press, 1939
43. L. Pauling, *"General Chemistry"* W.H.Freeman, 1947
44. F. Hund, *Z. Physik*, **37**, 742 (1927)
45. R. S. Mulliken, *Phys. Rev.*, **32**, 186 (1928)
46. J. E. Lenard-Jones, *Trans Faraday Soc.*, **25**, 668 (1929)
47. E. Hückel, *Z. Physik*, **70**, 204 (1931)
48. P. Debye, *Physik. Z.*, **13**, Nr, 3, 97(1912)
49. P. Debye, *"Polar Molecules"* Dover Publishing, 1929
50. W. L. Bragg, *Camb. Phil. Soc. Proc.*, November (1912), Proc. Roy. Soc. London A, **89**, 248 (1913)
51. W. H. Bragg, W. L. Bragg, *Proc. Roy Soc. London* A, **88**, 428 (1913)
52. S. Nishikawa, *Proc. Tokyo Math-Phys. Soc.*, **8**, 199 (1915)
53. I. Nitta, Bull. *Chem. Soc. Jpn.*, **1**, 62 (1926)
54. K. Lonsdale, *Nature*, **122**, 810 (1928)
55. J. D. Bernal, R. H. Fowler, *J. Chem. Phys.*, **1**, 515 (1953)
56. J. D. Bernal, D. Crowfoot, *Nature*, **133**, 794 (1934)
57. R. Wierl, *Physik. Z.*, **31**, 366 (1930)
58. A. Brown, *"J. D. Bernal The Sage of Science"* Oxford Univ. Press, 2005
59. A. Einstein, *Physik. Z.*, **18**, 121 (1917)

60. C. V. Raman, K. S. Krishnan, *Nature*, **121**, 501 (1928)
61. I. I. Rabi, J. R. Zacharias, S. Millman, P. Kusch, *Phys. Rev.*, **53**, 318 (1938)
62. E. K. Zavoisky, J. *Phys. USSR*, **9**, 447 (1945); **10**, 197 (1946)
63. R. Marcelin, *Compt rendus*, **151**, 1052 (1910)
64. W. C. McC. Lewis, *J. Chem. Soc.*, **113**, 471 (1918)
65. R. Marcelin, *J. de chim. phys.*, **12**, 451 (1914)
66. K. F. Herzfeld, *Ann. der Physik*, **4**, 59, 635 (1919), *Z. Elektrochem*, 25, 301
67. H. Eyring, M. Polany, *Z. Physikal. Chem.* B. **12**, 279 (1931)
68. H. Peltzer, E. Wigner, *Z. Physikal. Chem.*, B, **15**, 445 (1932)
69. H. Eyring, *J. Chem. Phys.* **3**, 107 (1935)
70. M. G. Evans and M. Polany, *Trans. Faraday Soc.*, **31**, 875 (1935)
71. H. Eyring, J. Walter, G. Kimball, "*Quantum Chemistry*" John Wiley and Sons, New York, 1944
72. G. Glasstone, K. J. Laidler, H. Eyring, "*The Theory of Rate Processes*" McGraw-Hill, New York, 1941
73. M. Smoluchowski, *Physikal. Z.*, **17**, 557, 589 (1916)
74. H. A. Kramers, *Physica*, 7, 284 (1941)
75. M. Bodenstein, S. C. Lind, *Z. phsikal. Chem.*, **57**, 168 (1907)
76. J. A. Christiansen, Det. Kgl. Damske Vid. Selskab., *Math. Phys. Medd l*, **14**, 1 (1919), Ph. D. Thesis, University of Copenhagen (1921)
77. F. A. Lindeman, *Trans. Faraday Soc.*, **17**, 598 (1922)
78. C. N. Hinshelwood, *Proc. Roy. Soc.*, A, **113**, 230 (1927)
79. O. K. Rice, C. H. Rampsberger, *J. Am. Chem. Soc.*, **49**, 1617 (1927)
80. L. S. Kassel, *J. Phys. Chem.*, **32**, 225 (1928)
81. N. N. Semenov, *Z. Physik*, **46**, 101 (1927)
82. C. N. Hishelwood and H. W. Thompson, *Proc. Roy. Soc. A.* **118**, 170 (1928)
83. J. Stark, *Physik. Z.*, **9**, 88, 84, (1908)
84. A. Einstein, *Ann. Phys.*, **37**, 832 (1912), **38**, 881 (1912)
85. M. Bodenstein, W. Dux, *Z. Physikal. Chem.*, **85**, 297 (1913)
86. W. Nernst, *Z. Elektrochem.*, 24, 335 (1918)
87. G. N. Lewis and M. Kasha, *J. Am. Chem. Soc.*, **66**, 2100 (1944)
88. I. Langmuir, *J. Am. Chem. Soc.*, **39**, 1848 (1917)
89. I. Langmuir, *J. Am. Chem. Soc.*, **38**, 2221(1916)
90. E. Rutherford, Phil. *Mag. Ser. 6*, **37**, 581 (1919)
91. I. Curie, F. Joliot, *Compt Rendus*, **158**, 254 (1934)
92. O. Hahn, F. Strassman, *Naturwiss.*, **27**, 11 (1938)
93. L. Meitner, O. R. Frisch, *Nature*, **143**, 239 (1939)
94. L. A. Turner, *Rev. Mod. Phys.*, **12**, 1 (1940)
95. R. L. Sime, "*Lise Meitner, A Life in Physics*" Univ. of California Press, 1996

96. E. McMillan, P. H. Abelson, *Phys. Rev.*, **57**, 1185 (1940)

97. G. T. Seaborg, *Nucleonics*, **5**, 16 (1949)

98. G. de Hevesy, F. A. Paneth, *Z. Anorg. Chem.*, **82**, 322 (1913)

99. S. Ruben, M. Kamen, *Phys. Rev.*, **59**, 349 (1941)

100. F. Pregl, *"Die Quantitativ Organische Mikuroanalyse"* Springer Berlin, 1917

101. Y. Heyrovsky, *Phil. Mag.*, **45**, 303 (1923)

102. Y. Heyrovsky, M. Shikata, *Rec. Trav. Chim. Pay-Bus*, **49**, 469 (1925)

103. M. Tswett, *Ber. Deutsch. Bot. Gesel.*, **24**, 5, 316 (1906)

104. A. J. Martin, R. C. Synge, *Biochem. J.*, **35**, 1358 (1941)

105. W. F. Libby, *"Radiocarbon Dating"* University of Chicago, Chicago, 1952

106. D. Coster, G. de Hevesy, *Nature*, **111**, 252 (1923)

107. K. Yoshihara,*"Kagakushi Kenkyu"*, **34**, 137 (2007)

108. A. Werner, *"Neuere Anschaung auf dem Gebiete der Anorganische Chemie"* F. Vieweg und Sohn, Braunschweig, 1905

109. N. V. Sidgwick, *"The Electronic Theory of Valency"* Clarendon press, Oxford, 1927

110. H. Bethe, *Ann. Physik*, **3**, 135 (1929)

111. L. Pauling, *J. Am. Chem. Soc.*, **53**, 1386 (1931)

112. J. H. van Vleck, *J. Chem. Phys.*, **3**. 803, 807 (1935)

113. C. J. Ballhausen, *"Introduction to Ligand Field Theory"* McGraw-Hill, 1962

114. C. Longet-Higgins, Le Bell, *J. Chem. Soc.*, 250 (1943)

115. H. Kragh, *"From Geochemistry to Cosmochemistry"* in ref. 12

116. F. W. Clark, *United States Geological Survey Bulletin*, No. 770 (1924)

117. V. M. Goldschmidt, *Videnskapsselskapets Skrifter, I. Mat. Naturv. Klasse*, No. 31(123)

118. R. Robinson, *"Two lectures on anOutline of an Electronical Theory of the Course of Organic Reactions"* Institute of Chemistry of Great Britain and Ireland, London 1932

119. K. Ingold, *Chem. Rev.*, **15**, 225 (1934)

120. K. Ingold, *"Structure and Mechanism in Organic Chemistry"* Cornell Univ. Press, Ithaca, N. Y. 1953

121. L. Hammett, *Chem. Rev.*, **17**, 125 (1935)

122. M. Gomberg, *J. Am. Chem. Soc.*, **23**, 757 (1900)

123. J. D. Kemp, K. S. Pitzer, *J. Am. Chem. Soc.*, **59**, 276 (1937)

124. S. Mizushima, Y. Morino, *Bull. Chem. Soc. Japan*, **17**, 94 (1942)

125. H. Sache, *Ber.*, **23**, 1363 (1890)

126. E. Mohr, *J. Prakt. Chem.*, **2**, 98, 315 (1918)

127. O. Hassel, *Tidsskr. Kjemi, Bergvesen Met.*, **3**, 32 (1943)

128. D. H. R. Barton, *Experientia*, **6**, 316 (1950)
129. V. Grignard, *Compt rend.* **130**, 1322 (1900)
130. O. Diels, K. Adler, Liebigs *Ann. Chem.*, **460**, 98 (1928)
131. H. Staudinger, *Ber.*, **53**, 1073 (1920)
132. Y. Furukawa, p.228–245 in ref. 15
133. E. Fischer, *Ber. Chem. Ges.*, **40**, 1754 (1907)
134. A. R. Todd, *J. Chem. Soc.*, 693 (1946)
135. R. Willstätter, *Nobel Lecture*, 1920
136. H. Fischer, *Nobel Lecture*, 1930
137. P. Karrer, "*Leherbuch der Organische Chemie*"
138. P. Karrer, *Nobel Lecture*, 1937
139. A. Harden, W. J. Young, *Proc. Roy. Soc. Ser. B*, **77**, 405 (1906)
140. H. von Euler-Chelpin, *Nobel Lecture*, 1930
141. R. Willstätter, "*Problem and Method in Enzyme Research*" Cornell Univ. Press, 1927
142. J. Sumner, *J. Biol. Chem.*, **69**, 435 (1926)
143. J. Northrop, *Nobel Lecture*, 137
144. W. Stanley, *Science*, **81**, 644 (1935)
145. O. Warburg, *Biochem. Z.*, **152**, 479 (1924)
146. O. Warburg, *Biochem. Z.*, **214**, 64 (1929)
147. D. Keilin, *Proc. Roy. Soc. B.*, **98**, 312 (1925)
148. T. Thunberg, *Skad. Arch. Phigiolo.*, **40**, 1 (1920)
149. O. Warburg, W. Christian, *Biochem. Z.*, **254**, 438 (1932)
150. O. Warburg, W. Christian, *Biochem. Z.*, **298**, 368 (1938)
151. O. Myerhoff, *Z. Physiol. Chem.*, **101**, 165 (1918)
152. C. H. Fiske, Y. Subbarow, *Science*, **70**, 381 (1929)
153. K. Lohman, *Nturwiss.*, **17**, 624 (1929)
154. A Szent- Györgyi, "*Studies on Biological Oxidation and Some of its Catalysis*" Eggenberg and Barth, Budapest and Leipzig (1937)
155. H. A. Krebs, W. A. Johnson, *Enzymologia*, **4**, 148 (1937)
156. G. T. Cori, C. F. Cori, C. Schmidt, *J. Biol. Chem.*, **129**, 629 (1939)
157. C. B. van Niel, *Archiv fur Mikrobiologie*, **3**, 1 (1931)
158. R. Shoenheimer, D. Rittenberg, *Science*, **82**, 156 (1935)
159. R. Shoenheimer, S. Ratner, D. Rittenberg, *J. Biol. Chem.*, **130**, 703 (1939)
160. M. Kaji, "*Kagakusi kennkyu*" (*Studies of History of Chemistry*), **38**, 173 (2011)
161. Y. Furukawa, "*Kagakushi kennkyu*" (*Studies of History of Chemistry*), **37**, (2010)
162. S. Tomonaga, H. Ezawa ed. "*Kagakusya no jiyuna Rakuen*" (*Free Paradise of Scientists*) Iwanamibunko, 2001

Chronology of Modern Chemistry and Science Technology (first half of the 20th century)

Year	Physics	Physical chemistry, inorganic chemistry, analytical chemistry, radiochemistry	Organic chemistry, industrial chemistry	Biochemistry, pharmacy
1895	Discovery of X-rays (Röntgen, 1895) Discovery of radioactivity (Becquerel, 1896) Zeeman effect (Zeeman, 1896) Discovery of electrons (Thomson, 1897)	Discovery of radium and polonium (Marie and Pierre Curie, 1898) Discovery of neon, krypton, and xenon (Ramsay, 1898)		Cell-free fermentation (Buchner, 1897)
1900	Quantum theory of radiation (Planck, 1900) Saturnian model of the atom (Nagaoka, 1903) Theory of special relativity (Einstein, 1905) Photon hypothesis (Einstein, 1905) Theory of the specific heat of solids (Einstein, 1906) Analysis of anode lines (J.J. Thomson, 1907) Liquefaction of helium (Kamerlingh-Onnes, 1908) Proof of the existence of atoms and molecules (Perrin, 1908) Discovery of α-particles (Rutherford, 1908) Measurement of the charge of the electron (Millikan, 1909)	Microscope observation of colloidal particles (Zsigmondy, 1903) Chromatography (Tsvet, 1903) Decay of radioactive elements (Rutherford, Soddy, 1903) Theory of Brownian motion (Einstein, 1906) Third law of thermodynamics (Nernst, 1906) Glass electrode (Cremer, Haber, 1906) Geochemical data (Clark, 1908) Concept of pH (Sørensen, 1909)	Discovery of the triphenylmethyl radical (Gomberg, 1900) Grignard reagents (Grignard, 1901) Discussion of reaction mechanisms (Lapworth, 1903) Hydrogenation by nickel catalyst (Sabatier, 1905) Discovery of chlorophyll-a and -b (Willstätter, 1906) Synthesis of Bakelite (Baekeland, 1906) Synthesis of octapeptide (E. Fischer, 1907) Synthesis of ammonia from nitrogen and hydrogen (Haber, Le Rossignol, 1909)	Adrenaline crystal (Takamine, 1900) Peptide theory of protein structures (E. Fischer, 1902) Discovery of hormones (Bayliss, Starling, 1902) Involvement of coenzymes and phosphoric acid in fermentation (Harden, Young, 1904) Salvarsan (Ehrlich, Hata, 1909)

1910	Existence of the atomic nucleus (Rutherford, 1911) Discovery of superconductivity (Kamerlingh-Onnes, 1911) X-ray diffraction by crystals (Laue, 1912) Determination of crystal structures by X-ray diffraction (William and Lawrence Bragg, 1913) Quantum theory of the atomic structure (Bohr, 1913) Element-specific relationships between X-rays and atomic number (Moseley, 1913) Stark effect (Stark, 1913) X-ray diffraction by powder crystals (Debye, Scherrer, 1916) Theory of absorption and emission of electromagnetic waves (Einstein, 1917) Destruction of the atomic nucleus by α-particles (Rutherford, 1919)	Concept of the isotope (Soddy, 1910) Synthesis of borane (Stock, 1912) Photochemical equivalents (Einstein, 1912) Dipole moment of the molecule (Debye, 1912) Founding of the tracer method (Hevesy, 1913) Law of displacement of radioactive elements (Soddy, Fajans, 1913) Chemical bonding by electron pairs (Lewis, 1916) Theory of ion bonding (Kossel, 1916) Diffusion-controlled reaction rates (Smoluchowski, 1916) Organic micro-analysis (Pregl, 1916) Absorption isotherm (Langmuir, 1916) Mechanism of chain reactions (Nernst, 1918) Mass spectrometry (Aston, 1919)	Liquefaction of coal by hydrogenation (Bergius, 1913) Synthesis of tropinone (Robinson, 1917)	Extraction of oryzanine (Suzuki, 1910) Naming of the vitamins (Funk, 1912) Dehydrogenase reaction (Wieland, 1912) Discovery of vitamin A (McCollum, Davis, 1913) Kinetics of enzymatic reactions (Michaelis, Menten, 1913) Methylglyoxal theory in glycolysis mechanisms (Neuberg, 1913) Dehydrogenation theory of respiration (Thunberg, 1917)
1920	Magnetic moment of the atom (Stern, Gerlach, 1921) Compton effect (Compton, 1922) Concept of matter waves (de Broglie, 1923)	Hydrogen bonds (Latimer, Rodebush, 1920) Mechanism of unimolecular reactions (Lindemann, Christiansen, 1920–1921) Development of the ultracentrifuge (Svedberg, 1923) Theory of electrolyte solutions (Debye, Hückel, 1923) Extension of the concept of acids and bases (Brønsted, Lowry, Lewis, 1923) Branched chain reactions (Semenov, Hinshelwood, 1923)	Proposal of polymer theory (Staudinger, 1920) Beginning of electronic theory of organic chemistry (Robinson, 1922) Synthesis of aliphatic hydrocarbons from water gas (Fischer, Tropsch, 1923)	Extraction of insulin (Banting, Best, 1921) Chemical transmission in nerves (Loewi, 1921) Oxygen activation theory of respiration (Warburg, 1923)

Year	Physics	Physical chemistry, inorganic chemistry, analytical chemistry, radiochemistry	Organic chemistry, industrial chemistry	Biochemistry, pharmacy
1920	Introduction of electron spin (Goudsmit, Uhlenbeck, 1925)	Proposal of adiabatic demagnetization (Giauque, 1924)	Projection formula of sugar structures (Haworth, 1925)	Discovery of cytochromes (Keilin, 1925)
	Pauli exclusion principle (Pauli, 1925)	Development of the polarograph (Heyrovsky, Shikata, 1925)		
	Matrix mechanics (Heisenberg, 1925)			
	Wave mechanics (Schrödinger, 1926)		Systematization of electronic theory of organic chemistry (Ingold, 1926)	Crystallization of urease (Sumner, 1926)
	Fermi statistics (Fermi, 1926)			
	Uncertainty principle (Heisenberg, 1927)			
	Electron beam diffraction (Davisson, Germer, G.P. Thomson, 1927)	Description of coordination bonds (Sidgwick, 1927)		
	Theory of ferromagnetism (Heisenberg, 1928)	Quantum theory of covalent bonds (Heitler, London, 1927)	Diels-Alder reaction (Diels, Alder, 1928)	
	Electron theory of metals (Bloch, 1928)	Molecular orbital theory (Hund, Mulliken, 1927)	Synthesis of hemin (H. Fischer, 1929)	Discovery of penicillin (Fleming, 1929)
	Discovery of the Raman effect (Raman, Krishnan, 1928)		Introduction of conformations (Haworth, 1929)	Discovery of ATP (Fiske, Subbarow, Lohmann, 1929)
	Relativistic electron equation (Dirac, 1928)	Crystal fields Theory (Bethe, 1929)	Isolation of estrone (female sex hormone) (Butenandt, Doisy, 1929)	
1930	Cyclotron (Lawrence, 1930)	Electronegativity (Pauling, 1931)	Chair structure of cyclohexane (Hassel, 1930)	Crystallization of pepsin (Northrop, 1930)
	Electron microscope (Knoll, Ruska, 1931)	Hybrid orbitals (Pauling, Slater, 1931)	Development of organic glass by polymerization of methyl/methacrylate (Bauer, 1931)	Generalization of photosynthesis reactions (van Niel, 1931)
	Discovery of the neutron (Chadwick, 1932)	Deuterium and heavy water (Urey, 1932)		Discovery of yellow enzymes (Warburg, Christian, 1932)
	Artificial nuclear transformation by high-voltage accelerator (Cockcroft, Walton, 1932)	Ligand field theory (van Vleck, 1932)	Development of synthetic rubber (Carothers et al., 1931)	
			Development of the Hückel method (Hückel, 1931)	
				Establishment of the glycolytic pathway (Embden, Meyerhof, 1933)
	Phase-contrast microscope (Zernike, 1934)	Discovery of artificial radioactivity (Frederic and Irene Joliot-Curie, 1934)	Synthesis of vitamin B$_2$ (Karrer, Kuhn, 1934)	C4 dicarboxylic acid theory of metabolism (Szent-Györgyi, 1935)
	Meson theory (Yukawa, 1934)	Transition state theory (Eyring, Polanyi, Evans, 1935)	Synthesis of androsterone (male sex hormone) (Butenandt, 1934)	Use of tracers in metabolic research (Schoenheimer, 1935)
		Radioactivation analysis (Hevesy, 1936)	Hammett's rule (Hammett, 1935)	Crystallization of tobacco mosaic virus

	Theoretical description of uranium fission (Meitner, Frisch, 1938)	Development of electrophoresis (Tiselius, 1937)	Synthetic nylon fiber (Carothers, 1936)	(Stanley, 1936)
		Discovery of uranium fission (Hahn, Strassmann, 1938)	Synthesis of vitamin A (Kuhn, 1937)	Citric acid cycle (Krebs, 1936)
		Abundance ratios of the elements (Goldschmidt, 1938)		Origin of life (Oparin, 1938)
				Insecticidal effects of DDT (Müller, 1938)
	Description of stellar heat sources by nuclear reactions (Bethe, 1939)		Structural determination of vitamin K (Karrer et al., 1939)	
	Magnetic resonance method (Rabi, 1939)			
1940		Plutonium (Seaborg, 1941)	Structure of chlorophyll (H. Fischer, 1940)	Study of photosynthesis by ^{18}O (Ruben, Kamen, 1941)
				Concept of high-energy phosphate bonds (Kalckar, Lipmann, 1941)
		Bridged structure of diborane (Longuet-Higgins, 1943)		
	Electron spin resonance (Zavoisky, 1944)	Partition chromatography (Martin, Synge, 1944)		Transformation of pneumococcus (Avery, 1944)
		Triplet sate theory of excited molecules (Lewis, Kasha, 1944)		

Part 3

Contemporary Chemistry

Chapter 5: Chemistry in the second half of the 20th century (I)

Advances in the observation, analysis, and fabrication of molecules

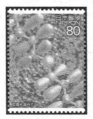

A postage stamp featuring polyacetylene, a conducting polymer discovered by Hideki Shirakawa. (Stamp: Japan, 2004)

The Second World War was a major turning point after which the environment surrounding the natural sciences changed in major ways. The world war and the ensuing U.S.-Soviet cold war had a significant impact on the development of the natural sciences. The scientific and technological superiority of the allied powers was a key factor behind their victory in the war; even more importantly, the emergence of the atomic bomb offered a clear demonstration of the ultimate authority of science and technology. The success of the Manhattan Project to build the atomic bomb vividly illustrated the magnitude of the achievements that were possible when science and engineering efforts were organized on a large scale. Before the war, scientific research had been conducted on small scales, primarily at universities; after the war, research began to receive major government support, spurring the emergence in many fields of big science—large-scale research conducted by organized teams of researchers. Before the war, most scientists were seen as educated intellectuals affiliated with universities; after the war, this situation was thoroughly transformed as the number of scientists increased significantly.

Whereas Europe was the unquestioned global center of science in the prewar era, the war undermined this position of dominance, and the center of science shifted from Europe to the U.S. The decline of German science—which suffered both from Germany's defeat in the war and from the expulsion of leading Jewish scientists—was particularly acute. The U.S. had already begun to produce its own top-notch scientists before the war; as their ranks were joined by first-rate scientists emigrating from Europe, the development of the natural sciences came to proceed largely in the U.S., undergirded by the nation's overwhelming economic power. England, despite its economic struggles, also produced much innovative research in basic science; particularly noteworthy here are the pioneering contributions of English science to the structural analysis of DNA and proteins.

The science of modern chemistry—which began in the early twentieth century as an attempt to understand chemical phenomena based on the structure of atoms and molecules, with particular focus on the behavior of electrons—became an increasingly mature and precise academic discipline, aided by the incorporation of progress in physics. Meanwhile, the chemistry of life phenomena fused with the field of molecular biology—launched in 1953 by the determination of the structure of DNA—to form one of the major drivers of the impressive development of the life sciences in the second half of the 20th century. In this way, chemistry developed into a broad field that overlaps with physics and biology. Indeed, the fruits of the development of chemistry in the latter part of the 20th century are so vast that we cover them in two separate chapters; the chemistry of life phenomena will be largely relegated to the following chapter. In researching this chapter we have relied primarily on references [1]–[13] and on a variety of Internet resources, particularly the website NobelPrize.org.

5.1 Overall trends

The societal backdrop underlying science in the post-war era[1, 2]

First, we briefly survey the evolution of the environment in which science is conducted. In prewar America, financial support for basic research was mostly provided by private organizations or private-sector corporations. In July 1945, MIT electrical engineer Vannevar Bush—who had overseen the mobilization of American scientists—wrote Science: The Endless Frontier, a report for the President that discussed postwar science and technology policy. This report was based on the notion of a linear model of science and technology development, in which progress originates from basic research leading to new concepts and principles, from which new technologies, new products, and new industries arise. Thus, the U.S. should take the lead in supporting the development of basic science in the new postwar era. Based on this philosophy, the U.S. government began to provide generous funding for basic scientific research in the years after the Second World War. The National Science Foundation (NSF) was established in 1950, and the federal government channeled large amounts of investment through this organization and through the National Institutes of Health (NIH). Even basic research with no immediate ties to practical applications was supported as long as its scientific value could be demonstrated. The military and the Atomic Energy Commission also supported research plans with no direct relationship to military objectives. The U.S.–Soviet cold war that began immediately after the Second World War further accelerated these trends. In the postwar era, as Western nations recovered, many countries came to appreciate the importance of science and technology for their own security and prosperity and devoted effort to scientific advancement. In this way, government support for science and technology in many countries rose to levels unheard of in the pre-war era. In addition, technologies developed during the war for military purposes were used in the postwar era for basic science research, where they spurred major progress. After the war, feelings of disillusionment with physics—the science that produced the atomic bomb—were not entirely absent; nonetheless, faith in the boundless progress in science remained the dominant sensibility, and there was a general belief that science, as long as it was not put to nefarious purposes—could make major contributions to the progress and well-being of mankind.

In 1957, the Soviet Union successfully launched Sputnik, the first man-made satellite; this further spurred the design of government policies to promote science and technology in the U.S. and ensured continuing support for science overall, particularly in the fields of space science and atomic energy. Thus science became a government-supported enterprise and began to make major progress as an organized endeavor. Many universities expanded their science and engineering faculties; the number of graduates grew explosively, and many large corporations established research laboratories to pursue basic science. The nations of Europe, and Japan, learned from this example, stepping up efforts to promote science and technology as their postwar economic recovery progressed. The Soviet Union and the Eastern-bloc countries also supported science and technology as crucial for national survival and development, albeit with various biases depending on the particular branch of science. In the late 1960s, the impact of the Vietnam War, and of pollution and other societal problems, began to erode society's unquestioning faith in scientific and technological progress. However, throughout the era of the U.S.–Soviet cold war (1946–1990), science continued to be regarded as crucial for national security, development, and confidence, and for these reasons there was continual support for science, which consequently enjoyed major progress.

The 1980s saw the rise of neoliberal government in the U.S. and England; meanwhile, the collapse of the Soviet Union in 1990 brought the cold war to an end. These societal transformations presaged a transformation in the philosophy underlying support for science and technology. The final 15 years of the 20th century witnessed the advance of economic globalization; in an era of increasing global economic competition, the notion that science and technology were important for economic growth received intense new emphasis. Ties between universities and industry became further entrenched, and research that could lead to patents came to be encouraged at universities as well as in the private sector. Meanwhile, the progress of science ensured that it was increasingly a large-scale, high-cost endeavor; this created a more urgent need to justify to taxpayers the enormous financial outlays required to fund science projects. Traditionally, basic science supported industry and was itself supported as a wellspring from which new industries emerged; however, in the present era it seems the overall tendency to emphasize applied science will only continue to grow.

The situation in Japan

As we saw in the previous chapter, up to around 1940, the progress of Japanese science had been steady, and the nation had made significant strides toward approaching the level of the advanced nations of Europe and the U.S. However, Japan suffered a devastating blow in the Second World War, and in the immediate aftermath of this defeat, the nation's scientific status had fallen far behind global standards in all but a few areas. Nonetheless, as the postwar recovery proceeded, the level of Japanese science rapidly caught up with the rest of the world, and the nation soon began to produce world-leading research. Among the factors that enabled this progress, we may cite the following:

1. Japan's system of higher education continued to function before and during the war, ensuring that talented individuals capable of driving the progress of science received the training they needed.
2. The education system was reformed in 1948; this led to a rapid increase in the number of universities and bolstered the ranks of researchers as workplaces for scientists grew.
3. The Fulbright system and other foreign-study programs created more opportunities for young researchers to study abroad in the U.S., where it was easier to learn cutting-edge disciplines.
4. The rapid economic growth that began around 1960 enabled the expansion and upgrading of scientific and technical facilities at universities; private-sector corporations established basic research laboratories, the numbers of scientists and engineers increased dramatically, and financial support for the sciences also gradually improved.

Thus, by the 1970s Japan had become an economic powerhouse with legitimate claims to global leadership in many fields of science.

Characteristic features of the science of chemistry in the second half of the 20th century

The science of chemistry underwent radical transformation and development in the second half of the 20th century. While some fields exhibited a continuation of the progress they had enjoyed in the first part of the century, other fields witnessed dramatic new developments spurred by unexpected discoveries and inventions. Here we offer a quick overview of some of noteworthy features characterizing the progress of chemistry during this time.

1. Revolutionary advances in electronics and computer technology, and corresponding progress in observational techniques exploiting these technologies, enabled the gathering of detailed information on the structure and reactions of matter at atomic and molecular levels. Chemistry increasingly became an exact science, and the boundary between physics and chemistry grew more difficult to discern.

2. Theoretical chemistry advanced to a point at which it could offer reliable descriptions and predictions of many experimental results. Explosive progress in computational capacity made it possible to perform calculations that could hardly have been imagined in previous eras, giving rise to the new sub-discipline of computational chemistry.

3. Technological progress enabled the analysis and synthesis of increasingly complex materials, spurring intense research efforts to synthesize and understand the properties of new materials with useful characteristics and features, including both organic and inorganic compounds. The discovery of unexpected new materials further entrenched the importance of chemistry as a cornerstone of materials science.

4. The elucidation of the structure of DNA in 1953 enabled a wide range of phenomena in the life sciences to be explained on a molecular level in the language of chemistry and physics. The borders between biology, chemistry and physics became increasingly indistinguishable, and studies of life phenomena became a vital field of chemistry research.

5. The various specialized subfields of chemistry burrowed to increasing depths, and the fragmentation of the discipline as a whole proceeded; nonetheless, the era was marked by vigorous research collaborations between different subfields of chemistry, as well as interdisciplinary research straddling the boundaries between fields such as chemistry and physics, chemistry and biology, and chemistry and medicine.

These developments also spurred major changes in the way chemical research was conducted. Up until the first half of the 20th century, research in chemistry was a relatively small-scale enterprise, dependent primarily on the creativity, ingenuity, and hard work of individual chemists. However, the second half of the century witnessed an increasing emergence of research conducted by large groups or dependent on costly, large-scale equipment.

The progress of chemistry in the latter part of the 20th century was nothing short of astonishing in any number of areas. To describe all of these advances would be not only impossible in any finite number of pages, but also well in excess of this author's capabilities. For these reasons, rather than discussing here the progress of individual subfields as we did in previous chapters, in this chapter we instead select a number of particular topics to discuss in depth. These are:

1. Progress in observational, measurement, and analytical techniques and the maturation of structural chemistry
2. Progress in theoretical and computational chemistry
3. The increasing precision of chemical reaction studies
4. The development of new synthesis techniques and the discovery of new substances
5. Foundations of materials science: the chemistry of the properties and functionality of substances
6. Chemistry of the Earth, the environment, and outer space
7. The birth of, and chemistry of, molecular biology and structural biology
8. The development of biochemistry (I): The chemistry of DNA and RNA
9. The development of biochemistry (II): Enzymes, metabolism, and molecular physiology

Topics (1)–(6) above are discussed in this chapter, while topics (7)–(9) are discussed in the following chapter. With such a survey we cannot hope to convey the full complexity of the development of chemical science in the latter half of the 20th century, but we can offer a fair outline of the flow of progress in fundamental chemistry. Even the mere task of assembling such a broad perspective on this range of topics proved a challenging task for a researcher trained in the single specialized subfield of physical chemistry, such as this author. For this reason, in assembling the content of this chapter, we have placed particular emphasis on research recognized by the Nobel Prize. Needless to say, the depth in which various topics are addressed reflects this author's particular taste and expertise.

Although the range of applications to which chemistry was put to use in the second half of the 20th century is unfathomably vast, and although the growth of the chemical industry transformed daily human life in this era, in this chapter we hardly touch on these developments. A detailed discussion of the postwar development of the worldwide chemistry industry may be found in the book by Aftalion[3].

5.2 Progress in observational, measurement, and analytical techniques and the maturation of structural chemistry

Techniques for observing atoms and molecules underwent explosive growth after the Second World War. These techniques were made possible first and foremost by advances in electronics and computer technology. Before the war, progress in wireless and radio engineering, as well as in communications applications and radio, had centered on vacuum-tube technology. However, the wartime development of radar methods for military applications spurred advances in electrical engineering and saw the birth of techniques for producing electromagnetic waves spanning a wide frequency range—from shortwave to microwave—as well as techniques for detecting and amplifying weak signals. The field of electrical engineering, as a discipline firmly rooted in physics, continued to use vacuum tubes through the 1950s, but from the 1960s onward the vacuum tube was replaced by the transistor, a device based on the semiconductor germanium and invented in 1948 by Bardeen, Brattain, and Shockley at Bell Laboratories in the U.S. Later, the development of the integrated circuit (IC) allowed the miniaturization of electronic circuitry and produced devices of high sensitivity and high reliability. In 1960, Maiman invented the ruby laser, and later other types of lasers were developed; these devices offered intense monochromatic beams of light with outstanding directivity—over a wide range of wavelengths of light—and were quickly adopted as light sources. These new technological achievements were immediately introduced into chemistry as tools for observation and analysis and led to revolutionary changes in methods of experimental chemistry.

Although various attempts to construct computers were made starting around the 1940s, it was in 1945 that the stored-program model of computer commonly used today was proposed by von Neumann. EDSAC, the first computer, was built at the University of Cambridge in 1947. At first these machines used vacuum tubes in the electronic circuits, and thus were large, consumed large amounts of energy, and had limited computational speed. The switch to transistor-based circuits starting in the early 1960s allowed miniaturization and higher speeds. The subsequent development of ICs and microprocessors allowed the further miniaturization and acceleration of computers, yielding rapid and dramatic improvements in computational speed and reliability. The trends of miniaturization, increased speed, and cost reductions continued in the 1980s, and computers came to be ubiquitous in chemistry laboratories, used both to control observational and measurement equipment and to process data. Analyses requiring enormous numbers of computations were made possible by the use of computers, enabling chemistry research to become even more precise.

In this section we survey advances in a variety of observational and analytical methods made possible by these technological developments in the second half of the 20th century. In many cases, new techniques of observation and analysis were invented and developed by physicists, but then immediately adopted by chemists for research in chemistry. Instrumental analysis had become important in analytical chemistry already by the early 20th century, but in the later part of the century this area incorporated many new methods of measurement and analysis and matured to a remarkable degree of sophistication. In the first part of the 20th century, most physical and analytical chemists were still using hand-made equipment in their research; however, the second half of the century witnessed the emergence of many industrial firms specializing in the development, manufacture, and sale of research instrumentation, upon which researchers have come to depend.

Progress in methods of structural analysis: Structural determination via diffraction techniques

Accurate determination of the atomic-level structure of molecules and solids is perhaps the most fundamental topic in modern chemical research, and the single structural-analysis technique that has made the greatest contributions to this subject is the diffraction of X-rays, electron beams, neutron beams, and other energy beams by the substance in question. As noted in the previous chapter, these analytical methods emerged in the first half of the 20th century, but it was not until the second half of the century that general-purpose machines—allowing convenient X-ray structural analysis—became available for sale, whereupon such machines quickly became widely used and developed into essential research tools spanning a wide range of subfields of chemistry. Such machines eventually came to be used to determine the structures of complicated biological polymers, exerting a major influence on the development of the life sciences.

X-ray diffraction

Thanks to progress in analytical techniques, improvements in detector sensitivity, and technological advances such as the use of computers for data analysis, it gradually became relatively easy to perform X-ray structural analysis on simple molecules. Determining the structure from the diffraction pattern requires information on both the intensity and phase of the diffracted X-rays; determining the phase in particular was a difficult problem, but it gradually came to be solved in the 1950s and thereafter by a number of clever innovations such as the direct method, the multi-wavelength anomalous diffraction method, and the heavy-atom substitution method. Herbert Hauptman and Jerome Karle established the direct method—in which mathematical methods are applied to the intensity distribution of the scattered X-rays to determine their phase—which allowed automated X-ray structural analysis of relatively simple molecules.

In the structural analysis of complicated molecules, the deciphering of data from large numbers of diffraction points remained an extremely difficult challenge even in the postwar era. However, improvements in the performance of computers gradually made it possible to solve this problem. In addition, synchrotron radia-

Herbert Hauptman (1917–2011)
An American crystallographer, Hauptman graduated from City College of New York and joined the U.S. Naval Research Laboratory after the Second World War. Together with Karle, he established methods for mathematically analyzing X-ray diffraction images of crystals to determine their molecular structure. He received the 1985 Nobel Prize in Chemistry.

Jerome Karle (1918–2013)
Karle was an American chemist and crystallographer. After graduating from City College of New York, he received a doctorate in physical chemistry from the University of Michigan. At the Naval Research Laboratory he collaborated with Hauptman to establish methods for mathematically analyzing X-ray diffraction images of crystals to determine their molecular structure. He received the 1985 Nobel Prize in Chemistry.

tion[1] became available as a source of X-rays to complement traditional methods that utilized the collisions of electrons with metal surfaces. The use of synchrotron radiation requires large-scale research facilities equipped with electron accelerators; in the 1980s, such facilities became increasingly available in advanced nations, providing X-ray sources with intensities far in excess of what had been possible before. Thus the structural analysis of complex organic compounds and biological macromolecules made remarkable strides, and a new domain of research known as structural biology was born, which was to make major progress as an interdisciplinary field straddling the boundary between physics, chemistry and biology. A major problem facing this field was the difficulty of obtaining crystals of sufficiently large sizes and high quality to enable analysis. Although much ingenious effort has been directed toward solving this problem, the challenge of growing crystals in a precisely controlled fashion remains unsolved, and the process still requires significant experience and relies heavily on trial and error.

Progress in the structural analysis of organic compounds proceeded apace, including the determination by Hodgkin and collaborators of the three-dimensional structure of penicillin in 1949 (Figure 5.1). Hodgkin and coworkers also determined the complicated structure of vitamin B_{12} in 1956 (Figure 5.14), and subsequent years witnessed a string of impressive successes in the field of the structural determination of natural organic compounds[14]. Attempts at X-ray

---

1. Synchrotron radiation: When high-energy electrons execute circular motion in a magnetic field, their acceleration—directed toward the center of the orbit—produces electromagnetic waves known as synchrotron radiation. Synchrotron accelerators offer powerful sources of light over a wide range of wavelengths, from the vacuum ultraviolet to the X-ray regime.

Dorothy Crowfoot Hodgkin (1910–1994)
After graduating from Oxford University, Dorothy Crowfoot began research in X-ray crystallography with Bernal at Cambridge in 1932; she became a lecturer at Oxford in 1934. In 1947 she became the first female member of the Royal Society. From 1960 to 1977 she was professor at Oxford. She achieved extraordinary accomplishments in the structural determination of molecules relevant to medicine and biology using X-ray diffraction. She received the 1964 Nobel Prize in Chemistry. (Column 17)

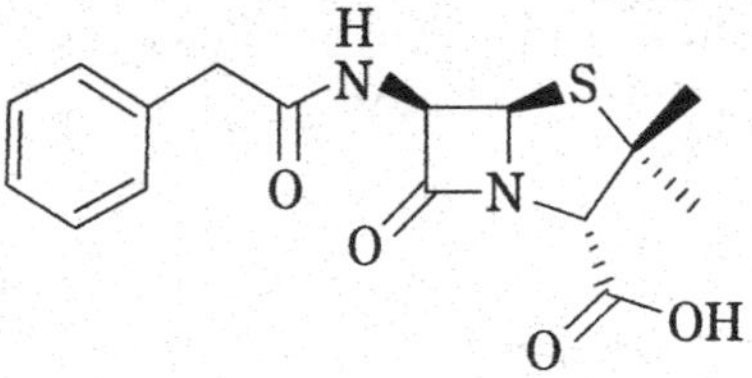

Figure 5.1: Structural formula of penicillin

structural analysis of globular proteins began in 1935 with the discovery of the diffraction pattern of pepsin crystals by Bernal and Crowfoot (whose married name was Hodgkin) at Cambridge[15]. Max Perutz, a student of Bernal, began in 1937 to analyze the structure of hemoglobin crystals. After the war, Perutz continued research on hemoglobin at Cambridge, where John Kendrew also joined in the effort and began studies of myoglobin. However, the problem of determining the phase remained unsolved, ensuring that research proceeded slowly. In 1953, Perutz discovered that replacement of mercury or silver atoms at specific points in the hemoglobin molecule left its structure unchanged (heavy atom isomorphous replacement). By comparing the diffracted intensity from such heavy atom replaced molecules with that from molecules that were not replaced, it was possible to solve the problem of the phase[16, 17]. Perutz's colleague Kendrew used similar isomorphous replacement method to obtain the low resolution structure of myoglobin in 1958, and again with a resolution of 2 Å in 1960[18]. The structural analysis of hemoglobin was more difficult, but Perutz announced the results of a low-resolution analysis in 1959 (Figure 5.2). Perutz finally announced the results of a high-resolution structural analysis of hemoglobin in 1968, some 30 years after he had begun research in this area. Perutz and Kendrew received the 1962 Nobel Prize in Chemistry for their research on the structure of globular proteins.

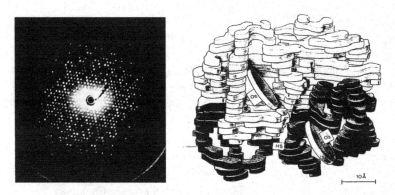

Figure 5.2 X-ray diffraction pattern of single-crystal hemoglobin (left) and the first low-resolution three-dimensional structure of hemoglobin obtained by Perutz (right).

The remaining years of the 20th century witnessed more Nobel Prizes awarded for research in the field of X-ray structural analysis. Oxford's Dorothy Hodgkin (Column 17) was awarded the Prize in 1964 for achievements in the structural determination of biologically relevant complex organic molecules. She was the third female recipient of the Nobel Prize in Chemistry, following the mother-daughter pair of Marie Curie and Irene Joliot-Curie. William Lipscomb, who determined the structure of many boranes and carboranes and pioneered new fields of inorganic chemistry, received the Prize in 1976, while Herbert Hauptman and Jerome Karle, who made major contributions to solving the phase problem and established the direct method of crystal structure determination, were awarded the Prize in 1985. Johann Deisenhofer, Robert Huber, and Hartmut Michel received the Prize in 1988 for determining the three-dimensional structure of a protein complex that plays a central role in photosynthesis reactions, thus contributing to our understanding of the initial stages of the photosynthesis process.

X-ray structural analysis also played a decisive role in the 1953 proposal by Watson and Crick of the structure of DNA. Later, the 1980s saw rapid progress in X-ray structural analysis of nucleic acids, proteins, and enzymes, giving rise to yet another batch of Nobel Chemistry laureates. We will discuss this work in the following chapter.

Diffraction of electron and neutron beams
Beams of electrons or neutrons have a de Broglie wavelength given by $\lambda=h/mv$ (Chapter 3, see page 215); they display diffraction phenomena similar to those of X-rays, and provide an important source of structural-analysis information,

Max Perutz (1914–2002)

Born in Austria, Perutz moved to England after graduating from the University of Vienna, where he studied at the Cavendish Laboratory in Cambridge, first with Bernal (from 1937 to 1939) and with William Lawrence Bragg (after 1939), focusing on the X-ray structural analysis of hemoglobin. In 1958 he reported a low-resolution analysis of the three-dimensional structure of the hemoglobin molecule. In 1962 he became the first director of the molecular biology research laboratory at Cambridge. From 1963 to 1969 he was chairman of the European Molecular Biology Laboratory. He received the 1962 Nobel Prize in Chemistry.

John Kendrew (1917–1997)

After graduating from Cambridge University, Kendrew studied military strategy at Royal Air Force Headquarters during the Second World War. After the war he returned to Cambridge, where he studied under Bragg at the Cavendish Laboratory, working with Perutz on the structural analysis of proteins using X-ray diffraction. He pioneered the methodology and the theoretical basis of X-ray analysis methods for crystallized proteins. In 1958 he reported a three-dimensional model of myoglobin. He received the 1962 Nobel Prize in Chemistry.

William Lipscomb (1919–2011)

After receiving his doctorate from the California Institute of Technology in 1946, Lipscomb worked as a professor at the University of Minnesota before joining the faculty at Harvard as a professor. From the X-ray structural analysis of boranes, he clarified the properties of two-electron bonds among three atoms, thus contributing to solving outstanding problems in chemical bonding. He received the 1976 Nobel Prize in Chemistry.

complementary to that obtained from X-ray diffraction. The phenomenon of diffraction of an electron beam was discovered in 1927 by the Americans Davisson and Germer and by the British scientist G. P. Thomson; however, the difficulty of the techniques involved ensured that application of this method to structural analysis initially progressed slowly. In the 1960s, high-vacuum technology advanced, and electron diffraction became a widely-used tool in structural analysis. In comparison to X-rays, electrons are more strongly scattered and absorbed by atoms, making them appropriate for studies of thin films and solid surfaces. In particular, low-energy electron diffraction (LEED), which observes the diffraction

Clifford Shull (1915–2001)
Shull was an American physicist. After graduating from the Carnegie Institute of Technology, he received his doctorate from New York University. He worked at Oak Ridge National Laboratory from 1946 to 1955 and was a professor at the Massachusetts Institute of Technology (MIT) between 1955 and 1986. At Oak Ridge he established the technique of neutron diffraction, in which the atomic structure of a substance is investigated by analyzing the scattering of a monoenergetic neutron beam by that substance. He received the 1994 Nobel Prize in Physics.

of relatively low-energy electrons, is a powerful tool for research on the structure of solid surfaces and structures adsorption structure, and it is widely used. Electron-beam diffraction is a tool that provides information complementary to that obtained by X-ray structural analysis of complex biological macromolecules, and since the 1970s has become one of the most important techniques in the study of such molecules.

Studies of the diffraction of neutron beams were begun in 1945 by Wollan at Oak Ridge National Laboratory in the U.S. He was later joined by Clifford Shull, and together they developed the basic techniques. Observation of diffraction phenomena requires high-density neutron beams, and for this purpose pulsed neutron beams obtained from atomic reactors or accelerators are used. Neutrons are scattered by electrons or atomic nuclei with magnetic moments; in contrast to the scattering of X-rays, the scattering of atomic nuclei is not proportional to the atomic number, but is dependent on the isotope. This makes the tool useful for determining the position of light atoms—particularly hydrogen—and relevant for studies not only of crystals but also non-crystalline solids, fluids, and biological macromolecules. In addition, phenomena related to the magnetic scattering of neutrons constitute powerful tools for determining the magnetic structure of magnetic crystals. Shull, and the Canadian physicist Brockhouse, received the 1994 Nobel Prize in Physics for their pioneering achievements in neutron-scattering studies, some 50 years after the initial discovery of the diffraction of neutron beams.

Dramatic progress in microscopy techniques: Direct observation of cells and surface atoms and molecules

The most direct method of obtaining knowledge of structures on the microscopic scale is to use observational techniques that allow atoms and molecules to be visualized by eye. This goal was achieved due to revolutionary advances in microscopy techniques in the second half of the 20th century. According to the theory of Abbe, established in the late 19th century, the

COLUMN 17

Dorothy Hodgkin and the structural determination of complex molecules[19]

Molecular structure of monomeric insulin. The sphere denotes a zinc ion.

Dorothy Crowfoot Hodgkin, a pioneer in the use of X-ray diffraction methods for the structural determination of proteins and other biologically important molecules, followed Marie and Irene Curie to become the third woman to receive the Nobel Prize in Chemistry. Of all the molecules whose structure she determined, those considered most important are penicillin, vitamin B_{12}, and insulin; at the time she embarked on the problem of determining the structure of these substances, the project was believed by other researchers to be essentially impossible. But Hodgkin, who was possessed of a brilliant imagination and an iron determination, invented new methods that pushed back the limitations of X-ray crystallographic techniques of her day and surmounted difficulties with a remarkable degree of resilience. Although her achievements were undoubtedly made possible by her extraordinary natural talents, the environment in which she grew up most likely played an additional role in cultivating her creativity and doggedness.

Dorothy Crowfoot was born in 1910 in Cairo, Egypt, then under British rule. Her father, a classicist and archaeologist, was the director of a school and of an excavation site in Egypt. Although her mother had not received academic training, through self-study she had taught herself the science of botany and had become an expert in ancient tapestries. When Dorothy was 4, the First World War broke out while the entire Crowfoot family was on vacation in England; her parents later returned to the Middle East, but Dorothy and her sisters were entrusted to the care of their tutor, and for the entire ensuing 4-year duration of the war she only saw her mother once. One cannot help but wonder if this unusual upbringing helped to foster Dorothy's independent spirit.

At the age of 10, Dorothy studied elementary chemistry at school, and it was here that she first grew crystals and became entranced with their beauty. Developing a fascination with chemistry, she joined her male classmates in attending chemistry classes that female students typically did not take. For her 16th birthday, she received from her mother the book *Concerning the Nature of Things* by Henry

Bragg; this further stoked her interest in crystals. At the age of 18 she rejoined her parents in Palestine, where she participated in survey excavations with her father and developed an interest in archaeology. Upon entering university she was torn between choosing to study chemistry or archaeology, but she eventually entered Oxford's Somerville College and majored in chemistry and physics.

At Oxford, after taking a special course on crystallography, Dorothy was encouraged to take up research on X-ray crystallography. However, Somerville College at that time was not a conducive atmosphere for such a pursuit. In 1932, the year she graduated, Dorothy's childhood friend Joseph paid a visit to Oxford. On the train, he had accidentally bumped into Cambridge physical chemistry professor Lowry, with whom he discussed Dorothy's future. Lowry suggested that Dorothy begin research with Bernal at Cambridge. In this way, Dorothy came to embark on a two-year course of research at Cambridge. In 1934, she captured X-ray diffraction images of pepsin, and she and Bernal thus became the founders of the field of X-ray crystallography of globular proteins. Dorothy later became a lecturer at Oxford, and in 1937 she received her doctorate from Cambridge for the structural analysis of sterol. That year she married the historian Thomas Hodgkin, with whom she later had three children.

As her impressive skills as a crystallographer became known, new types of crystals made their way to her door. From the famous organic chemist Robinson she received crystals of zinc insulin, which she began to study; however, the structure of insulin is particularly complex, and she would not successfully unravel it until some 30 years later, after she had already received the Nobel Prize. When demand for penicillin spiked during the Second World War, the determination of the structure of this molecule became urgent; in the summer of 1942, Dorothy received crystals of penicillin from Chain and began work to analyze its structure. This was a more complex structure than any of the organic molecules that had been analyzed by X-ray diffraction up to then; nonetheless, the basic structure was clear by 1945, and the full three-dimensional structure was reported in 1949. The next project she undertook—the structure of vitamin B_{12}, a problem related to malignant anemia (see page 531)—was even more complicated, and its calculations required the use of computers; the structural analysis was completed in 1957. For these achievements, Dorothy Crowfoot Hodgkin received the 1964 Nobel Prize in Chemistry.

Greatness as a scientist and basic decency as a human being do not always coincide, but Dorothy was loved and appreciated as a person by many people. Max Perutz wrote that "she will be remembered as a great chemist, a saintly, gentle and tolerant lover of people, and a devoted protagonist of peace." Incidentally, the British Prime Minister Margaret Thatcher was a student of Hodgkin at Oxford; although the two disagreed on political matters, they appear to have had great respect for one another.

resolution of an optical microscope is determined by the spread of the image point due to diffraction, and is typically around one-half the wavelength. For visible light this is approximately 0.2 µm, some 1000 times larger than the resolution required for direct observation of atoms and molecules. However, in the second half of the 20th century, techniques in optical microscopy made major advances, aided by progress in lasers, in methods for detecting ultra-weak light signals, and in computers. In particular, techniques of hyper-resolution fluorescence microscopy for observing single molecules using fluorescence microscopy underwent impressive development and had a major impact on biochemistry, cellular biology, and medicine.

In electron microscopy—in which electron beams are used in place of light—the fact that the de Broglie wavelength of the electron shrinks as its energy increases allows electron beams to achieve wavelengths sufficiently short to observe atoms. The electron microscope was first developed in the 1930s (see page 257), but it was not widely used until after the Second World War. During the 1950s and thereafter, electron microscopy came to be used in a wide range of fields spanning physics, chemistry, materials science, biology, and medicine, and technological advances increased the resolution to the point that large molecules could be directly observed. In the last 20 years of the 20th century, new microscopes based on new physical principles were developed, allowing atomic-level observation of solid surfaces and other targets. These developments had an enormous impact on chemistry, and recent advances in nanoscience and nanotechnology have been made possible by progress in microscopy techniques.

Advances in optical microscopy

The 20th century witnessed a variety of clever innovations in optical microscopy that further enhanced its usefulness. Phase differences arise when light passes through materials with different indices of refraction. In 1932, Fritz Zernike utilized this phenomenon to develop the phase-contrast microscope. Undyed cells and microorganisms are generally transparent and offer essentially no contrast, making them difficult to observe; taking the phase difference as the contrast allows observation. Zernike was awarded the 1953 Nobel Prize in Physics for this achievement. Other types of microscopy—including polarization microscopy, which uses the polarization of light, and differential interferometry, which uses both polarization and interference phenomena—became available and found uses in various types of application.

Tools such as fluorescence and Raman scattering came to be used for microscopy observations. Fluorescence microscopy was an old technique whose history dated back almost a century, but it became more widely used in the second half of the 20th century. In particular, use of confocal laser microscopy

Fritz Zernike (1888–1966)
A Dutch physicist, Zernike taught at Groningen University from 1920 to 1958. Based on his research on diffraction lattices, in 1934 he established the principle of phase contrast, and in 1938 he worked with Zeiss to develop the phase-contrast microscope, which enabled observation of the internal structure of cells. He received the 1953 Nobel Prize in Chemistry.

spread in the 1980s and thereafter. In one method, tiny regions of a sample are excited with laser light, their fluorescence is observed, and a computer is used to reconstruct the observational data, enabling high-resolution three-dimensional imaging. This technique significantly enhanced the usefulness of fluorescence microscopy measurements. In addition to using the inherent fluorescence of a sample, other methods for observing fluorescence were also discovered and put to use, including staining samples with fluorescent elements, labeling samples with fluorescence probes, and gene replacement to produce fluorescent proteins.

Single-molecule spectroscopy
What we observe in a typical spectroscopy experiment is the statistical average behavior of an aggregate (ensemble) of atoms or molecules. By directly observing single atoms and molecules, we glean information about the behavior and properties of individual molecules that are hidden in the statistical average. Since the late 1980s, it has been possible to use single-photon techniques to observe fluorescence from individual molecules in condensed systems; the processes of light emission by individual molecules have been clearly distinguished and measured. The first report of single-molecule spectroscopy was that of Moerner and collaborators in 1989, who investigated fluorescence from pentacene[2] in an ultra- low-temperature matrix and observed sharp single-molecule emissions embedded in a non-uniform band. The authors studied the details of spectral dispersion and linewidths due to variations in the local environment[20]. Later, similar techniques were applied to dendrimers (see page 551) and polymers containing different numbers of chromophores; this enabled the study of phenomena such as the transition from single-molecule fluorescence to ensemble fluorescence, and energy transport in the initial stages of photosynthesis.

Subsequently it became possible to label the chromophores of proteins and other biological macromolecules and observe the resulting fluorescence, thus

2. Pentacene: An aromatic hydrocarbon consisting of 5 benzene rings connected in a straight-line configuration.

Stefan Hell (1962–)
A German physicist, Hell received his PhD from the University of Heidelberg and is currently a Director of the Max Planck Institute for Biophysical Chemistry in Göttingen; he also leads a department at the German Cancer Research Center at Heidelberg University.

Eric Betzig (1962–)
An American physicist and engineer, Betzig received his PhD from Cornell University in 1988. After working at Bell Laboratories, he briefly worked outside of academia; since 2006 he has been a group leader at the Janelia Research Campus of the Howard Hughes Medical Institute.

William E. Moerner (1953–)
An American physical chemist, Moemer received his PhD from Cornell University, then worked at IBM and was a professor at the University of California at San Diego before becoming a professor of chemistry and applied physics at Stanford University in 1988.

yielding molecular-level information not available from observations of bulk samples. This spurred interest in using such information to understand changes in the structure and function of enzymes. Observations of fluorescence in biological macromolecules at room temperature developed into the method of three-dimensional fluorescence imaging of single molecules using a confocal microscope; this technique was to have a major impact on the life sciences.

Super-resolved fluorescence microscopy
Optical microscopy techniques advanced in significant ways in the late 20th and early 21st centuries. Ingenious new techniques—including stimulated emission depletion (STED) and photoactivated localization microscopy (PALM)—were invented to allow observations in the nanometer regime, exceeding the 0.2 µm limit derived by Abbe for the resolution of an optical microscope; this led to the development of ultra-high-resolution microscope technology and ushered in the possibility of nano-scale microscopy. STED utilizes both an exciting laser and an extinguishing laser—which depletes the population of excited molecules via stimulated emission—to observe only a portion of an excited region; the full sample is then scanned to yield high-resolution images.

In PALM, the sample is excited by a weak optical excitation and a snapshot image is captured, indicating fluorescence from well-separated individual molecules distributed throughout the sample; by repeating such measurements a large number of times and superposing the images thus obtained, one yields a high-resolution image of the full sample. These techniques, which have made it possible to observe the motion of molecules in cells at the level of individual molecules, have become extremely important tools for cutting-edge research in the life sciences. The 2014 Nobel Prize in Chemistry was awarded to Hell (the developer of STED), Betzig (the developer of PALM), and Moerner (who developed the concepts and techniques underlying single-molecule spectroscopy) for "the development of super-resolved fluorescence microscopy."

Green fluorescent protein (GTP) and fluorescence imaging
One technique is particularly useful for fluorescence imaging of biological macromolecules: green fluorescent protein (GFP)—which was discovered, isolated, and refined from Aequorea victoria by Osamu Shimomura (Column 18) in 1962[21]. In the 1990s, Prasher and collaborators identified the GFP gene, while the groups of Chalfie and Tsien introduced and observed GFP in cells of different species. In addition to the naturally-occurring forms of GFP, genetic engineering was used to construct mutant forms that offered different fluorescence characteristics and wavelengths. In addition, as the details of GFP fluorescence excitation was revealed, controlling GFP by fluorescence excitation became possible and resulted in the widespread use of GFP and related molecules in cellular biology, developmental biology, nerve-cell biology (neurobiology), and medicine. GFP allows non-invasive and real-time detection and can fuse with other proteins and operate as a fused protein, offering an extremely valuable tool for tracing the motion, position, and function of proteins.

Shimomura, who discovered GFP, together with Martin Chalfie—who demonstrated the value of GFP for applications to biology—and Roger Tsien, who developed fluorescent proteins with different fluorescence characteristics, were awarded the 2008 Nobel Prize in Chemistry for the discovery and development of green fluorescent protein. The evolution of GFP from discovery to development is an excellent example of how research born solely out of intellectual curiosity can bear fruits that develop into a wide range of useful applications.

Electron microscopy: TEM and SEM
As noted in the previous chapter, the first successful attempt to use an electron beam instead of light to obtain magnified images of structures was made in 1931 by the electrical engineers Ruska and Knoll at the Technical University of Berlin (see Section 4.3)[23]. However, it was not until after the Second World

Osamu Shimomura (1928–)

Shimomura was born in the city of Fukuchiyama, near Kyoto, Japan, and studied at the College of Pharmaceutical Sciences of Nagasaki Medical College. After working as a trainee instructor in the pharmacology department of Nagasaki University, in 1955 he became a researcher in the laboratory of Yoshimasa Hirata at Nagasaki University. In 1956 he succeeded in refining and crystallizing luciferin from sea fireflies (Vargula hilgendorfii). In 1960 he moved to the U.S. and became involved in research on the light-emission mechanisms of the jellyfish Aequorea victoria at Princeton University. He discovered green fluorescent protein (GFP) in 1962. From 1982 to 2001 he was a researcher at Woods Hole Oceanographic Institution. He received the 2008 Nobel Prize in Chemistry.

Martin Chalfie (1947–)

After receiving a doctorate in neurobiology from Harvard in 1977, Chalfie joined the faculty at Columbia, where he became a professor in 1982. He developed methods for inserting GFP genes into other types of cells, and those cells then synthesized GFP. He received the 2008 Nobel Prize in Chemistry.

Roger Tsien (1952–)

After studying at Harvard, Tsien received his doctorate in physiology from Cambridge. He worked at the University of California at Berkeley before joining the faculty of UC San Diego in 1989. He identified the relationship between the structure of GFP and the color emission resulting from modifications of its gene, and he developed proteins that fluoresce in colors other than green. He received the 2008 Nobel Prize in Chemistry.

War that the electron microscope would come to be widely used in research. The first electron microscopes were called transmission electron microscopes (TEMs), and their foundation was laid in the late 1930s by Ruska and Borries. After the war, Ruska continued to develop the TEM at Siemens, where he worked on improving the performance of the apparatus. In 1954, the "Emiskop 1" machine, which would come to be used in over 1,200 laboratories around the world, was first produced.

In TEM, a beam of electrons emitted by an electron gun is accelerated by a high voltage, focused by lenses formed by static electric and magnetic fields, and applied to a sample; the transmitted beam is enlarged by an electron lens and observed. Typical electron acceleration voltages are in the range 100—200 kV, although high-voltage machines using voltages between 500 kV and 3 MV

COLUMN 18

The life of Osamu Shimomura and the discovery of GFP: A wealth of serendipitous blessings[22]

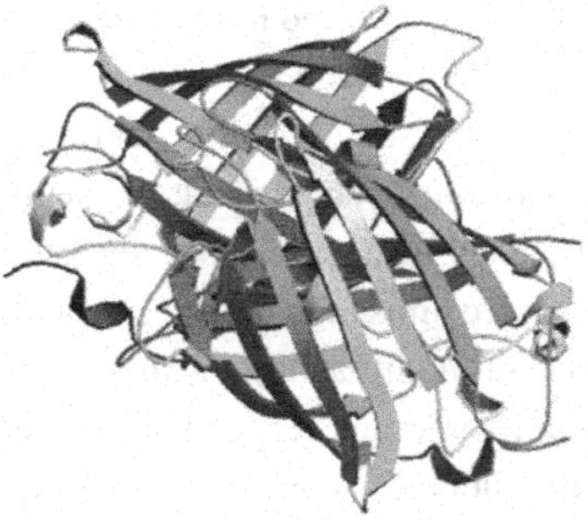

Molecular structure of GFP (ribbon model)

Fluorescence imaging using green fluorescent protein (GFP) and related molecules has evolved into a method for temporally- and spatially- resolved observation of the behavior of biological molecules at the single-molecule level. This approach has become ubiquitous across a wide range of fields in biochemistry, biology, and medicine. GFP was discovered by Osamu Shimomura in the course of research to determine the mechanism of light emission by the jellyfish *Aequorea victoria.* A look back at the process that led to this discovery is fascinating as a prototypical example in which research begun purely to answer questions in basic science wound up developing into large-scale applied research. Moreover, the life of Shimomura was itself filled with a series of fortuitous events, ranging from a boyhood turned upside down by war to his good fortune in becoming a researcher and making a major discovery. The story is a vivid reminder of how frequently scientific pursuits—and human lives—are dramatically impacted by chance events. We begin by surveying Shimomura's life.

Shimomura, the son of a professional soldier, was born in 1928 in Fukuchiyama City in Kyoto Prefecture, Japan. His father's career took the family to the former Manchuria (then a Japanese colony), to Sasebo near Nasagaki, and to various destinations in the Osaka area. As a middle-school student in Japan's prewar education system, Osamu evacuated to his mother's family home in Isahaya City in Nagasaki Prefecture, where he experienced the atomic bombing of the city from a distance of 20 km. After completing middle school, he attempted to proceed to high school, but was rejected after failing to receive any recommendations—due to his complete refusal to study in middle school. Instead, he entered the College of Pharmaceutical Sciences at Nagasaki Medical College, which had relocated near his home due to the atomic bomb blast. After graduation, he applied for a job at Takeda Pharmaceutical, but was rejected by his interviewer, who informed him that he was ill-suited for corporate work.

After failing his job-application exams, Shimomura found employment with Professor Shungo Yasunaga at Nagasaki Medical College, where he worked for four years as an instructor of experimental practice. In the spring of 1955, Professor Yasunaga visited Nagoya University in the hope of securing a place for Shimomura as a research student under Professor Fujio Egami, a biochemist; however, Professor Egami was absent that day. The man who took his place—Professor Yoshimasa Hirata—invited Shimomura to work with him instead. As a result of this unplanned circumstance, Shimomura became a researcher under Hirata, and it was from this entirely serendipitous day forward that Shimomura's luck began to change.

The research topic assigned by Hirata was "to refine and crystallize luciferin from sea fireflies (*Vargula hilgendorfii*.) This was an extremely difficult problem, and one which the group of Professor Frank Johnson at Princeton University had been trying to solve for 20 years. Shimomura immersed himself in his research and crystallized the protein in 10 months. The fact that Shimomura, despite his lack of training and having only a fundamental education in organic chemistry, was able to pull off this remarkable accomplishment testifies not only to his natural talents but also to his diligence. His results were published in an English-language paper in 1957, where they were read with great interest by Professor Johnson at Princeton—who invited Shimomura to work with him. Shimomura arrived at Princeton in 1960 as a foreign-study student supported by a Fulbright scholarship, and it was here under Johnson's guidance that he began research on the light-emission mechanism of the jellyfish *Aequorea victoria*.

Shimomura and Johnson gathered large numbers of jellyfish on the Pacific coasts of North America. They cut off the light-emitting regions of the jellyfish and compressed and filtered these to obtain a liquid. They extracted a small quantity of protein from this liquid that emitted blue light, which they purified and named *aequorin*. It became clear that aequorin emitted light in the presence of Ca^{2+} ions. Later, the two researchers conducted further studies during the course of which they isolated a different protein which, under irradiation with ultraviolet light, emitted green fluorescence. This was later termed *green fluorescent protein* (GFP). In the 1970s, Shimomura found that GFP contained special chromophores, and that aequorin formed a complex with GFP; light absorbed by aequorin excited the GFP chromophores via migration of resonant energy, causing fluorescence. Thus the mechanism by which the jellyfish emitted light was identified. This was already a spectacular result in the study of biological light emission, but the discovery was later to have an even larger impact on a wide range of fields spanning the natural sciences.

In the 1990s—as discussed elsewhere in this book (see page 427)—research on GFP made major strides, and its range of applicability broadened significantly. In a development that could hardly have been imagined at the time of its discovery, the method evolved into a revolutionary new tool allowing molecular-level tracking of biological processes inside living cells.

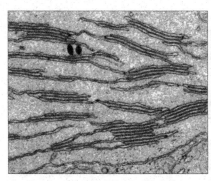

Interior of the *Arabidopsis* chloroplast   Tobacco mosaic virus

Figure 5.3: Examples of electron microscope images

now exist. The resolution is determined by the aberration of the objective lens and the wavelength of the electron beam; recent technological advances have made it possible to achieve resolutions of 0.2—0.3 nm, close to the theoretical limits and sufficient to observe individual atoms. Ruska, the developer of the TEM, received the 1986 Nobel Prize in Physics (see page 257).

An apparatus well-suited to the observation of surfaces is the scanning electron microscope (SEM). SEM images were first observed in 1935 by Knoll, but the technology did not become widely used for research until the 1960s. In SEM, a focused electron beam is scanned across the surface of a sample, the electrons emitted from each scanned point are detected, and these are amplified and synchronized with the scanner to produce an image.

Electron microscopy had a major impact on biology, allowing direct observation of viruses and cellular organelles (Figure 5.3). However, one problem was that observation of biological samples typically yielded poor image contrast because the molecules that compose biological structures are formed from light atoms such as hydrogen and carbon. In the late 1960s, Aaron Klug and collaborators developed a method in which two-dimensional electron microscope images are obtained from different angles, then merged and rearranged by computer to obtain three-dimensional images[24]. Using this method, these researchers successfully determined the structure of the tobacco mosaic virus, chromatins consisting of complexes involving proteins with tRNA or DNA, and other complicated biological macromolecules for which structural determination by X-ray diffraction would be difficult. Klug received the 1982 Nobel Prize in Chemistry for the development of electron-beam crystallography and for contributions to the understanding of the structure of biologically important complexes of proteins and nucleic acids.

Aaron Klug (1926–)

Klug is a British chemist who was born in Lithuania and grew up in South Africa. After graduating from university in South Africa, Klug went to England, where he completed his doctorate at Cambridge. Later, he collaborated with Franklin at Birkbeck College at the University of London and eventually became a researcher at the MRC Laboratory of Molecular Biology at Cambridge University, where he was named head of the laboratory's division of structural studies in 1978. In 1995 he became President of the Royal Society. He achieved numerous successes in understanding the three-dimensional structure of nucleic acids in viruses and the proteins that surround them. In particular, he developed crystallographic electron spectroscopy using electron microscope images. He received the 1982 Nobel Prize in Chemistry.

Scanning probe microscopy: STM and AFM

Scanning probe microscopy is a technique in which a sharp tip is affixed to a piezoelectric element—offering position control with atomic-scale resolution in all three directions—and the tip is scanned over the surface of a sample to provide information on the atomic structure and electronic state of the surface. The two prototypical examples of this technique are scanning tunneling microscopy (STM) and atomic force microscopy (AFM). The STM was invented in 1981 by Gerd Binnig and Heinrich Rohrer at IBM Laboratories in Zurich[25]. In STM, a sharp metallic tip is maintained at a distance of approximately 1 nm from the surface of a metallic or semiconducting sample and the tunneling current[3] flowing between the tip and the sample is measured; by using a piezoelectric element to scan the tip over the surface of the sample, one observes the structure and electronic state of the surface. STM offers resolutions of 0.1 nm in the lateral directions and 0.01 nm in the vertical (depth) direction, thus yielding atomic-level information on surface structure. By further performing spectroscopy (tunneling spectroscopy) at fixed atomic positions, it is possible to obtain information on the electronic structure of the surface. STM allows measurements not only in high vacuum, but also in air or in liquids. For these reasons, STM rapidly developed into a powerful tool for surface science research.

In AFM, the magnitude of the atomic-scale forces acting between the tip and the sample surface are read off either from the deflection of a cantilever to which

---

3. Tunneling current: A current that arises at the surface of a metal, semiconductor, or insulator when electrons with kinetic energies lower than a potential barrier nonetheless pass through the potential barrier and escape.

Gerd Binnig (1947–)
Binnig is a German-born Swiss physicist. After completing his studies at the University of Frankfurt in 1978, he joined the IBM Zurich Research Laboratory, where in 1981 he worked with Rohrer to develop the scanning tunneling microscope (STM). The STM allows observations of individual surface electrons and has made major contributions to the development of surface science. In 1986, Binnig achieved  further success with the development of the atomic force microscope. He shared the 1986 Nobel Prize in Physics with Rohrer and Ruska.

Heinrich Rohrer (1933–2013)
Rohrer was a Swiss physicist. After completing his doctorate at the Swiss Federal Institute of Technology with research on superconductors, in 1963 he joined the IBM Zurich Research Laboratory, where he first studied magnetoresistance and critical phenomena such as magnetic phase transitions. In 1978 he recruited Binnig, with whom he collaborated to design the STM, achieving success in 1981. He received the 1986 Nobel Prize in Physics.

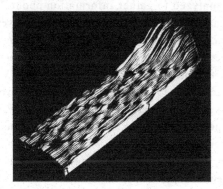

Figure 5.4: A silicon surface observed with an early STM
Note that steps with single-atom heights are visible.

the tip is mounted or from the shift in the resonance frequency of this cantilever. AFM was invented in 1986 by Binnig, Quate, and Gerber. Whereas STM is only applicable to the surface of conductors, AFM may be applied to insulators as well, and it is widely used today in research on surfaces. Binnig and Rohrer, the developers of the STM, shared the 1986 Nobel Prize in Physics with Ruska.

Scanning near-field optical microscopy

The resolution of an optical microscope is diffraction-limited to one-half the wavelength of the light. However, surface electromagnetic field (evanescent field) due to polarization fields induced on the surface of a body by incident light may be used to image and observe the optical properties of materials with resolutions of a few nanometers. This technique, known as scanning near-field optical microscopy (SNOM), was developed in 1984 by Pohl and colleagues. When a sharp tip on the order of several nanometers is brought to within a few dozen nanometers of a sample surface, the evanescent fields may be converted into propagating light or scattered light. The distribution of evanescent fields is determined by the material's electrical conductivity and surface structure. Thus by scanning the tip over the surface and measuring the propagating or scattered light, one obtains an image of the optical characteristics of the material surface. It is possible to achieve resolutions of 20 nm in the direction parallel to the surface; for this reason, the technique became commonly used to study nanoscale surface structure.

The emergence of lasers and the development of molecular spectroscopy: Observations of molecular structure and electronic state

Spectroscopic techniques—in which the spectrum of light emitted or absorbed by a material is analyzed to gather information about the structure and properties of the material—made major advances in the years following the Second World War. Spectroscopy generally requires a light source, a light detector, and a spectrometer for separating light into different frequencies. The light sources used for spectroscopy span a wide range of wavelengths, from the vacuum ultraviolet to the far infrared; on the short-wavelength end of the spectrum there was a surge of interest in X-ray spectroscopy, while on the long-wavelength end, spectroscopy began to be conducted at radio and microwave frequencies. The performance of various lamps and electron tubes used as light sources progressed, while the improved performance of photoelectric amplifiers and the development of sophisticated electronic techniques led to substantial increases in the sensitivity with which light signals can be detected. Progress in computer technology facilitated the signal processing of the detected light signal, spurring the introduction of new spectroscopic methods such as Fourier-transform spectroscopy. In addition, the increasing sophistication of analyses based on quantum mechanics allowed the field of molecular spectroscopy to enjoy major progress in the 1950s and thereafter.

However, the watershed event in the history of molecular spectroscopy was the emergence of the laser. Upon its invention in 1960, the laser was immediately adopted for chemical research, and progress in the application of lasers to a variety of problems in molecular spectroscopy continued through the 1970s and beyond. The arrival of the laser not only enabled high-resolution spectroscopy of complex molecules, but also allowed spectroscopic studies of short-lived radicals and excited states, clarifying their structure. In addition, the availability of intense monochromatic laser light gave birth to the new field of nonlinear spectroscopy. The pulse width of pulsed lasers shrunk from the regime of nanoseconds (10^{-9} s) in the 1960s, to picoseconds (10^{-12} s), femtoseconds (10^{-15} s), and finally to the attosecond (10^{-18} s) pulses available today, giving rise to the field of ultra-fast spectroscopy. In this section we survey progress in spectroscopy over a wide range of frequencies, from the microwave to the vacuum ultraviolet. Other techniques such as magnetic resonance—in which one observes the absorption or emission of electromagnetic waves associated with transitions between spin states in nuclei and electrons—and electron spectroscopy, in which one detects electrons, will be discussed in other sections.

Masers and Lasers

The term LASER is an acronym for Light Amplification of Stimulated Emission of Radiation; as the phrase suggests, the device works by amplifying the light arising from the stimulated emission of electromagnetic waves. The basic theoretical foundations were laid out in a 1917 paper by Einstein[26], but the device was not realized in practice until many years later. To achieve stimulated emission of light requires creating a population inversion, in which the occupation numbers of high-energy levels exceed the occupation numbers of low-energy levels (Figure 5.5). In 1953, Charles Townes and collaborators at Columbia University succeeded in creating a population inversion using a beam of ammonia molecules and in amplifying microwaves produced by stimulated emission; this was the invention of the maser, an acronym standing for Microwave Amplification by Stimulated Emission of Radiation[27]. However, the two-level maser developed by these researchers was not capable of continuous operation. Around the same time, Nikolay Basov and Aleksandr Prokhorov in the Soviet Union independently conceived of the idea of a maser, and in 1955 they invented a three-level maser capable of continuous operation. In later years the maser would come to be used for high-precision atomic clocks and frequency standards. For their fundamental contributions to the field of quantum electronics, Townes, Basov, and Prokhorov were awarded the 1964 Nobel Prize in Physics.

The theoretical possibility of obtaining similar amplification in the optical wavelength regime was presented in 1958 by Townes and Bell Laboratories' Arthur Schawlow. That same year, Prokhorov independently proposed the idea

Charles Townes (1915–)

Townes is an American physicist. Born in South Carolina, he completed his master's degree at Duke University and his doctorate at California Institute of Technology. In 1936 he joined Bell Laboratories, where he worked on radar research. In 1948 he became a professor at Columbia University, where he worked on microwave spectroscopy. In 1951 he conceived of the principles of the maser, which he successfully demonstrated in 1954. He further proposed the principles of lasers in 1958. He received the 1964 Nobel Prize in Physics.

Nicolay Basov (1922–2001)

Basov was a Russian physicist. During the Second World War he fought on the front lines in the Ukraine. After graduating from the Moscow Engineering Physics Institute in 1950, he joined the USSR Academy of Sciences, where he worked under Prokhorov. In 1952 they jointly proposed the principles of masers, and in 1955 they constructed a 3-level maser. He received the 1964 Nobel Prize in Physics.

Aleksandr Prokhorov (1916–2002)

Born in Australia to a family of revolutionaries who had fled imperial Russia, Prokhorov returned to Russia in 1923. After graduating from the Leningrad State University, in 1939 he entered the Lebedev Physical Institute in Moscow, where he conducted research on radio-wave propagation in ion layers. He was an infantry soldier in the Second World War. Around 1950, he began research on electromagnetic spectroscopy of molecular rotations and vibrations, and he worked with Basov to develop the maser. He also proposed optical resonators and contributed to the development of the laser. He received the 1964 Nobel Prize in Physics.

of the optical resonator[4]. The first successful laser was achieved in 1960 by the American, Theodore Maiman[28]. Maiman constructed a resonator from a ruby crystal with silver evaporated onto the surfaces of both ends, excited it by irradiation from a flash lamp to achieve a population inversion, and produced red laser light at a wavelength of 694.3 nm. The essential constituents of a laser are an optical resonator and a source of light excitation to produce a population

---

4. Optical resonator: A resonator for electromagnetic waves with frequencies in the optical regime. It is formed by arranging two reflecting surfaces to face one another to create a standing-wave configuration of light between them.

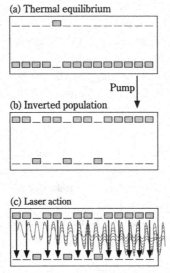

Figure 5.5: A schematic depiction of the principles underlying the generation of laser light

(a) The Boltzmann distribution that describes a thermal equilibrium state.

(b) A state in which a population inversion is present.

(c) A single emitted photon stimulates many subsequent emissions that amplify the light, as suggested by this cartoon.

inversion; these may be constructed using a variety of different media. In the years that followed various types of lasers were developed, including four-level lasers in which it was relatively easy to achieve population inversion (Figure 5.6). In 1961, Javan at Bell Labs developed the He-Ne gas laser (632.8 nm); this was followed in 1962 by the Nd:glass laser and the semiconductor laser, in 1964 by the CO_2 laser (10.6 μm) and the Nd:YAG laser (1064 nm) (Figure 5.7), and subsequently by many new types of lasers. 1966 saw the first dye laser with continuously tunable wavelength, while the first gaseous ultraviolet laser—the excimer laser—appeared in 1970. These instruments were promptly adopted for use in chemical research, where they quickly began to have a major impact. Excimer lasers in particular have found application in a wide range of areas including the modern semiconductor industry and vision correction (Lasik surgery).

By the 1970s it was possible to produce, over a wide range of wavelengths, laser light exhibiting several useful characteristics: 1) it was monochromatic with high output power, 2) it was phase-aligned with outstanding coherence, 3) its wavelength could be continuously tuned, 4) it was produced in ultra-short

Arthur Schawlow (1921–1999)

Schawlow was an American physicist. Though born in New York, he was raised in Toronto, Canada; after receiving his doctorate from the University of Toronto, he went to Columbia University, where he worked under Townes. From 1951 to 1961 he worked at Bell Laboratories, and from 1961 he was a professor at Stanford University. He received the 1981 Nobel Prize in Physics for his contributions to laser spectroscopy.

Nicolaas Bloembergen (1920–)

Bloembergen is a Dutch-born American physicist. After graduating from the University of Utrecht, he completed a doctoral degree at Harvard with research related to NMR relaxation; in 1957 he became a professor of physics at Harvard. From 1960 onward he conducted pioneering research in nonlinear spectroscopy. He received the 1981 Nobel Prize in Physics.

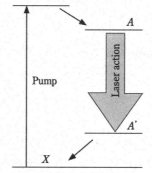

Figure 5.6 Energy levels in a 4-level laser

In this four-level laser, A and A' are both excited states, and it is easy to create a population inversion. Nd:YAG lasers are of this type.

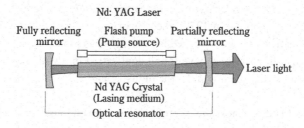

Figure 5.7: Schematic diagram of an Nd:YAG laser and an optical resonator

The lasing medium is yttrium (Y) to which neodymium (Nd) has been added pulses. These advances left the field of molecular spectroscopy poised to make major progress. We will next discuss the resulting developments. For their contributions to the development of laser spectroscopy, Nicolaas Bloembergen and Arthur Schawlow received the 1981 Nobel Prize in Physics.

Microwave spectroscopy

Techniques for working with microwaves emerged from research on radar during the Second World War, and after the war these gave birth to the field of microwave spectroscopy. At first, klystrons and other types of electron tubes were used to generate microwaves, but later these were replaced by solid-state devices such as Gunn diodes that provided a stable source of microwaves. The use of frequency synthesizers allowed continuous coverage of a wide range of frequencies. As discussed in Section 4.3 (see page 261), the energy gaps between the rotational energy levels of molecules correspond to the energies of microwaves, making microwave spectroscopy a powerful tool for precise structure determinations of simple molecules. In typical microwave spectroscopy, the sample gas is introduced into a cell forming one portion of a waveguide and its microwave absorption is detected by modulating a high-voltage electric field. In Fourier-transform spectroscopy, low-temperature sample gas is introduced through a pulse nozzle into a vacuum cavity resonator terminated at both ends by spherical mirrors. Pulsed microwave radiation induces transitions among rotational energy levels, and the resulting transient signal is detected and Fourier-transformed to yield the absorption spectrum. This method uses low-temperature samples and thus yields high-resolution spectra. Microwave spectroscopy not only gives detailed information about the structure of diatomic molecules, but also offers information on electric dipole moments, internal rotational energies, interactions between vibrational and rotational motions, magnetic moments, electric quadrupole moments, and more. However, for molecules consisting of larger numbers of atoms, the structure cannot be determined from the three moments of inertia; instead, to derive precise information in this case requires methods such as isotopic substitution and comparison.

Fourier-transform microwave spectroscopy allowed studies of the structures of short-lived free radicals (such as OH, CN, NO, CF, and CCH) and molecular ions (such as CO^+, HCO^+, and HCS^+). Microwave spectroscopy also proved a decisive technique for the structural determination of van der Waals complexes (such as $C_6H \cdot HCl$, $Kr \cdot HF$, and $SO_2 \cdot SO_2$)[5].

---

5. van der Waals complex: An aggregate of two or more neutral atoms or molecules bound together by weak van der Waals force.

Figure 5.8: *The 45-meter-diameter radio telescope at Nobeyama, Japan, used to detect interstellar molecules by observing electromagnetic waves in the millimeter wavelength range coming from outer space.*

Microwave spectroscopy is also tightly intertwined with radio astronomy observation of interstellar molecules (Figure 5.8). The existence of microwaves arriving from outer space was known before the Second World War, but the academic discipline of radio astronomy was not born until after the war. The new discipline confirmed the existence of a variety of molecules in space, and the spectra obtained from laboratory microwave spectroscopy proved to be valuable for identifying these molecules. (See 5.7.2)

Vibrational spectroscopy: infrared and Raman
Vibrational spectroscopy, which observes transitions among vibrational energy levels of molecules, exists in two main flavors: infrared and Raman spectroscopy (see page 262). In general, the position of a multi-atomic molecule containing N atomic nuclei is specified by 3N coordinates, with all nuclei exhibiting vibrations about an equilibrium point. Assuming that all nuclei exhibit harmonic vibrations at a common set of vibrational frequencies, we have 3N-5 normal vibrations for linear molecules and 3N-6 normal vibrations for nonlinear molecules. In infrared spectroscopy, one observes absorption due to transitions from the ground state to the first excited state of the vibrational energies, typically due to changes in molecular dipole moments (Figure 4.8). Before the war, infrared spectroscopy was used primarily by physical chemists as a tool for studying the structural chemistry of simple molecules. After the war, physical chemistry research—including structural studies of biologically-relevant molecules and other complex molecules—progressed; meanwhile, the fact that infrared spectrometers were also becoming readily available to chemists in other subfields

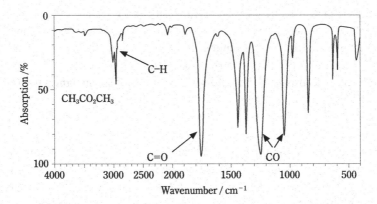

Figure 5.9: An example of an infrared absorption spectrum (for methyl acetate)

The absorption band near 3000 cm^{-1} is due to stretching vibrations of the C-H bond, while the peak near 1735 cm^{-1} is due to stretching vibrations of the C=O bond. The peaks at 1250 cm^{-1} and 1050 cm^{-1} arise from vibrations of the C-O bond.

ensured that infrared spectroscopy would become widely used as a convenient tool for the identification and analysis of compounds. The normal vibrations involve vibrations by all nuclei in a molecule; however, absorption spectra are peaked at particular wavenumbers that are strongly dependent on specific functional groups within the molecule (Figure 5.9). For example, a C=O group will exhibit characteristic absorption in the vicinity of 1700 cm^{-1}, while OH groups absorb in the vicinity of 3000 cm^{-1}. Thus infrared absorption provides a sort of "fingerprint" of the molecule. For this reason, infrared absorption spectra are widely used in organic chemistry and the chemistry of complexes, where the technique yields information complementary to that obtained from NMR and visible or ultraviolet absorption.

Until the 1970s, infrared spectrometers used beam splitters to divide the light obtained from infrared sources in two; one beam was passed through the sample, while the other beam was passed through a reference cell, and a comparison of the two beams allowed measurement of the absorption from the sample. Later, Fourier-transform infrared spectroscopy became the dominant technique. In this method, a Michelson interferometer[6] is used to obtain an interferogram containing information on infrared light absorption; this is then Fourier-transformed to provide the spectrum. In traditional methods, the spectrum had

---

6. Michelson interferometer: An interferometer invented by Michelson that allows precision measurements of light wavelengths and spectral lines and is widely used in high-resolution spectrometers.

been obtained by plotting the absorbed intensity on the vertical axis against the frequency on the horizontal axis; with Fourier-transform methods, information on all frequencies is obtained simultaneously, and accumulating spectra of this type yields an impressive improvement in sensitivity.

In Raman spectroscopy, a sample is illuminated by incident light at frequency v_0, and the scattered light at frequency $v_0 \pm v_i$ is observed; here v_i is a frequency associated with a vibrational or rotational transition in a molecule. Because Raman scattering arises from variations in molecular polarizability, its selection rules differ from those of infrared absorption (see page 263), and it allows observation of vibrations that cannot be observed by infrared spectroscopy. In Raman spectroscopy, the sample is irradiated by intense monochromatic light and the scattered light is observed after separation by a spectrometer. In general, Raman-scattered light is far weaker than Rayleigh-scattered light, so the technique requires high-resolution spectrometers and detectors capable of detecting weak light. With the emergence of the laser it became possible to obtain intense monochromatic light, and the field of Raman spectroscopy enjoyed major progress in the 1970s and thereafter. Raman scattering comes in both resonant and non-resonant cases. By utilizing the resonance Raman effect, in which the intensity of the Raman-scattered light increases dramatically as the frequency of the incident light approaches the frequency of absorption due to electronic transitions in the molecule, it is possible to measure tiny samples or dilute solution, making resonance Raman scattering a powerful tool for studying biological macromolecules or molecular species with short lifetimes. Only the vibrations of certain chromophores are visible in Raman spectra; this offers detailed information on the structure in the vicinity of chromophores and makes Raman scattering a powerful tool for studying the structure of chromoproteins such as heme protein.

A technique that significantly enhances the intensity of Raman-scattered light is surface-enhanced Raman scattering (SERS). This method utilizes the fact that the intensity of Raman scattering from a molecule adsorbed on a metal surface may be much greater than that of the same molecule in solution. This phenomenon was discovered in 1974 by Fleischmann and coworkers; in 1977, Van Duyne proposed that the intensity amplification could be attributed to local amplification of electromagnetic fields due to the excitation of plasmons[7] in the metal surface.

As laser spectroscopy techniques advanced, more sophisticated methods of Raman spectroscopy—such as stimulated-emission Raman spectroscopy and

7. Plasmon: A phenomenon in which free electrons in a metal, semiconductor, or other material oscillate collectively in a way that mimics particle-like behavior.

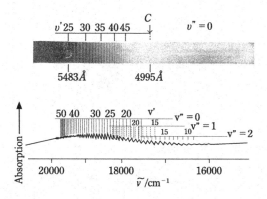

Figure 5.10: Visual absorption spectrum of I_2 vapor

Upper: As seen in a photograph.
Lower: As recorded by a spectrograph (spectrophotometer).
The absorption corresponds to transitions from the vibrational ground state v" to the first vibrational excited state v'.

coherent anti-Stokes Raman spectroscopy (CARS)—were developed, and these also were put to vigorous use in various fields of chemistry.

Ultraviolet and visible absorption, fluorescence and phosphorescence
In the 1930s, electronic states of molecules became understood through the concept of molecular orbitals, and the absorption and emission of ultraviolet and visible light was understood to arise from electronic transitions between molecular orbitals. Research on the electronic spectra of diatomic molecules and other simple molecules began before the Second World War and continued after the war, with extensive efforts contributed by Herzberg and other spectroscopists. The electronic spectra of simple molecules could be obtained as high-resolution spectra in the gas phase; they show clear splitting due to vibrational and rotational energy levels, thus yielding detailed information on molecular structure. These results were compiled into a monograph by Herzberg (Figure 5.10)[29].

Before the war, spectroscopists had used photographs for the observation of spectra. In the 1950s, spectrophotometers that recorded the intensity of absorption and emission as a function of frequency became commercially available and were widely adopted. Low-resolution ultraviolet and visible spectra came to be commonly used by chemists to identify and analyze compounds.

The absorption spectra of more complicated molecules such as aromatic compounds and metal complexes also came to be discussed in terms of molecular orbitals in the 1950s. By around 1950, chemists had succeeded in recognizing and distinguishing absorption processes arising from a variety of different

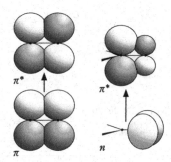

Figure 5.11: Schematic depiction of ππ* and nπ* transitions

Left: A ππ* transition in C=C in which an electron in a bonding π orbital is excited into an anti-bonding π* orbital.

Right: An nπ*transition in C=O in which an electron in an n orbital of the O atom is excited into a π* orbital.

transitions in aromatic compounds; these included ππ* transitions (transitions between π electronic orbitals), nπ* transitions (in which an electron in an nonbonding n orbital is excited into a π* orbital; see Figure 5.11), σπ* transitions (in which an electron in a σ orbital is excited into a π* orbital) and the reverse process of πσ* transitions; dd* and dπ* transitions (involving d electrons in metal complexes), and CT transitions (involving charge transfer (CT) among molecules). In each case, the characteristic features of the absorption spectra came to be understood, thus advancing qualitative knowledge of electronic spectra. Below we consider CT absorption in more detail.

Ethanol solutions of iodine have a brown color that differs from the purple color characteristic of elemental iodine. The appearance of new features in the absorption spectra of molecules that are not present in the spectra of the constituent molecules had been known previously, but it was not until 1952 that Mulliken explained this type of absorption in terms of CT spectra, due to transitions from the ground state to excited states in a complex (charge transfer complex) formed when an electron donor (D) transfers some of its electrons to an electron acceptor (A) to yield a more stable configuration[30]. This type of spectrum was observed around the same time in Japan, by the group of Saburo Nagakura at Tokyo University, in compounds such as quinhydrone, formed from quinone and hydroquinone. They also used the concept of intramolecular charge transfer to interpret absorption spectra of some substituted benzenes. Subsequently, similar phenomena were observed in metal complexes, including ligand-metal charge transfer (LMCT) absorption—associated with charge transfer from a ligand to a metal—or the reverse phenomenon of metal-ligand charge transfer (MLCT) absorption.

Saburo Nagakura (1920–)
After graduating from the University of Tokyo in 1943, Nagakura became a professor at the Institute for Solid State Physics at the University of Tokyo in 1959, rising in 1981 to Director of the Institute for Molecular Science. In 1988 he became Director of the Okazaki National Research Institutes. His accomplishments spanned many fields of research, including the electronic structure and reactivity of molecules—particularly charge-transfer states and photochemistry—and he played a leading role in the development of Japanese molecular science. He received Japan's Order of Culture in 1990.

In 1944, Lewis and Kasha proposed that fluorescence was due to radiative transitions from an excited singlet state (S_1) to the ground state (S_0), while phosphorescence was associated with transitions from a triplet excited state (T_1) to the ground state. However, this proposal was not generally accepted until the early 1950s. In 1949, Kasha proposed the existence of rapid non-radiative processes from the S_1 state to the T_1 state[31]. He also proposed "Kasha's Rule," which held that even molecules excited to highly excited states relax to the S_1 state via non-radiative relaxation processes and that fluorescence occurs from the S_1 state. Fluorescence and phosphorescence of organic compounds in condensed systems were a topic of active study from the 1950s through the 1960s, and during this time a detailed understanding of the radiative and non-radiative processes by which excited molecules relax gradually emerged.

High-resolution gas-phase spectroscopy of relatively large molecules began around 1970 with studies of benzene and similar molecules; with the emergence of wavelength-tunable dye lasers, it became possible to excite a single vibronic energy level and observe the resulting fluorescence. By fixing the wavelength of the laser light and performing spectroscopic measurements of the light emitted by the sample, one obtains a fluorescence spectrum; by varying the wavelength of the exciting laser light and observing the intensity of the resulting fluorescence one obtains a fluorescence excitation spectrum. From the 1970s onward it was possible to use techniques for cooling gas molecules to low temperatures by diluting samples with noble gases and spraying the gas into vacuum as an ultrasonic jet stream. This enabled spectra to be obtained with sufficiently high resolution to allow separation of the rotational structure even for relatively large molecules. Fine-grained analysis of these spectra then gives detailed information on the structure and dynamics of excited electronic states.

Before the emergence of the laser, spectroscopic studies could only observe single-photon processes, in which an atom or molecule absorbs only one photon. The availability of intense laser beams made it possible to obtain high-density optical fields by focusing laser light, creating the possibility of multi-photon processes in which an atom or molecule absorbs two or more photons simultaneously. Two-photon absorption is governed by selection rules that differ from those relevant for one-photon absorption, allowing the observation of absorption phenomena that are not observable in the one-photon case. In addition, molecules are readily ionized by two-photon or multi-photon processes, and the resulting ions and electrons may be detected with high sensitivity, leading to advances in spectroscopic methods utilizing multi-photon ionization detection. Another technique that came to enjoy widespread use was the optical-optical double resonance method, which uses two lasers.

Nonlinear spectroscopy

As discussed in the previous section, the use of lasers as light sources led to many new developments in the field of spectroscopy, and a wide variety of new spectroscopic methods exploiting the particular properties of laser light emerged in the 1960s and thereafter. Among these was a class of methods utilizing intense, monochromatic, and phase-matched coherent light that came to be known generally as techniques of nonlinear spectroscopy. In addition, by exploiting the ability to produce short light pulses it became possible to study processes occurring on extremely short time scales—ranging from nano- (10^{-9}) to femto- (10^{-12}) seconds—giving rise to the field of ultra-fast spectroscopy. Here we will consider representative examples of both techniques.

The phenomena observed in spectroscopy may be understood by considering the polarization P induced in matter by the electric field E of the incident light. If the incident light is weak, this polarization is proportional to E (i.e., it is given by a first order expression in the field strength E), but under illumination by strong laser light, second and third order terms in the field strength E become important, allowing the observation of nonlinear optical phenomena. This is the effect that is exploited by nonlinear spectroscopy.

Nonlinear processes arising from second order terms in E have been used for purposes such second-harmonic generation (producing light with twice the frequency of the incident light); more recently, these terms have formed the basis for *sum-frequency-generation spectroscopy*, a powerful tool for studying surfaces and interfaces that yields information such as the orientation of molecules at surfaces. However, these terms vanish in systems possessing inversion symmetry, and thus typical gases and liquids do not exhibit nonlinear processes due to second-order terms. For this reason, most nonlinear spectroscopy techniques used in chemistry to date have utilized third order terms in E; examples of such techniques include coherent anti-Stokes Raman spectroscopy (CARS), photon echo, and the transient grating method. The processes of multi-photon absorption and optical-optical double resonance, which we discussed in the previous section, may also be understood as applications of third-order nonlinearity effects. In CARS, the sample is illuminated at particular angles by laser light at two frequencies, v_1 and v_2 (where $v_1 > v_2$ corresponds to transitions from the ground state to an excited state), and scattered light at frequency $2v_1 - v_2$ is observed. When the difference frequency $v_1 - v_2$ lies in the Raman-active frequency band of the sample, a strong spectrum is obtained. This technique is useful for obtaining high-resolution Raman spectra in systems with strong background scattering. Photon echo is a spectroscopic method with similarities to magnetic-resonance spin echo in the optical regime (see page 456); here, the sample is excited by two laser pulses whose time delay is adjusted and the resulting echo light signal is measured. In the transient grating method, the sample is illuminated by two exciting beams of the same wavelength that intersect at a particular angle, and the phenomena excited by the resulting transient interference pattern are detected by a probe beam. All of these methods are useful for observing dynamic processes such as high-speed molecular motion and relaxation phenomena.

Ultra-fast spectroscopy

Methods in which short light pulses are used to excite molecules and the resulting short-lived absorption is observed were pioneered in 1949 by English scientists Porter and Norrish, who developed a flash-photolysis method using flash discharge tubes[32]; such methods later came to be widely used for studies of fast reactions and short-lived excited states and free radicals. The flash photolysis method is restricted to observations on time scales on the order of microseconds (10^{-6} s), but the development of lasers made it possible to observe phenomena on time scales ranging from nanoseconds (10^{-9} s) to femtoseconds (10^{-15} s), giving rise to the field known as ultra-fast laser spectroscopy.

The most frequently used techniques in this field use pump-probe methods to observe transient absorption. A sample is excited by an ultra-short-pulse pump laser, and the resulting short-lived absorption processes are observed as the absorption of a white-light probe beam. By varying the delay time between the pump and probe pulses, it is possible to track the dynamics of the induced short-lived processes. Short-lived species induced by laser pulses may also be detected by other methods, including fluorescence and photo-ionization. Because ultra-fast laser spectroscopy uses light pulses that achieve high instantaneous intensities, they involve nonlinear optical effects in an essential way, and the fields of ultra-fast spectroscopy and nonlinear spectroscopy are closely related. For a survey of applications of ultra-fast spectroscopy to studies of chemical reactions, see Section 5.4.2.

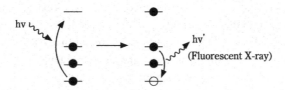

Figure 5.12: Principles underlying the emission of fluorescent X-rays

An inner-shell electron is excited by a high-energy X-ray or radiation beam; this creates a hole, which is filled by an electron from another orbital, and the excess energy is emitted in the form of a fluorescent X-ray.

X-ray Spectroscopy

Fluorescent emissions from matter excited through irradiation by high-energy X-rays or electron beams, as well as X-ray absorption by matter, also offer valuable information. X-ray fluorescence arises when an outer-shell electron in an atom falls into a hole created by ionization of an inner-shell electron, emitting an X-ray in the process. The energy of the emitted X-ray corresponds to the energy difference between the two atomic shells and is characteristic of the atom in question, and X-ray spectroscopic methods that exploit this fact have become widely used for elemental analysis and related purposes. (Fig. 5.12)

With the advent of synchrotron radiation, it became possible to make use of intense and wavelength-tunable X-rays, advancing the state of absorption spectroscopy in the X-ray regime. X-ray absorption spectra exhibit sharp, discontinuous absorption edges superimposed on continuous absorption bands at lower frequencies. These correspond to processes in which specific inner-shell orbitals absorb or emit electrons. In the vicinity of these absorption edges, a characteristic oscillatory structure known as *X-ray absorption near edge structure* (XANES) is observed; because this structure is due to transitions of inner-shell electrons, analyzing it yields information on the electronic state of the X-ray-absorbing atoms. The structure observed at higher energies, known as *extended X-ray absorption fine structure* (EXAFS), is thought to arise from interference between the waves of electrons emitted from X-ray-absorbing atoms and waves reflected from adjacent atoms; this yields information on the structure in the vicinity of the atom being studied.

γ-Ray spectroscopy: The Mössbauer effect

The *Mössbauer effect*—a resonant absorption phenomenon in which an atomic nucleus in a solid emits, without recoil, a γ-ray which is subsequently absorbed, also without recoil, by another atomic nucleus of the same species—was discovered in 1958 by the German scientist Rudolf Mössbauer[33]. Spectroscopic

> **Rudolf Mössbauer (1929–2011)**
> Mössbauer was a German physicist. After receiving his doctorate in 1958 from the Technical University of Munich, he worked as a professor at the California Institute of Technology before returning in 1964 to the Technical University of Munich as a professor. He discovered the Mössbauer effect in 1957 and received the 1961 Nobel Prize in Physics.

methods utilizing this phenomenon are known as *Mössbauer spectroscopy*. The energy levels of atomic nuclei vary slightly depending on the electronic structure of the solid, and by inducing motion in either the emitter or the absorber and observing the resulting Doppler shift, it is possible to obtain spectra. These spectra exhibit shifts due to interactions with electrons and splittings due to internal magnetic fields and nuclear quadrupole moments, making the method a powerful tool for obtaining information on properties such as bonding states, atomic configurations, and magnetic states of more than 20 atomic species—particularly Fe—which is used in a wide range of fields of physics, chemistry, biology, and mineralogy. Mössbauer received the 1961 Nobel Prize in Physics.

The development of electron spectroscopy techniques: Observing the inner shells of atoms and the configurations of surfaces

The spectroscopic methods discussed thus far have all been based on measuring the intensity of absorbed or emitted electromagnetic radiation (light). In electron spectroscopy, one instead measures the kinetic energies of electrons emitted from a sample due to irradiation by X-rays, light, or electron beams; this allows analysis of the structure and electronic configuration of the substance. Electron spectroscopy underwent major development in the postwar era, driven primarily by advances in vacuum technology and in methods for measuring electron energies. Electron spectroscopy is a method for studying the microscopic states of surfaces, and it has become a powerful tool for clarifying the absorption structure of surfaces and for understanding the details of catalytic processes; in recent years the technique has made major contributions to the development of surface science and catalysis science. Methods of electron spectroscopy include X-ray and UV electron spectroscopy, Auger electron spectroscopy, and electron energy-loss spectroscopy.

Photoelectron spectroscopy

Photoelectron spectroscopy measures the kinetic energies of electrons emitted from matter due to optical excitation; the technique yields information on the binding energies of electrons and other physical quantities (Figure 5.13).

Research on the energies of electrons emitted due to the photoelectric effect was already in progress as early as the 1910s, but research before the Second World War succeeded only in obtaining broadband spectra. The first successful reports of high-resolution X-ray photoelectron spectroscopy (XPS) came in 1957 from the group of Kai Siegbahn in Sweden[34], who used an apparatus designed by nuclear physicists to make precision measurements of the energies of electrons emitted in β-decay. By plotting the kinetic energies of the electrons emitted under monochromatic X-ray illumination on the horizontal axis versus the number of electrons emitted on the vertical axis, these researchers discovered an extremely sharp peak that had not been previously observed (Figure 5.13). Recognizing that this effect arose from electrons kicked out of atomic shells within the sample, they proceeded to conduct systematic studies of electron binding energies in different shells across a variety of elements, thus establishing XPS as a solid experimental method. In XPS, it is possible to induce the emission of inner-shell electrons; moreover, ionization energies are affected by the surrounding chemical environment. This makes XPS a useful tool for surface elemental analysis and surface state analysis, in which context it is sometimes known as electron spectroscopy for chemical analysis (ESCA). Experimental instruments became available for purchase in 1969, after which use of the method as an analytical tool became widespread. For his achievements in the discovery of XPS, Siegbahn shared the 1981 Nobel Prize in Physics with Schawlow and Bloembergen, the pioneers of laser spectroscopy.

Whereas XPS studies electrons emitted from the surfaces of solids, the method known as *ultraviolet photoelectron spectroscopy* (UPS) uses a He lamp with a wavelength in the vacuum ultraviolet (58.4 nm) to perform photoelectron spectroscopy on gas-phase molecules. This method was developed in the early 1960s by the physical chemist Turner at Oxford. In UPS, only the valence electrons of atoms are emitted; however, the method offers high resolution and can experimentally measure the energies of molecular orbitals, and for this reason it has become a useful tool for studying the electronic structure of molecules. Denoting the energy of the exciting photon by hv, the ionization energy needed to emit an electron from orbital i by I_i, and the energy of the emitted electron by E_k, we have the relation $E_k = hv - I_i$. UPS spectra may be interpreted on the basis of an approximation known as Koopmans' theorem, which holds that I_i is equal to the orbital energy of the emitted electron; this allows the method to determine the energies of molecular orbitals, a purpose for which it has become widely used.

In recent years, the use of synchrotron radiation as a light source has made it possible to cover a continuous range of incident photon energies, allowing sophisticated measurements, including observations of resonance effects, angle-

Kai Siegbahn (1918–2007)
Siegbahn was a Swedish physicist. After receiving his doctorate in 1944 from the University of Stockholm, he became a professor at the Royal Institute of Stockholm before becoming a professor at Uppsala University in 1954. From the 1950s onward he worked to improve analytical instruments using particle beams and to establish high-resolution methods of photoelectron spectroscopy. He received the 1981 Nobel Prize in Physics.

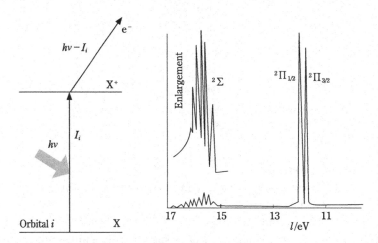

Figure 5.13: Principles of photoelectron spectroscopy (left) and the photoelectron spectrum of HBr (right).

When an atom or molecule is excited from state X to an ionized state X^+ by a photon with energy $h\nu$ in the X-ray or vacuum ultraviolet regime, an electron with energy $h\nu - I_i$ is emitted. In the spectrum the band at right corresponds to the ionization of an electron from an isolated electron pair (lone electron pair) in the Br atom; the band at left corresponds to the ionization of a bonding electron.

resolved measurements, and spin-resolved measurements. This has spurred progress in applying the technique to studies of the electronic structure of materials, photo-ionization, and photochemical reactions. In addition, angle-resolved photoelectron spectroscopy—in which the photoelectrons emitted in particular directions are detected—yields information on surface band structure and the orientation of adsorbed molecules. Because photoelectrons are only emitted from depths of a few hundred pm to a few pm beneath the surface, photoelectron spectroscopy methods yield detailed information on the electronic structure of solid surfaces and their adsorbed atoms and molecules, making the method a powerful tool in surface science.

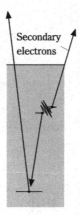

Figure 5.14: A schematic depiction of the principles of Auger spectroscopy

When an electron falls into a hole, liberating a certain quantity of energy, an electron from a different orbital is emitted with that energy (secondary electron emission).

Auger electron spectroscopy

Auger electron spectroscopy is based on the *Auger* effect, in which irradiation of a sample by a high-energy radiation beam or electron beam causes emission of an inner-shell electron; an electron from an outer shell then falls into the resulting hole, and a secondary electron is subsequently emitted with energy equal to the difference between orbital energies (Figure 5.14). The phenomenon of the Auger effect was in fact first reported by Meitner in 1922, but it is named for the French scientist Auger, who discovered it independently in 1925. The energy of the secondary electron is characteristic of the element in question and is affected by bonding state, and analyzing secondary-electron energies thus yields information on the chemical composition of surfaces. The technique has been actively used for surface analysis since the 1960s. In typical experiments, a sample is irradiated by a beam of electrons with energies of 1–5 keV and the energies of the emitted electrons are analyzed.

Electron energy loss spectroscopy (EELS)

The method of electron energy loss spectroscopy (EELS) was pioneered in the mid-1940s by Hillier and Baker; starting around 1980, the method began to be used as a spectroscopic technique for studying clean surfaces with no adsorbed material. In this method, a beam of electrons with fixed kinetic energies is directed at a sample surface, and the intensity of the reflected electron beam is expressed as a function of the energy lost due to inelastic scattering at the surface. Electrons lose energy through a variety of mechanisms—including the excitation of surface phonons[8], excitation of surface electronic bands or vibrational energy levels, and inner-shell ionization—and detailed analysis yields useful information about the surface at the atomic and molecular levels. In applications to chemistry, the method is particularly useful as a method of analyzing the vibrational spectra of adsorbed molecules. Technological advances have made it possible to collect high-resolution (HR) spectra, a method known as HREELS.

Magnetic resonance methods: Spectroscopic techniques that use spin as a probe

One experimental technique that experienced rapid growth after the Second World War and significantly transformed chemical research is the method of *magnetic resonance*, in which one observes absorption associated with transitions among energy levels of the spin system that are Zeeman-split by a magnetic field. As discussed in the previous chapter, the first observation of nuclear magnetic resonance (NMR)—in which the spins of atomic nuclei are observed—was made by Rabi in 1938 using a molecular beam of lithium chloride, while successful observations of electron spin resonance (ESR, also known as electron paramagnetic resonance or EPR) were made in 1944 by Zavoisky in the Soviet Union. Dutch scientist Gorter had attempted in the pre-war years to observe NMR in condensed systems, but was unsuccessful; the first successful observations in condensed systems were not made until after the war. Magnetic resonance underwent major development in the postwar years, eventually coming to play a major role in a wide range of fields spanning the natural sciences, including physics, chemistry, biology, and medicine. Other magnetic resonance techniques include nuclear quadrupole resonance (NQR)[9] as well as muon spin rotation, relaxation, and resonance (μSR)[10].

---

8. Phonon: A quasiparticle corresponding to a quantized lattice vibration or (in a broader sense) sound wave.

9. Nuclear quadrupole resonance (NQR): In atomic nuclei with nuclear spin greater than 1, the electric charge distribution of the nucleus may deviate from spherical symmetry to develop a nuclear quadrupole moment, whose interaction with an electric-field gradient creates splittings of the nuclear spin energy levels. Resonances due to these splittings are known as nuclear quadrupole resonances.

10. μSR: A method in which a substance is irradiated by μ particles (a type of elementary particle, also known as muons) and the rotation, relaxation, and resonance of the electrons resulting from the decay of these muons is observed to yield information on the structure of the substance.

Nuclear magnetic resonance (NMR)

In early 1946, the groups of Edward Purcell at Harvard University and of Felix Bloch at Stanford University independently reported successful observations of NMR signals in condensed systems[35, 36]. Purcell's group observed resonance phenomena involving protons in paraffin, while Bloch's group observed protons in an aqueous solution of ferric nitrate. The initial research objectives were to determine nuclear magnetic moments, but the discovery had a major impact on the physics world, and Purcell and Bloch shared the 1952 Nobel Prize in Physics. At first the interests of physicists were directed primarily toward understanding the fundamentals of resonance phenomena and relaxation phenomena[11], but the field quickly experienced a sequence of discoveries of unanticipated phenomena. First, the fact that the same nucleus can exhibit slightly different resonance frequencies depending on the surrounding chemical environment was observed in metals by Knight in 1949. Next, 1950 brought reports of *chemical shifts*—different resonance frequencies for the same nucleus in different compounds—for ^1H (the proton), ^{14}N, and ^{19}F, a development which stimulated strong interest in NMR among chemists[37]. The following year, methyl, methylene, and hydroxyl protons were separated in observations of ethyl alcohol spectra, showing that the same nucleus in the same molecule can exhibit different resonance frequencies depending on the type of compound (chemical shifts) (Figure 5.15).

Chemical shifts arise from the influence on the nucleus of the internal magnetic field due to magnetic moments induced in the electronic system in the vicinity of the nucleus in an atom or molecule by an external magnetic field. These are proportional to the strength of the external magnetic field, typically on the order of 10^{-5}–10^{-4} times the external field strength. Further splittings due to interactions between nuclear spins (spin—spin coupling) were subsequently observed even for the same type of nucleus in the same molecule. Thus it became clear that NMR spectra contain a wealth of information on molecular structure and electronic state, and their applications to chemical research grew rapidly beginning in the 1950s. Theoretical understanding of chemical shifts and spin—spin coupling also advanced, and based on this progress NMR became a powerful tool for determining the structure of organic compounds. By observing variations in spectra and linewidths and analyzing relaxation phenomena, the technique could also be used to study dynamical effects such as bond rotation

---

11. Relaxation phenomena: Phenomena that arise when a system in an equilibrium state is jolted into a new equilibrium state by an external force, and then the external force is removed to allow the system to return to its initial equilibrium. The time needed for the system to relax is known as the *relaxation time*.

Edward Purcell (1912–1997)

After studying electrical engineering at Purdue University and physics at Harvard University, Purcell became a lecturer and assistant professor at Harvard before becoming a full professor there in 1949. During the Second World War he worked on radar research. In 1946 he established nuclear magnetic resonance methods for measuring the magnetic moments of atomic nuclei in liquids and solids. He also made contributions to radio astronomy, successfully observing microwave emissions from interstellar hydrogen atoms in 1952. He received the 1952 Nobel Prize in Physics.

Felix Bloch (1905–1983)

Born in Switzerland, Bloch studied at the Swiss Federal Institute of Technology in Zurich before becoming an assistant to Heisenberg in Leipzig. In 1928 he proposed a form for the wavefunction of electrons in solids (the Bloch wavefunction). In 1934 he moved to Stanford University in the U.S., where in 1939 he collaborated with Alvarez to measure the magnetic moment of the neutron. During the Second World War he worked on atomic energy and radar. In 1946 he successfully measured nuclear magnetic moments using nuclear magnetic induction methods, and he contributed to developing the theory of magnetic resonance. He received the 1952 Nobel Prize in Physics.

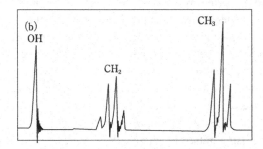

Figure 5.15: NMR spectrum of protons in ethyl alcohol

Chemical shifts split the spectrum into three groups: OH, CH_2, and CH_3. Within each group, there are further splittings due to spin-spin coupling. The horizontal axis is the magnetic-field shift and the vertical axis is the signal strength.

within molecules, changes in conformations, and reaction rates. Thus NMR quickly became an indispensable method for chemical research.

Applications to chemistry required high-resolution NMR observations that could resolve chemical shifts and spin—spin couplings. In the late 1950s, high-resolution NMR spectrometers became available for purchase, and these machines came to be used for chemical research spanning a wide range of fields. NMR spectrometers of this era used the *continuous-wave NMR (CW-NMR)* approach, in which the sample is continuously irradiated by radio waves at a fixed resonance frequency and the magnetic field is scanned to obtain spectra. However, the inadequate sensitivity of this method proved problematic, causing difficulties in measuring small quantities of naturally-occurring nuclei such as ^{13}C. Increasing the magnetic field strength and the frequency proved an effective means of improving sensitivity and spectral resolution; however, until around 1970, NMR instruments generally used electromagnets, which for protons effectively limit the frequency to around 100 MHz. Eventually, superconducting magnets capable of providing the stable, uniform magnetic fields required for high-frequency NMR were developed; thus it became possible to conduct NMR measurements at high frequencies, and by the end of the 20th century the frequencies of NMR measurements had risen to a few hundred MHz. However, high-resolution NMR spectra for solid samples—in which the directivity (anisotropy) of chemical shifts and of spin-spin couplings are not averaged—remained out of reach. In 1958, Andrew proposed a method for eliminating anisotropy by rotating the sample at high speed about an axis forming an angle of 54.2 degrees (the so-called "magic angle") to the direction of the magnetic field. This approach was later combined with cross-polarization techniques to pave the way toward high-resolution NMR studies of solids.[12]

Pioneering studies of the use of pulsed radio waves to obtain NMR signals were conducted by Torrey in 1949 and by Hahn in 1950. Torrey observed the decaying oscillations of the transient signal under pulsed radio-wave excitation[38], while Hahn illuminated the sample with two pulsed radio waves separated by a certain time interval and investigated the behavior of the resulting echo signal[39]. Until the 1960s, pulsed NMR was used primarily for studies of relaxation phenomena, but in 1966 Richard Ernst and collaborators developed a Fourier-transform (FT) NMR method using pulsed irradiation, thus achieving an epochal advance in NMR methodology (Figure 5.16). Rectangular wave pulses at the resonance frequency include components at all frequencies in the vicinity of the resonance

---

12. Cross-polarization: A measurement technique in which the magnetic moment of a ^1H spin, which has a relatively short relaxation time, is transferred to a nuclear spin with a relatively long relaxation time (for example, ^{13}C).

Richard Ernst (1933–)

Ernst studied chemical engineering at the Swiss Federal Institute of Technology in Zurich; after receiving his doctorate he worked in NMR research and development at Varian Associates. In 1976 he returned to his alma mater as a professor. In 1966 he developed Fourier-transform NMR techniques, thus succeeding in dramatically improving NMR sensitivity; in 1976 he developed the two-dimensional NMR method, thus paving the way toward the widespread adoption of NMR methods across many fields of science, including not only physics and chemistry, but also biology and medicine. He received the 1991 Nobel Prize in Chemistry.

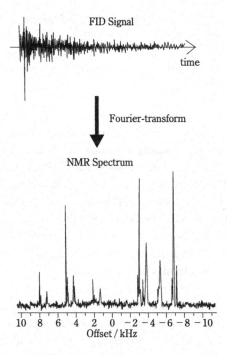

Figure 5.16: Fourier-transform of an NMR spectrum

The time variation of the NMR signal (FID signal) obtained by pulsed radio-wave illumination of a sample, as shown at top, is Fourier-transformed to obtain the spectrum shown at bottom.

frequency, and irradiating a sample with pulsed radio waves thus excites the entirety of the spin system, yielding a superposition of transient time-domain NMR signals (free induction decay) containing information spanning the full

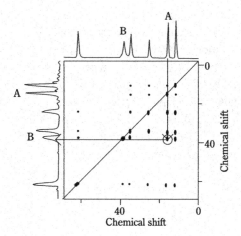

Figure 5.17: An example of a two-dimensional NMR spectrum indicating correlations between signals at two frequencies

The rectangular figure in the center is a contour plot of the spectral magnitude. Signals enclosed by ○ indicate the presence of a relationship between the nuclei responsible for the A and B peaks in the one-dimensional spectra.

spectrum. Fourier-transform analysis of this time-domain signal then gives the frequency-domain spectrum. Signals obtained from repeated pulse illumination may be added to achieve dramatic improvements in the signal-to-noise ratio, allowing FT-NMR methods to realize significantly greater sensitivities than were possible with the CW-NMR techniques used previously. In this way it became possible, and even easy, to study complicated molecules by observing the NMR spectra of ^{13}C and other nuclei that exist naturally in trace amounts. However, as the molecule grows more complex, its NMR spectrum becomes exceedingly complicated, and without further refinements the method is not able to analyze spectra from complex molecules such as proteins.

Two-dimensional NMR, proposed in 1971 by Jeener and developed in 1976 by Ernst, opened the door to the NMR analysis of complicated molecules (Figure 5.17)[40]. When a sample is irradiated by multiple pulses and the intervals between pulses are varied, the transient signals that result are influenced by spin-spin couplings and other magnetic interactions. By Fourier-transforming the temporal variation of the signal, treating the spectrum as a two-dimensional function of the two frequencies, and observing correlations between peaks in the 1-dimensional spectra, one obtains information on spin-spin couplings and the spatial distance between spins. This technique was subsequently extended to 3 or 4 dimensions. The general method of multidimensional FT-NMR has found uses across a wide range of fields in chemistry and biology. In two-dimensional

Kurt Wüthrich (1938–)
After receiving his doctorate from the University of Basel, Wüthrich worked at the University of California and Bell Laboratories before moving to the Swiss Federal Institute in Zurich, where he became a professor in 1980. He developed methods for determining the structures of biological macromolecules using NMR, formulating principles for computing the distances between atoms in proteins; he paved the way toward the use of NMR for the structural analysis of proteins. He received the 2002 Nobel Prize in Chemistry.

NMR, by observing ^{13}C and ^{15}N nuclei through resonances of the ^1H nucleus, it is possible to investigate questions such as "which hydrogens are bound to which carbons or nitrogens?" and "which hydrogens exist near each other?" This enables applications of NMR to three-dimensional structural analysis of biological macromolecules such as proteins and nucleic acids[41]. Indeed, with NMR it is possible to observe biological macromolecules in solution, and the technique thus serves to complement X-ray structural analysis; however, to observe ^{13}C and ^{15}N nuclei, one must use biological macromolecules tagged with ^{13}C and ^{15}N nuclei.

Pioneering research on NMR led to more Nobel prizes beyond those awarded to the field's founding fathers, Rabi, Purcell, and Bloch. Ernst, the developer of FT-NMR and two-dimensional NMR, received the 1991 Nobel Prize in Chemistry. Later, Kurt Wüthrich, who developed NMR methods for determining the three-dimensional structure of biological macromolecules in solution, was one recipient of the 2002 Nobel Prize in Chemistry.

Electron spin resonance (ESR or EPR)

Electron spin resonance (ESR) was discovered by Zavoisky in the Soviet Union during the Second World War (see page 264), but the first major developments were achieved between the late 1940s and early 1950s by a group of physicists led by Bleaney at Oxford University. Microwave techniques developed during the war for radar purposes were often put to use for basic science research after the war, and ESR research was one of the fields that benefited from this technological transfer. Because ESR focuses on electron spins, it is a powerful tool for studying the electronic states of paramagnetic molecules such as free radicals or transition-metal complexes with unpaired electrons. Detailed studies of metallic complexes were conducted at Oxford, where the foundations of ESR spectral analysis—including considerations of spin-orbit interactions, interactions between electron spins, and interactions between electronic and nuclear

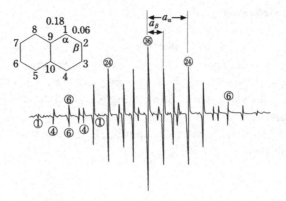

Figure 5.18: ESR spectrum of negative ion radical of naphthalene

The protons at sites α and β in naphthalene exhibit hyperfine splittings a_a and $a_β$, from which the density of unpaired electrons in the carbon atoms at sites α and β are determined to be 0.18 and 0.063 respectively.

spins—were laid. Applications of ESR in chemical research began in the early 1950s. Stable organic free radicals were the first to be studied extensively, and it became clear from the work by Weissman and others that analysis of hyperfine splittings (HFS) due to interactions between electron and nuclear spins could be used to determine the distribution of unpaired electron densities in molecules (Figure 5.18). In particular, so-called McConnell's equation[13] was proposed in 1956 to describe correlations between the density of unpaired π-electrons in molecular orbitals and hyperfine splittings due to protons in radicals of aromatic molecules, and this development spurred major progress in the use of ESR for chemical studies[42]. After this, ESR came to be widely used to study chemical species containing unpaired electrons, including transient species induced by radiation beams or light illumination—such as short-lived free radicals or excited triplet states—as well as transition-metal complexes. The 1950s also witnessed the first analyses of linewidths and relaxation times to understand reaction rates and variations in the conformation of radicals. New techniques—including the matrix isolation method, in which unstable free radicals were trapped in low-temperature solids for observation, and the spin-trap method, in which unstable free radicals are trapped in a spin-trap agent and detected after being converted to stable free radicals—came to be widely used in the 1960s and thereafter

---

13. McConnell's equation: Denoting the density of unpaired electrons in a 2p carbon orbital by $ρ$ and the magnitude of the hyperfine splitting due to protons by a_H, McConnell's equation shows that the relation $a_H = ρQ$ is approximately satisfied.

for studies of short-lived free radicals. The 1970s saw the development of the time-resolved ESR method, in which large transient spin moments induced by pulsed laser light or electron-beam excitation are used for detection purposes; this technique came to exert a commanding influence in studies of paramagnetic molecules of short-lifetimes on the order of nanoseconds, including radicals and excited triplet states.

In applications of ESR to chemistry, continuous-wave (CW) ESR using X-band (9 GHz) microwaves remained the primary method for many years, but in the 1970s pulsed ESR gradually came to enjoy more widespread use. In addition, electron nuclear double resonance (ENDOR)—which combines ESR with nuclear spin excitation—was developed in the late 1950s, while optically-detected magnetic resonance (ODMR), in which ESR is detected by studying variations in light emission, emerged in the late 1960s; both of these techniques soon came to be used in various specific applications. In the 1980s, high-frequency ESR techniques using high-frequency microwaves and superconducting magnets were developed, and these methods gradually came to be adopted as well.

Applications of ESR to biologically-relevant molecules began already in the 1950s, with initial targets including photosynthetic systems and myoglobin. In the early 1960s, the spin-labeling method—in which stable nitroxide radicals are used as tags to study the states and dynamic behavior of biologically-relevant molecules and biological membranes—was developed, and this technique soon came to enjoy extensive application to various fields of biochemistry. A variety of advanced ESR technologies—including ENDOR, pulsed ESR, and high-frequency ESR—have been actively applied since the 1980s to studies of diverse systems such as photosynthetic molecules and metalloenzymes.

Magnetic Resonance Imaging (MRI)

An application of nuclear magnetic resonance that emerged to play an immensely important role in the field of medicine is the technique of *magnetic resonance imaging* (MRI). Developed by physical chemist Paul Lauterbur (Column 19) at the State University of New York at Stony Brook in 1973, this technique superimposes a magnetic field gradient on top of a static background magnetic field and obtains NMR signals from various regions of a body, thus yielding cross-sectional images[43]. The promise of this technique for medical applications was immediately recognized, and attempts to obtain cross-sectional images of living organisms soon began. Thanks to advances in computer technology and manufacturing methods for superconducting magnets, by the 1980s it was possible to obtain images of the human body; as is well known, today MRI technology stands with X-rays and CT scans as one of the most important technologies for obtaining cross-sectional images of the human body, and applications of the

Paul Lauterbur (1929–2007)
After studying at the Case Institute of Technology, Lauterbur joined the Mellon Institute, where he worked on NMR studies of ^{13}C. In 1966 he moved to the State University of New York at Stony Brook, where he became a professor in 1969. In 1985 he moved to the University of Illinois to lead the Biomedical Magnetic Resonance Laboratory. In 1973 he proposed the principles of MRI, in which NMR is used to obtain cross-sectional images, thus laying the groundwork for the development of MRI. He received the 2003 Nobel Prize in Physiology or Medicine.

Peter Mansfield (1933–)
Mansfield is a British physicist. After receiving his doctorate from Queen Mary College at the University of London in 1962, he became a professor at the University of Nottingham in 1979. In 1978 he contributed to the development of MRI by inventing a high-speed image scanning method. He received the 2003 Nobel Prize in Physiology or Medicine.

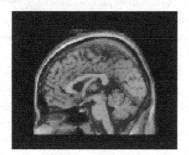

Figure 5.19: An MRI image of a human head

method in the medical field have become commonplace. Because the relaxation times of protons—which exist in great numbers in biological organisms—differ depending on their local environment, proton MRI yields images with outstanding spatial resolution and contrast (Figure 5.19). In recent years, MRI has also been applied to neuroscience research in the form of functional MRI (fMRI). When nerve cells are excited due to brain activity, blood flow surges in their vicinity; this causes a decrease in reduced hemoglobin in blood vessels and increases the MRI signal. This technique may thus be used to investigate brain activity.

The 2003 Nobel Prize in Physiology or Medicine was awarded to Paul Lauterbur and British physicist Peter Mansfield for their contributions to the invention and development of MRI.

COLUMN 19

Lauterbur and the birth of MRI

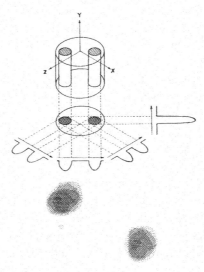

A: A schematic depiction of the basic principles of MRI[43]
B: Image of water in a capillary tube

Although NMR techniques were originally invented to achieve a simple goal of basic science—the determination of nuclear magnetic moments—within just a half-century they had spawned the method of MRI, a crucial modern tool for observing the interior of the human body. This author had the good fortune to spend over ten years with Professor Lauterbur, the developer of MRI, in the chemistry department at the State University of New York—and thus has the great advantage of intimate familiarity with the environment in which MRI methods were pioneered. In this column we will get to know Dr. Lauterbur and survey the historical trajectory that resulted in the birth of MRI.

Paul Lauterbur was born in 1929 in the rural town of Sydney, Ohio; he grew up on a farm, removed from city life and surrounded by the wonders of nature. From his earliest years he was interested in science and nature and dreamt of unraveling the mysteries of the natural world. After graduating with a degree in engineering from the Case Institute of Technology in Cleveland, he chose not to proceed to graduate school, and instead joined the Mellon Institute in Pittsburgh. His stated reason for not attending graduate school was that he found studying and attending lectures boring, and that he preferred instead to enter an environment in which he could conduct his own experimental research; still, while working at the Mellon Institute he enrolled at the University of Pittsburgh and received a doctorate in 1962. At Mellon he was exposed to NMR during research on the states of motion of polymers, and he conducted NMR studies of a variety of nuclei, including the first observations of NMR spectra for ^{13}C isotopes—work which quickly became famous. In 1963, his skills and his research track record earned him an invitation to join the State University of New York at Stony Brook—which had only just been founded—as an associate professor, which he accepted, thus joining the academic track after all. This career trajectory was somewhat unique among American academic scientists, a factor which may well have been related to Lauterbur's essential creativity. In his new position, Lauterbur first studied applications of NMR to chemistry, but toward the end of the 1960s his interest shifted to applications of NMR to biological macromolecules and living systems.

In the summer of 1971, Lauterbur came to be involved in the management of a small NMR company. It was in this company's laboratory that he became acquainted with NMR observations of mice, and where he had the first-hand opportunity to observe that NMR signals from mice with tumors exhibited clearly distinct behavior from signals from healthy mice. This discovery had a profound impact on Lauterbur: it occurred to him that many incredible opportunities could be realized if only it were possible to observe an organism alive and *in situ*—without removing any sample tissue—and to determine which parts of the organism were responsible for specific MRI signals. He then began to turn over in his mind a variety of techniques for enabling such measurements. It was at this point that he hit on the idea of using a magnetic field gradient to investigate the interior of three-dimensional bodies.

In his 1973 manuscript—which later became the centerpiece of his Nobel Prize award—Lauterbur schematically illustrates the operating principles of MRI by depicting samples of water (H_2O) in two capillary tubes placed in a glass vessel with an inner radius of 4.2 mm filled with heavy water (Figure A). He also states the possibility of using differences in relaxation times to enable selective imaging. The key insight in his proposal was the idea of using a magnetic field gradient to achieve spatially-resolved information; one hears many stories of scientists who read the paper and dejectedly wondered "Why didn't I think of that!!" And yet the problem is surely among those whose solution is easy to see only in hindsight. Moreover, executing the idea and proving its validity by obtaining actual images was certainly no easy task. The first publication only included 2-dimensional images, but Lauterbur quickly developed an algorithm for assembling three-dimensional images by capturing image data from a range of different angles, and the technique was used almost immediately to observe the three-dimensional distribution of water in a living organism. The first test subjects to be imaged were clams taken from a nearby beach; the next images were of living field mice. Already at this time the Chemistry department at Stony Brook was aflutter with high hopes that this work would lead to a future Nobel Prize.

Hopes of applying the technique for medical applications were high from the very start and are mentioned in the first publications; however, this required the development of magnets large enough to enclose a human being, and thus the first applications to clinical medicine did not come until the 1980s. After this time, many corporations joined the fray, and the field began to develop rapidly. Comparing the early image of the chest of a field mouse—circa 1975—to the modern image shown on page 462 (taken in the early 21st century) suggests an entirely different world of sensitivity and resolution, and yet the entire field started with the simple image in Figure B.

One aspect of the history of MRI that this author has always found particularly striking is that, throughout the initial phases of the research, Dr. Lauterbur did not rely on the assistance of his graduate students or postdoctoral researchers, performing instead to achieve everything on his own. Thus we see here an example of research born essentially out of serendipitous coincidence, neither planned nor supported by large budgets, which nonetheless spurred the development of an entirely new field. This history offers a perfect case study in the ability of small-scale research—based on the ideas of a single individual of surpassing ingenuity—to lead to revolutionary progress. The story of the development of MRI speaks powerfully to the importance of fostering and encouraging precisely this sort of research.

Progress in methods of separation and analysis

The various methods of spectroscopy and magnetic resonance that we have discussed thus far have also enjoyed widespread adoption as important methods of instrumental analysis. For example, fluorescent X-rays have been widely used for analytical purposes as characteristic X-rays emitted by elements in a substance. Two other methods that have proven particularly important as techniques of separation and analysis in modern chemistry are mass spectrometry (MS) and chromatography. Although both of these techniques were invented before the Second World War (see Section 4.6), in the postwar era they underwent significant development and came to be used for research in a wide range of fields of chemistry. Combinations of the techniques, such as GC/MS and LC/MS, have also proven to have important analytical applications. Chromatography made major contributions to the development of biochemistry, while mass spectrometers played an important role in determining the structures of proteins and have become essential tools for biochemical and medical research.

Advances in methods of mass spectroscopy

The mass spectrometer developed by Aston and Dempster between 1918 and 1919—an apparatus that ionized a sample and measured its mass/charge ratio (see page 193)—was used before the Second World War primarily to study isotopes. The resolution of the apparatus was significantly improved by the introduction of a double-focusing mass analyzer by the American, Dempster, and the Germans, Mattauch and Herzog, in 1935 and 1936. After the war, the device experienced dramatic improvements in both resolution and sensitivity, allowing the analysis of molecules; this made it an important research tool across a wide range of fields in chemical research. The apparatus also came to play an important role as a detector for spectroscopic measurements and for studies of reactions. These advances were made possible by revolutionary progress in vacuum engineering, electronics, and computer technology.

A mass spectrometer consists of three main parts: an ionizer, an ion analyzer, and an ion detector. We first survey the evolution of sample ionizers, for which a variety of ingenious methods have been invented. One simple method that was long used is electron ionization (EI), in which energized electrons emitted from a heated filament collide against atoms or molecules in a sample to induce ionization. However, this technique suffers from a number of drawbacks, including the fact that it tends to cause fragmentation of the sample molecules and that it places restrictions on the molecular weights of those molecules. Methods developed to replace EI include the chemical ionization method, which utilizes ion-molecule reactions, and the field desorption method, in which a whisker electrode is coated with the sample and ionization is induced by strong

John Fenn (1917–2010)
Fenn was an American analytical chemist. After completing his doctorate at Yale University in 1940, he worked in the corporate world before beginning research at Princeton University in 1952; from 1967 onward he was a professor at Yale University. In 1988 he developed the electron-spray ionization method, a pioneering technique in polymer mass analysis. He received the 2002 Nobel Prize in Chemistry.

Koichi Tanaka (1959–)
After graduating from Tohoku University in 1983 with a degree in Electrical Engineering, Tanaka joined Shimadzu Corporation. At Shimadzu's central laboratories he conducted applied research on the mass analysis of polymeric substances. In 1987 he developed a technique for using laser light to irradiate a polymer mixed with a matrix to produce ionization without destroying the polymer. He became a Fellow of Shimadzu Corporation in 2002. He received the 2002 Nobel Prize in Chemistry.

electric fields near the electrode. However, both of these methods are limited in the types of samples they can address. The breakthrough development that made possible the successful soft ionization of proteins and other polymer substances—thus making mass spectrometry an important analytical tool in biochemistry—was achieved in the mid-1980s by Fenn, Tanaka, Hillenkamp and others.

In 1984, the group of John Fenn and collaborators formed charged liquid droplets by spraying a sample dissolved in solution through a capillary to which a high voltage was applied. By distilling solvent molecules from the products of this step it was possible to analyze the liquid drops, ultimately yielding ions of the original sample molecules; this was the development of *electro-spray ionization*. Unlike other ionization techniques, this was a "soft" method that could be used to ionize proteins and other polymers[44].

In 1985, Hillenkamp, Karas and collaborators mixed the amino acid alanine in a tryptophan matrix and demonstrated that ionization could be easily achieved via irradiation with 266 nm laser light. It was subsequently confirmed that mixing with a matrix in this way allowed ionization even of high-molecular-weight peptides; the technique was named *Matrix-Assisted Laser Desorption/Ionization* (MALDI). In 1987, Koichi Tanaka and a group at Shimadzu Corporation used a glycerol matrix blended with fine cobalt particles and illumination with a 337 nm laser to achieve successful ionization of a protein, carboxy-peptidase A,

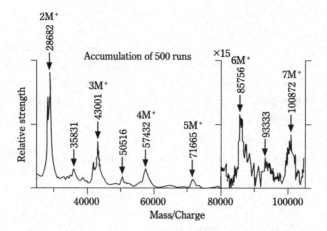

Figure 5.20: Mass spectrum of cluster ions from lysozyme (molecular weight 14.3 kDa)

which has a molecular weight of 34.5 kDa; this opened the door to applications of MALDI for protein analysis[45]. Hillenkamp and coworkers also immediately proceeded to demonstrate ionization of albumen (67 kDa) using a nicotine acid matrix and 266 nm laser light[46]. In this way, mass spectrometry by MALDI became an extremely powerful method in biochemical research (Figure 5.20). John Fenn and Koichi Tanaka received the 2002 Nobel Prize in Chemistry for the development of soft desorption/ionization methods for mass spectroscopy of biological macromolecules. Tanaka's receipt of the Prize was viewed as an example of how a corporate researcher—blessed with a good research idea and a little bit of luck—can conduct high-impact research.

A method of ion analysis that has been used since before the Second World War is to pass ions through a magnetic field and utilize the deviation in their flight trajectories due to the Lorentz force. Using double-focusing techniques—in which both electric and magnetic fields are used to reduce the spread in the initial angles and energies of the ions—high resolution can be achieved.

The velocity of an ion traveling through an electric field is dependent on the mass of the ion; therefore, the time required for ions to reach a detector can be used to separate ions by mass. This idea existed before the Second World War, but there had been no successful demonstrations of the concept. In the postwar era, the technology for manipulating extremely short electrical pulses advanced, leading to the development of time-of-flight (TOF) mass spectrometers. These devices release ions at a given instant and record the time elapsed before they reach a detector.

In 1953, German physicists Paul and Steinwedel developed a quadrupole mass spectrometer. In this device, ions are passed between four parallel rod-shaped electrodes to which a fixed voltage and a high-frequency voltage are simultaneously applied; only certain ions succeed in passing through the gap between the electrodes. By varying the applied voltage as an ion beam is passing through the device, it is possible to measure the mass/charge ratio of the ions, thus separating the ions by mass. This technique is convenient for use in small-scale detectors. Quadrupole electrodes are also used as ion traps. In this case, by varying the voltage, ions are isolated by their selective release. In 1974, a Fourier-transform ion cyclotron resonance[14] mass spectrometer was developed; this device achieves extremely high resolution. These methods may be used as appropriate for the particular purpose at hand; it is also possible to combine the methods to produce a tandem method.

Tools for ion detection include electron multiplier tubes and microchannel plates. By plotting experimental data on a graph with the mass/charge ratio on the horizontal axis and the detected intensity on the vertical axis, one obtains a mass spectrum. Mass spectra for many organic compounds and biological molecules exist in databases, and comparison with these allows compounds in samples to be identified.

Chromatography

The modern methods of chromatography pioneered by Martin and Synge developed into a variety of new techniques in the postwar era. These included gas chromatography (GC), high-performance liquid chromatography (HPLC), thin-layer chromatography (TLC), size-exclusion chromatography (SEC), and supercritical fluid chromatography (SCFC).

The possibility of GC had been proposed already in the 1941 paper of Martin and Synge, but it was overlooked for most of the following decade. In 1949, Martin collaborated with James to develop this technique; in his acceptance speech for the 1952 Nobel Prize in Chemistry he announced the successful separation of many organic acids and amines using gas chromatography[47]. Martin and Synge injected a sample of a mixture into a heated glass tube containing small-particle carriers coated with a non-volatile liquid, and then detected the well-separated sample components by forcing them out of the

---

14. Cyclotron resonance: In a magnetic field of flux density B, a particle of mass m and charge Ze executes circular motion at the cyclotron frequency, $\omega_c = ZeB/mc$. Resonant absorption of electromagnetic energy at that frequency increases the radius of the orbit; this is known as *cyclotron resonance*.

glass tube using compressed gas. This technique was immediately adopted by petroleum chemists as an easy method for separating mixtures of hydrocarbons. In the 1950s, heat conduction detectors, hydrogen flame ionization detectors, and other high-sensitivity detectors were developed; these were rapidly adopted for widespread use in analyzing organic compounds.

The research by Martin and Synge also demonstrated the possibility of high-performance liquid chromatography (HPLC) using small-particle adhesives and high pressures. The HPLC method—using a high-pressure pump to force liquids through a substrate consisting of a porous silica gel doped with alkyl groups—has been used since the late 1960s and has come to enjoy widespread use as a simple and highly reliable technique in analytical chemistry and biology.

In 1938, the Russian pharmacologists Izmailov and Shraiber reported that a glass substrate coated with a thin layer of aluminum oxide could be used in the same way as paper chromatography; however, at the time this research did not attract significant attention. Although there were numerous developments in chromatography from the late 1940s into the 1950s, it was in 1956 that German chemist Stahl established the method of thin-layer chromatography (TLC) with a silica-gel substrate. TLC was then rapidly adopted as an analytical technique. In 1973, Halpaap developed the method of high-performance TLC (HPTLC), which uses a substrate coated with a thin layer of silica gel formed from small-diameter particles; this method yielded significant improvements in analytical precision.

Porous molecular sieves can separate samples based on molecular sizes, and chromatography based on these tools is known as *size-exclusion chromatography.* (SEC) The method of *supercritical fluid chromatography* (SCFC) uses a supercritical fluid—a fluid maintained at high pressure above its critical temperature—as the moving phase. Both of these methods have found wide application in a number of specific areas.

5.3 Advances in theoretical and computational chemistry: Understanding and predicting chemical phenomena

Efforts to apply quantum mechanics and statistical mechanics to develop an understanding of chemical phenomena were initiated in the first half of the 20th century by a number of chemists and physicists of the day, and it was during this time that theoretical chemistry emerged as an important subfield within chemistry as a whole (see Section 4.3). The birth of quantum mechanics provided the essential framework within which the essence of chemical phenomena could be understood; however, the wave equation for atoms and molecules with many electrons could not be rigorously solved, and thus the central question became how to construct approximate solutions that describe experimental reality as accurately as possible. Even for approximate solutions, the computational cost grows exponentially as the number of electrons in the problem increases, a fact which restricted calculations to simple molecules. When computers were developed in the postwar era, they were immediately applied to molecular calculations, but computer performance remained poor throughout the 1950s, and many arguments remained qualitative, based on methods such as molecular orbital theory using empirical parameters or valence bond theory based on chemical intuition. The 1960s saw the development of transistor-based computers, while integrated circuits (ICs) and microprocessors were introduced in the 1970s; these developments spurred the astonishing growth in computer performance that has continued through to the present. Between the 1960s and the present day, the computational speed of computers has grown in rough accordance with Moore's Law, which predicts a doubling of speed roughly every 2 years; by the year 2000, the computational speed of computers had grown by a factor of around 10^6 compared to the computers of 40 years prior. This progress enabled gargantuan numbers of computations to be performed, and molecular-orbital calculations with no dependence on empirical parameters eventually came to dominate quantum chemistry; the reliability of these calculations also improved remarkably. Indeed, even biologically relevant molecules and other complex molecules became accessible to computational studies. Thus, by the end of the 20th century computational quantum chemistry had become an essential tool for analyzing many experimental results.

For theoretical treatments of chemical phenomena in complex systems such as polymers, solutions, solids, and surfaces, molecular dynamics (MD) simulations—made possible by the progress of computers—came to play a key role in analyzing experimental results and obtaining information at microscopic scales that cannot be easily obtained from experiment. As the performance of computers improved, the approximations in the models used could be refined, yielding a sequence of ever-more-reliable computational results.

Another area that witnessed major progress is the theory of chemical reactions. Here quantum chemistry calculations made it possible to gather reliable information on phenomena such as transition states and reaction pathways that cannot be gleaned from experiments. In this way it became possible to reproduce theoretically a number of experimental results regarding the reactions of simple molecules in the gas phase; the theoretical understanding of complicated reactions in solution or catalytic reactions has also made impressive progress.

By 1950, there were still only a small number of theoretical chemists in the world. Today the number of chemists involved in theoretical and computational chemistry has grown immensely, and these scientists also play important roles in the chemical and pharmaceutical industries.

In this section we survey progress in theoretical computational chemistry—placing particular emphasis on the development of methodologies—in two particular areas: (1) quantum chemistry computations and (2) computations in thermodynamics and statistical mechanics. The evolution of computational chemistry in reaction theory will be discussed in Section 5.4.

Computational quantum chemistry[48]

As the 20th century progressed into the second half, chemists used the concepts of molecular orbitals and valence bond theory to acquire an increasingly sophisticated qualitative understanding of chemical bonds, molecular structure, and reactivity. The electronic structure of simple molecules was explained on the basis of molecular orbital diagrams, while the properties of aromatic hydrocarbons were explained using Hückel molecular orbitals (HMOs, see page 250). HMOs were originally restricted to treatment of π electron orbitals, but in 1963 Hoffmann proposed an extended Hückel method that included σ orbitals, which could explain the stability of many molecules. Successful examples of theoretical studies based on these early quantum-chemistry methods include Fukui's theory of frontier orbitals and the Woodward-Hoffmann rules, which successfully explained many experimental results in organic chemistry and had a major impact on the field. These developments are discussed in Section 5.4. Still, calculations at the level of HMOs were unable to explain many other experimental results, particularly those concerning electronic transitions or electronic spectra. Efforts to achieve higher-precision quantum chemistry calculations began in earnest around 1950.

The Hartree-Fock (HF) Model

Attempts to use the Hartree-Fock method (see page 219)—which had been proposed to treat many-electron atoms—to treat electrons in molecules began around the end of the 1940s. In 1951, Roothaan at the University of Chicago proposed a method using the mean-field approximation to express the N-electron wavefunction of a molecule as a Slater determinant involving N single-electron wavefunctions consisting of molecular orbitals Ψ_i[49]. The Slater determinant expressed as a product of molecular orbitals was antisymmetric with respect to electron exchange. The molecular orbitals Ψ_i were expressed as linear combinations of basis functions χ_μ which primarily represented atomic orbitals χ_μ:

$$\psi_i = \Sigma c_{\mu i} \chi_\mu$$

This was called Roothaan's equation; in matrix notation it takes the form $FC = SCE$ and represents the Hartree-Fock approach to molecules[49]. Here F is known as the *Fock matrix*, C is the matrix of expansion coefficients, S is the matrix of overlap integrals, and the entries of E are the orbital energies. As in the case of atoms, this equation is solved self-consistently (for a description of the self-consistent field method, or SCF, see page 219); the resulting technique is known as the *Hartree-Fock* method.

Roothaan's initial equation assumed the *restricted Hartree-Fock* (RHF) configuration, appropriate for closed-shell molecules in which each orbital is occupied by 2 paired electrons. In 1954, this was extended by Pople and Nesbet to the *unrestricted Hartree-Fock* (UHF) method, capable of treating open-shell molecules containing unpaired electrons. Solving the Hartree-Fock equations for a molecule required the calculation of many types of integrals that enter the matrices; this posed a major computational challenge before the advent of high-speed computers. The original choice of basis functions were atomic orbital functions of the Slater type (Slater-type orbitals, STOs), but the calculation of multi-center integrals for STOs was difficult. For this reason, a variety of semi-empirical methods were proposed; in these techniques, computationally expensive integrals were treated as parameters to be determined by fitting to experimental data. A representative example of this type of semi-empirical technique was the Pariser-Parr-Pople (PPP) method, proposed in 1953 to treat π-electron systems; this method was effective for explaining the spectrum and electronic transitions of conjugated compounds[50,51]. Semi-empirical methods were extended in the mid-1960s into methods capable of treating all valence electrons; techniques such as CNDO (which completely neglected differential overlap within electron repulsion integrals), INDO (which retained some of those contributions), and MNDO (which adjusted the parameters in molecular integrals to fit experimental data) were all widely used. However, the accuracy and reliability of the results predicted by semi-empirical methods of this sort were questionable in view of the approximations used and the introduction of empirical parameters.

The development of *ab-initio* methods, which do not use experimental data, also began in the 1950s. In 1950, Boys at Cambridge University showed that using Gaussian-type orbitals (GTOs) as basis functions allowed the integrals needed for SCF calculations to be computed analytically. As approximations of atomic orbitals, Gaussian-type orbitals are inferior to Slater-type orbitals; however, the approximation improves as more basis functions are included. In the early days of computers, both Slater-type and Gaussian-type orbitals were used as basis functions; however, by the late 1960s most chemists used programs based on Gaussian-type orbitals. In 1969, John Pople's group presented a program using a linear combination of one STO and 3 GTOs—the so-called STO-3G basis[52]—and this was improved in 1971 to 6-31G. In the ensuing years, programs performing *ab-initio* quantum chemistry calculations using Gaussian-

John Pople (1925–2004)
Pople received his degree in mathematics from Cambridge University in 1951. He was interested in theoretical chemistry and developed methods for quantum-chemistry calculations. In 1964 he became a professor at Carnegie-Mellon University in Pittsburgh. His *Gaussian* program, first released in 1970, has been subject to continual improvements and remains widely used worldwide for molecular calculations. He made leading contributions to the development of quantum-chemistry calculation methods, including both semi-empirical methods in the 1950s and 60s and *ab-initio* methods from the 1970s onward. He received the 1998 Nobel Prize in Chemistry.

type orbitals and outfitted with numerous improvements were publicly released; these became widely used to provide the necessary level of precision.

Treatment of electron correlation

A major drawback of the Hartree-Fock (HF) model is that it ignores correlations between the motion of electrons with antiparallel spins. This problem is intrinsic to efforts to express the many-electron wavefunction as a single Slater determinant, and improving it requires the addition of further determinants. Thus the method of *configuration interaction* (CI), in which the ground-state wavefunction obtained in the HF method is augmented by additional wavefunctions describing excited-state electron configurations, came to be used in the 1970s and thereafter. In this method, the ground-state wavefunction Ψ is expressed in the form

$$\Psi = a_0 \Psi_0 + \sum a_i \Psi_i + \cdots\cdots$$

Here Ψ_0 is the wavefunction corresponding to the ground-state electron configuration in the HF approximation, while Ψ_i describes an excited-state wavefunction in which electrons occupy other orbitals; the coefficients $\{a_0, a_i\}$ are determined by the variational method. The technique exists in multiple variants, including CISD (in which only singly-excited or doubly-excited electronic configurations are retained) and CISDT (which additionally retains triply-excited configurations). As the performance of computers improved, it became possible to consider other extensions, including the full-CI method and the *multiple-reference*-CI (MR-CI) method, in which the configuration that serves as the basis for excitations is taken to be not only the HF ground-state

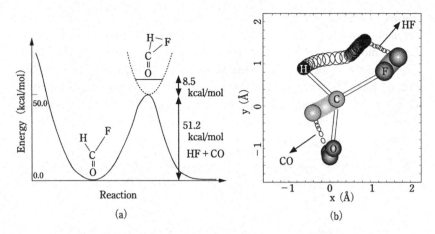

Figure 5.21: Results of a quantum-chemical calculation of the reaction in which an HFCO molecule dissociates into HF and CO

(a) Variation in the potential energy associated with the reaction.
(b) Variation in the molecular structure associated with the reaction. The hydrogen-carbon and fluorine-carbon bonds break, resulting in bonding between hydrogen and fluorine atoms.

configuration, but also includes multiple additional electronic configurations. Other high-precision CI methods have also been developed; these include cluster expansion methods—which describe multi-electron excitations as products of excitations involving fewer electrons—and multiple-configuration SCF methods, which optimize the molecular orbitals themselves together with their linear-combination coefficients in the CI wavefunction.

Chemists attempted to improve computational accuracy using methods other than CI. Moller-Plesset perturbation theory, a technique proposed in 1934 to treat many-electron atoms, was applied to molecular computations in the 1970s. In this approach, electron correlations are captured by retaining perturbative terms of orders 2 or greater; the procedure with perturbations of order n is known as MPn (where n=2,3,4,...).

Thus quantum-chemical calculations gradually became more precise and eventually attained a degree of reliability. These results not only helped to explain experimental results but also yielded information on reaction intermediates and transition states that are not accessible by experiment. Figure 5.21 shows one example of such a computational result.

John Pople, who played a leading role in developing quantum-chemical computation methods based on molecular orbitals, shared the 1998 Nobel Prize in Chemistry with Walter Kohn, who invented the technique of density functional theory discussed below.

Density Functional Theory (DFT)

The method of *density functional theory* (DFT), which calculates the electronic states of molecules and solids in terms of the electron density, has been widely used since the 1970s as a tool complementary to *ab-initio* molecular orbital theory for handling many-electron systems. In this theory, the electron density replaces the electron wavefunction as the central object of study. The foundations of density functional theory are based on the following two theorems, established in 1964 by Hohenberg and Kohn[53]. For a non-degenerate system of N electrons,

> *Theorem 1:* The ground state energy, E_g, is uniquely determined by the single-electron density $\rho(r)$. Here $\rho(r)$ may be determined from the solution of the Schrödinger equation in an external potential V.

> *Theorem 2:* Among all N-normalized trial electron densities $\rho'(r)$, the ground-state energy E_g assumes its minimal value for the true ground-state density $\rho'(r)$.

These theorems represent properties that were known to hold for the ground-state wave function, but express these properties in terms of the ground-state electron density; they demonstrate that ground-state wave functions and ground-state electron densities exist in one-to-one correspondence. In 1965, Kohn and Sham introduced the notion of orbitals into DFT; noting that the electron density $\rho(r)$ may be expressed in terms of N single-electron orbitals $\{\Psi_i(r)\}$ in the form $\rho(r)=\Sigma|\Psi_i(r)|^2$, they derived an equation known as the *Kohn-Sham* equation that resembles the Schrödinger equation for a single-electron orbital[54]. To solve this equation requires an approximate representation of the exchange/correlation interaction E_{xc} to the total electron-electron interaction energy. For this purpose Kohn and Sham used an approximation known as the *local density approximation* (LDA), in which one uses the expression for E_{xc} appropriate for the homogeneous electron gas. For applications to molecular problems, a variety of improvements were later developed; these include the local spin density approximation (LSDA), the generalized gradient approximation (GGA), and hybrid functionals that make use of the exact Hartree-Fock exchange term.

Because density-functional theory allows calculations that retain electron correlation to be completed in a short amount of time, it is well-suited for application to large molecules, surfaces, and solids, and it has been widely applied to such problems; Figure 5.22 shows an example of computational results for large systems. However, in comparison to the MO method, the DFT technique suffers from several drawbacks:

1. At present, the E_{xc} functional is only known approximately or empirically, and there are no standard criteria for assessing the accuracy of a density functional.

Walter Kohn (1923–)
Kohn was born in Vienna but fled to England when the Nazis occupied Austria. After receiving his undergraduate education at the University of Toronto in Canada, he received a doctorate in theoretical physics from Harvard University, working with Schwinger. From 1960 to 1979 he was a professor at the University of California at Santa Barbara. His interest in the electronic state of alloys led him to develop density functional theory, which came to be widely used for chemical calculations on systems such as large molecules and surfaces. He received the 1988 Nobel Prize in Chemistry.

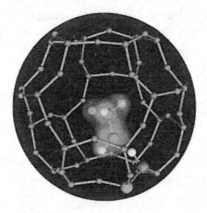

Figure 5.22: Structure of methanol inside a zeolite lattice, as computed using density functional theory

2. Calculation of physical quantities is difficult in some cases.
3. The accuracy of computations for excited states is poor.
4. Treatment of weak interactions such as van der Waals forces is generally difficult.

Molecular mechanics and computational methods for complicated molecules
Molecular mechanics (MM) is a technique in which (a) the atoms that comprise molecules are treated as spheres, (b) physical quantities—such as the force constants associated with the stretching of atomic bonds or the variation of bond angles, as well as the strengths of interactions between unbound atoms—are taken as parameters, and (c) the molecular potential energy is determined as a function of the atomic positions and used to predict quantities such as molecular

Martin Karplus (1930–)
Karplus is an Austrian-born American theoretical chemist. After receiving his PhD under Pauling at the California Institute of Technology, he taught at the University of Illinois and Columbia University before becoming a professor at Harvard in 1967. His accomplishments span many fields of theoretical chemistry; he is particularly well known for molecular-dynamics simulations of biological macromolecules and for the Karplus equations describing spin-spin coupling in nuclear magnetic resonance. He received the 2013 Nobel Prize in Chemistry.

Michael Levitt (1947–)
Born in South Africa to Jewish immigrants, Levitt studied physics at Kings College, received his PhD in computational biology from Cambridge, and today is a professor of structural biology at Stanford University. He is known for his pioneering research on molecular-dynamics calculations of DNA and proteins. He received the 2013 Nobel Prize in Chemistry.

Arieh Warshel (1940–)
Warshel is an Israeli-American theoretical chemist born in Palestine. After receiving his PhD in chemical physics from the Weizmann Institute of Science in Israel, he worked at Harvard, at the Weizmann, and at Cambridge before becoming a professor at the University of Southern California in 1976. He is known for his pioneering accomplishments in computer simulations of the functional properties of biological systems. He received the 2013 Nobel Prize in Chemistry.

structure or conformation in equilibrium, vibrational frequencies, and heat of formation. Because the method requires little computation time, it may be used to simulate very large molecules or molecular aggregates. In the early 1970s, Allinger and colleagues developed the first MM program, named simply MM1; since that time, a variety of programs offering different choices of parameters have been developed, and these have been widely adopted for purposes such as predicting the structures of complicated organic molecules. MM considers various contributions to the potential energy between atoms—including stretching vibrations, bending vibrations, contortions of dihedral angles, out-of-plane vibrations, electrostatic interactions, and van der Waals interactions—and determines the optimal structure that minimizes the total potential energy. Compared to calculations in *ab-initio* molecular orbital theory, MM calculations require far less computation time; still, computations on giant molecules such as proteins require huge numbers of parameters. Various techniques have been proposed for reducing the number of parameters, but these frequently result in reduced computational accuracy.

Although MM is capable of treating large systems, it is not well-suited to studies of chemical reactions. On the other hand, quantum chemical methods such as *ab-initio* computations and DFT techniques cannot treat large systems. For this reason, the years since the 1970s have witnessed the development of hybrid methods that combine quantum-chemical calculations with MM and other methods of classical mechanics to treat complicated systems such as chemical reaction systems or biomolecular systems. In recent years, the QM/MM method and the ONIOM technique[15]—methods which exploit the advantages of both quantum-mechanical and classical calculations—have been particularly fruitful. Three scientists who played pioneering roles in developing computational methods for complicated molecular systems of this type—Martin Karplus, Michael Levitt, and Arieh Warshel—received the 2013 Nobel Prize in Chemistry for "the development of multiscale models for complex chemical systems".

Thermodynamics and Statistical Mechanics

Applications of statistical mechanics to complicated systems such as solutions made solid progress in the second half of the 20th century, and the new fields of non-equilibrium thermodynamics, non-equilibrium statistical mechanics, and polymer solution theory were developed. The increasing speed of computers since the 1970s has made molecular dynamics (MD) calculations an important tool for the structural determination of proteins and other complex systems. As it became possible to compute the structure and macroscopic properties of complex molecular systems, these new tools came to stand alongside quantum-chemistry calculations as key pillars of computational chemistry, and they have found wide application.

Non-equilibrium thermodynamics and dissipative structures

A new development in the statistical mechanics of non-equilibrium processes was the advent of *linear response theory,* a subject to which major contributions were made by physicists in the 1950s, particularly Ryogo Kubo of Tokyo University. This theory considers the application of an external field—such as an electric or magnetic field—to a system in a state of thermal equilibrium, and attempts to discern the resulting change in the state of the system; it would

---

15. ONIOM: A computational method for studying proteins and other complicated molecular systems in which the core regions of the system are treated using high-accuracy quantum chemistry calculations, while more distant portions are treated using simpler, less accurate methods. Developed by Keiji Morokuma and others.

Ryogo Kubo (1920–1995)
After graduating in 1941 from Tokyo Imperial University, Kubo was a professor at the University of Tokyo from 1954 to 1980. From 1981 to 1992 he was a professor at the Faculty of Science and Technology of Keio University. He was known for his work on the elasticity of rubber and on magnetic resonance, and for developing linear response theory for irreversible processes. He received Japan's Order of Culture in 1973 and the Boltzmann Medal in 1977.

Ilya Prigogine (1917–2003)
Prigogine was born in Russia but fled with his family, who was critical of the post-revolutionary government; they passed through Germany and settled in Belgium in 1929. Prigogine studied chemistry at the Free University of Brussels, where in 1951 he became a professor of physical chemistry and theoretical physics. At Brussels he led a research group in statistical mechanics and thermodynamics and made major contributions to the theoretical understanding of non-equilibrium thermodynamics, particularly to dissipative structures. He is also known for his monographs, including *Order out of Chaos* and *The End of Certainty*. He received the 1977 Nobel Prize in Chemistry.

go on to play a key role in interpreting relaxation phenomena in fields such as magnetic resonance and laser spectroscopy. In this theory, the effects of time-dependent external fields are treated as perturbation, and the response of physical quantities to external fields is computed approximately—to first-order accuracy in the external field strengths—using the equations of motion for density matrix describing the system.

Many phenomena of interest in the natural world involve non-equilibrium states. The field of non-equilibrium thermodynamics—for which the groundwork was laid by Onsager in the 1930s—was taken up in the immediate postwar era by Ilya Prigogine, who ushered in key new developments. Prigogine's group at the Free University of Brussels in Belgium made dazzling contributions to a wide variety of fields, including the statistical thermodynamics of solutions, nonlinear thermodynamics, and non-equilibrium statistical thermodynamics; in particular, the group advanced the statistical theory of nonlinear thermodynamics for systems in states far removed from equilibrium and emphasized the notion of dissipative structures as mechanisms for imposing order in non-equilibrium systems[55]. The term *dissipative structures* refers to macroscopic structures that arise in physical systems undergoing dissipative processes in which mechanical

Paul Flory (1910–1985)
After completing his doctorate at Ohio State University, Flory joined DuPont Laboratories, where he worked in polymer research under Carothers. In 1948 he was invited to become a professor at Cornell University, and in 1957 he became director of the Mellon Institute; from 1961 to 1975 he was a professor at Stanford University. He laid the physicochemical foundations for many areas of polymer chemistry, using both theoretical and experimental approaches; he made major contributions to developing the thermodynamics and statistical mechanics of polymer solutions and chain molecules. He received the 1974 Nobel Prize in Chemistry for his fundamental contributions to understanding the physical chemistry of polymers.

or electrical energy is converted irreversibly into heat. One well-known reaction that furnishes a canonical example of a dissipative structure in a chemical reaction system is the Belousov-Zhabotinsky (BZ) reaction, in which the density varies periodically in both space and time. Prigogine's group proposed a mathematical model of a simple reaction network known as the *Brusselator*, which helped to illuminate oscillatory reactions of this type and clarified the conditions under which they occur.

Polymer solution theory
One of the many postwar advances in the statistical thermodynamics of solutions was the development of *polymer solution theory*. Experiments on freezing-point reduction and osmotic pressure in polymer solutions had highlighted the strong deviations of these solutions from ideal behavior. In 1942, Flory and Huggins independently used lattice models[16] to compute thermodynamic quantities, which they used to explain the vapor pressure and osmotic pressure of polymer solutions. Starting around 1948, Flory introduced the notion of excluded volume into the theory of polymer solutions; this opened the door to new theoretical developments[56]. This paradigm begins by observing that each structural element in a molecular chain of linear polymers has a volume; this volume must thus be excluded from the space that may be occupied by the other structural

---

16. Lattice models: A technique in which one imagines the system under consideration to be covered by a three-dimensional lattice such as a crystal lattice, on whose sites exist solvent molecules and solute molecules; thermodynamic quantities are then calculated by computing the partition function with respect to the configurations of these molecules.

elements. This effect causes the spreading of a polymer chain in solution to be greater than would be expected were the volume-exclusion effect neglected. Based on this concept, Flory was able to explain many problems regarding the properties of polymer solutions. In actual polymers there are interactions between the structural elements; the temperature at which the effect of these interactions precisely cancels the effect of the volume-exclusion phenomenon is defined as the theta temperature θ. At this temperature, volume-exclusion effects are negligible and properties arising solely from interactions between neighboring elements become particularly noteworthy; this observation pointed up the importance of measuring the physical properties of polymer solutions. In later years the polymer solution theory pioneered by Flory was further developed by many scientists.

Computer simulations: Molecular Dynamics and Monte Carlo methods

In many cases, detailed information on the microscopic structure and properties of assemblies of atoms and molecules is difficult to obtain experimentally. The field of *molecular simulation*—which attempts to use computer simulations, based on models of the structure and interactions of atoms and molecules, to glean information that cannot be obtained from experiment on the behavior of ensembles—emerged as a major area of research as the performance of computers improved. Typical examples of such methods include molecular dynamics (MD) and Monte Carlo methods; of the two, MD is more frequently applied to problems in chemistry.

In MD methods, one hypothesizes an appropriate set of potentials describing interactions between the molecules that make up a complex system. Given these potentials, one computes the total force f on each molecule due to all other molecules in the system, then integrates the Newtonian equation of motion from classical mechanics, $f=ma$, to solve for the dynamics of the system. These integrals are impossible to evaluate analytically; instead, one solves them numerically to determine the motion of each individual molecule, thus determining the time variation of the molecular ensemble as a whole. The results may be used to simulate the stable structures, thermodynamic properties, and dynamics of systems. The origins of this method may be traced to the 1957 work of Alder and Wainwright, who modeled an ensemble of rigid spheres to simulate solid—liquid phase transitions[57]. The method was extended by Rahman in 1964 to a system of mass points with a continuous potential, and it gradually became possible to study increasingly complex systems of molecules. Advances in the development of computational algorithms, innovations in the treatment of boundary conditions, and the availability of high-speed computers in the 1970s and thereafter allowed the treatment of large numbers of molecules, and the

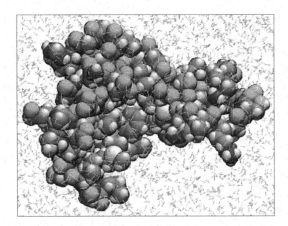

Figure 5.23: An example of a MD simulation

A protein and solvent water molecules, as modeled using MD simulations. Thin lines indicate water molecules.

method came to be widely adopted in many fields of chemistry. As computational speeds improved, the reliability of calculations improved in parallel. Studies based on MD simulations have made noteworthy contributions to answering many questions, including the structure and function of water and of proteins; in addition to the usefulness of the technique for basic research in chemistry, it has also come to enjoy widespread adoption for applied problems such as the design of new drugs. Figure 5.23 shows a snapshot of a system during a MD calculation.

Because MD calculations allow calculations of statistical ensembles under a variety of conditions (including constant temperature, constant pressure, constant volume, and constant energy), they may be used to study a wide variety of systems, including bulk systems, clusters, surfaces, and interfaces. Although in principle the potentials involve sums of interactions between all atoms and molecules that comprise a system, in practice they are typically approximated by combinations of pairwise potentials acting between atoms. The potentials acting between atoms within molecules are frequently modeled using empirical or semi-empirical parameters similar to those used in MM studies; the interactions between non-bound molecules may be modeled using the electrostatic Coulomb interaction or potentials of the Lennard-Jones type, which include both dispersion forces and short-distance repulsion. All of these potentials cause problems due to inaccuracy. This difficulty is addressed by *ab-initio* molecular dynamics, in which potentials are computed using quantum-mechanical methods; although this method was developed by Car and Parrinello in 1984, its applicability to practical systems remains limited.

Monte Carlo methods use random numbers to conduct numerical simulations of randomly-occurring natural processes, thus obtaining approximate numerical solutions to a variety of problems; although the basic ideas were first proposed by von Neumann and Ulam during the Second World War, it was not until the advent of high-speed computers that these methods came to be widely used in statistical physics. One commonly-used technique for calculations on systems in equilibrium states is the Metropolis algorithm, which is used for sampling states, proposed by Metropolis in 1953. In this method, one makes a small change to the state of a system and computes the infinitesimal change dE in the energy E of the state; if $dE<0$ then the new state is selected with probability 1, while if $dE>0$ then the new state is selected with probability $e^{-dE/kT}$. The large number of states selected in this way will constitute a normal distribution, and the macroscopic properties of the system can be obtained from the ensemble average. However, this method does not allow studies of the dynamic properties of systems, and for this reason its applications to chemistry are somewhat limited compared to the MD approach.

5.4 The increasing precision of chemical reaction studies

One of the central problems of chemistry is to understand, on a microscopic level, the essence of why chemical reactions take place. In the second half of the 20th century, advances in observation and measurement technologies and progress in theoretical and computational chemistry led to major developments in the study of reactions and molecular dynamics, and the molecular-level details of chemical reactions came to be understood. In this section we survey the nature of these developments. First, on the experimental side, techniques such as chemical relaxation methods and flash photolysis became available in the immediate postwar years, opening the door to studies of fast reactions. These methods allowed direct observations of short-lived reaction intermediaries and excited states, enabling dramatic progress in studies of reaction mechanisms and excited states. The development of the laser further facilitated research on short-lived chemical species, and by the end of the 20th century it had even become possible to observe transition states in some cases, giving rise to the field of femtosecond chemistry. In studies of gas-phase reactions, crossed molecular beams allowed elucidation of the details of two-molecule reactions. There were major advances in all fields related to chemical reactions, including photochemistry, surface chemistry, and catalytic chemistry.

Theoretical understanding of reactions advanced in parallel with this experimental progress. Among the most dazzling successes to arise from early applications of quantum chemistry was the emergence of organic chemical reaction theory, based on empirical models of molecular orbitals such as Fukui's frontier electron theory and the Woodward-Hoffmann rules for the conservation of orbital symmetry. Thereafter, progress in *ab-initio* computational quantum chemistry enabled the elucidation of transition states not accessible by experiments, allowing detailed studies of reaction mechanisms. One important class of reactions taking place in solution is electron transfer reactions; a general theory of these based on thermodynamics was proposed by Marcus. The theory of reactions in solution later advanced by incorporating the latest progress in statistical mechanics.

Experimental studies of reaction rates and reaction intermediaries

Oxidation/reduction reactions involving metal complexes

Oxidation/reduction/ (redox) reactions and substitution reactions are among the most important types of chemical reactions. Electron transfer plays a fundamental role in redox reactions, and electron-transfer reactions are thus important for these processes, but the details of reaction mechanisms and reaction rates were poorly understood until after the Second World War. At that time, radioactive isotopes came to be widely used in reaction studies, and there was progress in research on simple electron-transfer reactions such as

$$*Fe^{2+}(aq) + Fe^{3+}(aq) = *Fe^{3+}(aq) + Fe^{2+}(aq) \text{ (where } * \text{ denotes a radioactive isotope)}$$

Similarly, the rates of transition-metal complex oxidation/reduction reactions involving electron transfer were eventually understood; an example of such a reaction is

$$Co^{III}(NH_3)_5Cl^{2+} + C_r^{II}(H_2O)_6^{2+} \rightarrow Co^{II}(NH_3)_5 H_2O^{2+} + C_r^{III}Cl(H_2O)_5^{2+}$$

In the early 1950s it was recognized that the properties of the metal ions involved in the formation of complexes are important factors determining the rates of such reactions. In addition, it became clear that reaction rates were heavily dependent on the electronic state of the central metallic ion. In the 1950s, methods such as the rapid-flow method and the stopped-flow method were developed for the determination of reaction rates, and these yielded rapid progress in research on fast reactions.

One pioneer of research on electron transfer reactions and ligand substitution reactions involving transition-metal complexes was Henry Taube. In research starting around 1953, he clarified the concepts of *inner-sphere* and *outer-sphere* reactions as mechanisms for electron-transfer reactions in metal complexes[58]

Henry Taube (1915–2005)
Born in Saskatchewan, Canada, Taube completed his master's degree at the University of Saskatchewan and obtained his PhD from the University of California at Berkeley. He later worked as a professor at Cornell University, the University of Chicago, and Stanford University. He was a pioneer in research on the reactivity of transition-metal complexes, and from 1950 through the 1960s he was a leading researcher of reactions involving metal complexes, including electron-transfer reaction mechanisms and ligand substitution reactions. He received the 1983 Nobel Prize in Chemistry.

 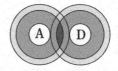

Figure 5.24: Schematic depiction of the outer-sphere and inner-sphere models of transition states

Left: Outer sphere
Right: Inner sphere
A is an electron acceptor, and D is a donor.

(Figure 5.24). The reaction involving CO^{III} and Cr^{II} complexes is a typical example of an inner-sphere reaction; here chloride ions exist within the ligand spheres of both the Co^{III} and Cr^{II} complexes, and electron transfer takes place via intermediates bridging the two ions. In contrast, outer-sphere electron transfer reactions take place without such intermediates; it was believed that electron migration takes place more rapidly than ligand substitution. The rate of electron-transfer processes in transition-metal complexes attracted the interest of theorists, eventually culminating with the Marcus theory of electron transfer.

Organic reaction theory and reaction intermediates
In the field of organic reaction theory, various relations between reaction rates and free energies—exemplified by the Hammett equation (see page 320)—were verified and refined in multiple reaction systems; at the same time, progress in observational techniques made it possible to experimentally detect certain reaction intermediates that had been hypothesized by Ingold and others on the basis of electronic theory of organic reaction.

George Olah (1927–)
Born in Budapest, Hungary, Olah studied at the Budapest University of Technology and Economics, where he later taught. He fled Hungary after the 1956 revolution, passing through England before moving to Ontario, Canada, where he worked at Dow Chemical. He later returned to academia, becoming a professor at Case Western Reserve University in 1965, and after 1977 he became a professor at the University of Southern California. He showed that strong acids could be used to produce carbocations with 5-fold coordination, and he developed reactions using strong acids that were important for industrial applications. He received the 1994 Nobel Prize in Chemistry.

In Ingold's electronic theory (see page 319), the crucial intermediates in reactions such as nucleophilic substitution or dissociation reactions were carbocations in which the carbon atoms were positively charged, such as $(CH_3)_2CH^+$, which arises from the dissociation of $(CH_3)_2CHCl$. Many experimental results on the rates and stereochemistry of organic reactions were explained by assuming the existence of carbocations of this type, but many points regarding the structure of such cations remained unclear, and furious controversy as to their nature continued through the 1950s and 60s. Under ordinary conditions, the carbocations are short-lived and cannot be directly observed; the discovery that such short-lived species can, under certain conditions, be *long*-lived was made by George Olah, who pioneered the use of spectroscopic methods for understanding their structure. In the mid-1960s, Olah succeeded in using superacids—such as the extremely strong acid FSO_3H—SbF_5—to produce stable carbocations at low temperatures; he then used high-resolution NMR techniques to study their structure and stability[59]. In particular, the research by Olah resolved a longstanding controversy by showing that the 2-norbornyl cation was a non-classical ion (carbonium ion) possessing five-fold coordinated carbon atoms, as had been proposed by Winstein (Figure 5.25). Olah received the 1994 Nobel Prize in Chemistry.

The existence of *carboanions* such as CH_3^-, in which negative charge exists primarily on the carbon atom, had also been hypothesized as intermediates in dissociation reactions and other organometallic chemical reactions; eventually, the existence of stable carboanions was confirmed. For example, the diphenylmethyl carboanion was produced at low temperatures in the form of lithium and crown ether salts.[60]

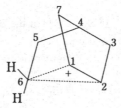

Figure 5.25: The non-classical 2-norbornyl cation

In the years after the Second World War, free radicals serving as reaction intermediaries became a subject of extensive study by ESR and spectroscopic methods. Among these, the chemistry of *carbenes*—which contain carbon atoms possessing two unpaired electrons—made significant progress beginning in the mid-1950s. Carbenes exist in many derivative forms, all with the base molecule methylene (H_2C:); questions of interest included whether the two unpaired electrons exist in a singlet or triplet state and what was the corresponding geometric structure of the molecule. For methylene and simple carbohydrate carbenes it eventually became clear that the ground state is a triplet state, while the singlet state is an excited state lying some 30 kJ/mol above the ground state. Because singlet and triplet carbenes exhibit different reactivities, these molecules were also of interest from the standpoint of organic reaction theory. For example, in addition reactions involving alkenes, singlet carbenes behave concertedly and react in a single step, whereas triplet carbenes exhibit a two-state process of radical addition. The chemistry of carbenes became an active topic of research spanning many subfields of chemistry, including organometallic chemistry, and proved to have applications in synthetic organic chemistry (see page 520).

Two chemists who helped to clarify the mechanisms of complicated organic reactions and enzymatic reactions—particularly from the perspective of stereochemistry—were Cornforth and Prelog. Cornforth used the method of isotopic labeling to elucidate the stereochemistry of enzymatic reactions in which a non-chiral molecule is transformed into a chiral molecule by the action of a chiral enzyme; in particular, he clarified the process by which the molecular backbone of cholesterol is formed. Prelog's work focused on the relationship between the stereochemistry of molecules involved in organic reactions and the progress of those reactions. He studied the stereochemistry of molecules consisting of 8 to 12 carbon atoms that formed highly mobile ring structures, as well as the stereochemistry of reactions between chiral molecules. The 1975 Nobel Prize in Chemistry was awarded to Cornforth "for his work on the

John Cornforth (1917–2013)
Cornforth was an Australian-born British organic chemist.
After receiving his BSc from the University of Sydney, Cornforth completed his PhD at Oxford in 1941, working under Robinson. In 1946, Cornforth joined the Medical Research Council, where he worked on organic synthesis, and the stereochemistry of enzyme reactions. He was a professor at the University of Warwick from 1965 to 1971 and at Sussex University from 1971 to 1975. He was completely deaf. He received the 1975 Nobel Prize in Chemistry.

Vladimir Prelog (1906–1998)
Born in Sarajevo in Bosnia-Herzegovina, Prelog received a ScD in chemical engineering from the Czech Technical University at Prague in 1929. Beginning in 1935, he worked as a researcher at the University of Zagreb, but in 1941 he fled to Switzerland to escape the Nazis. He began research at the Swiss Federal Institute in Zurich, where he became Director of the Organic Chemistry Laboratory upon the retirement of Ruzicka. His research spanned a wide range of fields. He received the 1975 Nobel Prize in Chemistry.

stereochemistry of enzyme-catalyzed reactions" and to Prelog "for his research into the stereochemistry of organic molecules and reactions."

Observation of short-lived species and studies of fast reactions

Many standard reactions, such as neutralization reactions, appear to occur instantaneously. But what is the *actual* rate at which these reactions proceed? This question was to remain unclear until the second half of the 20th century. In 1923, British scientists Hartridge and Roughton developed a rapid-mixing technique, which was later improved by Chance and others and applied to studies of enzyme reactions. However, in the first half of the twentieth century it was not possible to measure the rates of reactions occurring on timescales below milliseconds (10^{-3} s). In the early 1950s, the development of chemical relaxation methods and flash photolysis improved time resolution by a factor of almost 1000 essentially overnight, allowing measurements of processes with time scales on the order of microseconds (10^{-6} s). Then, beginning in the late 1960s, the emergence of lasers enabled observation of phenomena on timescales of nanoseconds (10^{-9} s), which later shrunk to picoseconds (10^{-12} s) and femtoseconds (10^{-15} s). Thus, over the span of 50 years, the timescales on which physical processes could be observed shrank by a factor of 10^{-12}. This technological progress led to revolutionary changes in the study of chemical reactions. In addition, studies using molecular beams yielded detailed information specific to the instant at which a reaction

Manfred Eigen (1927–)
Eigen studied physics and chemistry at the University of Göttingen and received his PhD in 1951. In 1953 he became a researcher at the Max Plank Institute for Biophysical Chemistry, where he rose to the rank of Director in 1964. He developed the chemical relaxation method and pioneered the study of fast reactions in solution. From the mid-1960s onward he worked on the application of the relaxation method on biochemical reactions. After receiving the 1967 Nobel Prize in Chemistry, he conducted research on molecular evolution and the origins of life.

occurs. Thus the combination of laser and molecular-beam techniques led to a dramatic increase in understanding transition states.

Chemical relaxation methods
When a system in an equilibrium state of the form $A \rightleftharpoons B$ is subject to a sudden change in one or more of the physical parameters governing the equilibrium—such as the temperature or pressure—the concentrations in the system evolve toward new equilibrium values. Techniques that follow these evolutions to determine the rates of the reactions $A \rightarrow B$ and $B \rightarrow A$ are known as chemical relaxation methods; they were pioneered in the 1950s by Manfred Eigen at the Max Planck Institute in Germany[61]. Eigen, together with his colleagues Tamm and Kurtze, measured the absorption of ultrasonic waves by aqueous solutions of various types of salts; in 1953 they showed that an analysis of the absorption data allowed determination of the reaction rates of rapid processes occurring in the solution. Eigen and collaborators went on to develop other techniques—including the temperature-jump method, in which a large discharge current is passed through a cell to induce a rapid temperature increase via Joule heating and the ensuing relaxation is observed, and the pressure-jump method, in which the pressure is rapidly varied—and used them to determine rate constants for many rapid reactions. For example, they determined the rate constant of the neutralization reaction $H^+ + OH^- \rightarrow H_2O$ to be 1.4×10^{11} M^{-1} s^{-1} at a temperature of 25°C. Eigen's group also measured, over a wide range of conditions, the rate of exchange between bulk water and water molecules hydrated to metal ions in aqueous solution; they concluded that this rate differed significantly depending on the metal in question. They also conducted pioneering studies of the applications of chemical relaxation methods to organic processes and enzymatic reactions such as keto-enol conversion. Chemical relaxation methods later came to be used in a wide range of fields of chemistry, including biochemistry.

George Porter (1920–2002)
After studying at the University of Leeds, Porter worked on radar development during the Second World War; in 1945 he moved to Cambridge, where he worked under Norrish, studying gas-phase reactions. In 1950 he successfully used flash photolysis to observe free radicals. He also achieved many results related to the research on energy levels and reactivity of lowest excited triplet states. In 1955 he became a professor at the University of Sheffield, and became a professor at the Royal Institution in 1966. He received the 1967 Nobel Prize in Chemistry.

Ronald Norrish (1897–1978)
Norrish was a British physical chemist. He studied at Cambridge University, where he received his doctorate and became a lecturer and then a professor in 1937. He conducted research on photochemical reactions related to combustion and polymerization reactions. Together with Porter, he developed the method of flash photolysis, which allowed studies of short-lived chemical species. He received the 1967 Nobel Prize in Chemistry.

Flash photolysis and the chemistry of short-lived species
The existence of short-lived free radicals had been postulated to explain the mechanisms of many chemical reactions, but no direct experimental confirmation had been obtained. The method of flash photolysis, introduced just after the Second World War by George Porter and Ronald Norrish, allowed the identification of short-lived chemical species and became an extremely powerful tool for studies of rapid reactions. George Porter, who was involved in the development of radar by the British Royal Navy during the Second World War, became a graduate student in the laboratory of Ronald Norrish at Cambridge after the war. Porter developed the method of flash photolysis—in which the flash produced by pulse discharge in a noble gas is used to induce photolysis, and the spectra of the resulting short-lived chemical species are tracked—and co-authored the first report of this new technique with Norrish in 1949[62]. Porter was the sole author on the second report[63]. Norrish was a well-known chemist who had already achieved many accomplishments in the study of chemical reactions, focusing primarily on photochemistry, and together these two scientists applied the new technique to the study of many photochemical reactions. In flash photolysis, information on the structure of the short-lived species is obtained from the vibrational and rotational bands in the absorption spectra, while information on reaction rates is obtained from the time dependence of the spectrum following photolysis.

The first object of study was the ClO radical produced by flash photolysis of a gaseous mixture of oxygen and chlorine; this yielded information on the structure of the radical and on the reaction itself. Later, Porter and Norrish detected many radicals in the gas and liquid phases—including benzyl and anilino radicals produced by photolysis of aromatic compounds—and determined their structure and properties.

Among the short-lived species studied by Porter and his collaborators using flash photolysis methods were the lowest excited triplet states of aromatic compounds in solution. It had been proposed by Lewis and Kasha in 1944 that phosphorescence in low-temperature rigid media of organic molecules was due to the lowest excited triplet state; however, phosphorescence was not observed in the gas or liquid phases, and thus the role of excited triplet states in photochemistry was not fully recognized. In 1952, Porter and Windsor used flash photolysis to study aromatic compounds in an ordinary solution; they succeeded in measuring the spectra of short-lived triplet states with lifetimes on the order of milliseconds, thus confirming the general existence of triplet states[64]. Porter's group went on to study various phenomena in triplet states—such as reactivity, energy transfer, and triplet-triplet annihilation—thus clarifying the role of triplet excited states in photochemical processes in solution. For their achievements in the development of the flash photolysis method, Porter and Norrish shared the 1967 Nobel Prize in Chemistry with Eigen.

The development of pulsed lasers in the 1960s made it possible to achieve photo-excitation using light pulses with durations on the order of nanoseconds (10^{-9} s), and laser flash photolysis quickly developed into a tool for studying fast chemical phenomena and short-lived chemical species. In recent years, it has become possible to trace the evolution of picosecond (10^{-12} s) or sub-picosecond phenomena, allowing detailed investigations of short-lived excited singlet states and reaction intermediates.

One scientist who made major contributions to studying the structure of short-lived radicals using flash photolysis was Herzberg. As noted in the previous chapter (see page 262), he escaped Nazi Germany and joined the National Research Council in Ottawa, Canada; in 1949 he became Director of the Division of Pure Physics at this institution, where he conducted research on spectroscopy. His laboratory became a world center of spectroscopic studies. At Ottawa, Herzberg—who had been interested since the early 1940s in identifying molecules and ions existing in interstellar space and in comets—pursued precision studies of the structure of free radicals and ions. His group conducted detailed spectroscopic studies of more than 30 free radicals, including species of chemical interest such as methyl and methylene radicals; in addition to determining structures with high precision, these studies showed

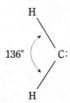

Figure 5.26: Structure of the ground state of the methylene radical

these structures changed significantly as the radical went from its ground state to an excited state.

For the methylene (:CH$_2$) radical obtained from flash photolysis of diazomethane, it was initially believed that the ground state was a linear molecule in a triplet state, but subsequent detailed studies revealed the true structure to be a bent molecule possessing a 136° bond angle (Figure 5.26)[65]. Herzberg received the 1971 Nobel Prize in Chemistry for his contributions to the structure and electronic configuration of molecules, particularly free radicals.

Another technique that was developed at the end of the 1950s was *pulse radiolysis*, which replaced light pulses with a high-energy electron beam obtained from a particle accelerator. This method was developed in the late 1950s and came to be used for the study of short-lived radicals, and it allowed measurements of the visible and ultraviolet absorption spectra, and the ESR spectra, of many radicals and ions. This method was also used in 1962 to observe the absorption spectrum of hydrated electrons. The time resolution of pulse radiolysis was later improved, and in recent years the technique has been used for picosecond pulse studies.

A powerful method for studying short-lived unstable chemical species is the *matrix isolation* method. In this technique, the unstable species is separated using a chemically inactive matrix—such as the low-temperature solid form of a noble gas—and studied using spectroscopic methods. The precursors of this method were studies of active chemical species in glasses, conducted already in the prewar era. However, the modern matrix isolation method was first developed in the mid-1950s by George Pimentel and many other researchers. This method was subsequently used as a tool for studying reactive chemical species and remains widely used to this day.

Dynamics of elemental reactions

The reaction rates observed in typical reactions are statistical averages over the ensemble of molecules in different energy states. To investigate the details of elemental reaction processes, it is desirable to study state-selective reactions, in which one investigates the detailed states of the reaction products arising from reactant molecules in a particular state. Studies of this sort—known as *state-to-state chemistry*—made major progress starting around 1960. The major tools used in this research were crossed molecular beams and energy analyses of product molecules based on infrared light emission. These studies clarified the detailed time variation of the motion and energy of molecules as the molecules collide and react. Attempts to glean information on transition states directly from experiments were also conducted using ultra-fast lasers and molecular beam techniques, leading to the emergence of the field known as *femtochemistry*.

Studies using crossed molecular beams and infrared light emission
Research based on molecular-beam techniques advanced in tandem with the progress of vacuum technology. In the 1920s, physical chemist Otto Stern at the University of Frankfurt in Germany conducted the famous Stern-Gerlach experiment with beams of silver atoms, while Rabi at Columbia University used molecular beams to measure nuclear magnetic resonance in 1936. Although studies of chemical reactions using molecular beams were in progress before the Second World War[17], the first attempts to study elemental processes in chemical reactions using two crossed molecular beams began in 1954. In that year, Datz and Taylor at Oak Ridge National Laboratory in the U.S. used crossed molecular beams to investigate the reaction $K+HBr \rightarrow KBr+H$. They demonstrated the possibility of using the different surface ionization detection efficiencies of K and KBr to distinguish KBr particles scattered in the reaction from the background of K particles arising from elastic scattering[66].

Starting in the early 1960s, detailed research on chemical reactions using crossed molecular beams was advanced by the group of Dudley Herschbach at the University of California at Berkeley. Their first subjects were the differential cross section and angular distribution of scattering in the reactions (A) $K+CH_3I \rightarrow KI+CH_3$ and (B) $K+Br_2 \rightarrow KBr+Br$. They analyzed experimental data using a dynamic analysis based on classical molecular collision theory and showed that reaction (A) was a recoil reaction in which the reaction product KI emerged in the direction opposite that of the incident K atoms, while (B) was a stripping reaction in which the scattered direction of the product molecules agreed with the direction of the relative velocity of the product molecules[67]. They also demonstrated that most of the energy produced in the reaction was converted into internal energy of the reaction products. Herschbach's group systematically

---

17. In the 1930s, the group of Nobuji Sasaki at Kyoto University conducted pioneering studies of chemical reactions using molecular beams. For example, they studied the reaction of Cl_2 gas with a molecular beam of Na atoms and attempted direct measurements of reaction rates.

Dudley Herschbach (1932–)

After receiving his doctorate in chemical physics from Harvard University, Herschbach began research on molecular beams at the University of California; he pioneered studies of reaction dynamics using molecular beams, and laid the foundations for the so-called "alkali age" of molecular-beam research. He became a professor at Harvard University in 1963, where he worked with Lee to develop a general-purpose molecular beam apparatus; he conducted pioneering research on reaction dynamics using molecular beams. He received the 1986 Nobel Prize in Chemistry.

investigated a large number of reactions involving alkali metals; however, these were all reactions that occur instantaneously upon collision of the molecules. In 1966, they discovered an example of a reaction that takes place via a long-lived reaction complex in the form of an exchange reaction of the type $A+X^-B^+$ $\rightarrow A^+X^-+B$ (where X is a halogen atom). In cases like this, the complex rotates many times before the reaction takes place, and Herschbach and coworkers showed that the angular momentum at this time plays an important role in the scattering.

The early studies using crossed molecular beams were restricted to reactions involving alkali atoms. However, in the second half of the 1960s, the group of Herschbach—now at Harvard—used supersonic jets as molecular beams to develop an apparatus capable of using time-of-flight measurements to resolve the velocities of the reaction products. The scientist Yuan Lee played a key role in the development of the instrument, which uses an ultra-high-vacuum mass spectrometer as the detector (Figure 5.27)[68]. This apparatus was used to study reactions such as $Cl+Br_2$, $H+Cl_2$, and $Cl+HI$, in each case yielding detailed information on reaction dynamics.

Methods for using crossed molecular beams to conduct research were developed by many researchers, but Yuan Lee would go on to make particularly significant contributions. He developed increasingly sophisticated instruments involving crossed molecular beams and conducted many brilliant studies at Chicago and Berkeley. These included (1) studies of reactions of F or O atoms with a variety of molecules, including simple organic molecules; (2) precision measurements of the scattering potentials of noble gas atoms, and (3) studies of the mechanism of multi-photon dissociation reactions. Lee identified the vibrational states of the DF molecule resulting from the reaction of F and D_2; the results of this study lent insight into the workings of DF chemical lasers.

Yuan Lee (1936–)

Born in Taiwan, Lee studied at National Taiwan University and completed his master's degree at National Tsing Hua University before completing his doctorate at the University of California at Berkeley. In 1965 he worked with Herschbach and his research group, playing a central role in the development of a general-purpose molecular-beam apparatus; he later worked at the University of Chicago and the University of California, becoming a leader in research on reaction dynamics using molecular beams. He received the 1986 Nobel Prize in Chemistry.

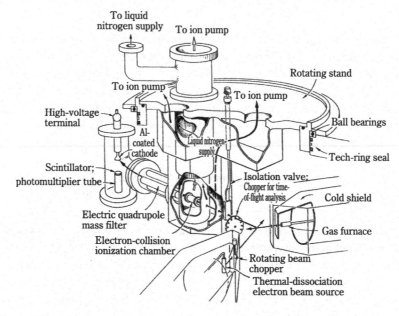

Figure 5.27: The apparatus designed by Lee, Herschbach, and collaborators for scattering experiments involving crossed molecular beams (1969)

His studies of large numbers of reactions demonstrated the generality of both instantaneous direct reactions—as discovered in early research involving alkali metals—and of reactions involving the formation of long-lived complexes.

Meanwhile, research involving chemical light emission in the infrared region, initiated by John Polanyi in 1958, would prove complementary to studies involving crossed molecular beams. Polanyi's methods, which involved the observation and analysis of extremely weak light emissions from reaction

John Polanyi (1929–)
The son of Michael Polanyi (see page 269), John Polanyi was born in Berlin and grew up in England; after studying at the University of Manchester, he worked at the Canadian National Research Council in Ottawa and at Princeton University before transferring to the University of Toronto in 1956, where he became a professor in 1962. He used chemical emissions in the infrared region to study reactions of simple atoms and molecules, and he made major contributions to the development of reaction dynamics. He received the 1986 Nobel Prize in Chemistry.

products, yielded information on the details of energy transfer and the energy distribution of molecules before and after the reaction. The excess energy produced in reactions is stored as internal energy within the reaction products, and is eventually emitted as infrared light. A typical example of the type of reaction first studied by Polanyi's group is $H+Cl_2 \rightarrow HCl+Cl$; Polanyi's work showed that the HCl molecule produced in this reaction exists in an excited vibrational state[69]. Through a series of investigations, Polanyi's group derived the general relationship between the shape of the potential energy surface and the energy distribution of the reaction products. In reactions that involve potential energy barriers in the initial stage of the reaction, the excess energy released in the reaction tends to take the form of internal vibrational energy of the reaction products; the reaction of H and Cl_2 discussed above is an example of this type of reaction. Thus, Polanyi demonstrated that chemical reactions can produce molecules in highly-excited vibrational states; this created the possibility of lasers based on chemical reactions. The first such *chemical laser* was realized in 1965 by Kasper and Pimentel[70].

Herschbach, Lee, and Polanyi received the 1986 Nobel Prize in Chemistry for their contributions concerning studies of the dynamics of elemental chemical processes.

Femtochemistry and observations of transition states
Thanks to progress in laser technology, in the late 1980s it became possible to investigate the detailed processes through which chemical reactions occur. The time required for a reaction to proceed through a transition state is typically on the order of femtoseconds (10^{-15} s), far shorter than a picosecond (10^{-12} s). Using light pulses with durations on the order of femtoseconds, it became possible to observe the processes by which chemical reactions proceed and to perform spectroscopic investigations of transition states and reaction intermediates,

Ahmed Zewail (1946–)

Born in a suburb of Alexandria, Egypt, Zewail completed his master's degree at the University of Alexandria before coming to the U.S. to obtain his doctorate from the University of Pennsylvania. From 1976 onward he was a professor at the California Institute of Technology, where he used ultrafast lasers to advance studies on excited states and chemical reactions; in the late 1980s he conducted pioneering research in femtochemistry. He received the 1999 Nobel Prize in Chemistry.

giving rise to the field known as *femtochemistry*. One widely-used technique is the *pump-probe* method, in which a reaction is initiated by exciting a molecule using a pump pulse and the state of the reaction system at some later time is investigated using a probe pulse. This technique was pioneered by Ahmed Zewail at the California Institute of Technology. The first experiment, reported in 1988, studied the photolysis of cyanogen iodide, ICN→I+CN. This study observed a transition state formed by the breaking of the I-C bond and showed that the reaction occurred over a time of 200 femtoseconds[71].

Another study by Zewail investigated the details of the dissociation reaction of sodium iodide (NaI). In this experiment, a pump pulse was used to excite the ion pair ($Na^+ I^-$) into the covalently-bonded entity [NaI]*, and the researchers observed the variation in the properties of the molecule as it vibrated. They showed that the probability for the molecule to return to its ground state and dissociate into Na and I was greatest at the precise point in the [NaI]* molecular vibration cycle at which the nuclei are separated by a distance of 6.9 Å[72] (Figure 5.28). Zewail's group also investigated the reaction of hydrogen and carbon dioxide ($H+CO_2 \rightarrow CO+OH$) and discovered that this reaction proceeds through the relatively long-lived (1000 fs) intermediate HOCO.

In later work, Zewail's group used femtochemistry methods to study transition states of a variety of reactions, including the dissociation of tetrafluorodiiodomethane ($C_2I_2F_4$) to form tetrafluoroethylene (C_2F_4), the reaction converting cyclobutane into ethylene, and the isomerization reaction by which cis-stilbene changes properties to become trans-stilbene. In the ensuing years many other researchers began to work in femtochemistry, and today it remains an active field of research. Early-stage research investigated individual molecules in molecular beams, but subsequent work broadened the range of the technique to reactions in solutions, on surfaces, and within living organisms.

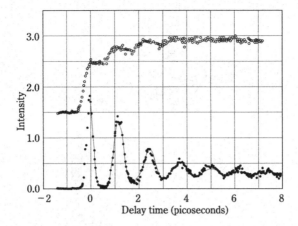

Figure 5.28: A figure illustrating the time evolution of bound Na and dissociated Na during a dissociation process induced by optical excitation of NaI

The bottom curve corresponds to bound Na, while the upper curve corresponds to dissociated Na.

Dynamics of excited molecules

Advances in spectroscopy research and in the quantum-chemical theory of molecules, both of which were underway before the Second World War, had succeeded by the mid-20th century in uncovering a wealth of information on atoms and molecules; however, it was only with the advent of the laser that the production of excited molecules, and studies of their structure and properties, made explosive progress. The detailed processes by which excited molecules relax and react with other species were clarified, and the field of photochemistry—which might equally well be termed *excited-molecule chemistry*—matured into a major subfield within chemistry. A molecule in an excited state may relax through a variety of processes, including light emission (fluorescence or phosphorescence), vibrational relaxation, internal conversion[18], inter-system crossing[19], energy transfer, and chemical reaction. The decay of the molecular excitation thus depends on rates at which these various processes proceed (Figure 5.29). The second half of the 20th century saw major progress in the understanding of all of these processes. In this section we survey these

---

18. Internal conversion: A process by which an atom or molecule in an excited state transitions non-radiatively to another excited state of the same spin multiplicity.

19. Inter-system crossing: A process by which an atom or molecule in an excited state non-radiatively changes its spin multiplicity. In most organic molecules, optical excitation produces singlet excited states, so it is common to observe transitions from such a state to an excited triplet state.

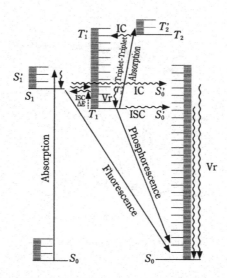

Figure 5.29: Several processes involving excited states

A molecule excited to the S_1 state emits fluorescence (fluoresces) and either (a) returns to the S_0 state, (b) transitions to the S_0' state—a highly vibrationally-excited version of the S_0 state—through the process of internal conversion (IC), or (c) transitions to the T_1' state through the process of inter-system crossing (ISC). The S_0' and T_1' states undergo vibrational relaxation (Vr) to relax non-radiatively to the S_0 or T_1 states, respectively. The T_1 states returns to the S_0 state by emitting phosphorescence or non-radiatively through the S_0' state.

developments, focusing on studies of the processes of excitation, energy relaxation, and transfer.

Relaxation of excited molecules

How is vibrational energy transferred between molecules—or within a single molecule? The question is important for understanding chemical reactions. When two molecules collide in the gas phase, there is a transfer of vibrational energy from the molecule with the greater vibrational energy (the hotter molecule) to the molecule with the lesser energy (the colder molecule). In addition, the vibrational energy of a molecule excited to a given vibrational mode changes over time into energy associated with other vibrational modes. Since the 1970s, it has been possible to use tunable-wavelength lasers to excite specific molecular vibrations and thus study the detailed processes by which vibrational energy migrates and relaxes.

Two traditional methods for studying the relaxation of vibrational energy within a molecule are (a) to use infrared light to excite a molecule from its ground state to a specific highly excited vibrational state and observe the processes by which it relaxes, and (b) to use ultraviolet or visible light excitation to create an electronic excited state in a specific excited vibrational state and observe the spectrum of the resulting fluorescence. This technique was exemplified by studies that began with cooled alkyl benzene in the S_1 state, excited vibrations in the benzene ring, and investigated the rate at which this vibrational energy is redistributed into vibrational energy of the alkyl group. These studies revealed that the most important factor governing the rate of these *intramolecular vibrational energy redistribution* (IVR) processes was the density of the vibrational state.

Before the advent of lasers, relaxation processes in electronically excited molecules were studied in condensed systems; subjects of particularly active investigation included interactions between states due to vibronic interactions or spin-orbit coupling, radiative transitions, and non-radiative transitions, and a qualitative understanding of their mechanisms was obtained. For intersystem crossings between excited singlet states and excited triplet states, transitions between excited states of the same symmetry were known to be slow, while transitions between states of different symmetries, such as $n\pi^*-\pi\pi^*$, were known to be fast; this was known as *El Sayed's rule*. The importance of perturbations inducing non-radiative transitions, and of the Franck-Condon factor[20], were also recognized. It was against this backdrop that the theory of non-radiative transitions was formulated in the early 1960s by Lin, Jortner, and Bixon. The emergence of high-resolution tunable-wavelength lasers and supersonic jet spectroscopy allowed individual molecules to be excited to specific vibrationally-excited states, whereupon their relaxation processes could be studied in detail; this further enriched understanding of relaxation processes in excited molecules. Particularly noteworthy among these studies was the observation of quantum beats[21] in the decay of fluorescence from relatively large molecules such as anthracene; this phenomenon indicated quantum-mechanical interference between a small number of energy levels relevant to vibrational relaxation within the molecule.

Energy transfer
Phenomena in which energy is transferred from a molecule in an excited state to another molecule in its ground state are important for photochemical processes

---

20. Franck-Condon factor: The transition probability due to light absorption or emission by a molecule is approximately given by the square of the transition moment times the square of the overlap integral between the vibrational wave functions of the ground and excited states. This latter factor is known as the *Franck-Condon factor*.

21. Quantum beats: A phenomenon in which the intensity of fluorescence from an atom or molecule excited for a short time oscillates with a frequency proportional to the energy gap ΔE between two nearby energy levels, indicating coherent quantum evolution.

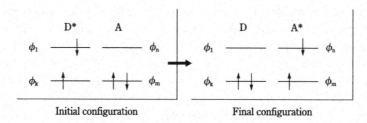

Figure 5.30: Schematic depiction of energy transfer

D: Donor, A: Acceptor, *: excited state
In the initial configuration, D is in an excited state and A is in its ground state. In the final configuration, D is in its ground state and A is in an excited state.

such as photosensitization or quenching reactions and photosynthesis (Figure 5.30). Mechanisms for the transfer of excitation energy differ depending on whether the energy is transferred between singlet states (such as reactions of the form $^1D^*+^1A\rightarrow^1D+^1A^*$, where D and A denote energy donors and acceptors and * indicates an excited state) or between a triplet state and a singlet state ($^3D^*+^1A\rightarrow{}^1D+^3A^*$). The former process is explained by the *Förster mechanism*, proposed in 1948 by Theodor Förster, in which energy transfer is mediated by the dipole-dipole interaction between the D and A transition moments[73]. In the Förster mechanism, the rate of energy transfer is proportional to the product of the squared magnitudes of the D and A transition moments and to the overlap between the D fluorescence spectrum and the A absorption spectrum; it varies as the inverse 6th power of the distance between the two transition moments. For molecules with large transition moments, this picture suffices to explain energy transfer at long distances; however, as the distance between molecules is reduced it becomes important to consider effects due to the overlap of molecular orbitals. Energy transfer between triplet and singlet states is explained by the *Dexter mechanism*, proposed by Dexter in 1953, which is based on the effect of exchange interactions between molecules[74]. This also explained the phenomenon of triplet-triplet annihilation, in which collisions between molecules in excited triplet states produce molecules in singlet excited states and the ground state.

Photochemistry[75]

The science of photochemistry developed at a remarkable pace in the second half of the 20th century. In the first 25 years of this period, basic photochemical processes were studied primarily using the method of flash photolysis. The following 25 years witnessed the advent of high-speed laser spectroscopy, which served not only to clarify many fast light-induced reaction processes, but also to advance the state of research on complex systems such as polymers and biomolecules. This period also saw tremendous growth in research with a more application-oriented focus, including applications to organic synthesis, solar energy conversion, artificial photosynthesis, and photocatalysis. In this section we can survey only the tip of this vast iceberg.

Basic Research

Because excited triplet states are relatively long-lived, it is easy for photoreactions to take place in such states. Several reactions were targets of active study from the late 1950s through the 1960s, including reactions in which hydrogen atoms are removed from aromatic carbonyl molecules such as benzophenone in excited triplet states. In these molecules, the $^3n\pi^*$ and $^3\pi\pi^*$ states (see page 444) are nearly degenerate, and the properties of the lowest excited triplet state differ depending on the substituent group and the solvent. For hydrogen-removing reactions, the reactivity of the $^3n\pi^*$state was shown to be vastly greater than that of the $^3\pi\pi^*$state.

It had been known since the work of Weller and collaborators in the 1950s that excimers—complexes formed from two identical molecules, with one in the ground state and one in an excited state—as well as *exciplexes*, charge-transfer complexes involving distinct molecules, can be produced and gave rise to broad fluorescence phenomena. With the advent of laser spectroscopy, it became possible to generate these complexes in excited states, whereupon the mechanisms of light emission came to be studied in greater detail.

Experimental studies of electron-transfer reactions began with research on metal complexes, but the emergence of fast laser spectroscopy led to progress in studies of electron-transfer reactions in excited molecules; as a large body of experimental data accumulated, it became possible to compare observed results to the predictions of Marcus theory (see page 509). Major contributions to this field were made by the group of Noboru Mataga at Osaka University. In particular, these researchers conducted a detailed investigation of the "inverted region" in the Marcus theory, contributing to the validation and refinement of the theoretical formulas.

One research subject that developed alongside electron-transfer reactions as an active target of ultra-fast laser spectroscopy studies was the field of *proton*-transfer reactions. The prototypical example of a proton-transfer reaction within a molecule occurs in the case of methyl salicylate. There are hydrogen bonds in the ground state of this molecule; however, in the excited state, an OH proton adjacent to a C=O bond may migrate, producing an isomer. This proton migration is an ultra-fast phenomenon, occurring on timescales on the order of femtoseconds, and many quantities—such as the rate of migration and its dependence on the excitation energy—have been investigated in detail. Indeed, the general subject of proton transfer—both within and between molecules—has been the topic of many studies.

Other precision studies have investigated the radicals formed by photo-excitation. The radicals formed when a molecule dissociates upon photoexcitation in solution are first produced as pairs of radicals in a cage configuration in the solution. Singlet-state radical pairs will recombine, but triplet-state radical pairs do not recombine, and thus transitions between singlet and triplet states in radical pairs are important for the recombining process. These transitions are caused by magnetic interactions and are influenced by external magnetic fields and by microwave irradiation; studies of magnetic-field effects and ESR led to detailed information on the dynamics of radical pairs. Japanese researchers made numerous contributions to this field[76].

Many photo-isomerization reactions have been studied using ultra-fast laser spectroscopy; one particularly representative example is the cis-trans isomerization of stilbene. This reaction was first studied in the 1950s using triplet sensitization reactions; since the 1970s it has been extensively studied using picosecond spectroscopy, and the reaction mechanism has been investigated in detail. However, the precise nature of the mechanism remained a topic of controversy even at the end of the 20th century.

Application-oriented research

Two topics that have been extensively studied for a variety of promising applications—but which are also interesting from the perspective of basic science—are photochromism and photocatalysis. In photochromism, a phenomenon known since antiquity, a chemical substance changes color upon the absorption of light, returning to its original color when placed in darkness. Although the effect has been observed in some inorganic compounds, it is particularly pronounced in the case of organic compounds, where it has applications to optical memory materials and optical switches; the phenomenon was a topic of active research in the second half of the 20th century. The change in color may be due to any number of light-induced reactions, including ring-opening reactions, cis-trans isomerization, proton migration within molecules, production of free radicals due to dissociation, and electron migration. One particular family of interesting substances are the diarylethene species—studied by Masahiro Irie and collaborators at Kyushu University—which are thermally stable.

The term *photocatalyst* is used in general to describe substances which, when irradiated by light, exhibit catalytic effects; among such substances, titanium oxides have attracted particular attention for their light-induced catalytic properties. In 1972, Kenichi Honda and Akira Fujishima at the University of Tokyo published a paper in *Nature* reporting the photo dissociation of water into hydrogen and oxygen using a powder of titanium dioxide (TiO_2); this paper attracted significant attention[77]. In this process, valence-band electrons in titanium oxide are excited into the conduction band by ultraviolet light, creating electron-hole pairs; the *holes* have extremely powerful oxidizing effects, while the *electrons* have extremely powerful reducing effects, and together they facilitate the dissociation of water into hydrogen and oxygen. This phenomenon—known as the *Honda-Fujishima effect*—attracted significant attention as a novel use of light energy; however, the poor photolytic efficiency meant that the process was not useful for practical applications. In contrast, the use of titanium oxide as an photocatalyst also attracted attention for its sterilizing effects and its ability to decompose organic substances—both of which properties follow from its strong oxidizing and reducing behavior—and the compound has since found broad application.

Photosynthesis reactions are perhaps the most important photoreactions in the natural world and have been the subject of extensive research, as discussed in Section 6.3.5. In the second half of the 20th century, many studies attempted to mimic the process of photosynthesis in plants by inducing *artificial* photosynthesis; to date, none of these efforts has been successful, and the topic remains an open challenge for the 21st century.

Advances in reaction theory
Theory of organic reactions
Electronic theory of organic chemistry, which began to be formulated in the 1930s, held that nucleophilic species would react in regions of high electron density, while electrophilic species would react in regions of low electron density. The success of Hückel molecular-orbital (MO) theory in explaining the stability of aromatic compounds and other phenomena spurred efforts to understand reactivity in terms of π electron density; nonetheless, the thinking of chemists remained dominated by the notion that reactivity was determined by electron density, which itself was determined by the full set of occupied electronic orbitals. However, many experimental results could not be explained by this picture. In 1952, Kenichi Fukui at Kyoto University proposed that electrophilic substitution reactions in aromatic hydrocarbons took place in whichever site had the largest coefficient of the highest occupied molecular orbital (HOMO), while nucleophilic reactions were governed by the coefficient of the lowest unoccupied molecular orbital (LUMO)[78] (Figure 5.31). Fukui and collaborators used the term *frontier orbitals* to refer to the HOMO and the LUMO, and in subsequent years they attempted to expand the theoretical underpinnings and the range of

Kenichi Fukui (1918–1998)

Fukui studied under Gen'itsu Kita at Kyoto University and began his career as an experimental organic chemist, but he was naturally gifted in mathematics, and he taught himself quantum mechanics and contributed to theoretical studies of organic reactions. He was a professor at Kyoto University from 1951 to 1982. In 1952 he proposed the frontier orbital theory of chemical reactions, and he later achieved numerous successes in the study of chemical reactions, including HOMO-LUMO interaction theory and analysis of reaction pathways. He also trained many successors at Kyoto University, making major contributions to the development of theoretical chemistry in Japan. He received the 1981 Nobel Prize in Chemistry.

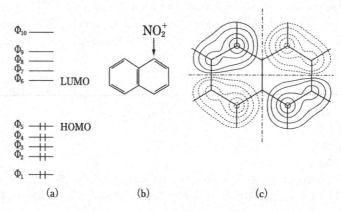

$$(a) \qquad\qquad (b) \qquad\qquad (c)$$

Figure 5.31: Frontier orbitals of naphthalene

(a) Energy level scheme of π electron orbitals of naphthalene. The highest-energy orbital among all molecular orbitals occupied by electrons is known as the *highest occupied molecular orbital* (HOMO); similarly, the lowest-energy orbital among all unoccupied molecular orbital is known as the *lowest unoccupied molecular orbital* (LUMO). Together, the HOMO and LUMO are known as *frontier orbitals*. The above figure illustrates the case of naphthalene.

(b) Electrophilic attack of NO_2^+ to naphthalene. (c) HOMO of naphthalene indicating the electron density.

applicability of this concept. Later, in 1964, these scientists further developed the frontier orbital theory by noting the important role played by the symmetry and phase of the HOMO and LUMO of the reactant molecules in cycloaddition reactions such as the Diels-Alder reaction (see page 328). However, at that time frontier orbital theory had not yet attracted much attention.

Roald Hoffmann (1937–)
Born to a Jewish family in Poland, Hoffmann was sent to a forced-labor camp in 1941, but he escaped with his mother and survived the remainder of the war in hiding. After the war he left Poland, passing through Austria and Germany before settling in the U.S. in 1949. After graduating from Columbia University, he studied with Lipscomb at Harvard University, obtaining his doctorate in 1962. In 1963 he  proposed the extended Hückel method. Thereafter he began collaborative research with the organic chemist Woodward, with whom he wrote the 1965 paper that formed the basis of the Woodward-Hoffmann rules. He received the 1981 Nobel Prize in Chemistry.

In 1965, Robert Woodward and Roald Hoffmann explained electrocyclic reactions, cycloaddition reactions, sigmatropic rearrangements[22], and a series of other examples of reactions with stereoselectivity—so-called *pericyclic reactions*[23]—on the basis of general principles of conservation of orbital symmetry in reactions[79]. Their principles, which came to be known as the *Woodward-Hoffmann rules*, went on to have a resounding impact across chemical disciplines. Thermally- and photochemically-induced reactions such as ring-closing reactions involving butadiene or ring-opening reactions involving cyclobutene—exhibit different stereoselectivities depending on whether the reaction is thermally induced or photochemically induced; the Woodward-Hoffmann rules explained these differences in terms of the symmetry of the molecular orbitals involved in the reactions and the conservation of orbital symmetry in these reactions (Figure 5.32). The emergence of the Woodward-Hoffmann rules demonstrated the importance of frontier orbitals in reactions and led to an increased appreciation for Fukui's frontier orbital theory.

Kenichi Fukui and Roald Hoffmann received the 1981 Nobel Prize in Chemistry for their achievements in theoretical studies of chemical reaction processes. Because Woodward passed away in 1979, he could not be named a

---

22. Sigmatropic rearrangement: When a σ bond adjacent to a π bonded system moves to a new position in a molecule in a concerted manner, the resulting reconfiguration of the π-bonded system is known as a *sigmatropic rearrangement*.

23. Pericyclic reaction: A reaction characterized by the following features. 1) At least one of the reactants or one of the productt substances is unsaturated. 2) The formation or breaking of a σ bond is accompanied by the disappearance or formation of a π bond. 3) The reorganization of electrons takes place within a cyclical chain. These are concerted reactions that are neither ionic in character nor radical reactions.

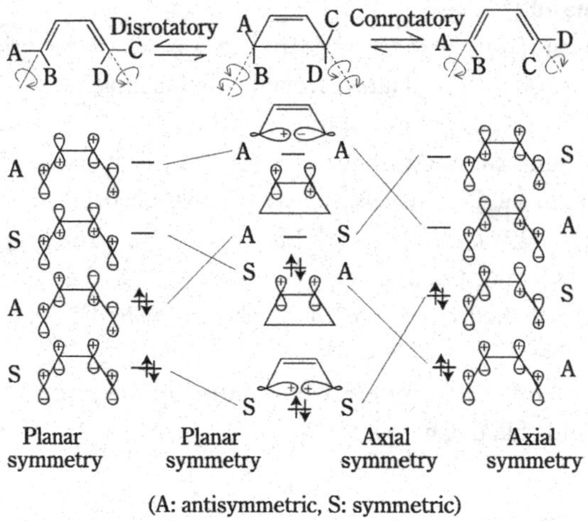

Figure 5.32: Stereoselectivity in ring-closing involving butadiene and ring-opening involving cyclobutene

Reactions proceed while conserving the symmetry of the orbitals involved. Thermal reactions tend to involve conrotatory rotation; such reactions preserve axial symmetry and have low activation energy. On the other hand, photochemical reactions tend to involve disrotatory rotation; these reactions preserve planar symmetry and exhibit different selectivity with respect to three-dimensional configuration.

co-recipient of the 1981 Nobel Prize; however, he had already received the Prize in 1965 for his contributions to organic synthesis (see Section 5.5.3 and column 19). All of this research furnished excellent examples of how new concepts based on molecular orbitals could offer explanations of complex chemical reaction processes. At the time Fukui proposed his theory, the notion that frontier orbitals, or any other specific orbitals, could determine chemical reactivity was quite novel—so novel, in fact, that its announcement elicited extensive criticism. However, in 1952 Fukui was inspired by a paper by Mulliken on charge transfer interaction to develop a theoretical foundation for his approach. Thereafter, Fukui continued to work at Kyoto University for many years, collaborating with many other researchers to develop and apply the theory of frontier orbitals.

Theory of electron-transfer reactions

Electron transfer reactions, which necessarily involve oxidation and reduction reactions among the reactant species, are among the most fundamental of all chemical reactions. The foundations of the microscopic understanding of these reactions were established by Marcus. Around the mid-1950s, Rudolph Marcus was inspired by the work of Taube and coworkers on electron-transfer reactions

Rudolf Marcus (1923–)
Born in Montreal, Canada, Marcus studied at McGill University and received a PhD in 1946. After conducting research at the University of North Carolina and the Polytechnic Institute of Brooklyn, he became a professor at the University of Illinois, and after 1978 was a professor at the California Institute of Technology. In addition to pioneering research on the theory of electron-transfer reactions, he is known for his contributions to chemical reaction theory, including the Rice-Ramsperger-Kassel-Marcus (RRKM) theory of unimolecular reactions. He received the 1992 Nobel Prize in Chemistry.

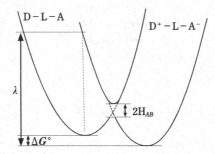

Figure 5.33: The evolution of free energy in an electron-transfer reaction

The plot shows both the reference free energy ΔG^0 and the reorganization energy λ. The horizontal axis is a reaction coordinate indicating the progress of the reaction, while the vertical axis is the free energy.

in metal complexes (see page 485) to launch his own research. He noted that the rate of electron exchange between Fe_2^+ and Fe_3^+ ions in water was slow, and designed a theory around the notion that this slowness was due to significant changes in the hydration structure surrounding the ions before and after the electron transfer (Figure 5.33). In a series of papers beginning in 1956, Marcus derived the following equation for the reaction rate constant k:[80]

$$k = A\exp(-\Delta G^*/k_B T)$$

Here the quantity ΔG^* is defined by $\Delta G^*=(1/4)(1+\Delta G^0/\lambda)^2$, where ΔG^0 is the standard free energy of the reaction and λ—the *reorganization energy*—includes contributions from the reorganization of the solvent (λ_0) and from vibrations (λ_i). In his first paper, Marcus approximated the solvent as a continuous dielectric body, but in later work he offered an improved treatment based on molecular theory and statistical mechanics. Marcus went on to apply this theory to the treatment of electron transfer reactions at electrodes. In the ensuing years, efforts to compare the predictions from Marcus theory to experimental results focused initially on electron-transfer reactions in complexes and the quenching of fluorescence, and obtained generally satisfactory agreement. The success of the theory was definitively established by experiments conducted in 1984 by Miller and coworkers. One important prediction of the Marcus equation was the existence of an *inverted region*. In a series of reactions with similar values of λ but differing values of ΔG^0, the value of ΔG^* was observed to decrease as ΔG^0 shrinks toward zero; at the particular point $\Delta G^0 = -\lambda$, one finds $\Delta G^* = 0$, and as ΔG^0 is further reduced one finds that ΔG^* starts to *increase*. Marcus referred to the case $-\Delta G^0 < \lambda$ as the normal region and to $-\Delta G^0 > \lambda$ as the *inverted region*. The experiment by Miller and colleagues measured the rate of electron transfer between biphenyl negative ions and systems consisting of four cyclohexane rings connected by various electron acceptor molecules; upon studying the relationship between the transfer rate and the quantity $-\Delta G^0$, the experimental data were in good agreement with the predictions of Marcus theory, thus confirming the existence of the inverted region[81]. Later studies compared the predictions of Marcus theory to experimental results in a wide variety of systems—including recombination reactions of charge-transfer complexes produced by photoexcitation and reactions between electrodes—and established the essential correctness of the theory.

The pre-exponential factor A in the initial equation was later made more precise by inclusion of detailed quantum-mechanical considerations. Today Marcus theory is widely used to explain electron-transfer reactions in many fields of biochemistry and biology, including studies of proteins and photosynthetic systems. Marcus received the 1992 Nobel Prize in Chemistry for his contributions to the theory of electron-transfer reactions.

Theory of elemental reactions

Reaction-rate theory had taken an important first step toward theoretical predictions of rate constants with the emergence of Eyring's transition-state theory, but accurate predictions remained challenging. In the second half of the 20th century, detailed investigations in the theory of reaction rates utilized approaches based on statistical theories and approaches based on theories of collisions and scattering. Here we briefly touch on these developments.

The theory of unimolecular reactions based on the RRK theory (see page 272) was improved by Marcus in 1952 by the introduction of transition-state concepts and became generally known as the RRKM theory[82]. This theory assumed equilibrium between a state A* with energy E^* and a transition state A‡; the reaction rate constant is computed from the number of A* and A‡ states. Consequently, calculations required information on the structure, potential energies, and vibrational frequencies of transition states. Because information on transition states cannot be obtained from experiment, reliable quantum-chemical calculations were essential. Moreover, this theory was based on the assumption that all states accessible to a system would be realized in equivalent ways; this required the assumption that transfers of energies among vibrational energy levels and redistributions of energies within molecules were rapid compared to the reaction rate. The validity of this assumption would later become a question of both theoretical and experimental interest after the emergence of ultra-fast laser spectroscopy.

Transition-state theory scored major successes in explaining reaction rates, but the underlying assumptions of the theory were difficult to justify, and the theory was met with much criticism and controversy. Efforts to overcome these doubts continued for many years after the Second World War; one such effort took the form of variational transition-state theory. Transition-state theory had assumed that, as long as a reacting system passed through a transition state, the reaction was guaranteed to occur; however, in reality, the system may pass back through the transition state, in which case the reaction will not occur. To avoid this problem, a method was developed in which the position on the potential surface separating the reactants and the reaction products was varied, minimizing the one-directional flow crossing this surface to arrive at the reactants. The relationship between transition-state theory and Kramers' theory (see page 271) was also considered in detail, and a unified understanding of the two paradigms was developed. In addition, the significance of active complexes and of activation energies was subject to a detailed reassessment, and these concepts were given new meanings.

Surface reactions and catalytic reactions

Chemical reactions at surfaces are an important field of research and are relevant to the study of catalytic reactions in heterogeneous systems. Modern surface chemistry was pioneered by Langmuir in the early 20th century (Section 4.4), but atomic and molecular-level studies of chemical reactions at surfaces made little progress for many years thereafter. Reasons for this include the difficulty of controlling the composition and state of a surface and the absence of direct methods for observing phenomena occurring at surfaces. This situation began to change in the decade between 1950 and 1960, as progress in semiconductor technology led to advances in techniques for handling surfaces under high vacuum. Several methods for atomic-scale observation of surfaces under high vacuum were developed during this time, and by the end of the 1960s a broad new field of *surface science* had emerged, encompassing aspects of physics, chemistry, and engineering. In this new environment, atomic- and molecular-level studies of chemical reactions at surfaces became active areas of research.

In the mid-1960s, the technique of low-energy electron diffraction (LEED, see page 452) was introduced as a tool for surface structure research; this method allowed studies of the atomic-level structure of clean single-crystal surfaces, and the relationship between reactivity and the surface structure of platinum and other metals used as catalysts became a key area of interest. As a result, it became clear that surface steps (atomic-level gradations) and lattice defects played important roles in governing reactions at surfaces. In addition, it became clear that many other types of structures were present at surfaces; these included surface reconstructions—producing arrays of atoms differing from the configuration present in the interior of solids—as well as surface-layer structures arising when molecules adsorb to surfaces, and *surface alloys* formed at metal surfaces by the addition of other types of metals.

The bonding strength of molecules adsorbed to surfaces may be measured from the ease with which the molecules detach as the temperature is raised. These phenomena may be broadly classified into physical adsorption, in which the bonding is due to weak forces such as van der Waals forces, and chemical adsorption, in which the molecules form covalent bonds with surface atoms. The introduction of electron energy loss spectroscopy (EELS) allowed observation of the vibrational spectrum of surface-adsorbed molecules. In this way, the bonding states and structures of adsorbed molecules gradually came to be understood. It also became possible to track the nature of the changes experienced by molecules reacting at surfaces as the reaction proceeded. Observations of surfaces were further advanced by the emergence of scanning tunneling microscopy (STM) in the 1980s, which allowed the visualization of fine gradations in the charge density of arrays of surface atoms. In addition, photoelectron spectroscopy and similar methods allowed detailed investigation of surface electronic structures. Thus, studies of reaction processes at surfaces became an extremely active area of research.

Against this backdrop, two pioneering scientists who advanced the state of modern surface chemistry research were Gerhard Ertl at the Fritz Haber Laboratory of the Max Planck Institute and Gabor Somorjai at the University of California at Berkeley. Ertl used iron catalysts—well known in the context of the Haber-Bosch method—to conduct detailed studies of the mechanism of ammonia synthesis from nitrogen and hydrogen (Figure 5.34). In this reaction, Ertl showed that N_2 molecules dissociate on an iron surface to produce surface-bound N atoms;

Gerhard Ertl (1936–)

After studying at the Technical University of Stuttgart and receiving his doctorate from the Technical University of Munich, Ertl worked as an assistant and a lecturer before becoming a professor at the University of Hanover in 1968. In 1986 he became a professor at the Free University of Berlin, a professor at the Technical University of Berlin, and director of the Fritz Haber Institute. He exploited the latest observational techniques—including low-energy electron diffraction, ultraviolet photoelectron spectroscopy, and scanning tunneling microscopy—to conduct pioneering research on surface catalytic reactions. In particular, he is known for clarifying the molecular mechanisms of ammonia synthesis reactions and for research on the oxidation of carbon monoxide using platinum catalysts. He received the 2007 Nobel Prize in Physics.

Gabor Somorjai (1935–)

Somorjai is a Hungarian-born American chemist. Born to a Jewish family, he escaped the Nazis and studied chemical engineering at the Budapest University of Technology and Economics. After the 1956 uprising in Hungary, he moved to the U.S., obtaining a doctorate from the University of California at Berkeley. He worked temporarily at IBM before becoming a professor at Berkeley, where he conducted pioneering research in surface chemistry.

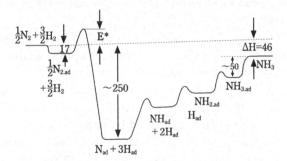

Figure 5.34: Enthalpy changes in each step of the reaction process by which N_2 and H_2 react at an iron surface to produce NH_3. (Units: kJ / mol)

these react in stages with H atoms produced by the dissociation adsorption of H_2, producing reaction products NH, NH_2, and finally NH_3, which escapes into the gas phase. Ertl was able to determine the change in enthalpy accompanying each of these processes. He also studied catalytic oxidation reactions of CO at platinum surfaces and discovered that these processes include oscillatory reactions that produce CO_2[83]. Ertl received the 2007 Nobel Prize in Chemistry for his contributions to the study of chemical processes at solid surfaces.

Somorjai discovered that dissociation reactions at platinum surfaces occur at surface steps or defects. He also discovered that ethylidyne (CH_3C_2) is produced at the surfaces of platinum catalysts in the hydrogenation reaction of ethylene, and he determined its structure. It was later demonstrated that the structure of ethylidyne is similar to that of organometallic molecules, involving three atoms pulled from the hexagonal structure of the platinum surface to bind to carbon atoms.

5.5 Discovery and synthesis of new substances

By 1945, all naturally-occurring elements had been discovered, and it was believed that essentially all simple compounds had been discovered as well. However, between the end of the Second World War and 1984, elements 97 through 109 were synthesized as man-made elements. As for simple inorganic compounds, in 1962 it was discovered that noble gases—which had been believed to be inert and not form compounds—were in fact able to form stable compounds. Furthermore, in 1985, fullerenes such as C_{60} and C_{70} were discovered. These findings—which showed that as-yet-unknown structures remained to be discovered even among ubiquitous elements such as carbon—sent a shockwave through the world of chemistry. In addition, discoveries of new, complicated molecules with relevance to living organisms continued unabated throughout this period.

The synthesis of useful new substances—and of substances that exist in the natural world—is one of the central objectives of chemistry. Synthetic chemistry developed as one of the central supporting pillars of the science of chemistry, and—as noted in the previous chapter—the field made major progress in the first half of the 20th century. In the second half of the century, synthetic chemistry continued its impressive development, and many complicated natural organic compounds and metallic complexes were synthesized. Synthetic chemists worked tirelessly to develop new methods of synthesis, exploiting all tools at their disposal—including infrared, visible, and ultraviolet spectroscopy, NMR, X-ray diffraction, mass spectrometry, and chromatography—and, when necessary, making use of the results of theoretical analyses based on methods such as molecular orbital theory or molecular mechanics. As the relationship between the structure and properties of molecules became increasingly well understood, methods for designing and synthesizing molecules for particular purposes became increasingly widely adopted. As a result, in today's world we are progressively approaching the point at which any arbitrary desired molecule can be synthesized at will. The synthesis of molecules with particular properties is also a goal of basic science, where it is used to investigate theoretical concepts. In this section, we survey a variety of topics in this area: the creation of new elements, new inorganic compounds, and new organometallic compounds; the development of new methods of synthesis, including methods for synthesizing complex naturally-occurring organic compounds; and research on supramolecules and molecular aggregates.

New elements and new material groups
Synthesis of artificial elements and the periodic table
In 1945, the number of known elements stood at 96. In the years after the Second World War, nuclear reactions were used to synthesize many new man-made elements. By 1984, the elements up to element 109—meitnerium (Mt)—had been added to the periodic table. Among these, einsteinium (Es, element 99) had a half-life of 20 days, while mendelevium (Md, element 101) had a half-life of 766 minutes and that of nobelium (No, element 102) only 2.3 seconds; after these elements, the half-lives progressively shrunk, with meitnerium having a half-life of only 3.4 milliseconds. The scientist Glenn Seaborg was involved in the discovery of elements 94 through 102. He showed that all of these elements belonged to the sequence of actinides, which starts with thorium (element 90), and that their trivalent oxidized states were stable and exhibited similar properties; this was the founding of the chemistry of actinides. Among these elements, the most intensively studied atoms were uranium—produced in large quantities for nuclear reactors—and plutonium, whose long half-life makes it useful as an atomic fuel and for use in nuclear weapons.

Noble gas compounds
The noble gases—for which the s and p orbitals of the outermost shell are fully occupied by electrons—were initially believed to be stable and to not form compounds. However, in 1933 Pauling anticipated that heavy noble gases could form compounds with fluorine or oxygen. In 1962, Neil Bartlett at the University of British Columbia in Canada studied the reaction of Xe and PtF_6 and reported the production of $Xe^+[PtF_6]^-$[84]. In the same year, Claassen obtained XeF_4 by reacting Xe and fluorine (F) at high temperature[85]. Subsequent studies produced other related compounds, including XeF_2, XeF_6, $XeOF_2$, $XeOF_4$, XeO_3, and XeO_4. Other noble gas compounds such as ArF,

KrF_2, $XeCl_2$, and Xe_2 exist as short-lived metastable compounds and are widely used in excimer lasers.

Clusters

The term *cluster* was first introduced in the early 1960s to describe metallic cluster compounds involving bonds between metal atoms. However, the term later became widely used to describe nanoparticles (with sizes on the order of 1 nm) consisting of aggregates of a few to a few thousand atoms or molecules. Clusters are particularly interesting as intermediate forms of matter existing somewhere between molecules and condensed-matter systems. A subject of interest in the field of physical chemistry was the variation in the physical properties and reactivity of clusters with the number of constituent atoms or molecules, and this became a topic of active research starting around 1980.

Clusters, which form due to van der Waals forces or hydrogen bonds between atoms or simple molecules, are produced as isolated gas-phase molecules by devices that emit supersonic molecular beams. This method simultaneously produces clusters of various sizes, but mass separation or similar techniques may be used to select clusters of specific sizes, whose structures and properties may then be analyzed using a variety of spectroscopic methods. For example, detailed studies were conducted on the structure of clusters consisting of various aromatic compounds—such as benzene or phenol—hydrated by water molecules. Other studies traced the variation in solvation structure and reactivity as individual solvent molecules were added, one molecule at a time.

Interesting size-dependent phenomena of clusters of metal atoms, discovered during investigations of the properties of clusters, include (a) a metal-insulator transition in clusters of mercury atoms at a cluster size in the range 200–300 atoms, and (b) a size-dependent ferromagnetic-antiferromagnetic transition in clusters of manganese atoms. In addition, the rate of dissociation reactions for methanol molecules atop nickel cluster ions was observed to depend on the cluster size, an interesting finding because of its relation to catalytic function.

The structures that received the most attention in relation to catalytic reactions were metal clusters consisting of aggregates of atoms built up around a central core formed by several metal atoms bound to one another. Typical examples of such clusters are structures of the form M_xL_y (where M is a metallic atom and L is a ligand), including metal-carbonyl clusters such as $Ir_4(CO)_{12}$ and $Fe_3(CO)_{12}$. Metal clusters of this form serve as catalysts for the CO hydrogenation reaction known as the *Fischer-Tropsch* reaction.

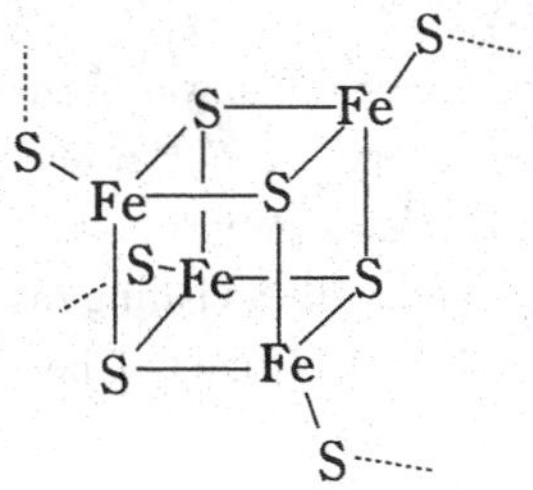

Figure 5.35: 4Fe-4S cluster structure of ferredoxin

Cluster molecules built up around a central cube consisting of 4 iron atoms and 4 sulfur atoms were discovered in the early 1960s in the biological enzyme ferredoxin and were also observed to form the functional portions of iron proteins that promote electron transfer reactions in living systems (Figure 5.35). Metal clusters containing cubes of this sort are also known to exist for nickel, tungsten, zinc, manganese, and chrome, and have been discovered among naturally-occurring enzymes in living organisms.

Quasicrystals

It was long believed that solids exist in either crystalline or non-crystalline (amorphous) phases, and thus the discovery in the 1980s of substances that exist in a third type of solid phase—neither crystalline nor amorphous—caused quite a stir. In 1984, Israeli chemist Daniel Shechtman showed that an Al-Mn alloy cooled rapidly from the liquid phase gave rise to electron-beam diffraction spots resembling crystals, but nonetheless exhibited five-fold rotational symmetry, which cannot exist in a crystalline phase; Shechtman reported that this substance had the symmetry of an icosahedron[86]. Although this finding contradicted the conventional wisdom of the time, similar solids were subsequently discovered; these came to known as *quasicrystals*. It became clear that quasicrystals, although not possessing the translational symmetry (periodicity) of crystals, nonetheless exhibit highly ordered structures. The first quasicrystals to be discovered were thermodynamically unstable, and upon heating reorganized into more stable crystal phases; however, stable quasicrystals were later discovered by Tsai and collaborators at Tohoku University. Quasicrystals attracted much interest for their curious physical properties, which differ from those of crystalline solids. For example, aluminum, copper, and iron are all good conductors of electricity, but Al-Cu-Fe quasicrystals formed from these three elements have extremely high electrical resistivity. Moreover, the non-crystalline nature of bulk quasicrystals makes it difficult for cleaved surfaces to form in these materials, making them relatively rigid and strong. Shechtman was awarded the 2011 Nobel Prize in Chemistry for his discovery of quasicrystals.

Organometallic compounds

Organometallic compounds, in which carbon atoms bond directly with metal atoms, had been known since the 19th century, and organic synthesis in the first half of the 20th century produced many useful compounds of this type, including Grignard reagents and alkali-lithium compounds. In the second half of the 20th century, organometallic compounds featuring new types of bonds—including ferrocene and other metallocenes—were discovered; in addition, the discovery of Ziegler-Natta catalysts underscored the importance of organometallic compounds as catalysts, and the field of organometallic chemistry underwent rapid development as a field lying at the nexus of organic chemistry and inorganic chemistry (complex chemistry).

Ferrocene [$(C_5H_5)_2Fe$] was first synthesized in 1951 by Paulson and Kealy and independently by Miller and collaborators[87], but the correct structure was initially not identified. The following year, in 1952, Woodward and Geoffrey Wilkinson at Harvard University used measurements of infrared absorption, magnetic susceptibility, and dipole moment, as well as reactivity, to propose a sandwich structure in which a divalent iron ion was surrounded by two cyclopentadienyl(CP) ions; Woodward named this compound ferrocene (Figure 5.36)[88]. Around the same time, Ernst Fischer and W. Pfab at the Technical University of Munich used X-ray structure analysis to confirm the sandwich structure of ferrocene and began research on other bis-cyclopentadienyl transition-metal compounds such as nickelocene and cobaltocene[89]. These scientists later used NMR and X-ray structural analysis to verify the sandwich structure. Wilkinson further developed this line of research by synthesizing similar compounds of ruthenium and cobalt. This was the origin of a sequence of research efforts on *metallocenes*—transition-metal compounds with the structure [$(C_5H_5)_2M$].

Geoffrey Wilkinson (1921–1996)
Wilkinson studied at Imperial College of London and received his doctorate in 1946. He worked under Seaborg at the University of California, studying radioactive isotopes of lanthanide metals. Later, he moved to the Massachusetts Institute of Technology where he switched focus to inorganic chemistry; in 1951 he began research on ferrocenes with Woodward at Harvard University. In 1956 he returned to England, where he became a professor of inorganic chemistry at Imperial College. In 1965 he discovered the Wilkinson complex, which is useful as a catalyst. He received the 1973 Nobel Prize in Chemistry.

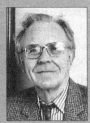

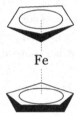

Figure 5.36: Structure of ferrocene

Compounds of this form with Fe replaced by other metal atoms are generally known as *metallocenes.*

The structure and electronic configuration of ferrocene introduced new concepts in organometallic chemistry. In ferrocene, the iron atom is in the +2 oxidized state, while the CP ring is a negative ion with a 6π electronic configuration; the ring stabilizes and exhibits aromaticity. Because of these π electrons, the CP ring forms a covalently-bonded π complex with the metal ion; together with the 6 electrons from Fe^{2+}, the complex has a total of 18 electrons with the electronic configuration of a noble gas and was thus believed to be stable. It was also demonstrated that the chemical properties of this compound exhibited reactivity characteristics of aromatic species.

Fischer and Wilkinson went on to make many contributions to organometallic chemistry. Fischer succeeded in synthesizing $Cr(C_6H_6)_2$, a compound consisting of a benzene-chrome sandwich. Wilkinson discovered CP_2ReH, a sandwich compound involving a bond between hydrogen and rhenium atoms, in an NMR spectroscopy study. In the 1960s he developed the Wilkinson catalyst, Rh(Cl)(PPh_3), which is effective in the hydrogenation of olefin.

Ernst Fischer (1918–2007)
Fischer studied at the Technical University of Munich, where he earned a doctorate in 1952 for research in the field of organometallic chemistry. He remained at that university as an assistant, where he began research on ferrocenes. In 1959 he became a professor at the Technical University of Munich, rising in 1964 to head its inorganic chemistry laboratory. Independently of Wilkinson, Fischer identified the structures of a variety of sandwich compounds formed by organometallic compounds. He received the 1973 Nobel Prize in Chemistry.

Fischer's next major contribution was to the chemistry of carbene (M=CRR') and carbyne (M≡CR) complexes. These compounds were thought to be important for reactions in organic chemistry involving transition metal catalysts, but they were not previously known as stable compounds. In 1964, Fischer succeeded in isolating stable carbene, and later his group synthesized and studied many compounds of the form M=C(ER)R' (where E is O, S, or NR''). These became known as "Fischer-type carbene." In Fischer-type carbene, the metal atom is an acceptor of π electrons and the carbon atom in the carbene is electrophilic. In contrast, in the carbene catalysts developed by Schrock for metathesis catalysis (see page 526), the carbon atom in the carbene is nucleophilic.

Fischer and Wilkinson received the 1973 Nobel Prize in Chemistry for research on metal-organic complexes with sandwich structures.

New synthetic methods for organic compounds
Synthetic organic chemistry made remarkable progress in the second half of the 20th century, with new methods of synthesis yielding an enormous bounty of research results. Because we lack the space here to discuss all of these in detail, we restrict our discussion to just those achievements that earned Nobel Prizes in Chemistry up to the year 2011.

Polymer synthesis
Polymer synthesis gave rise to products like nylon and plastics in the 1930s and developed particularly rapidly after the Second World War, laying the foundations for the explosive growth of the polymer industry in the postwar era. A typical example of the new methods that were developed is the catalysis method devised by Ziegler and Natta (Ziegler-Natta catalysis). German chemist Karl Ziegler had been involved in research on organometallic compounds in the prewar era at the University of Halle in Germany, and after the war at the Max

Karl Ziegler (1898–1973)
Ziegler studied at the University of Marburg, then worked at the Universities of Frankfurt and Heidelberg before becoming a professor at Halle in 1936. From 1943 onward he directed the Kaiser Wilhelm Institute for Coal Research. He researched the synthesis of organometallic compounds and polymerization methods for ethylene that used these compounds as catalysts. In 1952 he discovered that ethylene can be made to polymerize at room temperature and pressure by using a catalyst consisting of triethyl aluminum and titanium tetrachloride; this laid the foundations for many developments in the polymer industry. He received the 1963 Nobel Prize in Chemistry.

Giulio Natta (1903–1979)
After receiving his doctorate from Milan Polytechnic, Natta worked at the Universities of Pavia, Rome, and Turin before becoming a professor at Milan Polytechnic in 1938. He researched the applications of X-ray and electron-beam diffraction, from solids to catalysts and polymer compounds. He used Ziegler catalysts to synthesize polymers, and by studying the relationship between the stereo-regularity of these polymers and their properties, he laid the foundations for many developments in the polymer industry. He received the 1963 Nobel Prize in Chemistry.

Planck Institute and elsewhere; his research advanced the systematic study of new reactions. In 1953 he discovered that a catalyst formed by combining ethyl aluminum with titanium tetrachloride could be used to fuse ethylene into highly crystalline polyethylene at normal temperature and pressure; this represented the development of a low-pressure synthesis method for polyethylene[90]. Meanwhile, the chemist Giulio Natta, a professor at the Milan Polytechnic in Italy, had been using X-ray diffraction to study the structure of polymer compounds since 1938; using Ziegler's catalysts he synthesized polymers and showed that they exhibited three-dimensional regularity and that changes in their three-dimensional structure gave rise to changes in their properties[91]. He also succeeded in polymerizing polypropylene to a high degree of polymerization using triethyl aluminum and titanium trichloride, showing that this substance exhibits a stereospecific structure. Thereafter, he combined metal alkyls from groups 1, 2, and 3 with metals from groups 4–8 and showed that these were effective in polymerizing olefins and similar species; these became known as Ziegler-Natta catalysts. The development of highly active and stereospecific catalysts went on to make

Herbert Brown (1912–2004)

Born in London, Brown's family moved to the U.S. when he was 2. He lost his father at the age of 14 and was forced to drop out of school for three years to work to support his family. He received a scholarship to study at the University of Chicago, where he received a B.S. in 1936 and a PhD in 1938 in inorganic chemistry. He worked at Wayne State University before becoming a professor at Purdue University in 1947. Starting with studies of reactions between diboranes and carbonyl compounds, he discovered sodium borohydride, a compound with outstanding reducing ability; he went on to discover many new synthetic reactions. He received the 1979 Nobel Prize in Chemistry.

Georg Wittig (1897–1987)

Wittig was a German organic chemist. After receiving his doctorate from the University of Marburg in 1926, he worked at universities in various regions of Germany before becoming a professor at Heidelberg in 1956. He is known for the Wittig reactions, and he developed synthetic reactions using organic phosphorus as a base. He received the 1979 Nobel Prize in Chemistry.

dramatic advances and spurred major developments in the petrochemical industry. Ziegler-Natta catalysts also made major contributions to the development of organometallic chemistry through research on their reaction mechanisms. Ziegler and Natta received the 1963 Nobel Prize in Chemistry.

Brown and Wittig

Among the synthetic methods developed between the mid-war to early postwar years, two that are particularly useful are the method using organic boron compounds developed by Brown and the method using organic phosphorus compounds developed by Wittig. Herbert (H.C.) Brown began research on diboranes (see page 310) in Schlesinger's laboratory at the University of Chicago in the years before the Second World War. Starting with studies of reactions between diboranes and carbonyl compounds, he discovered that hydrogenated sodium boride ($NaBH_4$) exhibited outstanding reducing ability. Subsequently, at Purdue University, Brown made a systematic study of reactions involving hydrogenated boron and organic boron compounds, discovering hydroboration reactions in which a diborane or an alkyl borane reacts with olefin. These reactions turned out to be extremely useful for organic synthesis, and Brown was a pioneer in opening this new domain of synthetic organic chemistry. He also conducted research on the structure of carbonium ions and on the relationship between reactivity and steric strain, and he made many contributions to the theory of organic reactions[92]. At Purdue, Brown trained a large number of successors; these included a large number of Japanese researchers, including Negishi and Suzuki, the recipients of the 2010 Nobel Prize in Chemistry. In 1979 Brown and Wittig were awarded the Nobel Prize in Chemistry for their development of new organic synthetic methods.

Georg Wittig became a professor at the University of Tübingen in 1944 and a professor at the University of Heidelberg in 1965, where he made numerous contributions spanning a wide range of subjects in organic chemistry. The most famous of these was his discovery of the *Wittig reaction*, an olefin synthesis reaction using an organic phosphorus compound; this reaction went on to be widely adopted in organic synthesis[93].

Asymmetric synthesis

After 1979, for a couple of decades there were no Nobel Prizes awarded for methods of organic synthesis. Nobel Prize awards for this type of work resumed in the early 21st century, reflecting the remarkable progress of organic synthetic methods during the final quarter of the 20th century and the significant impact of these methods for practical applications. The achievements for which these Nobel Prizes were awarded were all synthetic methods that made use of new catalysts.

Many naturally-occurring compounds are *chiral*, possessing distinct enantiomers. Nucleic acids, proteins, and other biological macromolecules are all chiral, and one isomer of a chiral compound exhibits interactions distinct from those of its mirror-image counterpart. In other words, living organisms are able to recognize a biologically-active enantiomer and its mirror-image counterpart as two distinct compounds. For this reason, methods of *asymmetric synthesis*—which selectively synthesize only one of two possible enantiomers—are important for synthesizing products such as pharmaceuticals or agrichemicals. This situation spurred significant progress in the development of methods for selective asymmetric synthesis. In 1968, American scientist William Knowles at Monsanto Corporation used an asymmetric catalysis process in which a Wilkinson complex ligand (a rhodium complex that catalyzes reduction reactions) was replaced with a chiral phosphine; this demonstrated the possibility of selectively synthesizing one particular enantiomer[94]. Knowles and his colleagues later improved this catalyst for applications to the industrial production of L-DOPA, a treatment for Parkinson's disease.

Studies of reduction reactions involving asymmetric catalysts were significantly advanced by the group of Ryoji Noyori at Nagoya University in Japan. In 1980, these scientists used a rhodium complex, whose ligand was a chiral diphosphine known as BINAP, as a catalyst for the asymmetric

William Knowles (1917–2012)
After graduating from Harvard, Knowles received his
doctorate from Columbia University in 1942, then began
life as a researcher in corporate laboratories; until 1986 he
was a researcher at Monsanto. In 1968 he used a rhodium
complex to demonstrate for the first time the possibility
of asymmetric synthesis of an organic compound using a
hydrogenation reaction. This was used in 1974 to synthesize
L-DOPA, a drug for treating Parkinson's disease, and was the first industrial
application of asymmetric synthesis. He received the 2001 Nobel Prize in
Chemistry.

Ryoji Noyori (1938–)
Noyori completed his master's degree at Kyoto University in
1963, then worked as an assistant there and as an assistant
professor at Nagoya University in 1968 before accepting
a foreign-study opportunity in the Corey laboratory at
Harvard University. In 1972 he became a professor at
Nagoya University. His specialties were precision organic
chemistry and molecular catalytic chemistry, and he worked
on asymmetric synthesis reactions. He used chiral catalysts—particularly
Ru-BINAP—to achieve high-efficiency asymmetric hydrogenation reactions,
making major contributions to the industrialization of asymmetric synthesis.
In 2003 he became director of RIKEN. In 2000 he received Japan's Order of
Culture, and in 2001 he received the Nobel Prize in Chemistry.

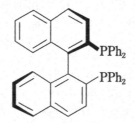

Figure 5.37: Structure of BINAP

hydrogenation of amino acids; they demonstrated a nearly 100% excess of one
chiral isomer over the other[95]. Noyori et al. went on to use BINAP complexes
of rhodium and ruthenium to achieve high-efficiency asymmetric reduction of
a variety of functional groups, contributing significantly to the development of

Barry Sharpless (1941–)
Sharpless graduated in 1963 from Dartmouth College and received his doctorate from Stanford University in 1968. He began work at the Massachusetts Institute of Technology in 1970, and in 1990 became a professor at the Scripps Research Institute. He studied oxidation reactions involving chiral catalysts, and he developed catalysts to assist in various asymmetric oxidation reactions. His methods are used to synthesize various pharmaceuticals. He received the 2001 Nobel Prize in Chemistry.

the technique (Figure 5.37). This method is also used for industrial production of pharmaceuticals and other products, including the synthesis of L-menthol.

Meanwhile, Barry Sharpless and Tsutomu Katsuki at the Scripps Research Institute were advancing the study of asymmetric oxidation reactions using chiral catalysts. In 1980, they used a titanium complex as an asymmetric catalyst to convert aryl alcohol into a chiral epoxide[96]. Epoxide is a useful intermediate product in many synthetic reactions, and this reaction opened the door to the synthesis of many asymmetric compounds. Knowles, Noyori, and Sharpless received the 2001 Nobel Prize in Chemistry for their contributions to the development of catalytic asymmetric synthesis; the award reflected an appreciation not only of the academic quality of the work in question, but of its practical value as well.

Metathesis

The term *metathesis* derives from the Greek meaning "to switch positions;" in the context of organic synthesis, metathesis refers to catalytic reactions such as that depicted in Figure 5.38, involving a crossover recombination in the bonding between two types of olefins. Reactions of this type were discovered as early as the 1950s in the polymer industry, but metathesis reactions were not recognized as useful methods of synthesis until 1967, when Calderon and colleagues reported that a tungsten chloride-triethyl aluminum catalyst could be used to convert

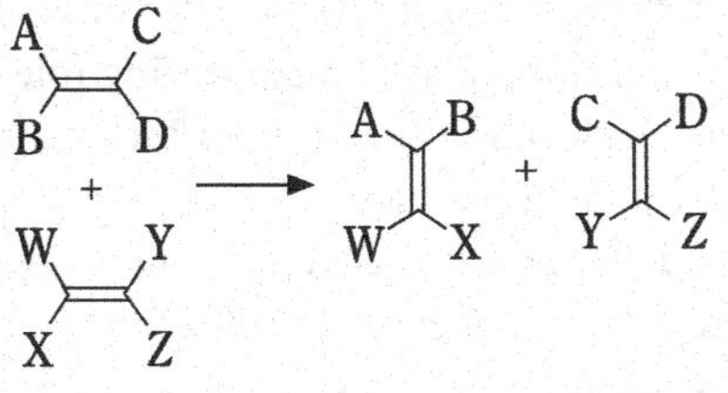

Figure 5.38: A metathesis reaction

Yves Chauvin (1930–)
After graduating in 1954 from the Lyon School of Chemistry, Physics, and Electronics, Chauvin entered the Institut Français du Pétrole as a researcher in 1960, and served as its director from 1991 to 1995. In 1971 he explained the mechanism by which metal carbene complexes promote methathesis reactions, and he later laid the foundations for the development of catalysts for metathesis reactions. He received the 2005 Nobel Prize in Chemistry.

Richard Schrock (1945–)
Schrock is an American organic chemist. After receiving his doctorate from Harvard in 1971, he worked at Cambridge University and at DuPont Corporation before joining the Massachusetts Institute of Technology in 1975, where he became a professor in 1980. In 1990 he discovered that metathesis reactions proceed efficiently through catalysis by molybdenum complexes. He received the 2005 Nobel Prize in Chemistry.

Robert Grubbs (1942–)
Grubbs is an American organic chemist. After receiving his doctorate from Columbia University in 1963, he worked at Stanford University and the University of Michigan before becoming a professor at the California Institute of Technology in 1978. He studied catalytic reactions involving organometallic substances and developed numerous catalysts, including the Grubbs catalysts used for metathesis reactions. He received the 2005 Nobel Prize in Chemistry.

2-pentene into 3-hexane and 2-butane. In 1971, French chemist Yves Chauvin suggested a reaction mechanism in which, after a metal cyclobutane complex in the form of a four-site ring is formed from a carbene metal complex (see page 520) and olefin, other olefins and other carbene complexes could be produced[97]. This proposal opened the door to the design of new catalysts, and the search for effective catalysts came to constitute the core of research on metathesis reactions. Two researchers who played a pioneering role in this search were Richard Schrock at the Massachusetts Institute of Technology and Robert Grubs at the California Institute of Technology.

In 1980, Schrock and colleagues discovered that a carbene complex of tantalum can catalyze metathesis reactions[98]. In 1990, the same group reported that molybdenum complexes exhibit high activity. However, these complexes suffer

Richard Heck (1931–)
Heck studied at the University of California at Los Angeles, receiving a BS degree in 1952 and a PhD in 1954. After working as a postdoctoral researcher, in 1957 he joined Hercules Inc. in Delaware. Between 1971 and 1989, he was a professor at the University of Delaware. The development of Heck reactions began in the late 1960s and was furthered in the 1970s by Mizoroki and by Heck himself; their importance gradually came to be recognized. He received the 2010 Nobel Prize in Chemistry.

from the drawback that they are unstable in water or oxygen. In 1992, Grubs and coworkers reported a ruthenium carbene complex that was relatively stable in water and oxygen and was effective in catalyzing metathesis reactions[99]. Since then, ruthenium carbene complexes developed by many researchers have been widely used, and metathesis reactions have become important components of many synthetic processes, particularly for the synthesis of naturally-occurring compounds. Metathesis reactions have also found widespread use in the pharmaceutical and chemical industries. Chauvin, Schrock, and Grubs shared the 2005 Nobel Prize in Chemistry for their contributions to the study of metathesis.

Cross-coupling and similar reactions
Bonds between carbon atoms are the basis of organic compounds, and thus the development of techniques for selectively creating new single C-C bonds between two reactant molecules is vital in synthetic organic chemistry. This challenge was taken up by many chemists in the second half of the 20th century, and major progress was achieved. Such reactions are known as *coupling* reactions; the term *homocoupling* is used if the two molecules that bind are identical; otherwise the term *cross-coupling* is used. Research on cross-coupling reactions began to make major progress in the late 1960s, and Japanese researchers made significant contributions to this effort.

In the second half of the 20th century, methods of synthetic organic chemistry that use transition-metal complexes as catalysts experienced major advances. In cross-coupling reactions, elements such as palladium (Pd), nickel, and copper are used as catalysts; palladium is particularly effective, a discovery that merited the 2010 Nobel Prize in Chemistry. In 1968, American chemist Richard Heck discovered that phenylstyrene could be produced by adding ethylene to halogenated phenyl palladium (PhPdX). In 1972, he used Pd as a catalyst to

Eiichi Negishi (1935–)

Born in Hsinking in Japanese-controlled Manchuria, he was raised in Seoul while Korea was under Japanese rule. After the Second World War, he moved to Yamato City in Japan's Kanagawa prefecture, where he spent his teenage years. After graduating from the Department of Applied Chemistry at the University of Tokyo in 1958, he worked at Teijin, then studied at the University of Pennsylvania on

a Fulbright scholarship, where he received his PhD in 1963. In 1966 he left Teijin to become a researcher in the Brown laboratory at Purdue University. In 1972 he became a professor at Syracuse University, then became a professor at Purdue University in 1979. He received the 2010 Nobel Prize in Chemistry.

Akira Suzuki (1930–)

Born in the northern Japanese island of Hokkaido, Suzuki lost his father at an early age, but studied hard to graduate from the Faculty of Science at Hokkaido University. After completing his doctorate there, in 1959 he was employed as an assistant, and in 1961 he became an assistant professor. After a three-year period in the laboratory of H. C. Brown at Purdue University, he returned to his professorship in 1973. In 1979, he and his assistant Miyaura discovered

the Suzuki-Miyaura coupling reaction. He received the 2010 Nobel Prize in Chemistry and Japan's 2010 Order of Culture.

develop the Heck reaction (1), in which an organic halogen compound (RX) is coupled to olefin[100]:

$$RX + H_2C=CHR' + Pd \text{ catalyst} \rightarrow RHC=CHR' \qquad (1)$$

(Here R and R' are aryl, vinyl, or alkyl groups and X is a halogen).

In this reaction, the electrophilic agent RX first reacts with Pd to form a palladium complex (RPdX); next, the nucleophilic olefin bonds with this complex, and then R migrates to the olefin to form a new C-C bond. Finally, both hydrogen and palladium are removed to produce substituted olefin.

Meanwhile, in 1976 Purdue University's Eiichi Negishi developed the following reaction—the *Negishi Reaction*—in which an organic zinc compound (RZnY) is used as a nucleophilic agent to induce coupling to a halogen compound R'X[101]:

Teruaki Mukaiyama (1927–)
After graduating in 1948 from the Tokyo Institute of Technology, Mukaiyama became a professor there in 1963; he became a professor at Tokyo University in 1974 and at the Tokyo University of Science in 1987. In addition to developing the Mukaiyama aldol reaction and other new types of reactions in synthetic organic chemistry, he succeeded in the total synthesis of taxol, an anti-cancer drug. He received Japan's Order of Culture in 1987.

$$RZnY + R'X + Pd \text{ catalyst} \rightarrow R - R' + MX$$

This reaction may be regarded as an improvement of the reaction reported in 1972 by Kumada and Tamao at Kyoto University, who used a Grignard reagent (see page 326) and a nickel catalyst to achieve a coupling reaction, but it was easier to carry out than the previously reported reaction. Further progress was achieved in 1979 by Akira Suzuki and Norio Miyaura at Hokkaido University, who used organic boron compounds (RBY) as nucleophilic agents to develop similar coupling reactions that proceed in the presence of base[102]. The use of organic boron compounds made it easier to carry out coupling reactions under milder conditions, thus enhancing their usefulness.

Cross-coupling reactions allow selective bonding of molecules with high efficiency; this technique was frequently used to synthesize complicated compounds, including full synthesis of naturally-occurring compounds. In addition, on the applied side, the method has become widely adopted for the manufacture of pharmaceuticals and liquid crystals, thus exerting a major impact on industry. Heck, Negishi, and Suzuki received the 2010 Nobel Prize in Chemistry for their research on cross-coupling of palladium compounds in organic synthesis.

As noted in this chapter, Japanese organic chemists made many contributions to the field of synthetic organic chemistry. During the postwar recovery period, Hitoshi Nozaki at Kyoto University developed new reactions using the properties of various metallic elements. Noyori and many other talented chemists were trained under Nozaki. Teruaki Mukaiyama at Tokyo University developed numerous new synthetic reactions—including the Mukaiyama aldol reaction—and trained many outstanding organic chemists.

Synthesis of naturally-occurring organic compounds
Research on complicated naturally-occurring organic compounds has been a major branch of chemistry ever since the early 20th century. In the first half of the 20th century, structural determination using methods of organic

Robert Woodward (1917–1979)
Born in Boston, Woodward had a strong interest in chemistry from an early age. He studied at the Massachusetts Institute of Technology and earned his doctorate at the age of 20. In 1938 he became a junior fellow at Harvard, where he worked his way up through the ranks to become a professor in 1950. In 1941 he discovered the rules describing ultraviolet absorption by conjugate unsaturated compounds. He determined the structures and synthetic strategies for various antibiotics, alkaloids, terpenoids, and other species; he established new synthetic methods, and achieved the total synthesis of compounds such as quinine, strychnine, chlorophyll, and vitamin B12. He received the 1965 Nobel Prize in Chemistry. (Column 20)

chemistry was a major goal; the years after the Second World War witnessed the introduction of X-ray diffraction, NMR, spectroscopic methods (including infrared, visible, and ultraviolet absorption) and other techniques that facilitated structural analysis. Meanwhile, the field of synthesis of naturally-occurring organic compounds underwent major development; at first this approach was used merely as a means of confirming structures, but it evolved gradually into a method whose objective was the synthesis of the compound itself. The goal of *total synthesis*, in which naturally-occurring organic compounds are synthesized from compounds that may be easily purchased—arose as the supreme challenge in organic chemistry. This field not only pioneered many sophisticated and refined techniques, but also met the practical needs of areas such as medicine, agriculture, and engineering.

Woodward and total synthesis
Many crucial innovations in the development of synthetic methods for naturally-occurring organic compounds were contributed by one chemist, the ingenious Robert Woodward (Column 20). Woodward, who worked at Harvard University, made major contributions spanning the entirety of the field of natural-products chemistry, including structure determination; however, his achievements in the total synthesis of naturally-occurring organic compounds were peerless. In 1944, he collaborated with W. Doering to synthesize the antimalarial agent quinine[103]. This synthesis was an epochal advance for the time. Thereafter, from the late 1940s through the 1950s, Woodward achieved the synthesis of a series of compounds believed at the time to be difficult to synthesize[104]. These included the steroids cholesterol (1951) and cortisone (1951) and the alkaloids reserpine (1956) and strychnine (1954)

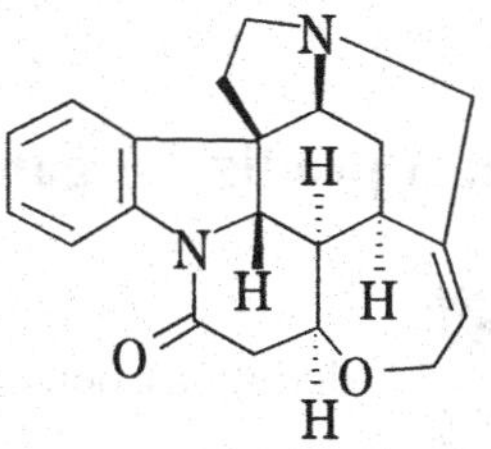

Figure 5.39: Strychnine

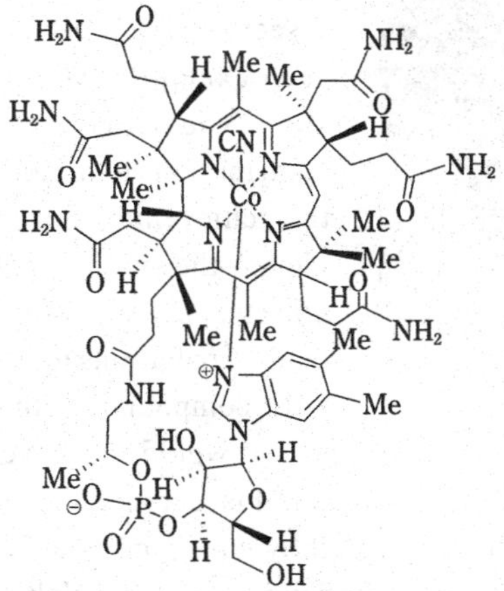

Figure 5.40: Vitamin B$_{12}$

(Figure 5.39). Woodward's synthetic methods, based on his deep insights into reaction mechanisms and exhaustive knowledge of physical organic chemistry, involved thorough planning from start to finish; they marked a decisive shift in organic synthesis from a set of techniques relying on empirical observation into a highly precise style of chemistry evoking art as much as science. Woodward went on to succeed in synthesizing the photosynthetic dye chlorophyll in 1960, and the antibiotic tetracycline in 1962; although he received the 1965 Nobel Prize in Chemistry for his contributions to organic synthetic methods, the accomplishment that must surely rank as the highlight of his research career came *after* this Prize award, with his total synthesis of vitamin B$_{12}$ in 1972 (Figure 5.40). This project, conducted in collaboration with the group of the Swiss chemist Albert Eschenmoser, involved the effort of over 100 researchers spanning more than a decade; the synthesis, which

Woodward: A prodigal organic chemist

Woodward giving a lecture

Some fields of scientific endeavor are indelibly associated with their prodigies: in physics, one has Einstein; in mathematics, Ramanujan. Are there any such geniuses in chemistry? Although his style differed entirely from those of Einstein or Ramanujan, it seems quite clear that Woodward is one such example. Although he received the 1965 Nobel Prize in Chemistry for his contributions to organic synthetic methods, the achievement considered his crowning glory—the synthesis of vitamin B_{12}—came *after* that prize award. He also made important contributions to the study of organometallic compounds with sandwich structures, the subject of the 1973 Nobel Prize; it would not have been inappropriate for Woodward to join Wilkinson and Fischer as a co-recipient of that prize. Meanwhile, if he had lived just slightly longer, he would almost certainly have joined Fukui and Hoffmann in receiving the 1981 Nobel Prize.

Woodward, the grandson of a pharmacist, was born in 1917 in a suburb of Boston. When he was just 1 year old his father died of Spanish influenza, and he was raised by his mother; he developed a strong interest in chemistry from an early age and frequently studied on his own. While attending public elementary and middle schools, he immersed himself in chemistry experiments in his home laboratory, and by the time he entered high school he is said to have completed most of the experiments in the textbook on organic chemistry experiments by Gattermann that was then used in college courses. He skipped grades in school and enrolled at the Massachusetts Institute of Technology at the age of 16. However, he ignored all formal study and never attended lectures, attempting to pass his courses by taking the exams alone. For this reason he was eventually expelled, but he was later re-admitted and graduated in 1936. The following year, at the age of 20, he received a PhD in chemistry. His doctoral thesis addressed the synthesis of the female hormone estrone. In 1937 he became a Junior Fellow at Harvard University, where he continued to work as a researcher for many years; he became a full professor in 1950.

In the early 1940s, Woodward became well known for pioneering the application of ultraviolet spectroscopy to the structural determination of naturally-occurring compounds. Based on extensive experimental data, he discovered a correlation between chemical structure and ultraviolet absorption, and he discovered the laws that would later be known as the Woodward rules. This was the first instance in which he used an analytical tool that was just in the process of being developed to determine the structure of complex organic compounds; he later went on to pioneer a new approach in which infrared spectroscopy, NMR, optical rotary dispersion, and other spectroscopic methods were applied to problems of structural determination and synthesis.

In the mid-1940s he began research on the synthesis of naturally-occurring organic compounds. As discussed elsewhere in this book, he proceeded to synthesize several naturally-occurring organic compounds whose synthesis was then thought to be impossible, thus laying the foundations for a new era in synthetic organic chemistry that may well be termed the Woodward Era. His success in achieving the total synthesis of vitamin B_{12} is one of the most important milestones in the history of organic synthesis. It was during the course of this synthesis that he hit on the idea of the Woodward-Hoffmann rules, which would become the most important rules in organic reaction theory. His ideas were verified through theoretical calculations by Hoffmann, and the two scientists jointly published the rules; Hoffmann received the 1981 Nobel Prize for this work. Woodward was one of the greatest chemists of all time, with many major accomplishments both within and beyond the field of synthetic organic chemistry.

Woodward's quirky habits were the source of many an anecdote. He liked the color blue, wearing exclusively navy-colored suits and sky-blue neckties; he even painted his parking space blue. His Thursday evening laboratory seminars were legendary, often continuing until late in the night. His lectures were long, often lasting 3 to 4 hours. In most of his lectures he used no slides, instead drawing complex chemical formulas on the blackboard with colored chalk. During his lectures, Woodward would align pieces of new colored chalk on one side of a table, and a series of cigars on the other side; he would use one cigar to light the other. He disliked exercising, and survived on little sleep; he was a heavy smoker and enjoyed scotch and martinis. However—perhaps because of this less-than-austere lifestyle—he died of a heart attack in his sleep in 1979, at the age of 62.

Although he was unable to receive a second Nobel Prize, he did receive numerous honors from nations around the world. These included the U.S. National Medal of Science, the Davy Medal from Britain's Royal Society, the Lavoisier Metal from the Société Chimique de France, and the Order of the Rising Sun, Second Class, from Japan.

COLUMN 21

Research on the *fugu* pufferfish and the competition surrounding its structural determination[109]

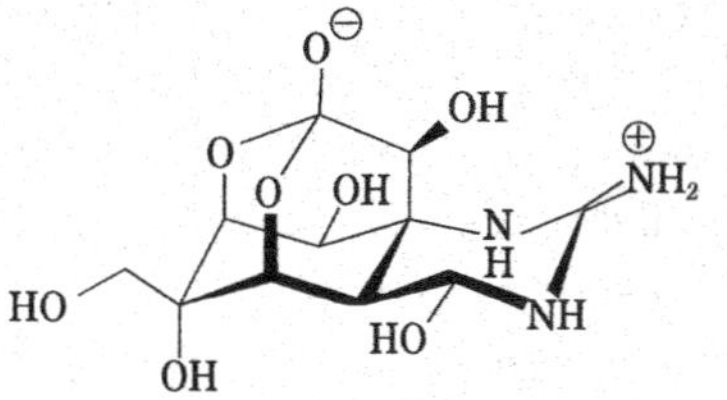

The molecular structure of tetrodotoxin, the poison in fugu pufferfish

The *fugu* pufferfish has long been a favorite delicacy of the Japanese people, and yet poisoning due to fugu poison was for many years a frequent occurrence. Perhaps for these reasons, many Japanese scientists have conducted research on this subject. Although fugu poison is mentioned in an entry in Captain Cook's 1774 ship's log, the poison is not found in fugu in the Atlantic, which may explain why interest in fugu poison was relatively weak in the western hemisphere. Here we survey the historical trajectory from the discovery of fugu poison to the elucidation of its structure and eventually its total synthesis.

Scientific research on fugu poison began in 1887 with studies conducted at Tokyo Imperial University by Juntaro Takahashi, a professor of pharmacology, and Kichindo Inoko, an assistant professor. They showed that fugu poisoning does not occur via decay, but rather by a poisonous agent present inside the body of the fugu itself.

Fugu poison was first isolated in 1907 by Yoshizumi Tahara, head of the Institute of Hygienic Sciences (an office of Japan's Ministry of Home Affairs). Tahara graduated from the Department of Pharmacology at the Tokyo Medical School and worked at the pharmaceutical office of the Institute of Hygienic Sciences where he supervised the analysis of foodstuffs. He reorganized his workspace into a laboratory and became its director of research. Focusing on the analysis of compounds in animals—a field in which European progress was sluggish—he began research on fugu poison. In 1890 he embarked on a three-year foreign-study course in Germany, where he conducted research at the Universities of Munich and Freiburg. At Munich he worked under Adolf von Baeyer (see page 147). Upon returning to Japan he resumed his studies on fugu poison; after more than 10 years of difficult research, in 1907 he successfully isolated the poisonous component. From the academic name for fugu—*tetraodontidae*—and the word *toxin* he constructed the name *tetrodotoxin*. However, the purity of his isolated samples was extremely low.

In 1950, Akira Yokoo at Okayama University succeeded in isolating crystals of tetrodotoxin. This demonstrated that high-purity samples could be obtained, thus opening the door to structural determination. The 1950s was an era in which the structural determination of complex organic compounds was becoming a popular endeavor around the world; Japanese chemistry had made it through a difficult period after the end of the Second World War, and was seeking to catch up to the rest of the world. The structural determination of tetrodotoxin was a tempting research topic, and researchers at Okayama University, Tokyo University, and Nagoya University took up the challenge.

In 1964, an international conference on naturally-occurring compounds was convened in Kyoto. At this conference, three research groups—led by Yoshimasa Hirata and Toshio Goto at Nagoya University, Kyosuke Tsuda at Tokyo University, and Robert Woodward at Harvard University—independently reported a determination of the structure of tetrodotoxin. The research methods used by the three groups differed slightly, but the structures they reported were all basically identical. This circumstance sent a huge wave of energy coursing through the Japanese chemistry establishment. It demonstrated that the level of Japan's research in the chemistry of naturally-occurring compounds rivaled that of the top-level research conducted elsewhere in the world, and that the groups of Hirata and Tsuda were competitive with that of Woodward. Further precise structural determinations by X-ray diffraction were conducted by the group of Isamu Nitta at Osaka University in 1970.

Once the structure had been determined, the next challenge was synthesis. Hirata's student, Yoshito Kishi, succeeded in the total synthesis of tetrodotoxin in 1972. This synthesis was a complicated endeavor involving a total of 29 steps. Kishi later became a professor at Harvard, where he worked in the field of synthesizing naturally-occurring compounds found in the ocean; he succeeded in the total synthesis of palytoxin (a type of poison found in marine life) and other compounds whose complexity exceeded even that of tetrodotoxin. The tetrodotoxin synthesized by Kishi was a racemic mixture containing two distinct optical isomers; the asymmetric total synthesis was only completed in 2003 by the groups of Minoru Isobe at Nagoya University and J. du Bois at Stanford University.

Yoshimasa Hirata, who made major contributions to the structural determination of tetrodotoxin, graduated from Tokyo University in 1941 and became an assistant professor at Nagoya University in 1944. He continued his research under difficult circumstances during and after the Second World War; he succeeded in the structural determination and synthesis of 3-oxykynurenine, a substance discovered in mutated eggs of silkworms. In 1952 and 1953 he was a foreign-study student in the group of Louis Fieser at Harvard; in 1954 he became a professor at Nagoya University and achieved significant results in the chemistry of naturally-occurring compounds. He also excelled as an educator; his students included Yoshito Kishi (professor at Harvard), Koji Nakanishi (professor at Columbia), Toshio Goto (professor at Nagoya University), Daisuke Uemura (professor at Nagoya University), and Osamu Shimomura (2008 Nobel Laureate in Chemistry). He is also known for recognizing the talents of the young Ryoji Noyori and inviting him to Nagoya University.

Elias Corey (1928–)

After receiving his doctorate from the Massachusetts Institute of Technology in 1951, Corey worked at the University of Illinois before becoming a professor at Harvard in 1959. He studied synthesis methods for naturally-occurring organic compounds; he established the method of retrosynthesis and combined it with computer analysis to achieve new developments in organic synthesis. He is known for synthesizing physiologically relevant compounds, including prostaglandins and erythromycin. He received the 1990 Nobel Prize in Chemistry for his development of theory and methodology in organic synthesis.

involves nearly 100 steps, demonstrated that—with appropriate techniques and extensive effort—even extraordinarily complicated organic compounds could be synthesized[105].

In addition to this work, Woodward was involved in two other Nobel-worthy research projects, including research on ferrocenes and the discovery of the Woodward-Hoffmann rules. Although Woodward's Nobel Prize was in 1965, by the time the Woodward-Hoffmann Rule was awarded the Prize in 1979 he had unfortunately passed away and thus could not become a two-time Nobel Laureate. Without question, Woodward must be considered the greatest organic chemists of the second half of the 20th century.

Retrosynthesis and solid-phase polymerization

In the second half of the 20th century, the total synthesis of naturally-occurring organic compounds developed into an active area of research within synthetic organic chemistry; to date, many complicated compounds have been synthesized. A chemist who played a leading role in these developments from the 1960s to the 1990s was Elias Corey at Harvard University. He introduced the idea of *retrosynthesis* and developed a methodology for synthesizing complicated biologically active compounds with selectivity in position and stereospecifisity[106]. In this approach, a multi-stage synthesis proceeds by dividing the target compound into a number of precursors with simpler structures and identifying an optimal synthetic pathway. Corey combined this technique with computer analysis to realize new possibilities in organic synthesis. Using these methods, his group succeeded in the total synthesis of many biologically active compounds[107], including prostaglandin-F_2 (1971), erythronolide (1975), and leukotriene (1981). Corey made major contributions to transforming organic

Vincent du Vigneaud (1901–1978)

An American biochemist, du Vigneaud completed his master's degree at the University of Illinois and his PhD at the University of Rochester. He worked at the University of Illinois and at George Washington University before becoming a professor of biochemistry at Cornell University in 1938. He received the 1955 Nobel Prize in Chemistry for the structural identification and total synthesis of biologically active compounds containing sulfur.

Bruce Merrifield (1921–2006)

Born in Texas, Merrifield studied at the University of California at Los Angeles, where he received his PhD in biochemistry in 1949. He joined the Rockefeller Institute for Medical Research, and from the 1950s through the 1960s he developed solid-state synthetic methods for peptides; he made major contributions to studies of hormones and enzymes. He received the 1984 Nobel Prize in Chemistry.

synthesis from something of a black art into a precise science that anyone could replicate. He received the 1990 Nobel Prize in Chemistry for the development of the theory and methodology of organic synthesis. In the second half of the 20th century, the total synthesis of naturally occurring compounds enjoyed explosive growth due to the involvement of many outstanding chemists, with Woodward and Corey chief among them; the contributions of Japanese researchers to this field have also been particularly noteworthy.

The synthesis of polypeptides—compounds which are important in biochemistry—was initiated by Emil Fischer in the early 20th century. In 1953, Vincent du Vigneaud synthesized oxytocin (pituitary-gland hormone), a biologically active polypeptide, and in the ensuing years many other polypeptides have been synthesized. The first syntheses were conducted in solution, but in this case there is some loss associated with the removal and purification of the reaction products at each stage, leading to low overall yield. In 1962, Bruce Merrifield developed the solid-phase polymerization process (Figure 5.41), which avoids this difficulty[108]. In this method, an amino acid is bound to an inactive substrate such as particles of polystyrene resin; the amino acid and the reactant are then added in a step-by-step process to synthesize a polypeptide. Eventually it became possible to automate the synthesis process; this made it easy to synthesize polypeptides with high yield. du Vigneaud received the 1955 Nobel Prize in Chemistry for his work on sulfur-containing biological compounds, and particularly for the structural determination and total synthesis of oxytocin and vasopressin. Merrifield received the 1984 Nobel Prize in Chemistry for his development of peptide synthesis methods based on

Kyosuke Tsuda (1907–1999)
After graduating from the Department of Pharmacology at the Faculty of
Medicine, Tokyo University in 1929, Tsuda taught at Tokyo University as an
assistant professor and then as a professor at Kyushu University before becoming a professor at the Tokyo University Institute of Applied Microbiology
in 1955. He contributed to developing the chemistry of naturally-occurring
organic compounds, including research on the isolation and structural
determination of fugu poison and studies of alkaloids in legumes. In 1982 he
received Japan's Order of Culture.

Koji Nakanishi (1925–)
Born in Hong Kong, Nakanishi graduated from Nagoya University in 1947.
In 1958 he became a professor at Tokyo University of Education, then moved
to Tohoku University in 1963. In 1969 he became a professor at Columbia
University. He isolated and determined the structures of many naturally-
occurring organic compounds and is particularly well-known for novel
applications of NMR and CD spectral methods. He received Japan's Order
of Culture in 2007.

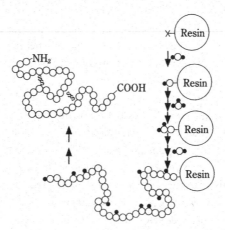

Figure 5.41: Schematic depiction of solid-state synthesis

A functional group (denoted by X in the figure) is attached to beads of inactive resin (denoted by
large circles ○); this functional group and a monomer unit (denoted by small circles ○) are bound
to another group (denoted by filled circles ●) to achieve binding of the monomer to the beads.
This is then bound to the next monomer unit, and the process is repeated to assemble a polymer.

solid-phase reactions. Solid-phase synthesis later became a powerful tool for the
synthesis of nucleic acids, where it contributed significantly to the development
of genetic engineering techniques.

The contributions of Japanese researchers to the organic chemistry of naturally-occurring compounds have been numerous and substantial. As just a few examples we may cite Kyosuke Tsuda, Yoshimasa Hirata, and Yoshito Kishi—who contributed to studies of tetrodotoxin, the poisonous compound in the pufferfish *fugu*—and Koji Nakanishi, who introduced new physical techniques of structural identification for biologically active compounds and clarified the structure of ginkgolide, a component present in gingko leaves. We have already discussed the work of Teruaki Mukaiyama, who attracted attention for using an independent method to rapidly synthesize the anti-cancer drug Taxol.

Supramolecular chemistry (guest-host chemistry)[110]

As the study of typical covalently-bonded molecules matured, research on supramolecules or molecular aggregates—arising from weak interactions among molecules—made major progress in the second half of the 20th century. The fact that weak interactions play an important role in chemical reactions within living organisms was recognized already by the end of the 19th century. Indeed, O'Sullivan and Thompson proposed in 1890 that the enzyme invertase and a substrate sugar can form a temporary complex, and that this was the mechanism by which inversion reactions of sugars proceeded. In 1894, Emil Fischer likened the interaction between an enzyme and a substrate to a lock-and-key mechanism, emphasizing the selectivity of bonding. In the years that followed, the importance of a variety of weak interactions and bonds of various characters—and their importance in chemistry—gradually became clear; these included ionic bonds, hydrogen bonds, van der Waals bonds, and donor-acceptor bonds. However, it was not until the 1970s that supramolecular chemistry (also known as guest-host chemistry) emerged as a new independent subfield. This development was presaged by advances in biochemistry and the emergence of molecular biology, which stimulated the interest of many chemists in elucidating the nature of biological functions involving proteins, enzymes and nucleic acids. However, the direct impetus for the birth of supramolecular

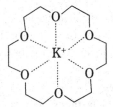

Figure 5.42: An 18-crown ether-6 incorporating a K+ ion

Charles Pedersen (1904–1989)

The child of a Norwegian nautical engineer and a Japanese mother, Pedersen was born in Busan, Korea, and received a middle-school education in Yokohama, Japan. He later graduated from the University of Dayton and completed a master's degree at the Massachusetts Institute of Technology. He then joined DuPont Corporation, where he worked for 42 years. His accomplishments in applied research were widely recognized, and in 1947 he was promoted to a position in which he could conduct research freely; he became involved with a variety of different studies, and discovered crown ethers while investigating oxidation reactions mediated by metallic catalysts. He received the 1987 Nobel Prize in Chemistry.

chemistry was Charles Pedersen's serendipitous discovery of crown ethers, which he reported in 1967.

Crown Ethers

In 1962, Pedersen, who was studying oxidation reactions with metallic catalysts at DuPont Corporation, attempted to synthesize pentadentate ligands from monoethers of catechol and di(2-chloroethyl)ether; as a result, he isolated white fiber-shaped crystals with a yield of 0.4%. He found these crystals interesting and made a detailed study of their structure and properties; he discovered that they were hexadentate ligands in the shape of large rings and that they had the ability to capture sodium and other metal ions. Pedersen went on to study a series of large ring-shaped ethers, which he called *crown ethers*; in 1967 he published a paper discussing their synthesis and their ability to capture metal cations (Figure 5.42)[111].

Denoting the number of atoms forming the rings as x and the number oxygen atoms by y, crown ethers have the general form x-crown-y-ether; the size of the cation that may be captured varies with the size of the ring. This paper attracted immediate interest from researchers across a wide range of subfields of chemistry. In 1960, Pedersen had begun to study the influence of multidentate phenol ligands on the catalytic activity of vanadyl groups (VO), and it was in the course of this investigation that he discovered crown ethers; the story forms a prototypical example of the role of serendipity in major discoveries.

Cryptands and host molecules

Following the work of Pedersen, the chemistry of molecules with multiple multidentate ligands was studied by other researchers—particularly Jean-Marie

Jean-Marie Lehn (1939–)

After completing his doctorate at the University of Strasbourg, Lehn was a postdoctoral researcher in the group of Woodward, then became a professor at Strasbourg and at the College de France in Paris. He developed cryptands that corresponded to three-dimensional versions of crown ethers, and he defined and developed the field of supramolecular chemistry based on intermolecular forces. Lehn is blessed with a fertile imagination and a genius-level intellect; he is a culturally-savvy European who loves music and fine art. He received the 1987 Nobel Prize in Chemistry.

Donald Cram (1919–2001)

After graduating from the University of Nebraska, Cram received his PhD from Harvard. He became a professor at the University of California at Los Angeles, where he achieved numerous accomplishments spanning a wide range of fields in organic chemistry. In particular, he developed the chemistry of crown ethers, pioneering the science of guest-host chemistry. He also excelled as an educator; the organic chemistry textbook he co-authored with Hammond was translated into 12 languages and achieved widespread adoption. He received the 1987 Nobel Prize in Chemistry.

Lehn at the University of Strasbourg in France and Donald Cram at the University of California at Los Angeles—who significantly broadened and developed the field (Figure 5.43). Lehn recognized that cage-shaped molecules containing two or more rings bond more strongly to cations; a characteristic example of this phenomenon is the molecule named [2.2.2] cryptand, which exists in a cage structure consisting of 2 nitrogen atoms, 6 oxygen atoms, and 18 carbon atoms. This molecule is a strong binder of potassium ions. Cryptand forms complexes with cations such as NH_4^+, alkali metals, alkali earths, and lanthanides; its selectivity with respect to ion radius is higher than that of crown ether[112]. Cram designed host molecules with even higher selectivities, which he used to develop advanced methods of organic synthesis. In this way, host molecules with selectivity not only for metallic cations but also for organic cations and anions and for small neutral molecules were developed. Cram also developed optically active crown ethers that could bind selectively to just one of two enantiomers; these were man-made compounds that mimicked the capabilities of enzymes[113].

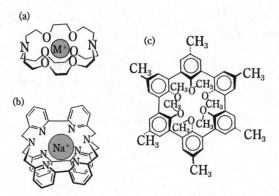

Figure 5.43. (a) and (b) are examples of cryptands that have internally captured metal ions

(a) is a [2.2.2] cryptand. (c) is an example of a host molecule synthesized by Cram.

These studies were connected to research on many related subjects—including ion transport in biological membranes, selective isolation and identification of ions, ion-selective electrodes and cation sensors, catalytic activity, and the isolation of enantiomers—and were extensively developed. The field of supramolecular chemistry (or "host-guest chemistry") pioneered by Lehn and Cram made rapid progress in the 1980s and thereafter, and the molecules synthesized using these techniques grew increasingly refined and complicated. These advances spurred developments across a wide range of fields—including molecular identification, artificial enzyme systems, artificial photosynthesis, molecular sensors, and molecular machines—and the field of supramolecular chemistry developed into a broad interdisciplinary field of research. Cram, Lehn, and Pedersen received the 1987 Nobel Prize in Chemistry for their contributions to the development and use of molecules with structure-specific interactions of high selectivity. The stark differences of personality and background among these three researchers make them an unusually interesting trio of Prize laureates; Pedersen in particular is noteworthy for being a Nobel-Prize winner despite having never received a doctorate.

Fullerenes: A new type of carbon compound[114]
Fullerene
Fullerene is a general term for molecules consisting of hollow spheres, ellipsoids, or cylinders comprising carbon atoms. Isolated carbon was known to exist only in the forms of graphite, diamond, and amorphous carbon; it thus caused quite a stir when, in 1985, the spherical molecule C_{60} was discovered by Robert Curl,

Robert Curl (1933–)

An American spectroscopist, Curl studied at Rice University and received his doctorate from the University of California at Berkeley; he worked as a researcher at Harvard before becoming a professor at Rice University. He used infrared laser spectroscopy to study the dynamics of radicals; he collaborated with Kroto and Smalley on an important series of studies relating to the discovery of fullerenes. He received the 1996 Nobel Prize in Chemistry.

Harold Kroto (1939–)

A British physical chemist, Kroto studied at Sheffield University; after receiving his PhD, he worked at Canada's National Research Council and at Bell Laboratories before obtaining a position at Sussex University in 1967, where he became a full professor in 1985. In 1985, in an attempt to synthesize linear molecules identified in interstellar formations, he collaborated with Smalley on an experiment using laser ablation of graphite; this led to the discovery of fullerenes. He received the 1996 Nobel Prize in Chemistry.

Richard Smalley (1943–2005)

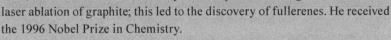

An American physical chemist, Smalley studied at the University of Michigan and received his doctorate from Princeton. After working for Shell Oil, he was a researcher at the University of Chicago before moving to Rice University, where he became a professor in 1981. He was a pioneer of cluster spectroscopy studies using laser ablation methods, and the apparatus he developed was used by him in collaboration with Kroto and Curl to perform the research that led to the discovery of fullerenes. He received the 1996 Nobel Prize in Chemistry.

Harold Kroto, and Richard Smalley. This discovery was a blessing of serendipity and was due to the collaboration of researchers originating from different fields. Kroto, who was interested in red giant stars, believed that long-chain carbon molecules must exist in the vicinity of red giants, and he sought to produce these molecules in the laboratory and study their properties. Curl, a spectroscopist, suggested that Kroto contact his colleague Smalley—who had developed an apparatus for producing and observing clusters by laser evaporation—and thus the three researchers came to collaborate. In 1985, their groups succeeded in

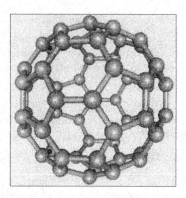

Figure 5.44: Structural Model of C_{60}

producing clusters consisting primarily of C_{60} and C_{70} through laser evaporation of graphite[115]. C_{60} was particularly stable, and the researchers believed it to have the structure of a soccer ball; inspired by the similarity of this shape to the geodesic domes designed by American architect Buckminster Fuller, they named the molecule *buckminster-fullerene* (Figure 5.44). The existence of molecules with this structure had been deduced as early as 1970 by Eiji Osawa; sadly, Osawa's paper appeared in a Japanese-language journal and did not attract wide attention at the time[116].

Curl, Kroto, and Smalley subsequently continued their research, gradually accumulating evidence supporting their hypothesized structure; in 1990, the German physicists Krätschmer and Huffman succeeded in producing gram-sized samples of C_{60} and C_{70} via arc discharge between graphite pillars in a helium environment, confirming the proposed structure[117]. Later work produced a series of other carbon clusters—including C_{76}, C_{78}, and C_{84}—as well as carbon clusters enclosing metal atoms; all of these carbon clusters were termed fullerenes. Thus a new field of chemistry and physics was established focusing on the study of C_{60} and similar carbon clusters.

C_{60} readily accepts electrons to become a negatively-charged ion; it was discovered that crystals of C_{60} potassium salts (K_3C_{60}) became superconducting at a temperature of 19 K. C_{60} and its derivatives were also widely adopted in studies of electron-transfer reactions and catalytic reactions. However, the next truly major breakthroughs in the subsequent development of fullerene science were the discoveries of carbon nanotubes and graphene.

Curl, Kroto, and Smalley received the 1996 Nobel Prize in Chemistry for their discovery of fullerenes. Although the efforts of Krätschmer and Huffman in synthesizing large quantities of fullerenes were crucial to the development of

Sumio Iijima (1939–)
After completing his PhD in physics at the Graduate School of Tohoku University in 1968, Iijima became a research assistant at the Research Institute for Scientific Measurements of Tohoku University. In 1970 he became a researcher at the University of Arizona; in 1982 he joined the Research Development Corporation of Japan, and in 1987 he became a principal researcher at NEC Corporation. In 1991 he used high-resolution electron microscopy to discover carbon nanotubes. He received Japan's Order of Culture in 2009.

Andre Geim (1958–)
A Russian-born Dutch physicist, Geim received his PhD in 1982 from the Institute of Solid State Physics at the Russian Academic of Sciences. After 1990, he conducted research at various universities across Europe, and in 2001 he became a professor at the University of Manchester. He received the 2010 Nobel Prize in Physics for his revolutionary research on graphene.

Konstantin Novoselov (1974–)
A Russian physicist currently living in England, Novoselov studied at the Moscow Institute of Physics and Technology before earning his PhD from the University of Nijmegen in the Netherlands. At Manchester University he worked with Geim to conduct ground-breaking research on graphene. He received the 2010 Nobel Prize in Physics.

fullerene science, the Nobel Prize committee chose to recognize only the work of the initial three discoverers.

Carbon Nanotubes and Graphene

The existence of carbon nanotubes was known as early as the 1970s, but it was not until the 1990s that the subject began to attract major attention. In 1991, Sumio Iijima at the Tsukuba Research Laboratory of Japan's NEC Corporation inspected transmission electron microscope images of the detritus accumulated (deposited sediment) on the negative electrode when fullerenes were produced by arc discharge between carbon electrodes. He discovered *carbon nanotubes* (CNTs)— tubes with radii on the order of nanometers that seemed to be constructed from networks of 6-site carbon rings—whose structure he determined using electron-beam diffraction[118]. These CNTs exhibited a multilayered structure, but later single-layered CNTs were also discovered (Figure 5.45). CNTs are structurally equivalent to monatomic graphite films (graphene sheets) rolled into cylinders;

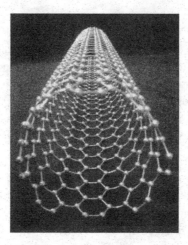

Figure 5.45: A model illustrating the structure of a carbon nanotube (CNT)

the geometrical structure of graphene sheets allows this rolling operation to be carried out in three distinct ways, and the resulting nanotube exhibits different properties—including metallic or semiconducting behavior—depending on how the sheet is rolled. CNTs are expected to have many applications in electronics due to their unusual electronic properties; meanwhile, their outstanding mechanical properties—including lightness, tensile strength, and elasticity—have also raised interest in the possibility of applications as a structural material.

In 2004, Andre Geim and Konstantin Novoselov at the University of Manchester succeeded in using adhesive tape to peel off monoatomic graphene films from graphite crystals. In the ensuing years the remarkable physical properties of graphene have attracted enormous attention. Graphene exhibits extremely high electrical and thermal conductivity. Because it consists of just a single layer of carbon atoms, it is highly transparent, yet nonetheless extraordinarily strong. The unusual properties stemming from its two-dimensional structure have captivated the theoretical curiosity of physicists, while high hopes remain for its practical usefulness as a material.

Geim and Novoselov received the 2010 Nobel Prize in Physics for their pioneering research on the two-dimensional substance graphene.

5.6 The chemistry of functional and physical properties: The foundations of materials science

The study of the physical properties of matter has long been a subject of chemical research, largely as a subdomain of physical chemistry and inorganic chemistry.

However, through the first half of the 20th century the study of physical properties within chemistry focused primarily on liquids, solutions, phase equilibria, and colloids; studies of the physical properties of solids—based on quantum mechanics and statistical mechanics—was largely confined to the domain of physics, with few chemists conducting research in this area. In the second half of the 20th century this situation was significantly transformed, with chemists becoming remarkably active in research of material properties. Underlying this evolution was the development of materials science, a foundational discipline for the electronics and polymer industries. Materials science research enjoyed significant progress due in part to enthusiastic support from industry, which held high hopes for practical applications.

With the 1948 invention of the transistor by the group of Brattain, Bardeen, and Shockley at Bell Laboratories, the field of electronic engineering evolved from reliance on vacuum tubes to a grounding in semiconductors; the electronics industry enjoyed explosive growth and developed into a major industrial sector. These developments were supported by modern solid-state physics, which studied the electrical and magnetic properties of solids on the basis of quantum mechanics and statistical mechanics; at first, the contributions of chemists to this endeavor were few. However, starting around 1970, interest began to grow in the electrical and magnetic properties of a wider variety of materials beyond traditional metals and semiconductors—including polymeric materials and organic compounds—and from this point onward chemists became actively involved in studies of physical properties. Collaborative research between physicists and chemists proved essential for researching the physical properties of widely varying substances, and the study of physical properties underwent major development as a highly colorful interdisciplinary field of research.

In the field of polymeric materials, which provided new chemical substances in the 1930s (see Sections 4.8 and 4.11), the development of Ziegler-Natta catalysts in 1953 enabled easy synthesis of polypropylene and other plastics. The strength and heat-resistance of plastics were significantly improved, and these products came to be widely used; research on the physical properties of polymeric materials became a topic of interest among chemists and physicists alike. In addition, functional polymers exhibiting special capabilities were fabricated and their properties were studied in detail.

In recent years chemists have studied the physical properties of an increasingly broad array of substances—including liquid crystals, surfaces, interfaces, photonic materials, ceramics, and carbon materials—and the study of physical properties is developing into one of the major subfields of chemistry. In this chapter we will consider some highlights of this rapidly growing field, restricting our focus to basic research only.

New functional materials

The second half of the 20th century witnessed the development and study of a wide variety of new materials, including the supramolecules and novel forms of carbon discussed in the previous section. Many of these new substances drew attention for the promise they offered for practical applications thanks to their novel functional properties. In this section we briefly touch on a few of these interesting new forms of matter.

Self-assembled films and nanoparticles

The science of chemistry is replete with examples of natural phenomena in which a system of molecules spontaneously and reversibly transforms from a disordered to an ordered configuration. As the 20th century drew to a close, the fields known as *nanoscience* and *nanotechnology* became the focus of increasing attention, and attempts to exploit these types of self-assembly behavior proceeded in earnest. One such phenomenon, of interest to pure and applied scientists alike, was the existence of *self-assembled monolayers* (SAMs) formed by the chemical adsorption of organic molecules. SAMs are single-molecule films in which one end of an amphiphilic organic molecule chemically adheres to a substrate; intermolecular interactions between the adsorbed molecules then cause them to self-assemble into an aligned configuration. Many molecules exhibiting self-assembly behavior contain long-chain alkyl groups, with the self-assembled film stabilized by van der Waals forces or hydrophobic interactions.

The substrate used to fabricate a SAM may be a metallic surface—such as gold, silver, copper, platinum, or mercury—or a semiconducting surface such as gallium arsenide or indium phosphide; in these cases, the self-assembling molecules are typically organic sulfur molecules [such as thiols (-SH) or disulfides (-S-S-)] which allow exploitation of the affinities of sulfur atoms for metal atoms. When non-metallic oxide surfaces are used as substrates, organic silane molecules are used as the self-assembling molecules. SAMs have found application in many areas of chemistry and engineering as surface-quality improvements.

Tiny particles in the size range 1–100 nm are known as *nanoparticles.* Nanoparticles—which exhibit large ratios of surface area to volume—have attracted attention for their remarkable characteristics, such as the dramatic dependence of their physical properties on particle size, attributable to quantum-mechanical effects. In particular, nanoparticles of gold and silver exhibit unusual absorption profiles due to surface plasmon resonances; for these particles there is a unique dependence of the *color* of the particle on the particle size. The striking colors of metallic nanoparticles have in fact been used since antiquity for applications such as stained-glass art, but it was only in the second half of the

20th century that various ingenious techniques for manufacturing nanoparticles were developed, and since that time the study of nanoparticles has become an active area of research in both basic and applied science. It was subsequently discovered that gold nanoparticles exhibit catalytic effects for many reactions, and this too became a topic of active research.

Meanwhile, semiconducting nanoparticles of materials such as cadmium selenide (CdSe) and zinc sulfide (ZnS) have the ability to trap electrons within them, thus giving rise to unique particle-size-dependent properties due to quantum-mechanical effects. These structures—known as *quantum dots*— have been active topics of study since the early 1980s, with a wide variety of applications envisioned. This novel form of matter offers particularly promising potential for applications to optical materials.

Liquid crystals and supramolecules

Certain substances composed of rod-shaped or planar molecules can—in particular temperature ranges—exist in states in which the *orientation* of the molecules exhibits a regular, ordered pattern, but the overall substance simultaneously exhibits the fluidity of a liquid. Such states, intermediate between liquids and solids, are known as *liquid crystals* (Figure 5.46) Liquid crystals formed from rod-shaped molecules exist in three forms: *nematic* liquid crystals, exhibiting longitudinal order in which the long axes of the molecules are aligned; *smectic* liquid crystals, which form layers exhibiting additional regularity in the positions of the molecular centers of mass; and *cholesteric* liquid crystals, in which the overall arrangement of the molecules forms a helical structure. The transition of these substances from ordered crystals to random liquids happens in stages, a fact first discovered in 1888 by the Austrian botanist Reinitzer in studies of the

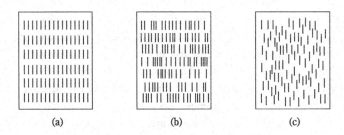

(a)   (b)   (c)

Figure 5.46: Crystals and liquid crystals

(a) *Crystals* exhibit fixed orientations and spatial periodicity.
(b) *Smectic states* exhibit fixed orientations and arrange themselves into evenly-spaced planes, but exhibit no in-plane periodicity.
(c) *Nematic states* exhibit fixed orientations but no spatial periodicity.

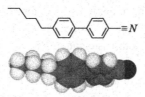

Figure 5.47: Molecular model of 4-alkyl-4-cyanobiphenyl

aromatic acid cholesteryl. Almost immediately after this finding, the German physicist Lehmann recognized that the transition was one from a crystal to a liquid crystal, and that the liquid-crystal phase exhibited optical anisotropy (birefringence). However, research on liquid crystals then made little progress for several decades; it was not until the 1960s and thereafter that the subject began to attract significant attention. Liquid crystals became a particularly popular subject of study after 1969, when Kelker synthesized molecules of the room-temperature nematic liquid crystal N-(4-methoxybenzylidene)-4-butylaniline (MBBA). Phase transitions, light scattering, and other phenomena in liquid crystals were extensively researched by the French physicist de Gennes and his coworkers, who received the 1991 Nobel Prize in Physics. Later, liquid crystals based on cyanobiphenyls were developed; these had low melting points and were chemically stable, and they found application in technologies such as liquid-crystal displays (Figure 5.47)

Polymer chemistry made dramatic advances in the years following the Second World War, and this progress stimulated the development of polymeric materials with a variety of functional properties. By varying reactants and reaction conditions, it became possible to synthesize polymers exhibiting a wide range of properties, including hardness, tensile strength, elasticity, optical sensitivity, viscosity, thermal stability, solubility, affinity or repulsion to a solvent, and optical response. *Copolymers,* obtained by combining two or more different monomers, were effective in producing polymers with desirable functional properties; in particular, *block polymers* consisting of chains of two distinct polymers connected by chemical bonds attracted significant attention [Figure 5.48(a)]. Tri- block polymers are formed from two polymers A and B, with B sandwiched between two polymers of A. If A and B do not exhibit chemical affinity for each other, one polymer will attempt to avoid the other polymer, leading to repulsion. In this case, the edges of polymer A will round themselves to form spheres, yielding a polymer in which spheres of molecule A are distributed throughout a continuous matrix of molecule B. Some block

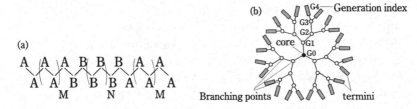

*Figure 5.48: Schematic illustrations of the structures of block polymers (a) and den-
drimers (b)*

polymers formed in this way adopt regular structures with two phases separated
by nanometer-scale distances. Efforts to confer such polymers with desirable
mechanical, optical, electrical, magnetic, or hydrodynamic properties became
an active area of research.

Among the types of polymer that became subjects of active research in the
1980s and thereafter were the polymers known as *dendrimers* [Figure 5.48(b)].
Dendrimers consist of a central molecule—known as the *core*—from which
side-chains known as *dendrons* emanate. The index of a branch of dendrons is
known as its *generation*. The concept of dendrimers was proposed by Tomalia in
1985. Because the core molecule is surrounded by dendrons and thus is isolated
from the external environment, these polymers exhibit unusual reactivities and
light-emission behaviors; they offer significant promise for applications to new
functional materials, and many types of dendrimers have been fabricated and
studied for their properties.

Optical materials
The remarkable progress since the 1980s of optical communication technologies
based on optical fibers was made possible by the use of a method known as
chemical vapor deposition (CVD) to develop quartz fibers of extraordinarily
high transparency. In this process, the burning of a silicon compound in a stream
of oxygen gas produces a "soot" consisting of pure oxidized silicon, which
adheres to the inner wall of a glass tube. The glass tube—with the oxidized
silicon still affixed to its inner surface—is then melted and pulled to produce a
glass-membrane quartz fiber with low optical losses. These fibers are essential
for practical optical communication systems. Optical amplifiers based on erbium
(Er)-doped glass were also developed. This was the birth of photonics, a field
which offers replacements for traditional electronics technologies; the field is
made possible by the progress of optical material engineering and its foundations
in the science of chemistry.

Hiroo Inokuchi (1927–2014)
After graduating from the School of Science at the University of Tokyo in 1948, Inokuchi became an assistant and then an assistant professor there before becoming a professor at the Institute for Solid State Physics at the University of Tokyo in 1967. In 1975 he also became a professor at the Institute for Molecular Science, which he directed from 1987 to 1993. He studied the electrical conductivity of organic materials, discovered organic semiconductors, and established the conceptual foundations for the field of organic semiconductors. He also conducted pioneering research on the electrical conductivity of charge-transfer complexes. He received Japan's 2001 Order of Culture.

Electrically conducting substances

Until the 1970s, research on electrically conducting substances had primarily focused on metals and inorganic semiconductors. Although the study of such materials ranked among the most important domains of solid-state physics, the fact that semiconductor applications routinely required silicon and germanium samples with purities of one part in a million or better ensured that chemical research was an important part of the story as well, essential for the development of high-purity fabrication methods. The subsequent emergence of compound inorganic semiconductors formed from two distinct components—such as indium antimonide and gallium arsenide—created further opportunities for research collaborations between chemists and physicists. It was later discovered that even amorphous silicon exhibits semiconducting properties, and this finding gave rise to a new domain of research beyond the realm of traditional solid-state physics, with its emphasis on the properties of crystals; this was a domain in which chemists would play an increasingly prominent role. Thus, the second half of the 20th century witnessed research on the electrical conductivity of materials becoming a topic of active interest among chemists, with studies focusing primarily on the electrical conductivity of organic compounds. The 1985 discovery of high-temperature oxide superconductors further inspired many chemists to begin searching for other substances exhibiting high-temperature superconductivity, and research on the electrical conductivity of materials enjoyed rapid progress as an interdisciplinary field straddling the boundaries of physics and chemistry.

Research on conducting organic compounds[119]

Studies of organic conductors began as early as the 1940s in Japan, England, and the Soviet Union, and compounds consisting of large numbers of compressed

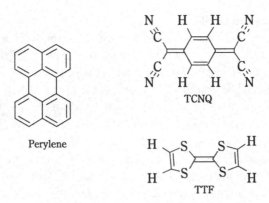

Figure 5.49: Left: Perylene. Right, upper: TCNQ (Tetracyanoquinodimethane). Right, lower: TTF (Tetrathiafulvalene)

benzene rings were known to exhibit semiconducting behavior. In 1954, a group at Tokyo University, including Hideo Akamatsu, Hiroo Inokuchi, and Yoshio Matsunaga, discovered that the complex obtained by applying bromine to perylene exhibited high electrical conductivity[120]. This demonstrated that organic substances could serve as electrical conductors, and research on organic conductors took off in earnest in the 1960s (Figure 5.49). It was thought that the conductivity of these materials was mediated by the highly mobile π electrons they contained. In 1973, it was discovered by Ferraris and collaborators that charge-transfer (CT) complexes (see page 444) of tetracyanoquinodimethane (TCNQ) and tetrathiafulvalene (TTF) exhibited conductivities similar to those of metals; after this, the conduction properties of organic substances became a topic of major interest[121]. Crystals of TTF and TCNQ complexes have charges of +0.56 and −0.56 respectively; some types of molecules form a comb structure consisting of parallel molecular planes, and it became clear that electrons flow easily in the direction of the molecular overlap, thus clarifying the one-dimensional nature of this electrical conductor. This complex exhibits metallic behavior in the sense that its conductivity increases with decreasing temperature; however, below 54 K it becomes an insulator. This is an instance of the metal-insulator transition predicted for low-dimensional conductors by the British physicist Peierls. After the discovery that TTF-TCNQ was a metal (a so-called *molecular metal*), research on molecular metals became an active field; the conductivity of many complexes was studied, and the conditions under which conduction phenomena emerge in these complexes were investigated.

If the transition to insulator at low-temperature in low-dimensional conductors could be suppressed, then it might be possible to create new types

Hideki Shirakawa (1936–)

After graduating from the School of Science at the Tokyo Institute of Technology in 1961, Shirakawa worked as an assistant in the Chemical Resources Laboratory there before becoming a researcher at the University of Pennsylvania; he later became an assistant professor and then a professor at the University of Tsukuba, Japan. In 1967 he discovered that high-density catalysts could be used to produce thin-film polyacetylene, and he conducted detailed studies of its structure and properties. Thereafter, in joint work with MacDiarmid and Heeger, he discovered polymers whose electrical conductivity rivaled that of metals. He received the 2000 Nobel Prize in Chemistry.

Alan MacDiarmid (1927–2007)

Born in New Zealand, MacDiarmid graduated from a local university before embarking on a foreign-study course at the University of Wisconsin in the U.S., where he received a PhD; later, he attended Cambridge University on a scholarship, where he earned a second PhD. In 1955 he began working at the University of Pennsylvania, where he became a professor in 1964. During his first 20 years as a professor, he studied the chemistry of silicon, but after learning of Shirakawa's polyacetylene, MacDiarmid invited Shirakawa to his laboratory, where they collaborated with Heeger—a professor of physics—to conduct the research that earned them the Nobel Prize. He received the 2000 Nobel Prize in Chemistry.

Alan Heeger (1936–)

After majoring in physics and mathematics at the University of Nebraska, Heeger obtained his doctorate in physics at the University of California at Berkeley. From 1962 to 1982 he worked at the University of Pennsylvania, and he was later a professor at the University of California at Santa Barbara. In collaboration with Shirakawa and MacDiarmid, he contributed to the theoretical basis for the field of conducting polymers, and he later worked on light-emission phenomena in polymer semiconductors. He received the 2000 Nobel Prize in Chemistry.

of superconductors; this realization spurred progress in research along these lines. Superconductivity in complexes of tetramethyltetraselenafulvalene (TMTSF)—which differs from TTF by the replacement of S atoms with Se

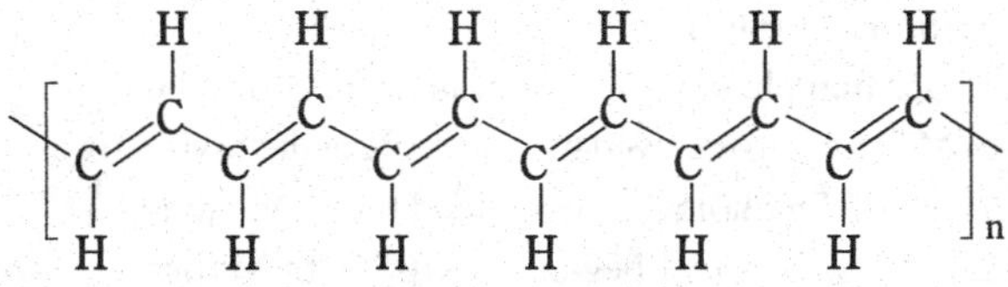

Figure 5.50: Trans-polyacetylene

atoms (which have larger atomic radii) and the replacement of H atoms with methyl groups—was discovered in 1980, stimulating interest in superconducting phenomena in molecular metals. The discovery of superconductivity in the inorganic polymer $(SN)_x$ at ultra-low temperatures (0.3 K) also attracted significant attention. Transition temperatures that have been achieved to date for organic superconductors include 12.3 K in charge-transfer complexes and 33 K in C_{60} complexes.

Organic polymers had been used as insulators since time immemorial, but in the 1970s the electrical conductivity of such polymers became a topic of interest. In 1963, the group of Weiss and collaborators in Australia discovered that the black-colored substance obtained by doping polypyrole with iodine exhibits high electrical conductivity, and similar high conductivity was later observed in polyanyline and related substances. However, these early studies attracted little attention; the field of conducting polymer research did not began to cause a major stir until 1977, when Hideki Shirakawa, Alan MacDiarmid, and Alan Heeger showed that doping polyacetylene with iodine produced a conducting polymer with conductivity rivaling that of metals[122]. In the early 1970s, Hideki Shirakawa—at the Tokyo Institute of Technology—developed a method of synthesizing polyacetylene in such a way that the ratio of cis- to trans-polyacetylene growth could be controlled (Figure 5.50). The electrical conductivity of trans-polyacetylene was some 10^5 times greater than that of cis-polyacetylene, but was still lower than that of typical metals. Meanwhile, at the University of Pennsylvania, Alan MacDiarmid and his physicist colleague Alan Heeger had learned of Shirakawa's polyacetylene; they collaborated with Shirakawa to study polyacetylene doped with halogens, and found that the doping could instantly increase the electrical conductivity by a factor of some 10^3 [123]. It was later discovered that the conductivity of cis-polyacetylene increases by a factor of 10^{11} upon doping with AsF_5. Following these studies, many highly conducting polymers—including polypyrole and polyanyline—were studied; although conducting polymers offer particular promising potential for applications, they are also topics of great interest in

Johannes Bednorz (1950–)
Bednorz is a German physicist. After graduating from the University of Münster, he received his doctorate at the Swiss Federal Institute of Technology, then joined the IBM laboratories at Zurich. In 1983 he began working with Müller to study the possibility of superconductivity in oxides; in 1986 he discovered that an oxide of barium, lanthanum, and copper became superconducting at a temperature of 35 K. He received the 1987 Nobel Prize in Physics.

Karl Alexander Müller (1927–)
Born in Switzerland, Müller studied physics at the Swiss Federal Institute of Technology, where he obtained his doctorate in 1957. In 1963 he joined the IBM laboratories, where he studied the physical properties of SrTiO3 and related perovskites. In the early 1980s he began to explore the possibility of oxide superconductors, and this led to the 1986 discovery of high-temperature superconducting oxides. He received the 1987 Nobel Prize in Physics.

basic science research. Upon doping with halogens, electrons migrate from the polyacetylene to the halogens, producing cation radicals or *polarons*[24]. The behavior of polarons and the mechanisms of conductivity were also subjects of interest to solid-state physicists. Indeed, research on conducting polymers is an excellent demonstration of the dramatic progress that can be achieved through interdisciplinary collaborations between chemists and physicists. Shirakawa, MacDiarmid, and Heeger received the 2000 Nobel Prize in Chemistry.

High-temperature superconductors
The phenomenon of superconductivity in metals was first discovered in mercury in 1911 by Kamerlingh-Onnes. In the ensuing years, scientists searched—with little success—for substances which became superconducting at significantly higher transition temperatures; the transition temperature of 23 K found for an NbGe alloy in 1973 was the highest that had been observed to that time, and the search for higher-temperature superconductors over the next 13 years was fruitless, leading many to believe that superconductivity could not occur at significantly higher temperature. Moreover, it was believed that superconducting

---

24. Polaron: A configuration in which conduction electrons move through a crystal accompanied by a local deformation of the crystal lattice.

phenomena could only occur in metals. It thus came as a surprise when, in 1985, Georg Bednorz and Alexander Müller at the IBM research laboratory in Zurich observed that the electrical resistivity of a certain copper oxide ($La_{1.85}Ba_{0.15}CuO_4$), which is not metal, decreased at temperatures below around 30 K and fell strictly to zero below 10 K[124]. This result was quickly verified in various other laboratories around the world, and in 1986 it was confirmed that ceramic materials of this type (La-Ba-Cu-O system) were high-temperature superconductors with transition temperatures of 35 K. These findings turned the conventional wisdom surrounding superconductivity on its head and sparked an intense global search for other high-temperature superconductors—a pursuit which continued for many years and involved many solid-state physicists and solid-state chemists around the world. Searching for new materials is of course a task at which chemists excel, and many chemists were inspired by this new challenge. In 1987, the yttrium-based copper oxide $YBa_2Cu_3O_7$ was discovered to have a transition temperature of 92 K[125], and subsequent studies revealed certain mercury-based copper oxides with transition temperatures of 160 K. The primary driver of this intense pursuit of high-temperature superconductors was the huge promise they offer for practical applications, but these substances—being ceramics, which differ completely from traditional superconducting metals or alloys—are also of intense academic interest. The first question was whether or not the mechanism of superconductivity in these newly discovered superconductors could be understood within the framework of traditional BCS theory[25]. This question spurred an enormous amount of experimental and theoretical research; for experimental results, one requires high-precision measurements on high-purity samples, and tight collaboration between physicists and chemists proved invaluable for this purpose. The magnitude of the impact of the discovery of high-temperature superconductors is highlighted by the fact that Bednorz and Müller received the Nobel Prize in Physics in 1987, just one year after their discovery. Research on high-temperature superconductors continued throughout the 1990s and beyond, and many new types of such materials were discovered.

Magnetism and magnetic materials

Although the study of magnetic materials has profited from important contributions from chemists—such as the prewar (1930) invention of ferrites ($FeO \cdot Fe_2O_3$)

---

25. BCS theory: A theory proposed by Bardeen, Cooper, and Schrieffer in 1956 to explain the phenomenon of superconductivity. In the BCS theory, electrons form pairs due to an attractive interaction mediated by the electron-lattice interaction; the ground state of a superconductor then consists of a condensate of these pairs, leading to superconducting behavior.

by Yogoro Kato and Takeshi Takei at the Tokyo Institute of Technology—research on magnetism has traditionally been conducted primarily by physicists. Although the essential nature of the main types of magnetic behavior in materials—including diamagnetism, paramagnetism, ferromagnetism, and antiferromagnetism—was established in the prewar era thanks to the emergence of quantum mechanics, research on magnetism expanded in the postwar era to encompass a wide range of substances, including metals, alloys, inorganic compounds, and organic compounds. In the second half of the 20th century, the range of applications for magnetic materials steadily broadened: magnetic materials were used as components of communication devices and in magnetic memory elements such as magnetic tapes and discs, while strong magnets such as superconducting magnets were used in magnetic resonance experiments and MRI machines. This was the societal backdrop against which basic research on magnetism continued to progress in the postwar era. In the first quarter-century after the Second World War, research on magnetic materials was conducted primarily by physicists; however, as magnetism studies eventually broadened to encompass a wider variety of material systems, more and more research was conducted by chemists, and the study of magnetism became an increasingly interdisciplinary pursuit. Toward the end of the 20th century, the field of *spin-tronics*—which uses both the charge and spin of the electron—became a topic of interest, spurring further interest in magnetic materials.

Varieties of magnetic materials[126]

In 1948, Louis Neel discovered crystals containing two types of magnetic ions with oppositely-directed spins; because the magnitudes of their magnetic moments differed, the overall crystal was a magnetic body, termed a *ferrimagnet*. It became clear that magnetite, a substance known to be magnetic since ancient times, was ferrimagnetic. Beginning in the 1950s, neutron diffraction was used to investigate the magnetic structures of many types of inorganic compounds, and the existence of *helical magnets*—in which the direction of magnetization rotates gradually in a curious spiral pattern as one passes through successive layers of atoms in the crystal structure—was discovered. Thus the existence of various new types of magnetic materials came to be recognized alongside traditionally known classes such as paramagnetic, ferromagnetic, and antiferromagnetic materials (Figure 5.51).

The ions responsible for magnetism in crystal lattices typically experience three-dimensional magnetic interactions. However, in some cases the interactions in one or two dimensions may be much weaker than those in other dimensions; it is then possible to treat the material as a low-dimensional magnet, in which the magnetic interactions of ions are approximated as two-dimensional or one-

Louis Neel (1904–2000)

Neel was a French physicist. After graduating from the Ecole Normale Superieure in Paris, Neel studied magnetism in metals and alloys at the University of Strasbourg, where he became a professor in 1937. From 1945 to 1967 he was a professor at the University of Grenoble. He achieved many remarkable results in the theory of antiferromagnetism, ferrimagnetism, and magnetic domains. He received the 1970 Nobel Prize in Physics.

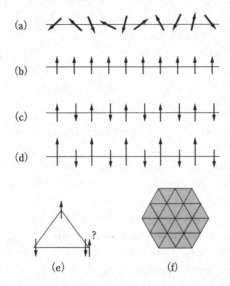

Figure 5.51: A schematic depiction of the spin configurations in various types of magnetic materials

(a) A paramagnet. (b) A ferromagnet. (c) An antiferromagnet. (d) A ferrimagnet. (e) Frustration shown by three antiferromagnetic spins. (f) A triangular lattice.

dimensional in nature. Low-dimensional magnets exhibit magnetic properties that differ from those of traditional three-dimensional magnets, and the magnetic behavior of these systems has been an active subject of research since the 1960s, particularly in conjunction with questions of order-disorder transitions in low-dimensional systems. Materials that may be approximated as one-dimensional magnetic material include $CsCoCl_3$, $CsCuCl_3$, and $KCuCl_3$, while materials that may be approximated as two-dimensional ones include $CoCl_2$, $NiCl_2$, and $FeCl_2$.

One type of antiferromagnetic material that has attracted significant attention in recent years is the *frustration magnetic material*. As an example, consider an equilateral triangle with one spin on each vertex, and suppose that an

Peter Grünberg (1939–)
Grünberg is a Czech-born German physicist. After receiving his PhD from the Darmstadt University of Technology in 1969, Grünberg worked from 1972 to 2004 as a researcher and professor at the Institute for Solid State Physics in Julich. He discovered the giant magnetoresistance effect and received the 2007 Nobel Prize in Physics.

Albert Fert (1938–)
Fert is a French physicist. He studied mathematics and physics at the Ecole Normale Superieure and received his doctorate from the University of Paris-Sud in 1970, where he became a professor in 1976. In 1988 he discovered the phenomenon of giant magnetoresistance in a material consisting of alternating thin films of chrome and iron. He received the 2007 Nobel Prize in Physics.

antiferromagnetic interaction is present between all pairs of spins. If two of the spins antialign (one pointing upward and the other pointing downward), then the energy of the third spin is the same regardless of whether it points up or down, and thus a competition arises between the two possible directions. This type of spin competition is known as *spin frustration*. Spin configurations consisting of triangular lattices (formed by connecting equilateral triangles) or Kagome lattices exhibit large spin system fluctuations due to frustration, and they offer promising potential for the realization of novel magnetic properties.

In typical metallic magnetic materials, the inner-shell electrons involved in magnetism are bound to the atomic nucleus and make essentially no contribution to the electrical conductivity. However, in systems for which these inner-shell electrons lie close to the Fermi level[26], magnetic electrons can move between atoms and are involved in electrical conduction. In systems of this type— examples include cerium compounds—magnetic electrons experience strong interactions and can lead either to magnetic behavior or to superconducting behavior depending on various conditions; these systems are fascinating cases in which magnetism and superconductivity are intimately linked.

In the mid-1980s it was discovered that magnetic films formed by vacuum evaporation coating of two or more metals exhibit interesting physical properties. In 1987, German scientist Peter Grünberg and French scientist Albert Fert, with collaborators, discovered that a multilayer film consisting of thin ferromagnetic

---

26. Fermi level: In a system consisting of electrons, protons, or other fermions (particles with half-integer spin), the highest energy level occupied at a temperature of absolute zero is known as the *Fermi level*.

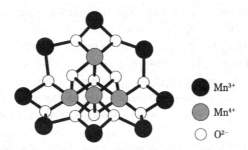

Figure 5.52: Structure of a Mn$_{12}$ cluster, an example of a single-molecule magnet
The large black circles denote Mn ions

and antiferromagnetic layers exhibited *giant magnetoresistance* (GMR)—an enormous variation in electrical resistance depending on the external magnetic field. GMR later attracted attention for its application to magnetic heads in hard disk drives, and Grünberg and Fert received the 2007 Nobel Prize in Physics. The magnetic properties of thin films are also of interest for their relation to spintronics.

Molecular magnets[127]
The field of magnetochemistry, which studies the magnetism of compounds on a macroscopic level, dates back to the prewar era; however, it was not until the late 1960s that research on molecular magnetism—which exploits the properties of molecules—began in earnest, spurred by advances in the study of metal complexes and organic radicals and by the progress of magnetism research in materials physics. In 1967, Koichi Ito discovered the lowest quintuplet state of a carbene molecule[128], and subsequently the groups of Ito, at Osaka City University, and Hiizu Iwamura, at the Institute for Molecular Science and Tokyo University, conducted detailed studies of magnetism in high-spin states of synthetic carbene molecules. In 1987, Miller and collaborators discovered bulk ferromagnetism in TCNE [Fe^{III}{C$_5$(CH$_3$)$_5$}$_2$], a charge-transfer complex containing a metal complex. In 1991, Minoru Kinoshita's group at the University of Tokyo made the first discovery of a ferromagnetic transition in a pure organic radical crystal. These developments help to ensure that research on molecular magnetism became an active topic of study among chemists.

The field of molecular magnetism later expanded to encompass studies of a wide variety of materials; these included attempts to understand the mechanisms of ferromagnetic and antiferromagnetic interactions on the basis of molecular theory, efforts to use molecular design principles to synthesize molecules with custom-designed magnetic properties, and research into phenomena in which magnetic properties are significantly affected by light, temperature, pressure, or other conditions. Organic magnetic materials—composed of organic molecules and complexes but nonetheless exhibiting ferrimagnetism or spin frustration— were discovered as well. One system that attracted particular attention was the prussian-blue-like complex, $A[B(CN)_6]$ (where A and B are transition-metal atoms), in which magnetism is governed by the choice of A and B. Optically-induced magnetism was also observed in complexes of this type.

One discovery that was heralded as the fruit of interdisciplinary research collaboration between physicists and chemists was that of the high-spin (S=10) state of the Mn_{12} acetate complex, discovered in the 1990s (Figure 5.52)[129]. Magnetic hysteresis with molecular origins was observed in this complex, and it was termed a single-molecule magnet. Magnetization curves at a temperature of 2 K display a step due to quantized magnetization tunneling. Investigations on single-molecule magnets went on to become an active area of research.

Optical properties

The study of the optical properties of solids was a well-established field of materials physics already in the prewar era. Once the basic nature of the spectra of atoms and simple molecules had been explained by quantum mechanics, physicists turned to focus on the optical properties of solids, and a wave of pioneering research began. One subject of study was the existence of color centers, formed by irradiating an ionic crystal (such as an alkali halide) with a beam of electrons or radiation. Research in this area gave rise to the concept of *excitons* in insulators and semiconductors; these are states in which an excited electron binds to a hole due to Coulomb forces. The exciton concept may be broadly divided into two categories: *Frenkel excitons* and *Wannier excitons*. The former, first proposed by Frenkel in 1931, are "small" excited states in which the extent of the wavefunction is much smaller than the lattice constant; the latter, proposed later by Wannier, are large excitations in which the extent of the wavefunction is much larger than the lattice constant. The exciton concept has been used to explain many processes associated with optical excitations in insulators and semiconductors, including light emission, energy relaxation, energy transport, charge recombination, and photoconductivity.

The recombination of electrons and holes excited by an impressed voltage in a semiconductor or related material liberates excitonic energy; the phenomenon in which this energy is emitted in the form of light is known as *electroluminescence* (EL), and a device based on this process—the light-emitting diode (LED)—has found widespread application in illumination and display technologies. In an LED, an electron from an n-type semiconductor and a hole from a p-type semiconductor recombine in an emission region to produce light. Red LEDs were

Isamu Akasaki (1929–)

A Japanese applied physicist and chemical engineer, Akasaki studied at Kyoto University, where he did undergraduate work in chemistry and earned a doctorate in engineering at Nagoya University. He worked at Matsushita Research Institute Tokyo, Inc. and was a professor at Nagoya University before becoming a professor at Meijo University. He was awarded the 2014 Nobel Prize in Physics.

Hiroshi Amano (1960–)

A Japanese engineer, Amano completed both undergraduate and doctoral work in engineering at Nagoya University, where he is currently a professor of engineering research. He was awarded the 2014 Nobel Prize in Physics.

Shuji Nakamura (1954–)

A Japanese-born American engineer, Nakamura graduated from Tokushima University and completed doctoral work in engineering. After working at Nichia Corporation, he became a professor at the University of California at Santa Barbara. He was awarded the 2014 Nobel Prize in Physics.

invented by Holonyak in the late 1960s using the semiconductor gallium arsenide (GaAs); by the end of the 1970s, LEDs of other colors—including yellow and green—had been developed. However, the difficulty of obtaining high-quality crystals of the semiconductor gallium nitride (GaN) prevented progress toward the development of blue LEDs. In 1986, the group of Isamu Akasaki and Hiroshi Amano at Nagoya University succeeded in using evaporation techniques to produce a thin film of high-quality GaN crystal on a sapphire substrate; this paved the way to the development of blue LEDs. Shuji Nakamura at Nichia Corporation pioneered the techniques that enabled mass-production of blue LEDs. With this final step, LEDs of all three primary colors could be realized, achieving the long-sought goal of constructing white LEDs. Both the efficiency and the lifetime of LEDs vastly exceed those of incandescent bulbs or fluorescent lights, and the widespread deployment of this technology is expected to make major contributions to the reduction of power consumption. Three scientists who contributed to the development of blue LEDs—Akasaki, Amano, and Nakamura—received the 2014 Nobel Prize in Physics for "the invention of efficient blue light-emitting diodes which has enabled bright and energy-saving white light sources."

Among the various types of solids, one class of materials of particular interest to chemists is *molecular crystals*. These materials were a subject of active research in the 1960s and 70s, during which time various phenomena in molecular crystals were investigated, including absorption and emission spectra, relaxation of excited states, energy transport, and photoelectric conductivity.

5.7 Chemistry of the Earth, the atmosphere, and outer space

The domain of chemistry—which seeks to understand the physical world around us on the basis of molecules and atoms—encompasses not only the entirety of the Earth but of the universe as well, and the branches of the field that address these realms include geochemistry and cosmochemistry. Although these are subfields of the broader multidisciplinary endeavors of earth science and space science, chemistry plays important roles in these disciplines. In this chapter we can only touch on a portion of their full breadth.

In the second half of the 20th century—amidst trends such as the rapidly increasing global population, the emergence of mass-consumption societies on a worldwide scale, and the concentration of populations in cities—the problem of the Earth's deteriorating environment, and how to improve it, became an urgent challenge. These difficulties may be viewed as unfortunate negative consequences of the progress of science and technology, inseparable from the enormous improvements in human living standards made possible by that very progress. For example, chemistry made major contributions to the prosperity and convenience of daily human life through the invention of polymer products such as plastics; however, this very progress also created the problem of how to dispose of massive quantities of plastic waste. Similarly, the invention of DDT as a pesticide made major contributions to the eradication of malaria and other diseases, and the widespread availability of agrichemicals enabled increases in the production of foodstuffs; however, these chemical substances accumulate in the bodies of animals through the food chain, creating serious concerns that they may lead to degradation of our natural environment. Environmental problems involve a complex interplay between various contributing factors, and in many cases their solution requires multidisciplinary approaches that go beyond the realm of chemistry alone; nonetheless, there are numerous ways in which chemistry can play a decisive role. In this section we will discuss some of these, focusing primarily on the role of chemistry within the problem of the destruction of the ozone layer.

The chemistry of the Earth and its environment
Geochemistry
As discussed in the previous chapter, geochemistry began as a field of research primarily concerned with the structure, chemical composition, and

circulation of the material constituents of the Earth. The field studied the elements, isotopes, and chemical species found within and at the surface of the planet—investigating both temporal and spatial variations—and attempted to discern the laws and principles that governed their distribution, movement, and evolution. Thus the field exhibited significant overlap with neighboring disciplines, including mineralogy, lithology, geology, and planetary physics; by the late 20th century the science had evolved beyond the normal boundaries of chemistry to constitute one subdomain within the larger multidisciplinary field of earth science. By incorporating chemical techniques and viewpoints based on the latest developments in analytical chemistry, inorganic chemistry, physical chemistry and organic chemistry, this subdomain played an important role in the discipline as a whole; nonetheless, understanding the transformation and circulation of matter on a global scale required considerations of complex interactions on large scales rather than the molecular-level considerations of typical chemistry research.

Chemistry and the problem of ozone-layer destruction

One planetary-scale problem that chemists made particularly significant contributions toward clarifying was the problem of the destruction of the Earth's ozone layer[130]. The atmosphere surrounding the Earth contains small quantities of ozone (O_3). This ozone absorbs the majority of the sun's ultraviolet radiation, protecting life on Earth from the damaging effects of ultraviolet rays. Ozone is produced from oxygen molecules (O_2) in the atmosphere via the reaction sequence

$$O_2 + hv \rightarrow 2O, O + O_2 + M \rightarrow O_3 + M \text{ (where M is either } N_2 \text{ or } O_2\text{).}$$

In the 1930s, the British geophysicist Chapman proposed a theory for the production and destruction of atmospheric ozone; according to this theory, the ozone concentration varies with altitude, with an *ozone layer* existing between 15 km and 50 km. However, subsequent observations showed deviations from Chapman's predictions, indicating that the presence of other chemical species was impacting the concentration of ozone in the ozone layer. In 1970, Paul Crutzen showed that O and NO_2 react catalytically with ozone to reduce the ozone concentration[131]. NO and NO_2 are supplied by soil bacteria at the Earth's surface and by N_2O emissions from seawater. In 1971, Johnston conducted a detailed study of the reactivity of nitrogen oxides with ozone, in which he noted the danger of destruction of the ozone layer due to accumulation of nitrogen oxides caused by supersonic aircraft flying in the stratosphere.

Paul Crutzen (1933–)
Born in Amsterdam, Crutzen completed elementary school
under difficult circumstances amidst the German occupa-
tion during the Second World War. After graduating from
technical school, he was employed as a civil engineer, but in
1959 he was hired as a computer engineer in the meteorology department of Stockholm University, where he began
climate research. In 1973 he received a doctorate from the
University of Stockholm, where he became a research professor. In the 1970s
he noted the possibility that human activity may impact the ozone layer of
the stratosphere. He received the 1995 Nobel Prize in Chemistry.

In 1974, Mario Molina and Sherwood Rowland pointed out that chlorofluorocarbons (CFCs) such as $CFCl_3$ and CF_2Cl_2—stable, non-toxic gases used
for coolants and aerosol sprays—posed a threat to the ozone layer due to their
chemical non-reactivity[132]. As CFCs rise through the atmosphere without
reacting, they enter the stratosphere, whereupon they are dissociated by
ultraviolet light to produce chlorine atoms; these are then thought to enter the
catalytic cycle and destroy the ozone layer, just as nitrogen oxides do. This report
caused serious and widespread concern regarding the ozone layer and initiated
a phase of comprehensive research on stratosphere ozone chemistry, in which
the reactions of many molecular species were studied in detail. 1985 brought
observation of an actual thinning of the ozone layer above the South Pole—the
so-called *ozone hole*, promoting the abstract threat of ozone-layer destruction
to the status of imminent reality. The 1987 Montreal Protocol established strict
international regulations on the use of chlorofluorocarbons and other dangerous
gases. This case represents an example of major contributions by chemists
toward the resolution of a global-scale environmental problem. Crutzen, Molina,
and Rowland received the 1995 Nobel Prize in Chemistry for their pioneering
work in elucidating the chemistry of the ozone layer.

Chemistry and the problem of Earth's environment
The relationship between chemistry and environmental problems is complex and
multifaceted. First we have the problem of environmental pollution by harmful
substances. Next there is the problem of waste treatment. Other problems include
global-scale variations in climate or the oceans. All of these are interrelated,
and there are no easy solutions; in all cases, chemistry plays a major role in
identifying potential resolutions.

Mario Molina (1943–)
Born in Mexico City, Molina graduated from the National Autonomous University of Mexico and later studied at the University of Freiburg in Germany and the University of California at Berkeley before obtaining his doctorate in 1972. In 1974, as a postdoctoral researcher at the University of California at Irvine, he worked with Rowland to point out the possibility of damage to the ozone layer from chlorofluorocarbons. In 1989 he became a professor at the Massachusetts Institute of Technology. He received the 1995 Nobel Prize in Chemistry.

Sherwood Rowland (1927–2012)
After completing his doctorate in 1952 with Libby at the University of Chicago for research on radiation chemistry, Rowland worked at Princeton University and at the University of Kansas before becoming a professor at the University of California at Irvine in 1964. In 1974 he collaborated with Molina to predict the possibility of damage to the ozone layer due to chlorofluorocarbons;
he also explained the mechanism of this process. He strove to ensure that scientists, politicians, and the general public understood the problems caused by the thinning of the ozone layer, and he contributed to the advancement of international initiatives to ban the use of chlorofluorocarbons. He received the 1995 Nobel Prize in Chemistry.

The problem of environmental degradation due to the disposal of harmful waste products grew dire in the second half of the 20th century as the global population soared and human consumption increased concomitantly. Water at or beneath the Earth's surface is a precious resource, and protecting water from contamination is a crucial challenge. Eutrophication—a problem caused by phosphoric acid and other nutrients in fertilizers—has exposed many bodies of water to the risk of biological death.

Protecting water from contamination requires identifying the source of the contaminants and clarifying the processes by which the contaminants migrate and transform. Many waste products are buried underground; to ensure that underground sites are safe places for waste disposal requires sufficient knowledge of their physics, chemistry, and biology, and the migration and transformation of waste products must be accurately predicted. Here the application of analytical chemistry is an essential tool. However, in the same way that chlorofluorocarbons once believed harmless turned out to be responsible for ozone-layer destruction, in the second half of the 20th century it was recognized that many unexpected substances were causing environmental damage, increasing the severity of environmental challenges. A typical example of this phenomenon was the problem of pesticides, as exemplified by DDT.

DDT was used as an effective pesticide that is harmless to humans, and its use made major contributions to eradicating malaria and other diseases carried by insects. However, because DDT is an extremely stable compound, large-scale dispersal of DDT leads to large accumulations of DDT in the environment; through the food chain, these accumulate in the bodies of birds, where they prevent the birds from reproducing, a problem noted in Rachel Carson's book *Silent Spring* published in 1962[133]. Similar concerns arose regarding herbicides that were once believed to be important tools for eliminating weeds and improving the production of foodstuffs. The later part of the 20th century witnessed many similar examples in which substances once believed to be safe, stable, and effective turned out in fact to pose serious and unexpected problems. A crucial step in resolving these problems will be to acquire a detailed understanding of the processes affecting the chemical substances that impact the environment and living creatures.

Toward the end of the 20th century, the possibility of climate change attributable to increased atmospheric levels of CO_2 and CH_4 due to human activity—the problem of "global warming"—came to be recognized as a serious problem. Over the 100-year period between 1906 and 2005, the average temperature at the surface of the Earth rose 0.74 °C. The average atmospheric concentration of CO_2 rose from 300 ppm to 381 ppm during this period, and in recent years has been growing at a rate of approximately 2 ppm per year. Of the energy received by the Earth from the sun, some fraction is re-radiated back into space and the remainder is converted into thermal energy at the Earth's surface; the surface temperature of the planet is determined by the balance of these two processes. Because greenhouse gases like CO_2 and CH_4 absorb infrared radiation, their presence blocks the re-radiation of infrared rays away from the Earth, promoting temperature increase at the planetary surface. The opinions of experts differ on the question of precisely *how much* of the observed recent increases in global temperature and CO_2 concentration may be attributed to human activity, versus how much are attributable to natural causes. Still, by the end of the 20th century the notion that mass consumption of fossil fuels and other human activities were important contributing factors to global warming and to increased concentrations of greenhouse gases such as CO_2 and CH_4 had become a mainstream view among a majority of observers. Moreover, the serious problems associated with increased atmospheric CO_2 concentrations are not limited to increased temperatures. Indeed, atmospheric CO_2 dissolves in seawater and makes it more acidic. If the atmospheric CO_2 concentration continues to grow at the present rate, it is expected to have a significant impact on ocean-based life systems in the near future. The task of responding to global warming, halting the increase in concentrations of CO_2 and other greenhouse gases, and preventing further degradation of the Earth's environment is a challenge that demands much from the science of chemistry, including but not limited to the development of alternative energy sources to replace fossil fuels.

The chemistry of outer space

Outer space remains the final frontier of human exploration—and the target of limitless curiosity. In the first half of the 20th century, the field of astrophysics—which lies at the nexus of physics and astronomy—made major strides and established itself as a proper academic discipline; the second half of the 20th century similarly witnessed the emergence of astrochemistry (cosmochemistry) as an interdisciplinary field lying at the nexus of astronomy and chemistry. One factor underlying this development was the remarkable progress in space observation technologies, particularly telescopes collecting electromagnetic waves of wide ranges of wavelengths; another factor was the increasingly sophisticated understanding of molecules made possible by advances in molecular spectroscopy. Thus the science of chemistry came to play an increasingly important role in a variety of subdomains of space science, including the discovery and identification of interstellar molecules and the study of chemical processes underlying the formation of stars, the evolution of the universe, and the origins of life. Here we touch briefly on a few of these fascinating topics.

Observation of interstellar molecules[134]

The existence of molecules in large regions of space between stars was confirmed in 1940 using optical telescopes to observe the absorption of CH, CH+, and CN. The emergence in the 1960s of telescopes capable of observing electromagnetic waves in the millimeter and centimeter wavelength regime further demonstrated that in fact there were many different molecular species in space. Temperatures in interstellar regions are low, on the order of 10—100 K, and thus electromagnetic wave emissions due to electronic transitions or vibrational transitions cannot be observed; instead, telescopes typically can detect only radio-wave emissions associated with transitions among rotational states. Moreover, the waves observed are extremely weak, and thus it was only after the development of radio telescopes with large receiving antennas (see page 440) that it became possible to observe many molecular species. Later, advances in techniques for detecting signals in the infrared regime allowed the observation of infrared spectroscopy.

Because the density of molecules in space is low—on the order of $1-10^5/$ cm^3—the probability of reactions induced by collisions between molecules is also low, allowing the existence of unstable radicals and molecular ions. Following the 1963 observation of the OH radical, discoveries of unstable molecules of this type proceeded in rapid succession; by the end of the 20th century, the existence of over 100 types of molecules had been confirmed. These included: diatomic molecules such as H_2, C_2, CH, CO, CO^+, CN, CP, CS, FeO, HCl, HN, HO, N_2, O_2, SiC, SiO, PN, PO, and NaCl; triatomic molecules such as H_3^+, C_3, CH_2, H_2O, HCN, HNC, and HCP; four-atom molecules such as C_2H_2, H_2CO, C_3N, C_3O, H_3O+, and NH_3; five-atom molecules such as C_5, CH_4, HCOOH, and SiH_4; the six-atom molecule methanol; the seven-atom molecule methyl amine; the 8-atom molecule acetic acid; the 9-atom molecule ethanol; the 10-atom molecule acetone; the 12-atom molecule benzene; and the 18-atom molecule naphthalene, to name just a few of the molecules with up to 18 atoms that were discovered during this time. Gradually it became clear that interstellar space contained these gaseous molecules in addition to atoms like hydrogen and helium and their associated ions; also present was *interstellar dust*, consisting of small particles of silicon, carbon, magnesium, iron, and other materials.

The discovery of interstellar molecules that could not be observed using visible light gave rise to studies of *interstellar molecular clouds*. As an interstellar molecular cloud evolves, gravitational compression causes its temperature and pressure to increase, eventually giving rise to a star. The star begins its life in a state known as a *protostar*, during which it emits infrared radiation; eventually, nuclear fusion reactions begin, and the star becomes an ordinary star. As the star's evolution continues, it eventually becomes a *red dwarf*, with the elements synthesized inside the star forming various molecules and solid particles. These forms of matter are then emitted into interstellar space, where they again begin to form interstellar molecular clouds. Chemists' grasp of the chemistry of these interstellar molecules played a crucial role in working out the birth-death sequence of stars, a compelling galactic-scale detective story (Figure 5.53).

Chemical reactions in outer space[134]

Hydrogen accounts for the majority of the atoms in interstellar space. Next is helium (around 15%), while other atoms such as C, N, and O exist in concentrations of only 0.01% that of H. Consequently, reactions in interstellar

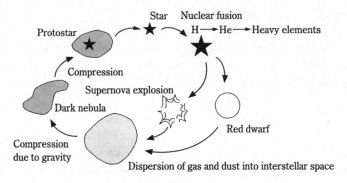

Figure 5.53: The birth-death cycle of a star

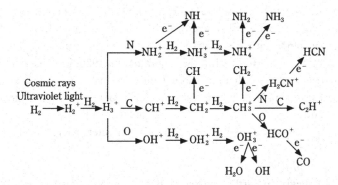

Figure 5.54: Early-stage processes in the formation of molecules in space

space are dominated by reactions involving hydrogen; the low-temperature, low-pressure environment of space ensures that the reactions that occur commonly in space differ from those on earth (Figure 5.54). In a reaction such as A+B →AB, the product AB molecule is in a highly excited vibrational state; however, because this molecule cannot release excess energy through collisions, it dissociates. In addition, the low temperature ensures that very little activation energy is available, and hence reactions of the form AB+C → AC+B do not occur. In 1973, W.D. Watson, and independently E. Herbst and W. Klemperer, proposed a model in which reactions in interstellar space are due to reactions between ions and molecules[135, 136]. In molecular clouds with hydrogen densities of 100/ cm^3, hydrogen exists in primarily molecular form, and hence reactions involving H$_2$ are dominant. High-energy cosmic rays or ultraviolet radiation may ionize

Aleksandr Oparin (1894–1980)
After studying plant physiology at Moscow State University, Oparin studied in Germany with Neuberg, Kossel, and Willstätter. In 1929 he became a professor of plant physiology at Moscow State University. In 1935 he became a professor at the Bakh Institute of Biochemistry, and became the director in 1946. In the 1920s he began research on the origins of life on Earth, and in 1923 he published a short volume on this subject. In 1937 he published a more full account, entitled The Origin of Life; revised versions of this work appeared in 1957 and 1966. His accomplishments also included work on the effects of enzymes in cells, and on industrial biochemistry.

H_2 into H_2^+, which then reacts with H_2 to form H_3^+ and H. H_3^+ readily transfers a proton to other atoms or molecules. This then leads to chain reactions of the type shown in the figure below, leading to the formation of molecules.

The most important molecular species in this reaction process is H_3^+. The spectrum of H_3^+ was observed in the laboratory in 1980, but observations in interstellar clouds were difficult; it was not until 1998 that T. Geballe and Takeshi Oka successfully made such observations[137]. The density of H_3^+ in interstellar clouds was inferred from the signal strength, and these data spurred a discussion of the detailed processes of molecule formation in outer space.

The origins of life

How did life begin on Earth? The question is among the most formidable of all remaining problems in science, and the pursuit of an answer is a major theme of chemistry and biology; the challenge was taken up by many scientists in the second half of the 20th century. Although comprehensive answers still seem far off, it is interesting to survey 20th-century progress in this area as one of the great open questions in the natural sciences. Life on Earth is estimated to have begun some 3.8 billion years ago. How, on a planet containing only inorganic compounds, did the organic compounds that comprise living organisms come into existence—and how did they go on to develop into living things?

The origins of life and chemical evolution

In 1924, the Russian biochemist Aleksandr Oparin published a book entitled *The Origin of Life*, in which he proposed a theory of chemical evolution. Oparin believed that, in the distant past of the cosmos, the Earth's atmosphere was a reductive system of hydrogen, ammonia, methane, and water vapor. These compounds received energy from cosmic rays, solar light, lightning, and other sources; this induced chemical reactions producing small organic molecules,

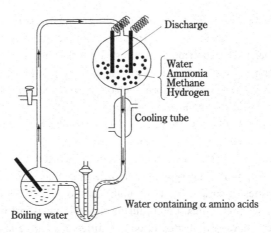

Figure 5.55: Schematic depiction of the experimental apparatus of Miller and Urey

which were absorbed by the oceans. Small-molecule organic compounds then combined with each other to form organic polymers, which condensed to form a primordial soup of polymer compounds. From this, micellar aggregates known as *coacervates* then formed, and it was these entities that led to the origins of life. Around the same time, the English biologist J. B. S. Haldane independently proposed similar ideas, and this theory of chemical evolution became known as the *Oparin-Haldane hypothesis*.

In the postwar era, chemical evolution theory developed from this hypothesis. In 1951, Bernal proposed his "clay theory," which held that the adhesion of small organic molecules to a clay-like substrate was important in the process by which those molecules combined to form polymers. For reactions on surfaces, the German scientist Gunter Wachtershauser proposed in 1988 a "surface metabolism theory," on the basis of which he discussed polymer-formation reactions of organic substances on pyrite (FeS_2) surfaces. The formation of organic compounds from inorganic compounds, which constitutes the first stage in the Oparin hypothesis, was verified by an experiment conducted by Miller and Urey at the University of Chicago in 1953 (Figure 5.55)[138, 139]. These scientists placed a gaseous mixture of methane, hydrogen, ammonia, and water vapor—the constituents thought to make up the Earth's primordial atmosphere—in a glass vessel, then introduced spark discharges to mimic lightning. They continued electrical discharges for one week, all the while cycling the gaseous mixture; upon analyzing the liquid contained in the apparatus at the end of the experiment, they found that α amino acids and several other organic compounds had formed. Because α amino acids are molecular components of proteins, this experiment demonstrated that the molecules necessary for the foundations of life could have been produced on the primordial Earth; the finding created a major stir. However, the primordial atmosphere later came to be seen as a more oxidative environment, consisting primarily of carbon monoxide, carbon dioxide, nitrogen, and water. In such an environment, the creation of amino acids would be more difficult, and thus the questions surrounding the formation of the organic molecules necessary for the foundations of life remained unanswered.

Another possibility that came to be considered is the idea that perhaps α amino acids and similar compounds came from outer space. The primordial Earth was a severe environment which suffered constant bombardment by meteorites; among these was a particular type of meteorite known as a *chondrite*, which contains quantities of carbon. In 1980, many amino acids were identified in extracts from such meteorites. Later, during the 1986 near-Earth approach of Halley's Comet, surveys indicated large quantities of organic matter present in this comet. Although amino acids have not yet been confirmed to exist in interstellar matter, advances in radio telescopes have created high hopes for such a discovery. These findings lend credence to the thought that perhaps the organic molecules that gave rise to life on Earth somehow arrived from outer space. A further piece of evidence in support of this theory is the problem of amino acid chirality. Although amino acids are optically active molecules that exist in an L-form and a D-form, the amino acids that comprise living organisms on earth are exclusively L-form. Why is this? The L-form of some amino acids detected in meteorites is present in excess, and it is possible that the origin of this phenomenon somehow lies in outer space. Using polarized light, it is possible to achieve selective dissociation of either D-type or L-type amino acids. Perhaps the interstellar dust from which meteorites formed was exposed to polarized light in space, resulting in an abundance of D-type or L-type amino acids. In any event, the reality is that even just the *origins* of the organic molecules present in the primordial Earth remain unclear to this day.

The second half of the 20th century also witnessed the proposal of various hypotheses regarding the pathways by which organic matter developed into living organisms. Life on earth requires both the self-replicating abilities of nucleic acids and the catalytic functions of proteins. But which of these two essential ingredients came first? The question has provoked a long-simmering debate in the scientific community. The discovery of RNA with catalytic activity (the ribozyme, Section 6.2.4) lent support to the *RNA world* theory, which postulates that the first living organisms were composed of RNA; however, definitive evidence of this picture is still lacking. Thus the scientific questions surrounding the origins of life remain as mysterious as ever, and their resolution is a task that has carried over to the science of the 21st century.

References

1. S. Nakayama,"*Kagakugijyutu no Kokusai Kyousouryoku*"(*International Competitiveness of Science and Technology*) Asahi Sinbunsha, 2006
2. T. Hiroshige, "*Kagaku no Shakaisi*" (*Social History of Science*) Iwanami Gendaibunnko, 2002
3. F. Aftalion, "*A History of International Chemical Industry*" Chemical Heritage Foundation, 1991
4. L. K. James "Nobel Laureates in Chemistry 1901–1992" Amer. Chem. Soc./ Chem. Her. Found. 1994
5. "Iwanami Rikagaku Jiten"(Iwanami Dictionary of Physics and Chemistry), 5th ed. Iwanami Shoten, 1998
6. Royal Society of Chemistry , N. Hall ed. "*The Age of the Molecules*" London, UK, 1999
7. N. Hall ed. "*The New Chemistry*" Cambridge Univ. Press, 2000
8. C. Reinhardt ed. "*Chemical Sciences in the 20th Century*" Wiley-VCH, Weinheim,
9. National Research Council "*Opportunities in Chemistry*" National Academy Press, Washington, 1985
10. G. C. Pimentel, J. A. Coonrod, "*Opportunities in Chemistry*" National Academy Pres, Washington, D. C., 1987
11. National Research Council, "*Beyond the Molecular Frontier*" National Academies Press, Washington, D.C. 2005
12. N. Hirota, K. Kajimoto ed. "*Gendaikagaku eno Shoutai*" (*Invitation to Contemporary Chemistry*) Asakura Shoten, 2001
13. W. J. Moore, "*Physical Chemistry*" 4th ed. Prentice Hall, New Jersey, 1972
14. D. Hodgkin, Nobel Lecture 1964, Noberprize.org
15. J. D. Bernal, D. Crowfoot, *Nature,* **133**, 794 (1934)
16. D. W. Green, V. M. Ingram, M. F. Perutz, *Proc. Roy. Soc. London*, A **225**, 287 (1954)
17. M. Perutz, Nobel Lecture 1962, Nobelprize.org
18. J. C. Kendrew, R. E. Dickerson, B. E. Strandberg, R. G. Hart, D. R. Davies, D. C. Phillips, V. C. Shore, *Nature*, **185**, 422 (1960)
19. S. B. McGrayne, "*Nobel Women in Science*" Birch Lane Press, 1922, p.225–254
20. W. E. Moerner, L. Kador, *Phys. Rev. Lett.*, **62**, 2535 (1989)
21. O. Shimomura, F. H. Johnson, Y. Saiga, *J. Cell. Comp. Phisiol.* **59**, 223 (1962)
22. ja.wikipedia.org/wiki/Osamu Shimomura

23. E. Ruska, Nobel lecture 1986, Nobelprize.org.
24. D. J. DeRosier, A. J. Klug, *Nature*, **217**, 130 (1968)
25. G. Binnig, H. Rohrer, Ch. Gerber, E. Weibl, *Phisica*, B **109/110**, 2075 (1982)
26. A. Einstein, *Ann. Physik*, **18**, 121 (1918)
27. J. P. Gordon, H. J. Ziegler, C. H. Towns, *Phys. Rev.* **95**, 282 (1954)
28. T. H. Maiman, *Nature*, **187**, 493 (1960)
29. G. Herzberg, "*Molecular Spectra and Molecular Structure, I Diatomic Molecules*, 2nd Ed. 1950, *II Infrared and Raman Spectra*, 1945, *III Spectra of Polyatomic Molecules*", 1966, Van Nostrand Co. Toronto, Canada
30. R. S. Mulliken, *J. Am. Chem. Soc.*, **74**, 811(1952)
31. M. Kasha, *Discussions Faraday Soc.*, **9**, 14 (1950)
32. R. G. W. Norrish and G. Porter, *Nature*, 164, 658 (1949)
33. R. L. Mössbauer, Zeit. Physik, A **151**, 124 (1958)
34. C. Nording, E. Sokolowski, K. Siegbahn, *Phys. Rev.*, **105**, 1676 (1957)
35. E. M. Purcell, H. C. Torrey, R. V. Pound, *Phys. Rev.*, **69**, 37 (1946)
36. F. Bloch, W. W. Hansen, M. E. Packard, *Phys. Rev.*, **69**, 127 (1946)
37. J. T. Arnold, S. S. Dharmatti, M. E. Packard, *J. Chem. Phys.* **19**, 507 (1951)
38. H. C. Torrey, *Phys. Rev.*, **76**, 1059 (1946)
39. E. L. Hahn, *Phys. Rev.*, **77**, 297 (1950)
40. W. P. Aue, E. Bartholi, R. R. Ernst, *J. Chem. Phys.*, **64**, 2229 (1976)
41. K. Wüthrich, Nobel Lecture, 2002, Noberprize.org.
42. H. M. McConnell, *J. Chem. Phys.*, **26**, 764 (1956)
43. P. C. Lauterbur, *Nature*, **249**, 190 (1973)
44. M. Yamashita, J. B. Fenn, *J. Phys. Chem.* **88**, 4451 (1984)
45. K. Tanaka, Y. Ido, S. Akita, Y. Yoshida, T. Yoshida, *Rapid. Commun. Mass Spectrom.*, **36**, 59 (1988)
46. M. Karas, F. Hillenkamp, *Annal. Chem.* **60**, 2299 (1988)
47. J. Martin, Nobel Lecture, 1952, Nobelprize.org.
48. Y. Harada, "*Ryousikagaku*" (*Quantum Chemistry*) Shoukabo, 2007
49. C. C. Roothaan, *Rev. Mod. Phys.*, **23**, 69 (1951)
50. R. Pariser, R. Parr, *J. Chem. Phys.*, **21**, 466 (1953), **21**, 767 (1963)
51. J. Pople, *Trans, Faraday Soc.*, **4**, 1375 (1953)
52. W. Hehre, R. F. Stewart, J. A. Pople, *J. Chem. Phys.*, **51**, 2657 (1969)
53. P. Hohnberg, W. Kohn, *Phys. Rev.*, **136**, B 864 (1964)
54. W. Kohn and J. Sham, *Phys. Rev.*, **140**, A1133 (1965)
55. I. Prigogin, Nobel Lecture, 1977
56. P. Flory, *J. Chem. Phys.*, **17**, 303 (1949)
57. B. J. Adler, T. E. Wainwright, *J. Chem. Phys.*, **27**, 1208 (1957)
58. H. Taube, *Chem. Rev.*, 50, 69 (1952)

59. P. Schleyer, W. E. Watts, R. C. Fort, M. B. Comisarpe, G. A. Olah, *J. Am. Chem. Soc.*, **86**, 5679 (1964)

60. M. M. Olmstead, M. M. Power, *J. Am. Chem. Soc.*, **107**, 2174 (1985)

61. M. Eigen, G. Kurtze, K. Tamm, *Z. Elektrochem*, **57**, 103 (1953)

62. R. G. W. Norrish, G. Porter, *Nature*, **164**, 658 (1949)

63. G. Porter, Proc. Roy. Soc. London, A 200, 284 (1949)

64. G. Porter, M. W. Windsor, *J. Chem. Phys.*, **21**, 2088 (1953)

65. G. Herzberg, J. W. C. Johns, *J. Chem. Phys.*, **54**, 2276 (1971)

66. E. Taylor, S. Datz, *J. Chem. Phys.*, **23**, 1711 (1955)

67. D. R. Herschbach, *Adv. Chem. Phys.*, **10**, 319 (1966)

68. Y. T. Lee, J. D. McDonald, P. R. Lebreton, D. R. Herschbach, *Rev. Sci. Instr.*, **40**, 1402 (1969)

69. J. C. Polanyi, *J. Chem. Phys.* **31**, 1338 (1959)

70. J. V. V. Kasper, G. Pimental, *Phys. Rev. Lett.*, **14**, 352 (1965)

71. A. H. Zwail, *Science*, **242**, 1645 (1988)

72. T. S. Rose, M. J. Rosker, A. H. Zwail, *J. Chem. Phys.*, **88**, 6672 (1988), **91**, 7415 (1989)

73. T. Förster, *Ann. Phys.* (Leipzig), **2**, 55 (1948)

74. D. L. Dexter, *J. Chem. Phys.*, **21**, 836 (1953)

75. M. Itoh, *"Reza- Hikarikagaku"* (*Laser Photochemistry*) Shoukabo (2002)

76 S. Nagakura, H. Hayashi, T. Azumi,ed. "Dynamic Spin Chemistry" Kodansya, 1998

77. K. Honda, A. Fujishima, *Nature*, **238**, 37 (1972)

78. K. Fukui, T. Yonezawa, C. Nagta, H. Shingu, *J. Chem. Phys.*, **22**, 1433 (1954)

79. R. B. Woodward, R. Hoffman, *J. Am. Chem. Soc.*, **87**, 395 (1965)

80. R. Marcus, *J. Chem. Phys.*, **24**, 966, 979 (1956)

81. J. R. Miller, L. T. Calcaterra, G. L. Closs, *J. Am. Chem. Soc.*, **106**, 3047 (1984)

82. R. Marcus, *J. Chem. Phys.*, **21**, 359 (1952)

83. *"Chemical Processes on Solid Surface"* 2007, Nobelprize. org

84. N. Bartlett, *Proc. Chem. Soc. London*, 1962, 218

85. H. H. Classen, H. Selig, J. G. Malm, *J. Am. Chem. Soc.*, **84**, 3593 (1962)

86. D. Schechtman, I. Bleach, D. Gratias, J. Cahan, *Phys. Rev. Letters*, **53**, 1951 (1984)

87. T. J. Kealy, P. L. Pauson, *Nature*, 168, 1039 (1951)

88. G. Wilkinson, M. Rosenblum, M. C. Whiting, R. B. Woodward, *J. Am. Chem. Soc.*, **74**, 2125 (1952)

89. E. O. Fischer, W. Pfab, *Z. Naturforsch.*, **76**, 377 (1952)

90. K. Ziegler, E. Holzkamp, H. Breil, H. Mirtin, *Angew. Chem.*, **67**, 541 (1955)

91. G. Natta, Nobel Lecture, 1963, Nobelprize.org
92. H. C. Brown, Nobel Lecture, 1979, Nobelprize.org
93. G. Wittig, U. Schollkopf, *Chem. Ber.* **87**, 1318 (1954)
94. W. S. Knowles, M. J. Sabacky, Chem. Commun. 1445 (1968)
95. A. Miyashita, A. Yasuda, H. Takaya, K. Toriumi, T. Itoh, T. Souichi, R. Noyori, *J. Am. Chem. Soc.*, **102**, 7932 (1980)
96. T. Katsuki, K. B. Sharpless, *J. Am. Chem. Soc.*, **102**, 5974 (1980)
97. J. –L. Herisson, Y. Chauvin, Makromol. Chem. **141**, 161 (1971)
98. R. R. Schrock, S. M. Rocklage, J. H. Wengrovius, G. Rupprecht, J. Fellman, *J. Mol.Catal.* **8**, 73 (1980)
99. S. T. Nguyen, L. K. Johnson, R. H. Grubbs, *J. Am. Chem. Soc.*, **114**, 3974 (1992)
100. R. F. Heck, J. P. Nolley, *J. Org. Chem.*, **37**, 2320 (1972)
101. E. Negishi, A. O. King, N. Okukado, *J. Org. Chem.* **42**, 1821 (1977)
102. A. Miyaura, A. Suzuki, *Chem. Commun.* 866 (1979)
103. R. B. Woodward, W. E. Doering, *J. Am. Chem. Soc.*, **66**, 849 (1944)
104. R. B. Woodward, M. P. Cava, W. D. Ollis, A. Hunger, H. U. Daeniker, K. Schenker, *J. Am. Chem. Soc.*, 76, 4749 (1954)
105. R. B. Woodward, *Pure and Applied Chemistry*, 33, 145 (1973)
106. E. J. Corey, X. –M. Cheng, "*The Logic of Cemical Synthesis*" John Wiley, New York, 1989
107. E. J. Corey, N. H. Andersen, R. M. Carlson, J. Paust, E. Vlattas, R. E. K. Winter, *J. Am. Chem.*, **90**, 3245 (1968)
108. R. B. Merrifield, *J. Am. Chem. Soc.*, **85**, 2149 (1963)
109. J.-M. Lehn, "*Supuramolecular Chemistry*" VCH, Weinheim, 1995
110. K. Takeuchi, "*Jinbutu de Kataru Kagakunyuumon*" (*Introduction to Chemistry by stories about characters*) Iwanami Shinsho, 2010
111. C. Pedersen, *J. Am. Chem. Soc.*, 89, 2495, 7019 (1967)
112. J.- M. Lehn, *Structure and Bonding*, **16**, 1 (1973)
113. E. P. Kyba, R. C. Helgeson, K. Madan, G. W. Gokel, T. L. Tarnowski, S. S. Moore, D. J. Cram, *J. Am. Chem. Soc.*, 99, 2564 (1977)
114. H. Shinohara, "*Nanoka-bon no Kagaku*" (*Sceience of nano carbon*) Kodansha, 2007
115. H. W. Kroto, J. R. Heath, S. C. O'Brien, R. F. Curl, R. E. Smally, *Nature*, **318**, 162 (1985)
116. E. Osawa, *Kagaku* (Chemistry), **25**, 854 (1970)
117. W. Krätschmer, K. Fostiropoulos, R. Huffman, *Chem. Phys. Lett.*, **170**, 167 (1990)
118. S. Iijima, *Nature*, **354**, 56 (1991)

119. G. Saito ed. *"Bunsierekutoronikusu no Hanashi"* (*Story about molecular electronics*) Kei Di Neobukku, 2008

120. H. Akamatsu, H. Inokuchi, Y. Matsunaga, *Nature*, **173**, 168 (1954)

121. J. P. Ferraris, D. O. Cowan, V. Walaska, J. H. Perlstein, *J. Am. Chem. Soc.*, **95**, 498 (1973)

122. The Nobel Prize in Chemistry 2000,Advanced Information, *"Conductive Polymers"* Nobelprize.org

123. H. Shirakawa, E. J. Louis, A. G. MacDiarmid, A. J. Heeger, *J. Chem. Soc. Chem. Comm.* (1977), 579

124. J. G. Bednorz, K. A. Mueller, *Z. Phys.* B**64** (2), 189, (1986)

125. K. M. Wu et. al., *Phys. Rev. Lett.* **58**, 908 (1987)

126. M. Mekata, *Pariti*-(Parity), **25**, No. 7, 50 (2010)

127. K Itoh ed. *"Bunsijise"* (*Molecular magnetism*), Gakkai Shuppan Senta-, 1996

128. K. Itoh, *Chem. Phys. Lett.* **1**. 235 (1967)

129. A. Caneschi, D. Gateschi, R. Sessoli, A.-L. Barra, L. C. Brunel, M. Gillot, *J. Am. Chem. Soc.*, **113**, 5873 (1991)

130. The Nobel Prize in Chemistry 1995, Press Release, Nobelprize.org

131. P. J. Crutzen, Quart. *J. Roy. Meteor. Soc.*, **96**, 320 (1970)

132. M. Molina, F. S. Rowland, Nature, **249**, 810 (1974)

133. R. Carson, *Silent Spring*, Penguin Modern Classics, 2000.

134. The Chemical Society of Japan, ed. *"Sentan kagaku siri-zu IV"* (*Cutting edge chemistry series*) III *Supe-su kemisutori-*(*Space chemistry*), Maruzen, 2004

135. W. D. Watson, *Astrophys. J.* 183, L17 (1973)

136. E. Herbst, W. Klemperer, *Astrophys. J.* **183**, 505 (1973)

137. T. R. Geballe, T. Oka, *Nature*, **384**, 334 (1996)

138. S. Miller, *Science*, **117**, 528 (1953)

139. S. Miller, H. C. Urey, *Science*, **130**, 245 (1959)

Chapter 6: Chemistry in the second half of the 20th century (II)

An understanding of the phenomena of life based on molecules

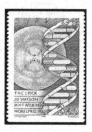

Stamp depicting a model of DNA and Franklin's X-ray diffraction data for DNA

Life is based on chemical phenomena. The developments in biochemistry that began in the early 20th century brought about major advances in how chemistry shed light on life processes, especially metabolism. However, in the first half of the 20th century, chemistry had not yet directly engaged in addressing the important issues in biology, such as genetics, development, and evolution. This situation was completely altered when Watson and Crick elucidated the structure of DNA in 1953. In addition to acting as an impetus for the development of new paths that increased our understanding of life at the molecular level, the discovery of the structure of DNA catalyzed the emergence of molecular biology as a cutting-edge discipline. This is a place where biology and chemistry are entirely contiguous with one another, and the principal components of biology came to be understood on the basis of structures of molecules, interactions between molecules, and processes of conversion between molecules. Molecular biology had previously been regarded as a field of biology, but looking at it from the standpoint of being a discipline based on molecules, it is itself a form of chemistry, and chemistry came to be regarded as encompassing many aspects of biology. In fact, there is virtually no boundary between biochemistry and molecular biology, and typical modern biochemistry textbooks deal with a broad range of issues, including those related to genes.[1] Looking at the list of Nobel Prizes in Chemistry and Nobel Prizes in Physiology or Medicine from the second half of the 20th century to the present, one sees that the overwhelming majority of researchers were active in the fields of biochemistry and molecular biology. These research efforts unified conventional biochemistry and molecular biology into one major field, which could be called life molecular science, and represent many of the most significant advances in chemistry and biology.

The second half of the 20th century also saw remarkable progress in traditional aspects of biochemistry such as proteins, enzymes, and metabolism. A precise discussion based on the structures of biopolymers became possible in biochemistry, as empirical chemistry had transformed into a sophisticated academic discipline thanks to advances in structural chemistry in the first half of the 20th century. What made this possible were the advances in the techniques of structural analysis, such as X-ray crystallography and NMR, which in turn gave birth to the field of structural biology, which is concerned with the elucidation of the structure and function of biopolymers.

This chapter reviews developments in the second half of the 20th century that facilitated our understanding of the molecular theories of life phenomena. Indeed our understanding of the chemistry of life underwent major developments in the second half of the 20th century, and considerable advances were made. Not being an expert in this field, the author is unable to cover the entire picture. As such, the summary provided here focuses on achievements that resulted in the awarding of a Nobel Prize. In writing this chapter, the author referred primarily to information in references [1] to [8] and the Internet, especially Nobelprize.org.

6.1 Birth of molecular biology and structural biology

When Watson and Crick elucidated the structure of DNA in what is widely regarded as the greatest scientific discovery in the second half of the 20th century, they indirectly spawned the new field of molecular biology; this chapter first follows the path leading up to the structural elucidation of DNA, explores how this discovery came to be, and reflects on early research developments on nucleic acids. Next, the developments in structural biology, which had a major impact on biochemistry, will be described.

Road to structural analysis of DNA

How the structure of DNA was elucidated is a drama that is filled with many interesting anecdotes. There are four leading roles in this drama: James Watson, Francis Crick, Maurice Wilkins, and Rosalind Franklin. The first three were awarded the Nobel Prize in Physiology or Medicine in 1962; however, as Franklin died in 1958, she was not considered for the Nobel Prize. The structural elucidation of DNA was the culmination of the efforts of numerous workers. Here, these are referenced and summarized to the best of the author's ability.[5,9–12]

Many of the scientists who tried to understand the physical basis of the biological phenomena of genetics were originally physicists. However, since genes are composed of a large macromolecule called DNA, this endeavor became a matter of chemistry, seen from the viewpoint of elucidating the molecular structure. One scientist whose research was based on this recognition is Pauling, who is widely considered to be one of the foremost figures in structural chemistry. However, Pauling was not responsible for correctly elucidating the structure of DNA, but rather ironically the team of Crick, who had turned away from physics and disliked chemistry, and Watson, a biologist. The elucidation of the structure of DNA drove the emergence of molecular biology and biophysics as new disciplines in academia. Biochemistry, which had long been a field of study of the molecules involved in life phenomena, had played only a supporting role in the birth of these new disciplines, but in later developments in molecular biology, biochemists would also make major contributions.

Avery and Chargaff's research

By the 1940s, it was generally believed that the substance responsible for genetics was a protein (see page 335), but experimental results suggesting that nucleic acid is the bearer of genes also began to appear. In 1923, the British microbiologist, Frederick Griffith, discovered that *Diplococcus pneumoniae* comprises two strains, one that is pathogenic and one that is not. In 1928, he made the startling discovery that when mice were injected with the non-pathogenic bacteria (R strain) along with a form of the pathogenic bacteria (S strain) that had been heat-killed and thus was no longer pathogenic, the S-strain bacteria were found in the mice and the S-strain bacteria replicated repeatedly.[13] This meant that the R-strain bacteria had been transformed into the S-strain bacteria.

Intrigued by this discovery, Oswald Avery of the Rockefeller Institute in New York initiated research into isolating and identifying the substance that was responsible for this transformation. After many long years spent in careful experimentation, Avery, MacLeod, and McCarty published a paper in 1944 claiming that this substance is DNA.[14] Their conclusion was as follows:

> The evidence presented supports the belief that a nucleic acid of the deoxyribose type is the fundamental unit of the transforming principle of Pneumococcus Type III.

However, their conclusion was not generally accepted immediately, and numerous scholars doubted the conclusions of Avery et al., saying among other things that protein would not have been completely removed by their purification method. One of his fiercest opponents was Mirsky, an eminent biochemist who also worked at the Rockefeller Institute. Nonetheless, in response to this report, some scientists began to advance the study of nucleic acid and the latter half of the 1940s saw considerable attention being paid to nucleic acid research. Avery wrote this groundbreaking paper when he was 67 years old. Passing away in 1955,

Oswald Avery (1877–1955)

The child of a British clergyman, Avery was born in Halifax, Canada. In 1887, the family moved to New York, and he earned a Bachelor of Arts at Colgate University, but he aspired to become a doctor and went on to study at the Columbia University College of Physicians and Surgeons. He began microbiology research at the Hoagland Laboratory in 1907, and in 1923 became a researcher at the Rockefeller Institute, where he studied the toxicity and antigenicity of Diplococcus pneumoniae. Inspired by Griffith's research, he engaged in experiments to shed light on how genes work, and in 1944, at the age of 67, he published his definitive paper showing that DNA is responsible for genetic inheritance. However, at the time, his discovery garnered little attention.

Erwin Chargaff (1905–2002)

An Austrian-born biochemist in the U.S., Chargaff majored in organic chemistry at the University of Vienna, earning his degree in 1928; in 1933, he worked at the Pasteur Institute. He emigrated to the U.S. to serve as an assistant professor starting in 1938, and in 1950, as a professor in the Department of Chemistry at Columbia University. Prompted by Avery's research, in 1944 he turned to the study of nucleic acid, and discovered Chargaff's rules, which led to the discovery of the double helix structure of DNA.

he was never awarded the Nobel Prize. Today, however, he is widely regarded as someone who should have been awarded a Nobel Prize.

One person who paid attention to Avery's research and provided the next major step was Erwin Chargaff, a biochemist at New York's Columbia University. Using ultraviolet spectroscopy and the technology of paper chromatography, both of which were just beginning to be used in biochemistry at the time, Chargaff and his collaborators analyzed the base components of DNA from various species. Their results showed that the ratio of base components in DNA from the same species was always the same, but that DNA from different species had different component ratios. This finding implied that DNA had the same level of diversity as the number of species in existence. They also discovered Chargaff's rules, which state that, of the four kinds of bases that constitute DNA, 1) the amount of purine bases (adenine (A) and guanine (G)) is the same as the amount of pyrimidine bases (thymine (T) and cytosine (C)), and 2) the ratio of A and T and the ratio of G and C are both always 1:1.[15] Chargaff was unable to clarify the important significance of these rules, but Chargaff's research did dictate that Levene's DNA tetranucleotide theory (see page 335) was incorrect.

Research advancing the notion that DNA is the substance responsible for genetic inheritance included experiments by Hershey and Chase, published in 1952.[16] They conducted their research using viruses and bacteriophages that infect bacteria. The phages they used were made of DNA and a protein shell enclosing the DNA. They labeled proteins with ^{35}S and DNA with ^{32}P. They found ^{32}P, but not ^{35}S, in the phages that were produced. Their findings categorically showed that DNA was the material of genes, prompting many researchers to become interested in DNA.

Physicists and genetics
With the realization that phenomena in chemistry could be understood from the viewpoint of physics, the 1930s saw the appearance of physicists who believed that biology might also be understood from the perspective of physics. In his 1930 speech entitled "Light and life", however, Bohr inferred that life cannot be reduced to atomic physics, but rather the process of life involve some kind of complementarity, like the physical complementarity seen between waves and particles. A young theoretical physicist who listened to Bohr's lecture, Max Delbrück, believed that genes, too, might have complementarity, and considered investigating this. He opened a private society for study at his home, where he held discussions with geneticists. In 1935, they proposed a quantum theoretical model for mutations. Their model states that when radiation strikes a gene, the gene undergoes a quantum leap and undergoes a mutation. In 1936, collaboration between physicists and geneticists culminated in a biophysics conference, and in the years following 1938 the Rockefeller Foundation actively supported a biophysics program. Delbrück's paper in 1937 caught the attention of Salvador Luria, who had been contemplating changing course from radiology to biophysics at Fermi's laboratory in Italy.

With support from the Rockefeller Foundation, Delbrück moved to the U.S. in 1937, where he established himself at the laboratory of T. H. Morgan, a major figure in genetics at the California Institute of Technology. There, he used bacteriophages as a material to begin researching genes. In 1940, he became a lecturer at Vanderbilt University. In the summer of 1941, at a biology laboratory in the New York suburb of Cold Spring Harbor, 80 geneticists and nucleic acid researchers convened to begin a symposium on the topic of "genes and chromosomes—structures and functions". It was here that Delbrück and Luria met and began collaborative research. Having fled Europe in 1940, Luria found work as a lecturer in bacteriology at Indiana University through Columbia University. Later, Luria would instruct Watson at Indiana University. Delbrück and Luria used Cold Spring Harbor as a base to launch their research group

Max Delbrück (1906–1981)
A German-born molecular biologist working in the U.S., Delbrück studied physics at the University of Göttingen and began as a theoretical physicist, but he became interested in genetic phenomena at the Kaiser Wilhelm Institute for Chemistry. Moving to the U.S. in 1934, he was a professor at Vanderbilt University before he became a professor at the California Institute of Technology in 1947. He was a leader of the phage research group, and was awarded the Nobel Prize in Physiology or Medicine in 1969.

Salvador Luria (1912–1991)
Luria was an Italian-born geneticist in the U.S. who studied medicine at the University of Turin and learned experimental techniques for studying phages at the Pasteur Institute. He emigrated to the U.S. and served as a professor at Indiana University and then, beginning in 1950, at the Massachusetts Institute of Technology. Along with Delbrück, he formed a phage research group and promoted phage research, revealing that mutations occur in bacteria and phages. He was a recipient of the 1969 Nobel Prize in Physiology or Medicine.

on bacteriophages. There, every summer, they held phage workshops and endeavored to expand the horizons of phage research.

In 1944, one of the founders of quantum mechanics, Schrödinger, published a booklet called "What is life?", based on the contents of a public lecture given in Dublin.[17] This was a discussion of life as viewed by a physicist, on the basis of the knowledge of genetics at the time. Today we know that much of its content was incorrect, but it had a major impact on young scientists of the time. A considerable number of the young people involved in physics research in Europe were also considering a move to biophysics, feeling disillusioned with the fact that physics had been used as a tool of war, including the atomic bomb. Watson, Crick and Wilkins were all affected by this book and aspired to study DNA.

Studying biopolymers by X-ray structural analysis and structural chemistry
As noted in Chapter 5, attempts to use X-rays to elucidate the structure of biopolymers had already begun in the 1920s. William Astbury, who was conducting X-ray structural analysis at the Royal Institution of London under the supervision of Henry Bragg, moved to the University of Leeds in 1928 and began researching fibrous proteins such as keratin and collagen. In the early 1930s, he discovered that large structural changes take place when wet wool or hair is stretched. X-ray data showed that fibers before stretching had a helical structure that repeated at intervals of 5.1 Å. Astbury proposed that unstretched

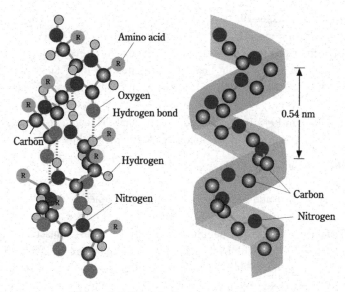

Figure 6.1: The α-helix structure of protein

The right-side figure is a schematic diagram where the structure of the main chain is represented with a ribbon coil.

protein molecules take a helical structure (α-type), but when stretched, the coil of the helix is broken and they enter a stretched state (β-type).

In the summer of 1937, Pauling considered a helical structure for polypeptides consistent with Astbury's X-ray data, but there was insufficient data to definitively support this structure immediately. Therefore, in the early1940s, Pauling ran detailed X-ray structural analyses of amino acids and polypeptides alongside his colleague, Corey, and determined their bond angles and bond lengths; they found evidence that the peptide bond is a planar structure, and that polypeptides adopt a structure where the dihedral angles between the planes are not distorted. Based on this, Pauling developed an α-helix model and in 1950 proposed a model where the pitch of the helix is 5.4 Å, and each turn comprises 3.6 peptide bonds (Fig. 6.1).[18] Pauling believed that fibrous proteins have a double-helix structure. Pauling did not have experimental results to support this, but he was convinced that this structure was correct based on his knowledge of structural chemistry. Pauling also believed that hemoglobin, myoglobin, and other globular proteins were also composed of this kind of α-helix.

At the Cavendish Laboratory, where Bragg served as director, Perutz and Kendrew were continuing their X-ray structural analyses of proteins. Shortly before Pauling's paper was published, they too published a paper about the α-helix structure, but it had an error. As soon as Pauling's paper was published, Perutz ran experiments and found evidence that Pauling's proposal was correct. Thus, the competition to elucidate the α-helix structure of polypeptides ended in a victory for Pauling, dealing a severe blow to Bragg's self-esteem.[19] By this time, it was generally considered that DNA was responsible for transmitting genetic information, so the next target was DNA, and it was obvious that Pauling would also work towards determining its structure.

The chemical structure of nucleic acid was clarified by the beginning of the 1950s, thanks chiefly to the research of Levene and Todd. Nucleic acid is a linear polymer consisting of nucleotides; sugar residues at the 3'- and 5'-positions are cross-linked with phosphoric acid (see page 336). The phosphodiester groups of a polynucleotide are acidic, and nucleic acid is a polyvalent anion at physiological pH. The question was thus, 'what kind of three-dimensional steric structure is adopted by nucleic acid, and how is the structure related to its functions?'.

The pioneer of X-ray structural analysis of DNA in the U.K. was Astbury. In 1938, he reported that DNA has a 2.7-nm repeating structure, where the bases are stacked at intervals of 0.34 nm in a plane.[20] However, he did not go any further than this. Instead, it was the group led by John Randall of King's College at the University of London that focused on the X-ray structural analysis of DNA. Randall had researched luminescence at the University of Birmingham before the war. During the war he helped develop the cavity magnetron for radar, and after the war he launched a biophysics project. Upon being nominated for the position of chief professor of physics at King's College, he acquired a large sum of research funding from the Medical Research Council (MRC) and the Rockefeller Foundation, and established a group to begin intensive research. Maurice Wilkins, who before the war was studying luminescence under Randall and during the war participated in developing the atomic bomb, was also added to this group in 1946, and by around 1950 Wilkins had started studying DNA using X-ray diffraction. At this time, many investigators had started to realize that genetic information is stored in DNA and that it was important to determine its structure. In early 1951, Wilkins obtained X-ray diffraction photographs of DNA that were much more clearly defined than what had been previously reported.

Rosalind Franklin joined Randall's group in January 1951. She had majored in physical chemistry at Cambridge where she studied coal, and later was involved in the study of graphite and carbon particles in Paris. Franklin was familiar with the techniques of X-ray structural analysis. Randall hoped that she would be skilled at X-ray structural analysis and tried to recruit her to work on DNA at King's College. However, Wilkins, who had already begun his own X-ray diffraction studies on DNA at King's College and was obtaining results, was not told about this, resulting in a misunderstanding between Wilkins and

Maurice Wilkins (1916–2004)

Born in New Zealand, Wilkins graduated from the University of Cambridge, and during the war was involved in the atomic bomb development program in the U.S. In 1946, he was teaching at King's College in London. His research first focused on studying viruses using polarized light microscopes, but by around 1950 he had begun X-ray structural analysis of DNA, and contributed to the creation of the Watson-Crick model of DNA. In 1962 he was awarded the Nobel Prize in Physiology or Medicine, and became a professor at King's College in 1970.

Rosalind Franklin (1920–1958)

Franklin majored in physical chemistry at the University of Cambridge and earned a PhD for her structural study of coal; in 1947 to 1950, she was engaged in X-ray structural analysis of graphite at the Laboratoire Central des Services Chimiques de l'Etat in Paris. In 1951, she became a researcher at King's College London in the biophysics institute, where she began X-ray analysis of DNA. Franklin contributed to the creation of the Watson-Crick model by taking what were the best X-ray photographs at the time. From 1953 onwards, she made achievements in the X-ray structural analysis of tobacco mosaic virus at Birkbeck College, but in 1958 died at the young age of 38 from cancer.

Franklin. Franklin felt that Wilkins was abandoning X-ray diffraction studies on DNA, so the relationship between the two was strained from the beginning. They both decided to continue studying DNA but did not have any discussions between themselves.[12]

Meanwhile, at Bragg's Cavendish Laboratory, Perutz and Kendrew were continuing their structural analysis of hemoglobin and myoglobin. Structural analysis of proteins was very challenging work, and the likelihood of getting any useful results could not be foreseen at that time. Bragg and Randall had a British-style gentlemen's agreement, however, and Bragg did not engage in DNA structural analysis, recognizing it as work being done at King's College.

In 1949, Francis Crick joined Perutz's group, and began research for his dissertation on the structural analysis of polypeptides and proteins. Despite being 33 years old, he was still a graduate student with no achievements. While he was widely regarded as being intelligent, he was also considered as being too talkative and rude. Experimental research that required patience, such as X-ray structural analysis of proteins, did not agree with his personality.

Francis Crick (1916–2004)
Crick studied physics at University College London and
began research at graduate school in Cambridge, but his
work was interrupted by the war. During the war he was
engaged in researching mines for the Royal Navy. After the
war, he became disillusioned with physics and turned to
biology, beginning structural studies of proteins by X-ray
diffraction under Perutz. In 1951 he began researching the

structure of DNA by modeling, alongside Watson, and the next year they
proposed a double helix structure. He also played a leading role in later
developments in molecular biology, and between 1962 and 1977 he was a
research member of the Medical Research Council Laboratory of Molecular
Biology in Cambridge. He became a member of the Salk Institute in the U.S.
in 1972 and remained there until 2004. In 1962 he was awarded the Nobel
Prize in Physiology or Medicine along with Watson and Wilkins.

James Watson had earned a degree at the age of 21 in bacteriophage
research under Luria at Indiana University before studying at the University of
Copenhagen. Watson planned to study the biochemistry of viruses in 1950–51,
but in May 1951 he saw Wilkins' X-ray diffraction photographs of DNA crystals
at a symposium in Naples and thought he should also get involved in X-ray
structural analysis. With Luria's assistance, he became a post-doctoral researcher
with Kendrew at Cavendish, and in October 1951 he came to Cambridge. Thus,
the four leading roles appearing in the drama of the structural analysis of DNA
took their places in Cambridge and London.

The Watson-Crick model and structural analysis of DNA
Despite 12 years of age difference between them, Watson and Crick got along
well from the start. The two liked being talkative, and were fiercely passionate
in their desire to elucidate the structure of DNA above all else. They were
complementary in their research skills. Crick was a physicist at heart and
excelled at theory, while Watson the biologist was equipped with biological
knowledge about genes. Nonetheless, because there was a relationship with
King's College, DNA research was not something that Bragg, as director, would
admit to. Crick's duty was protein research and Watson's was tobacco mosaic
virus. Still, they began modeling with the belief that if they constructed models
based on the experimental data obtained at King's College, they might determine
the correct structure for DNA. Their approach, indeed, was the one used by
Pauling when elucidating the structure of the α-helix. Crick had become close

James Watson (1928–)
A precocious genius, Watson was fascinated by birds and majored in zoology, graduating from the University of Chicago in 1947. He became interested in genetics and researched the impact of X-rays on bacteriophages at Indiana University, earning a degree at the age of 21. In 1951, he went to the University of Cambridge to study, researching alongside Crick at the Cavendish Laboratory, and proposed  a double helix structure for DNA. Later, after serving as a professor at the California Institute of Technology, he went on to serve as a professor at Harvard University beginning in 1955, and then from 1968 he was a director at the Cold Spring Harbor Laboratory. His "Molecular Biology of the Gene", published in its first edition in 1965, is famous as a superb textbook. His book "The Double Helix" looks back at his DNA model-making days and is interesting as a frankly worded statement of personal opinion, but it has many problems in objectivity. In 1962 he, together with Crick and Wilkins, was awarded the Nobel Prize in Physiology or Medicine.

friends with Wilkins and therefore was able to obtain the latest X-ray diffraction data. In this regard, they were much more favorably positioned than Pauling, who had to rely on Astbury's old data.

Using data that Watson learned of from Franklin's research seminar at King's, the two immediately made a model, but in this model the nucleic acid bases protruded out. They showed this model to Wilkins, Franklin, and other researchers at King's, but no one supported the model. In the summer of 1952, Watson and Crick began to think that the bases assemble to link DNA molecules together. In July, Chargaff visited the laboratory, and drew attention to the fact that A and G were present in the same quantity in DNA, as were C and T. The idea emerged that pairs of A and T, and of C and G, linked by hydrogen bonds, served as bridges to form a double helix. For A and G, and C and T, to assemble together through hydrogen bonding required knowledge about tautomers[1]; advice from a physical chemist named Donohue who was working at the laboratory at the time proved useful.

In December 1952, Peter Pauling, who was a graduate student in Cambridge and the son of Linus Pauling, received a letter from his father claiming that he had elucidated the structure of DNA. Hearing this news, Watson and Crick rushed to complete their model. One month into the new year, Peter received a preprint of a paper from his father and showed it to Watson and Crick. Surprisingly, it was

---

1. Tautomer: When there are two different isomers that readily interconvert.

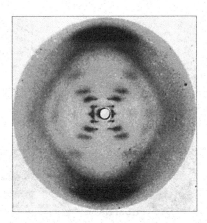

Figure 6.2: Franklin's X-ray diffraction photographs of DNA showing the double helix structure

an incorrect triple helix structure in which the bases come outward, as in their previous model. Several days later, Watson showed a copy of Pauling's paper to Wilkins. Wilkins, without Franklin's permission, showed Watson the best X-ray photographs of DNA that Franklin had taken. These photographs clearly showed that DNA has a double helix structure (Fig. 6.2). These photographs convinced Watson and Crick that the double helix structure linked by base pairs was correct. Their model was completed in early March 1953, and the paper was immediately sent to the journal *Nature*.

X-ray structural analysis at King's College was not going smoothly, due in part to the discord between Wilkins and Franklin. Wilkins believed in a helical structure of DNA, based on his data, but Franklin had the best data. She rejected the helical structure at first, but in early 1953, analysis of her own data led her independently to propose a double helix structure, and she was in the process of drafting a paper. Thus, the edition of *Nature* published on 25 April 1953 had three different papers about the structure of DNA. The first paper was Watson and Crick's proposal of a double helix model (Fig. 6.3),[21] the second paper was by Wilkins and his collaborators, suggesting that DNA has a helical structure based X-ray diffraction results,[22] and the third paper was by Franklin and Gosling discussing the results of X-ray diffraction to determine the double helix structure[23] In 1953, Franklin left King's College, where she had felt ill at ease, and moved to Bernal's laboratory at Birkbeck College in London, where she began structural analysis of viruses. She died of cancer at the young age of 38 in 1958.

The 1953 papers did not necessarily result in all biochemists and biologists immediately believing the Watson-Crick model. Some were still skeptical, and

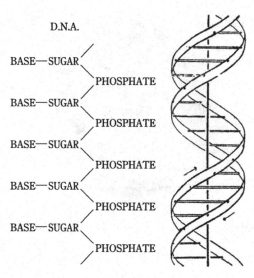

Figure 6.3: Chemical structure of DNA in Watson and Crick's paper (left) and model of the double helix structure (right)

experimental verification was needed. Over the next ten years, various types of experimental data were accumulated, and the correctness of the model was verified, earning Watson, Crick, and Wilkins the Nobel Prize in Physiology or Medicine in 1962. (Column 22)

Replication of DNA
The end of the Watson-Crick paper on the double helix model concludes with the following sentence.

It has not escaped our notice that the specific pairing we have postulated immediately suggests a possible copying mechanism for the genetic material.

Their next paper pointed out that DNA strands could serve as templates that direct synthesis of the complementary strand. This was experimentally proven by a tour-de-force experiment in 1958 by Meselson and Stahl, who were in Pauling's laboratory.[25] They noticed that if labeled parent DNA unravels and becomes a template for daughter DNA to be semiconservatively replicated, then only one of the two strands is labeled. They grew *E. coli* in medium containing ^{15}N and then cultured it in regular medium containing ^{14}N atoms, then used density gradient ultracentrifugation to see how the mass of the DNA changed as the *E. coli* replicated repeatedly. The experimental results they

COLUMN 22

Linus Pauling's successes and failures

Pauling's triple helix DNA molecule model

Linus Pauling was one of the greatest chemists of the 20th century. From the 1920s to the 1930s, he became a leader in structural chemistry, with outstanding successes in the field of chemical bond theory. Armed with structural chemistry, he turned to elucidating the structures of biopolymers, beginning in the latter half of the 1930s. He enjoyed brilliant success in the discovery of the α-helix structure of proteins, but failed at elucidating the structure of DNA.

In the 1930s, proteins were generally thought to be the most important molecules that held the key to life phenomena, but their structures were not understood. Pauling was the first to understand the structure of the peptide bond and the amino acids that constitute proteins, and used this knowledge to infer the structures of proteins. In 1937 he was working with Corey on the X-ray structural analysis of amino acids. (see page 586)

Pauling temporarily left the structural study of proteins. In February 1948, he received an honorary doctorate from Cambridge, and visited the U.K. in order to give Friday lectures at the Royal Institution. After two weeks of these lectures, he was in bed with a cold. Out of boredom, he took out paper and a pencil and began thinking of models for the three-dimensional structure of keratin, with the idea that it adopts a helical structure. Fixing the peptide bond moiety to a plane, he made a model by folding the paper to represent where he thought rotation around bonds was possible. He obtained a helical structure that was consistent with the bond lengths and bond angles that were known for amino acids and peptide bonds. This was the birth of the α-helix model.

In the spring of 1950, the Cambridge group of Bragg, Perutz, and colleagues reported their paper entitled "Polypeptide chain configurations in crystalline proteins". This paper, however, showed that they did not understand that the structure of the peptide bond is planar. Pauling and Corey rushed to send a preliminary report on their α-helix model to the *Journal of the American Chemical Society* (*JACS*) in October 1950, and their full report was published in May 1951 in the *Proceedings of the National Academy of Sciences* (*PNAS*).

Upon reading the Pauling-Corey paper, Perutz hurried to verify whether their model was correct, and confirmed that Pauling's model was right.

Around 1950, the importance of DNA in genetic phenomena was gaining recognition. Around the summer of 1951, Pauling also became interested in the structural elucidation of DNA. In November 1952, inspired by a seminar at the department of biology, he gathered information about nucleotides and DNA the next day and began working to construct a model of DNA. He arrived at a triple helix structure in which a phosphate group is at the center and three strands of DNA are entwined together. Pauling searched through the literature looking for information about DNA structures obtained from X-ray diffraction, but found only Astbury's old data. In the absence of any data to verify the correctness of the model, he continued thinking about a structure where each part of the DNA fits well, and completed a triple helix model in which most of the atoms fit nicely. Pauling and Corey sent their triple helix paper to *PNAS* on 31 December. However, the Watson-Crick paper and Franklin's X-ray diffraction results supporting it soon made it apparent that the triple helix was wrong. Pauling was defeated in the race to elucidate the structure of DNA. One wonders why was he so fixated on the triple helix and failed to arrive at the double helix. The reason lies in the density of DNA. DNA samples contain large amounts of bound water, and the actual DNA represents only two thirds of the mass of hydrated DNA; this bound water is why Pauling thought the double helix was a triple helix.

One also wonders why, in the race to elucidate the structure of DNA, Pauling, who was the leading structural chemist of the time, ended up losing to Watson and Crick, who were still novice researchers. There are various opinions about this.[24] Pauling did not know the latest X-ray analysis results for DNA, but some are of the opinion that the U.S. government is to blame for this. Some say it is because Pauling, who was opposed to nuclear testing in the U.S. at the time, was not issued a passport and was unable to attend academic conferences in the U.K., which meant he could not access the latest data being obtained at King's College.

Pauling himself attributed his defeat to the fact that he relied on X-ray analysis data that was unsatisfactory, the fact that he relied too much on the density of hydrated DNA, and the fact that he lacked information on DNA subunits. There is a large difference, however, between his success in elucidating the structure of the α-helix and his failure with the structure of DNA. Research on the α-helix took place over many long years, whereas his study of DNA was finished in a hurry, in barely six weeks. Even the genius Pauling, it seems, needed more than six weeks to be able to pursue such a major quarry.

Arthur Kornberg (1918–2007)
Studying chemistry at the City University of New York and medicine at the University of Rochester, Kornberg worked from 1943 to 1953 at the National Institutes of Health (NIH) before serving as a professor of microbiology at Washington University (1953 to 1959) and later as a biochemistry professor at Stanford University. During his years at NIH,  he studied the synthesis and decomposition of coenzymes (especially pyridine nucleotides); in 1956 he succeeded in isolating and purifying DNA polymerase from E. coli, then synthesized DNA in vitro. He was a recipient of the 1959 Nobel Prize in Physiology or Medicine.

obtained were consistent with the Watson-Crick model. They also demonstrated that DNA is a double helix. Cells containing ^{15}N-DNA were cultured for one generation in a ^{14}N-medium, then the DNA was heated and denatured and the two strands were separated. Subjecting this denatured DNA to density gradient ultracentrifugation produced two bands, corresponding to a ^{15}N strand and a ^{14}N strand. Meselson and Stahl showed that these DNA strands were half the mass of the DNA before denaturation.

How does DNA replication occur? Later research showed that this is a complex process involving various enzymes. The first researcher to show this was Arthur Kornberg of Washington University (St. Louis) in 1957.[26] His group added ^{14}C-labeled thymidine (the nucleotide containing thymine as its base) to DNA-free supernatant extracted from E. coli, then added ATP and DNA taken from bovine thymus and studied the reaction. The results revealed that the ^{14}C was incorporated into the DNA. This signifies that DNA was synthesized by the enzymes in the supernatant. Kornberg worked to purify the enzymes involved in DNA synthesis by E. coli and succeeded in purifying DNA polymerase, publishing his work in 1958. The enzyme replicated DNA faithfully, even with template DNA from organisms other than E. coli, reproducing the compositional ratio of AT vs. GC in the template DNA. This research earned Kornberg the Nobel Prize in Physiology or Medicine in 1958, along with Ochoa who had succeeded in synthesizing RNA. This achievement was a major contribution of biochemists in early DNA research.

DNA and RNA
By around 1960, the double helix structure of DNA was established. However, the question remained, 'what is the role of RNA, and what is the relationship between DNA and RNA? The function of DNA is to direct its own replication

and the transcription of RNA molecules, but RNA itself has various biological functions. These important functions include what is called transcription and translation. Translation is the process by which proteins are synthesized under the direction of template DNA.

The involvement of RNA in protein synthesis had already been made evident by Caspersson and Brachet's research in the second half of the 1930s. They discovered that in eukaryotic cells, DNA is inside the nucleus and RNA is mainly in the cytoplasm. Brachet discovered that RNA-containing particles in the cytoplasm are also rich in protein. They focused on the fact that the concentration of these RNA-protein particles, which were later named ribosomes, is related to the rate of protein synthesis by the cells, suggesting an association between RNA and protein synthesis. Brachet used amino acids labeled with radioactive isotopes to show that most amino acids incorporated into protein are bound to ribosomes. It was also shown that in eukaryotes, DNA does not come into direct contact with ribosomes, and protein synthesis is not directly directed by DNA.

Structural analysis of proteins and the birth of structural biology

In the 1930s and 1940s, many biochemists believed that elucidating the structures of proteins held the key to understanding the chemistry underlying the phenomena of life. However, protein structures are complex, and the 1930s ended without any established theory regarding their structures. Protein structures are classed as primary, secondary, tertiary, or quaternary. The primary structure is the amino acid sequence of the polypeptide chain, the secondary structure is the partial steric structure of the polypeptide main chain, the tertiary structure is the three-dimensional structure of the polypeptide chain overall, and the quaternary structure is the steric configuration of the assembled subunits. The primary structure of proteins was investigated by Frederick Sanger who, in 1953, first determined the complete amino acid sequence of insulin. Later, the amino acid sequences of several thousand different proteins were determined. Sanger's work had a major influence on later developments in biochemistry.

The secondary structure, as stated in the preceding section, was first understood based on the results of Pauling and Corey in the 1930s and 1940s regarding polypeptide bonds, but detailed research on the α-helix, and another important structure, the β-sheet, began in the 1950s.

Elucidating the three-dimensional structure of a protein was an extremely difficult task, but as stated in Section 5.2.1, the approximate structures of myoglobin and hemoglobin were determined at the end of the 1950s. This research began in the 1930s, but was interrupted because of the war; thus, it took nearly 30 years until the first protein structure was determined. Success was due

Frederick Sanger (1918–2013)
Sanger was raised in the household of a Quaker private doctor, and studied at the University of Cambridge, earning a B.A. in natural sciences in 1939. He then majored in biochemistry to earn a PhD for his study of lysine metabolism. From 1944 to 1951, he was a fellow of medical research at Cambridge, and from 1951 until his retirement in 1983 he was on staff at the Medical Research Council. In 1962, he  began research at the Laboratory of Molecular Biology at Cambridge. His lifetime involvement in studying the primary structure of biopolymers, and his success in 1951 in determining the first amino acid sequence of a protein, insulin, earned him his first Nobel Prize in Chemistry in 1958. In 1975, he developed a new method for sequencing DNA, and was awarded his second Nobel Prize in Chemistry in 1980.

to tenacious patience, supported by astounding optimism and indomitable spirit at the University of Cambridge. Thus, structural biology was born at Cambridge.

Determining the first primary structure of a protein
Frederick Sanger began studying the amino acid sequence of bovine insulin after earning a degree at Cambridge for his research on lysine metabolism. In 1945, he developed a method for confirming the N-terminal amino acid of a polypeptide chain by labeling it with fluoro-dinitrobenzene (his DNP method). Using this method, Sanger showed that insulin is composed of two polypeptide chains. Next, he devised a method for cleaving the –S–S– bonds connecting the two strands, to obtain two different polypeptide chains. These were further cut by acid and enzyme action into short pieces, which were then separated and identified. He made full use of chromatography and electrophoresis in the processes of separation and identification, but the use of paper chromatography, only recently developed, was especially effective. A method of determining the sequence in order from the N-terminus was used to determine the amino acid sequences of the fragments. The sequence of the entire chain was reconstructed from the sequences of the fragments thus determined, and then the positions of the –S–S– bonds were determined. In this manner, the complete chemical structure of insulin, which is composed of 51 amino acids, was determined in 1955.[27] This was a major step towards elucidating the structures of proteins. He also determined the structure of insulin from other animals, including pigs, sheep, and horses, and discovered that the sequence was specific for different animals. These brilliant results were obtained by painstaking work

Stanford Moore (1913–1982)
An American biochemist, Moore earned a degree at the University of Wisconsin in 1938 before joining the Rockefeller Institute and, in 1952, become a professor at Rockefeller University. He was committed to improving chromatography, and with Stein determined the primary structure of ribonuclease. He and Stein were awarded the 1972 Nobel Prize in Chemistry.

William Stein (1911–1980)
Stein was an American biochemist who joined the Rockefeller Institute after earning his degree at Columbia University. At Columbia, he conducted structural studies of proteins, improved methods for analyzing amino acids and peptides, and with Moore determined the primary structure of ribonuclease. He and Moore were the recipients of the 1972 Nobel Prize in Chemistry.

and ingeniously resolving a number of difficulties. He was awarded the Nobel Prize in Chemistry in 1958 for this achievement. Sanger then devoted himself to research at Cambridge, and 22 years later, in 1980, he was awarded his second Nobel Prize in Chemistry for sequencing the DNA of an *E. coli* phage. (Column 24)

Sanger's work was followed by Stanford Moore and William Stein's determination of the chemical structure of the enzyme ribonuclease, composed of 124 amino acids. This protein has more amino acid residues than insulin, but they developed a new form of chromatography that used an ion exchange resin and an automatic sorting machine that made amino acid analysis easier. Later, determining the primary structure of a protein became easy due to many technological developments, and by the end of the 20th century proteins from several thousand different species were sequenced. It was Sanger, however, who determined the fundamental approach for sequencing proteins. Moore and Stein were awarded the Nobel Prize in Chemistry in 1972 for their contribution to the understanding of the relationship between the structure of ribonuclease and the catalytic activity of its active center..

The primary structure of a protein is the result of the expression of genetic information, so elucidating the primary structure is fundamental to the study of protein structure. In 1949, Pauling and Itano discovered that there is an electrophoretic difference between the hemoglobin of healthy adults (HbA) and the hemoglobin of patients with sickle cell anemia (HbS), and inferred that sickle cell anemia is a molecular disease caused by a hemoglobin mutation. In 1956, Ingram demonstrated that this electrophoretic shift is due to the substitution of a specific amino acid in hemoglobin. It became clear that with this mutation, the mutated hemoglobin (HbS) binds oxygen poorly and causes HbS to more readily turn into fibers, which changes the shape of the red blood cells.

It became evident that amino acid sequencing of protein is also helpful in elucidating the classification and evolution of organisms. Comparison of cytochrome c (see page 352) from many eukaryotes shows that there are many non-variable residues, or residues that replace only similar amino acids. These studies showed that the sequences of cytochrome c in organisms belonging to the same family were similar. The evolutionary difference of two homologous proteins can be obtained by counting the amino acid differences between the two proteins. The phylogenetic tree of species, which is based on protein structures, was made by analyzing such data. It was found that when protein mutations are plotted with respect to the time of branching of species (obtained by radiometric dating of fossils), then a linear relationship is obtained, showing that allowable mutations occur at a constant rate.

Three dimensional structure of proteins

One person who made a major contribution to elucidating the secondary structure of proteins was Pauling. As stated previously, the fact that the peptide bond is a planar structure was shown by Pauling and Corey's research in the 1930s and 1940s, and it was established that the amino acids in polypeptides are connected through planar peptide bonds. It follows that the structure of a polypeptide would be determined by the angle of twist (dihedral angle) between peptide bonds, but steric hindrance limits the possible values to a given range. Another factor that determines a protein structure is stabilization by hydrogen bonding. The fact that steric hindrance and hydrogen bonding produce the stable α-helix structure was derived by Pauling and Corey in 1951 from their study of molecular models. This was a very important discovery in biochemistry, but it also had a major impact on the elucidation of the structure of DNA.

In the same year, Pauling and Corey proposed another secondary structure, the β-pleated sheet structure (Fig. 6.4). In this structure, peptide bonds with the same dihedral angle are repeated, and stabilization comes from hydrogen bonding between the backbones of the polypeptides. The two hydrogen-bonded peptide chains can have the opposite direction or they can have the same direction.

In keratin, silk fibroin, collagen, and other fibrous proteins, the two-dimensional structure is the basis for the entire structure of the protein. Because these proteins do not crystallize, X-ray diffraction studies of crystals are not possible, but the long axes of the molecules are aligned with the direction of the fibers. X-ray diffraction studies of these fibrous proteins was begun by Astbury in the early 1930s and lay the groundwork for experiments following Pauling's proposed models of the α-helix and β-sheet structures.

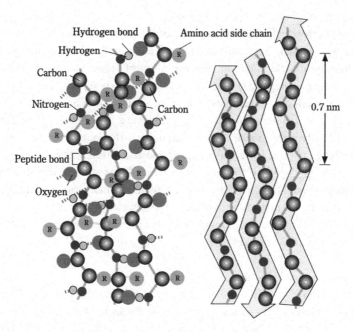

Figure 6.4: Protein β-sheet structure

Adjacent single strands are connected by hydrogen bonding to make a sheet. The right-side figure shows a schematic view of the strands displayed as ribbons.

X-ray analysis of the three-dimensional structures of proteins was a driving
force that created the new field of structural biology. Perutz and Kendrew's
structural analyses of hemoglobin and myoglobin (see page 418) were followed
by rapid advancements from around 1980, helped by dramatic developments in
computer technology. Two-dimensional nuclear magnetic resonance (NMR)
also became available in the mid-1980s for the structural analysis of smaller
proteins. Many life phenomena thus came to be understood on the basis of the
structures of proteins and other macromolecules, and on the basis of intermolecular interactions.

Structures and functions of proteins
As the structures of proteins became more apparent, the relationships between
their structures and functions also came under scrutiny. Hemoglobin and
myoglobin were the most important proteins during this early developmental
process. Myoglobin has an ellipsoid structure in which eight helical structures
are linked by short peptide structures. A heme is located inside a hydrophobic

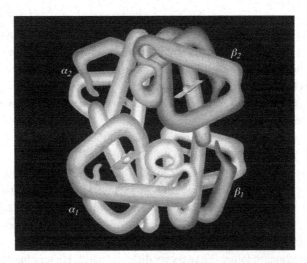

Figure 6.5: Structures of deoxyhemoglobin, composed of four units (α_1, α_2, β_1, β_2), made using computer graphics. Cylinders indicate α-helices

pocket created by two helices. Hemoglobin has four subunits—α_1, α_2, β_1, and β_2—where two α- and two β-subunits are arranged in two-fold symmetry. The heme is found between two globins of each subunit (Fig. 6.5). Oxygen binds to hemoglobin with positive cooperativity because when oxygen binds to one heme, the binding affinity of other heme is increased. However, the heme–heme distance in hemoglobin molecules remains between 25 to 37 Å whether oxygen is bound or not, so understanding this cooperativity was a problem. Perutz proposed explaining cooperativity based on X-ray structural analysis.[28]

At the beginning of the 1960s, Monod, Jacob, and Changeux of the Pasteur Institute in Paris focused their attention on cooperative binding between the hemoglobin molecule and oxygen, and the similarities with the catalytic behavior of enzymes involved in biosynthesis in certain species. Their results showed a shared phenomenon, where the plot of the amount of ligand bound to the protein vs. the concentration of ligand or substrate in solution has a sigmoidal shape. This sigmoid saturation curve is observed for many different proteins; Monod, Jacob, and Changeux named this the allosteric effect.[29] They hypothesized that this curve was due to cooperative behavior in proteins composed of several subunits that adopted at least two reversible states. One of the two states was termed the T-state (tense), where the subunits are confined and have low affinity for the substrate, and the other was termed the R-state (relaxed), where this confinement is released and the protein exhibits a high affinity for the substrate.

In the Perutz structure, oxygen binds to the Fe(II) of the heme, and this causes the Fe(II) to fit tightly in the heme plane, which prompts the hemoglobin to change state from T to R. The T-state is stabilized by a series of ionic bonds that are broken in the R-state. Thus, the binding of O_2 to Fe(II) in one subunit triggers a change in the structures of the other subunits, resulting in an increase in the affinity of the Fe(II) for O_2. The binding between hemoglobin and oxygen was thus representative of the allosteric effect.

Max Perutz and John Kendrew were awarded the Nobel Prize in Chemistry in 1962. Because Watson, Crick, and Wilkins won the Nobel Prize in Physiology or Medicine that year, it was a year in which Nobel prizes were awarded for the structural analysis of compounds important for the elucidation of life phenomena, namely, DNA and proteins. The awarding of these prizes showed the broad-sweeping victories of structural chemistry. However, as noted in this chapter, it is very interesting that elucidating the structures of these two families of compounds meant following two very different paths. History has shown that there is not necessarily only one road to an important discovery.

Folding and structural changes

As the structures of protein molecules became increasingly clear, one lingering problem that caught the interest of many researchers was the problem of the folding of protein molecules. Protein molecules spontaneously fold under physiological conditions, which means their three-dimensional structure is determined by the primary structure. However, even at the end of the 20th century it was still impossible to predict the tertiary structure from the primary structure, and solving the problem of folding was carried over into the 21st century as a critical issue for theoretical chemistry.

In 1957, Christian Anfinsen and his collaborators showed that when ribonuclease is reduced by β-mercaptoethanol in urea, the S–S bond of the disulfide is broken and the structure completely unravels, but when urea is removed and the system is exposed to oxygen, the protein returns to its original state and enzymatic activity is also restored.[30] This result supported the notion that the formation of the three-dimensional structures of globular proteins is a spontaneous process that is determined by the primary structure of the protein. However, subsequent detailed research would reveal that in many instances, the folding process also requires an auxiliary protein, such as a molecular chaperone[2]

---

2. Molecular chaperone: When a protein is partially folded and forms a multimer, another protein, called a molecular chaperone, binds temporarily to the partially folded protein to help it reach its correct, final structure.

Christian Anfinsen (1916–1995)
An American biochemist, Anfinsen earned a degree at Harvard in 1943 before studying at the University of Pennsylvania, Harvard, and in Stockholm. Anfinsen worked at the National Institutes of Health (NIH) in the USA and studied the relationship between the structure and physiological function of proteins. He was a recipient of the 1972 Nobel Prize in Chemistry.

Anfinsen was awarded the Nobel Prize in Chemistry in 1972. He was given this award for "his work on ribonuclease, especially concerning the connection between the amino acid sequence and the biologically active conformation".

6.2 Development of biochemistry (I): Chemistry of DNA and RNA

The fundamental correctness of the three-dimensional structure of DNA was established by around 1960, and the general mechanism of DNA replication was elucidated. The next major task was to elucidate how the code of life written as the base sequence of DNA is decrypted and how cells read the messages in DNA molecules. Other tasks were to determine the base sequence of DNA, investigate the details of the structures of genes, and shed light on the relationship between structure and function. From the mid-1950s to the 1960s, it became increasingly apparent that there is a mechanism by which DNA information is transcribed into RNA and then amino acids are joined to synthesize a protein. This understanding brought about rapid advances in the study of transcription, translation, replication, repair, recombination, gene expression, and other principal functions of nucleic acid.

Determining the base sequence of nucleic acid was even more difficult than determining the amino acid sequences of proteins, but in 1975 a new method for quickly sequencing DNA was developed which led to rapid advances. A major problem in biochemical research was the difficulty of obtaining sufficient enzyme, protein or other samples, but in 1972 recombinant DNA techniques were developed that helped overcome this hurdle, and genetic engineering emerged. Development of the polymerase chain reaction (PCR) for quickly amplifying fragments of DNA in 1985 led to progress in reading genetic information, and at the end of the 20th century the project to read the human genome was close to completion. With the help of such technological advancements, the study of nucleic acid saw explosive growth in the second half of the 20th century. This section reflects on progress in the field which forms the center of molecular biology, namely, the expression and transmission of genetic information. Many

George Gamow (1904–1968)
An American physicist born in Russia, Gamow studied in Leningrad before conducting research in Göttingen, Copenhagen, and Cambridge; in 1934, he emigrated to the U.S. and was a professor at George Washington University and the University of Colorado. He is known for proposing the theories of α decay and β decay, and the Big Bang theory for the origin of the universe.

of the achievements that were prominent in this field were recognized by Nobel Prizes in Chemistry and in Physiology or Medicine; these achievements form the center of this review.

Transcription and translation of DNA information

One might ask how proteins, which are composed of a series of amino acids, are made from information that is encoded in the base sequences of DNA. Central to this are the processes of "transcription", which is when DNA serves as a template for the synthesis of RNA, and "translation", which is when a protein is synthesized based on the information encoded in RNA. From the mid-1950s to the 1960s, molecular biologists were focused on understanding these processes.

Genetic information and RNA

In 1954, the theoretical physicist George Gamow was impressed by the Watson-Crick double helix model and became interested in the genetic code. He noted that there are 20 different kinds of main amino acids that constitute proteins, and defining these with four different kinds of bases means that a sequence of at least three bases must be read for each kind of amino acid.[31] However, it was already known that proteins are not made in the nucleus, which is where DNA is located, and that instead RNA is important in the synthesis of proteins. Therefore, research first focused on RNA. Gamow and Watson formed the "RNA Tie Club"[3], and advanced a "decipherment" movement. In 1955, Crick believed that for every amino acid there is a specific adapter molecule, which transports the amino acid to the site of protein synthesis. He expected that these "adapters" were RNA molecules. Crick's idea was shown to be correct by Zamecnik in Boston. Zamecnik developed a cell-free system for synthesizing proteins. He used rat liver tissue components to reproduce a simplified cell interior inside a

---

3. RNA Tie Club: A club formed by Watson and Gamow, where membership was limited to 20 members so as to match the 20 amino acids, and each person was allocated a tie pin corresponding to a particular amino acid.

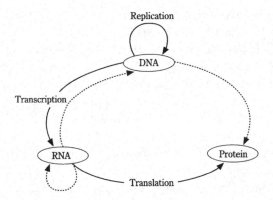

Figure 6.6: Central dogma of molecular biology

The solid arrows show the flow of information. The dotted arrows show flow of information that only occurs rarely.

test tube, and tracked the manner in which amino acids labeled with radioactive isotopes were incorporated into proteins. This allowed him to show that proteins are synthesized on the ribosome. Then, Zamecnik and Hoagland discovered that before being incorporated into a polypeptide chain, amino acids are joined to a small RNA molecule. This discovery of transfer RNA (tRNA) proved Crick's theory that there is a specific RNA adaptor for each amino acid. This finding also showed that there are different types of RNA.

In 1958, Crick made a single flowchart summarizing the relationship between DNA, RNA, and protein, which at the time had been recognized only vaguely; he called this his central dogma of molecular biology (Fig. 6.6).[32] It states that "DNA directs its own replication and its transcription to yield RNA which in turn, directs its translation to form proteins.". However, there was no concrete information regarding RNA directing protein translation. Prior to the discovery of tRNA, it was believed that all cellular RNA acted as a template, but it was becoming increasingly apparent that this way of thinking was problematic. RNA strands in ribosomes are of constant length, but whenever an RNA strand is the template for protein synthesis, its length would have to vary depending on the size of the protein. It was also odd that the base sequence of ribosomal RNA strands had no correlation with the base sequence of chromosomal DNA molecules.

By 1960, a third form of RNA, called messenger RNA (mRNA), was discovered, and these discrepancies were resolved. This discovery involved many scientists, including Nomura, Watson, Meselson, Jacob, and Brenner. They

Robert Holley (1922–1993)
Holley was born in Illinois, studied chemistry at the
University of Illinois, and earned a PhD majoring in organic
chemistry at Cornell University. He became an assistant
professor of organic chemistry at Cornell, but turned to
biochemistry and became a professor of biochemistry in
1962. Around 1957, he started research to determine the
structure of alanine tRNA from yeast. In 1964 his group
determined the overall structure by using two different ribonucleases to break
tRNA into its components and then comparing the degraded components.
This achievement would be key in explaining the mechanism of protein
synthesis, and earned him the Nobel Prize in Physiology or Medicine in 1968.

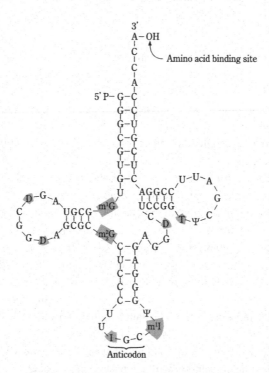

Figure 6.7: Cloverleaf model of the yeast t-RNA base sequence

The hatched portions contain derivatives of uracil, cytosine, adenine, and guanine.

Martin Temin (1934–1994)

Temin was an American virologist who earned a degree at the California Institute of Technology in 1959, then studied oncogenesis by Rous sarcoma virus at the University of Wisconsin; in 1970, he discovered reverse transcriptase, which reverse transcribes viral RNA into DNA. He was awarded the 1975 Nobel Prize in Physiology or Medicine.

David Baltimore (1938–)

An American virologist, Baltimore studied medicine at the Rockefeller University and earned his degree in 1964, then became a professor at the Salk Institute and later in 1972 became a professor at the Massachusetts Institute of Technology. He discovered the mechanism by which an RNA tumor virus reverse-transcribes its RNA genetic information into the DNA of the host cell and makes the cell cancerous. Also, independent of Temin, he discovered the enzyme reverse transcriptase by which RNA information is copied into DNA. Together with Temin, he was awarded the 1975 Nobel Prize in Physiology or Medicine.

revealed a schema whereby mRNA transcribes DNA information to provide a faithful template for protein synthesis, then tRNA bound to a specific amino acid binds to the mRNA inside the ribosome. This results in the amino acids lining up and binding to each other in the proper order to form the polypeptide chain.

In 1965, after seven years of hard work, Robert Holley was the first to determine the sequence of a biologically significant nucleic acid, yeast tRNA.[33] This tRNA has a base sequence structured into a cloverleaf-like shape (Fig. 6.7). This discovery was followed by the determination of the structures of many different kinds of tRNA, and all had the cloverleaf form proposed by Holley. These tRNAs are alike in having a stem of seven base pairs with a phosphate group at the 5'-terminus, and three arms that have loops. The loop of the arm with five base pairs opposite the stem contains the anticodon, which is a sequence of three bases that is complementary to the codon (see next section). X-ray structural analysis revealed the three-dimensional structure of tRNA in 1974. Holley was awarded the Nobel Prize in Physiology or Medicine, together with Nirenberg and Khorana, in 1968.

In 1970, it became clear that there were deviations from the flow of DNA→ RNA → protein in the central dogma of molecular biology proposed by Crick. Studying sarcoma in chickens, Martin Temin thought that viral RNA might be reverse-transcribed into DNA inside the host cell. Mizutani, working in Temin's laboratory, discovered an enzyme that synthesizes DNA with RNA as a template in virus extract. David Baltimore of the Massachusetts Institute of Technology (MIT) around the same time discovered RNA-dependent DNA polymerase in mouse leukemia virus extract. Temin and Baltimore were awarded the 1975 Nobel Prize in Physiology or Medicine for the discovery of "reverse transcriptases".

Deciphering of the genetic code
The next problem was to identify the rules by which the DNA base sequence is changed into an amino acid sequence: deciphering DNA. Encoding 20 amino acids with only four bases requires sequences of at least three bases (codons). This means that $4^3=64$ combinations of three base sequences are possible. In 1961, Sydney Brenner and Crick used chemicals that trigger mutations to add or remove single base pairs to/from bacteriophage DNA. They then investigated the nature of mutations that occur when base pairs are inserted or deleted, and came to the following conclusions. 1) Inserting or deleting a base pair shifts the reading frame where sequential base pairs are read as a codon. 2) Encryption is with three base sequences. 3) All 64 combinations of three base sequence encryptions encode an amino acid. These findings revealed that encryption is by three base sequences, but the enormous task of decryption still remained.

The effort to decrypt DNA was led by Marshall Nirenberg, a young researcher at the National Institutes of Health (NIH). He and his colleague, Matthaei, investigated whether RNA synthesized in vitro functions the same as naturally occurring messenger RNA when proteins are synthesized in a cell-free system. They used a method developed by Grünberg-Manago six years earlier to make polyuracil (UUUUU ...) and added it to the cell-free system. The results they obtained were surprising in that only simple proteins composed entirely of one kind of amino acid, phenylalanine, were synthesized. This showed that the cipher UUU codes for phenylalanine.[34]

Nirenberg's discovery prompted the start of a competition to decipher the remaining 63 three-base sequences. Gobind Khorana, an Indian-born biochemist at the University of Wisconsin, successfully decrypted many of the ciphers by developing a method of accurately synthesizing RNA with a simple repetitive sequence. Thus, all 64 codons except for three were decoded to designate one of twenty amino acids by 1966. The remaining three were later found to be stop codons which terminate deciphering. Nirenberg and Khorana were awarded the Nobel Prize in Physiology or Medicine in 1968.

Regulation of gene expression and operon theory
The flow of information DNA $\rightarrow$ RNA $\rightarrow$ protein had been revealed by 1965. However, some of the proteins thus formed are present in large amounts and

Sydney Brenner (1927–)
Brenner is a British molecular biologist who finished his bachelor's and master's degrees at the University of Witwatersrand in South Africa before earning a doctorate at the University of Oxford in 1954. He served as a director at the Medical Research Council Molecular Genetics Unit in Cambridge and then went on to be director at The Molecular Sciences Institute in La Jolla, California, in 1966. In 2002, he was awarded the Nobel Prize in Physiology or Medicine for his research on "organ growth and the process of programmed cell death".

Marshall Nirenberg (1927–2010)
An American biochemist, Nirenberg studied at the University of Florida and the University of Michigan; in 1957, he joined the National Institutes of Health (NIH), where he succeeded in synthesizing a protein in vitro in 1961, providing clues for deciphering genetic information. Later, he collaborated with Khorana to synthesize 64 different kinds of trinucleotides and determine which sequence encodes which amino acid. He was a recipient of the 1968 Nobel Prize in Physiology or Medicine.

Gobind Khorana (1922–2011)
An Indian-born biochemist in the U.S., Khorana studied chemistry at the University of the Punjab and earned a degree at the University of Liverpool. After engaging in research in Canada, he went on to serve as a professor at the University of Wisconsin and then in 1970 as a professor at the Massachusetts Institute of Technology. He showed which combinations of DNA bases encoded which specific amino acid, thus helping decipher the genetic code. He shared the 1968 Nobel Prize in Physiology or Medicine with Nirenberg.

others are not. This raises the question of why. Many genes are switched on to begin making a particular protein only in specific cells and at a certain time, but what mechanism causes this to happen? The first to focus their research on gene switches were Francois Jacob and Jacques Monod of the Pasteur Institute in Paris in the early 1960s.[35] They focused their study on the fact that *E. coli* can utilize lactose. *E. coli* synthesizes an enzyme called β-galactosidase in order to digest lactose, but this enzyme is not produced when lactose is absent from the *E. coli* culture medium. When lactose is added to the *E. coli* culture medium, however, the *E. coli* cells begin producing this enzyme. Believing that the presence of lactose somehow triggers the production of β-galactosidase, they set out to clarify this mechanism. From clever experiments, they found evidence that

Francois Jacob (1920–2013)

Born in Nancy as the only son of a Jewish family, Jacob was known as a child prodigy. Aspiring to be a doctor, he was admitted to the Sorbonne, but during the war he moved to the U.K. and volunteered for de Gaulle's Free French Forces, and was seriously injured in the North African campaign. He graduated from university in 1947, but no longer desired to be a surgeon and sought instead to become a researcher.  He joined Lwoff and Monod's group at the Pasteur Institute even though, past the age of 30, he still had almost no basic knowledge of biochemistry. He and Monod worked very well together and they jointly conducted research on gene expression and regulation in E. coli. Together, they lay the foundation of molecular genetics, for example, with their operon theory. In 1964, Jacob was appointed a professor at the College de France. His book "The Statue Within"[36] was acclaimed as an autobiography with high literary value. He was a recipient of the 1965 Nobel Prize in Physiology or Medicine.

Jacques Monod (1910–1976)

Monod had many abilities, and he was torn between becoming a musician and becoming a biologist, but in the end he chose biology. After graduating from the Sorbonne, he became an assistant and then an assistant professor there. During the war, he joined the resistance against the Germans. At the end of the war he became a staff member at the Pasteur Institute. He discovered that changing the composition of the culture medium causes changes in  enzyme synthesis in bacteria. This discovery led him, along with Jacob, to propose their operon theory. Monod also discovered the allosteric effect in enzymes. He was active during the period when molecular biology was gaining momentum. Monod was excellent at philosophical speculation, and wrote a book "Chance and Necessity"[37] in which he showed a materialistic worldview founded on knowledge of molecular biology. The book created a sensation and become a worldwide bestseller. He shared the 1965 Nobel Prize in Physiology or Medicine with Jacob and Lwoff.

there is an inhibitory factor (repressor) molecule that hinders transcription of the β-galactosidase gene when lactose is not present. The repressor no longer works when bound to lactose, so they believed lactose is needed for the β-galactosidase gene to be transcribed.

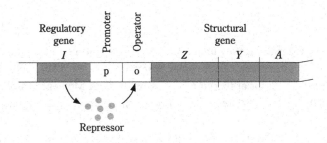

Figure 6.8: Schematic diagram of an E. coli operon

Expression of the structural gene is suppressed when the repressor binds to the operator.

Advancing this idea, Jacob and Monod proposed the following operon theory about the expression and regulation of genes (Fig. 6.8). *E. coli* possesses structural genes[4], called the Z gene, which produces β-galactosidase, and the Y gene, which produces an enzyme called β-galactoside permease. As long as a repressor involving regulatory gene I is bound to operator gene o, expression of the structural genes for producing the enzymes needed to degrade lactose is inhibited. When lactose is present, however, the structure of the repressor changes and can no longer bind to the o gene, allowing the adjacent structural gene group to be transcribed into mRNA. This operon theory greatly stimulated later research on gene expression and regulation. Despite the difficulty of the problem of revealing the identity of the repressor, Gilbert and Müller-Hill showed at the end of the 1960's that the repressor is a protein that binds to DNA. With the nature of the repressor molecule revealed, the activity of the gene was shown to be directly regulated by the protein binding to the DNA molecule.

Jacques Monod and Francois Jacob were awarded the Nobel Prize in Physiology or Medicine in 1965, together with Andre Lwoff.

Replication, repair, and lifetime of DNA
DNA replication
The mechanism of DNA replication was predicted immediately from the 1953 Watson-Crick model, but it was not until about 20 years later that details of the replication mechanism became clearer. Kornberg's research in 1957 showed that DNA replication requires an enzyme called DNA polymerase, and that the replicated DNA strands extend in a direction going from the 5' end to the 3'

---

4. Structural gene: A structural gene is the term for a DNA region that corresponds directly to a protein such as an enzyme; a set of structural genes regulated by the same operator is called an operon.

Reiji Okazaki (1930–1975)
Born in Hiroshima City, Okazaki graduated from Nagoya University in 1953 and left Japan to study under Arthur Kornberg in 1960. He returned to Japan in 1963 to become an assistant professor at Nagoya University, then becoming a professor at that university's Institute of Molecular Biology in 1967. In 1966, he discovered Okazaki fragments, and in 1972 he completed his DNA non-continuous synthesis model, but his exposure to the atomic bomb at Hiroshima caused him to develop leukemia, and in 1975 he sadly passed away at the young age of 44.

end. In 1963, Cairns used *E. coli* DNA to show that replication proceeds with the helix unwinding from a single starting point (which he called the replication fork). However, in double-stranded DNA, the two strands were thought to be oriented in opposite directions, making it unclear how both strands could extend in the 5'→3' direction. In 1967, Reiji Okazaki of Nagoya University proposed a semi-discontinuous replication model, which states that the two parental strands are replicated by different methods: one strand extends in the 5'→3' direction, which is the direction of movement of the replication fork, and the other strand is replicated by short strands (Okazaki fragments) that are discontinuously synthesized and then joined together by DNA ligase.[38] RNA acts as a primer when DNA strands are produced by DNA polymerase[5].

DNA replication is a complex process involving various enzymes. DNA polymerase encompasses not only what is called DNA polymerase I (Pol I), which was discovered by Kornberg, but also DNA polymerase II (Pol II) and DNA polymerase III (Pol III), which were discovered and characterized later. Recent advances in X-ray crystallography have enabled more detailed investigation of the structures of these enzymes and a better understanding of the process of DNA replication at the molecular level.

PCR method
The polymerase chain reaction (PCR) technique is a method by which a specific DNA strand is artificially replicated to produce large amounts of that DNA strand. PCR was first developed in the mid-1980's by Kary Mullis of Cetus Corporation (Fig. 6.9).[39] The technology on which this method was founded came from Khorana and Kleppe, but it was Mullis who proposed the

---

5. Primer: A short fragment of nucleic acid that supplies the 3'OH and serves as the starting point for DNA synthesis by DNA polymerase.

Kary Mullis (1944–)

Mullis was born in North Carolina in the U.S., studied chemistry at the Georgia Institute of Technology in Atlanta, and earned his PhD at the University of California, Berkeley. After conducting research at several universities, he joined Cetus Corporation, a biotechnology company, in 1979 and developed PCR. He left the company in 1986, and from 1988 onward was engaged in freelance consultancy. He was a recipient of the 1993 Nobel Prize in Chemistry.

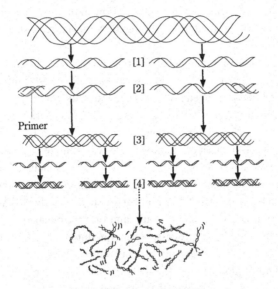

Figure 6.9: Mechanism of the PCR method

[1] DNA strands are heated and unraveled.[2] The synthesis-initializing primer is added.[3] DNA polymerase replicates the DNA, producing a copy of the same double-stranded DNA.[4] The above operation is repeated to obtain large amounts of the DNA.

ideas that improved the technique, and a large number of researchers at Cetus were involved in making PCR into a revolutionary technique that transformed molecular biology. In this method, a solution containing fragments of DNA to be replicated, as well as different nucleoside triphosphates and thermophilic bacterial enzymes, is heated to 90°C. This causes the double-stranded DNA to separate into single strands that become templates for DNA replicative enzymes at 70°C. The amount of DNA is doubled in one operation; thus, for every n times the operation is repeated, 2^n-times the amount of DNA can be obtained. This method is simple, and has found wide-reaching applications in DNA testing and genome analysis.[40] It was thanks to this technology that large-scale genome analysis such as the Human Genome Project[6] became possible. Its developer, Kary Mullis, was awarded the Nobel Prize in Chemistry in 1993.

Mullis, it should be noted, is unlike other Nobel laureates and holds controversial opinions, and has been criticized for his opinions and behavior. A surfer playboy, he has said that the idea for PCR came to him during a drive with his girlfriend. He denies that AIDS is caused by the HIV virus, is against the majority opinion on climate change, and has even professed belief in astrology. It is a far cry from the image one usually holds of a typical important chemist, but such interesting personalities do number among the Nobel laureates (see Column 23).

DNA repair and lifetime: telomeres and telomerase

Intracellular DNA can be damaged in several ways. The most significant cause is believed to be reactive oxygen[7], which is generated during normal metabolism. Other causes include irradiation with ultraviolet rays, X-rays, and radiation, and carcinogens such as hydrocarbons in tobacco smoke. For example, when the oxygen at the 8-position of guanine in DNA is oxidized by reactive oxygen into OH, producing 8-hydroxyguanine, the G-C pair is changed to a T-A pair during replication because the DNA synthesizing enzyme mistakes G for T. When such an error occurs, DNA replication stops and the DNA is repaired. DNA replication is not 100% accurate, and on rare occasion a mistake can occur during the replication process. Normal maintenance of life requires that the errors in DNA be repaired; the mechanism involves specific enzymes. A damaged strand of DNA in the double helix can be repaired in situ; there is another mechanism that removes the damaged nucleotides and replaces them with the correct bases. If both strands of the double helix are damaged, there is an additional mechanism for fixing this.

DNA cannot self-replicate infinitely, even if DNA damage is repaired. The distal ends of chromosomes in eukaryotic cells contain a linear chain of DNA called telomere, where a characteristic base sequence (which in mammals is TTAGGG) is repeated. DNA is replicated by the action of DNA polymerase, but

---

6. Human Genome Project: A project to analyze the entire base sequence of the human genome in order to elucidate the functions of genes in health and disease. It was initiated in 1990 by the U.S. Department of Energy and the National Institutes of Health. Later, the U.K., Japan, France, Germany, and China also participated. The project was completed in 2003.

7. Reactive oxygen: General term for highly reactive oxygen-containing molecules, radicals, atoms, or singlet oxygen. O_2^-, H_2O_2, OH, HO_2, and other species are extremely reactive.

COLUMN 23

Mullis, an unusual chemist, and the development of PCR[40,41]

Mullis the surfer

PCR is a revolutionary technology that makes it possible to exponentially amplify trace amounts of DNA fragments in a short period of time. PCR has been used for basic research on DNA structure and function, and also in vital areas such as the diagnosis of genetic diseases, DNA testing, and the Human Genome Project. PCR has even been called the greatest invention in biotechnology. However, the man awarded the Nobel Prize in Chemistry as the developer of PCR, Kary Mullis, is perhaps the most eccentric person in the history of Nobel laureates.

The career of Kary Mullis is itself unusual. He earned a degree in biochemistry at the University of California, Berkeley, in 1972, but after earning his degree he set out to write a novel. He abandoned this almost immediately and became a researcher at the University of Kansas Medical School, after which he returned to Berkeley and spent two years running a restaurant/coffee shop. Then, he worked as a postdoctoral fellow in pharmaceutical chemistry at the University of California, San Francisco, but in 1979 he joined Cetus Corporation, a biotechnology venture. An unparalleled philanderer and a surf bum, he took LSD while in graduate school.

His work at Cetus was on the synthesis of oligonucleotides. At first, he was interested and put effort in his work, but he soon tired of the tedious and repetitive tasks, and his irreverent and uncooperative manner resulted in friction with his colleagues. In 1981, he became general manager of the DNA synthesis lab, and devised innovative ways to speed up the production of oligonucleotides, but his relationship with the other laboratory staff only worsened.

Mullis recounted how he got the idea for PCR. One night in the spring of 1983, he went for a drive toward his mountain villa along with his girlfriend. In the middle of this drive, the idea came to him in a flash while he was thinking about DNA sequencing. When a given portion of DNA is nipped with a pair of oligonucleotide primers and then the DNA fragment is copied using DNA polymerase, two fragments are generated. Repeating this operation would make it possible to synthesize DNA fragments in large quantities. Mullis was very excited about this idea. In August, he presented this idea during one of his company's seminars. However, the reaction was lukewarm. Although the idea seemed good, it had to be shown that it actually works well. Repeated experiments lead to promising results in December of that year. However, opposition to his behavior intensified in the company. Due in part to the fact that his relationship with his girlfriend had also ended, Mullins became emotionally unstable, and his presence in the company was increasingly problematic.

In 1984, Cetus formed a group to focus on the PCR project, and also began to investigate applications of PCR. Plans were made to amplify the sequence of the β globin gene responsible for sickle cell anemia. In the spring of 1985, the group obtained data showing that DNA fragments of the genome could be amplified several hundred-thousand-fold. This result was to be announced as a paper, and Mullis was to write an "idea" manuscript while an "application" paper by the whole group would be written separately. However, Mullis struggled to write the paper, and the "application" manuscript was submitted first to *Science* and published in the 20 December 1985 edition. Mullis submitted his "idea" manuscript to *Nature*, but it was not accepted, and his re-submission to *Science* was also rejected. Greatly delayed, his paper was finally published in *Methods in Enzymology* in 1987, but by this time PCR was already widely known.

One problem in the development of PCR was the fact that DNA polymerase was unstable, and was degraded by the high temperature used during amplification. Therefore, fresh enzyme had to be added every time the replication cycle was repeated. Mullis in 1986 began using polymerase from heat-resistant bacteria, with the result that DNA polymerase only had to be added once. In this way, PCR became a technology that revolutionized biotechnology. Mullis left Cetus Corp. in 1986.

The Nobel Prize for the development of PCR was awarded solely to Mullis, a fact that did not go without criticism. PCR was developed through the collaboration of several superb technicians at Cetus, where it was first developed as a technology. Without Mullis, however, it is unlikely that PCR would have been developed at Cetus. Another criticism of the Nobel Prize award was that PCR is not a new idea at all, but instead pulled together known techniques into an optimized methodology. However, it is not uncommon for innovative technologies to include known techniques. Though he has often been an eccentric troublemaker, there is no doubt that Mullis is also a richly creative and uniquely talented person.

Elizabeth Blackburn (1948–)

Blackburn is an Australian-born biologist in the U.S. who graduated from the University of Melbourne before earning a degree at Cambridge University in 1975. She served as an assistant professor and then a professor at the University of California, Berkeley, and then as a professor at the San Francisco campus in 1990. She discovered telomerase in 1984, and was a recipient of the 2009 Nobel Prize in Physiology or Medicine.

Jack Szostak (1952–)

A British-born biologist in the U.S., Szostak was raised in Canada and graduated from McGill University; in 1977, he earned a degree at Cornell University. He moved to Harvard Medical School in 1979 and become a professor there. He is famous for elucidating the function of telomeres, for artificially synthesizing yeast chromosomes, and for other achievements. Together with Balckburn and Greider, he was awarded the Nobel Prize in Physiology or Medicine in 2009.

Carol Greider (1961–)

Greider is an American molecular biologist. In 1983, she graduated from the University of California, Santa Barbara, and then earned a degree at the Berkeley campus; in 1997, she began working as a professor at the Johns Hopkins University School of Medicine. In 1984, she discovered telomerase working in collaboration with Blackburn; she was the third recipient of the 2009 Nobel Prize in Physiology or Medicine.

the ends of linear DNA are not replicated, so the DNA is shortened each time the chromosome divides and replicates. This means that replication progressively shortens the chromosome, and ultimately the chromosome should disappear, but in reality this is not the case. At the beginning of the 1970s, this contraction was recognized as the "end replication problem"; during the 1980s, increased under-standing of telomeres and their function advanced and the problem was solved.

In 1980, Elizabeth Blackburn was analyzing the telomeres in DNA from tetrahymena, a unicellular eukaryote, and discovered that it comprises a repeating base sequence of CCCCAA. Blackburn and Jack Szostak, who had been studying the degradation of linear DNA in yeast cells (minichromosomes), collaborated in their research and discovered that when tetrahymena telomeres are attached to the ends of yeast minichromosomes, the chromosomes do not degrade.[42] This finding showed that telomeres, which have a characteristic base sequence, protect DNA. In the course of this research, they also found that telomeres stretch or shorten. Blackburn and her graduate student, Carol Greider, studied how telomeres are synthesized and discovered telomerase, an enzyme involved in the extension of the specific telomere nucleotide sequence.[43]

Telomerase is composed of protein and RNA, and the RNA has the base sequence characteristic of telomeres. The action of telomerase makes it possible to extend the telomeric DNA, and to compensate for the shortening of telomeres that occurs during replication.

Later research would show that telomeres and telomerase are involved in senescence and the onset of cancer, further highlighting their importance. When telomeres are shortened, cell proliferation is inhibited and ultimately division ceases. Telomerase prevents shortening of telomeres and thus extends cell life. Senescence in organisms is a complex process that cannot be attributed to any one cause, but telomere shortening is inferred to be one of these causes. Cancer cells have higher telomerase activity, which is believed to be partially responsible for the unrestricted proliferation of cancer cells. Inhibiting telomerase activity may be a promising approach to beating cancer.

Blackburn, Greider and Szostak contributed greatly to the study of telomeres and were together awarded the 2009 Nobel Prize in Physiology or Medicine. Their research was initiated by curiosity and was not aimed at applications to medicine.

Manipulating and sequencing nucleic acid
Manipulating nucleic acid

Because of the large size of nucleic acid molecules, sequencing and otherwise investigating them requires large amounts of material. After 1970, techniques for manipulating nucleic acid advanced, and the study of nucleic acid entered a new period of development. In the 1960s, the Swiss biochemist Werner Aber discovered "restriction enzymes", which cleave DNA at a given point in the base sequence, because proliferation of phages was limited with certain strains of *E. coli*. He also discovered "DNA modifying enzymes", which modify the base sequence recognized by a restriction enzyme so that the sequence can no longer be cleaved. In 1970, Hamilton Smith of Johns Hopkins University purified different types of restriction enzymes that cut specific base sequences. Using a restriction enzyme obtained from the influenza virus, Smith's colleague Daniel Nathans succeeded in cleaving the circular DNA in monkey tumor virus SV40 into 11 fragments and in creating a cleavage map. It was thus shown that restriction enzymes are a kind of chemical "scissors" that cleave DNA at specific nucleotide sequences.

Paul Berg of Stanford University began studying the expression of genes by incorporating bacterial genes into SV40 DNA and then infecting animal

Werner Arber (1929–)
A Swiss microbiologist, Arber studied at the Swiss Federal Institute of Technology, the University of Geneva, and the University of Southern California; after serving as an instructor at the University of Southern California and the University of Geneva, he was appointed as a professor at the University of Basel in 1971. From the late 1950s to the early 1960s, he studied the phenomenon of host-controlled variation, which led to the discovery of restriction enzymes. In 1978, he was awarded the Nobel Prize in Physiology or Medicine.

Hamilton Smith (1931–)
A microbiologist in the U.S., Smith graduated from the University of California, Berkeley, then earned a degree in medicine at Johns Hopkins University in 1956. In 1967 he became an assistant professor there, and in 1973 he became a professor. From 1967 onward, he studied the influenza virus and determined how to remove the DNA from phage virus. In 1970 he discovered restriction enzymes. He was one of three recipients of the Nobel Prize in Physiology or Medicine in 1978.

Daniel Nathans (1928–1999)
Nathans was a molecular biologist in the U.S. who earned a degree at Washington University, was on staff at the National Cancer Institute and the Rockefeller Institute, then became a professor at Johns Hopkins University in 1962. He studied the tumor virus SV40, and used restriction enzymes to fragment its DNA and solve its gene structure. In 1978, he was awarded the Nobel Prize in Physiology or Medicine together with Arber and Smith.

cells with it. In his research, he made use of the restriction enzyme EcoR I, which was isolated from *E. coli*. This enzyme recognizes the base sequence GAATTC and cleaves between the G and A. The corresponding sequence of the complementary strand is CTTAAG, and this too is cleaved between A and G, so that a single strand of TTAA remains at the end. This is called a sticky end, and the two complementary ends easily anneal. Sticky ends act as a pasting margin when DNA is being cleaved to incorporate a desired gene. Berg, who was equipped with "scissors" for clipping DNA and "pasting margins" for joining DNA termini together, announced the technique of "genetic recombination" in 1972.[44] "Genetic recombination" technology in 1973 became much simpler thanks to the development of a method by Boyer and Cohen, making it possible to synthesize large amounts of DNA. Cohen, who had been studying *E. coli*

Paul Berg (1926–)
Berg graduated from Pennsylvania State University and
then earned his PhD at Case Western Reserve University;
after a fellowship at the University of Copenhagen in
Denmark and at Washington University, he became a
professor of biochemistry at Stanford University in 1959.
In the 1960s he began working on nucleic acid synthesis
in E. coli, the elucidation of regulatory mechanisms of
tRNA synthesis, and studying the tumor virus SV40. In 1972 he developed
a technique for connecting the λ phage gene of E. coli with an SV40 gene
in vitro, making him a trailblazer in gene recombination experiments and a
pioneer in genetic engineering. He was a recipient of the 1980 Nobel Prize
in Chemistry.

plasmids (extrachromosomal genes)[8] at Stanford University, and Boyer, who had
been studying restriction enzymes at the University of California (San Francisco
campus), collaborated in their research and succeeded in establishing a method
for using EcoR I to incorporate a desired DNA fragment into a plasmid and
multiplying it in *E. coli*. This method made it possible to clone the DNA of any
organism (where "cloning" means producing many copies of the same DNA
fragment following transfection into a bacterial cell).

Arber, Smith, and Nathans, who were instrumental in the discovery of
restriction enzymes, were awarded the Nobel Prize in Physiology or Medicine
in 1978, and Berg was awarded the Nobel Prize in Chemistry in 1980 for
his biochemical study of nucleic acid that laid the groundwork for genetic
engineering. However, despite contributing greatly to the development of gene
recombination technology, Boyer and Cohen were not awarded Nobel Prizes.

Sequencing nucleic acid
Advances in the sequencing of nucleic acid advanced using a similar strategy
as had been used to sequence proteins. 1) A polynucleotide chain is specifically
degraded and separated into small fragments. 2) Each of the individual
fragments is sequenced. 3) The order of the fragments in the original sequence
is determined. Until the mid-1970s, nucleic acid sequencing was slow, and
only a handful of RNA sequences had been determined. From 1975 onward,

---

8. Plasmid: A general term for a DNA molecule found outside the nucleus of yeast or bacteria
    such as *E. coli*. Plasmids are replicated and inherited by the next generation independent of the
    chromosome during cell division.

Walter Gilbert (1932–)
Gilbert is a molecular biologist in the U.S. who graduated from Harvard University and earned a degree studying mathematics and physics at the University of Cambridge. In 1968, he became a professor of theoretical physics at Harvard, but switched to molecular biology and developed a method of sequencing DNA, earning a Nobel Prize in Chemistry in 1980.

however, development was accelerated by 1) the discovery of restriction enzymes that specifically cleave double-stranded DNA, 2) the development of DNA sequencing methods, and 3) the development of molecular cloning techniques.

Walter Gilbert of Harvard University and Frederick Sanger of the University of Cambridge proposed two distinctly different sequencing methods at around the same time.[45,46] Gilbert, a theoretical physicist, had been appointed as an instructor at Harvard, but at Watson's prompting he turned to the study of nucleic acid. A chemical cleavage method proposed by Gilbert and Maxam entailed chemical treatment, where one end of single-stranded DNA is labeled with radioactive ^{32}P and cleavage takes place at the site of a particular base. This provides DNA of various lengths with one end labeled with ^{32}P. Sequencing involves separating the DNA fragments by electrophoresis on a gel, detecting their positions on the gel by autoradiography, and inferring the sequence from comparison of the fragment positions on the gel.[45] Sanger, who had been awarded a Nobel Prize for determining the primary structure of proteins, had developed a chain termination method in which the DNA to be sequenced serves as a template; a ^{32}P-labeled primer is added and bases complementary to the template are attached by DNA polymerase. If the sample is divided into four sequencing reactions and one of the four kinds of deoxynucleotide triphosphate (dNTP) and dideoxynucleotide triphosphates (ddNTP) are added to one reaction mix, the synthesis of the complementary strand stops when ddNTP is incorporated. The sequence is determined by separating the synthesized strands of various lengths by electrophoresis, visualizing the fragment bands by radiography, and reading the sequence by comparing the band pattern of the four electrophoresed reaction mixes.[46]

Determining long DNA sequences such as chromosomes required fast sequencing. The chain termination method was amenable to automation and computerization, and a rapid variation of the Sanger method has been used for genome analysis. Gilbert, Sanger and Berg were awarded the Nobel Prize in Chemistry in 1980. This was Sanger's second Nobel Prize in Chemistry; he is the only person to have won the Nobel Prize in Chemistry twice. (see Column 24)

COLUMN 24[7]

Sanger, the two-time Nobel laureate in Chemistry

Sanger (right) and the Swedish princess (left) at his first Nobel Prize award ceremony

Only three people have won two Nobel Prizes in science. They are Marie Curie (1903 Prize in Physics and 1911 Prize in Chemistry), John Bardeen (1956 and 1972 Prizes in Physics), and Frederick Sanger (1956 and 1980 Prizes in Chemistry). Linus Pauling did win the Nobel Prize twice, but once it was the Peace Prize, so Sanger is the only person to have been awarded two Prizes in Chemistry. One asks what kind of person Sanger was and how he was able to achieve such a monumental feat.

Frederick Sanger was born in 1918 in a small village in Gloucestershire, U.K., the son of a general practitioner and a Quaker. In 1936, he entered St. John's College at the University of Cambridge, where he studied natural sciences. In 1939, he graduated from university majoring in biochemistry. The war began, but being a Quaker, he was exempted from military service as a conscientious objector. In 1940, he began research for his PhD, earning his degree in 1943 for his studies on the metabolism of the amino acid lysine.

After earning his degree, he joined a group headed by Chibnall, a protein chemist who had just become a senior professor in biochemistry in Cambridge. At the time, nothing was known about the structure of proteins, or even whether proteins had a constant structure. Chibnall had been studying the amino acid composition of bovine insulin and asked Sanger to collaborate. Sanger started research to sequence the amino acids in insulin. Insulin is a low molecular weight protein that is readily available. Using the method described in this text (see page 597), Sanger systematically solved problems and in 1956 determined the primary structure of bovine insulin, which is composed of 51 amino acids. For this achievement, he was awarded his first Nobel Prize in Chemistry, in 1958. He was 40 years old. Tiselius, who was chairman of the selection committee, stated that the intention for the Nobel Prize is not only to reward achievement, but also to encourage future work; in his case, this was indeed quite true. Sanger was awarded his second Nobel Prize in Chemistry after 22 years of further diligent work.

After winning the Nobel Prize, everyone expected Sanger to start sequencing the amino acids of a larger protein, but he left this to other people, and instead challenged himself with a new problem: the sequencing of nucleic acid. He first attempted to sequence RNA, and encountered difficulties different from those encountered in determining an amino acid sequence. These were overcome, and Sanger's group succeeded in sequencing the 5S ribosomal RNA from *E. coli*. However, the race to sequence RNA was won by Robert Holley (1968 Nobel Prize in Physiology or Medicine).

Sanger next took on the task of sequencing DNA. He developed a method he called a Plus and Minus technique, and in 1975 he sequenced the nucleotides of the bacteriophage φX174. In his method, a single strand of DNA acts as a template for synthesizing various fragments using DNA polymerase. The fragments are fractionated by electrophoresis, and the sequence is inferred from the differences in length between the fragments. In 1977, his group introduced a new method called dideoxy chain termination. This method made it possible to rapidly and accurately sequence long DNA strands. His group sequenced human mitochondrial DNA (16,569 base pairs) and the genome of bacteriophage λ (48,502 base pairs). For this achievement, he was jointly awarded the Nobel Prize in Chemistry with Walter Gilbert in 1980. Later, his method was improved and automated, and used to sequence the human genome. Sanger thus developed methods for sequencing the components of the most important molecules involved in life phenomena: proteins and DNA. His major accomplishments were achieved by carefully and systematically solving the problems he encountered. Thus, his achievements stand in contrast to achievements made with a single spark of inspiration.

Having been raised in a Quaker household, he was a pacifist who abhorred violence. However, he gradually lost his faith and became agnostic. He stated, "I was brought up as a Quaker, and for them truth is very important... one is obviously looking for truth, but one needs some evidence for it. Even if I wanted to believe in God I would find it very difficult. I would need to see proof." A reserved and quiet person, he never promoted his own achievements, and declined an offer of a knighthood as he did not wish to be given the title "Sir". He retired in 1983 and enjoyed the rest of his days tending his garden at his home in the Cambridge suburbs. He died in 2013.

Richard Roberts (1943–)

A British-born molecular biologist in the U.S., Roberts earned a degree at the University of Sheffield. He was briefly at Harvard University and Cold Spring Harbor Laboratory, then became a chief science officer at New England Biolabs in 1992. In 1977, he demonstrated that the adenovirus genes are interrupted, and discovered that DNA includes both exon and intron regions. He was awarded the 1993 Nobel Prize in Physiology or Medicine.

Philip Sharp (1944–)

Sharp is a molecular biologist in the U.S. who earned a degree at the University of Illinois. He worked at Cold Spring Harbor Laboratory and in 1974 became director of the Massachusetts Institute of Technology's Center for Cancer Research. In 1977, he discovered that DNA includes both exon and intron regions. With Roberts, he was awarded the 1993 Nobel Prize in Physiology or Medicine.

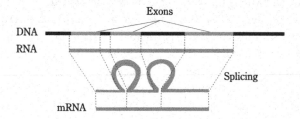

Figure 6.10: Schematic diagram of the splicing process

Introns and exons

An important discovery from DNA sequencing was that genes are interrupted. In 1977, Richard Roberts and Philip Sharp independently discovered that the genes in the adenovirus (which causes the common cold) are not a continuous base sequence. Instead, the portions (exons) that encrypt genetic information are interrupted by portions (introns) that do not carry genetic information. Using electron microscopy and biochemical approaches, they showed that during transcription into RNA, the intron portions are looped out and cut away from the exons, then the exons are spliced together to produce the translatable mRNA (Fig. 6.10). Interrupted genes were later found to be common in the DNA of many different eukaryotes. Subsequent research showed that in eukaryotes, first the complete structural gene comprising introns is transcribed into an mRNA precursor, then the introns are cut out and the exons are linked together to generate the mature mRNA. The majority of eukaryotic genes contain introns, and the discovery of interrupted genes is fundamental to understanding genes.

Roberts and Sharp were awarded the Nobel Prize in Physiology or Medicine in 1993.

Functions of RNA and synthesis and degradation of proteins

By the 1960s, the process by which proteins are synthesized based on information in DNA was understood, but the detailed mechanisms of transcription and translation remained unclear. Active research to shed light on these mechanisms began in the 1970s. Many new facts regarding the functions of RNA and degradation of proteins were discovered, and the chemistry underlying DNA and RNA became increasingly clear.

Details of the transcription machinery

RNA polymerases are enzymes that catalyze the synthesis of complementary RNA from the base sequence of the template DNA using the A, U, G, and C nucleotides and ribose triphosphate. Bacterial RNA polymerase was independently discovered by Hurwitz and Weiss in 1960. Studies during the 1970s revealed the structure of RNA polymerase and the mechanism of transcription in greater detail. In RNA polymerase from *E. coli*, subunits called σ factors bind to a core enzyme composed of four subunits; electron diffraction analysis revealed that RNA polymerase has a large channel to which DNA binds. Transcription begins with the σ factors recognizing a specific base sequence called the promoter; the double-stranded DNA unzips and transcription starts, with the strands extending in the direction from 5' to 3'. Transcription terminates at a specific site in the DNA.

Transcription in eukaryotes is far more complex. Three types of RNA polymerase (types I, II, and III) had been identified by 1969 by Roeder and Rutter. Each polymerase synthesizes different types of RNA. RNA polymerase II synthesizes mRNA. From the end of the 1970s to the 1980s, it was discovered that in eukaryotes, transcription requires a group of proteins called transcription factors (TFs) in addition to RNA polymerase. RNA polymerase II works in conjunction with transcription factors TFIIA, B, D, E, F, and H. Transcription factors have DNA binding sites and transcription activation sites, and recognize unique sequences of DNA.

Roger Kornberg from Stanford University made major contributions towards elucidating the mechanism of transcription in eukaryotes. He began working on this problem in the 1980s, using yeast as a model system. Kornberg made full use of advanced biochemical techniques, along with X-ray crystallographic analysis and electron microscopy observation to elucidate the steric structures of molecules involved in transcription (Fig. 6.11). In the 1990s, the structures of the proteins involved in transcription became increasingly apparent, and in

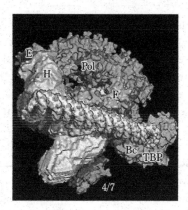

*Figure 6.11: Structure of the transcription initiating complex composed of RNA
polymerase II, determined by X-ray and electron diffraction studies of
crystals. The structure shows DNA associated with the polymerase.*

2001 the three-dimensional structure of the RNA polymerase II complex was
solved.[47] Roger Kornberg was awarded the Nobel Prize in Chemistry in 2006
for "his studies of the molecular basis of eukaryotic transcription". He is the
eldest son of Arthur Kornberg, who was awarded the Nobel Prize in Physiology
or Medicine in 1959 for the discovery of DNA polymerase. The second son,
Thomas Kornberg, discovered DNA polymerase II and III.

As mentioned earlier, Sharp and Roberts showed in 1977 that eukaryotic
RNA transcribed from DNA (i.e., the precursor of mRNA) is composed of
exon portions that encode genetic information relevant to protein synthesis and
intron portions that do not. The end of the 1970s and the 1980s saw progress in
understanding the process of RNA splicing, where the intron portions of RNA
are removed and mRNA is produced. The next problem was elucidating the
mechanism by which information is transferred from mRNA to tRNA in the
ribosomes in the cell during protein synthesis.

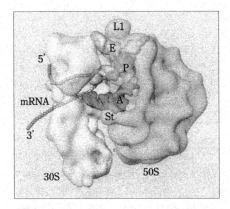

Figure 6.12: View of the structure of the E. coli ribosome at 25Å resolution

The left-side shows the 30 S subunit and the right-side shows the 50 S subunit. tRNA occupies the A, P, and E sites.

Protein synthesis and ribosomes

tRNA recognizes combinations of three bases (called a codon) as a code corresponding to a single α-amino acid. There are $4^3=64$ different codons, but only 61 initialize reading of one of the 20 amino acids. The other three codons are termination codons. A three nucleotide base sequence, called the anticodon, on tRNA recognizes the corresponding mRNA codon and binds, thus delivering the amino acid to a specific position on the mRNA. Because proteins are synthesized on ribosomes inside the cell, elucidating the detailed mechanism of protein synthesis required structural analysis of ribosomes. The general shape of the ribosome was inferred from electron microscopy studies, and by around 1980 it was known that prokaryotic ribosomes are composed of two subunits, 30 S and 50 S (where S is a unit of size). The 30 S subunit is composed of 20 different proteins and one RNA (rRNA) and the 50 S subunit is composed of 33 different proteins and three rRNAs (Fig. 6.12). In the 1980s, detailed studies of the mechanism of ribosomal protein synthesis were accompanied by progress in X-ray crystallographic structural analysis.

Prokaryotic protein synthesis is an extremely complex process that takes place in the ribosomes and involves mRNA, tRNA, rRNA, GTP (guanosine triphosphate) and other factors. The basic outline is as follows. (1) Activation of amino acids: Amino acids are linked through an ester bond to the 2' or 3' end of their specific tRNA to generate aminoacyl-tRNA. (2) Formation of the initiation complex: An initiation factor binds to the 30 S subunit of the ribosome and the ribosome dissociates. mRNA, initiation tRNA, and GTP bind to the dissociated 30 S subunit to form the 30 S initiation complex. The 50 S subunit binds to

Ada Yonath (1939–)

Born in Jerusalem, the child of Jews who had immigrated to Palestine from Poland, Yonath worked to support her schooling despite poverty and studied chemistry at Hebrew University. She earned a PhD in 1968 majoring in X-ray crystallography at the Weizmann Institute of Science. In 1970 she established a laboratory for the X-ray structural analysis of proteins, and between 1979 and 1984 she served  as group leader for the Max Planck Institute for Molecular Genetics in Berlin. In the 1970s she tackled the problem of elucidating the structure of the ribosome, which was believed to be impossible at the time. In 1980 she succeeded in crystallizing the ribosome, and after about 20 years of difficult research, she succeeded in determining its approximate crystal structure. Together with Steitz and Ramakrishnan, she was awarded the 2009 Nobel Prize in Chemistry.

Thomas Steitz (1940–)

An American biochemist, Steitz was born in Wisconsin and studied chemistry at Lawrence University; he studied under Lipscomb at Harvard University and earned his PhD in 1966 in biochemistry and molecular biology. Later, after serving as a postdoctoral fellow at the Laboratory of Molecular Biology at the University of Cambridge, he taught at Yale University and currently is a professor of molecular  biophysics and biochemistry. In the 1990s, he helped develop techniques that dramatically improved the resolution of X-ray structure diffraction. Together with Yonath and Ramakrishnan, he was a recipient of the 2009 Nobel Prize in Chemistry. The prize recognized his achievements in determining the structure and function of the ribosome, including elucidating the reaction mechanism of peptidyl transferase.

the initiation complex to form a 70 S initiation complex. (3) Generation and elongation of the polypeptide chain: A complex of protein elongation factors and aminoacyl-tRNA corresponding to the mRNA codon binds to the A-site of the 50 S subunit, and the growing peptide chain is bound to the P-site. The polypeptide chain detaches from the tRNA in the P-site and a peptide bond is formed between the C-terminal amino acid in the peptide and amino acid attached to the tRNA in the A-site. The A-site now contains the growing peptide. The empty tRNA

Venkatraman Ramakrishnan (1952–)
Ramakrishnan was born in the state of Tamil Nadu in India. He studied at a university in Vadodara, where he earned a BSc in physics. Emigrating to the U.S., he earned a PhD in physics at Ohio State University before studying biology at the University of California, San Diego. He began studying the ribosome as a postdoctoral fellow at Yale University, and continued researching the ribosome at Brookhaven National
Laboratory between 1983 and 1995. In 1995 he became a professor at the University of Utah, and in 1999 he moved to the Medical Research Council Laboratory of Molecular Biology at the University of Cambridge. His group succeeded in determining the high resolution structure of the 30 S subunit of the ribosome, and in 2009 he shared the Nobel Prize in Chemistry with Yonath and Steitz.

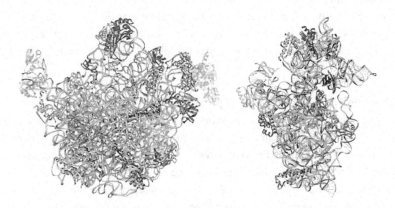

Figure 6.13: X-ray diffraction structure of the two subunits (50 S and 30 S) of the 70 S ribosome of the heat-resistant bacterium Thermus thermophilus

leaves the P-site and the A-site peptidyl-tRNA moves to the P-site. This process is repeated to elongate the polypeptide chain. (4) End of synthesis: An mRNA stop codon is recognized at the A-site and protein synthesis terminates.

Although very complex, protein biosynthesis is controlled very subtly, and is conducted accurately and without errors. To understand the process in greater detail, it was essential to understand the structure of the ribosome and the complexes produced during the reaction process. Crystal structure analysis of the ribosome started around 1980, and a high-resolution structure was available in 2000. The three major contributors to this effort were Ada Yonath, Thomas Steitz, and Venkatraman Ramakrishnan. Yonath succeeded in crystallizing heat-resistant bacterial ribosomes in the early 1980s, allowing X-ray crystal structural analysis.[48] Steitz solved the phase problem for the ribosome diffraction data using electron microscope images and heavy atom substitution, and in 2000 he succeeded in the high-resolution structural analysis of the 50 S subunit.[49] In the same year, Ramakrishnan obtained high-resolution images of the 30 S subunit (Fig. 6.13).[50] Thus, at the start of the 21st century, understanding the structure and function of the ribosome at the atomic level became possible. Yonath, Steitz, and Ramakrishnan shared the Nobel Prize in Chemistry in 2009 for their studies on the structure and function of the ribosome. Yonath, a woman scientist, was born in Israel and is now a professor of structural biology at the Weizmann Institute of Science. Steitz is a professor of molecular biophysics and biochemistry at Yale University in the U.S. Ramakrishnan was born in India and earned his PhD in physics in the U.S., and is currently group leader of the structural studies division at the Medical Research Council Laboratory of Molecular Biology in Cambridge. This team demonstrates the international and interdisciplinary nature of modern chemical studies.

Ubiquitin and the degradation of proteins
Pioneering research by Schoenheimer in 1942 using isotopes showed that proteins in the body are in a state of dynamic equilibrium: they are constantly being synthesized and broken down (see page 363). Cells are constantly synthesizing proteins from amino acids and simultaneously breaking proteins down into their components. This is how abnormal proteins that are harmful to the cell, and enzymes and control proteins that are no longer needed, are removed. Enzymes such as trypsin that hydrolyze proteins, and an organelle that breaks proteins down, called the lysosome, were well known. These processes do not require energy. Around 1950, however, scientists obtained experimental results indicating the existence of a separate proteolytic process that is dependent on ATP and does require energy. This proteolytic process was examined in greater detail near the end of the 1970s. From the 1970s to the 1980s, Aaron Ciechanover, Avram Hershko, Irwin Rose and colleagues showed that ubiquitin plays a critical role in intracellular proteolysis.[51–53]

Ubiquitin is a monomeric protein composed of 76 amino acids, and is abundant in eukaryotic cells. Ubiquitin binds through a thioester bond to ubiquitin-activating enzyme (E1) in the presence of ATP, then the ubiquitin

Aaron Ciechanover (1947–)

An Israeli biochemist, Ciechanover graduated from Hadassah Medical School in the Hebrew University and earned a degree at the Technion–Israel Institute of Technology. In 1977 he began working at the Technion, and in 2002 he was appointed as a professor. He, along with Hershko and Rose, studied proteolysis by ubiquitin starting in the late 1970s. With Hershko and Rose, he was awarded the 2004 Nobel Prize in Chemistry.

Avram Hershko (1937–)

Hershko is an Israeli biochemist who graduated from Hadassah Medical School in the Hebrew University and earned a degree in 1969. In 1972 he started working at the Technion–Israel Institute of Technology, and in 1998 he was appointed as a professor. He, along with Ciechanover and Rose, studied proteolysis by ubiquitin starting in the late 1970s, and with them was awarded the 2004 Nobel Prize in Chemistry.

Irwin Rose (1926–)

Rose is an American biochemist who earned a PhD at the University of Chicago and worked at Yale University's Fox Chase Cancer Center before joining the University of California, Irvine, in 1997. Together with Ciechanover and Hershko, he was awarded the 2004 Nobel Prize in Chemistry for his work on proteolysis by ubiquitin.

protein is transferred to ubiquitin-conjugating enzyme (E2). E2 complexes with ubiquitin ligase (E3), then ubiquitin is transferred from E2 to the side chain of a lysine on a target protein requiring proteolysis. A subsequent ubiquitin is added to the side chain of a lysine of the ubiquitin. This process is repeated until the protein is labeled with a chain of ubiquitin molecules. When this labeled protein binds to a large protein complex called a proteasome, the ubiquitin is removed and the protein is broken down. In addition to marking a protein for degradation, ubiquitin was demonstrated to be involved in other cellular processes such as cell cycle control, DNA repair, transcription regulation, and apoptosis[9].

Ciechanover, Hershko, and Rose were together awarded the 2004 Nobel Prize in Chemistry for the discovery of proteolysis via ubiquitin.

---

9. Apoptosis: The phenomenon whereby some cells "fall away" in programmed cell death by mechanisms determined in advance by genes.

Thomas Cech (1947–)

Cech was born in Chicago as the child of immigrants from Czechoslovakia. He grew up in Iowa and graduated from Grinnell College, then majored in biophysical chemistry at Berkeley, where he developed an interest in molecular biology. After earning his degree, he served as a postdoctoral fellow at the Massachusetts Institute of Technology (MIT) and then as instructor at the University of Colorado. His research focuses on the transcription process in the cell nucleus. He discovered that RNA can self-splice, and found that RNA also can exhibit catalytic function. He has also contributed to the study of the structure and function of telomeres. With Altman, he was a recipient of the 1989 Nobel Prize in Chemistry.

Sidney Altman (1939–)

Born in Montreal, Canada, as the child of immigrants from the Soviet Union, Altman studied physics at the Massachusetts Institute of Technology (MIT) and Columbia University. He became fascinated by molecular biology and majored in and earned a degree in biophysics at the University of Colorado. After postdoctoral fellowships at Harvard and Cambridge, he became a professor of biology at Yale University, and also served as dean of Yale College. He, independently of Cech, discovered that the catalytic activity in RNA-protein complexes is due to the RNA, and that the RNA molecule alone exhibits catalytic activity. With Cech, he was a recipient of the 1989 Nobel Prize in Chemistry.

Catalytic function of RNA and the RNA world

By the 1970s, it was believed that RNA acts as a mediator for the transfer of genetic information from DNA, and that compounds responsible for vital functions, such as enzymatic reactions, are proteins that are synthesized based on the information encoded in DNA. The development of genetic manipulation techniques such as cloning, genetic recombination, and DNA sequencing in the 1970s provided detailed information on the functions of RNA. Cech and Altman discovered that RNA itself has catalytic functions. This discovery led to increased awareness of the importance of RNA.

Thomas Cech studied rRNA genetic splicing in the flagellate protozoan tetrahymena in the early 1980s. He discovered that when an rRNA precursor isolated from this organism is cultured without protein, guanosine, or the guanine nucleotide, self-splicing occurred and the introns were spontaneously cut out and the exons linked together. In 1982, he showed that RNA functions as a catalyst in this reaction.[54] Sidney Altman, who had been studying RNase P, an enzyme composed of RNA and protein that cleaves RNA in *E. coli*, discovered that the RNA alone can act as an enzyme.[55] Later, it became apparent that various structures unique to RNA provide a catalytic function in self-splicing, and that RNA plays a central role in protein synthesis by the ribosome. RNA exhibiting functions similar to those of protein enzymes is called a ribozyme.

The discovery that RNA has catalytic actions inspired the 1986 proposal of the "RNA world hypothesis", which states that life originated from RNA that functioned as a gene and simultaneously had a self-replicating catalytic function. [56] According to this hypothesis, the primitive Earth supported self-replicating systems composed of RNA (the RNA world). This primordial system evolved, with the more stable DNA becoming the bearer of genetic information and the structurally more flexible proteins taking over enzymatic functions. This popular hypothesis presented interesting possibilities, but several problems have been noted. The RNA world remains a topic of debate due to insufficient experimental evidence supporting it.

For their discovery that RNA has catalytic actions, Altman and Cech were awarded the 1989 Nobel Prize in Chemistry.

6.3 Development of biochemistry (II): Enzymes, metabolism and molecular physiology

Traditional biochemistry also made significant advances in the second half of the 20th century due to advances in experimental techniques. First, progress in analytical techniques such as chromatography, ultracentrifugation, and electrophoresis made it easier to separate and analyze compounds involved in biochemical reactions. Next, radioactive isotopes, particularly ^{14}C, ^{15}N, and ^{32}P, became widely used in biochemical research, making it possible to track biochemical reactions and identify reaction intermediates. Moreover, with advances in X-ray structural analysis, the three-dimensional structures of enzymes were revealed, and the detailed mechanisms of enzymatic reactions were elucidated at the atomic level. Research using proteins modified by genetic manipulation also became possible, bringing about significant developments in the study of enzyme reactions and metabolism.

This section deals with enzyme structures and reaction mechanisms, metabolism and its regulation, biological membranes and membrane transport, electron transport, photosynthesis, and signaling. Progress in these fields was rapid, and there were many important developments. The overview presented here focuses on research recognized by Nobel Prizes in Chemistry and Physiology or Medicine.

Elucidating enzyme structures and reaction mechanisms
Advances in enzyme chemistry

Although Sumner and Northrop showed that enzymes are proteins, before World War II it was widely believed that low-molecular-weight prosthetic molecules are responsible for enzyme activity. It was only after the war that the structures of active sites of enzymes such as chymotrypsin, lysozyme, and pepsin were elucidated, and it was definitively shown that proteins are responsible for enzyme activity. After the war, spectroscopic methods such as ultraviolet and visible-light absorption spectroscopy and fluorescence spectroscopy came to be widely used; furthermore, the use of new methods such as chemically modified proteins led to significant advances in the study of enzymes. Other important developments contributing to an understanding of enzyme reaction mechanisms included new methods for studying reaction rates such as the stopped flow method, transition state theory proposed by Eyring, and physical organic chemistry kinetics developed by Ingold and Hammett. Beginning around 1980, structural analysis by X-ray diffraction and NMR advanced significantly. Thus, in the second half of the 20th century, the structures and functions of huge numbers of enzymes were being clearly explained.

Enzymes bind specifically to their substrate due to the shape of the enzyme and substrate molecules and the complementarity of their physical properties. Binding and reaction between enzyme and substrate is stereospecific. An efficient reaction requires that the substrate and enzyme approach one another with the appropriate orientation. Many enzymes involved in oxidation-reduction reactions or that catalyze the transfer of groups require a coenzyme, and many vitamins are precursors of coenzymes. Furthermore, many enzyme reactions are catalyzed by acid-base groups; examples include peptide and ester bond hydrolysis, phosphate group reactions, and tautomeric reactions. About one-third of enzymes require a metal ion for catalytic activity. Bound metal ions can neutralize a negative charge and act as Lewis acids.

Lysozymes and chymotrypsin

It is not possible to describe in detail the enormous number of enzyme reaction mechanisms elucidated thus far, but typical examples will be described briefly here. One example is lysozymes, which are enzymes that break down bacterial cell walls, and are widely distributed in the cells and secretions of vertebrates. The best-studied lysozyme is hen egg white (HEW) lysozyme, comprising a single-

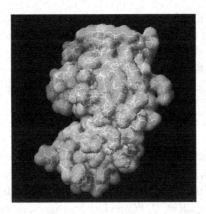

Figure 6.14: Surface structure of lysozyme generated by computer graphics
The cleft where the enzyme binds the substrate is clearly visible.

chain polypeptide of 129 amino acid residues and four disulfide bonds. The X-ray structure of HEW lysozyme was elucidated by Philips and colleagues in 1965; this was the first enzyme structure to be elucidated at high resolution (Fig. 6.14). The structure showed a cleft for substrate binding across one side of the enzyme. Lysozyme hydrolyzes the glycosidic linkage between polysaccharide components constituting the cell wall. Oligosaccharides of N-acetylglucosamine (NAG) inhibit lysozyme activity by binding to the active site. Substrate binding was investigated in greater detail from experiments using the model hexasaccharide, $(NAG)_6$, and led Philips to propose a catalytic mechanism.[57] According to this mechanism, (1) lysozyme binds to hexasaccharides in the cell wall, causing the ring of the fourth hexose to adopt a strained conformation. (2) A nearby glutamate donates a proton to the O of the strained glycosidic bond, the C–O bond is broken, and an oxonium ion is produced. (3) The oxonium ion is stabilized by electrostatic interaction with a nearby aspartate. (4) Water reacts with the oxonium ion and a product is produced, with the enzyme returning to its original state. This mechanism has been extensively studied, and Philips' theory is believed to be correct.

Another example is the reaction of serine proteases (proteolytic enzymes with a serine at the active center) such as chymotrypsin and trypsin. Chymotrypsin and trypsin are protein digestive enzymes secreted from the pancreatic duct. Research begun in the early 1950s had shown that the reaction mechanism of serine proteases involves the following steps. (1) The hydroxyl group of a serine at the active site of an enzyme nucleophilically attacks the carbonyl carbon atom of the peptide bond to be cleaved, forming a tetrahedral intermediate. (2) This intermediate is broken by migration of the electrons from the broken bond to the histidine in the active site. The polarity of the bond is increased through hydrogen bonding with aspartate, thus producing an acyl-enzyme intermediate; the water is replaced by deamination. (3) A tetrahedral intermediate is formed by the reverse process of (2), and a carboxyl product and the original enzyme are produced with the reverse process of (1). The X-ray crystal structure of bovine chymotrypsin determined by Blow in 1967 clarified the structure of the active site. Evidence for the tetrahedral intermediate was obtained from X-ray structural analysis of the complex of trypsin bound to soybean trypsin inhibitor.

Developments in metabolic research and their impact

Pioneering research on elucidating metabolic pathways was conducted by Schoenheimer using the isotopes ^2H and ^{15}H prior to World War II. After the war, radioactive isotopes such as ^3H, ^{14}C, and ^{32}P became more readily available, and research using molecules labeled with these nuclides greatly advanced metabolic study. The enzymes catalyzing the individual steps of a reaction were isolated, identified and crystallized, and the mechanisms underlying enzymatic reactions were investigated in greater detail. Around 1980, X-ray crystallographic analysis and two-dimensional NMR were used to determine the structures, functions and mechanisms of the action of enzymes.

Detailed mechanism of glycolysis

Metabolic substances such as sugars, proteins, and fats are first broken down into their constituent units (e.g., glucose, amino acids, fatty acids and glycerol) and are converted into a common intermediate, acetyl CoA. The acetyl group of this molecule is oxidized to CO_2 in the citric acid cycle, with the simultaneous reduction of NAD^+ and FAD. When these are re-oxidized by O_2 in the electron transport system, ATP is formed by oxidative phosphorylation. The outline of the glycolytic system by which glucose is broken down into pyruvic acid was revealed before World War II, as stated in the previous section, but the enzymatic reaction mechanism at each step was revealed in detail in the second half of the 20th century. In the glycolytic system, two molecules of pyruvic acid are generated from one molecule of glucose and delivered to the citric acid cycle, which in turn converts two molecules of ADP into ATP.

Although Hans Krebs had established the existence of the citric acid cycle before the war (see page 358), the mechanism by which citric acid is generated from pyruvic acid had not been elucidated, nor had the cycle been completely elucidated. The most important post-war contribution was the elucidation of the process by which a two-carbon compound, acetyl-CoA, is generated from pyruvic acid and condenses with oxaloacetic acid to produce citric acid.

In 1945, Lipmann and Kaplan discovered a new coenzyme that is effective in acetylating choline; Lipmann named this compound coenzyme A (CoA).

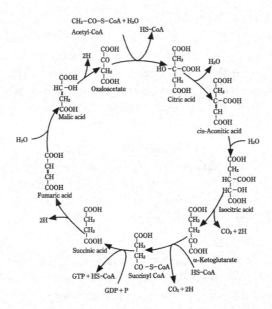

Figure 6.15: Citric acid cycle (completed figure)

Later, his group showed that this coenzyme comprises adenosine diphosphate, pantothenic acid, and a thiol (SH) group. In 1951, Ochoa and Lynen showed that acetyl CoA directly condenses with oxaloacetic acid when citric acid is produced, thus revealing the full picture of the citric acid cycle (Fig. 6.15). Each step in this cycle would later be extensively studied at the molecular and enzymatic level. Acetyl CoA is an important metabolic intermediate that links to an acetyl group via a terminal thiol group and transports the acetyl group to oxaloacetate.

In 1953, Hans Krebs and Fritz Lipmann were awarded the Nobel Prize in Physiology or Medicine, the former for the discovery of the tricarboxylic acid cycle and the latter for discovering coenzyme A and the significance of high-energy phosphate bonds in metabolism.

Metabolism of glycogen

Animals store excess glucose as glycogen mainly in their liver and muscles, and then break it down into glucose as needed. Glycogen is broken down by the action of glycogen phosphorylase to generate one molecule of glucose monophosphate (G1P) and glycogen shortened by one glucose. G1P is not immediately reconverted into glycogen and phosphoric acid. In 1957 the Argentinian biochemist Louis Leloir showed that G1P participates in the synthesis of uridine diphosphate glucose (UDPG) by reacting with uridine triphosphate glucose; subsequently, the glucosyl group is transferred to glycogen

Louis Leloir (1906–1987)

Leloir was an Argentinian biochemist who studied at the University of Buenos Aires and was a director of a biochemistry institution in Buenos Aires in 1947. He was awarded the 1970 Nobel Prize in Chemistry for his contributions to the elucidation of the biosynthesis of glycogen and the intermediates involved, and of the synthesis of sugars such as sucrose and lactose.

Edwin Krebs (1918–2009)

An American biochemist, Krebs earned a degree in medicine at Washington University in St. Louis and in 1957 became a biochemistry professor at the University of Washington in Seattle. He explained how the phosphorylation and dephosphorylation of glycogen phosphorylase are involved in glycogen metabolism. He was awarded the 1992 Nobel Prize in Physiology or Medicine.

Edmond Fischer (1920–)

Fischer is an American biochemist who was born in Shanghai to Swiss parents; he earned his degree at the University of Geneva and in 1961 became a professor at the University of Washington in Seattle. With Krebs, he explained how phosphorylation and dephosphorylation of phosphorylase are involved in glycogen metabolism. In 1992 he shared with Krebs the Nobel Prize in Physiology or Medicine.

by the action of glycogen synthase.[58] This discovery was a breakthrough and changed the conventional perspective of the biosynthesis of glycogen, earning Leloir the Nobel Prize in Chemistry in 1970.

Glycogen metabolism is a reaction controlled by the enzymes phosphorylase and synthase. The study of glycogen metabolism played a critical role in elucidating the control mechanisms by which the body maintains homeostasis. This control involves a variety of mechanisms, but only a brief discussion of research that led to the scientists being awarded the Nobel Prize in Physiology or Medicine is provided here.

Carl and Gerty Cori (see page 360) in 1938 discovered that there are two types of phosphorylase. The a-type is active even in the absence of adenosine monophosphate (AMP), but the b-type requires AMP in order to be activated. In 1949, Edwin Krebs and Edmond Fischer discovered that when a specific serine in the phosphorylase b protein is phosphorylated, the phosphorylase becomes the a-type, and dephosphorylation converts it again to the b-type. They showed also that in phosphorylation, a protein kinase acts as a catalyst, and dephosphorylation requires catalysis by a phosphatase. It thus became apparent that the

Earl Sutherland (1915–1974)
An American biochemist, Sutherland graduated as a doctor of medicine from Washington University and in 1963 became a professor at Vanderbilt University. He discovered that cAMP acts as a second messenger and studied its functions, helping elucidate the actions of hormones. He was a recipient of the 1971 Nobel Prize in Physiology or Medicine.

phosphorylases and synthases involved in glycogen metabolism are controlled with two enzymes that are reversibly interconvertible.

Glycogen metabolism in the liver is controlled by glucagon, whereas in other tissues it is controlled by insulin and by adrenaline and noradrenaline produced by the adrenal gland. These hormones convey stimuli via membrane receptor proteins in the plasma membrane. Different cells have different receptors and have different hormones that they respond to by releasing molecules called second messengers[10] into the cell; these second messengers transmit the hormone's message to the cell. In 1956, Earl Sutherland and Rall discovered cyclic AMP (cyclic adenosine monophosphate, cAMP) and determined that it mediates glucagon and adrenaline's hyperglycemic actions in the liver. Later, Sutherland showed that glucagon and adrenaline activate adenylate cyclase at the cell surface to produce cAMP, and to show that cAMP plays the role of a second messenger.

The study of metabolic control and regulation at the molecular level evolved greatly in the second half of the 20th century, and Sutherland, Krebs, Fischer and coworkers' pioneering research was particularly significant. The Nobel Prize in Physiology or Medicine was awarded to Sutherland in 1971 and to Krebs and Fischer in 1992. Both Sutherland and Krebs trained with Carl and Gerty Cori, who had an enormous impact on the study of metabolic control and regulation.

Biological membranes and membrane transport
Cells are surrounded by a cell membrane, and eukaryotic cells are partitioned by membranes into structures such as the nucleus, mitochondria, chloroplasts, endoplasmic reticulum and Golgi apparatus. Biological membranes are organized assemblies of lipids, proteins, and small amounts of sugars, and regulate the composition of the intracellular fluid by regulating the flow of

---

10. Second messenger: In information transfer in the body, "first messenger" is the name for compounds involved in direct information transfer between organs, cells, and tissues, whereas "second messenger" is the name for compounds involved in information transfer within the cell.

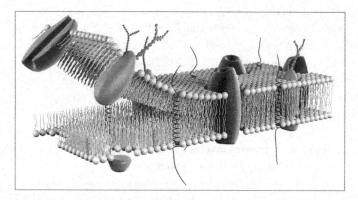

Figure 6.16: Schematic diagram of the plasma membrane

Membrane proteins are embedded in the phospholipid bilayer. Sugar molecules (indicated as beaded structures) are found on the outside of the membrane.

specific molecules and ions. Many biochemical reactions take place using a membrane as a scaffold.

Structure of biological membranes

Gorter and Grendel were the first to note the existence of lipid bilayer membranes, in 1925. They observed that when they extracted the cell membrane from red blood cells and spread the membranes over a surface to form a monomolecular film, the surface area obtained was double the original surface area. This result led them to propose that cell membranes are formed from a lipid bilayer. In 1935, Danielli and Davson proposed a model for membranes composed of lipids and proteins. The increasing use of electron microscopy to study cells after the war led to general acceptance of the idea that biological membranes are generally composed of a lipid bilayer, but understanding the structures of biological membranes in detail is a rather recent development.

Fluorescence, NMR, ESR and other spectroscopic methods shed light on the dynamic nature of artificial lipid bilayer membranes. Information on proteins in biological membranes also accumulated, and in 1972 a "fluid mosaic model" for biological membranes was proposed by Singer and Nicolson (Fig. 6.16). [59] According to this model, membrane proteins are like icebergs floating in a two-dimensional sea of lipids, and are free to move laterally unless hindered due to binding with other membrane components.

Membrane transport

Biological membranes are non-polar, so ions and polar substances are unable

Jens Skou (1918–)

Skou is a Danish chemist who studied medicine at the University of Copenhagen and graduated in 1944. He was trained in clinical practice, but became interested in the effects of local anesthetics and earned a PhD studying in a physiology laboratory at the University of Aarhus; in 1963 he became a professor of physiology at this university. In 1957, he discovered "Na+/K+-ATPase", a mechanism found in cell membranes that removes sodium ions and captures potassium ions. This was the first ion pump to be discovered. In 1997, he was awarded the Nobel Prize in Chemistry.

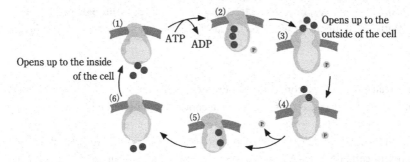

Figure 6.17: Schematic diagram of the mechanism of action of the Na⁺/K⁻-ATPase ion pump

(1) Na⁺ ions are incorporated into the enzyme. (2) ATP converts into ADP and one phosphate group binds to the enzyme. (3) The structure of the enzyme changes and the Na⁺ ions are released to the outside of the cell. (4) K⁺ ions are incorporated into the enzyme and phosphate groups are released. (5) K⁺ ions bind to the enzyme. (6) The K⁺ ions are released to the inside of the cell and the enzyme returns to its original state.

to passively cross the membrane. However, Na^+, K^+, Ca^{2+}, and metabolites such as pyruvic acid, amino acids, sugars, and nucleotides all pass through the membrane. Their transport is mediated by transport proteins. The permeation of ions through membranes had attracted interest since the 19th century as an important issue in physiology and biochemistry. At the start of the 1950s, Hodgkin and Huxley in the U.K. showed that the passage of Na^+ and K^+ ions through the membranes of nerve cells is important in nerve transmission. The structures of the proteins involved in ion transport, however, were not known, nor was the mechanism of membrane permeability. Both membrane protein structure and the mechanism of ion transport were important issues in biochemistry in the second half of the 20th century.

Roderick MacKinnon (1956–)
An American biochemist, MacKinnon graduated from the Tufts University School of Medicine; after his clinical training, he conducted basic research at Brandeis University and Harvard University. He was appointed a professor by The Rockefeller University in 1996. MacKinnon elucidated the structures and functions of ion channels that allow specific ions to pass through membranes. He was a recipient of the 2003 Nobel Prize in Chemistry.

Peter Agre (1949–)
Agre is an American biochemist who graduated from the Johns Hopkins University School of Medicine and trained and conducted research at the University of North Carolina, later returning to his alma mater and becoming a professor in 1993. While studying red blood cells, he discovered proteins called aquaporins, which selectively allow water to pass through membranes. With MacKinnon, he was a recipient of the 2003 Nobel Prize in Chemistry.

Transport along a concentration gradient is mediated by membrane channel proteins, whereas transport against the concentration gradient requires energy and occurs through ion pumps. Most ion pumps require the hydrolysis of ATP to supply the energy. The most researched transport system is a membrane protein called Na^+/K^+-ATPase, discovered by Jens Skou in 1957 (Fig. 6.17).[60] This enzyme protein is called the Na^+/K^+ pump because it pumps Na^+ out of the cell and pumps K^+ into the cell in conjugation with ATP hydrolysis. Three Na^+ ions are pumped out and two K^+ ions are pumped in, so a potential is generated, with the removal of one positive charge. This pump is involved in the electrical stimulation of nerve cells.

The study of ionophores (for example, valinomycin which is selective for K^+, and monensin, which is selective for Na^+), advanced after the war. Ionophores selectively bind to specific ions and facilitate their passage through the membrane. However, it is only recently that detailed structures of ion channels that selectively allow specific ions to pass through the membrane, and the mechanism of ion permeation, became clear. K^+ ions penetrate from inside the cell to outside the cell via membrane proteins called K^+ channels. This process is important in biochemical machinery, including for maintaining intracellular osmotic pressure and allowing neurotransmission. These channels are at least 10^4 times more selective to K^+ ions than to Na^+ ions, and have a large permeation rate. In 1998, Roderick MacKinnon of The Rockefeller University succeeded in his X-ray structural analysis of a K^+ channel called KcsA, and revealed the detailed mechanism of ion permeation.[61] His studies led to a model where oxygen atoms at the terminus of a protein helix form a cavity that selects K^+ ions; the K^+ ions are coordinated to oxygen atoms and pass through the channel.

Permeation of water through the cell membrane was a physiologically important problem, but one that was not solved until the second half of the 20th century. In the mid 1950s, it was found that water permeates quickly through a membrane by passing through water-specific pores in the cell membrane. Later research revealed the existence of a filter that lets only water through and not ions, but details regarding the filter remained a mystery. In the mid-1980s, Peter Agre, who had been studying the membrane proteins of red blood cells, discovered an unknown membrane protein by chance. He sequenced the protein and its corresponding DNA and inferred that it was a water channel. Then, in 1992, comparison of cells that contain this protein and cells that do not demonstrated that the protein was a membrane protein that forms a water channel.[62] This protein was later named aquaporin, and its high-resolution X-ray structure was reported in 2000. The presence of this kind of membrane protein has been widely confirmed in bacteria, plants and animals.

The Nobel Prize in Chemistry was awarded to Skou in 1997 for the discovery of Na^+/K^+-ATPase, and to Agre and MacKinnon in 2003 for the discovery of water channels and for studies on the structures and mechanisms of ion channels, respectively.

In vivo electron transport and oxidative phosphorylation

By the early 20th century it was clear that oxidation in the body requires catalysis by intracellular enzymes. Warburg and coworkers were instrumental in these studies. Glucose is ultimately oxidized to CO_2 by the actions of enzymes in the glycolytic process and the citric acid cycle. The oxidation of glucose by oxygen is represented by the following equation, and generates a large amount of free energy.

$$C_6H_{12}O_6 + 6O_2 \rightarrow 6CO_2 + 6H_2O \qquad \Delta G = -2823 \text{ kJmol}^{-1}$$

This reaction involves the transfer of 24 electrons and can be thought of as being divided into an oxidation reaction, where glucose carbon atoms are oxidized to CO_2 ($C_6H_{12}O_6+6H_2O\rightarrow6CO_2+24H^++24e^-$) and a reduction reaction, where an oxygen molecule is reduced to water ($6O_2+24H^++24e^-\rightarrow24H_2O$). In biological systems, the transport of electrons proceeds through a multi-stage process involving many different enzymes, and the free energy thus generated is stored in the form of ATP. In the first half of the 20th century, it was shown that pairs of electron pairs in in vivo oxidation processes are transferred to the coenzymes NAD^+ and FAD, generating 10 NADH and 2 $FADH_2$ (see page

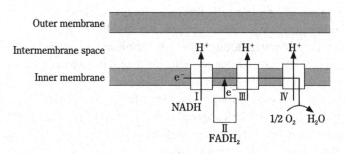

Figure 6.18: Schematic diagram of the electron transport system in mitochondria

353). The latter half of the 20th century saw detailed studies into the process by which these electrons drive stepwise oxidation and reduction along an electron transport chain, with oxygen ultimately being reduced to water. A relationship between electron transport and ATP generation by oxidative phosphorylation was thus revealed.

Electron transport mechanisms
In 1948, Lehninger and Kennedy showed that the mitochondria in cells contain many enzymes, including an enzyme needed for oxidation (pyruvic acid dehydrogenase), enzymes involved in the citric acid cycle, and various enzymes and proteins needed for electron transport and oxidative phosphorylation. Their findings revealed that oxidative metabolism occurs in the mitochondria in eukaryotic cells. After the war, electron microscopy revealed the internal structure of mitochondria, and details regarding transmembrane mitochondrial metabolism came to be discussed in detail.

In the 1950s, the radionuclide ^{32}P was used to study oxidative phosphorylation, and Lehninger's group showed a relationship between the phosphorylation of ADP and electron transport from NAD to oxygen. In 1957, coenzyme Q (CoQ) was discovered by Crane, and its role in the electron transport system was studied. Chance's group used spectroscopy to identify each of the steps of the electron transport system where oxidative phosphorylation takes place. Later research would reveal that electron transport involves four kinds of protein complexes embedded in mitochondria (Fig. 6.18). The complexes were named I, II, III, and IV in ascending order of standard reduction potential, and electrons flow in the direction NADH→complex I→CoQ→complex III→complex IV. This flow of electrons was determined from the effects of inhibitors specific for each complex and measurements of the standard reduction potential of each complex. From 1980 onward, advances in X-ray structural analysis revealed the structures of these protein complexes, allowing detailed discussion of the

Peter Mitchell (1920–1992)

After graduating from the University of Cambridge, Mitchell continued to conduct research at his alma mater and earned a degree in 1950. In 1955 he took up a research position at the University of Edinburgh, but he left the university in 1963 because he fell ill. Several years later, he built a laboratory in his home and resumed his work with a small research group to experimentally verify the "chemiosmotic theory" he had published in 1961. These studies showed that his hypothesis was correct, and his theory came to be increasingly accepted. In 1978, he was awarded the Nobel Prize in Chemistry, and he became a topic of discussion as a Nobel laureate not affiliated with a university. (Column 25)

mechanisms of electron transport. The change in standard free energy generated by the oxidation of NADH by oxygen is –218 kJmol⁻¹. The standard free energy needed to convert one ADP to one ATP is 30.5 kJmol⁻¹, so the oxidation of one molecule of NADH provides sufficient energy to produce several molecules of ATP. A major problem discussed during the 1950s was the mechanism by which the free energy obtained from the electron transport system is stored and used to synthesize ATP. Various hypotheses were suggested, but the theory that best explained the experimental facts was the "chemiosmotic theory" proposed by Peter Mitchell in 1961.[63]

Chemiosmotic theory

According to the chemiosmotic theory, the free energy obtained from electron transport causes H^+ to be pumped into the space between membranes from the mitochondrial matrix, resulting in an electrochemical H^+ concentration gradient across the inner membrane. This electrochemical potential gradient is used to synthesize ATP. As electrons are transferred sequentially to complexes I, III, and IV, H^+ is pumped to the outside from the inner membrane of the mitochondria, creating a concentration gradient of protons. Complex II also brings in electrons into the electron transfer system. Two electrons pass through complexes I, III, and IV, creating a proton concentration gradient that is used to synthesize one molecule of ATP at each of the complexes. The mechanisms of proton transport during electron transfer have been investigated in detail. Mitchell's chemiosmotic theory found increasing acceptance, and in 1978 he was awarded the Nobel Prize in Chemistry. (Column 25)

How is the free energy stored in the proton concentration gradient used to synthesize ATP? This is accomplished by an enzyme called ATP synthase. ATP

COLUMN 25[7]

Mitchell, the man who built a laboratory on his own

Glynn House at the time of construction in 1897

The achievements of many scientists who win a Nobel Prize are made by collaborating with many students and researchers in well-equipped laboratories at a large university. One amazing person, however, built a laboratory on his own property in the second half of the 20th century, conducted research there with just a few people, and made achievements that earned him the Nobel Prize. That person is Peter Mitchell.

Mitchell enrolled at Jesus College of the University of Cambridge in 1939, and studied the natural sciences and majored in biochemistry. Though neither his entrance examination results nor his performance while a student was good, he was fortunate enough to be at a graduate school where he could be tutored by great scholars such as Danielli and Keilin. In 1951, he earned a degree for his study of the actions of penicillin, and in the six years following 1950 he served as an assistant in the biochemistry department and studied the permeation of bacterial cell membranes by phosphorus. In 1955, he was invited to launch a research group in chemical biology at a zoology classroom in the University of Edinburgh. He published his "Chemiosmotic theory" see page 645) about oxidative phosphorylation in 1961 while at the University of Edinburgh. However, his theory went largely unnoticed because researchers in this field at the time believed that after a high-energy phosphate intermediate is formed, a phosphoric acid is transferred to ADP to generate ADP.

Taking time off from work in 1962 because of a gastric ulcer, he resigned from his position in 1963 and rested in a villa in the rural town of Bodmin in southwest England. At this time, he learned that a mansion called Glynn House, which had been built in the early 19th century, was being offered for sale, and so he invested his inheritance to buy it. He had always had an interest in architecture. Glynn House was timeworn and damaged, but he renovated it over two years and converted it into a home-turned-laboratory. During those two years, he took charge of the construction as his own architect and construction supervisor, putting his research aside. He recovered his health, and together with Jennifer Moyle, a collaborator during his days at Cambridge, he founded Glynn Research Institute in 1965. This was a small institute in which he and Moyle were the researchers, and a technician and secretary were hired. Here he set out to continue his research to prove his "chemiosmotic theory", which had been ignored by academia.

At that time, research into photosynthesis provided results supporting his hypothesis. Reports were published stating that the generation of ATP is driven by a proton concentration difference in chloroplasts, which supported his theory. He and Moyle also demonstrated that mitochondria export protons outside the membrane during oxidative phosphorylation. Data accumulated to support Mitchell's theory, and the theory transitioned from hypothesis to standard theory. The rapidity with which Mitchell's theory was accepted by leading researchers, shown in the figure, is truly impressive. Although he did not have a single supporter when he first proposed his theory in 1961, by 1977 everyone was a supporter except the distinguished scientist, Green. In 1978 Mitchell was the sole recipient of the Nobel Prize in Chemistry.

In his later years, Peter Mitchell became increasingly concerned about increased violence in densely populated societies, and he took an interest in the problem of communication between individuals in civilized society. From his own experiences in the world of science, he believed that lively, close relationships are easier to forge in small groups than in large groups, and that small groups often function more effectively. Given the current trend for research to be conducted on a larger scale and to be more collaborative than when Mitchell was conducting his research, I think many places are learning Mitchell's approach to research.

Researcher	1961	1965	1969	1973	1977	1980	Year
Green							
Lardy							
Chance							
Boyer							
Slater							
Ernster							
Good							
Lehninger							
Abron							
Papa							
Racker							
Skulachev							
Ramberg							
Witt							
Crofts							
Chapel							
Jagendorf							
Mitchell							

□ "Chemical theory" supporters   ■ "Chemiosmotic theory" supporters

Annual changes in "chemiosmotic theory" supporters

Paul Boyer (1918–)
Born and raised in Utah, Boyer studied chemistry at Brigham Young University, and in 1942 he earned a PhD in biochemistry at the University of Wisconsin. After engaging in wartime research on the stabilization of serum albumin, he taught at the University of Minnesota and studied enzymes. In 1963, he was appointed a professor of chemistry and biochemistry at the University of California, Los Angeles (UCLA). He continued his extensive studies of ATP synthase, which synthesizes ATP in the body, and proposed his "rotation theory", suggesting that synthesis is sustained by rotation of the enzyme. He was a recipient of the 1997 Nobel Prize in Chemistry.

John Walker (1941–)
Walker is a British biochemist who earned a degree at the University of Oxford and in 1982 became a professor at the MRC Laboratory of Molecular Biology in Cambridge. He worked on X-ray structural analysis of ATP synthase to verify Boyer's rotation theory. After 10 long years of research, he elucidated the mechanism of ATP synthesis. With Boyer, he was awarded the 1997 Nobel Prize in Chemistry.

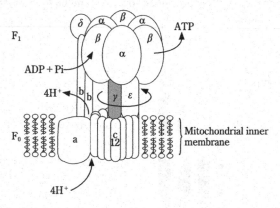

Figure 6.19: Schematic diagram of the structure of the mitochondrial membrane, F_1, F_0, ATP synthase

synthase is composed of an F_0 moiety, which is embedded into the membrane of mitochondria, and an F_1 moiety that extends into the matrix. In 1961, Racker succeeded in isolating the F_1 moiety, and showed that it is associated with the activity of ATPase. From the 1960s to the 1970s, Paul Boyer proposed a "binding change mechanism", claiming that rotation of the γ subunit in the center of F_1 alters the conformation (steric structure) of the α and β subunits, and that these conformational changes are associated with the synthesis of ATP (Fig. 6.19).[64]

Melvin Calvin (1911–1997)
Born in Minnesota, the child of Jewish immigrants from Russia, Calvin earned a degree studying the electron affinity of halogens at the University of Minnesota in 1935 before studying under Polanyi at the University of Manchester, where he developed an interest in photosynthesis. In 1937, he was invited by G. N. Lewis to join the staff at the University of California; from 1946 until 1980, he served as director of the biology and organics department of the Lawrence Radiation Laboratory and promoted research on photosynthesis. He was awarded the 1961 Nobel Prize in Chemistry for the discovery of the Calvin–Benson cycle, which relates to photosynthesis.

In this mechanism, the proton concentration gradient drives the rotation of the γ subunit, changing the conformations of the catalytic sites of the α and β subunits, and ADP is synthesized from ATP. In 1994, John Walker's group succeeded in the X-ray crystallographic structural analysis of the F_1 moiety,[65] and showed that Boyer's model was essentially correct. Boyer and Walker were awarded the Nobel Prize in Chemistry in 1997 for elucidating the enzymatic mechanism of ATP synthesis.

Photosynthesis

The study of photosynthesis is an important interdisciplinary area of research in chemistry, involving researchers in a wide range of fields, including many organic chemists, physical chemists, and biochemists. In the first half of the 20th century, photosynthesis was shown to comprise a light reaction that involves light, and a dark reaction that does not; in the former, light energy is used to synthesize ATP and NADPH, and in the latter. ATP and NADPH are used to synthesize water and carbohydrates from CO_2. (see section 4.10) Research after the war first elucidated the detailed mechanism of the dark reaction.

Dark reaction

Tracer experiments using radioactive isotopes had begun before the war, but when ^{14}C became more readily available after the war, M. Calvin's group at the University of California began research in 1946 to determine in detail how the radioactive label of $^{14}CO_2$ is incorporated into the series of photosynthetic reaction intermediates. First, they added $^{14}CO_2$ to a culture of the unicellular algae Chlorella for specified lengths of time under various light irradiation conditions, then killed the green algae to stop the photosynthesis reaction,

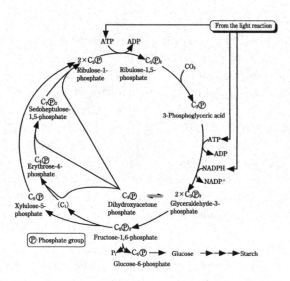

Figure 6.20: The Calvin–Benson cycle of photosynthesis

and studied the reaction pathway in detail. They used autoradiography and two-dimensional paper chromatography, which had just been developed at the time, to separate and identify the generated radioactive compounds. This work required enormous effort, but Calvin, Bassham, Benson and coworkers had achieved a good understanding of the dark reaction by 1953, and proposed a reaction pathway that today is called the Calvin-Benson cycle (Fig. 6.20). Experiments in which $^{14}CO_2$ was added to green algae and the algae were killed within 5 seconds resulted in the production and identification of the first stable compound produced in the reaction, 3-phosphoglyceric acid (3PG). In subsequent experiments, they observed that the concentrations of the photosynthesis intermediates remained constant when the green algae were exposed to sufficient $^{14}CO_2$ while being irradiated with light. However, they observed changes in the concentrations of the products when the $^{14}CO_2$ supply was cut off. CO_2 reactions with ribulose-1,5-bisphosphate (RuBP) were shown to produce two molecules of 3PG.[66] In the Calvin cycle, first ATP and NADPH form six molecules of glyceraldehyde-3-phosphate (GAP) from three molecules of RuBP and three molecules of CO_2, then one of these GAP molecules is used for biosynthesis while the remaining five GAP molecules pass through a cycle comprising C_3, C_4, C_5, C_6, and C_7 compounds to regenerate the RuBP starting material. RuBP carboxylase, the enzyme that catalyzes CO_2 fixation, accounts for 50% of leaf protein and is the most abundant protein in the living world.

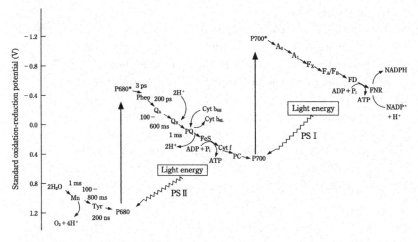

Figure 6.21: Z-scheme of photosynthetic reactions

A mechanism for the reaction catalyzed by this enzyme was proposed by Calvin. X-ray structural analysis showed that the enzyme has an $L_8 S_8$ structure composed of eight large subunits (L) and eight small subunits (S).

Melvin Calvin was awarded the Nobel Prize in Chemistry in 1961 for his research on photosynthesis in plants.

Light reaction and the initial process

The photochemical reaction in photosynthesis starts when light energy absorbed by antenna chlorophyll is moved from one chlorophyll molecule by excitation energy transfer and is trapped in the chlorophyll of the reaction center. In 1952, using purple photosynthetic bacteria, Duysens showed that chlorophyll is directly photo-oxidized during photosynthesis. The light reaction system of this bacterium is simpler than the light reaction systems of higher plants; therefore, purple photosynthetic bacteria were extensively used to study the initial processes of the reactions and the structures of the reaction centers.

Light reactions in plants take place inside the thylakoid membrane[11] in the chloroplasts, and include processes similar to electron transport and oxidative phosphorylation in mitochondria. In 1954, Arnon discovered a pathway for ATP synthesis that is dependent on light (photophosphorylation). In 1957, Emerson

---

11. Thylakoid membrane: A membranous structure arrayed in chloroplasts and cyanobacteria, where the photosynthetic reaction takes place.

Hartmut Michel (1948–)

Michel is a German biochemist who studied biochemistry at the University of Tübingen and in 1977 earned a PhD at the University of Würzburg. In 1979, he joined the Max Planck Institute of Biochemistry, and in 1987 became a deputy director of the Max Planck Institute for Biophysics. In 1982, he successfully extracted the membrane protein of photosynthetic bacteria and crystallized it, and along with  Deisenhofer and Huber elucidated the steric structure of the photosynthesis center by X-ray analysis. This achievement earned him, Deisenhofer and Huber the 1988 Nobel Prize in Chemistry.

Johann Deisenhofer (1943–)

Deisenhofer is a German biochemist who earned a degree at the Max Planck Institute in 1974, and between 1982 and 1985 determined the structure of the photosynthetic reaction center of photosynthetic bacteria together with Michel and Huber. In 1987 he moved to the Howard Hughes Medical Institute in the U.S. In 1988, he shared the Nobel Prize in Chemistry with Michel and Huber.

Robert Huber (1937–)

A German biochemist, Huber earned a degree at the Technical University of Munich before joining the Max Planck Institute of Biochemistry. Using X-ray diffraction, he, along with Michel and Deisenhofer elucidated the three-dimensional structure of the membrane protein of the photosynthesis center in photosynthetic bacteria and with them shared the 1988 Nobel Prize in Chemistry.

and coworkers discovered that the quantum yield of oxygen generated by the green alga Chlorella is increased significantly when the algae are irradiated simultaneously with yellow-green light and red monochromatic light with a long wavelength of at least 680 nm. They showed that the generation of oxygen by photosynthesis involves two different photochemical processes that are linked. Hill and Bendall in 1960 proposed a Z-mechanism in which a photochemical system I and a photochemical system II are arranged in series, and they showed an energy gradient of electrons flowing in the photochemical systems (Fig. 6.21). Photochemical system I (PSI) simultaneously produces a weak oxidizing agent and a strong reducing agent capable of reducing $NADP^+$, whereas photochemical system II (PSII) simultaneously produces a weak oxidizing agent and a strong oxidizing agent capable of oxidizing H_2O. PSI and PSII work at the same time,

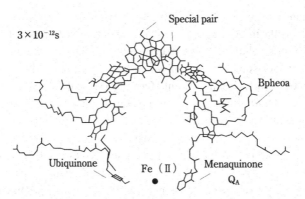

Figure 6.22: Structure of the reaction center in photosynthetic bacteria and the initial excitation process

Light excites the BChl*b* special pair to an excited state, then the excited electrons move to Bpheo*a* 3 picoseconds later; another 200 picoseconds later, they move to Q_A. After 100 microseconds, the electrons move to ubiquinone.

with NADP$^+$ being reduced with H_2O electrons to run photosynthesis. Detailed research conducted by many researchers from around 1960 onward revealed the detailed structures and mechanism of electron transport in photochemical systems. The electron transport system of the thylakoid membrane is composed of PSII, the cytochrome b_{6f} complex, and three kinds of protein complexes in PSI. Chlorophyll P680 (Chla) that has been photoexcited in the electron transport system PSII releases electrons, then an oxygen-generating complex comprising Mn compensates for this by extracting electrons from H_2O. The photoexcited P680 passes the electrons to Pheo by light-induced charge separation, and Pheo transports the electrons to the plastoquinone (Q) pool, converting the plastoquinone into plastoquinol. Plastoquinol reduces a cytochrome complex and then H$^+$ is pumped into the thylakoid membrane, thus reducing the plastocyanin. P700 in PSI also releases electrons by photoexcitation and becomes oxidized, but it is then reduced by plastocyanin and returned to its original state. The electrons released from P700 pass through a series of electron carriers before reducing NADP$^+$ to NADPH. ATP is produced by the action of ATP synthase through the proton concentration gradient created in this series of steps.

The study of the initial processes of photosynthesis, beginning around 1980, progressed rapidly due to advances in X-ray structural analysis, pico/femtosecond ultrafast spectroscopy, pulse ESR, and other biophysical techniques. The photosynthetic systems of purple photosynthetic bacteria were studied in the greatest detail. Hartmut Michel in 1982 succeeded in crystallizing the membrane protein of a reaction center of photosynthetic bacteria,[67] and in

1984 he, along with Johann Deisenhofer, Robert Huber and coworkers succeeded in determining the three-dimensional structure of the reaction center membrane protein at high resolution.[68] The structure showed that the reaction center is composed of four subunits, two of which have symmetrically-arranged dye molecules, with two molecules of Chl$a$ forming a special pair. The arrangement of Chl$a$, Pheo, Q, iron ions, and other moieties involved in the initial reaction became apparent, and it became possible to discuss the initial process in detail. Results from monitoring the electron states with ultrafast spectroscopy and ESR made it apparent that the special pair dimer creates a radical pair, $(Chla)_2^+$ +Pheo$^-$, by charge separation within 3 picoseconds of excitation, and that about 200 picoseconds later the electrons move to Q_A; 100 microseconds later, the special pair is returned to its original state (Fig. 6.22).

Deisenhofer, Huber, and Michel were together awarded the Nobel Prize in Chemistry in 1988. Their research is recognized as having not only contributed to major progress in photosynthesis research by showing the structures of the photosynthetic reaction centers, but also for paving the way to the crystallization and structural analysis of membrane proteins.

Signaling

To stay continuously active while maintaining harmony, a multi-cellular organism must be able to exchange information using chemical signals between its constituent cells. Hormones and other signaling molecules bind to cell surface receptors, then the receptors transmit signals to the cell interior. Pioneering research by Sutherland and colleagues showed that a signal (first messenger) used to transfer information between cells is converted at the cell membrane to cAMP. cAMP is a second messenger that transfers information within the cell. However, very little was known about the mechanism of intracellular signaling inside the cell. This question was addressed early in the second half of the 20th century.

Signaling mechanisms

From the end of the 1960s to the 1970s, Martin Rodbell and Alfred Gilman's group greatly expanded our understanding of this mechanism (Fig. 6.23). Rodbell discovered that signaling from the cell exterior to the cell interior requires moieties for mediating three different functions: signal identification, transduction, and amplification. They also showed that transduction is driven by guanosine triphosphate (GTP). Gilman's group isolated and purified the membrane protein (G protein) that binds to GTP and GDP (guanosinediphosphate), and showed that this protein is involved in transduction. G protein bound to GDP is inert, but when GDP is substituted with GTP, interactions with receptors stimulate G protein into an active state and it activates adenylate cyclase. G

Martin Rodbell (1925–1998)

An American biochemist, Rodbell graduated from Johns Hopkins University before earning a degree at the University of Washington in Seattle, and then later conducted research at the National Institutes of Health. In the 1960s, he proposed a theory that G-proteins act as a relay mechanism in the transduction of information from receptors to enzymes. He was awarded the Nobel Prize in Physiology or Medicine in 1994 together with Gilman.

Alfred Gilman (1941–)

Gilman is an America pharmacologist who graduated from Yale University and then earned his PhD at Case Western Reserve University; after appointments at the National Institutes of Health and the University of Virginia, he was appointed chairman of the Department of Pharmacology at the University of Texas Southwestern Medical Center at Dallas. He proved that G proteins are involved in signaling, and isolated and purified many different G proteins. He shared the Nobel Prize in Physiology or Medicine in 1994 with Rodbell.

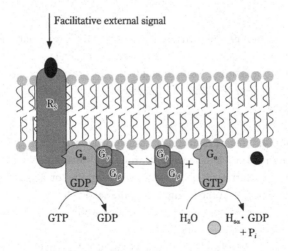

Figure 6.23: G protein structure and signaling mechanism

proteins are composed of three subunits, α, β, and γ (G_α, G_β, and G_γ), and the G_β and G_γ are bound to each other. G_α binds to either GDP or GTP. When G_α · GDP · G_β/G_γ binds to a receptor/hormone complex, the G_α replaces the GDP with GTP and dissociates from $G_\beta G_\gamma$, activating a signal amplification system. For example, in the metabolism of glycogen, adenylate cyclase generates cAMP with an amplifier. There are a number of different G proteins; each binds to a specific receptor and activates a specific amplification system. It became clear

Robert Lefkowitz (1943–)
Lefkowitz is an American biochemist who was born in New York and in 1966 earned his MD at Columbia University; after working in clinical medicine and conducting research at the National Institutes of Health (NIH), in 1973 he was appointed an associate professor and then in 1979 a professor at the Duke University Medical Center. In 2012, he and Kobilka were awarded the Nobel Prize in Chemistry for their pioneering work on G protein-coupled receptors.

Brian Kobilka (1955–)
An American biochemist, Kobilka graduated from the University of Minnesota, majoring in chemistry and biology before earning his MD at Yale University; after training as a physician, he was a post-doctoral fellow under Lefkowitz at Duke University. In 1989 he moved to Stanford University to be a professor of molecular and cellular physiology in the School of Medicine. In 2012, he was awarded the Nobel Prize in Chemistry for his study of G protein-coupled receptors, together with Lefkowitz.

that different stimuli produce different responses. G proteins play an important role in biochemistry and medicine, and were extensively studied in the final quarter of the 20th century. Rodbell and Gilman were awarded the 1994 Nobel Prize in Physiology or Medicine for the discovery of G proteins and their roles in cell signaling.

Structure and function of G protein-coupled receptors[69]
How can cells recognize stimuli that produce sights, smells, and tastes, or recognize transmitters such as adrenaline, histamine, or dopamine and then transmit their signals? At the beginning of the second half of the 20th century, it was known that receptors on the cell surface were critically important, but virtually nothing was known about them.

At the end of the 1960s, Robert Lefkowitz used radioactive isotopes of iodine to label various hormones such as adrenaline, then identified and purified the receptors. His group proposed a quaternary complex model to explain the activation of receptors (Fig. 6.24). His attention then turned to allosteric effects, where binding with adrenaline or an agonist[12] having a similar function strengthens the affinity between a receptor and G protein, and binding with a G protein strengthens binding between the receptor and agonist at the same time, thus activating

---

12. Agonist: Refers to a chemical that acts on receptor molecules in the body and functions similar to a hormone or neurotransmitter.

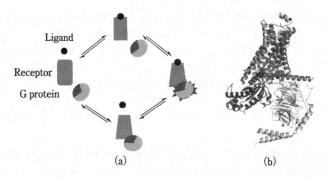

(a)   (b)

Figure 6.24: Structure and function of G protein-coupled receptors

(a) Thermodynamic cycle model of a quaternary complex composed of a ligand (●), receptor, and G protein.
(b) Structure of the quaternary complex based on X-ray structural analysis

the G protein. His group, together with Brian Kobilka and coworkers, analyzed the genes encoding adrenaline receptors and discovered that these receptors are similar to the receptors involved in vision. Furthermore, they showed that these receptors belong to a group of G protein-coupled receptors that all have similar structures and functions.

In 2011, Kobilka's group solved the X-ray structure of a receptor/G protein complex activated with a hormone and in the conformation required to transmit a signal, thereby revealing the details of structural changes that occur in receptors during activation.

Lefkowitz and Kobilka were awarded the Nobel Prize in Chemistry in 2012.

Neurotransmitters and nitric oxide (NO)

In 1921, the Austrian physiologist Otto Lewy was the first to show that a chemical is involved in neurotransmission at synapses; later, this chemical was identified to be acetylcholine [(CH_3COO CH_2 $CH_2N^+(CH_3)_3$]OH^-. Many neurotransmitters have been discovered since then, including amino acids, amines, and peptides, but in the 1980s the simple diatomic molecule NO was shown to be a signaling molecule, which attracted interest. In 1980, Furchgott discovered that acetylcholine dilates blood vessels by producing an unknown compound in the vascular endothelium; he named this compound endothelium-derived relaxing factor (EDRF). Around 1986, the research group of Robert Furchgott and Louis Ignarro spectroscopically identified EDRF as being NO. Ferid Murad discovered in 1987 that nitroglycerin and other nitrates release NO and activate guanylate cyclase in vascular smooth muscle, generating cyclic GMP and thereby triggering a relaxation reaction. Nitroglycerin was discovered by Nobel and had been used as an anti-anginal agent for more than 100 years; interestingly, Murad's studies revealed its mechanism of action and triggered a huge amount of research on NO. It became clear that NO is a signaling molecule important in many aspects of medicine. Furchgott, Ignarro and Murad were awarded the Nobel Prize in Physiology or Medicine in 1998 for the discovery of NO as a signaling molecule in the cardiovascular system.

Immunity and gene rearrangement

The elucidation of the mechanism of diversity expression in antibodies by Susumu Tonegawa was a major achievement in immunology and an important discovery showing the variability of genes. Humans and higher animals possess an immune defense system to defend themselves against intruders such as viruses and bacteria. This system includes B lymphocytes, which produce antibodies that can recognize and bind to diverse invading pathogens (antigens). The number of antigens is very large, but an antibody can be generated for every antigen. The humans genome comprises roughly 30,000 genes, but more than ten billion types of antibodies can be made. Tonegawa shed light on how this is possible by studying the mechanism by which mouse B lymphocytes make protein antibodies called immunoglobulins.

Immunoglobulins have a Y-shape and are composed of polypeptide chains called the L chains and H chains, as illustrated in Fig. 6.25. The tip of the Y-shape constitutes a moiety that recognizes the antigen and binds to it. This moiety is the variable section (V region) and varies depending on the antibody, whereas the structures of the other moieties remain constant. The V-region of the H chains is divided into three sections (V, D, and J); the specific combination of genes

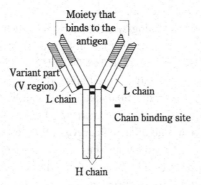

Figure 6.25: Schematic diagram of the structure of immunoglobulins

The inside Y-shape moieties are the H chains. The outside short moieties are the L chains. The hatched portions are the variable V regions.

Susumu Tonegawa (1939–)

A 1963 graduate of the Kyoto University Faculty of Science, Tonegawa received his Ph.D. from the University of California, San Diego in 1968. He was a post-doctoral researcher at the Salk Institute, then was a researcher at the Basel Institute for Immunology in Basel, Switzerland from 1971 to 1981. There, he elucidated the genetic-level mechanism responsible for the diversity of antibody molecules, which earned him a Nobel Prize. Later, he became a professor at The Howard Hughes Medical Institute in the Massachusetts Institute of Technology, and then turned his attention to the study of brain science and neuroscience. In 2009, he became the director of the RIKEN Brain Science Institute. He was awarded the 1987 Nobel Prize in Physiology or Medicine.

designating the amino acid sequence of each of these three sections determines the amino acid sequence of the V-region overall. The DNA of the hematopoietic cells that make B cells have about 100 gene candidates that designate these sections for V, 20 for D, and four for J. Selecting genes one at a time from these candidates generates approximately 8,000 different types of antibody molecules. The V regions of the L chains are likewise very diverse; by including the diversity produced from displacement of the connecting parts of the sections, an enormous number of diverse antibody molecules can be produced by B cells. A series of studies by Tonegawa and colleagues, begun in 1976, showed the mechanism by which genes are rearranged to create diverse antibodies.[70] This research not only elucidated the mechanism underlying immunity, but also was very important in terms of understanding gene rearrangement.

Reference

1. For example, D. Voet and J. Voet, *"Biochemistry"* *3rd ed.* John Wiley , 2004
2. J. Fruton, *"Molecules and Life Historical Essays on the Interplay of Chemistry and Life"* John Wiley and Sons, 1972
3. J. Fruton, *"Proteins, Enzymes, Genes"* Yale Univ. Press, New Heaven, 1999
4. K. Maruyama, *"Seikagaku wo tsukutta Hitobito"* (*People who made biochemistry*) Shokabo, 2001
5. M. Watanabe, *"DNA no Nazo ni idomu"* (*Challenge to the riddle of DNA*) Asahi Sensho, 1998
6. A. Ohno, *"137 Okunen no Monogatari"* (*Stories of 13.7 billion years*) Sankyo Syuppan, 2006

7. L. K. James "*Nobel Laureates in Chemistry 1901–1992*" Amer. Chem. Soc./ Chem. Her. Found. 1994

8. J. D. Watson with A. Berry, "*DNA: The secret of life*" Random House, New York, 2003

9. J. D. Watson, "*The Double Helix*" The New American Library, 1968

10. F. Crick, "*What Mad Pursuit: Personal View of Scientific Discovery*" Basic Books, New York, 1988

11. M. Wilkins, "*The Third Man of the Double Helix*" Oxford Univ. Press, 2003

12. B. Dixon, "*Rosalind Franklin: The Dark Lady of DNA*" Harper Collins, 2002

13. F. Griffith, *J. of Hygience,* **27**, 113 (1928)

14. O. T. Avery, C. M. Macleod, M. McCarty, *J. Experimental Medicine*, **79**, 137 (1944)

15. E. Chargaff, *Experientia*, **6**, 201 (1950)

16. A. D. Hershey, M. Chase, *J. Gen. Physiol.*, **36**, 39 (1952)

17. E. Schrödinger, "What is life ? The Physical Aspect of the Living Cell" Cambridge Univ. Press, 1944

18. L. Pauling, R. B. Corey, *J. Am. Chem. Soc.*, 72, 5349 (1950)

19. M. Perutz, "*I wish I had made you mad earlier*" Cold Spring Harbor Pres, 2002

20. W. T. Astbury, F. O. Bell, *Nature*, **141**, 747 (1938)

21. J. D. Watson, F. H. C. Crick, *Nature*, **171**, 737 (1953)

22. M. H. F. Wilkins, A. R. Stokes, H. R. Wilson, *Nature*, **171**, 738 (1953)

23. R. E. Franklin, R. G. Gosling, *Nature*, **171**, 740 (1953)

24. T. Hager, "*Force of Nature: The Life of Linus Pauling*" Simon & Schuster, 1995

25. M. Meselson, F. W. Stahl, *Proc. Natl. Acad. Sci.*, **44**, 671 (1958)

26. A. Kornberg, *Science*, **131**, 1503 (1960)

27. F. Sanger, *Adv. Protein Chem.*, **7**, 1 (1952), Nobel lecture 1958, Nobelprize. org

28. M. Perutz, *Nature*, **228**, 728 (1970)

29. J. Monod, J. P. Changeux, F. Jacob, *J. Mol. Bio.*, **6**, 306 (1963)

30. C. B. Anfinsen, *Science*, **181**, 223, (1973)

31. G. Gamow, *Nature*, **173**, 318 (1954)

32. F. H. C. Crick, *Symp. Soc. Exp. Biol.*, **12**, 138 (1958)

33. R. W. Holley, J. Apgar, G. A. Everett, J. T. Madison, M. Marquisse, S. H. Merrill, J. R. Penswick, A. Zamir, *Science*, **147**, 1462 (1965)

34. M. Nirenberg, J. H. Matthaei, *Proc. Natl. Acad. Sci.*, **47**, 1588 (1961)

35. F. Jacob, J. Monod, *J. Mol. Biol.*, **3**, 318 (1961)

36. J. Monod, *"Chance and Necessity: An Essay on the Natural Philosophy of Modern Biology"* Alfred Knopf, New York, 1971

37. F. Jacob, *"The Statue Within"*, Basic Books, 1995

38. R. Okazaki, T. Okazaki, K. Sakabe, K. Sugimoto, A. Sugino, *Proc. Natl. Acad. Sci.*, **59**, 598 (1968)

39. K. B. Mullis, *Sci. Am.* **262**, 56 (1990)

40. P. Rabinow, *"Making of PCR"* Univ. of Chicago Press, Chicago & London, 1996

41. K. Mullis, *"Dancing in the Mind Field"*, Vintage, 2001

42. J. W. Szostak, E. H. Blackburn, *Cell*, **29**, 245 (1982)

43. G. W. Greider, E. H. Blackburn, *Cell*, **43**, 405 (1985)

44. A. D. Jackson, R. H. Symons, P. Berg, *Proc. Natl. Acad. Sci. USA*, **69**, 2904 (1972)

45. W. Gilbert, A. Maxam, *Proc. Natl. Acad. Sci. USA*, **70**, 3581 (1973)

46. F. Sanger, S. Nicklen, A. R. Coulson, *Proc. Natl. Acad. Sci. USA*, **74**, 5463 (1977)

47. a) P. Cramer, D. A. Bushnell, R. Kornberg, *Science*, **292**, 1863 (2001)
    b) A. L. Gnatt, P. Cramer, J. Fu, D. A. Bushnell, R. D. Kornberg, *Science*, **292**, 1876 (2001)

48. A. Yonath, J. Mussig, B. Tesche, S. Lorenz, V. A. Erdmann, H. G. Wittmann, *Biochem. Int.* **1**, 428 (1980), A. Yonath, H. D. Bartunik, K. S. Bartels, H.G. Wittmann, *J. Mol. Biol.*, **177**, 201 (1984)

49. N. Ban, P. Nissen, J. Hansen, M. S. Capel, R. Sweet, P. B. Moore, T. A. Steitz, *Science*, **289**, 905 (2000)

50. W. A. Clemons, J. L. May, B. T. Wimberly, J. P. McCutcheon, M. S. Capel, V. Ramakrishnan, *Nature*, **400**, 833 (1999)

51. A. Ciechanover, Y. Hod, A. Hershko, *Biochem. Biophys. Res. Commun.*, **81**, 1108 (1978)

52. A. Hershko, A. Ciechanover, I. A. Rose, *Proc. Natl. Acad. Sci. USA*, **76**, 3107 (1979)

53. A. Ciechanover, H. Heller, S. Elias, A. L. Haas, A. Hershko, *Proc. Natl. Acad. Sci. USA*, **77**, 1365 (1980)

54. A. J. Zaug, T. R. Cech, *Nucleic Acids Res.*, **10**, 2823 (1982)

55. C. Gurrier-Takada, K. Gardiner, T. Marsh, N. Pace, S. Altman, *Cell*, **35**, 849 (1983)

56. W. Gilbret, *Nature*, **319**, 618 (1986)

57. C. C. F. Blake, L. N. Johnson, G. A. Mair, A. C. T. North, D. C. Philips, V. R. Sarma, *Proc. R. Soc, London*, Ser. B, **167**, 378 (1967)

58. L. F. Leloir, C. E. Cardini, *J. Am. Chem. Soc.*, **79**, 6340 (1957)

59. S. J. Singer, G. L. Nicolson, *Science*, **175**, 720 (1972)
60. J. C. Skou, *Biochim. Biophys. Acta.*, **20**, 394 (1957)
61. P. Doyle, J. Cabral, R. Pfuetzner, A. Kuo, J. Gulbis, S. Cohen, B. Chait, R. Mackinon, *Science*, **280**, 67 (1998)
62. G. M. Preston, T. Caroll, W. B. Guggino, P. Agre, *Science*, **256**, 358 (1992)
63. P. Mitchell, *Nature*, **191**, 144 (1961)
64. P. D. Boyer, R. L. Cross, W. Momsen, *Proc. Natl. Acad. Sci. USA*, **70**, 2837 (1973), P. D. Boyer, Nobel Lecture, 1997, Noberprize.org.
65. J. P. Abrahams, A. G. Leslie, W. Lutter, J. E. Walker, *Nature*, **370**, 621 (1994)
66. M. Calvin, J. Chem. Soc. (1956) 1895, *Science*, **135**, 879 (1962)
67. H. Michel, *J. Mol. Biol.*, **158**, 562 (1982)
68. J. Deisenhofer, O. Epp, K. Miki, R. Huber, H. Michel, *J. Mol. Biol.*, **180**, 385 (1984)
69. Nobel Prize in Chemistry 2012—*Advanced information*, Nobelprize.org
70. N. Hozumi, S. Tonegawa, *Proc. Natl. Acad. Sci. U. S. A.*, **73**, 3628 (1976)

Chronology of modern and contemporary chemistry (second half of the 20th century)

Year	Physics, applied physics, technology	Observation and analysis, reaction, theory and computation	New materials, synthesis, function and physical properties, environment	Biochemistry, molecular biology
1945	Neutron diffraction (Wollan, 1945) Nuclear magnetic resonance (Bloch, Purcell, 1946) Holography theory (Gabor, 1948) Invention of the transistor (Bardeen, Brattain, Shockley, 1948) Ferrimagnetism (Neel, 1948) Computer EDSAC (Cambridge University, 1949) Renormalization theory (Tomonaga, 1949)	^{14}C dating (Libby, 1946) Flash photolysis (Porter, Norrish, 1947) Excitation energy transfer (Förster, 1948) Polymer solution theory (Flory, 1948) Gas chromatography (Martin, James, 1949)	Promethium (Marinsky, Glendenin, Coryell 1945) Synthesis of ATP (Todd, 1948)	Discovery of coenzyme CoA (Lipmann, Kaplan, 1945)
1950	Invention of the maser (Townes, 1953) Quadrupole mass spectrometer (Paul, Steinwedel, 1953) Continuous oscillation maser (Basov, Prokhorov, 1955) Theory of superconductivity (Bardeen, Cooper, Schrieffer, 1957) Mossbauer effect (Mössbauer, 1957) Theory of irreversible processes (Kubo, 1957) Tunnel effect in semiconductors (Esaki, 1958)	Molecular HF formula (Roothaan, 1951) Frontier orbital theory (Fukui, 1952) Electron transfer reactions in metal complexes (Taube, 1952) Charge transfer spectrum (Mulliken, 1952), Chemical relaxation method (Eigen, 1953) Reaction research by crossed molecular beam (Datz, Taylor, 1954) Structural determination of vitamin B_{12} (Hodgkin, 1954) Theory of electron transfer reaction (Marcus, 1956) X-ray photoelectron spectroscopy (Siegbahn, 1957) MD calculation of hard spheres (Alder, Wainwright, 1957) Reaction research by chemiluminescence (Polanyi, 1958)	Synthesis of cholesterol (Woodward, 1951) Ferrocene (Wilkinson, Woodward, 1952) Generation of amino acids from inorganic matter (Miller, Urey, 1953) Catalytic polymerization of olefins (Ziegler, Natta, 1953) Wittig reaction (Wittig,1954) Discovery of the organic conductor (Akamatsu, Inokuchi, Matsunaga, 1954) Hydroboration reaction (Brown 1956)	Chargaff rule of nucleic acid (Chargaff, 1950) α-Helix structure of proteins (Pauling, 1950) Experiments showing that DNA is the carrier of genetic information (Hershey, Chase 1952) Double helix structure of DNA (Watson, Crick, 1953) X-ray analysis of the DNA double helix structure (Franklin, 1953) Enzymatic synthesis of RNA (Ochoa, 1955) Primary structure of insulin (Sanger, 1955) Fixation pathway of CO_2 in photosynthesis (Calvin, Benson, 1957) Discovery of cyclic AMP (Sutherland, 1957)

	Col A	Col B	Col C	Col D
	Mechanism of glycogen biosynthesis (Leloir, 1957) Na⁺, K⁺ transport ATPase (Skou, 1957) Enzymatic synthesis of DNA (Kornberg, 1958)	Structure determination of myoglobin (Kendrew, 1958)		Suggestion of the possibility of the laser (Townes, Schawlow, 1958) Integrated circuit (Kilby, 1958)
1960	Operon theory of gene regulation (Monod, Jacob, 1961) Chemiosmotic hypothesis (Mitchell, 1961) Deciphering of the genetic code (Nirenberg, 1961) Allosteric effect (Monod, Changeux, Lwoff, 1963) Sequencing of tRNA (Holley, 1965) Discovery of restriction enzymes (Arber, 1968)	Synthesis of chlorophyll (Woodward, 1960) Solid-phase polymerization method (Merrifield, 1962) Noble gas compound (Bartlett, 1962) Woodward–Hoffmann rules (Woodward, Hoffmann, 1965) The discovery of crown ethers (Pedersen, 1967) Asymmetric synthesis reactions (Knowles, 1968)	Discovery of green fluorescent protein (GFP) (Shinomura, 1962) Fourier transform NMR (Ernst, 1966) Density functional theory (Kohn, Sham 1968) Structure of hemoglobin (Perutz 1968) ab initio MO calculation (Pople, 1969) Identification of carbocation (Olah, 1969)	Invention of the laser (Maiman, 1960) Spontaneous symmetry breaking (Nambu, 1960) Tunneling effect in superconductors (Josephson, 1962) Quantum theory of optical coherence (Glauber, 1963) Quark theory (Gell-Mann et al., 1964) Fiber optic communication (Kao, Hockham, 1966) Amorphous electron theory (Mott, 1969)
1970	Fluid mosaic model of biological membranes (Singer, Nicholson, 1972) Gene recombination technology (Berg, 1972)	Mechanism of metathesis reaction (Chauvin, 1971) Prostaglandin synthesis (Corey, 1971) Heck reaction (Heck, 1972) Total synthesis of Vitamin B_{12} (Woodward, Eschenmoser, 1972) Metallic CT complexes (Ferraris,1973) Destruction of the ozone layer by fluorocarbons (Molina, Rowland, 1974)	Honda-Fujishima effect (Honda, Fujishima, 1972) NMR imaging (Lauterbur, 1973) Surface-enhanced Raman (Fleischmann, 1974)	Invention of CCD sensors (Boyle, Smith, 1970) ³He superfluid (Lee, Richardson, Osheroff, 1972) Third-generation quark and lepton prediction (Kobayashi, Masukawa, 1973)

Year	Physics, applied physics, technology	Observation and analysis, reaction, theory and computation	New materials, synthesis, function and physical properties, environment	Biochemistry, molecular biology
1970		Two-dimensional NMR (Ernst, 1976)	Negishi coupling reaction (Negishi, 1976) Conductive polymers (Shirakawa, Heeger, MacDiarmid, 1977) Suzuki coupling reaction (Suzuki, Miyaura, 1979)	Structure of antibody genes (Tonegawa, 1976) Sequencing of nucleic acid (Gilbert, Maxam, Sanger, 1977) Interrupted genes (Roberts, Sharp, 1977)
1980	Discovery of the quantum Hall effect (von Klitzing, 1980) Invention of the scanning tunneling microscope (Binnig, Rohrer, 1981) Discovery of high-temperature superconductors (Bednorz, Müller, 1985) Invention of the atomic force microscope (Binnig et al., 1986) Giant magnetoresistance (Grünberg, Fert, 1987)	Electrospray ionization (Fenn et al., 1984) MALDI method (Hillenkamp et al., 1985) Protein analysis by MALDI (Tanaka et al., 1987) Femtosecond spectroscopy (Zewail, 1987) Single-molecule spectroscopy (Moerner, 1989)	Development of asymmetric hydrogenation reaction (Noyori, 1980) Asymmetric oxygenation reaction (Sharpless, Katsuki, 1980) Discovery of quasi-crystals (Shechtman, 1984) Discovery of fullerenes (Curl, Kroto, Smalley, 1985)	Structure and function of telomeres (Blackburn, Szostak, 1980) Catalytic function of RNA (Cech, 1982) Development of PCR (Mullis, 1985) Structural analysis of photosynthetic reaction centers (Deisenhofer, Huber, Michel, 1985)
1990	Achievement of Bose-Einstein condensate (Cornell, Wieman, 1995)		Carbon nanotubes (Iijima, 1991)	

Chapter 7: Chemistry from the 20th century into the future

The logo of the International Union of Pure and Applied Chemistry (IUPAC)

The science of chemistry experienced major progress in the 20th century. What future can we expect for this discipline? In this chapter, based on our knowledge of 20th-century chemical history, we look forward to the chemistry of the 21st century; after all, although none of us is capable of *predicting* the future, to *ponder* the future is a task both interesting and important. We first survey the evolution of 20th-century chemistry, and the background underlying some of its most original research, through the prism of the Nobel Prize. We next study how the field of chemistry has changed from the late 20th through the early 21st centuries, and use this as input for speculating on the future directions of the field.

7.1 20th century chemistry and the Nobel Prize

The Nobel Prize is the most prestigious award in the world of science, and one which attracts the fascination of society at large. Because the prize was first awarded in 1901, the achievements it recognizes span the full range of the history of chemistry from the late 19th century through to the chemistry of the present day. For these reasons, it serves as a valuable milestone for tracking the evolving and transforming status of the field throughout the 20th century. Indeed, a glance at the roster of laureates over the past 114 years, and the accomplishments for which they were recognized, allows us to discern a significant variety of historical trends. Still, this pursuit is not without its shortcomings. In this section we first review the unique character of the Nobel Prize—including some of its problematic aspects—and then ask how the history and transformation of modern chemistry is reflected in the history of the Prize.

Figure 7.1: The Nobel Prize medal (for physics and chemistry)

The front side has an image of Alfred Nobel, while the back side depicts the goddess of science lifting a veil from the goddess of nature to reveal her face.

Finally, we ask how the most important discoveries in chemistry—and those which were to have the most profound influence on the development of the field—came to be made.

20th-century chemistry through the lens of the Nobel Prize in chemistry
The unique character of the Nobel Prize—and its problematic aspects[1,2]

The selection of Nobel Prize recipients over the 114-year stretch up to 2014— although not entirely without missteps, including cases in which the Prize was awarded for work later determined to be in error, or choices that now appear dubious for other reasons—has nonetheless been mostly judicious and appropriate overall, and this fact must surely be understood as the primary reason for the great prestige that the Prize has come to enjoy. Still, the Nobel Prize is not without problems as a benchmark for assessing the progress of science. Because the Prize is not awarded posthumously, one frequently encounters cases in which scientists responsible for extraordinary achievements are excluded from consideration. Particularly unfortunate examples with deep ties to chemistry include Moseley, who made major contributions to elucidating atomic structure; Franklin, who played a decisive role in determining the structure of DNA; and Avery, who first showed the importance of DNA in genetics. In all of these cases, the scientist in question passed away before the accomplishments came to be widely appreciated, thus forfeiting the opportunity to be awarded the Prize. Similarly, during the roughly 50 years that elapsed between the initial development of the electron microscope and the awarding of the Prize for that work to Ruska, his collaborator Knoll had passed way and was thus denied appropriate recognition.

Moreover, because the Prize may not be awarded to more than three recipients in any given year, one frequently finds cases in which a fourth deserving candidate missed a chance at eternal glory by a small margin. In cases in which a year's Prize is awarded to just a single recipient, one sometimes wonders if the chosen laureate's achievements really outshine so prominently those of other candidates. In the case of collaborative research, it is often difficult to apportion proper credit between a project's research leaders and the students and assistants who actually carried out the groundwork. Moreover, it is not easy to compare the relative importance of one subfield to that of another. The assessment of accomplishments invariably involves the preferences and subjective opinions of the selection committee; in this, as in any selection process carried out by a group of human beings, it is impossible to eradicate entirely the taint of mistakes, bias, or oversights. There are also cases in which the selection may be seen as unfair. For example, Mendeleev was not selected because of the opposition of an influential committee member. Another example is the Prize recognizing the discovery of nuclear disintegration, which was awarded to Hahn alone; many observers consider it unfair that Meitner was not named a co-recipient. In recent years the selection process has been carried out meticulously and with a great deal of effort, but still it cannot be perfect, as witness the frequent grumbling and dissent with which the decisions of the selection committee continue to be met to this day. In the present day of specialization, it may be particularly difficult for a committee with limited members to make proper decisions.

Because the Prize tends to emphasize initial discoveries and inventions, contributions that offer longstanding cumulative value tend to be overlooked. For example, G. N. Lewis, who made numerous contributions spanning a broad swath of physical chemistry, was a candidate for the Prize numerous times but never actually received it. Thus the roster of laureates omits even researchers who made major contributions to the advancement of their discipline. Nonetheless, even after accounting for all of these problematic aspects, a survey of the roster of Nobel Laureates and their accomplishments over the past 114 years gives a good sense of how the science of chemistry evolved and progressed over this interval. An Appendix at the end of this book lists all winners of the Nobel Prize in Chemistry, and the achievements for which they were recognized, between the years 1901 and 2014.

Changes in the fields for which the Prize was awarded
Table 7.1 presents a rough breakdown of the subfields of chemistry for which the Nobel Prize was awarded during the first and second halves of its 114-

year history. In the first 57 years, the Chemistry prize was roughly equally distributed between the areas of (1) physical chemistry, (2) organic chemistry, and (3) inorganic, analytical, and radiation chemistry. The large number of Prizes awarded for work related to radioactivity (see page 280) testifies to the importance of radioactivity and the major impact it had on the chemistry of the era; indeed, in the first half of the 20th century, radioactivity was a subject at the cutting edge of research in both physics and chemistry. Table 7.1 also indicates a relatively large number of Prizes awarded for work in physical chemistry, and particularly chemical thermodynamics, demonstrating not only the importance of this subject as a cornerstone of chemistry, but also the attractiveness of this new discipline as its development was stimulated by the incorporation of contemporaneous progress in physics. In the field of organic chemistry, we see many Prizes awarded for work related to the chemistry of living organisms, illustrating that the challenges of synthesizing and analyzing the structure of complicated organic compounds was a topic at the frontiers of chemistry research during this era. This field also played an essential role in laying foundations for the future rapid progress of biochemistry. However, relatively few Nobel Prizes in Chemistry were awarded for biochemical research during this era; those that were primarily recognized research on enzymes. The other important biochemical research of the day—including, for example, research on metabolism and reduction-oxidation reactions inside living organisms—tended instead to be recognized by the Prize in Physiology or Medicine. This fact undoubtedly reflects the increasing awareness of the importance of biochemistry in medicine, but may also indicate that the importance of biochemistry in chemistry was not yet fully appreciated.

In contrast, the situation in the second 57 years of the Prize is markedly different. One particular trend that stands out from Table 7.1 is the surge in the number of Prizes awarded for research in biochemistry and molecular biology; selections in these fields have been especially numerous in the past 30 years. To some extent this development is only a natural consequence of the explosive progress of the life sciences in the second half of the 20th century, but—in this author's opinion—it also reflects the increasing willingness of the Nobel Prize selection committee to adopt a broad-minded perspective in which molecular biology is understood as a subfield of chemistry. Groundbreaking achievements in biochemistry and molecular biology may be recognized either by the Prize in Chemistry or by the Prize in Physiology or Medicine, and the distinction between the two remains somewhat unclear. For example, in 1962, the Nobel Prize in Physiology or Medicine was awarded for the elucidation of the structure of DNA, while the Nobel Prize in Chemistry was awarded for the elucidation of the structure of proteins; it would not have been any less reasonable for the DNA work

Table 7.1: Breakdown of Nobel Prizes in Chemistry by Subfield

	1901 TO 1957	1957 TO 2014
Physical chemistry	14	37
Organic chemistry	19	25
Inorganic, Analytical, Radiation chemistry	16	8
Biochemistry (including molecular biology)	6	40
Other	3	2

*In cases where the work recognized by a Prize spans two or more subfields, we have counted the Prize in both areas.

to have been awarded the Chemistry Prize. Indeed, many achievements related to DNA and RNA which would typically be considered research in molecular biology—including the determination of DNA base sequences, the recombination of genes, the development of the PCR method, and the discovery of the catalytic effects of RNA—were later recognized by Nobel Prizes in Chemistry.

One also finds a relatively large number of Prizes awarded for research in physical chemistry, including theoretical chemistry. This is one measure of the major impact exerted on the wider field of chemistry by progress in the subfield of physical chemistry, aided both by the incorporation of new results from physics and by dramatic advances in measurement techniques and computer power. A significant number of Prizes have been awarded in the area of organic chemistry, including organometallic chemistry; awards for synthetic organic chemistry have been particular noteworthy in past 10 years. This development reflects the reality that the synthesis of materials is one of the central problems of chemistry—not only for its importance to the academic discipline itself, but also for its direct impact, through industrial applications, on the daily lives of humans around the world—and is also in keeping with recent trends in which the Nobel Prizes in Chemistry and Physics have begun to recognize research related to practical applications. Nobel's own bequest will states that the Prize is to be awarded to those who "have conferred the greatest benefit to mankind," so the notion of affording due consideration to applied research surely comports with Nobel's own wishes. Another recent trend is the awarding of the Chemistry Prize for interdisciplinary research that straddles multiple subfields—such as physical chemistry and biochemistry, analytical chemistry and biochemistry, or physical chemistry and organic chemistry—or that is difficult to classify given the traditional boundaries between subfields. In the opinion of this author, this phenomenon merely reflects the reality of present-day chemistry, in which new types of research that transcend traditional boundaries between physics, chemistry, and biology are constantly emerging.

The nationalities of Prize recipients

Table 7.2 presents a breakdown of prize recipients by nationality. In the first 57 years of the Prize, German recipients were by far the most numerous, but one finds also a fair number of recipients from Britain and the U.S. Indeed, in Britain—the birthplace of such luminaries as Newton and Darwin—the science of chemistry exhibited a certain independence and creativity during this time, while the U.S. had already achieved an impressive level of competence by the start of World War II. Other European nations—including the small countries of Switzerland and Sweden—are also healthily represented in this tally.

In the past 57 years of the Prize's history, the overwhelming majority of Prizes have been awarded to American chemists, reflecting the strength of American chemistry since World War II—a situation attributable not only to American economic superiority but also to the many outstanding universities in the U.S. and the nation's ability to attract extremely talented individuals from Europe and Asia. Among European nations, Britain boasts a number of Prize recipients that is disproportionately large in view of its population, although this trend may be somewhat diminished in recent years. Will British chemistry continue to produce innovative research in the 21st century? The decline in Prizes awarded to German chemists is readily discernible, reflecting both the nation's defeat in World War II and the Nazi persecution of Jewish scholars. The decline of France is also striking. On the other hand, the nation of Israel has recently boasted an impressive number of prize recipients for such a small country.

As for Nobel Prizes awarded to Japanese chemists, the long dry spell that followed the first such award—to Fukui, in 1981—was eventually broken by a string of recent successes that has served rapidly to increase the tally: Shirakawa in 2000, Noyori in 2001, Tanaka in 2002, Shimomura in 2009, and Suzuki and Negishi in 2011. Indeed, in the second 57 years of the Prize's history we find Japan pulling even with Germany, reflecting the impressive advances in Japanese chemistry made possible by postwar economic growth. Setting aside the U.S., this observation also testifies to the extent to which Japanese chemistry has risen to a level where it rivals the most advanced research conducted anywhere in the world. Of course, these numbers are not necessarily cause for unrestrained celebration. We must keep in mind that the work for which Shimomura and Negishi were recognized was carried out in the U.S. Moreover, Tanaka and Shimomura were almost unknown in the Japanese chemistry community before they received the Prize, testifying both to the unpredictable nature of important discoveries and to recent changes in the character of the science of chemistry. The Prizes awarded to Noyori, Suzuki, and Negishi recognized work in the area of synthetic organic chemistry, a field in which the level of Japanese research has

Table 7.2: Breakdown of Nobel Chemistry Prize recipients by nationality (only nations with two or more recipients are listed)

	1901 TO 1957	1958 TO 2014	TOTAL
U.S.	10	67	67
Germany	20	8	28
Britain	10	14	24
France	6	2	8
Japan	0	7	7
Switzerland	3	3	6
Israel	–	4	4
Sweden	3	0	3
Holland	2	0	2
Canada	0	2	2

traditionally been high. The Prize awarded to Shirakawa also recognized work in the synthesis of materials. In these Prize awards one sees hints that Japan's longstanding cultural tradition of "making things" seems to have seeped into the nation's scientific pursuits as well.

Serendipity[1] and the Nobel Prize

It is fascinating to ask precisely *how* Nobel-worthy research comes to be conducted. It is often noted[3] that many of the most important discoveries in the history of science were unexpected—arising instead from the sort of accidental coincidences described by the notion of *serendipity*—but to what extent is this really true? After all, one also finds many cases in which a topic thought to be important was chosen at the start of a research project and was then duly pursued, leading to significant results. Still, there are certainly many cases in which an accidental discovery led to the opening of a new field of research. Let us consider some particular examples of this type involving Nobel-Prize-winning research conducted in the late 20th century.

First, the discovery of ferrocene, which marked the beginning of the development of organometallic chemistry, could never have been predicted in advance. The same is true of Pedersen's discovery of crown ether, which spurred

---

1. *Serendipity:* A term coined by the English novelist Walpole taken from a fairy tale titled *The Three Princes of Serendip.* It refers to situations in which, in the course of searching for one item, one coincidentally happens to find something else of value. In the natural sciences it describes cases in which interesting and/or valuable discoveries are stumbled upon by accident.

the growth of supramolecular chemistry. Indeed, crown ether was discovered by accident as the byproduct of an attempt to make pentadentate ligands. Similarly, fullerenes such as C_{60} were identified in the course of research initially designed to search for long-chain carbon molecules in interstellar space. The discovery of ceramic materials that exhibit high-temperature superconductivity is yet another example of a fortuitous accidental breakthrough. A similar anecdote, perhaps of slightly different character, lies behind Shirakawa's discovery of conducting polymers: the new finding was made possible when a research student, attempting to conduct an experiment involving fabrication of thin polyacetylene films, prepared a catalyst whose concentration was incorrect by a factor of 1000. Thus the keys to major discoveries are frequently found lurking in unexpected places. In particular, in fields of chemistry that work with complicated materials, it seems reasonable to believe that many potential discoveries—the likes of which we cannot possibly imagine—remain to be uncovered.

In the fields of biochemistry and molecular biology, in which the central objects of study are the complex polymers that constitute living organisms, it goes without saying that many major discoveries have come about in an unexpected fashion. Indeed, the majority of the most important discoveries made at the early stages of research on DNA and RNA were entirely unforeseen, while many more recent discoveries—including the catalytic properties of RNA, and aquaporin, the protein which forms water channels in membranes—were unanticipated as well. In fields that are still developing, as well as in fields that involve complicated systems such as living organisms, it is only natural to expect many opportunities for unexpected discoveries. In fields such as medicine that treat complicated phenomena, it is probably safe to say that *most* important discoveries happen by chance; it has even been suggested that large-scale projects convened under the banner of grandiose goals such as "eradicating cancer" are utterly ineffectual at birthing innovative and original research results[3].

Even aside from the major discoveries of the examples above, surely any research chemist can tell stories of coming across an unanticipated result in the process of research—and being thus inspired to follow new directions in their research. However, although there is no question that good fortune plays a significant role in the process of chance discovery, it is equally important for chemists to have the ability to *notice* these fortuitous opportunities when they arise—to grab hold of them without overlooking them or letting them escape undetected—and to be equipped with the skills needed to develop them into new research projects; surely *this* preparation lies at the heart of scientific serendipity. After all, in the words of Pasteur himself, "chance favors only the prepared mind."

The importance of basic research

History is replete with examples in which the results of basic research, conducted solely for the purpose of satisfying intellectual curiosity, resulted in unexpected developments that yielded major progress, including on the applied side. Shimomura's discovery of green fluorescent protein (GFP) is a good example of this phenomenon. Nobody could have anticipated that GFP—which was discovered in the course of research begun to answer the question of how the jellyfish, *Aequorea victoria*, glows in the dark—would later turn out to be an important research tool in biology and medicine. Similarly, nuclear magnetic resonance techniques, which were initiated as a means of studying the magnetic properties of atomic nuclei, went on to become essential tools in physics, chemistry, and biology, and today have developed into the MRI methods that play an important role in medicine. The discovery of the laser arose out of basic research on stimulated emission from atoms and molecules, but today the use of lasers is ubiquitous across a broad swath of daily life, including applications such as optical communications, bar-code scanners, and laser discs. There are many examples like this in which the fruits of basic research went on to spur major progress in both basic and applied areas; and yet, at the time of the initial discovery, it would have been essentially impossible to anticipate the potential for such significant future developments. From a long-term perspective, there can be no question that providing adequate support for pure basic research—carried out solely to satisfy intellectual curiosity—is crucial for the advancement of science and technology.

In recent years, chemistry research has become a large-scale enterprise; we have witnessed a proliferation of large-scale research projects, and cases in which research funds are allocated preferentially to specific fields have become commonplace. Does this sort of research support lead to the sort of creative original research that gets recognized by Nobel Prizes? Project-based research is effective for the purpose of further developing fields in which some degree of insight has already been established, but it is not necessarily conducive to spurring the type of trailblazing research that opens doors to new fields. In the case of chemistry, the Nobel Prize has often been awarded for research that grew out of small-scale studies based on a creative new idea from a talented individual researcher. In the opinion of this author, it is crucial *not* to prioritize project-based research to the exclusion of other types of research, but rather to create research environments in which the seeds of promising future research can be broadly sown—and the ensuing green shoots can be properly cultivated. Those of us who take this view tend to survey today's environment—in which the tendency to emphasize research with short-term objectives seems to be stronger than ever—with some regret.

Pauling's predictions and chemistry in the second half of the 20th century

Given the reality that progress in chemistry is often driven in major ways by accidental or serendipitous discoveries, to what extent is it possible to predict the future of this field? Surely we can make *some* predictions regarding large-scale trends in the field as a whole. At the end of the first half of the 20th century, Pauling made a series of predictions regarding the state of chemistry in the year 2000[4], and it is a fascinating question to ask how accurately such a giant of chemistry succeeded in anticipating the future of his field. Pauling's predictions are summarized below.

1. A complete understanding of the forces that act between atoms and molecules will allow systematic predictions of the rates of chemical reactions. As a result, chemists will understand the ways in which catalysts operate and will learn to control reactions using custom-designed catalysts.

2. The chemists of the future may become skilled at using new methods—such as high-intensity radiation or extreme high temperatures and pressures—to induce chemical reactions in precisely the way they wish.

3. If a deeper understanding can be obtained of the relationship between molecular structure and the chemical and physical properties of materials, it should be possible to predict the requisite forms of new materials to be synthesized for a variety of specialized purposes.

4. Just as the study of new compounds of silicon and fluorine has made progress in recent years, there may well be new ways to utilize other elements. In particular, we can expect progress in the chemistry of elements that tend to form huge molecules, including phosphorous, vanadium, and molybdenum.

5. Research on metals, alloys, and intermetallic compounds has been somewhat neglected; however, as the theory of the structural chemistry of metallic materials progresses, it will become possible to express new alloys—with specialized properties for specialized applications—in terms of chemical formulas.

6. Progress will be made in the study of substances with physiological effects—particularly vitamins and pharmaceuticals—and the chemistry of physiologically active substances, grounded in an understanding of molecular structure, will advance.

7. The structure of proteins, nucleic acids and other large-molecule constituents of organic matter—including enzymes and genes—will be elucidated. The pharmaceutical effects of these substances will become clear, and chemists will make major contributions to preserving human health and combating disease.

Evidently Pauling's speculations were optimistic and full of hope. To what extent were they borne out by the actual evolution of chemistry by the end of the 20th century?

Predictions (1)–(3), regarding chemical reactions and synthesis, can surely be said to have been relatively prescient. Indeed, item (3) clearly anticipates the possibility of synthesizing molecules through molecular design technology. On the other hand, the predictions regarding catalysts were somewhat too optimistic. Even today we have not yet arrived at the point at which we can use custom-made catalysts to control reactions at will. Considering that Pauling was writing before computers had begun to be used in chemistry research, it is surprising to read such optimistic predictions of the authority of pure theory in chemistry and the fruits it would yield. Of course, while chemists have yet to obtain methods for controlling reactions precisely as they might wish, the emergence of high-power lasers represents a partial realization of prediction (2). Predictions (4) and (5)—regarding the chemistry of inorganic compounds, metals, and alloys—have not proven particularly accurate, perhaps illustrating the danger of excessively rose-colored thinking regarding complicated chemical phenomena. Meanwhile, the hopes of predictions (6) and (7)—suggesting major progress in the chemistry of life phenomena—have either been realized, or, indeed, exceeded. As far as Pauling was concerned, by 1950 the determination of the structure of DNA and proteins was only a matter of time, and it was surely this optimism that allowed him successfully to predict major progress to come in the life sciences. Although it is inevitably difficult to predict advances in individual areas of research, it is clear that Pauling was quite accurate in his predictions of overall future trends. In any event, Pauling's predictions abounded with hope and optimism for the orderly progress of chemistry, and we can safely say that most of them had been realized by the end of the 20th century.

7.2 Chemistry at the dawn of the 21st century

As we have seen, chemistry in the 20th century made major strides as the "science of atoms and molecules." However, the second half of the 20th century witnessed systematized forms of science and engineering being tightly interwoven into the very fabric of society—and being able no longer to avoid being impacted in major ways by the state of society itself. Indeed, the maturation of the traditional science of chemistry itself, and the transformation of the environment in which chemistry is done, have revealed signs of major changes in the direction of chemistry research. In this section, we discuss the transformations evident in the science of chemistry from the late 20th through the early 21st centuries, and then consider the future of the field. Although none of us can predict the future, in eras of transformation such the one in which we are living it is essential to anticipate what lies ahead in our subject. Needless to say, to survey the entirety of the enormous field of chemistry and project its future with any sort of accuracy is a task far in excess of this author's abilities; our intention here is merely to offer some fodder for speculation and debate.

The changing climate surrounding the practice of science

The end of the cold war in the early 1990s shifted the motivation for societies to support science and technology; in this new era, the aspects of science and technology that are relevant to economic development received particularly strong emphasis. For a brief time after the end of the cold war it was thought that capitalist economies were running smoothly; however, as we entered the 21st century, the world entered an era of major chaos and transformation. Indeed, since the September 11, 2001 terrorist attacks on the U.S., the world has witnessed an unceasing sequence of warfare and unrest, while the financial crisis that began with the collapse of Lehman Brothers in 2008 forced even the most advanced capitalist nations of the world—who, until recently, had been celebrating their victory in the cold war and the seemingly limitless prosperity that followed—to confront the various problems inherent in economic systems based on neoliberalism. The subsequent European financial crisis of 2011 continues to cast a dark shadow on the world economy. Broadly speaking, we face an era in which modern civilization—whose development depended on economic growth fueled by mass consumption and supported by the techno-logical innovations of the 20th century—must address a series of global-scale challenges facing mankind in the near future, including the worsening problem of global warming, the difficulty of accommodating the Earth's ever-growing population, the scarcity of resources and the insufficiency of foodstuffs, and the problems of energy sources. Although the progress of science and technology has brought material abundance to the developed nations of the world, it has also served to widen the gap between the world's rich and poor and to exacerbate the economic gulf between the northern and southern hemispheres. The progress of economic globalization will only further intensify the competition for techno-logical advances. In many nations around the world, the development of science and technology is seen as a necessary tool for winning economic competitions and ensuring national prosperity; although these nations do support scientific and technological research and development, they do so with a keen awareness of the *cost* of this support, and the importance of accountability—the duty to explain to taxpayers how their money is being used—has become widely acknowledged. This is one reason for the increasing trend toward the emphasis on research with applied orientations. Moreover, the number of scientists and engineers in advanced nations has grown to large proportions, and the situation surrounding the pursuit of scientific and engineering professions is undergoing major transformation[5]. We are living through an era in which the importance of *science for the sake of society*—seeking not only an expansion of human knowledge but also the continuing advancement of mankind—has become widely appreciated[6].

It was against this backdrop that Japan, in 1996, formulated its Science and Technology Basic Law, an effort to ensure Japan's status as a global leader in the creation of innovative science and technology by mandating that the promotion of science and technology be supported as a matter of government policy. When the Act was renewed for a second term in 2001, its mission was summarized in the following three principal goals[7]:

1. To ensure that Japan contributes to the world at large through the creation and application of knowledge—*the creation of new knowledge.*
2. To ensure Japan's international competitiveness and capacity for sustainable development—*the creation of new knowledge-based activities.*
3. To ensure that Japan offers a safe, secure, and high-quality standard of living to its people—*the creation of a plentiful knowledge-based society.*

These principles represent something of a hodgepodge of distinct values. Item (1) seeks to preserve the values of the traditional academic sciences. Item (2) embodies the values of economic nationalism; still, in view of the foundational role played by scientific and technological progress in modern-day capitalist economies, some sort of principle along these lines seems inescapable. Item (3) envisions a scientific and technological edifice that can contribute to protecting the environment and ensuring the health and welfare of the population. Thus the three aspirations reflect the full complexity of the relationship between science and society in the modern world, as must any set of principles that seeks to underpin support for science and technology. Indeed, the stated goals of support for science and technology in all advanced nations are similar to those espoused by Japan, though they may differ somewhat in the emphasis given to various aspects. In this chapter, we will study the particular science of chemistry in the context of this modern environment.

The current status of chemistry—and the challenges it faces

The direction of chemistry research exhibited clear changes between the late 20th and early 21st centuries. One reason for this is that the field of chemistry itself progressed, with its traditional subfields maturing as scientific disciplines; another reason is that the overlaps between chemistry and the other natural sciences became even more extensive. Moreover, the end of the cold war brought about a change in motivations for supporting science; societal demands for scientific endeavors that could be useful in promoting economic growth have without question exerted a major impact on the field.

The image of chemistry

As discussed in detail in Chapters 4 to 6, the science of chemistry made major advances in the 20th century. Broadly speaking, the impetus for this progress was twofold: human intellectual curiosity regarding the substances that exist in our universe and their transformations; and the basic human desire to lead healthy, pleasant, and abundant lives. The former motivation was the primary driver of academic research in basic chemistry, while the latter goal spurred major progress in technological applications of chemistry in engineering, medicine, pharmacology, agriculture, and the industries based on these fields, ultimately coming to exert a profound influence on human life. However, toward the end of the 20th century some observers began to point out that all of this progress had caused the science of chemistry to lose some of its identity—and to suffer *image problems* in which the world no longer recognized the fruits of chemistry research. This was the premise of a 2001 editorial in *Nature* entitled "A discipline buried by success," which is summarized in the following[8]

Chemistry is no longer restricted to its academic divisions of organic, inorganic and physical chemistry. In addition to territories such as catalysis, organic synthesis, polymers and materials science, chemistry has extended to new, adjacent boundary areas, and has begun to interact with various other fields of science.

But it is this diversity that has made it difficult to define chemistry. Because the chemistry 'brand' cannot be accurately defined, the discipline is easily misunderstood, and is not sufficiently appreciated. This has hurt the discipline's image.

To the general public, chemistry is synonymous with the industry with which it shares its name, and is associated with the negative aspects of the chemical industry that lead to pollution. To scientists in other disciplines, particularly young scientists, chemistry is seen as a mature discipline with its prime years already behind it. Chemists speak of many dreams, but in order for chemistry to achieve as much success in the 21st century as it has thus far, they must improve the image of chemistry, attract bright young scientists, and secure an adequate budget.

As the borders between scientific disciplines blur, fundamental chemistry skills such as synthesis and analysis will be all the more crucial for interdisciplinary subjects. Indeed, chemists are already flocking in increasing numbers to collaborate with biologists, physicists, engineers, computer scientists and other scientists. Indeed, cutting-edge interdisciplinary disciplines are full of opportunities for contributing to the enhancement of the image of chemists. In order for chemists to disseminate their contributions around the world, they must stop being modest and more aggressively promote awareness regarding their work.

This sense of apprehension is shared by many chemists. Indeed, the newsletter of the American Chemical Society has even considered the suggestion that the word "chemistry" should be replaced with "molecular science" to enhance the external image of the field.

The Breslow-Tirrell (BT) Report

The changes sweeping chemistry and the environment surrounding it may also be discerned from a report published in 2005 by the U.S. National Research Council. Every 20 years or so the Council releases a report on the current state and future prospects of chemistry[9,10]. The 2005 report, titled *Beyond the Molecular Frontier*[10], was prepared by a committee of 17 well-known chemists and chemical engineers led by Columbia University organic chemist Ronald Breslow and UC Santa Barbara chemical engineer Matthew Tirrell. This report (the "BT report") treated chemistry and chemical engineering—fields traditionally seen as distinct—together as a single field[2] known as "chemical science." This notion is premised on the observation that the basic and applied branches of chemistry have become tightly intertwined, with the gap between basic and applied research increasingly difficult to discern. In the cold-war era of the second half of the 20th century, support for basic research was justified on the basis of a linear model in which scientific progress was argued to proceed from basic research to applied research and from applied research to development. However, this model does not accurately reflect the actual practice of modern science research.

The BT report divided scientific research into three types: (1) pure basic research, (2) pure applied research, and (3) goal-oriented basic research. Type 1 was exemplified by Bohr's quantum theory, while type 2 was exemplified by research by Edison; characteristic examples of type 3 include Pasteur's work on microorganisms or Langmuir's research on surfaces. In today's world, more and more research of type 3 is being conducted, and the importance of this research is growing. Thus progress in science does not proceed simply from the basic to the applied; instead, the BT report suggests that the two types of research

---

2. Note that the distinction between chemistry and chemical engineering at U.S. universities is not the same as the distinction observed in Japan. A significant portion of the chemistry research conducted in the chemical engineering departments of Japanese universities would belong in the chemistry department of a U.S. university.

Ronald Breslow (1931–)
After completing undergraduate and graduate studies at Harvard, Breslow studied under Woodward, specializing in organic chemistry. Since 1956 he has been involved in research and education at Columbia University; he is well known for synthesizing compounds with interesting properties and as the founder of biomimetic chemistry. He received the Priestley Medal in 1999 and the Perkin Medal in 2010.

Matthew Tirrell (1950–)
After completing his doctorate in polymer science at the Massachusetts Institute of Technology, Tirrell worked from 1977 through 1999 at the University of Minnesota—where he served as a professor of chemical engineering and then department head. Later he was a professor of chemical engineering at the University of California at Santa Barbara and is currently Director of the Institute for Molecular Engineering at the University of Chicago. He achieved outstanding results in his research on polymer surfaces and in 2012 received the Polymer Physics Prize from the American Physical Society.

influence each other in a dynamic cyclic process. This observation is based on the maturation of basic chemistry and the increasing emphasis placed on applied research and development; indeed, it is undoubtedly true that cases in which pure and applied research influence one another to drive progress are on the rise. In the opinion of this author, robust progress in science depends on well-balanced support for all three of the aforementioned types of research; however, the trend in recent years seems to be toward emphasizing research of types (2) and (3).

The BT report makes the following observation regarding recent changes in chemistry and the environment surrounding it:

All of these factors began to change in the 1990s. Fundamental chemical research began to overlap with and penetrate chemical engineering to an unprecedented extent. This has been characteristic for interdisciplinary fields such as polymers, catalysis, electronic materials synthesis and processing, biological science and engineering, and computational science and engineering. These fields of research have become not just accepted but actually central to both chemistry and chemical engineering departments, and they cut across the traditional subdisciplinary boundaries. The nature of the efforts of chemists and chemical engineers in these areas are sometimes difficult to separate in a meaningful or useful way. Some research emphasizes fundamental curiosity or solving puzzles of nature, some

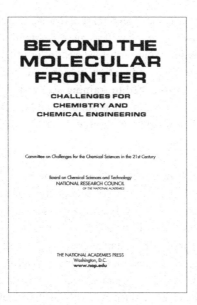

Figure 7.2: The title page of the report prepared by Breslow and Tirrell for the U.S. National Research Council

aim to test intriguing or provocative hypotheses, and some seek to improve our ability to address technological or societal problems.

Furthermore, chemists are increasingly involved in constructing, analyzing, and using complex systems and assemblies, from cells to clouds, from energy production to earth systems. This merges naturally with the systems approach of engineering. Approaching the chemical sciences from the traditionally chemical engineering end of the spectrum, we find chemical engineers increasingly entering, and in some cases leading, in more basic fields of chemistry because more science input is needed to solve technological problems or because the tools of the chemical engineer are more suited to discovery in certain areas.

The title of this report, "Beyond the Molecular Frontier," indicates that the committee took a broad view of the scope of chemistry, considering not only molecules but also domains such as materials, cells, life, and the Earth. The authors were most likely attempting to consider new possibilities for the future progress of chemistry. However, when the subject of chemistry is enlarged to encompass such a broad range of content, it becomes increasing difficult to define just what the field is about—whereupon we are quickly led back to the "image problems" noted by *Nature*. Moreover, to succeed in this sort of interdisciplinary research, it is essential for researchers to have deep knowledge of their own area of expertise, as well as an understanding of other fields and good skills for communicating with researchers in other fields. Present-day undergraduate and graduate education, which emphasize narrow fields of specialization, are not well suited to training scientists with these skills. The science of chemistry is transforming around us, and the BT report reminds us that the challenge of responding appropriately to this reality remains fraught with many difficult problems.

The challenges facing chemists and chemical engineers
The BT report, recognizing that new developments in chemistry are frequently spurred by unexpected discoveries, lists nine areas of particular importance for chemistry and chemical engineering today and in the near future. These are (1) synthesis, (2) transformation of substances, (3) isolation, identification, imaging, and measurement, (4) theoretical and computer modeling, (5) the boundaries separating chemistry from biology and medicine, (6) materials, (7) climate and environment, (8) energy, and (9) safety. In each of these areas, the authors survey future prospects—based on the achievements of chemistry and chemical engineering thus far—and outline some of the major challenges remaining to be addressed by future chemists and chemical engineers, examples of which include the following.

Synthesis and manufacturing
Developing new methods for fabricating any new substance of scientific or practical interest using compact methods of synthesis and manufacturing that offer high selectivity, low energy consumption, and minimal impact on the environment. Achieving this goal will require unceasing progress in the development of new methods of synthesis and manufacturing. The creation of new substances—including materials for pharmacology and other specialized purposes—will undoubtedly continue to offer benefits to mankind.

Materials and devices to ensure safety
Developing new materials and devices to detect and identify dangerous substances and organisms with high sensitivity and selectivity to protect civilians from terrorism, accidents, crime, and disease. Rapid and reliable detection of harmful disease vectors, extremely toxic chemical substances, hidden explosives, and other dangerous items is the first step in this direction. The next steps, which are challenges bestowed upon chemists, will require devising new methods to combat such threats.

Understanding and control of applications
Understanding and controlling how reactions occur on all time scales and for all molecules of all sizes. A fundamental understanding of reactions helps in designing new reactions and manufacturing methods and lends basic insight into chemistry as a science. Major progress toward this goal in the next 20—30 years will undoubtedly incorporate the following elements: realization of computational modeling on large-scale parallel-processing machines to predict molecular motion, development of techniques for studying and controlling individual molecules as well as aggregates of molecules, and advances in electron beams and ultra-high-speed light pulses, at wavelengths down to the X-ray region, for observing molecular structure. These are just a few examples of areas in which progress in the understanding of chemistry can also be useful for practical applications.

New materials and devices
Developing methods for designing and fabricating new substances, materials, and devices with properties custom-designed for certain specialized purposes. This will help to systematize the search for new materials, eliminating the trial-and-error process used in the past. Recent advances in chemical theories and computational techniques will help to enable this development.

Chemistry of living organisms
Acquiring a detailed understanding of the chemistry of living organisms. Understanding how many different types of proteins, nucleic acids, and other biologically relevant small molecules aggregate to form complexes with specific chemical functions; also understanding the complex chemical interactions that exist between the various constituents of living cells. Using chemistry to explain life processes is one of the major challenges that will continue into the future; it is particularly fascinating to ask what sort of chemistry underlies the processes of memory and cognition. As biology becomes increasingly a field of chemical science (and chemistry becomes increasingly one of the life sciences) we can expect major progress in this direction.

Pharmacology and therapies
Developing medicines and treatments for diseases that cannot presently be treated. Despite significant progress in the past—including development of many new pharmaceuticals by chemists and the development of new materials by engineers—this field contains many challenges remaining to be addressed. New medicines to treat cancer, diseases caused by viruses, and other illnesses will contribute in major ways to the well-being of mankind.

Utilization of self-assembly

Developing methods of self-assembly as an effective approach for the synthesis and manufacturing of complex systems and materials. Appropriately designed mixtures of chemical constituents can self-assemble—just as in living organisms—into complex aggregate structures from the nano- to the macro-scale. By expanding these techniques from small-scale laboratory structures into the realm of practical production, it may be possible to enact a revolution in chemical manufacturing.

Chemistry of the environment

Understanding the complex chemistry of the Earth—including the continents, the oceans, the climate, and the living world—and, in so doing, to ensure its adaptability for human life. This is a fundamental challenge to those of us who are active in the natural sciences; we hold the key to assisting in the formulation of policies to prevent environmental degradation. Moreover, chemists will be able to use this understanding to devise new methods of mitigating pollution and other threats to our planet.

Energy

Using new methods of energy creation, storage, and transport to develop unlimited and inexpensive energy sources to ensure a truly sustainable future. At present, the creation and use of energy is based on the consumption of limited resources, causing a variety of environmental problems. High hopes await future developments such as the creation of fuel cells—which enable the establishment of a new economy based on hydrogen created using a variety of non-fossil-fuel methods—as well as various new ways of using sunlight and superconductors to facilitate effective energy transport.

Self-optimized chemical systems

Designing and promoting chemical systems capable of self-optimization. Taking a page from the natural process of evolution—in which living systems optimize themselves—it should be possible for a system to create optimized substances, not by attempting to isolate the desired substance from a mixture of multiple constituents but rather by creating the optimal substance as the sole product of the fabrication process. The self-optimization approach will undoubtedly prove useful to insightful chemists as they seek to create new medicines, catalysts, and other important chemical substances.

A revolution in chemical manufacturing methods

Designing chemical processes to be safe, compact, flexible, energy-efficient, and environmentally friendly while enabling rapid commercialization of new products. Although much progress has already been made in the development of green chemistry, more work is needed before we will be able to meet global demand for the production of important chemical products through processes that inflict precisely zero damage to the Earth and its inhabitants.

Communication with citizens
Effectively educating the general public as to the contributions to society made by chemistry and chemical engineering. Chemists and chemical engineers must learn effective techniques for communicating with the general public, either directly or through the media. Chemists and chemical engineers must explain to the general public precisely what it is that they are doing—and how the goals of chemical sciences serve to improve society as a whole.

Education
Inspiring the brightest young scientists to pursue the chemical sciences and helping them meet their challenges. These future scientists will live exciting lives involving work at the molecular frontier or in domains beyond it and will surely make major contributions to meeting the pressing needs of mankind.

Overall one feels here a strong orientation toward applications, but this list communicates a clear understanding of the many challenges remaining in chemistry—both in *chemistry pursued to satisfy intellectual curiosity* and *chemistry pursued for the betterment of mankind.*

What are the big questions in chemistry?
Although the field of chemistry without question continues to offer many unsolved problems—such as those mentioned above—one must surely wonder if there remain any major mysteries in the field, of the sort that quicken the pulse and capture the intellectual curiosity of the human mind. This question was taken up by John Maddox, the longtime editor of *Nature*, who published in 1998 a book entitled *What remains to be discovered* in which he predicted the fields in which major discoveries could be expected in the 21st century[11]. Maddox discussed three topics in particular: matter, life, and the world around us. For Maddox, the problem of matter was rooted in the origins of matter, and the answers were to be found in outer space and in elementary particle physics. The problem of life was to be addressed by the science of biology; indeed, chemistry does not figure prominently in Maddox's book—and yet, in fact, chemistry is intimately related to the problems of "life" and "the world around us." In the area of "life" Maddox discusses the origins of life, the roles of cooperation and autonomy in biological phenomena, genes and their defects, and the theory of evolution. All

of these are questions traditionally associated with biology, and yet in each case the answers are believed to lie primarily in molecular-level research, marking them as challenges which—in this author's opinion—chemists are uniquely well-qualified to take on. In the domain of "the world around us," the problems to which chemistry can offer the most direct solutions are those "avoiding large-scale catastrophe due to environmental degradation." These include the problem of global warming due to increased atmospheric CO_2, but the contributions that chemistry can make toward "avoiding large-scale catastrophe" are not limited to the problem of global warming. Indeed, many of the issues raised by the BT report mentioned above are connected to the challenge of "avoiding large-scale catastrophe," and we can surely expect major progress to ensue from their pursuit.

In 2006, the journal *Nature* asked chemists to opine on the "big questions" in chemistry[12]. Physicists inquire as to the origin and structure of the universe, while biologists question the nature of life itself. Are there any similarly big questions remaining in chemistry? The responses of many well-known chemists to this question tend to agree on one point in particular: the chemistry of life processes. Richard Zare, a professor of physical chemistry at Stanford University, says that "To me, the big unanswered questions concern the chemistry of life processes," while Harvard's George Whitesides notes that "The nature of the cell is an entirely molecular problem." The folding of proteins, the genetic encoding of biomolecular functions, and high-level selective molecular recognition are all, in essence, chemistry problems. However, the big problems remaining in chemistry are not restricted to the domain of the living world. A crucial aspect of chemistry that is absent from other sciences is the possibility of "making things," and chemists are particularly skilled at manipulating atoms and molecules. Based on these insights, the *Nature* article lists 6 specific big questions.

1. How can we design molecules that exhibit specific features or dynamics?
2. What are the chemical foundations of cells?
3. How can we fabricate the materials that will be needed for the future development of fields such as energy, aerospace and space travel, and pharmaceuticals?
4. What is the chemical basis of cognition and memory?
5. How did life originate on Earth? Is it possible for life to begin in places other than our Earth?
6. How can we search the full space of all possible combinations of all elements?

Big questions such as these cannot fail to capture our intellectual curiosity. But is it truly essential to focus *exclusively* on such big questions? Indeed, by placing too much emphasis on the "big questions" might we not ultimately wind up

Richard Zare (1939–)

After undergraduate and graduate studies at Harvard, Zare received his doctorate in 1964. In 1969 he became a professor at Columbia University and has been a professor at Stanford since 1977. His areas of expertise include physical chemistry and analytical chemistry. He is known for his work in laser chemistry, and particularly for experimental and theoretical research on molecular-level chemical reactions. He received the 2005 Wolf Prize (Chemistry division)

George Whitesides (1939–)

After completing his undergraduate education at Harvard, Whitesides received his doctorate from Caltech in 1964. From 1963 to 1982 he was a professor at MIT, and since 1982 has been a professor at Harvard. He is known for contributions to a wide range of fields, including organometallic chemistry, self-assembled molecules, soft lithography, and nanotechnology. He received the Kyoto Prize (in the Advanced Technology division) in 2003 and the Priestley Medal in 2007.

deflating the spirit of creativity in chemistry? The history of chemistry provides ample evidence that groundbreaking discoveries often result from serendipitous findings during the course of ordinary research. Chemist and Nobel laureate Roald Hoffmann (see page 507) put it this way:

"There is no Holy Grail in chemistry… My natural philosophical disposition is not to work on big questions. I like working on many detailed small problems in this wonderful chemical garden, while keeping my eyes open for the connections."

This author feels a strong sense of kinship with Hoffman's sentiments. Many of the subjects raised by the BT report discussed above may not involve answering any "big questions," but they serve nonetheless to answer the *everyday* questions that we encounter in our daily lives—and to furnish responses to a host of problems for which our society is in increasingly dire need of solutions. Taking sure-footed steps to achieve solutions to these problems—while never losing sight of the "big questions" that inflame the intellectual curiosity of mankind—must be a crucial imperative for present-day chemistry.

7.3 The future of chemistry—and what we can expect from it

The physical world that surrounds us consists entirely of atoms and molecules, and the basic principles underlying the structure, kinetics, and interactions of atoms and molecules are now clear. However, atoms and molecules can combine in an essentially infinite variety of ways, and the structure and properties of the resulting materials are thus essentially infinitely complex and varied themselves. In the opinion of this author, the most profound hope we can have for the chemistry of the future is that it will "create new knowledge" by allowing us to understand the properties and reactions of materials that follow from this great diversity—and to use this understanding to create new materials. In today's world—in which chemistry has matured to the point at which detailed knowledge is already available at least for relatively simple molecules—the focus of research must surely shift toward "complex systems" in which a variety of constituents interact. This is one reason for the attractiveness of the chemistry of living systems, a field that abounds with unsolved problems. Even with the structures of proteins, DNA, and other biological polymers understood, the question of how these objects proceed, through intermolecular interaction networks, to underlie the complicated and yet miraculously precise workings of living organisms remains largely unanswered. Of course, needless to say, biological systems are not the only complex systems of interest. Understanding the occurrence of chemical reactions under varying conditions, explaining the properties and features of interesting substances, and synthesizing new materials are all challenges that, in one way or another, amount to complex-system problems. In addition, the chemistry of the world around us in the broadest sense—including not only atmospheric chemistry but also earth chemistry, the chemistry of the oceans, and the chemistry of outer space—also furnishes examples of complex systems in which many different constituents interact in complex ways. Moreover, systems composed of multiple disparate constituents are not the only complex systems of interest. Indeed, the new forms of carbon discovered toward the end of the 20th century—such as fullerenes and nanotubes—constitute new materials arising solely on the basis of *structural* diversity. For all the progress made thus far, the world around us boasts still an abundance of diversity far exceeding the capacity of our own feeble imaginations; in this author's opinion, it is here that we can expect to find the most promising future for the science of chemistry.

To date, mankind has derived a plentitude of material benefits from the incredible progress of chemistry, and yet modern civilization—which is itself supported by the advance of science and technology—now faces an increasingly urgent set of challenges. From environmental degradation due to the world's growing population, to global warming due to increased concentrations of greenhouse gases, shortages of energy, water, foodstuffs, and other resources, the emergence of new infectious diseases, pollution due to toxic substances, and more, the urgent global-scale problems expected to confront the human race continue to accumulate[13]. Although the solutions to these problems will require international cooperation, the present reality is that, in many cases, coordination among nations is exceedingly difficult due to conflicting national interests. Given the current global economic depression, the world's advanced and developing nations are seeking higher employment rates and economic growth. Some fear that, if economic growth proceeds worldwide—including in massively populous nations such as China and India—and the global upper and middle classes continue to grow, thus leading to significantly increased consumption standards for the world as a whole, then all of the aforementioned problems will grow only more pressing and the world will find itself on the brink of destruction. If we continue to hold on to the values we have embraced in the past, then—to this author, at least—it seems impossible to imagine that, merely by finding proper use for the progress of science and technology alone, we will be able find harmony with the global environment of our finite planet, solve the problems mentioned above, achieve sustainable growth, and avoid the demise of the human race. Instead, we must also find a way to recalibrate our imaginations and break out of the vicious cycle of societies dependent on perpetual mass consumption and economic growth. And yet no matter which question we consider—how do we solve the problem of global warming? How do we obtain safe and clean energy sources? How do we avoid the threat of shortages of resources and foodstuffs? How do we resolve the problem of pollution due to chemical materials?—the problems seem to grow more ominous every day, threatening the continued existence of mankind. Advances in chemistry and chemical engineering will be needed to solve these problems; if chemists can succeed in addressing the myriad challenges before us—such as those described by the BT report mentioned in the previous section—then there may yet be hope for solutions to some of the world's pressing difficulties.

2011 was the United Nations' "International Year of Chemistry,³" and to mark its start *Nature* compiled a special edition on the present and future of chemistry[14]. The issue noted that, although chemistry has achieved remarkable progress and has made major contributions to the development of other fields, it does not receive the recognition it deserves; still, the authors express great hope for future progress in the field, noting that many fascinating puzzles remain in the various branches of chemistry, and that chemistry is also poised to make major contributions toward the solution of several global-scale problems. The

---

3. International Year of Chemistry: 2011 was both the 100th anniversary of the Nobel Prize in Chemistry awarded to Marie Curie and the 100th anniversary of the founding of the International Union of Pure and Applied Chemistry (IUPAC), and a joint proposal by IUPAC and UNESCO to declare the year International Year of Chemistry was adopted by the United Nations. In a similar development, 2005—the 100th anniversary of Einstein's *annus mirabilis* (see page 200)—was named International Year of Physics.

Figure 7.3: The logo for International Year of Chemistry

special issue also included reviews of research on new carbon-based materials—exemplified by fullerenes, carbon nanotubes, and graphene—and of movements toward green chemistry in the chemical industry, as well as comments by many chemists on the present and future of chemistry. These included Harvard's George Whitesides and MIT's John Deutch, who contributed a commentary on the present and future of chemistry. An abridged excerpt of the article is given below.

Chemistry is the science that connects the relative simplicity of atoms and molecules to the complexity and function of macroscopic matter and of life. Some of the most interesting problems in science, and many of the most important facing society, need chemistry for their solution. Examples include: understanding life as networks of chemical reactions; interpreting the molecular basis of disease; global stewardship; the production, storage, and conservation of energy and water, and the management of carbon dioxide. Chemistry has been slow to exploit these research opportunities, and academic chemists are satisfied with the status quo and conservative in their outlook.

However, cracks in the 100-year-old structure of chemistry began to emerge in the 1990s. First, it became clear that chemistry's best intellectual opportunities lay outside its historical boundaries. The new frontiers of chemistry were the life sciences and materials science. Now others—energy, the environment, and health care—have joined the fray.

Next, "function" replaced "structure" as the objective of research in chemistry. Function is harder to handle than structure, and in particular it is difficult to design function, and usually emerges from empiricism and serendipity. (Society does not care if a molecule has a particular structure; it cares about its function.)

Third, academic chemistry is overpopulated, with an imbalance between the supply of and the demand for PhDs. The current PhD programs produced too few ideas and too many average scientists, and neither provided novel solutions to problems.

Finally, Balkanization of the field has led to specialization by young scientists, and we are producing chemists who lack the skills appropriate for the creation of new disciplines. A proper strategy for restructuring chemistry is needed. To address these problems, chemists must enact bold reforms, including the following.

1. Chemistry must reorganize to try to solve problems that are important and recognizable to the society that is paying for the research. To make fundamental discoveries, an approach that starts with practical problems, and uses them to reveal unsolved fundamental problems, will work at least as well as (and arguably better than) one that starts with the familiar questions of familiar disciplines.

2. Disciplines mature, and must be subsumed into others. Chemistry should cluster its teaching and research around the exciting and uncertain future rather than the ossified historical past. A first step is to merge chemistry and chemical-engineering departments. A second is to form broad new entities that address the most challenging problems that require the skills of chemists. Plausible topics could include functional materials, catalysis, complex dynamic networks, energy, the environment and sustainability, health, and out-of-equilibrium systems.

3. Chemistry has unique capabilities in many areas: complex kinetics, biological and environmental networks, synthesis of new molecules and forms of matter, examination of the properties of molecules, relating the properties of molecules to the properties of materials, and many others. A focus on these intellectual strengths avoids being second-best in someone else's game.

4. Many subdisciplines of chemistry still use an apprenticeship model in which a professor conceives the problem and strategy, and graduate students execute the bench work. It is hard to imagine a worse way to prepare tomorrow's chemists to work at the integration of many disciplines. Instead, professors should teach students the tools of curiosity. An independent, engaged student, exploring as a colleague in a promising area, will do better work than a simple apprentice.

Taken together, these statements propose a fairly radical break from the traditional practice of chemistry and the systems that support it, and many chemists may find themselves opposed to these ideas. They may also reflect the intensely competitive nature of American society and its philosophy of neoliberalism. However, if chemistry is to remain an attractive field into the future—and if it is to continue receiving support from society—then it must be prepared to confront new problems aggressively. In the present era of major changes sweeping both science and society, it is important for all of us to take these impassioned arguments to heart and to think carefully about the future of chemistry. After all, the future belongs to the young; if today's younger generation is willing to take on the challenges of steering new progress in chemistry, we have every reason to expect a healthy and robust future of exploration and discovery in our field—a hope that seems an appropriate note on which to end this book.

Reference

1. I. Hargitai, "*The Road to Stockholm*" Oxford Univ. Press, 2002
2. E. Norrby, "*Nobel Prizes and Life Sciences*" World Scientific, 2010
3. M. A. Meyers, "*Happy Accidents: Serendipity in Modern Medical Breakthroughs*" Arcade Publishing, 2010
4. J. R. Oppenheimer Ed. "*The age of Science, 1900–1950*" Scientific American, September 1950
5. F. Satoh, "*Shokugyou to siteno Kagaku*" (*Science as a profession*) Iwanami Sinsho, 2011
6. Y. Murakami, "*Ningen ni totte Kagaku towa nanika*" (*What is Science for human being* ?)
7. Science and Technology Basic Law (second period), Ministry of Education, Culture, Sports, Science and Technology, 2001
   National Research Council, "*Beyond the Molecular Frontier*" National Academy Press, Washington, D.C. 2005
12. J. Maddox, "*What remained to be discovered*" The Free Press, 1998 *Nature*, **442**, 500 (2006)
13. J. Martin, "*The Meaning of the 21st Century*" Riverhead Press, New York, 2007
14. *Nature*, **469**, 5 (2011)

Epilogue

In this volume we have surveyed the historical development of chemistry—with particular emphasis on the 20th century—all the while keeping in mind a simple underlying background question: What exactly *is* chemistry? As a definition of the field whose evolution we sought to trace, we took chemistry to be *the discipline that studies the structure and properties of atoms, molecules, and aggregates thereof, with particular emphasis on the creation and transformation of matter.* Of course, the research areas that lie at the cutting edge of any discipline change with the progress of time, and chemistry today exhibits significant overlap with other branches of science and technology, ensuring the exceeding difficulty of clearly defining which areas of research lie properly within the domain of chemistry. Nonetheless, the crucial importance of chemistry—as the core of all scientific and technological efforts that work with substances—only continues to grow in today's world. What is important for chemists is not to fuss over the precise definition of boundaries between fields, but rather to embrace the challenges of solving the problems that are truly important in science, of creating new knowledge, and of making useful contributions toward ensuring a future for mankind that is both materially and culturally rich.

At the beginning of the 20th century, the mere existence of atoms and molecules remained a matter of controversy and debate. By the end of the century it had become possible in practice to observe and manipulate individual atoms and molecules. Perhaps even more astoundingly, it had become possible to explain complex phenomena in living organisms in terms of the structure and transformation of molecules. These and other developments testify to the breathtaking progress of chemistry in the 20th century and the dramatic concomitant expansion of the world of human knowledge. The history of chemistry abounds with innumerable tales of fascinating discoveries, offering intense intellectual stimulation and giving great hope for the potential of mankind. Moreover, practical applications of chemistry have spurred major technological progress and have made significant contributions toward enabling the world's population to lead healthy and materially abundant lives.

And yet the progress of science and technology in the 20th century did not always serve to advance the well-being of mankind, and here chemistry was no exception. We need hardly point out the tragic misuse of science and technology

695

in warfare—a phenomenon crystallized by the emergence of the atomic bomb—while the peaceful use of atomic energy, a prospect that seemed to offer such great promise, has turned out to be problematic in many ways, for evidence of which we need look no further than the recent accident at the Fukushima nuclear power plant. The massive earthquake that struck Japan on March 11, 2011, and the accidents that followed, testify to the puniness of human capabilities in comparison with the raw power of nature, and expose the dangers inherent in controlling any large-scale technological endeavor. The economic growth that followed the end of World War II—which was driven by advances in science and technology—saddled the world with a variety of new problems, including the gap between rich and poor in advanced nations, the chasm in economic status between the northern and southern hemispheres, the worsening of global warming, and a coarsening of the human spirit. Meanwhile, the essential human character has not changed in the slightest since the days of ancient Greece; we continue to repeat the same foolish mistakes over and over again. It is a sad but true fact that mankind has yet to devise systems of government, economics, and society capable of directing scientific and technological advances entirely toward the betterment of human happiness and well-being. Indeed, to the contrary, one cannot help but fear that advances in science and technology are serving only to hasten the very destruction of mankind.

In the future, science and technology—which are powered by the intellectual curiosity and desire that lie at the very core of the human spirit—will continue to develop in startling ways, and not a single one of us is capable of standing in their way. The question is whether or not mankind can diverge from its present trajectory—in which civilization is dependent on economic growth and mass consumption—and erect a new type of society, in which advances in science and technology can be harnessed to address the challenging problems we face and to bring about a bright future for all. It seems clear that the 21st century will be the ultimate proving ground on which this question will find its answer. It is this author's most fervent and devout hope that the science of chemistry will contribute to the future well-being of mankind.

Afterword

The direct impetus for the writing of this book came in 2007, when my dear friend Professor Shigeki Kato, who was then Chairman of Kyoto University Press (KUP), suggested that I write a *"History of Modern Chemistry."* Back around 1995, when I was still at Kyoto University, I taught a course—offered to the first and second-year students in all departments—entitled *"Introduction to Modern Chemistry,"* in which I attempted around 10 hours' worth of lectures on the history of modern chemistry. At that time I was encouraged by the response of many students, who said things like "I had never heard that story before...and it was really interesting!" This made me think that, if I ever had the opportunity, I would love to write a textbook history of modern chemistry. And thus it was that I took Professor Kato's suggestion and began writing this book little by little at the end of 2007. Before that time I had read a few books discussing the history of chemistry, and the biographies of a few chemists, but I quickly discovered that the work of compiling a comprehensive textbook was more arduous than I had anticipated, and it was several years before I had completed a rough draft. Moreover, even after I had finally prepared this manuscript, I still felt my knowledge of the vast breadth of chemistry to be somewhat shallow, and there were many points in the book on which I lacked confidence in my treatment. Professor Kato graciously offered to review the book himself, but tragically took ill and passed away in 2010. However, many individuals kindly volunteered to read the book at its rough-draft stage and helped me to bring it to a state of completion. I would now like to express my heartfelt gratitude to all those who assisted in this effort.

First, Professor Tamejiro Hiyama, Chairman of KUP, read the entire manuscript and offered many valuable suggestions. In particular, Professor Hiyama helped to compensate for my inadequate knowledge concerning many topics in organic chemistry. Similarly, Professor Masaaki Baba of the Kyoto University School of Science also read the entire manuscript, pointed out many errors, and offered many valuable comments. Atsuyoshi Ono, Koa Kajimoto, Mitsuhiro Nozaki, Hiroyasu Nomura, Shigehiko Hayashi, and Tadashi Yamamoto all graciously agreed to read portions of the manuscript, correcting my mistakes and providing crucially important feedback.

Finally, the most helpful of all my editors were Tetsuya Suzuki, executive director of KUP, and Shoko Nagano, of the editorial department. Mr. Suzuki read the entire manuscript and proposed a large number of changes—ranging from the book's overall structure to details regarding particular passages—that greatly improved the quality of the book. Ms. Nagano meticulously studied the manuscript, pointed out many mistakes, and suggested many places in which the text could be expressed in a more easily-understood manner. Thanks to her masterful editorial work, the process of preparing this book proceeded smoothly. Moreover, my conversations with these two were among the most pleasant moments I experienced while working on this project. For the hard work and dedication they selflessly gave to the creation of this book, I thank them both from the bottom of my heart.

It is a terrible shame that Professor Kato is no longer with us to see this book come to fruition. Still, with the completion of this project I am delighted to have been able at last to keep the promise I made to him years ago. It is my devout hope that this book serves in some small way to ignite the passions of young readers—and to inspire their fascination with chemistry.

August 2013
Noboru Hirota

Appendix

The Lineage of Nobel Laureates with Connections to Liebig

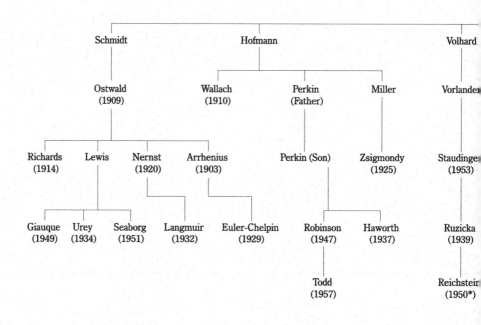

Numbers in parentheses indicate the year the Nobel Prize was awarded. * denotes prizes in physiology/medicine; all other prizes were in chemistry.
This figure was produced based on source material provided by Yusaku Ikegami.

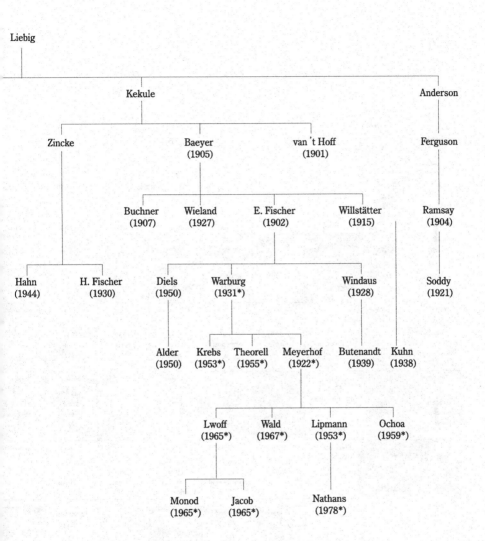

The History of the Discoveries of the Elements

Legend:

Hydrogen	→ Name of element
$_1$H	→ Atomic number, Element symbol
1766	
Cavendish	→ Year discovered
B	→ Discovered by
	→ Nationality

Color coding:

: Elements known prior to the publication of Lavoisie

: Elements discovered between 1790 and Me

: Elements discovered after 1870 that exist

☐ : Man-made elements

Nationalities:

B: Britain   G: Germany   F: Franc
I: Italy   R: Russia   U: U.S.
Ho: Holland   Sp: Spain   Hu: Hu
D: Denmark   USSR: Former Soviet Unic

1 (IA)								
Hydrogen $_1$H 1766 Cavendish B	2 (IIA)							
Lithium $_3$Li 1817 Arfwedson Swe	**Beryllium** $_4$Be 1797 Vauquelin F							
Sodium $_{11}$Na 1807 Davy B	**Magnesium** $_{12}$Mg 1808 Davy B							
		3 (IIIA)	4 (IVA)	5 (VA)	6 (VIA)	7 (VIIA)	8 (VIII)	9 (VIII)
Potassium $_{19}$K 1807 Davy B	**Calcium** $_{20}$Ca 1808 Davy B	**Scandium** $_{21}$Sc 1879 Nilson Swe	**Titanium** $_{22}$Ti 1791 Gregor B	**Vanadium** $_{23}$V 1830 Sefstrom Swe	**Chrome** $_{24}$Cr 1797 Vauquelin F	**Manganese** $_{25}$Mn 1774 Gahn Swe	**Iron** $_{26}$Fe	**Cobalt** $_{27}$Co 1737 Brandt Swe
Rubidium $_{37}$Rb 1861 Bunsen, Kirchhoff G	**Strontium** $_{38}$Sr 1808 Davy B	**Yttrium** $_{39}$Y 1794 Gadolin Swe	**Zirconium** $_{40}$Zr 1789 Klaproth G	**Niobium** $_{41}$Nb 1801 Hatchett B	**Molybdenum** $_{42}$Mo 1781 Hjelm Swe	**Technetium** $_{43}$Tc 1939 Perrier, Segre I	**Ruthenium** $_{44}$Ru 1844 Claus R	**Rhodium** $_{45}$Rh 1803 Wollaston B
Cesium $_{55}$Cs 1860 Bunsen, Kirchhoff G	**Barium** $_{56}$Ba 1808 Davy B	**Elements 57–71: Lanthanides**	**Hafnium** $_{72}$Hf 1923 Hevesy (Hu) Coster (Ho)	**Tantalum** $_{73}$Ta 1802 Ekeberg Swe	**Tungsten** $_{74}$W 1783 Elhuyar Brothers Sp	**Rhenium** $_{75}$Re 1925 Noddack, Tacke, Berg G	**Osmium** $_{76}$Os 1804 Tennant B	**Iridium** $_{77}$Ir 1804 Tennant B
Francium $_{87}$Fr 1939 Perey F	**Radium** $_{88}$Ra 1898 Pierre and Marie Curie F	**Elements 89–103: Actinides**	**Rutherfordium** $_{104}$Rf 1964 Flyorov et al. USSR	**Dubnium** $_{105}$Db 1968 Flyorov et al. USSR	**Seaborgium** $_{106}$Sg 1974 Soviet / American group	**Bohrium** $_{107}$Bh 1981 German group	**Hassium** $_{108}$Hs 1984 German group	**Meitnerium** $_{109}$Mt 1982 German group

Lanthanides:

Lanthanum $_{57}$La 1839 Mosander F	**Cerium** $_{58}$Ce 1803 Klaproth (G) Berzelius (Swe), et al.	**Praseodymium** $_{59}$Pr 1885 Welsbach A	**Neodymium** $_{60}$Nd 1885 Welsbach A	**Promethium** $_{61}$Pm 1947 Coryell et al. U	**Samarium** $_{62}$Sm 1879 Boisbaudran F

Actinides

Actinium $_{89}$Ac 1899 Debierne F	**Thorium** $_{90}$Th 1828 Berzelius Swe	**Protactinium** $_{91}$Pa 1918 Hahn et al. (G) Soddy (B)	**Uranium** $_{92}$U 1789 Klaproth G	**Neptunium** $_{93}$Np 1940 McMillan, Abelson U	**Plutonium** $_{94}$Pu 1940 Seaborg, McMillan et a U

18 (0)

Helium
$_2$He
1895
Ramsay
B

entary Treatise on Chemistry (1789)
's first periodic table (1869)
le form on Earth

Swe: Sweden
A: Austria
Swi: Switzerland

13 (IIIB)	14 (IVB)	15 (VB)	16 (VIB)	17 (VIIB)
Boron $_5$B 1808 Gay-Lussac, Thenard F	Carbon $_6$C	Nitrogen $_7$N 1772 Rutherford B	Oxygen $_8$O 1774 Priestley B	Fluorine $_9$F 1886 Moissan F	Neon $_{10}$Ne 1898 Ramsay, Travers B
Aluminum $_{13}$Al 1825 Oersted D	Silicon $_{14}$Si 1824 Berzelius Swe	Phosphorus $_{15}$P 1669 Brand G	Sulfur $_{16}$S	Chlorine $_{17}$Cl 1774 Scheele Swe	Argon $_{18}$Ar 1894 Ramsay, Lord Rayleigh B

10 (VIII)	11 (IB)	12 (IIB)						
Nickel $_{28}$Ni 1751 ronstedt Swe	Copper $_{29}$Cu	Zinc $_{30}$Zn	Gallium $_{31}$Ga 1875 Boisbaudran F	Germanium $_{32}$Ge 1886 Winkler G	Arsenic $_{33}$As	Selenium $_{34}$Se 1817 Berzelius Swe	Bromine $_{35}$Br 1826 Balard F	Krypton $_{36}$Kr 1898 Ramsay, Travers B
Palladium $_{46}$Pd 1803 Wollaston B	Silver $_{47}$Ag	Cadmium $_{48}$Cd 1817 Stromeyer G	Indium $_{49}$In 1863 Richter, Reich G	Tin $_{50}$Sn	Antimony $_{51}$Sb	Tellurium $_{52}$Te 1782 Müller A	Iodine $_{53}$I 1811 Courtois F	Xenon $_{54}$Xe 1898 Ramsay, Travers B
Platinum $_{78}$Pt 1748 le Ulloa Sp	Gold $_{79}$Au	Mercury $_{80}$Hg	Thallium $_{81}$Tl 1861 Crookes B	Lead $_{82}$Pb	Bismuth $_{83}$Bi	Polonium $_{84}$Po 1898 Pierre and Marie Curie	Astatine $_{85}$At 1940 Corson et al. U	Radon $_{86}$Rn 1900 Dorn G
rmstadtium $_{110}$Ds 1994 fmann et al. G	Roentgenium $_{111}$Rg 1994 German group	Copernicium $_{112}$Cn 1996 German group		Flerovium $_{114}$Fl 1998 Russian / American group		Livermorium $_{116}$Lv 2000 Russian / American group		
Europium $_{63}$Eu 1896 emarcay F	Gadolinium $_{64}$Gd 1880 Marignac Swi	Terbium $_{65}$Tb 1843 Mosander Swe	Dysprosium $_{66}$Dy 1886 Boisbaudran F	Holmium $_{67}$Ho 1879 Cleve Swe	Erbium $_{68}$Er 1843 Mosander Swe	Thulium $_{69}$Tm 1879 Cleve Swe	Ytterbium $_{70}$Yb 1878 Marignac Swi	Lutetium $_{71}$Lu 1907 Urbain F
Americium $_{95}$Am 1945 aborg et al. U.S.A.	Curium $_{96}$Cm 1944 Seaborg et al.	Berkelium $_{97}$Bk 1949 Thompson, Seaborg et al.	Californium $_{98}$Cf 1950 Thompson, Seaborg et al.	Einsteinium $_{99}$Es 1952 Seaborg et al.	Fermium $_{100}$Fm 1952 Seaborg et al.	Mendelevium $_{101}$Md 1955 Ghiorso et al. U.S.A.	Nobelium $_{102}$No 1957 Swedish / American / British team	Lawrencium $_{103}$Lr 1961 Ghiorso et al. U.S.A.

The History of Chemistry in Terms of Past Nobel Prize Winners

Year	Nobel Prize in Chemistry	Nobel Prizes in Physics / Physiology or Medicine (Selected)
1901	J. H. van 't Hoff (Holland): Discovery of the laws of chemical dynamics and osmotic pressure of solutions	W. C. Röntgen (Germany): Discovery of the remarkable rays subsequently named after him (Physics)
1902	E. Fischer (Germany): Sugar and purine synthesis	H. A. Lorentz (Holland), P. Zeeman (Germany): Research into the influence of magnetism upon radiation phenomena (Physics)
1903	S. A. Arrhenius (Sweden): Advancement of chemistry by electrolytic theory of dissociation	A. H. Becquerel (France):Discovery of spontaneous radioactivity. P. Curie (France), M. Curie (France): Joint research on radiation phenomena discovered by Becquerel (Physics)
1904	W. Ramsay (Britain): Discovery of inert gaseous elements in air and determination of their place in the periodic system	Lord Rayleigh (Britain): Investigations of the densities of the most important gases and for the discovery of argon in connection with these studies (Physics)
1905	J. F. W. A. von Baeyer (Germany): Research on organic dyes and hydroaromatic compounds	P. E. A. Lenard (Germany): Work on cathode rays (Physics)
1906	H. Moissan (France): Investigation and isolation of the element fluorine and invention of the electric furnace named after him	J. J. Thomson (Britain): Theoretical and experimental investigations on the conduction of electricity by gases (Physics)
1907	E. Buchner (Germany): Biochemical research and the discovery of cell-free fermentation	A. A. Michelson (U.S.): Optical precision instruments and the spectroscopic and metrological investigations carried out with them (Physics)
1908	E. Rutherford (Britain): Investigations into the disintegration of the elements, and the chemistry of radioactive substances	P. Ehrlich (Germany), E. Mechnikov (Russia): Work on immunity (Physiology / Medicine)
1909	F. W. Ostwald (Germany): Work on catalysis and investigations into the fundamental principles governing chemical equilibrium and rates of reaction	G. Marconi (Italy) and K. F. Braun (Germany) : Contributions to the development of wireless telegraphy (Physics)

1910	O. Wallach (Germany): Pioneering work in the field of alicyclic compounds	J. D. van der Waals (Holland): Work on the equations of state for gases and liquids (Physics)
		A. Kossel (Germany): Contribution to knowledge of cell chemistry through his work on proteins and nucleic substances (Physiology / Medicine)
1911	M. Curie (France): Discoveries of the elements of radium and polonium and studies of the properties of radium and its compounds	W. Wien (Germany): Discoveries regarding the laws governing the radiation of heat (Physics)
1912	V. Grignard (France): Discovery of Grignard reagent	
	P. Sabatier (France): Method of hydrogenating organic compounds in the presence of finely powdered metals	
1913	A. Werner (Switzerland): Work on the linkage of atoms in molecules that has opened up new fields of research in inorganic chemistry	H. Kamerlingh Onnes (Holland): Investigations on the properties of matter at low temperatures which led to the production of liquid helium (Physics)
1914	T. W. Richards (U.S.): Accurate determinations of the atomic weights of a large number of elements	M. von Laue (Germany): Discovery of the diffraction of X-rays by crystals (Physics)
1915	R. Willstätter (Germany): Research on plant pigments, especially chlorophyll	W. H. Bragg, W. L. Bragg (Britain): Analysis of crystal structure using X-rays (Physics)
1916	No Prize awarded	No Prize awarded
1917	No Prize awarded	C. G. Barkla (Britain): Discovery of characteristic X-ray emissions from elements (Physics)
1918	F. Haber (Germany): Synthesis of ammonia from its constituent elements (nitrogen and hydrogen)	M. Planck (Germany): Contributions to the advancement of physics by his discovery of energy quanta (Physics)
1919	No Prize awarded	J. Stark (Germany): Discovery of the Doppler effect in canal rays and the splitting of spectra lines in electric fields (Physics)
1920	W. H. Nernst (Germany): Work on thermochemistry	C. E. Guilaume (Switzerland): Precision measurements in Physics using his discovery of anomalies in nickel steel alloys (Physics)

Year	Nobel Prize in Chemistry	Nobel Prizes in Physics / Physiology or Medicine (Selected)
1921	F. Soddy (Britain): Contributions to the chemistry of radioactive substances and investigations into the origin and nature of isotopes	A. Einstein (Switzerland): Theoretical physics, and especially for his discovery of the law of the photoelectric effect (Physics)
1922	F. W. Aston (Britain): Discovery, using mass spectrometry, of isotopes of non-radioactive elements, and elucidation of the whole-number rule	N. H. Bohr (Denmark): Investigation of the structure of atoms and of the radiation emanating from them (Physics) A. V. Hill (Britain): Discovery of the production of heat in muscle O. Meyerhof (Germany): Discovery of the relationship between consumption of oxygen and metabolism of lactic acid in muscle (Physiology / Medicine)
1923	F. Pregl (Austria): Invention of the method of micro-analysis of organic substances	R. A. Millikan (U.S.): Work on the elementary charge of electricity and the photoelectric effect (Physics) F. G. Banting, J. J. R. Macleod (Canada): Discovery of insulin (Physiology / Medicine)
1924	No Prize awarded	M. Siegbahn (Sweden): Discoveries and research in the field of X-ray spectroscopy (Physics)
1925	R. Zsigmondy (Germany): Demonstration of the heterogeneous nature of colloid solutions and the establishment of modern colloid chemistry	J. Franck (Germany), G. Hertz (Germany): Discovery of the laws governing the impact of an electron upon an atom (Physics)
1926	T. Svedberg (Sweden): Work on dispersed systems	J. Perrin (France): Research on the discontinuous structure of matter, and particularly the discovery of sedimentation equilibrium (Physics)
1927	H. Wieland (Germany): Investigations of the constitution of bile acids and related substances	A. H. Compton (U.S.): Discovery of the Compton effect (Physics) C. T. R. Wilson (Britain): Work on the observation of charged particles using cloud chambers (Physics)
1928	A. Windaus (Germany): Research into the constitution of sterols and their connection with vitamins	O. W. Richardson (Britain): Research on the thermionic phenomenon and especially for the discovery of the Richardson effect (Physics)

1929	A. Harden (Britain), H. von Euler-Chelpin (Sweden): Investigations on the fermentation of sugar, and fermentative enzymes	L. V. de Broglie (France): Discovery of the wave nature of electrons (Physics) C. Eijkman (Germany): Discovery of the antineuritic vitamin (Physiology / Medicine) F. G. Hopkins (Britain): Discovery of growth-stimulating vitamins (Physiology / Medicine)
1930	H. Fischer (Germany): Research into the constitution of hemin and chlorophyll, and especially for the synthesis of hemin	C. V. Raman (India): Work on the scattering of light and for the discovery of the Raman effect (Physics)
1931	C. Bosch (Germany), F. Bergius (Germany): Invention and development of chemical high-pressure methods.	O. Warburg (Germany): Discovery of the nature and mode of action of the respiratory enzymes (Physiology / Medicine)
1932	I. Langmuir (U.S.): Discoveries and investigations in surface chemistry	W. Heisenberg (Germany): The creation of quantum mechanics and the discovery of para- and ortho-hydrogen (Physics)
1933	No Prize awarded	E. Schrödinger (Austria), P. A. M. Dirac (Britain): Discovery of new productive forms of atomic theory (Physics)
1934	H. C. Urey (U.S.): Discovery of heavy hydrogen	T. H. Morgan (U.S.): Discoveries concerning the role played by the chromosome in heredity (Physiology / Medicine)
1935	F. Joliot and I. Joliot-Curie (France): Synthesis of new radioactive elements	J. Chadwick (Britain): Discovery of the neutron (Physics)
1936	P. J. W. Debye (Holland): Determination of molecular structure using dipole moments and diffraction of X-rays and electrons in gases	H. H. Dale (Britain), O. Loewi (Germany): Discoveries relating to chemical transmission of nerve impulses (Physiology / Medicine)
1937	W. N. Haworth (Britain): Investigations on carbohydrates and vitamin C P. Karrer (Switzerland): Investigations on carotenoids, flavins, and vitamins A and B2	C. J. Davisson (U.S.), G. P. Thomson (Britain): Experimental discovery of the diffraction of electrons by crystals (Physics) A. Szent-Györgyi Nagyrapolt (Hungary): Discoveries in connection with biological combustion processes, particularly vitamin C and the catalysis of fumaric acid (Physiology / Medicine)

Year	Nobel Prize in Chemistry	Nobel Prizes in Physics / Physiology or Medicine (Selected)
1938	R. Kuhn (Germany): Work on carotenoids and vitamins	E. Fermi (Italy): Demonstrations of the existence of new radioactive elements produced by neutron irradiation, and for the related discovery of nuclear reactions induced by slow neutrons (Physics)
1939	A. F. J. Butenandt (Germany): Work on sex hormones L. Ruzicka (Switzerland): Work on the polymethylenes and higher terpenes	E. O. Lawrence (U.S.): Invention and development of the cyclotron and for results obtained with it, especially artificial radioactive elements (Physics) G. Domagk (Germany): Discovery of the antibacterial effects of prontosil (Physiology / Medicine)
1940–1942	No Prize awarded	No Prize awarded
1943	G. de Hevesy (Hungary): Work on the use of isotopes as tracers in the study of chemical processes	O. Stern (U.S.): Development of the molecular ray method and discovery of the magnetic moment of the proton (Physics) H. C. P. Dam (Denmark): Discovery of vitamin K E. A. Doisy (U.S.): Discovery of the chemical nature of vitamin K (Physiology / Medicine)
1944	O. Hahn (Germany): Discovery of fission of heavy nuclei	I. I. Rabi (U.S.): Resonance method for recording magnetic properties of atomic nuclei (Physics)
1945	A. Virtanen (Finland): Research and inventions in agricultural and nutrition chemistry, especially the fodder preservation method	W. Pauli (Switzerland): Discovery of the Pauli Principle (Physics) A. Fleming (Britain), E. B. Chain (Britain), H. W. Florey (Australia): Discovery of penicillin and its curative effects in various infectious diseases (Physiology / Medicine)
1946	J. B. Sumner (U.S.): Discovery that enzymes can be crystallized. J. H. Northrop and W. M. Stanley (U.S.): Preparation of enzymes and virus proteins in a pure form	P. W. Bridgman (U.S.): Invention of an apparatus to produce extremely high-pressures, and discoveries made in high-pressure physics (Physics) H. J. Müller (U.S.): Discovery of the production of mutations by means of X-ray irradiation (Physiology / Medicine)

Year		
1947	R. Robinson (Britain): Investigations on plant products of biological importance, especially alkaloids	C. F. Cori, G. T. Cori (U.S.): Discovery of the course of catalytic conversion of glycogen (Physiology / Medicine)
1948	A. W. Tiselius (Sweden): Research on electrophoresis and adsorption analysis, especially for discoveries concerning the complex nature of serum proteins	P. M. S. Blackett (Britain): Development of Wilson cloud chamber method and discoveries therewith in the fields of nuclear physics and cosmic rays (Physics) P. Müller (Switzerland): Discovery of the high efficiency of DDT as a contact poison against several arthropods (Physiology / Medicine)
1949	W. F. Giauque (U.S.): Contributions in the field of chemical thermodynamics, particularly concerning the behavior of substances at extremely low temperatures	H. Yukawa (Japan): Prediction of the existence of mesons based on theories of nuclear forces (Physics)
1950	O. P. Diels and K. Alder (Germany): Discovery and development of diene synthesis (the Diels–Alder reaction)	
1951	G. T. Seaborg and E. M. McMillan (U.S.): Discoveries in the chemistry of transuranium elements	J. D. Cockcroft, E. T. S. Walton (Britain): Pioneering work on the transmutation of atomic nuclei by accelerated atomic particles (Physics)
1952	A. J. P. Martin and R. L. M. Synge (Britain): Invention of partition chromatography	F. Bloch, E. M. Purcell (U.S.): Development of new methods for nuclear magnetic precision measurements and discoveries in connection therewith (Physics) S. A. Waksman (U.S.): Discovery of streptomycin (Physiology / Medicine)
1953	H. Staudinger (Germany): Discoveries in the field of macromolecular chemistry	F. Zernike (Holland): Demonstration of the phase contrast method, especially for the invention of phase-contrast microscopy (Physics) F. A. Lipmann (U.S.): Discovery of co-enzyme A and its importance for intermediary metabolism (Physiology / Medicine) H. Krebs (Britain): Discovery of the citric acid cycle (Physiology / Medicine)

Year	Nobel Prize in Chemistry	Nobel Prizes in Physics / Physiology or Medicine (Selected)
1954	L. C. Pauling (U.S.): Research into the nature of the chemical bond and its application to the elucidation of the structures of complex substances	M. Born (Germany): Fundamental research in quantum mechanics, especially for the statistical interpretation of the wave function (Physics) W. Bothe (Germany): Coincidence method and discoveries made therewith (Physics)
1955	V. du Vigneaud (U.S.): Work on biochemically important sulfur compounds, especially for the first synthesis of a polypeptide hormone	P. Kusch (U.S.): Precision determination of the magnetic moment of the electron (Physics) W. E. Lamb (U.S.): Discoveries concerning the fine structure of the hydrogen spectrum (Physics) H. Theorell (Sweden): Discoveries concerning the nature and mode of action of oxidation enzymes (Physiology / Medicine)
1956	S. C. N. Hinshelwood (Britain), N. N. Semenov (Soviet Union): Research into the mechanism of chemical reactions	W. B. Shockley, J. Bardeen, W. H. Brattain (U.S.): Research on semiconductors and the discovery of the transistor effect (Physics)
1957	A. Todd (Britain): Work on nucleotides and nucleotide co-enzymes	
1958	F. Sanger (Britain): Work on the structure of proteins, especially that of insulin	G. W. Beadle, E. L. Tatum (U.S.): Discovery that genes act by regulating definite chemical events (Physiology / Medicine) J. Lederberg (U.S.): Discoveries concerning genetic recombination and the organization of the genetic material of bacteria (Physiology /Medicine)
1959	J. Heyrovsky (Czechoslovakia): Discovery and invention of the polarographic method of analysis	S. Ochoa, A. Kornberg (U.S.): Discovery of the mechanisms in the biological synthesis of RNA and DNA (Physiology / Medicine)
1960	W. F. Libby (U.S.): Method to use carbon-14 for age determination in archaeology, geology, geophysics, and other branches of science	
1961	M. Calvin (U.S.): Research on carbon dioxide assimilation in plants	R. Mössbauer (Germany): Research on resonant absorption of gamma rays and discovery of the Mössbauer effect (Physics)

1962	M. F. Perutz and J. C. Kendrew (Britain): Studies of the structures of globular proteins	L. D. Landau (Soviet Union) pioneering theories for condensed matter, especially liquid helium (Physics)
		F. H. C. Crick (Britain), J. D. Watson (U.S.), M. H. F. Wilkins (Britain): Discoveries concerning the molecular structure of nucleic acids and its significance for information transfer in organisms (Physiology / Medicine)
1963	K. Ziegler (Germany) and G. Natta (Italy): Discoveries of the chemistry and practical development of high polymers	J. C. Eccles (Australia), A. L. Hodgkin, A. F. Huxley (Britain): Discoveries concerning the ionic mechanisms involved in excitation and inhibition in the peripheral and central portions of the nerve cell membrane (Physiology / Medicine)
1964	D. C. Hodgkin (Britain): Determination by X-ray techniques of the structures of important biochemical substances	C. H. Townes (U.S.), N. G. Basov, A. M. Prokhorov (Soviet Union): Fundamental work in quantum electronics and construction of oscillator and amplifiers based on the maser- laser principles (Physics)
		K. Bloch (U.S.), F. Lynen (Germany): Discoveries concerning the mechanism and regulation of cholesterol and fatty acid metabolism (Physiology / Medicine)
1965	R. B. Woodward (U.S.): Outstanding achievements in the art of organic synthesis	S. Tomonaga (Japan), J. Schwinger, R. P. Feynman (U.S.): Fundamental work in quantum electrodynamics, with deep consequences for the physics of elementary particles (Physics)
		F. Jacob, A. Lwoff, J. Monod (France): Discoveries concerning genetic control of enzyme and virus synthesis (Physiology / Medicine)
1966	R. S. Mulliken (U.S.): Fundamental work concerning chemical bonds and the electronic structure of molecules by the molecular orbital method	A. Kastler (France): Discovery and development of optical method for studying Hertzian resonance in atoms (Physics)

Year	Nobel Prize in Chemistry	Nobel Prizes in Physics / Physiology or Medicine (Selected)
1967	M. Eigen (Germany), R. G. W. Norrish (Britain), and G. Porter (Britain): Studies of extremely fast chemical reactions, effected by disturbing the equilibrium by means of very short pulses of energy	H. Bethe (Germany / U.S.): Contributions to the theory of nuclear reactions, especially the discoveries concerning energy production in stars (Physics) R. Granit (Finland / Sweden), H. K. Hartline, G. Wald (U.S.): Discoveries concerning the primary physiological and chemical visual processes in the eye (Physiology / Medicine)
1968	L. Onsager (Norway / U.S.): Discovery of the reciprocal relations bearing his name, which are fundamental for the thermodynamics of irreversible processes	R. W. Holley (U.S.), H. G. Khorana (U.S.), M. W. Nirenberg (U.S.): Interpretation of the genetic code and its function in protein synthesis (Physiology / Medicine)
1969	O. Hassel (Norway), D. H. R. Barton (Britain): Development of the concept of conformation and its application to chemistry.	M. Delbrück (Germany / U.S.), A. D. Hershey (U.S.), S. E. Luria (Italy): Discoveries concerning the replication mechanisms and the genetic structure of viruses (Physiology / Medicine)
1970	L. F. Leloir (Argentina): Discovery of sugar nucleotides and their role in the biosynthesis of carbohydrates	L. Neel (France): Fundamental work and discoveries concerning anti-ferromagnetism and ferrimagnetism which have led to important applications (Physics)
1971	G. Herzberg (Canada): Contributions to knowledge of the electronic structure and geometry of molecules, particularly free radicals	E. Sutherland (U.S.): Discoveries concerning the mechanisms of the action of hormones (Physiology / Medicine)
1972	C. B. Anfinsen (U.S.): Research on ribonuclease, particularly concerning the connection between amino-acid sequences and active conformation W. H. Stein, S. Moore (U.S.): Understanding of the relationship between the chemical structure of ribonuclease and the catalytic activity of its active center	J. Bardeen, L. N. Cooper, R. Schrieffer (U.S.): Jointly developed the theory of superconductivity, usually called the BCS-theory (Physics) G. Edelman (U.S.), R. R. Porter (Britain): Discoveries concerning the chemical structure of antibodies (Physiology / Medicine)

1973	E. O. Fischer (Germany), G. Wilkinson (Britain): Pioneering work on the chemistry of organometallic, so called sandwich compounds	L. Esaki (Japan), I. Giaever (Norway): Experimental discoveries regarding tunneling phenomena in semiconductors and superconductors (Physics) B. Josephson (Britain): Theoretical prediction of the properties of a supercurrent through a tunnel barrier, particularly Josephson effect (Physics)
1974	P. J. Flory (U.S.): Fundamental achievements, both theoretical and experimental, in the physical chemistry of macromolecules	
1975	J. W. Cornforth (Australia): Work on the stereochemistry of enzyme-catalyzed reactions V. Prelog (Switzerland): Research into the stereochemistry of organic molecules and reactions	R. Dulbecco (Italy), H. M. Temin, D. Baltimore (U.S.) Discoveries concerning the interaction between tumor viruses and genetic material of the cell (Physiology / Medicine)
1976	W. N. Lipscomb (U.S.): Studies on the structure of boranes, solving outstanding problems in chemical bonding	
1977	I. Prigogine (Belgium): Contributions to non-equilibrium thermodynamics, particularly dissipative structures	P. W. Anderson, J. H. van Vleck (U.S.), N. F. Mott (Britain) Fundamental theoretical investigations of the electronic structure of magnetic and disordered systems (Physics) R. Guillemin (U.S.), A. V. Schally (U.S.): Discoveries concerning peptide hormone production of the brain (Physiology / Medicine) R. Yalow (U.S.): Development of radioimmunoassays of peptide hormones (Physiology / Medicine)
1978	P. Mitchell (Britain): Contribution to the understanding biological energy through the formulation of the chemiosmotic theory	P. Kapitsa (Soviet Union) Basic inventions and discoveries in the area of low-temperature physics (Physics) D. Nathans, H. O. Smith (U.S.), W. Arber (Switzerland): Discovery of restriction enzymes and their applications (Physiology / Medicine)
1979	H. C. Brown (U.S.), G. Wittig (Germany): Development of the use of boron- and phosphorous-containing compounds, respectively, as important reagents in organic synthesis	G. N. Hounsfield (Britain), A. M. Cormack (U.S.): Development of computer-assisted X-ray tomography (Physiology / Medicine)

Year	Nobel Prize in Chemistry	Nobel Prizes in Physics / Physiology or Medicine (Selected)
1980	P. Berg (U.S.): Fundamental studies of the biochemistry of nucleic acids, with particular regard to recombinant-DNA F. Sanger, W. Gilbert (U.S.): Determination of the base sequences of nucleic acids	
1981	Kenichi Fukui (Japan), R. Hoffmann (U.S.): Theories, developed independently, concerning the course of chemical reactions	N. Bloembergen, A. L. Schawlow (U.S.): Contribution to the development of laser spectroscopy (Physics) K. Siegbahn (Sweden): Contribution to the development of high-resolution electron spectroscopy (Physics)
1982	A. Klug (Britain): Development of crystallographic electron microscopy and structural elucidation of biologically important nucleic acid-protein complexes	K. G. Wilson (U.S.): Theory of critical phenomena in connection with phase transitions of matter (Physics) S. K. Bergstrom, B. I. Samuelsson (Sweden), J. R. Vane (Britain): Discoveries concerning prostaglandins and related biologically active substances (Physiology / Medicine)
1983	H. Taube (U.S.): Work on the mechanisms of electron-transfer reactions, especially in metal complexes	B. McClintock (U.S.): Discovery of mobile genetic elements (Physiology / Medicine)
1984	R. B. Merrifield (U.S.): Development of methodology of chemical synthesis on a solid matrix	
1985	J. Karle, H. A. Hauptman (U.S.): Establishment of direct methods for the determination of crystal structures	K. von Klitzing (U.S.): Discovery of the quantized Hall effect (Physics)
1986	D. R. Herschbach (U.S.), Y. T. Lee (U.S., Taiwan), J. C. Polanyi (Canada): Contributions concerning the dynamics of chemical elementary processes	E. Ruska (Germany): Fundamental work in electron optics and the design of the first electron microscope (Physics) G. Binnig, H. Rohrer (Germany): Design of the scanning tunneling microscope (Physics)

1987	C. J. Pedersen, D. J. Cram (U.S.), and J.-M. Lehn (France): Development and use of molecules with structure-specific interactions of high selectivity	J. G. Bednorz (Germany), K. A. Müller (Switzerland): Break-through in the discovery of super conductivity by ceramic materials (Physics); S. Tonegawa (Japan): Discovery of the genetic principles for generation of antibody diversity (Physiology / Medicine)
1988	J. Deisenhofer, R. Huber, H. Michel (Germany): Determination of the three-dimensional structure of a photosynthetic reaction center	
1989	S. Altman, T. R. Cech (U.S.): Discovery of catalytic properties of RNA	N. F. Ramsey (U.S.): Invention of the separated oscillatory fields method and its use in the hydrogen maser and other atomic clocks (Physics); H. G. Dehmelt, W. Paul (Germany): Development of the ion trap technique (Physics)
1990	E. J. Corey (U.S.): Development of theory and methodology for organic synthesis	
1991	R. Ernst (Switzerland): Development of the methodology of high resolution nuclear magnetic resonance (NMR) spectroscopy	P. de Gennes (France): Mathematical research on phase transitions in complex polymers, liquid crystals, and superconducting materials (Physics); E. Neher, B. Sakman (Germany): Discoveries concerning the function of single ion channels in cells (Physiology / Medicine)
1992	R. A. Marcus (U.S.): Theory of electron transfer reactions in chemical systems	E. A. Fischer, E. G. Krebs (U.S.): Discoveries concerning reversible protein phosphorylation as a biological regulatory mechanism (Physiology / Medicine)
1993	K. B. Mullis (U.S.): Invention of the polymerase chain reaction (PCR) method; M. Smith (Canada): Contributions to the establishment of oligonucleotide-based, site-directed mutagenesis and its development for protein studies	P. A. Sharp, R. J. Roberts (U.S.): Discoveries of split genes (Physiology / Medicine)

Year	Nobel Prize in Chemistry	Nobel Prizes in Physics / Physiology or Medicine (Selected)
1994	G. A. Olah (U.S.): Contributions to carbocation chemistry	B. N. Brockhouse (Canada), C. G. Shull (U.S.): Development of neutron scattering techniques for materials research (Physics) M. Rodbell (U.S.), A. G. Gilman (U.S.): Discovery of G-proteins and the role they play in signal transduction in cells (Physiology / Medicine)
1995	F. S. Rowland (U.S.), M. Molina (Mexico), and P. J. Crutzen (Germany): Work in atmospheric chemistry, particularly concerning the formation and decomposition of ozone	
1996	R. F. Curl, R. E. Smalley (U.S.), H. W. Kroto (Britain): Discovery of carbon fullerenes (C_{60})	D. M. Lee, R. C. Richardson, D. D. Osheroff (U.S.): Discovery of superfluidity in helium-3 (Physics)
1997	P. D. Boyer (U.S.), J. E. Walker (Britain): Elucidation of the enzymatic mechanisms underlying in the synthesis of ATP J. C. Skou (Denmark): Discovery of the ion-transporting enzyme, Na+, K+-ATPase	S. Chu, W. D. Phillips (U.S.), C. Cohen-Tannoudji (France): Development of methods to cool and trap atoms with laser light (Physics)
1998	W. Kohn (U.S.): Development of the density-functional theory J. A. Pople (Britain): Development of computational methods in quantum chemistry	R. B. Laughlin, H. L. Stormer, D. C. Tsui (U.S.): Discovery of a new form of quantum fluid with fractionally charged excitations (Physics) R. F. Furchgott, L. J. Ignarro, F. Murad (U.S.): Discoveries concerning nitric oxide as a signaling molecule in the cardiovascular system (Physiology / Medicine)
1999	A. Zewail (U.S. / Egypt): Studies of the transition states of chemical reactions using femtosecond spectroscopy	G. Blobel (Germany): Discovery that proteins have intrinsic signals that govern their transport and localization in cells (Physiology / Medicine)
2000	A. J. Heeger, A. G. MacDiarmid (U.S.), H. Shirakawa (Japan): Discovery and development of conducting polymers	Z. I. Alferov (Russia), H. Kroemer (Germany): Development of semiconductor heterostructures used in high-speed- and opto-electronics (Physics) J. S. Kilby (U.S.): Invention of the integrated circuit (Physics)
2001	W. S. Knowles (U.S.), R. Noyori (Japan): Work on chirally catalyzed hydrogenation reactions B. Sharpless (U.S.): Work on chirally catalyzed oxidation reactions	E. A. Cornell, C. E. Wieman (U.S) and W. Ketterle (Germany): Achievement of Bose-Einstein condensation in dilute gases of alkali atoms, and for fundamental studies of the properties of the condensates (Physics)

2002	J. B. Fenn (U.S.), K. Tanaka: (Japan): Development of methods for identification and structural analysis of biological macromolecules (development of the soft ionization method for mass analysis) K. Wüthrich (Switzerland): Development of methods for identification and structural analysis of biological macromolecules (development of NMR spectroscopy for structural determination of biological macromolecules)	R. Davis (U.S.), M. Koshiba (Japan): Pioneering contributions to astrophysics, particularly the detection of cosmic neutrinos (Physics)
2003	P. Agre (U.S.): Discoveries related to channels in cell membranes (discovery of water channels) R. MacKinnon (U.S.): Discoveries related to channels in cell membranes (structural and mechanistic studies of ion channels)	A. A. Abrikosov, V. L. Ginzburg (Russia), A. J. Leggett (Britain, U.S.): Pioneering contributions to the theory of superconductors and superfluids (Physics) P. Lauterbur (U.S.), P. Mansfield (Britain): Discoveries related to nuclear magnetic resonance imaging (Physiology / Medicine)
2004	A. Ciechanover, A. Hershko (Israel), I. Rose (U.S): Discovery of ubiquitin-mediated protein degradation	R. Axel, L. B Buck (U.S.): Discovery of odorant receptors and the organization of the olfactory system (Physiology / Medicine)
2005	Y. Chauvin (France), R. H. Grubbs, R. R. Schrock (U.S.): Development of the metathesis method in organic synthesis	R. Glauber (U.S.): Contributions to the quantum theory of optical coherence (Physics) J. L. Hall, T. W. Hansch (US): Contributions to the development of precision laser spectroscopy methods, including the optical frequency comb technique (Physics)
2006	R. D. Kornberg (U.S.): Studies of the molecular basis of eukaryotic transcription	A. Z. Fire, C. C. Mello (U.S.): Discovery of RNA interference-gene silencing by double stranded RNA (Physiology / Medicine)
2007	G. Ertl (Germany): Studies of chemical processes on solid surfaces	A. Fert (France), P. Grünberg (Germany): Discovery of giant magneto-resistance (Physics)
2008	O. Shimomura (Japan), M. Chalfie, R. Y. Tsien (U.S.): Discovery and development of green fluorescent protein (GFP)	Y. Nambu (US): Discovery of the mechanism of spontaneous symmetry breaking in elementary particle physics and nuclear physics T. Masukawa and M. Kobayashi (Japan): Discovery of the origin of symmetry breaking that predicts the existence of at least three families of quarks in nature (Physics)

Year	Nobel Prize in Chemistry	Nobel Prizes in Physics / Physiology or Medicine (Selected)
2009	V. Ramakrishnan (India / U.S.), T. A. Steitz (U.S.), A. E. Yonath (Israel): Studies of the structure and function of the ribosome	C. K. Kao (Britain, U.S.): Achievements concerning light transmission in optical fibers for optical communications (Physics) W. Boyle (U.S., Canada), G. E. Smith (U.S.): Invention of the CCD sensor, an imaging semiconductor circuit (Physics) E. H. Blackburn, C. W. Greider, J. W. Szostak (U.S.): Discovery of how chromosomes are protected by telomeres and the enzyme telomerase (Physiology / Medicine)
2010	R. F. Heck (U.S.), E. Negishi, A. Suzuki (Japan): Palladium-catalyzed cross couplings in organic synthesis	A. Geim (Holland), K. Novoselov (Russia, Britain): Groundbreaking experiments regarding the two-dimensional material graphene (Physics)
2011	D. Shechtman: (Israel): Discovery of quasicrystals	
2012	R. J. Lefkowitz, B. K. Kobilka (U.S.): Research on G-protein-coupled receptors	S. Horoche (France), D. J. Wineland (U.S.): Experimental methods that enable measuring and manipulating individual quantum systems (Physics) J. B. Gurdon (Britain), S. Yamanaka (Japan): Discovery that mature cells can be reprogrammed to become pluripotent (Physiology / Medicine)
2013	M. Karplus (U.S.), M. Levitt (US, Britain, Israel), A. Warshel (U.S., Israel): Development of multiscale models for complex chemical systems	J. E. Rothman, R. W. Schekman, T. C. Sudof (U.S.): Discoveries of machinery regulating vesicle traffic, a major transport system in cells (Physiology / Medicine)
2014	E. Betzig (U.S.), S. W. Hell (Germany), W. E. Moerner (U.S.): Development of Super-resolved fluorescence microscopy	I. Akasaki, H. Amano (Japan), S. Nakamura (US): Invention of efficient blue light-emitting diodes which has enabled bright and energy-saving white light sources (Physics)

Figure Credits

Chapter 1

title page: Courtesy of the National Library of Medicine.

1.2: Hieronymus Brunschwig. Liber de arte distillandi. Strassburg: 1500. Page 39 verso. Courtesy of the National Library of Medicine.

1.3: Georgius Agricola; translated from the first Latin edition of 1556 by Herbert Clark Hoover and Lou Henry Hoover. De re metallica. London: The Mining magazine; 1912. Book VII, p.265. "A FIRST SMALL BALANCE. B SECOND. C THIRD, PLACED IN A CASE." Courtesy of Internet Archive.

1.4: Antoine Laurent Lavoisier. Traité élémentaire de chimie. Paris: Cuchet, 1789. Courtesy of HathiTrust.

1.5: Antoine Laurent Lavoisier. Traité élémentaire de chimie. Paris: Cuchet, 1789. p.192. Courtesy of HathiTrust

Paracelsus: Courtesy of the National Library of Medicine.

Robert Boyle: Courtesy of the National Library of Medicine.

Joseph Black: Courtesy of the National Library of Medicine.

Joseph Priestley: Courtesy of the National Library of Medicine.

Carl Scheele: Courtesy of the National Library of Medicine.

Henry Cavendish: Edgar Fahs Smith Collection, University of Pennsylvania Libraries

Antoine-Laurent de Lavoisier: Courtesy of the National Library of Medicine.

Column 1: Joseph Priestley. Disquisitions Relating to Matter and Spirit. London: J. Johnson; 1777. Courtesy of HathiTrust.

Column 2 (p.33): Courtesy of the National Library of Medicine.

Column 2 (p.34): Antoine Laurent Lavoisier. Traite elementaire de chimie. Paris: Cuchet, 1789, plate 4.

Chapter 2

title page: Courtesy of Masami Saitoh.

2.1: John Dalton. A new system of chemical philosophy. Lonodn: R. Bickerstaff; 1808. Volume: 1, p.218, Plate 4. Courtesy of HathiTrust. 742.

2.2: Alessandro Volta. On the Electricity Excited by the Mere Contact of Conducting Substances of Different Kinds. Philosophical Transactions of the Royal Society, 1800. v. 90, pt. 2. pp.403–431. Courtesy of Internet Archive.

2.3: E. Shimao, "Jinbutsu Kagakushi" (Characters in the history of chemistry) Asakura Shoten 2002, p.60.

2.4: Aaron J. Ihde, The Development of Modern Chemistry, Dover Publications, 1970, p.305, p.309.

2.5: Aaron J. Ihde, The Development of Modern Chemistry, Dover Publications, 1970, p.313.

2.7: Rene Dubos, Pasteur and Modern Science, ASM Press, Washington, D.C. 1998, p.18.

2.10 : Kirchhoff, G. and Bunsen, R. (1860), Chemische Analyse durch Spectralbeobachtungen. Ann. Phys., 186: 161–189. Copyright © 1860 WILEY-VCH Verlag GmbH & Co. KGaA, Weinheim.

2.11: The Special Collections Research Center, University of Chicago Library.

2.12: Lothar Meyer. Die Natur der chemischen Elemente als Funktion ihrer Atomgewichte. Annalen der Chemie und Pharmacie. 1870; Supplement 7: 354–364.

2.15: Based on Alfred Werner, New Idea of Complex Cobalt Compounds, 1911.

2.17: Based on W. J. Moore "Physical Chemistry"4th Ed. Prentice-Hall, 1972, p.143, Figure 4.11.

2.22: Photo courtesy of Tsuyama Archives of Western Learning.

John Dalton: Courtesy of the National Library of Medicine.

Jons Jacob Berzelius: Courtesy of the National Library of Medicine.

Joseph Gay-Lussac: Courtesy of the National Library of Medicine.

Amedeo Avogadro: Edgar Fahs Smith Collection, University of Pennsylvania Libraries.

Jean-Baptiste-André Dumas: Courtesy of the National Library of Medicine.

Humphry Davy: Courtesy of the National Library of Medicine.

Michael Faraday: Courtesy of the National Library of Medicine.

Friedrich Wöhler: Courtesy of the National Library of Medicine.

Justus von Liebig: Courtesy of the National Library of Medicine.

Auguste Laurent: Edgar Fahs Smith Collection, University of Pennsylvania Libraries.

Charles Gerhardt: Edgar Fahs Smith Collection, University of Pennsylvania Libraries.

Stanislao Cannizzaro: Edgar Fahs Smith Collection, University of Pennsylvania Libraries.

Friedrich August Kekule: Courtesy of the National Library of Medicine.

Louis Pasteur: Courtesy of the National Library of Medicine.

Jacobus Henricus van't Hoff: Edgar Fahs Smith Collection, University of Pennsylvania Libraries.

Marcellin Berthelot: Edgar Fahs Smith Collection, University of Pennsylvania Libraries.

Robert Bunsen: Courtesy of the National Library of Medicine.

Gustav Kirchhoff: Courtesy of the Library of Congress, LC-USZ62-133715.

Dmitri Ivanovich Mendeleev: Edgar Fahs Smith Collection, University of Pennsylvania Libraries.

Lothar Meyer: Edgar Fahs Smith Collection, University of Pennsylvania Libraries.

Ferdinand Henri Moissan: Reproduced courtesy of the Library of the Royal Society of Chemistry.

John Strutt (Lord Rayleigh): Science & Society Picture Library/ Aflo

William Ramsay: Edgar Fahs Smith Collection, University of Pennsylvania Libraries

Alfred Werner: ETH-Bibliothek Zurich, Image Archive.

Sadi Carnot: Edgar Fahs Smith Collection, University of Pennsylvania Libraries.

William Thomson: Edgar Fahs Smith Collection, University of Pennsylvania Libraries.

James Clerk Maxwell: ETH-Bibliothek Zurich, Image Archive.

Ludwig Boltzmann: © Archives of Graz University

Josiah Willard Gibbs: AIP Emilio Segre Visual Archives

Svante Arrhenius: Edgar Fahs Smith Collection, University of Pennsylvania Libraries.

Wilhelm Ostwald: Edgar Fahs Smith Collection, University of Pennsylvania Libraries.

Emil Fischer: Courtesy of the National Library of Medicine.

Friedrich Miescher: © Ralf Dahm, Mainz, Germany.

Eduard Buchner: Courtesy of the University of Würzburg.

William Perkin: Edgar Fahs Smith Collection, University of Pennsylvania Libraries.

Adolf von Baeyer: Edgar Fahs Smith Collection, University of Pennsylvania Libraries.

Alfred Nobel: © The Nobel Foundation

Yoan Udagawa: Courtesy of Takeda Science Foundation Kyo-U Library.

Mitsuru Kuhara: Courtesy of the Kuhara family, deposited at Tsuyama Archives of Western Learning.

Jyoji Sakurai: Photo courtesy of RIKEN.

Column 3 (p.56): Edgar Fahs Smith Collection, University of Pennsylvania Libraries.

Column 3 (p.57): James Gillray "New Discoveries in Pneumatics"

Column 4 (p.118): Based on P. Tans (2007) "Monthly mean atmospheric carbon dioxide at Mauna Loa Observatory, Hawaii". Global Monitoring Division, Earth System Research Laboratory, National Oceanic and Atmospheric Administration, U.S. Department of Commerce, U.S.A.

Column 4 (p.119): Courtesy of Masami Saitoh.

Column 5 (p.121): ©Technische Universität Braunschweig

Column 5 (p.122): ©Technische Universität Braunschweig

Column 6 (p.139): Courtesy of Liebig-Museum Giessen, Germany.

Column 6 (p.140): Liebigs Annalen. 1997. Volume 1997, Issue 12. Front cover. Copyright Wiley-VCH Verlag GmbH & Co. KGaA. Reproduced with permission.

Column 7 (p.157): Photo courtesy of Graduate School of Engineering, Kyoto University.

Column 7 (p.159): Hikorokuro Yoshida (1897) "Shinpen Kagaku Kyokasho"

Chapter 3

title page: Courtesy of the University of Würzburg.

3.1: Science Museum / Science & Society Picture Library

3.2: S. Honma, "Shinban Denshi to Genshikaku no Hakken" (S. Weinberg, "The Discoveries of Subatomic Particles") Chikuma Gakugei Bunko, 2006, p.7. Reproduced with permission.

3.3: Based on Linus Pauling, General Chemistry, Dover Publications, 1970, p.72, fig.3–25.

3.4: H. G. J. Moseley, M. A., The High-frequency spectra of the elements, Phil. Mag., 1913, p.1024.

3.5: Astons first Mass Spectrograph set up in the lab. C.1919. Courtesy of the Cavendish Laboratory, University of Cambridge.

3.6: Based on S. Honma, "Shinban Denshi to Genshikaku no Hakken" (S. Weinberg, "The Discoveries of Subatomic Particles") Chikuma Gakugei Bunko, 2006, p.251.

William Crookes: Courtesy of the National Library of Medicine.

Joseph John Thomson: Courtesy of the National Library of Medicine.

Wilhelm Röntgen: Courtesy of the National Library of Medicine.

William Henry Bragg: Courtesy of State Library of South Australia, B3991.

Lawrence Bragg: AIP Emilio Segre Visual Archives, Weber Collection

Henry Moseley: Edgar Fahs Smith Collection, University of Pennsylvania Libraries

Henri Becquerel: Courtesy of the National Library of Medicine.

Pierre Curie: Edgar Fahs Smith Collection, University of Pennsylvania Libraries

Marie Sklodowska-Curie: Edgar Fahs Smith Collection, University of Pennsylvania Libraries

Ernest Rutherford: Courtesy of the National Library of Medicine.

Frederick Soddy: Edgar Fahs Smith Collection, University of Pennsylvania Libraries.

Francis Aston: Edgar Fahs Smith Collection, University of Pennsylvania Libraries.

Harold Urey: Edgar Fahs Smith Collection, University of Pennsylvania Libraries.

Albert Einstein: ETH-Bibliothek Zurich, Image Archive.

Jean Perrin: Edgar Fahs Smith Collection, University of Pennsylvania Libraries.

Max Planck: Archives of the Max Planck Societey, Berlin.

Niels Bohr: ©The Niels Bohr Archive, Copenhagen

Arnold Sommerfeld: AIP Emilio Segre Visual Archives, Physics Today Collection.

Wolfgang Pauli: Science Photo Library/Aflo

James Chadwick: AIP Emilio Segre Visual Archives, Numeroff Collection.

Werner Heisenberg: Courtesy of Archiv of the University of Leipzig.

Louis de Broglie: AIP Emilio Segre Visual Archives, Physics Today Collection.

Erwin Schrödinger: AIP Emilio Segre Visual Archives, Physics Today Collection.

Max Born: Courtesy of The University of Edinburgh.

Paul Dirac: AIP Emilio Segre Visual Archives.

Column 8 (p.181): Courtesy of Wayne Boucher.

Column 8 (p.182): Courtesy of MRC Laboratory of Molecular Biology.

Column 9 (p.187): Wellcome Library, London

Column 9 (p.188): Courtesy of Masami Saitoh.

Chapter 4

title page: Courtesy of Masami Saitoh.

4.1: Valence and the structure of atoms and molecules. Gilbert Newton Lewis. New York: Chemical Catalog, 1923 (Monograph series/American Chemical Society). p.29. Fig. 3. Courtesy of HathiTrust.

4.2: Valence and the structure of atoms and molecules. Gilbert Newton Lewis. New York: Chemical Catalog, 1923 (Monograph series/American Chemical Society). p.33 Fig.4. Courtesy of HathiTrust.

4.3: Based on W. J. Moore, "Physical Chemistry" 4th ed. Prentice Hall, New Jersey, 1972.

4.8: Based on W. J. Moore, "Physical Chemistry" 4th ed. Prentice Hall, New Jersey, 1972.

4.10: Based on P. W. Atkins, "Physical Chemistry" 5th edition, p.952, fig. 27.15, 27.16.

4.11: Photo Courtesy of Beckman Coulter.

4.12: Reprinted by permission from Macmillan Publisher Ltd: Dent CE, Stepka W, Steward FC, Detection of the Free Amino-Acids of Plant Cells By Partition Chromatography, Nature 160; 1947:682–683. Copyright © 1946.

4.36: Photograph of a culture-plate showing the dissolution of staphylococcal colonies in the neighbourhood of a penicillium colony. (Fig.1, page 228–1). Alexander Fleming, On the Antibacterial Action of Cultures of a Penicillium, with Special Reference to Their Use in the Isolation of B. Influenzas. Br J Exp Pathol. 1929; 10(3): 226–236. Courtesy of PubMed Central.

Linus Pauling: AIP Emilio Segre Visual Archives, W. F. Meggers Gallery of Nobel Laureates.

Gilbert Newton Lewis: AIP Emilio Segre Visual Archives, photograph by Francis Simon.

Walther Nernst: ©UB der HU zu Berlin; Porträtsammlung; Richthofen, Nernst, Walter

William Giauque: AIP Emilio Segre Visual Archives, W. F. Meggers Gallery of Nobel Laureates.

Peter Debye: Archives of the Max Planck Societey, Berlin.

Walter Heitler: ETH-Bibliothek Zurich, Image Archive.

Fritz London: AIP Emilio Segre Visual Archives, Physics Today Collection.

Robert Mulliken: AIP Emilio Segre Visual Archives, Physics Today Collection.

Ernst Ruska: Archives of the Max Planck Societey, Berlin.

Gerhard Herzberg: Courtesy of National Research Council Canada Archives.

Chandrasekhara Raman: Edgar Fahs Smith Collection, University of Pennsylvania Libraries.

Isidor Rabi: AIP Emilio Segre Visual Archives, Physics Today Collection.

Henry Eyring: AIP Emilio Segre Visual Archives.

Michael Polanyi: Courtesy of Professor John Polanyi.

Theodor Svedberg: Edgar Fahs Smith Collection, University of Pennsylvania Libraries.

Irving Langmuir: AIP Emilio Segre Visual Archives.

Frederic Joliot-Curie: AIP Emilio Segre Visual Archives, W. F. Meggers Collection.

Irene Joliot-Curie: AIP Emilio Segre Visual Archives, W. F. Meggers Gallery of Nobel Laureates.

Lise Meitner: Archives of the Max Planck Societey, Berlin.

Otto Hahn: Archives of the Max Planck Societey, Berlin.

Glenn Seaborg: Courtesy of the U.S. Department of Energy.

Georg de Hevesy: AIP Emilio Segre Visual Archives, W. F. Meggers Collection.

Fritz Pregl: Photo: Graz University Archive.

Jaroslav Heyrovsky: Courtesy of Institute of Physical Chemistry of J. Heyrovsky of the AS CR.

Mikhail Tsvet: Photo courtesy of Itaru Matsushita.

Archer Martin: © National Portrait Gallery, London

Willard Libby: Courtesy of U.S. Department of Energy.

Victor Goldschmidt: Courtesy of University History Photobase, University of Oslo. Unknown photographer.

Christopher Ingold: Reproduced courtesy of the Library of the Royal Society of Chemistry

Victor Grignard: Mary Evans Picture Library/ Aflo

Otto Diels: © UB der HU zu Berlin; Porträtesammlung; Diels Otto

Kurt Alder: © The Department of Chemistry, the University of Cologne

Robert Robinson: "Robert Robinson, Professor of Organic Chemistry". Courtesy of the University of Sydney Archives, the reference number G3_224_1679.

Hermann Staudinger: ETH-Bibliothek Zurich, Image Archive.

Walter Haworth: Reproduced courtesy of the Library of the Royal Society of Chemistry

Richard Willstätter: Archives of the Max Planck Societey, Berlin.

Hans Fischer: Edgar Fahs Smith Collection, University of Pennsylvania Libraries.

Heinrich Wieland: Courtesy of Albert-Ludwigs-Universität Freiburg.

Adolf Butenandt: Archives of the Max Planck Societey, Berlin.

Paul Karrer: ETH-Bibliothek Zurich, Image Archive.

Richard Kuhn : Archives of the Max Planck Societey, Berlin.

Arthur Harden: Reproduced courtesy of the Library of the Royal Society of Chemistry.

Hans von Euler-Chelpin: Leopoldina-Archiv/MM 3470

James Sumner: HUP Sumner, James (1), Harvard University Archives

John Northrop: Courtesy of the Rockefeller University.

Wendell Stanley: Courtesy of the Rockefeller University.
Otto Warburg: Courtesy of the National Library of Medicine.
Otto Meyerhof : Archives of the Max Planck Societey, Berlin.
Albert Szent-Györgyi: Courtesy of the National Library of Medicine.
Hans Krebs: Reproduced courtesy of the Library of the Royal Society of
    Chemistry
Gerty Cori and Carl Cori: Becker Medical Library, Washington University
    School of Medicine.
Rudolph Schoenheimer: University Archives, Rare Book & Manuscript
    Library, Columbia University in the City of New York.
Umetaro Suzuki: Photo courtesy of RIKEN.
Jokichi Takamine: Photo courtesy of RIKEN.
Fritz Haber: Archives of the Max Planck Societey, Berlin.
Wallace Carothers: Photo courtesy of E. I. du Pont de Nemours and
    Company
Paul Ehrlich: Courtesy of the National Library of Medicine.
Kikunae Ikeda: Photo courtesy of RIKEN.
Riko Majima: Photo courtesy of RIKEN.
Gen'itsu Kita: Courtesy of the Kita family.
Column 10: The Bancroft Library, University of California.
Column 11: © Wolfgang Suschitzky / National Portrait Gallery, London
Column 12 (p.287): Photo: Deutsches Museum
Column 12 (p.289): Archives of the Max Planck Society, Berlin.
Column 13: Photo courtesy of Tohoku University Archives.
Column 14: PBD ID: 1E9Z
N.-C. Ha, S.-T. Oh, J.Y. Sung, K.-A. Cha, M. Hyung Lee, B.-H.Oh (2001)
    Supramolecular assembly and acid resistance of Helicobacter pylori
    urease. Nat.Struct.Biol. 8: 480.
Colunn 15: Archives of the Max Planck Societey, Berlin.
Column 16 (p.386): Courtesy of the Kondo family.
Column 16 (p.388): Photo courtesy of RIKEN.

Chapter 5

title page: Courtesy of Masami Saitoh.
5.2: Max F. Perutz — Nobel Lecture: X-ray Analysis of Haemoglobin. Nobel
    Lecture, December 11, 1962.© The Nobel Foundation 1962.
5.3: Left: Courtesy of professor Noriko Nagata.
Right: Courtesy of professor Yuichiro Watanabe.

Kenichi Fukui: Photo courtesy of Fukui Institute for Fundamental Chemistry, Kyoto University.

Roald Hoffmann: AIP Emilio Segre Visual Archives, Physics Today Collection.

Rudolf Marcus: AIP Emilio Segre Visual Archives.

Gerhard Ertl : Fritz-Haber-Institut

Geoffrey Wilkinson: © Liam Woon / National Portrait Gallery, London

Ernst Fischer: Technische Universität München

Karl Ziegler: Archives of the Max Planck Societey, Berlin.

Giulio Natta: Photo courtesy: Giulio Natta Archive.

Herbert Brown: Purdue University, Department of Chemistry

William Knowles: © 2005 National Academy of Sciences, U.S.A. PNAS is not responsible for the accuracy of this translation..

Ryoji Noyori: Photo courtesy of RIKEN.

Yves Chauvin: Reuters/ Aflo

Richard Heck: © The Nobel Foundation. Photo: Ulla Montan.

Eiichi Negishi: © The Nobel Foundation. Photo: Ulla Montan.

Akira Suzuki: © The Nobel Foundation. Photo: Ulla Montan.

Elias Corey: AP/ Aflo

Charles Pedersen: Photo courtesy of E. I. du Pont de Nemours and Company

Jean-Marie Lehn: Courtesy of Professor Jean-Marie Lehn

Donald Cram: UCLA Photography

Robert Curl: Photo: Thomas LaVerne

Harold Kroto: Photo: Margaret Kroto

Richard Smalley: AIP Emilio Segre Visual Archives, Physics Today Collection

Hideki Shirakawa: Photo Courtesy of University of Tsukuba.

Alan Heeger: Courtesy of Professor Alan Heeger.

Johannes Bednorz: © IBM Research-Zurich

Karl Alexander Müller : © IBM Research-Zurich

Paul Crutzen: Max-Planck-Gesellschaft

Sherwood Rowland: AIP Emilio Segre Visual Archives, Physics Today Collection.

Column17: PDBID: 4INS. E.N. Baker, T.L. Blundell, J.F. Cutfield, S.M. Cutfield, E.J. Dodson, G.G.Dodson, D.M. Hodgkin, R.E. Hubbard, N.W. Isaacs, C.D. Reynolds, K. Sakabe, N. Sakabe, N.M. Vijayan (1988) The structure of 2Zn pig insulin crystals at 1.5 A resolution. Philos.Trans.R.Soc. London,Ser.B 319: 369–456.

Column 18: PDBID: 1GFL. F. Yang, L.G. Moss, G.N. Phillips Jr. (1996) The molecular structure of green fluorescent protein. Nat.Biotechnol. 14: 1246–1251.

Column 19: Reprinted by permission from Macmillan Publishers Ltd: Lauterbur
   PC. Image Formation by Induced Local Interactions: Examples Employing
   Nuclear Magnetic Resonance. Nature 242; 1973: 190–91. Copyright ©1973.
Column 20: Courtesy of Harvard University Archives.

Chapter 6

title page: Courtesy of Masami Saitoh.
6.2: Reprinted by permission from Macmillan Publishers Ltd: Franklin R,
   Gosling RG. Molecular Configuration in Sodium Thymonucleate. Nature
   171; 1953: 740–41.
6.3: Reprinted by permission from Macmillan Publishers Ltd: Watson JD, Crick
   FHC. Genetical Implications of the structure of Deoxyribonucleic Acid.
   Nature 171; 1953: 964–7.
6.5: Science Photo Library/ Aflo
6.11: Reprinted from FEBS Lett., 579(4), Boeger H, Bushnell DA, Davis R,
   Griesenbeck J, Lorch Y, Strattan JS, Westover KD, Kornberg RD, Structural
   basis of eukaryotic gene transcription, 899–903, Copyright 2005, with
   permission from Elsevier.
6.12 : Voet, Biochemistry, 3rd Ed., Wiley, p.1313.
6.13: Reprinted from Cell, 107(5), Harms J, Schluenzen F, Zarivach R, Bashan
   A, Gat S, Agmon I, Bartels H, Franceschi F, Yonath A, High resolution
   structure of the large ribosomal subunit from a mesophilic eubacterium,
   679–88, Copyright 2001, with permission from Elsevier.
Reprinted from Cell, 102(5), Schluenzen F, Tocilj A, Zarivach R, Harms J,
   Gluehmann M, Janell D, Bashan A, Bartels H, Agmon I, Franceschi F,
   Yonath A, Structure of Functionally Activated Small Ribosomal Subunit at
   3.3 Å Resolution, 615–23, Copyright 2000, with permission from Elsevier.
6.14: Prepared with Jmol: an open-source Java viewer for chemical structures in
   3D. http://www.jmol.org/ (PDBID:5LYZ)
6.16: Science Photo Library/ Aflo
6.22 : Based on Voet, Biochemistry, 3rd Ed., Wiley, p.881, fig. 24–12.
6.24: PBD ID: 3SN6
Crystal structure of the β2 adrenergic receptor—Gs protein complex.
   Rasmussen, S.G., DeVree, B.T., Zou, Y., Kruse, A.C., Chung, K.Y., Kobilka,
   T.S., et.al, (2011) Nature 477: 549–555.
Oswald Avery: Courtesy of the Rockefeller University.
Maurice Wilkins: © National Portrait Gallery, London
Rosalind Franklin: Courtesy of Jenifer Glynn, from National Library of
   Medicine's Profiles in Science.

Francis Crick: Courtesy of the Salk Institute for Biological Studies.

James Watson: HUP Watson, James (2), Harvard University Archives

Arthur Kornberg: Courtesy of Arthur Kornberg, from National Library of
    Medicine's Profiles in Science.

Frederick Sanger: Courtesy of Genome Research Limited.

Robert Holley: Courtesy of the Salk Institute for Biological Studies.

Francois Jacob: Courtesy of Institut Pasteur.

Jacques Monod: Courtesy of Institut Pasteur.

Paul Berg: Courtesy of National Library of Medicine's Profiles in Science.

Roger Kornberg: Courtesy of Professor Roger Kornberg.

Ada Yonath: © Nobel Foundation. Photo: Ulla Montan.

Thomas Steitz: © The Nobel Foundation. Photo: Ulla Montan.

Venkatraman Ramakrishnan: © The Nobel Foundation. Photo: Ulla Montan.

Thomas Cech: Photo: University of Colorado, Glenn Asakawa

Sidney Altman: Photo: Michael Marsland, Yale University

Jens Skou: Photo: Lars Kruse, Aarhus University

Peter Mitchell: AP/ Aflo

Melvin Calvin: Berkeley Lab

Hartmut Michel: Max-Planck-Gesellschaft

Susumu Tonegawa: Photo courtesy of RIKEN.

Column 22: Science & Society Picture Library/ Aflo

Column 23: Courtesy of Nancy Cosgrove Mullis.

Column 24 : ZUMA Press/ Aflo

Column 25 (p.646): © The Francis Frith Collection

Column 25 (p.647): Based on K. Miura, "Noberusho no Hasso" Asahi
    Sinbunsha, 1985, p.108.

Chapter 7

title page: With kind permission of IUPAC.

7.1: © ® The Nobel Foundation. Photo: Lovisa Engblom.

7.2: Reprinted with permission from Beyond the Molecular Frontier:
    Challenges for Chemistry and Chemical Engineering, 2003 by the
    National Academy of Sciences, Courtesy of the National Academies Press,
    Washington, D.C.

7.3: With kind permission of IUPAC.

Name index

Subject index